STREET DESIGN

STREET DESIGN
The Secret to Great Cities and Towns

SECOND EDITION

John Massengale and Victor Dover

WILEY

SHARE
THE ROAD

CONTENTS

Chapter Three STREET SYSTEMS AND NETWORKS   256

Chapter Four NEW & RETROFITTED STREETS   336

FOREWORD TO THE FIRST EDITION

HRH Prince of Wales

CLARENCE HOUSE

I am delighted that an evolutionary change in thinking seems to be underway worldwide about how best we can make and sustain our villages, towns and cities. Streets are meant to be places for people and this is most evident in our historic centres – where comfortable streets have made it possible for neighbourhoods to adapt again and again, even as culture and technology have continually evolved. On the other hand, the harsh and unwelcoming streets of urban sprawl were made for moving cars at high speed; their very geometry is car-based and, not surprisingly, these streets are often unsafe for pedestrians.

A handful of fundamental principles seem to separate the streets where communities flourish from those that do not. These principles form the central message of this book.

Learning from observing is one of those fundamental principles. Observation is perhaps the street designer's most powerful tool. At Poundbury, in Dorset, where I have attempted to put these principles into practice for more than 20 years, we have applied lessons learnt from other market towns and villages. For example, we realized that if we wanted to create a walkable town, we would need to insist that the car was not the focus in the physical design, but rather that the street is a pedestrian space through which cars are permitted to move, albeit slowly and safely. There are virtually no traffic signs or painted lines on the ground; instead there are spatial changes and physical and architectural features in a sequence along the streets. These sorts of spatial sequences were normal features in historic towns. Designing the streets this way has lent safety to the town, but it has also created a built environment that relates to the human scale and to the local identity. It is certainly encouraging that all the efforts at Poundbury, in Dorset (England), and my Foundation for Building Community's work elsewhere, have been cited as models in the British Government's new manual "Designing Streets".

Learning from observing is at the root of all living traditions, especially art – and street design is indeed an art; it is the art of place-making and that can only be acquired through study and application. The best streets in the world's villages, towns and cities – whether modest or grand – continually remind one that *simplicity* is part of the recipe for success in this art. The advice of Victor Dover and John Massengale, their historic examples and their own designs, reflect that simplicity.

Great streets define great cities; great cities establish harmony with Nature, continually improve societies and stimulate economic progress in a genuinely sustainable way. There is a real urgency to apply these principles in street design. The rewards will be worth every bit of effort.

FOREWORD TO THE SECOND EDITION

Carlos Moreno

FOR TOO LONG, those of us who live in cities, both large and small, have accepted the unacceptable. We have accepted that in cities, our sense of time is warped because we have to spend so much of it adapting to the organization of the city and its absurd travel demands for daily life. Why should we adapt to transportation and degrade our quality of life? Why not adapt the city to our needs?

How do we do this? There are four guiding principles: ecology, for a green and sustainable city; proximity, to live with reduced distances for daily life; solidarity, to create contact and links between people; and participation, actively involving citizens in the transformation of their neighborhoods.

Like most cities, Paris has spent the past century adapting to the scale of the motor vehicle. Now, Mayor Hidalgo is blazing a new trail, developing a 15-minute city that returns our streets to human-scale public spaces for living. We are redesigning the city so that people can meet their daily needs in a satisfying and sustainable urban experience within the distance of a 15-minute walk or bike ride. This means proximity to work, housing, food, health, education, culture, and leisure.

To achieve this, we need streets where people want to walk and bike. Why should our streets be noisy and polluted when they can be quiet garden streets lined with trees where people can meet, walk to the baker, and children can walk to school? Our acceptance of the dysfunctions and indignities of the harsh modern city has gone on too long. We need to change that for the sake of justice, our well-being, and the climate.

Cities need streets where we can comfortably walk, bike, and meet. The book *Street Design: The Secret to Great Cities and Towns* shows how to create such streets for every conceivable situation. It is a practical guide, full of examples for reclaiming the streets in our cities and towns for walkable 15-minute neighborhoods with access to everything we need. I recommend it to everyone who wants to make their city, town, or neighborhood more equitable, walkable, and sustainable.

This transformation is not just about infrastructure; it is about fostering a sense of community and belonging. It is about creating spaces where people can connect, where local businesses can thrive, and where the environment can flourish. By prioritizing these principles, we can build cities that truly serve their inhabitants and create a model for urban living that is both humane and sustainable.

Figure 0.2: Paris, France. Aerial photograph looking southwest towards the Arc de Triomphe and the place Charles de Gaulle. © Domenico Convertini / CC BY-SA 2.0

PREFACE

WHY UPDATE *Street Design*? The book is only a little over 10 years old, but there have been many changes in the design of city streets during the last 10–15 years. We saw the beginnings of those changes in Complete Streets, Vision Zero, Walk Scores, bike lanes, bike share, and talk about "livable streets" and "streets for people." And in the last five years, European cities and towns have shot ahead of American-style traffic engineering in the making of streets that put walking first, cycling second, and driving third. We believe many Americans want streets for people, and we hope this new edition will help.

When we began working on the book in 2010, the context was very different than it is today. A few European cities like Amsterdam and Copenhagen started transforming their streets in the 1960s and 1970s, reducing the emphasis on cars and traffic flow and promoting the rights of people walking and cycling. By the time the twenty-first century arrived, cities and towns in the Netherlands, Denmark, and parts of northern Belgium had very different streets than most of the Western world, where cars came first.[1] But ideas and attitudes about street design were rapidly changing in many places.

As practitioners, we participated in many discussions, meetings, and conferences about transforming streets. With a grant from Richard Driehaus, we visited European towns and cities, including London, Paris, Barcelona, and Amsterdam, where 85% of the streets in the center were shared-space streets that put walking first, biking second, and driving third. We liked them so much that we put one on the cover of the first edition (Figure 5.1).[2]

London introduced congestion pricing in 2003, but when we visited London eight years later, the design of "streets for people" was still experimental and not always successful.[3] Riding around London on the shared "Boris Bikes" could be scary on many streets. Paris was different from London or Amsterdam (or Copenhagen or Milan). By the time we wrote *Street Design*, the City of Light had visually reclaimed historic plazas by moving the parking underground, making a more beautiful, pedestrian-friendly city. But as we wrote in the preface to the first edition (page xiv), traffic on many historic avenues and boulevards could still be overwhelming. Paris had some exemplary bike lanes (see https://street.design/parisbicycle), but not enough of them, while the city's bike-share program had many problems.

Here in America, New York City was leading the way. Mayor Bloomberg proposed the country's first Congestion Zone in 2007, and the next year the NYC DOT implemented a Tactical Urbanism plan next to Madison Square that is still one of the best in the country (page 444). But congestion pricing in New York City only began on January 5, 2025, and the "temporary" design for the intersection of Broadway and Fifth Avenue is almost the same today as it was 17 years ago. Chapter Four includes our sketch plan for a permanent design (Figure 4.141).

COVID-19 brought experimentation around the world. With streets empty everywhere, cities used the streets to make safe spaces for public life in a pandemic. European cities and towns took advantage of the crisis and the opportunity to permanently make their streets better places for city life. On the whole, however, American cities did not. DOTs in the United States went back to business as usual after the pandemic. Today, the difference between the quality of life on American and European streets is starker than ever. With few exceptions, our cities still put traffic engineering and traffic flow first. We talk about this in Chapter Five: "Whose Streets?"

> Many European cities used the crisis created by COVID-19 and the pandemic to permanently make their streets better places for public life. On the whole, however, American cities did not.

A few words about the new edition: There are almost 60% more photos and drawings, including the renderings for the new Catalog of Ten Essential Street Types at the end of Chapter One (interactive, three-dimensional models of the 10 types are online at street.design/

models). We reorganized the chapters—Chapter Two, "Historic Streets," and Chapter Four, "New and Retrofitted Streets" —have new case studies, written by us and by guest authors. In addition, we expanded on themes around different aspects of design that were in the first edition: some history of how we came to this point in the history of street design, how streets fit into a greater whole, how traffic engineers think, and how urban designers think (differently from Modern architects, on the whole). In the first edition, we argued for streets like Singelgracht and Worth Avenue where less is more. In the second edition, we say more about that and introduce the Monderman Rule, which says the best way to make safe streets is to "remove things" (the opposite of what traffic engineers do, according to Hans Monderman—see page 513).[4]

We wrote a little more about the importance of design, and how it can solve problems in unique ways. We made some extensive notes about how beauty affects us and why it's not just in the eye of the beholder (pages 2 and 622). We call urban designers "street whisperers" in the new book, able to consciously or intuitively pull together the elements that make people comfortable on the street.

Good city streets support public life, bad city streets do not. As we shall see, it should be quite easy to make a good street, but it is often quite difficult in America today.

Once we leave our house or apartment building, the first thing we see is the street. How we feel about the street, where the street takes us, and if we like the place it takes us, all contribute to how we feel about ourselves and our place in the world.

It's common on social media and in public planning meetings to hear someone exclaim that street design and urban design are "just aesthetics." We could not disagree more. If you want to see what a culture values, look at what it builds. In the last one hundred years, we have substituted bleak transportation corridors for civic space, where public life takes place. That is a societal mistake we discuss at the beginning of Chapter One.

We know that a young child's learning about the world outside their home is directly affected by how they move from the home onto the street. What is the character of that street? Are they allowed to venture on their own more than a few feet from the front door? Are they encouraged to explore the neighborhood on their own, or do they experience going into the world from the back seat of a car driven by a parent? Studies show these differences affect their feelings of independence and self-confidence.

The place where we live is a reflection of our self.[5] Imagine looking back at your home from the street, as though you are looking at yourself in the mirror. Do you like what you see? Does it make you confident, or do you feel that others have better places to live? Does your home look like your neighbors, or does it look like a "project," a place for people who couldn't afford something better? New Urban designer Ray Gindroz's long career redesigning housing projects and working in poor neighborhoods gave him data that showed children had more self-confidence, better grades in school, and happier lives.[6]

The street, the neighborhood, and the city or town where we live are also reflections of ourselves, fundamental to our self-image and how we fit into the world. They can welcome us or isolate us.[7]

PREFACE TO THE FIRST EDITION

WE—JOHN AND VICTOR—have been looking at and thinking about streets for decades. We've logged our favorites in sketchbooks and debated their many differences. At the same time, we've come to understand that what makes a good street is not as subjective or as complex as some might think. In fact, making good streets comes naturally to people, and has for thousands of years.

Studies show that when people are given maps of a town or city and asked to walk around and mark on the maps the places they like and don't like, their choices correspond to a great degree. Yes, some people have more formal tastes than others, and there are other preferences that might distinguish one person's favorites from another's. But increase the sample size, and the preferences become part of a predictable range with a lot of overlap. There is always a consensus about which places are the best and which are the worst, regardless of personal preferences. Practically everyone will say that the Piazza San Marco in Venice and the streets on the Left Bank in Paris are beautiful. Similar reactions are found in cities, towns, neighborhoods, and villages around the world.

Once we leave our house or apartment building, the first thing we see is the street. How we feel about the street, where the street takes us, and if we like the place it takes us, all contribute to how we feel about ourselves and our place in the world.

If there is so much consensus on what makes a good street, why are we still building so many bad and ugly ones? The reasons can be identified and addressed. Today, too few people bother to think about what makes them feel at home on the street in the first place. Take the time to look, and anyone will begin to notice the patterns of buildings, trees, and comfortable spaces that set the better streets apart from the rest.

For this book, we made lists of our favorite streets, and then examined what made them special. We asked our colleagues to tell us about the streets they admire, and we went into the library and looked online to find other lists of great streets. Then we went out to reexamine many of the streets in person—photographing them, taking measurements, and observing the way people behave and interact on them. We had the pleasure of visiting many great streets, and we were able to see how the experience of visiting the streets today compared to our memories or our colleagues' recommendations. We could see if there had been changes, and if those had made the street better or worse.

A problem we found everywhere was that the automobile has taken over our streets. Writers like Peter D. Norton have shown how Organized Motordom pushed everyone but the driver and his car to the side of the road—and then sometimes took the sidewalk, too.[8] Well-meaning authorities redesigned roads for "throughput" and removed obstacles like pedestrians, who were getting in the way of the cars. This emphasis on driving frequently undermined public spaces that were once wonderful for walking. Some of our favorite streets, when revisited, were no longer agreeable places to be: spending a few hours on the Boulevard St. Germain in Paris was exhausting, because of the never-ending noise, smell, and energy of the cars racing along it (Figure 0.3). The formerly pleasing broad High Street in Marlborough, England, no longer felt like the town center, due to the sheer volume of cars and trucks passing through it on their way from somewhere else to yet another place. Newer streets were often even more disagreeable.

Along with the success stories, therefore, we saw problems, and we also looked at new and old examples of streets commonly regarded as failures. We recorded some of those too. Our hope is that every reader will

Figure 0.3: Boulevard St. Germain, Paris, France. Once the center of Bohemian life in Paris, the boulevard St. Germain is now overwhelmed by traffic. *Flâneurs* today walk on other streets.

come away with a sharper sense of the elements that contribute to making a street a place that people seek out or avoid.

The good news is that today, all across the United States, we are in a period of rediscovering our old towns and cities and rebuilding our streets, and more and more people are seeing the need to curtail the radical influence of the car on our physical surroundings. One rallying term for this new vision of community is the Complete Street[9]—one where the pedestrian, the driver, the cyclist, and transit users all have a stake.

The bad news is that Americans are frequently ignoring the basic rules of placemaking in our attempts to create complete streets. Professionals of all stripes—often with competing agendas—are designing and building streets with specialized standards and criteria, which is one of the main reasons why our streets get worse and worse. The formulaic, seemingly ubiquitous use of yellow pedestrian crossings, red bus lanes, green bicycle tracks, ugly bumpouts, and uglier white plastic sticks make sense in the narrow focus of the specialist, but look at the world's best streets, as we have, and you will find they don't have these special things. What they do have is a limited palette of materials in the roadbed and on the sidewalk. When you visit them, what you notice is the beauty and the harmony of the place, not the details like the crosswalks or the bench selection.

In the following chapters, we will look at why streets matter. We will examine historic streets, retrofitted streets, new streets, and street networks. Our hope is that our readers will come to see that we can all envision better places—and then fix our streets, by design. We *must* make the most of the glorious new opportunities to build more walkable towns and cities by creating streets that are places *where people actually want to be.*

NOTES

1. The approaches in the Netherlands and Denmark were distinctly different: the Dutch put more emphasis on shared space, while the Danes created more multimodal transportation corridors and pedestrian malls. See pages 461 and 466. Italian cities also reduced the role of the car in their pedestrian-oriented *Centri Storici*, as documented by Bernard Rudofsky in *Streets for People: A Primer for Americans* (Doubleday, 1969). Also see John Massengale, "There Are Better Ways to Get Around Town," *New York Times*, May 15, 2018, https://www.nytimes.com/2018/05/15/opinion/there-are-better-ways-to-get-around-town.html.
America built many pedestrian malls in the 1960s and 1970s: many failed. See pages 77 and 523. The Departments of Transportation that controlled the design of American streets ignored the shared-space model.

2. See Figure 5.1 and the case study for the Spui on page 507. Today in Amsterdam, there are more bicycles now, plus e-bikes, electrified scooters, mopeds, etc. These make the conversation about bicycles more complicated, in the Netherlands and around the world. See page 587.

3. See our case studies for Kensington High Street (page 388) and Exhibition Road (page 513).

4. "The trouble with traffic engineers is that when there's a problem with a road, they always try to add something," Monderman said. "To my mind, it's much better to remove things." See James Surowiecki, "Roads Gone Wild," *Wired*, December 2004, https://www.wired.com/2004/12/traffic/. Also see the Coco Chanel Rule on page 400: "Before you leave the house, look in the mirror and take one thing off." "Simplicity is the keynote of all true elegance," Chanel said.

5. Clare Cooper Marcus, *House as a Mirror of Self: Exploring the Deeper Meaning of Home* (Nicolas-Hays, 1995). There is a free online copy at https://archive.org/details/houseasmirrorofs00marc.

6. Ray Gindroz, *The Place of Dwelling* (Prince's Foundation for the Built Environment, 2008): 11–15. Also see Rob Steuteville's review, "The Architectural Tuning of Settlements and The Place of Dwelling," CNU Public Square, January 1, 2009, https://www.cnu.org/publicsquare/architectural-tuning-settlements-and-place-dwelling.

7. The cul-de-sac filled with cookie-cutter houses that turn their back on the street, disconnected from the surroundings by an ugly, unwalkable arterial, isolates us in our houses, reflecting the deep divisions in contemporary society.

8. Peter D. Norton, *Fighting Traffic: The Dawn of the Motor Age in the American City* (Cambridge, MA: The MIT Press, 2011). Also see Tom Vanderbilt, *Traffic: Why We Drive the Way We Do (and What It Says About Us)* (New York: Vintage, 2009).

9. "Now, in communities across the country, a movement is growing to **complete the streets**. States, cities and towns are asking their planners and engineers to build road networks that are safer, more livable, and welcoming to everyone. Instituting a **Complete Streets policy** ensures that transportation planners and engineers consistently design and operate the entire roadway with **all users** in mind—including bicyclists, public transportation vehicles and riders, and pedestrians of all ages and abilities."—from the website of the National Complete Streets Coalition at http://www.smartgrowthamerica.org/complete-streets.

ACKNOWLEDGEMENTS

FIRST AND FOREMOST, we thank our wives, Maricé Chael and Melanie Hoffman, for their guidance and infinite patience: Melanie was also an indefatigable proofreader. We are deeply grateful to Emily Glavey and Kenneth Garcia at Dover, Kohl & Partners, who tirelessly performed indispensable research, drawing, and editorial tasks. And we thank our editor, Alice Truax, who somehow got us to rewrite most of the book. Lauren Poplawski and Kerstin Nasdeo at Wiley brought us along step by step, and graciously let us tinker with the book at every stage.

Richard Driehaus and the Driehaus Charitable Lead Trust gave us a crucial, generous grant to research the book. Paula McMenamin, Carol Wyant, Eric Alexander, and Vision Long Island signed on early to support this. Many others helped with information on specific places or tours on our travels. Robert Russell and Macky Hill taught us Charleston history, Christian Sottile and Thomas Wilson expanded our understanding of the Savannah grid, and Chris Gray at the Office for Metropolitan History in New York was, as always, a font of information about his city. Stephane Kirkland knows all there is to know about the history of streets in Paris. Matt Shannon introduced us to street types in Chicago. In two of our favorite towns—Great Barrington and Nantucket, in Massachusetts—David Scribner, Andrew Blechman, and Andrew Vorce helped us.

Besides the enormous debt we owe our guest essayists for broadening the insights in this book, some of them helped us in other ways as well. Hank Dittmar at the Prince's Foundation gave crucial guidance at the earliest stage. Paul Murrain took us on an astounding day-long walk in London. Gabriele Tagliaventi helped us see Bologna in new ways. Douglas Duany not only showed us Orvieto and the medieval streets of Rome, he put up with us for two weeks. We are also in debt to Rebecca Martin, Jim Evarts, Thomas Massengale, Elizabeth Plater-Zyberk, and Laura Heery Prozes for putting us up in Barcelona, London, Miami, and New York.

Our colleagues in the Congress for the New Urbanism contributed their wisdom in many discussions. Peter Katz coached and prodded. Rick Hall shared his insights and unflagging enthusiasm about engineering (and engineers). Billy Hattaway, Norm Garrick, Dewayne Carver, and Gary Toth also pitched in their engineering wisdom. Bob Gibbs shared his great knowledge of the rules of successful retail. Elizabeth Plater-Zyberk and Rusty Bloodworth pointed out London streets we didn't know, and the pro-urb, Urbanists, and Trad-Arch listservs helped us kick ideas around. Fred Kent, Dan Burden, Mike Lydon, Richard Layman, Michael Ronkin and others did the same in extensive email discussions. The CNU's Project for Transportation Reform under Marcy McInelly established a thought-provoking foundation for this work. At Smart Growth America and the Complete Streets Coalition, we owe a debt to Geoff Anderson, Ilana Preuss, Barbara McCann, and Roger Millar. Beth Osborne at the US DOT and the DC Director of Planning Harriet Tregoning are two of the wisest voices in Washington when it comes to street design. In New York, Janette Sadik-Khan and Jon Orcutt have been smart, inspiring, and generous.

Many people contributed original or historic photographs. Several deserve at least a mention here as well: Sandy Sorlien, Steve Mouzon, Steven Brooke, Joseph Ip, Peter Pennoyer, Anne Walker, David Dixon, David Fishman, and James Mercer. Stephanie Sayre at the Iowa State University Library, Marie Henke at the Nantucket Historical Society, Robert Peterson at the Ingham County DOT, Nilda Rivera at the Museum of the City of New York, Todd Gilbert at the New York Transit Museum, Larry Gould at the Metropolitan Transportation Authority, the staff at HistoryMiami, James Labey at the Royal Borough of Kensington and Chelsea, and Tallulah Morris at the Crown Estate in London all made our work easier. James Dougherty, Kenneth Garcia, Megan McLaughlin, and Andrew Georgiadis at Dover, Kohl & Partners contributed photographs, and James worked on our designs for the Yorkville Promenade, Jane Jacobs Square and Winslow Homer Walk, along with our collaborator Zeke Mermell. The entire crew at Dover-Kohl persistently helped, especially Kristen Thomas, Justin Falango, and, of course, Joseph Kohl.

FURTHER ACKNOWLEDGEMENTS FOR THE SECOND EDITION

WE MUST BEGIN by recognizing the years of diligent work by colleagues at Dover, Kohl & Partners who contributed to the new edition. This most especially includes Robin Crowder, who managed and improved every page of the manuscript, researched many topics, fixed myriad details, tracked down the sources for little-known photos, drew beautiful new cross sections and maps, and built the elaborate 3D computer models of each street in the catalog of street types. Eva Klovatskiy, Shriya Dhir, and Lee Dover refined those models and produced the animated video clips for the website. James Dougherty painted (and repainted, and repainted) the illustration on the cover until it was just right.

Our editors at Wiley, Todd Green and Kavin Shanmughasundaram, displayed extraordinary patience with the expanded scope of the book and the many revisions it therefore required.

You may notice New York City figures prominently in the book, as it should. We are grateful to the many New York friends and colleagues we interviewed, cited, or corresponded with, including DOT Commissioners Polly Trottenberg, Hank Gutman, and Ydanis Rodríguez, Deputy Commissioner Michael Replogle, and traffic engineers par excellence Sam Schwartz and Michael Flynn (named Commissioner just as we go to press). Streetsblog NYC continued to be an invaluable resource: in addition to founding editor Aaron Naparstek (who went on to be one of the founders of The War On Cars podcast) and the other Aarons, Aaron Short and Aaron Donovan, we should mention Gersh Kuntzman, David Meyer, Dave Colon, and Kevin Duggan. We can't forget their colleagues at Streetopia UWS, Lisa Orman and Carl Mahaney, or the leaders at other nonprofits in New York: Ben Furnas, Transportation Alternatives; Andrew Berman, Village Preservation; Anthony Wood, New York Preservation Archive; Sam Turvey, ReThinkNYC; Sean Khorsandi, Landmark West!; Lynn Ellsworth, Tribeca Trust; and Layla Law-Gisiko and John West at the City Club of New York. And thank you to inveterate New Yorkers of many stripes: the late great Christopher Gray, Roberta Gratz, Jeremiah Moss, Sam Stein, Gib Veconi, Kim Phillips-Fein, Alice Blank, George Janes, Nicole Gelinas, Heather Boyer, Damon Hemmerdinger, Jonathan Rose, Stephanie Azzarone, Martin Pedersen, Kurt Andersen, Tony Hiss, Ham Fish, Catherine Hughes, Kate Ascher, Alice Shays, and Michael King.

Colleagues from around the rest of the USA lent much-needed help, too, including advice, interviews, inspiration, and their own research, especially John Simmerman of Active Town Initiative, Prof. Bruce Stephenson, Scott Bernstein, Jessica Keller, Karen Christensen, Carie Penabad, Steven Semes, Mark Hewitt, Steve Wright, Andy Boenau, Jeffrey Tumlin, Angie Schmitt, and Michael Mehaffey of IMCL, Mallory Baches of CNU, members of the Urban Guild, and Vince Graham.

Thinkers from the United Kingdom from whom we received photos, assistance, and hours of interviews include Ben Bolgar of the King's Foundation, Paul Murrain, Matthew Hardy, Phil Jones, and Will Norman, London's Walking and Cycling Commissioner. Colleagues in the Netherlands who inspired ideas in the book include Melissa and Chris Bruntlett, Meredith Glaser of Urban Cycling Institute, Ronald Tamse, Henk Groenewegen, Ton Schaap, Peter Groenendaal, and Hans Karssenberg. Mikael Colville-Andersen of Copenhagen added a perspective both global and Danish to our understanding of streets. Peter Elmlund, Tigran Haas, and the Ax:son Johnson Foundation helped us in Sweden.

Finally, we thank Carlos Moreno, who wrote the new Foreword, and Elizabeth Plater-Zyberk, who wrote the new Afterword—both of us will be forever grateful for the many ways Lizz helped and encouraged each of us over the years. We also thank our colleagues from four continents for creating new case studies and essays for the Second Edition. Their writings frame the book, place it in a broader context, and point to new ways it can be used, now more than ever.

CHAPTER ONE

STREET DESIGN MATTERS

THE DESIGN OF CITIES begins with the design of streets. To make a good city, you need good public spaces, and that includes streets where people want to be. Streets should be safe and comfortable, useful and interesting,

◄ **Figure 1.1:** Broad Street, New York, New York. Looking north towards Wall Street around 1905. Broad Street has a long and varied history. Built in the seventeenth century by the Dutch West Indies Company, it once had a canal in the center and was named *Heere Gracht* (Lords' Canal). In 1676, the company filled the canal and created a public market space. George Washington's inauguration as the first President of the United States was at the head of the street, on Wall Street, and the infamous New York City Draft Riot filled Broad for four days in 1863. Four decades later, bankers and brokers traded stocks and bonds in the street, leading to the construction of the New Stock Exchange Building on the west side of the street (where it still stands). Today, Broad is car-free once again and one of the great public spaces in America. *Library of Congress / Detroit Publishing Company Photograph Collection*

and lively and attractive in every sense of the word. They need to be places for people.

People are social beings, who function best in societies with inviting public spaces. Over time, we have learned how to make cities and streets that do that. Think of the Roman forums, the New England town commons, or the boulevards of Paris.

In the last one hundred years, however, America led the way in changing the way we build and use streets, making them spaces where cars come first. The changes degraded what we call "the public realm," which James Howard Kunstler defines as "the physical manifestation of the common good." And, he adds, "When you degrade the public realm, the common good suffers."[1]

In this chapter, we explore the history of streets, how streets go together to make cities, how we experience streets, why we stopped making streets for people, and then present models for better streets in the twenty-first century. Chapters Two, Three, and Four are about historic streets, new streets, and retrofitted streets, organized by the ten essential street types we introduce at the end of Chapter One. In each chapter, we discuss context, principles, and step back for the bigger picture. Case studies present exemplary streets. Some case studies have pull quotes that highlight important principles or ideas.

1

In the final chapter, we ask "Whose Streets?" and talk about Streets for People.

If you want to know what a culture values, look at what it builds. Our cities and towns are some of the greatest works of human civilization. They all began with their streets.

THE CITY AND THE STREET

We often think of buildings when we think of urban design—as we should. Great streets require great buildings. Good streets can get by with merely good buildings. But most importantly, city streets are the spaces between the buildings, and those spaces need the art of *placemaking*. Placemaking turns public spaces into places where people want to be. A space is not a place unless there are people in it.

We will look at great streets in this book and explore what made them great places. We personally visited every street we wrote about. That is important. When you visit these streets, as we hope you will, you will find that some are exactly what you expect from the images and discussion here. But some might be different than you expect, and you won't fully understand many of them until you walk them yourself.

We hope you will understand them better after reading this book. Some urban designers have told us they keep the first edition close at hand, and that they always pull it out when they're designing a street. But having said that, we want to repeat that the experience of being on the street will give you sensory information you can only get in person.[2]

Many of the streets presented here are beautiful, so let's consider the idea of beauty for a moment. Scientists, neuroscientists, sociologists, artists, poets, and even New Urbanists (who use various types of community surveys during the public charrette process) have been gathering evidence the last few decades that refutes the Modernist idea that beauty is in the eye of the beholder. We all know beautiful places when we see them. If we walk into the Piazza San Marco in Venice, no one has to tell us it's beautiful. We experience places with our bodies, our senses, and our minds. We say more about this in the final chapter, and include some notes about beauty at the end of this chapter.[3]

Figure 1.2: Ponte Vecchio, Florence, Italy. Taddeo Gaddi, 1345; Giorgio Vasari, 1565. Looking north towards the Via Por Santa Maria. We are social beings: put interesting things in beautiful settings, and we enjoy gathering there.

People experience beauty on many types of streets, from small-town Main Streets to big-city boulevards and country lanes. We will talk about the full range of streets, but most of our examples will be in cities and towns. When the buildings and trees lining urban streets

give a sense of enclosure, and the proportions and details make a harmonious whole, these streets become places where people want to linger, sharing a common experience with their neighbors and fellow citizens.

Tragically, we rarely build streets like that today. The overwhelming majority of the streets in America have been built since World War II, and most were made for cars rather than people—like the seven-lane arterial road in the middle of nowhere lined with strip malls, shopping malls, big box centers, and the other detritus of modern suburban life (Figure 1.3). These cheaply built, poorly designed sites and buildings do not feel like authentic places. "There is no there there," Gertrude Stein famously wrote.[4] The roads are what Kunstler calls "auto sewers"—suburban "thoroughfares" sized by engineers to make the traffic flow like water in a pipe. But it feels more like sludge in a sewer pipe.

No one walks on these streets if they don't have to (Figure 1.4). The problem with the streets is not just their location, far from anything except other shopping centers and big box stores. Their design and construction are bad for people too. The scale is vast and frightening. Speeding cars roar by, there are large swales where the sidewalk should be, and crossing the street is difficult, with long expanses between traffic lights. Even when you reach your destination, you still must cross a large parking lot with no sidewalks or shade trees. It's all ugly, and it's all depressing.

Fortunately, after decades of fleeing cities and old towns, Americans have embraced walkable towns and neighborhoods again. There's a common understanding that the automobile-based building patterns made a physical environment inferior in many ways to the old pedestrian-based model: we need to remake our cities, towns, and streets for people. Accordingly, the federal and local governments appropriate billions of dollars in a well-intentioned—yet scattered and intermittent—effort to rebuild the nation's roads.

Figure 1.3: An "auto sewer" arterial that, except for the palm trees, could be Anywhere, USA. "The road is now like television, violent and tawdry. The landscape it runs through is littered with cartoon buildings and commercial messages. We whiz by them at fifty-five miles an hour and forget them, because one convenience store looks like the next. They do not celebrate anything beyond a mundane ability to sell merchandise. We don't want to remember them. We did not savor the approach, and we were not rewarded upon reaching the destination, and it will be the same next time, and every time. There is little sense of having arrived anywhere, because everyplace looks like no place in particular."— James Howard Kunstler, *The Geography of Nowhere*

Figure 1.4: A placeless cul-de-sac: a residential auto sewer in Anywhere, USA. Sprawl isolates us and contributes to the deep divisions in American society today. *Courtesy of Megan McLaughlin*

Figure 1.5: Steenstraat, Bruges, Belgium. Streets are the public spaces between the buildings. The harmonious whole is more important than any individual element. The great American landscape architect and urban designer Frederick Law Olmsted called this the Principle of Subordination.

> Our streets and squares make up what we call the public realm, which is the physical manifestation of the common good. When you degrade the public realm, the common good suffers.
>
> — James Howard Kunstler

Less encouraging is that many of the professionals involved in remaking our streets bring with them the criteria and biases of their specialties, which frequently prevents them from designing streets where people want to be. Bicycle specialists, pedestrian specialists, transit specialists, and even Complete Street specialists may understand the need to add a bike lane or a streetcar, but they rarely understand placemaking or the importance of the public realm. The professionals in charge usually continue the late-twentieth-century pattern of allocating most of the space in the street to the motor vehicle and its movement—now with bicycle and bus lanes. They introduce innovations that make the street safer for those riding bikes or even traveling on foot, but at the same time, they regularly diminish the experience of walking.

THE TRADITIONAL STREET

The history of urban design and street design in Western civilization has its roots in ancient Rome and Athens. For the Greeks and the Romans, the city was the place where men and women came together to make a good and civilized life. The words "civil," "civilization," and "citizen" come from the Roman word for city, "civitas." From the ancient Greek word for city, "polis," we get "polite," "political," and "police," which reflect the classical idea that the city was a political body of citizens, as well as the place where they politely came together to create civilization. For centuries, the first job of the architect when designing a new building was to make or reinforce a public realm that supported those ideals.

Figure 1.6: Via Appia Antica, Rome, Italy. Appius Claudius, 312 BC. "The Queen of the Long Roads." In the ancient world, street design became a hallmark of civilization. The Roman Empire was famous for the vast network of roads that tied it together. More than two thousand years later, we still say, "All roads lead to Rome"—reflecting the functional and symbolic importance of the ancient routes.

Figure 1.7: Old New York Police Headquarters, 240 Centre Street, New York, New York. Hoppin & Koen, 1905–1909. View looking south from Cleveland Place. A deflected vista in the late afternoon sun in the fall. The dome marks the old Headquarters as a civic building and visually connects it to the United States Courthouse (Cass Gilbert, 1929) and the New York Municipal Building (McKim, Mead & White, 1907) in the distance. "Architecture is the learned game, correct and magnificent, of forms assembled in the light." — Le Corbusier

Figure 1.8: Paris, France. An aerial photograph of the Right Bank from the early twentieth century. The Church of the Madeleine serves as the terminating object for four avenues. "The boldest conception of civic art makes it embrace not merely individual groups of buildings with their approaches and gardens but even entire cities. It is one thing to distribute fine groups of public buildings over the area of a city and to connect them effectively. It is a much more difficult thing to relate the entire city to such a scheme." — Werner Hegemann and Elbert Peets, *American Vitruvius: An Architects' Handbook of Civic Art*

Ancient Romans emphasized the importance of the *res publica* (the public realm) as the place where the citizens came together in the *polis*. It was shaped by the buildings in the private realm (*res privata*). In *The Architecture of Community*, the architect and urban designer Léon Krier uses diagrams to show that each realm is incomplete without the other, while the two combine to make the complete city (Figure 1.9).[5] Besides open space (streets, squares, and parks), the public realm includes public buildings such as churches and town halls. Much of the art of traditional urban design and town planning consists of two things: shaping and programming the public realm into a place where pedestrians want to be, and strategically placing public buildings (such as a market, a place

of worship, or a theater) so that they are understood to be more important than the private buildings.

In the modern world, we also have a large semipublic domain of stores, businesses, and places of entertainment, such as movie houses, restaurants, and nightclubs. Office buildings now frequently tower above the church steeples that used to be the tallest structures, and corporate headquarters like the Chicago Tribune Tower or the Woolworth Building in New York are distinguished from speculative office buildings by their monumentality and ornate architecture. All these buildings play a part in making urban places where people want to be. Some of these spaces were meant to inspire a sense of grandeur; others were designed to be intimate. Most importantly,

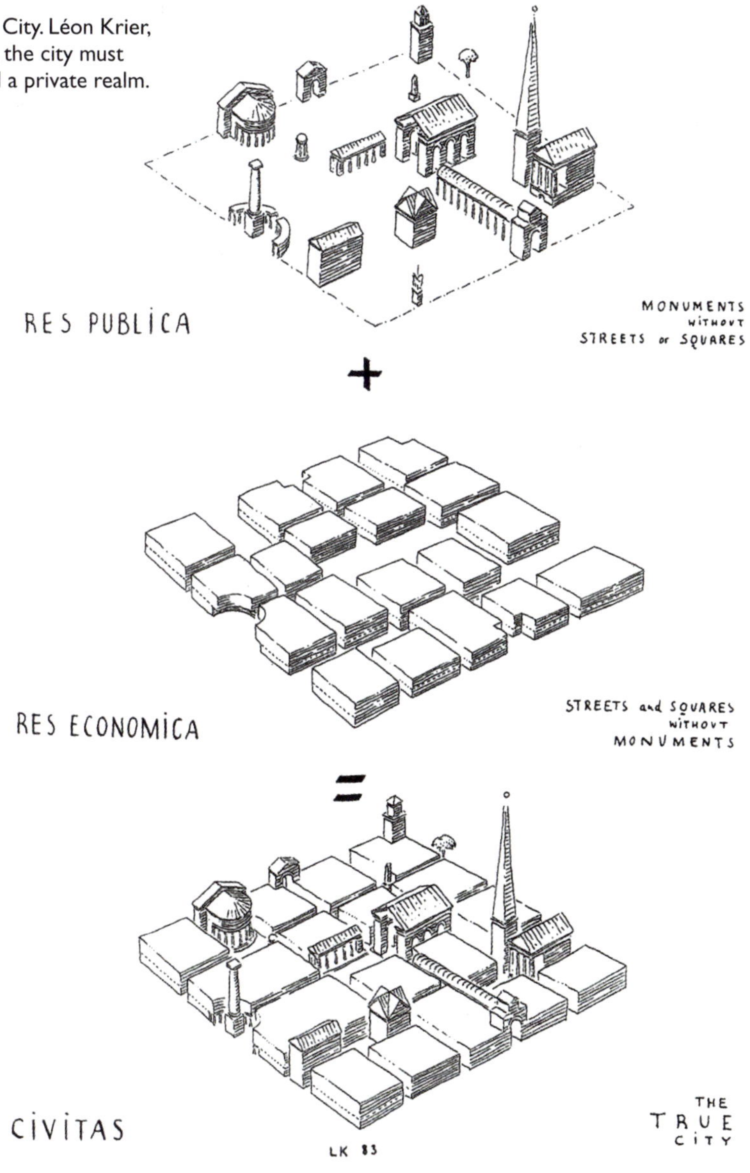

Figure 1.9: The True City. Léon Krier, 1983. To be complete, the city must have both a public and a private realm. *Courtesy of Léon Krier*

whether we are strolling through the ruins of the Roman Forum or exploring the streets of Back Bay Boston, all of these places were built to a human scale.

THE GRID

An American book about street design must mention the grid, however briefly. The rectilinear grid has been used to plan towns and cities since at least the fifteenth century BC, when the Chinese started a tradition of gridded plans they still employ today.[6] In the Western World, the use of rectilinear grids for town plans goes back to at least 2600 or 2500 BC, and the Romans institutionalized a standard gridded plan for the places colonized by the Roman Empire. Roman cities, fortified garrisons, and colonial outposts were designed around the famous *cardo* and *decumanus*. *Cardi* were north–south streets, and *decumani* east–west streets: the two central axes were the largest streets, known as the *cardo maximus* and the *decumanus maximus*. Where these crossed at the center of the town, there was normally a forum or public square. The most important streets in many European, Middle Eastern, and North African cities and towns today are still where the Romans built their *cardo maximus* and *decumanus maximus*.

Many early towns and cities in America were laid out by commercial interests that saw the grid as an efficient, simple way to divide open land into rectangular lots with clear boundaries that allowed the easy establishment of title. The seventeenth- and eighteenth-century settlements typically had level sites, often by a river or along the coast. There was not much topography on the flat sites to impede easy implementation of the plans, which frequently ended raggedly along the uneven shorelines. The grandest versions of these plans were in Philadelphia (1682) and Savannah (1733).

Philadelphia's influential plan (Figure 1.11) started a tradition for American gridded plans: the north–south streets were numbered, while the east–west streets were named after trees.

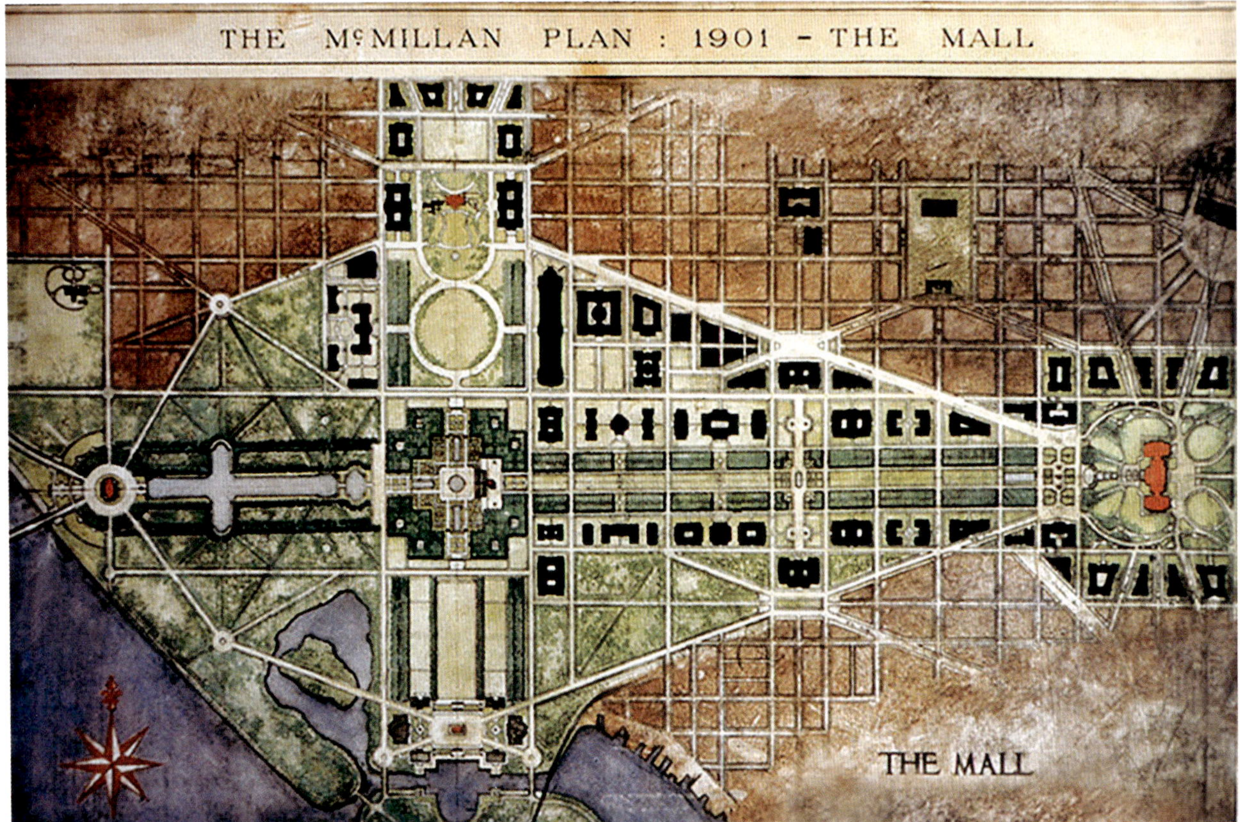

Figure 1.10: McMillan Plan, Washington, DC. Senate Park Commission (Daniel Burnham, Frederick Law Olmsted Jr., Charles F. McKim, *et al.*), 1901. At the peak of the City Beautiful movement, the Senate Park Commission hired leading architects and landscape architects to restore the clarity of Pierre Charles L'Enfant's 1791 plan for Washington. The Senate Park Commission was better known as the McMillan Commission, in honor of Senator James McMillan, whose Chief of Staff led the effort. *National Capital Planning Commission / Public Domain*

Savannah and Philadelphia started by the edge of rivers and took many decades to grow into their expansive plans. The plans for Philadelphia and Savannah (Figure 3.2 on page 258) included regularly repeated squares; in Philadelphia, one square was rented to a lumberyard until the city grew up around it. Savannah had a rich and varied plan that made it easy for the city to grow over time (see "The Streets of Charleston and Savannah" in Chapter Three).

The predictability of how a grid will shape development is another of its advantages. In 1811, New York's city fathers platted a grid across the island of Manhattan, which was still mainly covered by farmland, woods, and wetland. As in other American towns established by commercial interests, the grid simplified surveying and selling lots with clear titles. Land speculation started

immediately, and the population of Manhattan alone multiplied almost fourteen times before the end of the century. Speculation and growth were helped by the fact that the Commissioners' Plan of 1811 continued block sizes already in use in lower Manhattan, so that building types developed there could be easily used in new parts of the city. "A city is to be composed principally of the habitations of men, and that strait sided, and right-angled houses are the cheapest to build, and the most convenient to live in," a Commissioners' Plan report said.[7]

Grids were not only used in planning cities and towns. In 1785, the U.S. Congress passed a land ordinance that expanded on a similar one drafted by Thomas Jefferson the year before. The 1785 Land Ordinance drew a rectilinear National Grid across the country, dividing it into six-mile-square "townships." The squares were

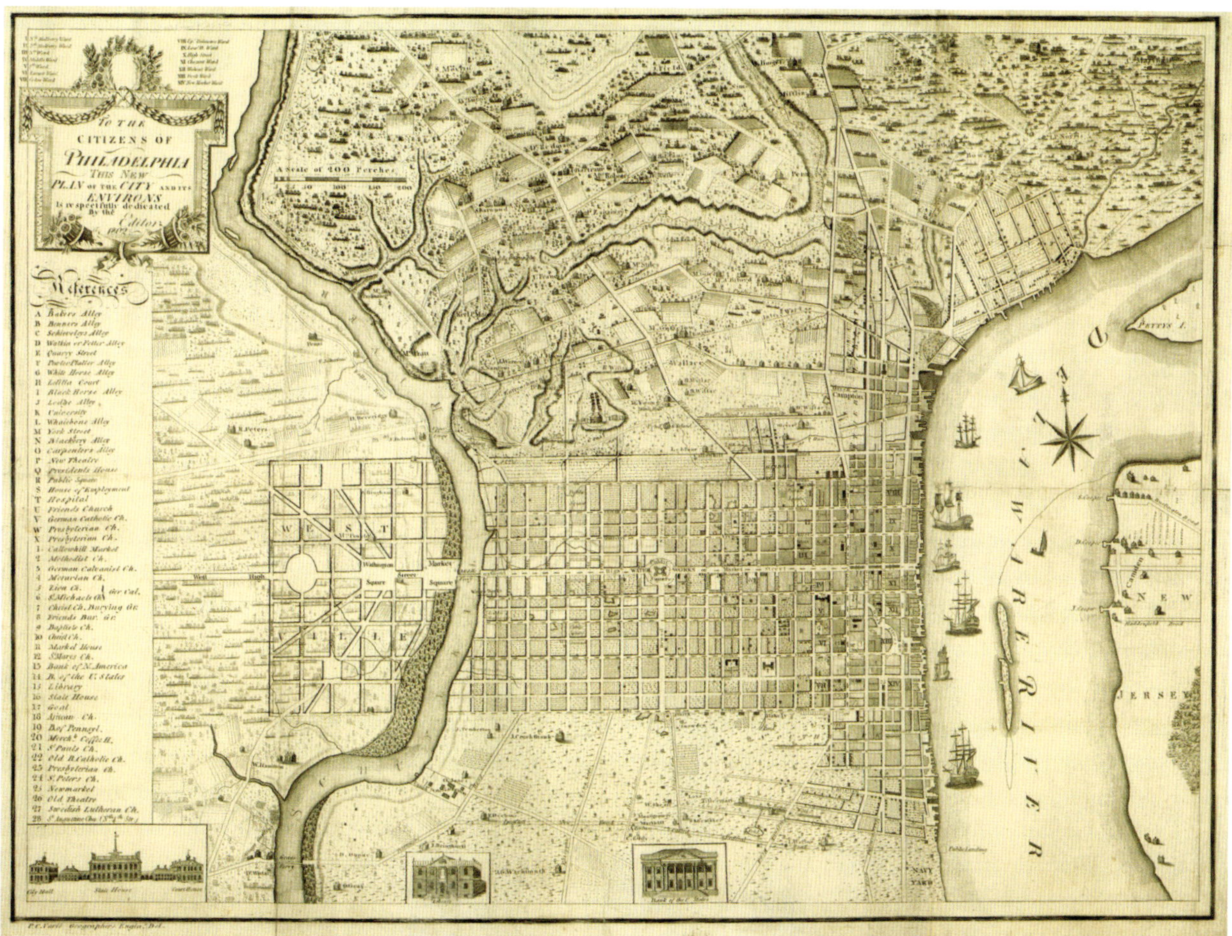

Figure 1.11: Philadelphia, Pennsylvania. William Penn, 1682. A plan of Philadelphia published in 1802. An early and influential American grid. *P. C. Varte / Public Domain*

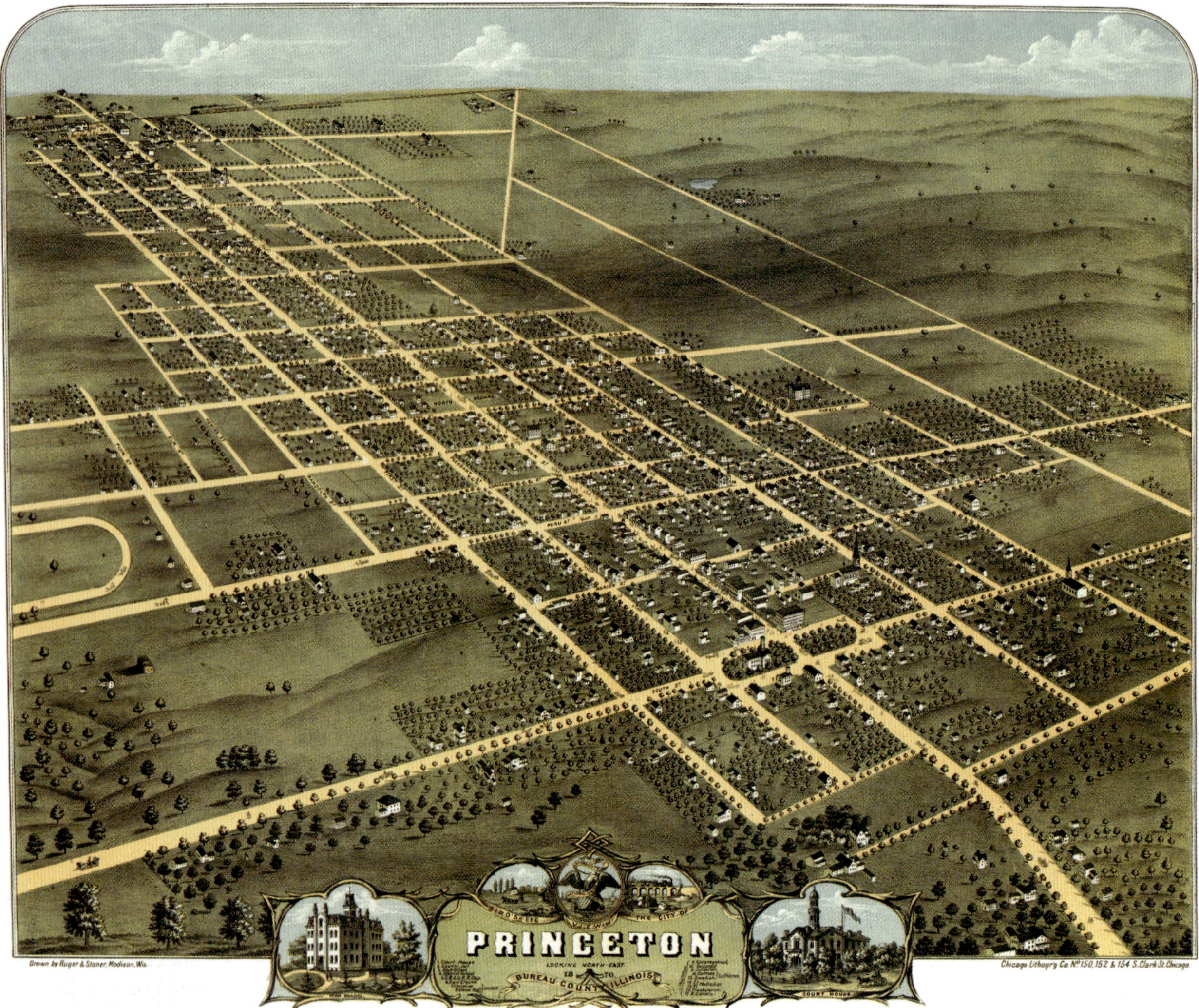

Figure 1.12: Princeton, Illinois. Bird's-eye view of Princeton in 1870. A perspective map looking northeast, not drawn to scale. The 1785 Land Ordinance laid the National Grid across undeveloped land in the Midwest and the Great Plains and led to the platting of gridded towns and cities. Companies across the country published and sold bird's-eye views like this in the nineteenth century. *Library of Congress / Ruger Map Collection*

further divided into thirty-six square sections of 640 acres each. Those sections could be sliced into half-sections, quarter-sections, etc., down to a dimension of sixty feet across (Figure 1.12). Congress established a plan for surveying and laying out the grid and then sold much of the land for a minimum of $1 an acre.

Satellite views of the Midwest and the Great Plains today clearly show the National Grid laid across the land. New towns and cities like Chicago were planned within the framework of the grid, and modern suburban and

exurban arterials frequently followed the National Grid. In the west and southwest, however, the grid sometimes ran into conflict with development patterns in lands formerly belonging to Spain. The royal regulations for Spanish colonies, the Laws of the Indies, governed the planning for new towns there. These included directions for siting and laying out towns with central squares in a rectilinear grid of twelve streets.

Santa Fe, New Mexico, was planned under the Laws of the Indies. It is one of the places in America where market demand for second houses for the rich has driven house prices ridiculously high, because of the beauty of the town. Like many classic American small towns, it has a simple grid. Wiscasset, Maine; Easton, Maryland; Marshall, Michigan; and Virginia City, Nevada, are a few of many exemplars across the country. Mayberry, North Carolina, and Bedford Falls, New York, are two fictional examples on television and in the movies. They all have good Main Streets where people enjoy walking, surrounded by simple grids with comfortable single-family houses on tree-lined streets (Figure 1.13).

Not all small towns with grids are classic places where we want to be, of course. In *American Notes*, Charles Dickens said Philadelphia is "a handsome city, but distractingly regular. After walking about for an hour or two I felt that I would have given the world for a crooked street."[8] A grid built without thought and attention to detail can be a rigid, unpleasant gridiron, but there are simple ways to soften the grid.

Taming the Grid

1. A beautiful tree canopy over a street is one of the easiest ways to soften a grid. When New Haven, Connecticut had so many American Elms that it was known as "Elm City", it was a more beautiful city.

2. A variety of street types and street widths combined with short blocks and squares enriches a grid. Savannah is the supreme example (see Figure 3.2 on page 258 in Chapter Three).

3. In a town or city with gentle slopes, slightly bending the streets to follow the topography softens the grid (Figure 1.14). A traditional design technique for city streets is to change the street's slope or direction only at an intersection with another street. As seen on Nassau Street in New York between Fulton Street and Wall Street, this produces a beautiful spatial effect.

4. When a town or city has dramatic hills, like San Francisco, running the grid up and down the hills with little deformation can be visually and experientially interesting. Even in Manhattan, where many hills were shaved down as the grid was built, the most pleasant parts of the long avenues frequently combine hills and plateaus to draw attention away from long, unterminated vistas. Madison Avenue in Carnegie Hill between 86th Street and 96th Street

Figure 1.13: Old South Road, Southport, Connecticut. Looking east from the intersection of Old South Road and Pequot Avenue. This classic American street near the Southport train station and the village center has large front yards and a single sidewalk on one side of the road.

Figure 1.14: Capitol Street, Charleston, West Virginia. The grid surrounding Capitol Street is almost orthogonal, but the streets bend slightly here and there, always in accordance with the city's gentle topography.

Figure 1.15: Avenue Foch, Paris, France. View looking west from the Arc de Triomphe. The broad axial avenue brings the green of the Bois de Boulogne into the heart of the city in a single, unified sequence. The French state used formal geometry and a monumental scale to communicate rationality and grandeur, symbolizing the order and power of imperial France.

and Lexington Avenue in Lenox Hill above and below 72nd Street are examples.

5. Shifting the grid can make shorter streets and add variety and richness to the plan. Sidney Place in Brooklyn, discussed in Chapter Two, is an example.

6. Laying diagonals across a grid, as in Pierre L'Enfant's plan for Washington, DC, adds interest if the design skillfully includes important buildings or monuments on the prominent sites created by the diagonals (Figure 1.10). Diagonals in a grid can also be disorienting, however, and produce awkwardly shaped lots that require custom building plans.

7. Streets that open to the surrounding area can draw the landscape into the town. Examples included streets in Santa Fe that point at the mountains and the crosstown streets in Manhattan that open to the Hudson and East rivers.

THE FORMAL, THE PICTURESQUE, AND THE HYBRID

City plans and street designs have historically fallen into three main types: formal plans, picturesque plans, and hybrid designs that combine the two. Preferences for the formal and the picturesque have alternated throughout history. As we have seen, the use of the simple grid goes back to at least the fifteenth century BC, but before the classical world of the ancient Greeks and Romans, many human settlements were planned and built in informal ways. In ancient Rome and Greece, Classical buildings were frequently used in both formal and informal plans, and we now know that during "the Dark Ages" between the fall of the Roman Empire and the beginning of the Italian Renaissance, planning was not exclusively what is now often called "medieval." Design tendencies became more formal as the Renaissance spread around the Western World, but in the nineteenth century, there was a swing towards eclecticism, romanticism, and the picturesque. And yet the nineteenth century also had formal designers, planners, and architects who consciously combined the two, as we shall see.

In the mid-nineteenth century, eclectic designers like the great landscape architect Frederick Law Olmsted used both formal and picturesque designs, depending on the context. Olmsted's plans for the streets of Llewellyn Park, New Jersey, and Riverside, Illinois, were winding and romantic, and his plans for Ocean Parkway and Eastern Parkway in Brooklyn, New York, were nineteenth-century American interpretations of the French *Grands Boulevards*. His designs for Prospect Park in Brooklyn and Central Park in Manhattan had informal and formal elements, like the Central Park Promenade. The Promenade, now known as the Mall or the Poet's Walk, is a long, straight, and level walk lined by majestic elms. The widest walkway in Central Park, the Promenade ends in a grand stair that descends to the Bethesda Terrace and Fountain. The stair is on axis with the Mall, another measure of formal design, and the statue of an angel that crowns the fountain is too. From the south, the angel visually "terminates" the axis—another hallmark of formal planning (Figure 1.16).

The Formal

The simple American grid was the work of the surveyor and the engineer. The grid has an obvious rectilinear formality, but it is frequently untouched by the principles of formal urban design. As we have seen, these Classical principles—order, harmony, balance, and legibility—shaped the form of great cities and buildings for thousands of years. In the late nineteenth century, the City Beautiful movement set out to reshape American towns and cities with the type of formal order found in the history of Classical architecture and urbanism in Europe. Most of the founders of this movement studied architecture and urbanism in Paris at the École des Beaux-Arts, where they saw firsthand the city's great avenues and boulevards from the eighteenth and nineteenth centuries. The students also studied and visited the landmarks of ancient Greece and Rome, the cities and buildings of the Italian Renaissance, and European Baroque design.

They brought all this back to America in the form of what they called Civic Art, which combined urban design, street design, building design, and the design and placement of monuments and sculptures in the city. They liked the American grid as a starting point but wanted to enrich it with a hierarchical range of streets and prominently sited civic buildings.

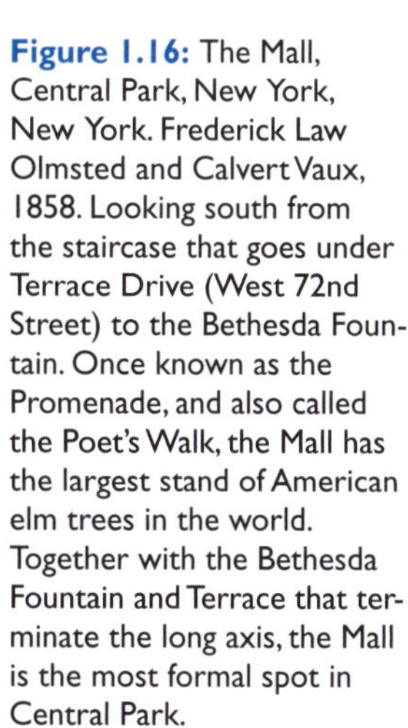

Figure 1.16: The Mall, Central Park, New York, New York. Frederick Law Olmsted and Calvert Vaux, 1858. Looking south from the staircase that goes under Terrace Drive (West 72nd Street) to the Bethesda Fountain. Once known as the Promenade, and also called the Poet's Walk, the Mall has the largest stand of American elm trees in the world. Together with the Bethesda Fountain and Terrace that terminate the long axis, the Mall is the most formal spot in Central Park.

CITY BEAUTIFUL

The first full-blown expression of the American City Beautiful movement was the World's Columbian Exposition, built in Chicago in 1893 with temporary white plaster buildings that transformed America with a Beaux-Arts-influenced Classical revival. City Beautiful philosophy included social and economic justice ideas that connected it to the Progressive movement in American politics. Master plans for American cities, like the 1909 Plan of Chicago, included housing reform and the planning of public parks, as well as formal interventions in simple grids, with new squares and tree-lined avenues visually terminated by important civic buildings.

Daniel Burnham—one of the founders of the City Beautiful movement, the organizer of the World's Columbian Exposition, and author of the 1909 Chicago Plan commonly known as the Burnham Plan—famously said, "Make no little plans." Many of the best American cities owe much of their present character to work built during the City Beautiful movement. In New York City, City Beautiful monuments include Grand Central Terminal; the Metropolitan Museum of Art; many of the great skyscrapers, like the Woolworth Building and the Municipal Building (Figure 1.17); the New York Public Library on 42nd Street and many of the branch libraries; the great public schools, parks, and playgrounds; and swaths of rowhouses and apartment buildings in all five boroughs.

Since this is a book about street design, however, it must be said that many of the street designs in the most ambitious City Beautiful plans were never executed, except in small parts here and there, even though American cities were undergoing enormous growth and had tremendous wealth. Of the great diagonals Burnham planned for Chicago, only one was constructed, while the City Beautiful streets designed for Manhattan by America's greatest architects exist only on paper. Two notable exceptions are the streets around the Mall in Washington, DC, redesigned as part of the McMillan Plan, and the Benjamin Franklin Parkway in Philadelphia, designed by the French landscape architect Jacques Gréber. The goals of the parkway design were broad enough to encompass slum clearance and the creation of sites for new civic buildings, including the Philadelphia Museum of Art and a new city library.

The construction of individual City Beautiful buildings and civic centers around the country (San Francisco and Cleveland have prominent examples) was more common. In retrospect, the City Beautiful civic centers were sometimes bad for street life, because they concentrated civic buildings around large plazas that drew pedestrians from the streets without filling the plazas. Almost a century after the Franklin Parkway opened in Philadelphia, the wide boulevard still has little street life most of the time.

The Picturesque

The canals and *calli* of Venice that disappointed McKim (see the City Beautiful box above) epitomize picturesque planning. The father of modern picturesque planning, Camillo Sitte, said about the Piazza San Marco, "So much beauty is united on this unique little patch of earth, that no painter has ever dreamt up anything surpassing it." After describing the elements and details that make the piazza, he continued, "However, it is the felicitous arrangement of them that contributes so decidedly to the whole arrangement. There is no doubt that if all these works of art were disposed separately according to the modern method, straight in line and geometrically centered, their effect would be immeasurably decreased."[10]

Sitte studied medieval towns and cities for his book, *City Planning According to Artistic Principles,* published in 1889. From his studies of medieval urbanism, he developed underlying principles of design for what we call picturesque urbanism (Figure 1.18). This was not medieval design but, rather, design based on the spaces that Sitte found pleasing in medieval streets, towns, and cities. He presented his ideas in words, plans, and perspective drawings, emphasizing that the goal was not to reproduce old places but to make new places with experiences as good as those produced by the best medieval designs. He also emphasized that understanding how to build engaging three-dimensional spaces was the key to creating picturesque urbanism. Irregular medieval squares and streets, he said, were more pleasing than the geometrical designs in

Figure 1.17: Manhattan Municipal Building, 1 Centre Street, New York, New York. McKim, Mead & White, 1907. Painting by Colin Campbell Cooper, 1922. Chambers Street, which is visually terminated by the Municipal Building, once continued through McKim, Mead & White's grand colonnade and arch. *The New-York Historical Society / Wikimedia Commons / CC-BY 2.5*

Infrequently visited parks separate the parkway from the low-density neighborhoods surrounding it, and the cultural institutions along the parkway don't draw much foot traffic. New Urbanists usually advocate locating civic buildings around the city to increase pedestrian traffic and maximize the use of civic buildings as civic monuments on important sites.

A story about three of the most influential founders of the City Beautiful movement demonstrates their fondness for the formal. In 1901, the U.S. Senate authorized the hiring of Burnham, Charles McKim, and Frederick Law Olmsted Jr. (the son of Frederick Law Olmsted, who with his half-brother continued their father's landscape architecture practice under the name Olmsted Bros.) to study the deterioration of the Mall in Washington, DC, and to make recommendations about how to address it. With a Senate aide, the three set off on a six-week tour of Europe to study "French planning" and "Italian architecture." After touring Paris and Versailles, they traveled to Venice, where McKim was unhappy because there was so little of the axial planning he enjoyed in Paris. Wandering the small, meandering passages of Venice, McKim became separated from the group. When Olmsted asked how they would find him, Burnham replied, "We will go to the Piazza San Marco and find him on the axis."[9] They did, and McKim was there.

Figure 1.18: Marché aux Poulets, Brussels. Drawing by Camillo Sitte, circa 1885. The composition of the gently curving street makes a space that deflects the view rather than formally terminating the vista with an object. The walk down the street presents a series of pictures and dynamic, unfolding events rather than a static scene.

later cities. He was particularly critical of contemporary engineering and formal designs like those favored by the City Beautiful movement.

Important to Sitte was the idea that the experience of one's surroundings was continuous through space and time. In a discussion of medieval street design in the Etruscan hill town of Orvieto, landscape architect and urban designer Douglas Duany talks about the vernacular mind of medieval Italy, which understood streets as a sequential experience (see the essay "Orvieto, Italy" in Chapter Two). For his era, Sitte developed his designs as a series of picturesque compositions, drawn in perspective at different points along the way. Thus, they were "picturesque" in two ways: as painterly compositions and as compositions that were asymmetrical and dynamically balanced rather than static, centered, and regular. But it is important to remember that his individual drawings were meant to be seen almost like a flip book. Picturesque designers today frequently "walk" their plans block by block, picturing the views at street level and adjusting angles and elements in the view to make them more pleasing and picturesque (Figure 1.18).

The Hybrid

In 1909, the influential English architect and planner Sir Raymond Unwin wrote one of the twentieth century's most important urban design books, *Town Planning in Practice*. After a discussion of plans from different cities and towns, he commented on the formal and the informal:

> We can hardly have examined the many different town plans referred to in the last chapter without realizing that in spite of their great variety, they fall into two clearly marked classes, which we may call the formal and the informal, and that there are to-day [*sic*] two schools of town designers, the work of one being based on the conviction that the treatment should be formal and regular in character, while that of the other springs from an equally strong belief that informality is desirable. From the views given of both types of towns, we should almost certainly agree that a high order of beauty has been obtained by each method, for although our personal preference may lean strongly to one or the other type, there will be few who will not admit great beauty in many of the examples of its opposite.[11]

Unwin was one of the town planners involved with the Garden City movement, which combined the formal and the informal in hybrid plans—usually for new towns with formal centers and streets that became increasingly

Figure 1.19: First Street, Gainesville, Florida. View looking south towards the Hippodrome State Theatre (Thomas Ryerson, 1911). A terminated street vista is perhaps the oldest, most dependable tool in the urban designer's kit; the axial geometry gives civic importance.

Figure 1.20: High Street, Oxford, England. William Wordsworth praised "the stream-like windings of that glorious street," one of the most beloved streets in England—and a wonderful example of a deflected vista. Contributing to the beauty of "The High" are the lean-in tree on the left and The Queen's College (Nicholas Hawksmoor, 1708–1710), which sits on a bend just beyond (for a plan of Oxford and the High Street, see Figure 2.113 on page 170). The planner Thomas Sharp described the sycamore as "one of the most important in the world: without it, the scene would suffer greatly."

informal as they radiated outward. The Olmsted Bros. plan for Forest Hills Gardens is an American example of the hybrid type (see Figure 3.53 on page 298). As used there, the hybrid plan illustrates what New Urbanists today call a Transect (discussed in Chapter Three in "The Transect Observed"), and it is common for New Urban designers to make similar, Transect-based hybrid designs. New Urban firms might also choose to do either a formal *or* an informal plan—based on the context or the client's preference.

> "There are two kinds of music," Duke Ellington famously said. "Good music, and the other kind." The same can be said about design.

Formal, picturesque, and hybrid designs can all be well or badly done, and *Street Design* does not advocate one over another. Some might prefer the music of Wolfgang Amadeus Mozart to Richard Wagner or vice versa, but they are both musical geniuses. "There are two kinds of music," Duke Ellington famously said. "Good music, and the other kind." The same can be said about design.

Many believe that beauty is in the eye of the beholder and that there can be no way to accurately define what individuals find beautiful. Recent studies, however, show that if individuals walk a prescribed city route with a map in hand and mark the places they like and don't like, there will be a high degree of correspondence in their preferences: the results show a consensus about what is beautiful, what is ugly, and how we respond to beauty and ugliness.[12] If the group sample is large enough, there will also be distinct patterns—some people will prefer more formal spaces, and some will like more picturesque places, for example. However, the favorite and least favorite places will still be consistent. For the purposes of urban design, street design, and this book, what is of greatest importance is that *all* the groups show a preference for the places made according to the principles of placemaking illustrated here, whether formal or informal.

As we wrote in the Preface, this is not "just aesthetics." Paraphrasing Winston Churchill, we shape our cities and thereafter they shape us. Cities, towns, and neighborhoods affect how we work, live, and play. They influence our happiness. They can make us feel lonely and sad. Cities are at the center of civilization and are among our greatest works of art. Compact, walkable cities are vital for the future of the planet. The design of cities begins with the streets. Let's look at one in New York City.

Figure 1.21: Mercato di Mezzo, Bologna, Italy. When a street has a sense of enclosure, and the proportions and details form a harmonious whole, it becomes a setting for sharing a common experience with our friends and neighbors.

EAST 70TH STREET: A BEAUTIFUL NEW YORK BLOCK

Legend has it that inveterate New Yorker Woody Allen calls the block of East 70th Street between Park Avenue and Lexington Avenue on Manhattan's Upper East Side the most beautiful block in New York.[13] That's interesting for several reasons, including that the repetitive nature of the Manhattan grid means the block is one of 150 or so similar blocks between Park Avenue and Lexington Avenue (and not unlike many other blocks in New York). Looking at what makes it better than many illustrates some basic principles of urban design. We will examine East 70th Street (Figure 1.22) the way that one would do that in person: by

Figure 1.22: East 70th Street, New York, New York. Looking west towards Park Avenue, in the section of the block that has the widest setback and the largest houses. Visible above the Asia Society building at the end of the block is 720 Park Avenue (also see Figure 2.34 on page 118). The apartment houses on the extra-wide avenue are 60 percent taller than those on the narrower side street. New York's great residential neighborhoods were built with lower height limits than the business districts, so the residential streets and the residences would have more light and air. See "A Short Discussion of Residential Building Heights in New York City" (page 31).

standing on the sidewalk, gazing up and down the street, and observing what it feels like. By the time we finish, we will also consider the size of the block and how it fits in the neighborhood and the Manhattan grid.

East 70th Street is a quiet street (not surprisingly, studies show we like quiet residential streets).[14] Standing under the trees on the sidewalk is comfortable. The width of the sidewalk allows plenty of space for the number of people walking there, and the width of the sidewalk and the height under the branches arching over the sidewalk are similar, so that the ratio of the horizontal dimension to the vertical is approximately 1-to-1—a proportion human beings find comfortable, as we shall see.

The parked cars and trees along the edge of the sidewalk shield us from cars and trucks driving by. That's also reassuring; these vehicles are often noisy and smelly, and they weigh four thousand to twenty thousand or more pounds, which can be both a physical and a psychological problem as they rush by at thirty to forty miles per hour. When we are on a sidewalk without a barrier, we know that a car could easily injure or even kill us if it were to veer onto the walk.*

When the traffic is quiet and we step out into the road—to cross it or to walk down the street—we find ourselves under a beautiful canopy of trees. The trees on East 70th Street, all in a line and regularly spaced, were used like architectural columns to provide visual order to the street—another quality humans find comforting. We say the trees "*were* used like architectural columns" because they have been significantly changed in the last two or three decades. Originally, the trees were all American sycamores. Sycamores are traditional American street trees that grow tall and form a canopy. Today, there are six or seven tree species on the short block, including ginkgo and locust trees that disrupt the canopy over the sidewalk and the street. The breaks and different tree types stand out like sore thumbs, diminishing the beauty of the street. If Woody Allen did indeed call this the most beautiful block in New York, he might have been talking about the old street, which had a fuller canopy. He once famously said that everything in New York used to be better (a belief we don't share). In any case, the role of trees in street design can be of critical importance.

East 70th Street is 60 feet across, like most of Manhattan's cross streets. That is to say, the distance from the property line on the north side of the street to the property line on the south side is 60 feet. The height of most of the buildings on the block averages 4½ to 5 generously sized stories, so the width-to-height ratio of the street is also approximately 1-to-1 (Figure 1.24). Traditional principles of urbanism say that the most comfortable streets are 1-to-1 or 1-to-1½, the width of the space between the buildings to the height of the buildings.[17] Many Italian piazzas are 1-to-3 (where the building height is one-third the width of the piazza). Once the proportions of an open space go beyond 1-to-5 or 1-to-6, though, the sense of spatial enclosure is lost.

The surveyors who laid out the Manhattan grid thought of it as a utilitarian network for future development, so many of the block's elements are simple. The question is, What was done to this block of East 70th Street that made it feel better than other, similar blocks? We've looked at some of the ways the street trees help. Urban designers also use trees to help define the space between the buildings on a block as an "outdoor room." In terms of the experience of the block, the canopy of the mature trees gives a "ceiling" to the room and limits how far we see in each direction, so that the space is visually contained. The slight slope of what is called Lenox Hill also helps, by bringing the canopy down into our view as we look towards Lexington Avenue.

Many of the elements of the block laid out by the surveyors are simple—but the principles of good street

---

* *Street Design* was first published in January, 2014, the same month Mayor Bill de Blasio announced new Vision Zero policies for New York (for more on Vision Zero, see pages 580, 582, 611–612 in Chapter Five).[15] Vision Zero led to the lowering of the speed limit on New York City streets to 25 miles per hour and the creation of Slow Zones with 20-mile-per-hour speed limits. When cars go 20 miles per hour or less, many streets will be more comfortable for walking without a wall of parked cars separating pedestrians from the cars. But East 70th Street is not in a Slow Zone, and the one-way street has two lanes of parking and only one traffic lane in a thirty-five-foot roadway, so some drivers speed in the wide space between the parked cars. Historically, almost all New York City streets were two-way, including East 70th Street (and the nearby avenues). Two lanes of parking and two lanes of traffic going in opposite directions slowed cars down, especially when they passed each other or when there was a double-parked car. Until 1950, cars could not park on New York City streets overnight.[16] In other words, as we say in Chapter Two, "Old streets are not what they used to be." We will look at this a little more in Chapter Five.

THE SEVEN ROLES OF THE URBAN STREET TREE

1. Define the space of the street.

This particularly applies to streets that are too wide for the height of the buildings, streets with holes in the streetwall, or suburban streets with buildings too far apart to contain the space of the street. Mature trees provide a canopy.

2. Define the pedestrian space.

A mature canopy hides the tops of tall buildings, giving the sidewalk a consistent human scale.

3. Calm traffic and protect pedestrians from cars.

4. Filter the sunlight.

Deciduous trees, unlike evergreens or palms, serve different functions in the summer and winter. Trees also lower city temperatures in the summer and change carbon dioxide into oxygen through photosynthesis.

5. Bring order to the street.

Trees should be laid out with regular geometries, repetition, consistent sizes, and alignment. On long, straight streets, trees that form canopies over the street limit the visual length of the street.

6. Visually soften the streetscape.

At some times of the day, the shadows are as beautiful as the trees.

7. Introduce the beauty of nature.

Living plants contrast with the buildings, and in many parts of the world introduce seasonal change, color, and fragrance.

Figure 1.23: East 70th Street, New York. A view of a block farther east on East 70th Street, looking back towards Second Avenue in the spring, with the trees in bloom. *Courtesy of Noel Y. Calingasan*

design *can* be simple. For example, the north and south "streetwalls" are parallel, and so are the sidewalks, the street trees, and the roadbed. Lining things up is a traditional design principle that brings a pleasing visual order. A street can certainly be more complex in shape, but the KISS Rule—"Keep It Simple, Stupid"—is often a good rule in street design, and "line things up" is a traditional principle of design that can work well too. Those are not the only good choices, of course. We will discuss these ideas in more detail in other parts of the book.

The Importance of the Building

When we walk down a street, we experience the space and its enclosure with our bodies and our senses. A good street is enclosed by trees or by buildings without many openings or interruptions. On a street shaped by buildings, we experience the design of the buildings that make the streetwall as essential elements of the street. Luckily for New York, the developers and designers of the early twentieth century—when New York was growing astonishingly quickly—believed that when they built, they had a responsibility to the city and the public realm, as well as to the inhabitants of the buildings. They created what are still the most popular neighborhoods and districts in New York City. Not coincidentally, they were saying to the world, "Look, we rival the great cities like London and Paris." One way they did that was to create streets where people wanted to live and work, lined by beautiful buildings.

The most beautiful block in New York boasts a high number of unusually distinguished houses. Built at a time when America produced a lot of wealth, they were designed by some of America's leading architects. The houses are in traditional styles, built with pleasing natural materials—marble, limestone, brownstone, brick, stucco, iron, copper, and wood. As an ensemble, they balance order and richness, which complements the simple New York City grid.

HARMONIC PROPORTIONS

I am every day more and more convinced of the truth of Pythagoras's saying, that nature is sure to act consistently and with constant analogy in all her operations: from whence I conclude, that the same numbers, by means of which the agreement of sound affects our ears with delight, are the very same which please our eyes and our minds.

— Leon Batista Alberti, *The Ten Books of Architecture*

During the Italian Renaissance (the "rebirth" of ancient Classical art, architecture, and learning), architects and painters adopted the principles of harmonic proportion they found in Classical designs. As described by the modern-day Classical architect Francis Terry, "If you take two violin strings under the same tension, with one half the length of the other, and then pluck them both, the difference in pitch is an octave which sounds beautiful."[18] Mathematically, the strings have a ratio of 1 to 2. If the strings have a relationship of 2 to 3, the result is a "fifth," and a relationship of 3 to 4 produces a "fourth." These mathematical ratios produce beautiful sounds. For centuries, urban designers have used ratios of harmonic proportion to design streets and squares.[19] The ratio of the height of the streetwall to the width of the street does not have to be a whole-number harmonic proportion: plus or minus a little works as well.[20]

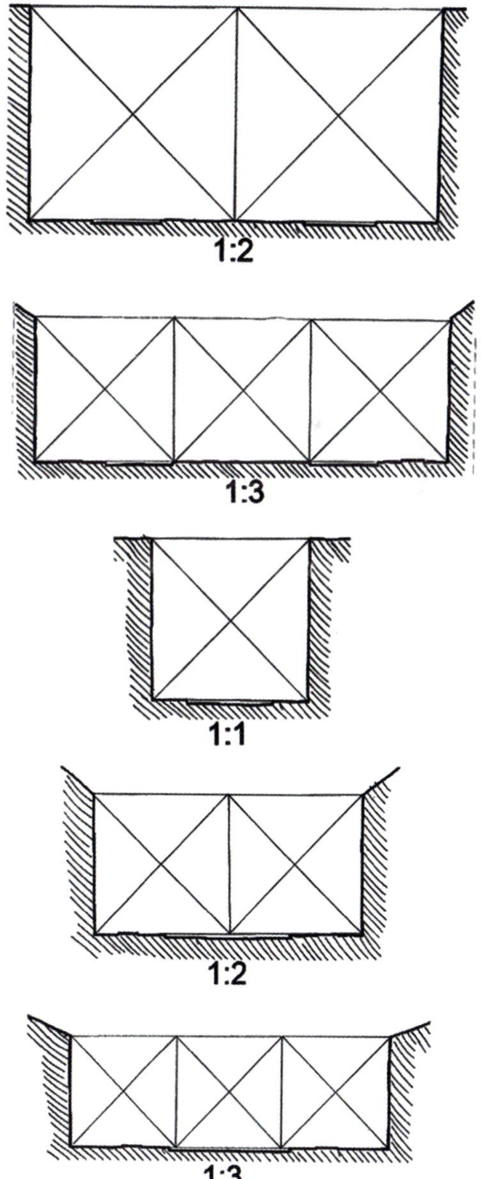

Figure 1.24: Street Sections with Harmonic Proportions. Harmonic proportions are the same as the numerical ratios used in music to create beautiful sounds. © *2007 Dover, Kohl & Partners*

The construction and the composition of the facades on the buildings that shape the street affect the feeling of enclosure. The solidity of the streetwall—the almost unbroken line of buildings—contribute to the making of an outdoor room where the pedestrian feels comfortable. The facades are well proportioned and, on the whole, simple and symmetrically composed. The elevations are divided into three parts (top, middle, and bottom) and have a level of detail and ornament that gives

scale and visual interest. As a group, the buildings are varied, and yet they achieve an overall harmony through their similarities in size, massing, proportion, scale, and windows.[21]

Most of the houses have similarly sized double-hung, vertical windows. Because the houses sit on a street, rather than in a piazza, we primarily see them obliquely, and the vertical openings set up a contrapuntal rhythm with the horizontal perspective. The windows are deeply set in the walls, increasing the play of light and shadow in a manner that adds a layer of visual interest.

The houses are taller than they are wide, which sets up a pleasing counter rhythm as we walk along the street. Most of the houses have similar but different arrangements of vertical windows: the patterns and the individual windows contribute to the counter rhythm, preventing our eyes from lazily looking down the street in a single gaze. The double-hung windows also contribute to the harmony of the whole, because most of the windows on the block and nearby on Park and Lexington avenues are comparable. The grand, twelve-story apartment houses on Park Avenue have similarly sized, double-hung, wooden windows and red-brick or limestone facades, just like most of the shorter, narrower houses on 70th Street. The shared family of materials and details bring order to the scene. The vertical rhythms, the beautiful ornament, and the play of light and shadow all add richness to the parallel rows of the buildings, the sidewalks, the trees, and the street (Figure 1.25).

A building on East 70th Street that interrupts the pattern is a Modernist-style house at number 124 East 70th Street designed by William Lescaze in 1941, with large horizontal windows (Figure 1.26). Looking back at the traditionally designed facades on the block, we see that they appear to respect the forces of gravity. For example, openings in brick walls are spanned by lintels that visually hold up the bricks above them, regardless of whether or not they actually support them (the real structure might be a steel frame). The exposed depth of the wall gives a mass that makes the wall seem to stand on its own, whether or not that is true.

In contrast, the brick on the Lescaze facade is cut like wallpaper, and the thinness of the wall is emphasized. If

Figure 1.25: Thomas W. Lamont House, 107 East 70th Street, New York, New York. Walker & Gillette, 1920. One of the large houses on the north side of the block affected by a covenant mandating a 10-foot setback.

Figure 1.26: Edward A. Norman House, 124 East 70th Street, New York, New York. William Lescaze, 1941. The thin column, the wide horizontal openings, and the shallow brick veneer of the Lescaze house emphasize that the facade is a curtain wall rather than a structural, load-bearing wall. That is one of the reasons why the house has a different effect on the streetscape than the neighboring houses. The others appear to respect the forces of gravity, with lintels, vertical openings, and thick walls. The Lescaze house appears to defy gravity. You don't need to be an architect to feel that in your body.

there were any doubt that the brick was in effect glued on, Lescaze cantilevered the wall and pulled back the glass entry wall of the first floor so that we see from below how thin the brick veneer is. Finally, the lightly framed glass wall across the width of the first floor visually creates a void, so that as far as we can tell, nothing holds up the weight of the house: the narrow corner column on one side is obviously not strong enough to bear the weight of the building above it, let alone the light curtain wall.

Modern architectural theory argues this is more honest construction: the design reveals that the brick is a veneer rather than load-bearing, while the real structure is inside the house. But in terms of placemaking, the degree to which Lescaze focused on the expression of construction and the creation of a unique object weakened the emphasis he gave to the larger context of the public realm and the making of a comfortable outdoor room. Compare the experience of walking on a block mainly flanked by vertical windows with the experience of walking on some nearby blocks with modern institutional buildings glazed with horizontal and ribbon windows: the rhythm of regular vertical windows in a horizontal streetwall creates a more well-defined space than a less visually interesting horizontal wall with horizontal windows.

Good streets, like good urbanism, require a balance of order and richness.

The facade Lescaze designed has a higher percentage of glass openings and a lower percentage of solid wall than the other buildings on the block. By itself, the twenty-foot-wide building does not break the streetwall, but a block lined with buildings like this would feel less contained. Some other blocks between Lexington and Park avenues have all-glass streetwalls: consequently, they feel less enclosed and are less pleasant to be on. The mechanical repetition of the glass curtain over a large area is monotonous, and when the eye is bored, the mind is bored. Only two blocks to the south, on 68th Street between Park and Lexington avenues, the architects of a large modern apartment building and the designers of a new synagogue across the street ignored some of the most basic principles while making the streetwall, so that a block that could be just as beautiful as the parallel block on 70th Street instead feels like an alley.

Compared to the other buildings on the street, the Lescaze facade has plenty of order, but little richness. That's not a problem on East 70th Street, because the other buildings offer so much richness. In fact, the Lescaze house provides interesting variety. But on any street where there are only boring glass facades, or when small simple modern buildings become bigger and more boring, that street suffers.

NEW YORK CITY'S UPPER EAST SIDE
A COMBINATION OF RICHNESS AND ORDER

- Blocks between Fifth Avenue and Third Avenue are short
  - New north-south avenues added to the repetitive Manhattan grid create short blocks
  - 200 feet by 400 feet ±
  - Gives convenient connections and a variety of routes for walking
  - Increases incidence of direct sunlight
  - The majority of the Manhattan grid has long crosstown blocks ranging from 610 to 920 feet (see Figure 1.87)
- Variety of Streets and Street Widths
  - Park Avenue, 140 feet wide with a planted median
  - Fifth and Third avenues, 100 feet
  - 79th and 72nd streets, 100 feet
  - Madison Avenue, 80 feet
  - Lexington Avenue, 75 feet
  - Side streets, 60 feet
  - Fifth Avenue faces Central Park
  - Short blockfronts face the avenues, only 200 feet long
  - Block of East 70th Street has extra setbacks and wider sidewalk
  - Wide sidewalks on Park and Fifth avenues
  - Wide sidewalks on East 72nd Street
  - Trees, topography, and Central Park create terminated vistas
- Building Heights
  - 1–1½ times the street width, with a maximum of 90 feet on East 70th Street and 150 feet on Park Avenue*
  - Taller buildings on wide streets, shorter buildings on narrower streets
  - Give daylight and sky view to the streets and residences

- Variety of Building Types, Sizes, and Residences
  - Large apartment houses
  - Small apartment houses
  - Mansions and rowhouses
  - Mansions and rowhouses converted to apartments
  - Private houses, condominium apartments, cooperative apartments, rental apartments
  - Stores and offices on Madison, Lexington, and Third avenues
  - Institutions like the Metropolitan Museum, Frick Collection, Asia Society, the New York Society Library
  - Hospitals, churches, temples, schools, libraries
- Strong Streetwalls
  - Well-designed, well-proportioned buildings
  - Consistent building heights and setbacks support the shaping of the blocks
  - Masonry facades with large wall areas create walls that look and feel solid
  - Vertical facades create counter-rhythm on the horizontal streets
  - Line Things Up: Buildings, sidewalks, trees, and roadways all support the order of the street while adding visual richness
- Street Trees
  - See "The Seven Roles of the Urban Street Tree" (page 20)
  - Soften the grid
  - Prevent long, unterminated vistas (hills also terminate vistas)
- Architecture
  - Historic District with well-designed buildings

---

*In 1929, a state-wide Multiple Dwelling Law allowed apartment buildings taller than 1½ times the width of the street, limited by two conditions. See "A Short Discussion of Residential Building Heights in New York City" on page 31.

- More Classical buildings than many parts of the city

- Regardless of size or style, buildings share similar materials and details, with restrained designs

- Facades divided into top, middle, base

- Facades have horizontal rhythms, made with vertical elements

- Masonry facades with details that suggest traditional construction

- Traffic & Transportation

  - No trucks or buses on Park Avenue

  - Side streets are typically low traffic**

  - Buses on Fifth, Madison, Lexington, and Third avenues

  - Crosstown buses on 72nd and 68th streets

  - Subway below Lexington Avenue

  - Taxis everywhere

**Modern DOT policies pump too much traffic through the neighborhood on the north–south avenues. Current DOT designs for the avenues are one-way, sub-urban Complete Streets that hurt walkability. The neighborhood would be helped if the city went back to the narrower, two-way avenues and streets that lasted in most cases until the 1960s. See page 581 Chapter Five.

The Lot and the Block

Rigid and autocratic, the Commissioners' Plan of 1811 was simultaneously loose and entrepreneurial. Its grid cut across hills and property lines while leaving unanswered the question of how the blocks would be filled. New York City famously introduced the first zoning in America in 1916, but long before that, a combination of planning regulations, building laws, development practices, building and development approvals, and sometimes deed restrictions determined lot sizes, as well as building sizes and use (office, retail, residential, and single or mixed use).[22] East 70th Street is part of Manhattan's Upper East Side, which owes its development to the planning of Central Park and the construction of multiple forms of mass transit,[23] combined with the phenomenal growth of New York in the nineteenth century.[24] New York envisioned both the Upper East Side and the Upper West Side, on the other side of Central Park, as large residential neighborhoods filled primarily by private development with rowhouses and apartment houses.

Building Types

American cities were built with a variety of locally developed building types. San Francisco is famous for its "Painted Ladies."[25] Charleston, South Carolina, developed the distinctive "Single House," discussed in Chapter Three (page 260). On the Upper East Side around East 70th Street, developers and builders continued the evolution of the single-family rowhouses and multifamily apartment houses already built in lower Manhattan and Brooklyn, deciding that the area between Central Park and Third Avenue would be filled with housing for the rich. (A loud elevated subway on Third Avenue made that a natural boundary for the wealthy neighborhood.)

The rowhouses were initially built with brownstone facades and a layout that by the end of the nineteenth century came to be known as the "American Plan." The ground floor was slightly below grade, in a raised basement. The parlor floor was one floor up, with a living room and a formal dining room. Exterior stairs known as stoops led to the main entrance and the parlor floor, which had the highest ceilings in the house. A service entrance was usually under the stoop. The stoop could extend four feet into the public sidewalk, and houses with stoops were set back an additional eight feet (unless the city approved different setbacks when the unimproved lots first sold). The city received annual rent for stoops that extended into the public right-of-way.

In the early twentieth century, many New Yorkers remodeled their brownstones, using the new "English Plan." English Plan houses had ground-floor entrances and no stoops. The new owners frequently replaced the brownstone facades with limestone or brick elevations and sometimes brought the front of the house closer to the street, since they no longer needed a setback for a stoop. Unusually, the houses on the north side of the block on lots perpendicular to 70th Street had restrictive

QUALITY AND QUANTITY: SOME THOUGHTS ON DENSITY AND BUILDING HEIGHT

Since we wrote the first edition of *Street Design*, frequent discussions about building heights and population density have become common across America. Affordable housing for all is an important issue in many places, and there is a large YIMBY movement—well funded by real estate interests—that argues all zoning and height limits are restrictive regulations that make it more difficult and expensive to build. If cities, towns, and suburbs would just let developers build what they want, where they want—the argument says—we would have all the affordable housing we need. The authors of *Street Design* think the truth is more complicated and nuanced. Here is a brief discussion of some of the reasons why we think that, and how the topic is closely related to street design.[26]

In the discussion of East 70th Street in the first edition of *Street Design*, we wrote that for thousands of years, good traditional streets frequently had a width-to-height ratio of 1-to-1 or 1-to-1½ (pages 19 and 21). We did not mention that for many years New York City limited residential building heights to 1½ times the width of the street (see "A Short Discussion of Residential Building Heights in New York," page 31). The physical character of the great residential neighborhoods like the Upper East Side and Harlem come from that time and those height limits. Today, the tallest "supertall" apartment tower on the new Billionaires' Row in mid-town Manhattan is 15.5 times as high as 57th Street is wide (and 25.8 times as tall as the width of 58th Street, on the north side of the building).[27] The residential tower is also 300 feet taller than the Empire State Building, which was the tallest building in the world from the time it was built in 1931 until the first World Trade Center tower surpassed it in 1971. But New Yorkers never wanted to *live* in the Empire State Building or the World Trade Center. We talk about some of the reasons for that in the discussion of height limits on residential buildings in New York (page 31).

In *The Death and Life of Great American Cities*, Jane Jacobs discusses "the kind of problem a city is", calling it "a problem in handling organized complexity."[28] Among the many interrelated issues, a successful business district with tall office towers has a different character than a neighborhood where people want to live. The YIMBY discussion, on the other hand, uses simplified talking points that emphasize one simple numeric quantity at a time: either building height, population density, the number of new building permits, or the residential unit count. Focusing on

individual numbers one at a time erodes the quality of cities and city life. Paraphrasing what Jacobs said about looking at street traffic in terms of pedestrians versus cars (discussed on pages 1, 458, and 574), reducing building design to simply height or unit count is "to go about the problem from the wrong end."[29] Urbanists must be generalists, balancing social, economic, environmental, transportation, and health issues with urban form and placemaking. If not, specialists like traffic engineers and luxury housing developers can erode cities and their quality of life.

Density can be reduced to a number, but that number does not tell us what it feels like to be in or at that place. East 70th Street is a part of the Upper East Side of Manhattan, which has over 100,000 people per square mile. That makes it one of the three or four densest residential neighborhoods in the Western World, more than twice as dense as the average neighborhood in Paris. Builders achieved that density 100 years ago with an average building height under 100 feet tall, and without an apartment building taller than 150 feet. And as we have seen, there are other qualities that make some blocks feel better than other blocks with almost identical dimensions.

Questions about urban form are essential for street designers. The height of the buildings on East 70th Street affects how we experience the space between the buildings, how the buildings enclose the street, and how they meet the street. The qualities that make East 70th Street feel good are missing in the photo of New Brooklyn (Figure 1.27), ironically on a street in the part of the borough widely known as Brownstone Brooklyn. The neighborhoods surrounding this section of Flatbush Avenue have the greatest collection of rowhouse streets in America: they create neighborhoods that are in great demand (Figure 1.28). Jane Jacobs was right when she observed that a greater variety of residential building types could have made the neighborhoods even better,[30] but these new apartment towers don't do that. Figure 1.27 shows what New Urbanists call "density without urbanism."

Please note: we are not saying that good streets must be expensive. Good streets are for everyone. They shape the public realm where public life takes place and city residents meet each other every day. The luxury towers in Figure 1.27 do not make good public spaces or create the great city where most New Yorkers want to live.

Figure 1.27: Flatbush Avenue, Brooklyn, New York. Looking south from Fourth Avenue. "New Brooklyn" is surrounded by some of the most loved places in New York City. Does anyone love the New York seen here? This is "density without urbanism." *Courtesy of Google*

Figure 1.28: Willow Street, Brooklyn, New York. The Brownstone Brooklyn neighborhoods of Brooklyn Heights, Fort Greene, Clinton Hill, Bedford-Stuyvesant, Crown Heights, Prospect Heights, Carroll Gardens, Cobble Hill, and Boerum Hill have an astounding collection of rowhouses and rowhouse streets. It is not hard to make a good street for people if you ignore the biases of specialists like traffic engineers and luxury housing developers. *Courtesy of Kenneth García*

covenants mandating 10-foot setbacks.[31] The sidewalk on that side of the street, therefore, is wider than the sidewalk on the southern side of the block. As a result, the limestone and brick mansions on the northern side get a bit more light and air than most New York houses, giving the street a subtle luxuriousness. Walking in the neighborhood, one can notice the difference, consciously or subconsciously.

The Commissioners' Plan specified that the north–south avenues platted in 1811 and the periodic wide cross streets like 72nd and 79th would be one hundred feet wide. Stores, restaurants, and other commercial spaces typically went on the avenues, sometimes on the wide cross streets, and rarely on the narrow cross streets.[32] Today, an urban designer preparing a form-based code begins by designing the street network and the streets, but they typically also choose the building types to line each street. Variables in that decision include the forms, sizes, and uses of the different types. Smaller buildings, like rowhouses, still go on the narrower streets like East 70th Street. Larger buildings, like apartment houses, still go on the broader streets like Park Avenue and East 72nd Street.

All the blocks, including the ones between the broader and narrower streets, are two hundred feet wide because that was the standard in New York by 1811. The two-hundred-foot block can be split down the middle to form back-to-back one-hundred-foot lots, each with its own street frontage, and a variety of building types had been developed to fit on those lots. A small house might have a lot sixteen feet wide by one hundred feet deep. A large house or an apartment building might have a lot twenty-four to forty-eight feet wide by one hundred feet deep. A very large apartment house might have a lot taking the entire two-hundred-foot length of the end of the block along the avenue. When a block fronted on both a wide cross street and a narrow cross street, larger buildings could go on the wider street, and smaller buildings on the narrower street. For example, apartment houses might sit on the lots along East 72nd Street, one of the wide streets, backing up to lots on East 71st Street with rowhouses. An important point for urban design and street design is that the larger buildings face each other across the wide street, appropriate to the shaping of that street, and smaller buildings face each other on the narrower street, appropriately sized

for that street. In urban design, this is known as the principle of "like faces like."

Thus, we have rowhouses of different widths but similar heights on the north and south sides of our block on East 70th Street and other building types on the east and west ends. At Park Avenue, there is a large institutional building—the Asia Society headquarters and museum—on the north corner and a large postwar apartment building on the south side. On the east end of the block, at Lexington Avenue, the architect Grosvenor Atterbury designed a house for himself: it has a storefront facing Lexington Avenue, the main retail street for the block (Figure 1.29).[33] There were once several brownstones on the block, but most were torn down and replaced by more fashionable designs or were thoroughly rebuilt with new facades. Because it faces the busy retail street, the brownstone on the corner of Lexington was given a storefront and converted to apartments instead.

The new apartment buildings at the Park Avenue end of the block share materials with the smaller houses; if they shared double-hung windows as well, the block would have more unity. One can argue that it is appropriate that the Asia Society, a civic building, be distinguished from the residential buildings by a more monumental scale and larger windows.

For decades, railroad trains ran in a trough north of East 26th Street at the center of Fourth Avenue, the original name for Park Avenue. The trains belched thick coal smoke that drove people away, but in the Commissioners' Plan of 1811, Fourth Avenue was the only north–south street between Fifth Avenue and Third Avenue, which are almost two thousand feet apart. To make walking in the neighborhoods around the railroad tracks more pleasant, two new avenues were cut in between Fifth Avenue and Fourth Avenue and between Fourth Avenue and Third Avenue, Madison Avenue being the former and Lexington Avenue the latter. What this means for pedestrians is that the east–west blocks between Fifth Avenue and Third Avenue are half the length of Manhattan's standard east–west blocks, which brings more sunlight and more freedom of movement to the experience of walking through the grid. As Jane Jacobs discusses in *The Death and Life of Great American Cities,* long blocks can be visually and even psychologically disagreeable.[34]

By 1913, when Grand Central Terminal opened, the trough in Fourth Avenue was covered and capped with planted medians, and all the trains under the cap were electrified. The newly named Park Avenue was no longer an ugly gash in the city fabric, filled with noisy, smoke-belching trains, but a broad, sunny street with a pleasant green center. The three-quarters of a square mile between Fifth Avenue and Third Avenue below 96th Street became one of the most desirable neighborhoods in Manhattan, soon claimed by the rich.

Central Park ran along the entire western edge of the Upper East Side: for sixty-one blocks, Fifth Avenue's apartment houses and houses looked out over one of the most beautiful parks in America. From midday on, the apartments facing the park get abundant natural light.

Figure 1.29: Lexington Avenue and East 70th Street, New York, New York. Looking west on 70th Street. The brownstone on the northwest corner of the intersection was renovated by the architect Grosvenor Atterbury for his own use in 1909.

Figure 1.30: East 72nd Street, New York, New York. Looking west from Park Avenue. The current fashion for planting multiple types of trees and choosing species that will never grow very high or have a wide spread prevents majestic allées like the old ones in Figures 1.16 and 2.73 on pages 13 and 142.

There are a few public buildings and museums on the residential avenue, but no offices or stores. Those are a few hundred feet away on Madison Avenue, which is narrower than the avenues in the Commissioners' Plan. Today, it is one of the most expensive international shopping streets, but for many years, Madison Avenue had all the stores residents needed.

The next street to the east is Park Avenue, the widest street in Manhattan, and another avenue with no stores allowed (except for a few small, grandfathered shops). Trucks and buses are banned too. Quiet and green, Park is lined with apartment houses that get more daylight than most Manhattan buildings. Then comes Lexington Avenue, another local shopping street. Like Madison Avenue, Lexington is narrower than the standard Manhattan

avenue, which is good for pedestrians. All the east–west blocks between Fifth Avenue and Third Avenue are equally short. The "most beautiful block in New York" benefits from all of this.

Our point is not that the Upper East Side of Manhattan is good because rich people live there, but that it was a place where people wanted to be. In that context, when Jane Jacobs published *Death and Life* in 1961, cities across America were suffering, and her city of New York was no exception. Industry moved to places where land and labor costs were lower, and the city's finances suffered. For a decade, the city's population shrank as New Yorkers moved to the suburbs. That prompted the urban renewal—called "urban removal" by Jacobs—that she studied in *The Death and Life of Great American Cities*.

Figure 1.31: West 74th Street, on the Upper West Side of Manhattan, looking towards Central Park. In this winter photo, the street is filled with sunlight in the late afternoon. The maximum height of the rowhouses and the apartment houses between the avenues is limited to 1 1/2 times the width of the street to let the light in. But the San Remo apartment house on Central Park West, seen at the end of the block, followed the 1929 Multiple Dwelling Law regulations that allowed residential towers up to 300 feet tall on very large lots along the avenues. The market, however, only built five apartment houses that tall: four faced Central Park, and the fifth sat directly on the East River. For another photo of the San Remo, see Figure 5.45.

A SHORT DISCUSSION OF RESIDENTIAL BUILDING HEIGHTS IN NEW YORK CITY

The topic of height limits for buildings is relevant in any discussion of urban design and street design. While discussing East 70th Street, let's briefly look at the planning regulations and building laws that contributed to the design of "the most beautiful block in New York." The towers of the New York City skyline are world famous, but few stop to think that the tall towers were all office buildings.

Until the nineteenth century, practicality limited buildings of all types to six or seven stories, because climbing to the top floor of taller buildings was too difficult as an everyday activity. By 1867, the increasing use of safety elevators and steel-frame construction* meant new office buildings might be as tall as eight to ten stories. New fire codes limited building heights, but New Yorkers also began to worry about the effect tall residential buildings could have on residential neighborhoods.

Common law traditions suggested that homeowners had a right to the sunlight that would "naturally" reach their land.[35] Many New York residents feared the shadows from tall buildings: in the 1880s private groups commissioned studies of "the high building question" and asked the city and the state to impose a moratorium on tall buildings. The New York State Legislature responded in 1885 with a bill limiting the heights of residential buildings in New York to seventy feet on side streets and narrow streets and eighty feet on avenues and wide streets.[36] Interestingly, seventy feet was the same number used in ancient Rome in 64 AD,[37] while Georges-Eugène Haussmann set lower height limits in Paris in 1859. The new Parisian boulevards were twenty meters wide, lined with grand apartment houses twenty meters tall (65 ½ feet). Buildings on narrower streets were limited to 57½ feet.

In New York City, a series of Tenement House Acts, building laws, and fire laws regulated the design and construction of apartment buildings. Over time, the laws and regulations increased residential height limits. With the rise of new technology and safety elevators, New York City raised residential building heights to 1½ times the street width or 150 feet, whichever was less (as mentioned previously in "Quality and Quantity, Some Thoughts on Density and Building Height"). But not until forty-nine years after New York first limited the height of residential buildings did the city famously pass America's first zoning in 1916, similarly regulating the design of tall office buildings.

In other words, New York City controlled the height of residential buildings long before it passed zoning regulations that controlled the height of commercial buildings, which could be larger and bulkier. The old laws and regulations for residential buildings continued to apply until New York State passed the Multiple Dwelling Law of 1929.[38]

There were three primary reasons New York wanted height limits for residential buildings and neighborhoods that were lower than the limits for office buildings in business districts: for better fire-safety conditions, for more natural light and ventilation in the apartments and on the streets, and to maintain a visible and psychological connection to the street and the sky from the apartments. Tall buildings on sixty-foot-wide side streets make residential streets feel dark and closed in. For residents in the apartments, keeping a feeling of connection to the street and the sky minimized the sense of peering into the neighbors' apartments across the street. Walking around Manhattan today, we see the wisdom of limiting the building heights to 1½ times the width of the streets. The spaces feel good, especially when they are filled with sunlight. One of the qualities that subtly contributes to our block on East 70th Street is the deed covenant that gives the north side of the block a little extra sunlight. Conversely, in the winter, some of the new supertall residential towers cast shadows across Manhattan and Central Park over 1½ miles long.

---

*Before steel-frame construction, "tall" buildings had load-bearing masonry walls that held up the building and everything in it. The taller a building was, the thicker the walls needed to be on the lower levels. The thickness of the wall on the lower floors became a limiting factor in construction.

When the New York State Legislature passed the state-wide Multiple Dwelling Law in 1929, the MDL allowed apartment houses taller than 150 feet, but only in limited circumstances. Residential buildings on lots that were 30,000 square feet or larger and on wide streets could have 150-foot towers on top of a 150-foot base. Buildings on smaller lots could have penthouses above the old height limits, but steep setbacks typically limited the additional height to one story behind a terrace. Notable exceptions were on Park Avenue, where the width of the street allowed multi-story rooftop additions that set back on each floor. A popular example is 720 Park Avenue, on the northwest corner of Park Avenue and East 70th Street (Figure 2.33).

New York City limited residential building heights to 1 ½ times the street width. That meant 150 feet tall on the avenues and wide cross streets, and 90 feet tall on the other cross streets.

The Great Depression, World War II, the move to the suburbs that followed the war, and perceptions of demand in the market meant that only five of the three-hundred-foot towers were ever built.[39] New York's 1961 Zoning Resolution changed the rules, giving height bonuses for towers set back from the street. Tower construction costs were too high for low-income buildings until post-war Modernism introduced bare-bones construction paid for by federal funding that dictated "towers in the parks" housing like the projects built by the New York City Housing Authority.[40]

Once New York had tall apartment towers, many New Yorkers found that although the views were exciting, they did not like the daily experience of living high above the ground, where they felt cut off from city life. People on the ground seemed as small as ants. Going down to the city could take a long time, particularly when there was a problem with the elevator service, and going up required fast elevators.

A former Chair of the New York City Planning Commission, Joe Rose, called the 1961 Zoning Resolution an ideological statement of Modern planning, particularly referring to the height bonuses for plazas between the buildings and the street. "We are in crisis because in many instances our zoning promotes an architectural vision that does violence to our urban fabric," Rose said in a talk in 1999. "The 1961 zoning changes imposed an aesthetic regime of "towers-in-the-park which has been proven to be a fundamentally flawed, anti-urban, and anti-New York concept."[41] He believed the new zoning was responsible for the creation of Historic Districts. The first Historic District came three years after the 1961 Resolution. Today, New York City has over 160 Historic Districts. In 2000, Rose proposed zoning changes that emphasized maintaining street walls and limiting tower heights. In his own telling, Rose placed the proposal on the desk of Mayor Rudolph Giuliani, who within half an hour had a phone call from the Real Estate Board of New York (REBNY) that killed Rose's plan.[42]

At the end of the twentieth century, New York developers discovered that there was a small but lucrative market for apartments that rose above the buildings around them. Apartments with unobstructed views could sometimes command a 30 percent premium in price. Thinking about the effect of many tall residential towers on New York City, Rose said, "Views have become so prized that we unleashed an intense desire for building height without regard for neighborhood character or scale. Each new building tries to achieve better views by being taller than the last. The consequence has been a powerful inducement to break away vertically as far as possible from the neighborhood pack. While there is nothing wrong with nice views, it is not necessary to have a city shaped by a desperate grab for them."[43]

In the following years, however, some of REBNY's biggest, richest, and most politically influential developers built the first supertall residential towers in New York and marketed the apartments to members of the global one percent who wanted to park money in real estate investments in the city. As we write the second edition, some of the supertall, super-luxury towers on Billionaires' Row in midtown Manhattan are the most profitable buildings in the history of New York. REBNY therefore convinced New York's Governor to remove impediments to residential height limits across the state.[44] If the powerful group triumphs in the end, East 70th Street between Park and Lexington might someday include a tower 20 or 30 times greater than the width of the street. But that is another story.[45]

One thing she discovered, usually overlooked today, was that when the city declined, residential neighborhoods with a variety of building types remained more popular and suffered less abandonment. Brooklyn Heights, which had a greater variety of dwelling types, had less decline and recovered more quickly, than Brownstone Brooklyn.[46] Rather than being a neighborhood of streets lined solely with rowhouses, it had rowhouses, small apartment buildings, mid-rise apartment buildings, and larger apartment houses up to 150 feet tall. These buildings contained a variety of rental apartments and cooperatively owned apartments. The apartments ranged from small one-bedroom apartments and studios to large apartments for families. Some of the single-family rowhouses were mansions, and some of the very rich, like Andrew Carnegie and Henry Clay Frick, owned freestanding mansions (both museums today). But the Upper East Side was also home to the most luxurious apartment houses in America.

Despite being built for the rich, the Upper East Side had a similar mix of less expensive buildings and apartments. By 1963, many brownstones and rowhouses had been converted into multifamily houses with small rent-regulated apartments. On the basis of market demand, the Upper East Side was the most popular neighborhood in New York City.

Sitting in the middle of the Upper East Side, the block on East 70th we have looked at both benefited from, and contributed to, the surrounding urbanism. An intrinsic part of that urbanism was that the wide avenues had taller buildings, while the narrower side streets had smaller, shorter, and less wide buildings. The neighborhood had shorter blocks than most Manhattan neighborhoods, and two of the avenues were among the sunniest in the city. It all contributed to the appeal of the block between Park and Lexington avenues on East 70th Street (Figure 1.29). In Chapters Three and Five, we will discuss why good streets are inseparable from the city or town around them.

A BRAVE NEW WORLD

For street design in America, everything changed after World War II, reflecting great changes in American business, society, and culture. The way we used streets, the way we designed streets, and the way we built our communities were all new and different.

In the first half of the twentieth century, America had walkable towns, cities, and neighborhoods connected by an efficient network of streetcars, trains, and boats. Before the Depression brought so much hardship, the majority of Americans could lead a comfortable life without owning a car. A group that called itself "Organized Motordom" successfully worked to change that, however. A consortium of car manufacturers, oil companies, road builders, and the like, Organized Motordom wanted to sell cars. Their early experiences made them realize they had to claim control of the streets as places for cars, so that car owners could quickly and easily drive wherever they wanted.[47]

Motordom had not yet invented traffic engineering and all the car-first detritus engineers created: traffic signals, stop signs, striped lanes, and the like. Everyone had a right to be anywhere on the street at any time, but cars quickly made that space dangerous for pedestrians (a word that was rarely used until the twentieth century).[48] The new pedestrians fought back. "The pedestrian," one New Yorker said, "as an American citizen, naturally resents any intrusion upon his prior constitutional rights." The Anglo-American legal tradition confirmed pedestrians' right to the street.[49] Both the United States and the United Kingdom once had red flag laws that required motor vehicles in cities be preceded by a man on foot carrying a flag to warn pedestrians of the approaching danger.

Bigger changes came after the war when Organized Motordom teamed up with planners, architects, developers, bankers, and the Federal government to change the way we lived. Owning a single-family house in the suburbs—which required owning a car—was sold as the American Dream. Since then, the majority of Americans have exchanged the city and the town for sprawl, by some estimates constructing more than three-quarters of the roads and buildings in America. Today, most Americans have to drive everywhere for everything: in most of the country, living, working, and shopping are all separated by roads built for the use of cars.

Several factors contributed to the desire for change. Soldiers coming home from Europe wanted better lives than the ones they had left behind. Most Americans lived on farms and in cities before 1940, but by 1945 those cities were frequently seen as dirty and crumbling, and old neighborhoods as socially constricting. There was a new sense of possibility and social mobility; after the war, the GI Bill encouraged veterans to go to college and aim for white-collar jobs instead of accepting the same blue-collar jobs their parents had. And when they graduated, the government gave them low-cost loans to buy single-family houses in the suburbs. Frequently unspoken was that the education bills and mortgage subsidies were primarily for white Americans. Redlining kept minorities in the cities.

Cars for the new life in the suburbs were more affordable after the war. Detroit switched from manufacturing tanks and planes for war to producing hundreds of thousands of cars. At the same time, the flight from the cities was just beginning, so suburban roads were empty, convenient to drive, and often beautiful. In addition, having won the war with technology and industry, we had faith in technology and industry. In 1956, the former Supreme Commander in Chief of the Allied Forces, President Dwight D. Eisenhower, signed the Federal-Aid Highway Act, which funded the largest public works program in the history of the world: our interstate highway system. The highways were the centerpiece and symbol of a new transportation system that substituted personal motor vehicles for public transportation. Organized Motordom had won the day.

Don't honk your horn. Raise your voice. We fought, and we won, our right of way. But now, two miles of road are wearing out for every one being built. Write your hometown officials and postcard your newspaper editors. Demand better highways and more parking space…. Give yourself the green light.

— General Motors propaganda film
Give Yourself the Green Light, by
Jam Handy Productions, 1954

Transportation engineers successfully convinced cities that if they failed to build modern highways, they would be unable to compete in the twentieth century, so cities rebuilt their roads to make it easier for suburbanites to drive in and out of the city. In the suburbs, planners developed a new system of automobile-based zoning that separated uses, creating single-use housing developments, shopping centers, and office parks. It was all part of a new world in which President Eisenhower's Secretary of Defense said, "I thought what was good for our country

Figure 1.32: St. Charles Avenue, New Orleans, Louisiana. America once had a network of trains, boats, and streetcars that made it easy to travel to most cities and towns without a car. New Orleans streetcars like the famous streetcar named "Desire" sometimes travel in broad grassy medians New Orleanians call "the neutral ground."

was good for General Motors, and vice versa."[50] Famously misquoted as "What's good for General Motors is good for the country," and frequently attributed to President Eisenhower, the remark expressed the dominant ethos of the time.

Corporate America saw specialization as one of the reasons why Detroit was able to produce all the planes, tanks, and munitions that beat the Nazis, and specialization became central to the creation of the new automobile-based way of life that promised economic and social mobility. Traffic engineers specializing in auto movement invented new road systems. Developers stopped building mixed-use towns and created single-use "products" like housing subdivisions and shopping centers. Loan officers at banks became specialists in lending for the single-use products: a housing lender would not lend money for shopping center construction or an office park.

> The street wears us out. It is altogether disgusting. Why, then, does it still exist?
>
> — Le Corbusier, *La Ville Radieuse*

Architects and urban designers shared in this reinvention of the way we live. Three decades earlier, the most influential architect and urban designer in the Western World, Le Corbusier, had written, "The street wears us out. It is altogether disgusting. Why, then, does it still exist?"[51] His solution was to design a new type of urbanism without ordinary streets. In the famous *Plan Voisin*, Corbusier proposed demolishing most of central Paris north of the Seine and building a new city with sixty-story towers in large parks. Surrounding the parks were multilevel, multimodal streets (called "machines for circulation"[52]) with cars, buses, trains, and even planes. "The cities will be part of the country," Corbusier wrote in a tribute to technology and futurism:

ORGANIZED MOTORDOM: STREETS FOR CARS

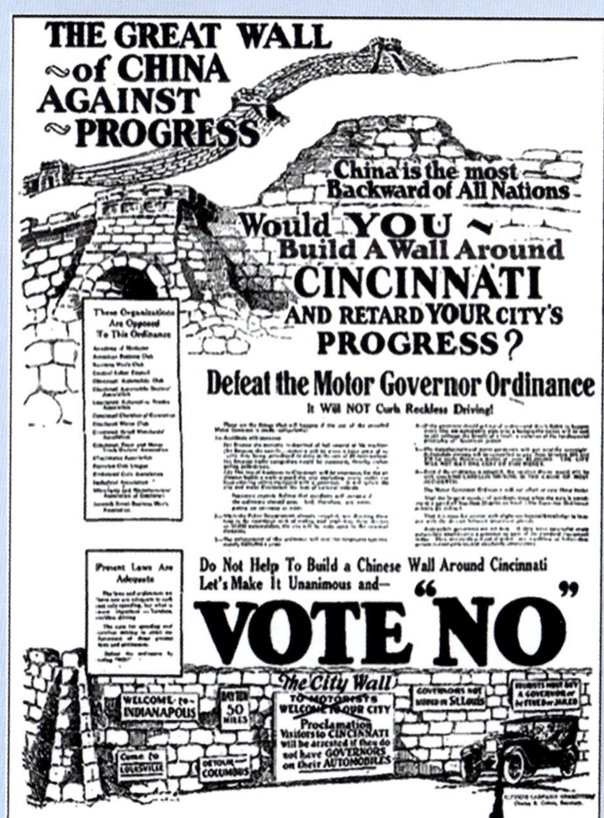

How did the American city become an automotive city? Why was much of the city physically destroyed and rebuilt to accommodate automobiles? The case presented in this book is that before the city could be physically reconstructed for the sake of motorists, its streets had to be socially reconstructed as places where motorists unquestionably belonged.

This social reconstruction was only one of several ways in which people tried to solve a new problem. New automobiles were incompatible with old street uses. Until the 1920s, under prevailing conceptions of the street, cars were at best uninvited guests. To many, they were unruly intruders. They obstructed and endangered street uses of long-standing legitimacy. As a Providence newspaper editor expressed the problem in 1921, "it is impossible for all classes of modern traffic to occupy the same right of way at the same time in safety."
—Peter D. Norton, *Fighting Traffic, The Dawn of the Motor Age in the American City*

Figure 1.33: Advertisement by Citizens' Committee, *Cincinnati Post*, November 5, 1923. When Cincinnati, Ohio, considered putting speed governors on cars, Motordom organized and paid for a Citizen's Committee that ran ads against the campaign, wrote letters to every household in Cincinnati, and stationed 400 people at the polls on election day to urge citizens to vote "NO." *Cincinnati Post, via Peter Norton, Fighting Traffic*

I shall live 30 miles from my office in one direction, under a pine tree; my secretary will live 30 miles away from it too, in the other direction, under another pine tree. We shall both have our own car. We shall use up tires, wear out road surfaces and gears, consume oil and gasoline. All of which will necessitate a great deal of work… enough for all.[53]

It was only in the 1940s, however, that Le Corbusier's ideas became standard in American architecture and urbanism. Supersized blocks (superblocks) with towers in parks became so accepted in professional thinking that the federal funding incentives to construct them were impossible to pass up—as Robert Moses learned.

While Moses was working with New York University on a postwar plan for the redevelopment of Washington Square South in Greenwich Village, he saw renderings by Eggers & Higgins—the successor firm to the office of the Classical architect John Russell Pope—that showed two sides of Washington Square rebuilt with Georgian-style buildings. Moses was so impressed that when the university subsequently chose to build only one Georgian building (the NYU Law School), he wrote a letter of complaint to the university's chancellor. During this same period, Moses oversaw slum clearance plans funded by the Federal Housing Act of 1949 for a forty-acre site southeast of the square to be developed as Washington Square Village (Figure 1.34). He asked Eggers & Higgins to design a Georgian-style development, but Eggers soon came back to Moses, reporting that the Federal funding and regulations required superblocks with residential-only towers in parks. The authors of the Slum Clearance section of the Housing Act, known as Title I, explained that an "unwise mixture of residential and commercial uses of land" and the "frozen patterns of street layouts" like the Manhattan grid were among the causes of slum conditions.

Always a pragmatist, Moses quickly accepted the new standards and built the conventional Modernist, tower-in-the-park project seen in Figure 1.34. In the following decades, he used federal funds to tear down block after block of "blight" in New York City and build new superblocks of towers-in-the-parking-lot housing

Figure 1.34: Washington Square Village, New York, New York. Rendering, circa 1950. An early site study for Robert Moses by the architects Eggers & Higgins. Committee on Slum Clearance Plans. *Courtesy of the Avery Architectural and Fine Arts Library*

projects. That is the Moses we know from Robert Caro's Pulitzer-Prize-winning biography *The Power Broker: Robert Moses and the Fall of New York.*[54] Similar postwar federal funding requirements changed the parkways Moses built. The regulations produced anti-urban highways like the Cross-Bronx Expressway, a central element in Caro's story.

The highways were part of the Dwight D. Eisenhower National System of Interstate and Defense Highways, initially funded by the Federal-Aid Highway Act of 1956. The highway spending and the Federal Functional Classification System (see below) that gave federal money to State Departments of Transportation for federally approved roads supported a new way of living we can reasonably call the largest and most expensive top-down social experiment in the history of the world. American sprawl wouldn't exist without our extensive highway system and millions of miles of Functional Classification roads that promote driving as the national transportation system.

FUNCTIONAL CLASSIFICATION

Functional Classification is the system American transportation planners and traffic engineers use today when designing roads. In addition to the limited-access highway, the system has three—and only three—road types (known as "facilities"): "Arterials," "Collectors," and "Local Roads." Planners and engineers developed Functional Classification in the United States in the years following World War II. It was mandated by law in the 1960s and 1970s and enshrined in the 1968 Federal-Aid Highway Act, which required the classification of all roads in the country to establish funding priorities. Functional Classification tells planners and engineers what types of roads to design and how they should connect (or not).[55]

As shown in Figure 1.35, Functional Classification is based on a philosophy of "mobility" versus "access." "Mobility" refers to mobility for motor vehicles: fast, easy-flowing traffic is at the top of the graph. "Access" represents how easy it is for vehicles to connect to and enter the thoroughfare: the higher the frequency and number of intersections and curb cuts along a road, the more access there is for motor vehicles. Figure 1.35 shows that in the world of Functional Classification increasing access decreases mobility, and vice versa. This leads to three guiding principles of Functional Classification:

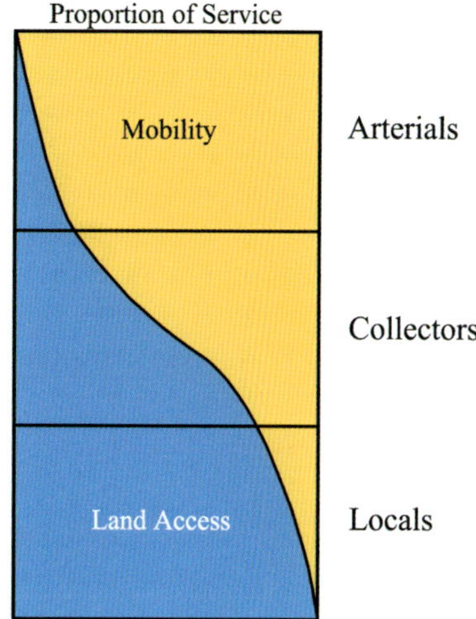

Figure 1.35: Proportion of Service Diagram for the Functional Classification System. *Courtesy of Dover, Kohl & Partners, redrawn from the 1969 Federal Aid Highways report*

1. The longer the trip, the faster the travel speed should be.
2. The faster the road speed, the more isolated the road should be from its surroundings.
3. Isolated roads can become bottlenecks, so they must grow wider as the "ADT" (the number of Average Daily Trips) grows.[56]

Functional Classification imposes a hierarchy of roads. Highways and freeways are at the top, followed by arterials, collector roads, and local roads, in that order. Highways have limited exits and entrances, arterial roads have fewer intersections and curb cuts than collector roads, and local roads are considered optimal when they are dead-end cul-de-sacs (Figure 1.36). The system channels traffic onto large, wide arterials and collector roads like the Coral Gables "stroad" in Figure 1.39 (for more on stroads, see pages 504 and 505).

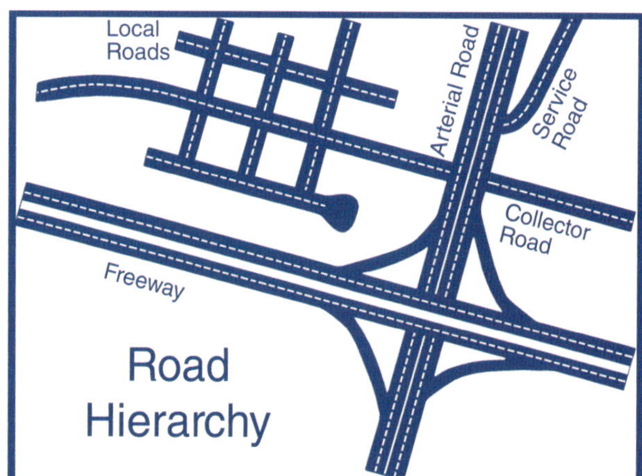

Figure 1.36: Road Hierarchy Diagram. *Courtesy of Dover, Kohl & Partners, based on the cover of a Wisconsin DOT manual.*

Figure 1.37: I-95 and I-395 Interchange, Overtown, Miami, Florida. Aerial view. Hundreds of houses, businesses, and institutions were displaced to build I-95 and the interchange, in a disastrous urban renewal scheme motivated by racism and automania. The city was left more divided than ever. © 2013 Pictometry International Corp.

Figure 1.38: Amsterdam Avenue (Tenth Avenue), New York, New York. Looking south from Morningside Heights towards Midtown Manhattan. New York's utilitarian Commissioners' Plan of 1811 has many examples of the "endless section" and the unterminated vista. However, the New York City Department of Transportation exacerbated the problem by removing everything but motor vehicles from the roadway and turning most avenues into one-way, suburban-style transportation corridors with synchronized traffic lights that encouraged fast driving.

Unlike a city grid, which disperses traffic by giving drivers multiple routes to choose from, a "dendritic" (tree-like) Functional Classification network concentrates traffic on a limited number of arterial roads and highways.

An organization called the American Association of State Highway and Transportation Officials (AASHTO) publishes a book entitled *A Policy on Geometric Design of Highways and Streets* ("The Green Book") that sets the standards for the design and construction of federally- and state-funded roads in America. Based on traffic flow, traffic speed, and traffic volume, the Green Book sets the standards for what AASHTO calls "Levels of Service" (LOS). The Green Book allows five LOS, denoted by a letter grade of A through F. LOS prioritizes the car over all other modes of transportation, including walking.

PROBLEMS WITH FUNCTIONAL CLASSIFICATION

All time-saving methods, to which alleviation of traffic congestion and other conveniences belong, do not, paradoxically, save any time, but simply fill the time available in such a manner that one has no more time at all. The result of this is inevitable, breathless haste, superficiality and nervous fatigue with all the related symptoms like nervous hunger, impatience, irritability, distractedness etc.

— Carl Jung, publictransit.us

Figure 1.39: Red Road, Coral Gables, Florida. Colored plastic sticks are everywhere on modern streets. We are so used to them that we subconsciously discount how alien (and alienating) they feel to the pedestrian.

In theory, Level of Service evaluations demonstrate which roads and intersections are creating bottlenecks and lead to their redesign. In reality, growth and density invariably make the Functional Classification system fail on its own terms, because when everyone drives everywhere for everything, it is impossible to build roads large enough to handle peak concentrations of cars. And while Functional Classification is designed to increase safety and lower traffic fatalities, it is a system that produces over forty thousand traffic deaths every year. Studies by rogue transportation engineers show that wide roads frequently have more fatalities than narrower streets, and that increasing the number of intersections rather than minimizing intersections also lowers fatalities.[57] In other words, two fundamental principles of Functional Classification—increasing road width and limiting access—can increase deaths caused by traffic collisions.

Unfortunately, Functional Classification and Level of Service standards also conflict on many levels with making streets where people want to be (Figure 1.37). In their efforts to make the car's journey ever more efficient, engineers create large, barren, noisy places that pedestrians avoid when they can. The level of alienation from ordinary human experience that has crept into this field is reflected in its language: transportation engineers refer to trees as Fixed Hazardous Objects, or FHOs. Even worse, people walking are referred to as Moving Hazardous Objects (MHOs) that lower the Level of Service and are banned from many roads. The criteria for Functional Classification obviously do not include making a public realm where people want to be.

The Green Books call the standards for the one-way arterials and couplets that have gutted so many American cities and towns "Urban." They are really suburban, or sub-urban. The other Geographic Definition is "Rural." Those standards are also for large over-engineered suburban roads. By the standards of street designers in Europe, AASHTO's design requirements for roads in cities, sprawl, and rural areas are all the same. The city streets people love, including every street in this book, get failing grades from AASHTO.

"The Functional Classification System and Level of Service need to be entirely reevaluated," says former Milwaukee Mayor and Congress for the New Urbanism President John Norquist. "In certain rural contexts, the system may sometimes make sense, but applying it to urban contexts doesn't. For example, Greenwich Village is rated F (lowest), based on congestion. It's congested with people who want to be there! They're buying stuff, and creating jobs, and creating art. It's an F that all good urbanism gets."

Even if all these accusations are true, the automobile is still an improvement on its principal alternative, the pedestrian. Pedestrians are easily damaged. Try this test: Hit a pedestrian with a car. Now have the pedestrian hit the car back.... Which is in better shape?

— P.J. O'Rourke, *Give War a Chance*

TRANSPORTATION AND TRAFFIC ENGINEERING

Street Design has many comments and complaints about the damage traffic engineering has done to our streets and the public realm. The problem, of course,

is the system, not the individuals who administer and wrestle it. Some of our best friends are traffic engineers and transportation planners. Rick Hall, Norman Garrick, Walter Kulash, Rick Chellman, Billy Hattaway, DeWayne Carver, Peter Swift, Ian Lockwood, Wade Walker, Dan Burden, Wesley Marshall, Chris McCahill, Sam Schwartz, Michael Flynn, Andy Boenau, Phil Jones, and Andrew Cameron taught us many of the lessons in this book. There are more, including the renowned author of *Confessions of a Recovering Engineer*, Charles Marohn. We've included an excerpt from his book (pages 41–43) that explains the process engineers follow when designing a street. The traffic manuals they use are like cookbooks, he says. "If you wish to make a certain type of chocolate cookie, a cookbook will provide the common ingredients found in cookies and the specific way to arrange them for a particular recipe. Likewise, if you wish to build a certain type of street, an engineering manual will explain how to assemble all of the components to get the desired outcome."

Engineers and transportation planners start the design of a street by choosing a design speed—how fast traffic should go on the street. Next, they decide on the volume of traffic they think is appropriate, usually with computer programs that forecast future traffic. (Note that the computer programs were devised by asphalt mongers for asphalt mongers; the default settings assume virtually all future trips will be driving trips, typically with only one person in each car, and they

apply absurd "growth rates" that become self-fulfilling prophecies.) Then they consult the traffic manual for the recipe—or formula—for the street, including the number of traffic lanes, the width of the traffic lanes, and so on. To a disappointed neighbor, the engineers and transportation planners can then say, "Sorry, the computer made me do it," and "the manual made me do it" (Figure 1.40).

We can understand the design by looking at a "section" of the road, a slice through the street that shows all the dimensions, including the overall width. When the new road is in sprawl or exurbia, where there is lots of land, it is a simple matter to extrude the section like spaghetti from a pasta machine. Urban designers sometimes call this "the endless section," because it never changes.

That's a problem when the design is for an existing road in a city or town. That's when the devil is in the details. Some sections of the road might be too narrow for all the elements in the formula, because sidewalks or trees get in the way. Some design speeds might be too fast because of Moving Hazardous Objects (people on foot) in the area.

The endless section works well for drivers on broad arterials in the middle of nowhere. When they are confident about what lies ahead, they can drive faster. But the endless section is boring for people walking. We like a combination of order and variety, ideally with human-scaled buildings and spaces along the way. Creating that combination and solving problems with design is why designers say God is in the details. Dealing with irregularities that interfere with the endless section is why engineers say the devil is in the details.

EXCERPT: *Confessions of a Recovering Engineer* / Charles L. Marohn Jr.[58]

Transportation professionals consider their texts, and by extension their entire profession, as being… value free. This is wrong. At the foundation of traffic engineering is a collection of deeply infused values. These values are so deep, and so core to the profession, that practitioners do not consider them values. They bristle at the suggestion. For practitioners, these values are merely self-evident truths—something like gravity that it is not necessary to believe in because it just is.

These values are expressed in the range of options that engineers consider, the way that they discuss different approaches, and the transportation systems that they build. This would not be a problem, and we could allow this entire profession to retain their sacred texts and practices unchallenged by heretical viewpoints, if they could find a way to address the damage traffic engineering is doing to our communities.

They cannot do this for a simple reason: The damage being done is the culmination of those values. The injuries and deaths, the destruction of wealth and stagnating of neighborhoods, the unfathomable backlog of maintenance costs with which most American cities struggle, are all a byproduct of the values at the heart of traffic engineering. Addressing the damage requires addressing the values, but you cannot address something that you deny even exists.

The underlying values of the transportation system are not the American public's values. They are not even human values. They are values unique to a profession that has been empowered with reshaping an entire continent around a new, experimental idea of how to build human habitat.

Let us identify those values.

The underlying values of the transportation system are not the American public's values. They are not even human values. They are values unique to a profession that has been empowered with reshaping an entire continent around a new, experimental idea of how to build human habitat.

The Design Process

When an engineer sits down to design a street, they begin the process with the design speed. I have been in countless meetings where engineers presented technical design sheets and even in-depth studies for a street project. Never, and I mean never, was any elected official or any member of the public asked to weigh in on the design speed.

Never once did I hear one of my fellow professional engineers say, "So, what are you trying to accomplish with this street in terms of speed?"

No. The design speed is solely the purview of the engineering professional, with a preference for accommodating higher speeds over lower.

Why?

Choosing a design speed is, by its nature, an application of core values. When we pick a speed, we are selecting among different, competing priorities. Is it more important that peak traffic move quickly, or is it more important to maximize the development potential of the street? Do we compromise the safety of people crossing on foot in order to obtain a higher automobile speed, or do we reduce automobile speed in order to improve safety for people outside of a vehicle?

These are policy decisions, and like all policy decisions, they should be decided by some duly elected or appointed collection of public officials. In a democratic system of representative government, representatives of the people should be provided the full range of options and be allowed to weigh them against each other. That rarely happens, and I have never heard of an instance where it has happened for a local street.

Many of my engineering colleagues will reply that they do not control the speed at which people drive—that travel speed is ultimately an enforcement issue. Such an assertion should be professional malpractice. It selectively denies both what engineers know and how they act on that knowledge.

For example, professional engineers understand how to design for high speeds. When building a high-speed roadway, the engineer will design wider lanes, more sweeping curves, wider recovery areas, and broader clear zones than they will on lower-speed roadways. There is a clear design objective—high speed—and a professional understanding of how to achieve it safely.

There is rarely any acknowledgment of the opposite capability, however: that slow traffic speeds can be obtained by narrowing lanes, creating tighter curves, and reducing or eliminating clear zones. High speeds are a design issue, but low speeds are an enforcement issue. That is incoherent, but it is consistent with an underlying set of values that prefer higher speeds.

Once the engineer has chosen a design speed, they then determine the volume of traffic they will accommodate. How many motor vehicles will this street be designed to handle? This is the second step of the design process, and the second instance where the design professional independently makes a decision that is, at its heart, a value decision.

Standard practice is to design the street to handle all of the traffic that routinely uses it at present, plus any increase in traffic that is anticipated in the future. There is no consideration given as to whether that is too much traffic for the street, and rarely is there a conversation of whether other alternatives should be considered. If traffic is present, it is the traffic engineer's calling to accommodate it. No nonprofessional is given an opportunity to suggest otherwise.

Now that they have identified the design speed and traffic volume, the traffic engineer consults one of the books of standards to determine how to assemble a safe street. Given a certain speed and volume, how does the design cookbook indicate the street's ingredients be assembled? Within the design process, the answer to that question is, by definition, safe. Any other design would generally be considered a compromise of safety.

The final step of the design process then is to take the "safe" design and determine how much it will cost. This dollar amount is the price for a responsible street design. Any questioning of this minimum effort would be considered a reckless endangerment of human life.

Now we have the traffic engineering profession's values as expressed in the design process. In order of importance, those values are traffic speed, traffic volume, safety, and cost.

I have presented the profession's values in this way to dozens of audiences, comprising thousands of people, across North America. I then ask them to identify their values. These are mixed audiences of professionals and nonprofessionals, people involved in local government decisions and those who are not. There is always a broad consensus.

I ask them to think about a street where they live, or one where they shop or like to go out to eat. I then ask them to shout out, in unison, which value they consider most important as applied to that street. The answer, overwhelmingly, is safety.

And of course, it is. Most humans, including most traffic engineers when they stop to consider what is being asked, would sacrifice much of the street's performance in terms of speed or volume in order to make it safer. Safety is the top value nearly all people apply to street design.

As we continue, I ask for them to shout out their second most important value. Again, there is no real ambiguity. Nearly everyone chooses "cost."

Again, most Americans today would sacrifice the ability of someone to drive at speed, and the capacity of a street to accommodate a specific volume of traffic, to have a more cost-effective design. I acknowledge that this collective response may differ from the preferences of the individual driving the street, but there are always competing interests between an individual and society. In public policy, we routinely ponder such tradeoffs.

While safety and cost are the top values for nearly everyone, the third value expressed by the groups with whom I have interacted is perhaps the most telling. I ask, "In a tradeoff between speed and volume, would you prefer a design that moves fewer vehicles at a higher speed, or one that moves more vehicles but at reduced speed? Would you emphasize speed or volume?" The answer, overwhelmingly, is "volume."

Table 1.1: Values Applied to the Design of Streets*

Current Practice	Most Humans
Design Speed	Safety
Traffic Volume	Cost
Safety	Traffic Volume
Cost	Design Speed

*In order of priority, highest priority first.

And that makes sense. To the extent that the street is used to convey traffic, sacrificing the number of cars that can pass through in a given time frame just so those drivers can go faster is counterproductive by any meaningful measure. If we can slow down traffic speeds, and it means that more vehicles can pass through and people arrive at their destinations sooner, why would we not do that? Most people would.

The values of the design process—the values applied to street design—are not values that most people would identify with. I would assert that this includes most traffic engineers, which suggests that design professionals are not morally deficient people but simply that they have accepted these underlying values without debate, internal or otherwise.

STREETS FOR PEOPLE

At the beginning of this chapter, we wrote that a street is not a place unless there are people there. Cities are for people, and good cities need places where people want to be. We've looked at the history of the street, how we experience streets, and how the Western World and particularly the United States transformed the street and its uses. With a combination of policy, regulation, federal funding, new lending practices at banks, corporate priorities, zoning, and changes in planning and urban design, America built a new transportation system and moved a large part of the population from cities and towns to an unprecedented form of automobile-based, unsustainable land use we now call sprawl.

Having built a way of life around the car, we must be practical and respond to peoples' needs and desires to use their cars, but there are many downsides. The way we use our cars is the biggest single reason why Americans have been the largest contributors to greenhouse gases in the atmosphere and therefore to climate change (you can't spell "carbon" without "car," they say). Over a third of Americans are too young, too old, or too poor to drive, and most live in places where this inability to get around is a real hardship. Those who do drive are part of an increasingly unhealthy population that suffers from all the fallout from inactivity: obesity, heart disease, high blood pressure, diabetes, and the like. The list is long, and the isolation of sprawl contributes to a well-documented epidemic of loneliness and division among Americans. In the name of practicality, people have created auto sewers around the world that diminish the use of the public realm for everyone but drivers.

We changed the nature of how we make and use streets. The building of a city begins with its streets, and we created sprawl by reimagining the street. Functional

Classification's catalog of streets—the highway, the arterial, the collector, and the local roads—was an essential and central part of the operating system for the reinvention of American life. Expanding automobile ownership and putting cars first on public roads in the 1920s was the first step towards land-use regulations that separated uses and created a world in which it is necessary to drive to work, to stores, and home.

Moving forward, we must build in ways that foster community and less driving. In the rest of this chapter, we look at the qualities of streets that make them better places for people.

SEVERAL SCENES OR ONE?

Part of the art of street design is choosing a balance of street types for a city. Some streets are monumental—long and straight, and often visually terminated by a monument or a monumental building. Historically, French designers masterfully plotted streets like that, coding the architecture precisely, so that the background buildings that make up the urban fabric have a majestic presence enhanced rather than diminished by repetition. In extreme cases like the rue Royale in Paris, all the facades between the place de la Concorde and the place de la Madeleine were designed by the Royal Architect Ange-Jacques Gabriel and built before the lots were sold (Figure 1.41). In addition, trees and light fixtures repeat regularly in long, straight rows.

A well-designed monumental street also has a balance in the relationships between its parts, however. Consider the avenue des Champs-Élysées, which begins at the very large place de la Concorde, passes through parkland with freestanding buildings and pavilions of various sizes, goes around a "rond point," followed by the most famous

Figure 1.41: Rue Royale, Paris, France. A hand-tinted view from the place de la Concorde, looking towards the Church of the Madeleine. Since the seventeenth century, the planners and designers of Paris have frequently added Classical street and building designs to the ancient city. Parisian architecture and urbanism have codes that promote formality and repetition, but no one comes home from Paris saying, "The streets were so boring." *Brooklyn Museum / Public Domain*

offsets, each part reinforces the sense that it is a legible, self-contained piece of the town. The sections establish an impression of "local-ness" and intimacy; every address along the street is unique.

The Importance of Physical Design

Urban designers must be generalists, able to bring together expertise and ideas from many fields, including engineering, retail, finance, law, architecture, and city

section of the avenue, and terminates at the Charles de Gaulle-Étoile and the Arc de Triomphe. The Champs-Élysées was once a long avenue in the woods, but over time, French planners made it a series of connected streets, each part with its own character.

Less monumental streets also combine individual segments created over time, perhaps only a block or a few blocks long, laid end-to-end. The character can vary in each segment: the total width and building-height-to-street-width proportion changes, the details adjust, and a mix of buildings old and new unfolds as we travel along the road. When the segments are defined at their ends by intersections, terminated vistas, curves, cranks, or

Figure 1.42: New College Lane, Oxford, England. View looking west under the Hertford Bridge (Sir Thomas Jackson, completed 1914) towards the Sheldonian Theatre (Christopher Wren, 1668) and the Bodleian Library, with the Old Bodleian (1600–1619) on the left and the Clarendon Building (Nicholas Hawksmoor, 1715) on the right. Intersections, terminated vistas, curves, cranks, offsets, and spans can divide a street into spatial segments, making each part a complete scene and a legible piece of the town. The spatial variety creates "local-ness" and orientation, making every address and vantage point unique.

planning. However, regulators who do not design increasingly dominate the field, turning city planning departments into permitting departments. To make our streets places where people want to be, we must focus on what planners call "physical planning," designing or coding the physical form of the city, town, or neighborhood. Distilled down, this is the most crucial professional leadership work involving the built and natural environment today: configuring beautiful, durable, and sustainable places for people. If the urban designer successfully does all those things, his or her design will necessarily include information from many specialized areas of expertise.

For more than half a century, we steadily increased our extravagant dependence on the car. In the present era, our task is to modify and sometimes reverse this financially and environmentally costly experiment. We must pay closer attention to the lessons of history, to see what works and what doesn't. Unlike the global sprawl project and the transformation of ordinary city streets into auto sewers, the new urbanism must take its cues from five thousand years of human experience in building successful, robust human settlements.

WORK OF MANY HANDS

There should be no reason, finally, why the decisions taken by elected authority cannot be larger ones, disciplining anarchy to make the city what it has always been, the ultimate work of human art: making possible the effective action not only of the group but the individual citizen, so liberating what Sophocles called "the feelings that make the town."

— Vincent Scully, *The Death of the Street*

An urban design cannot be an artist's personal vision unless that vision is sufficiently broad and meaningful to attract the general public now and in the future. The public realm is for all citizens. Every architect or developer who designs a building that shapes the public realm should recognize their responsibility to contribute positively to the common good. Urban design and the design of urban architecture are public arts. That differentiates urban design from painting or sculpture designed to be displayed in a gallery or in a collector's private home—where one may choose to look at it or not.

The city is an assembly of public works of art that are never finished, produced by many hands collaborating over time. That said, there are many opportunities for personal vision in placemaking. One distinction between the urban designer and the painter or sculptor working alone is that the urban designer rarely starts with a truly blank canvas or plain piece of marble; there are always existing conditions that constrain and unlock the artistic response, as each successive round of work builds upon the layers that precede it. In one of the best-known examples, Michelangelo took hold of the incoherent public space of the Campidoglio in Rome and used geometry and a common architectural vocabulary to establish a legible order. The masterpiece is an intervention of finite size connected to the surrounding spaces and the fabric of the whole city (Figure 1.43).

This way of seeing cities and streets—as a family of spaces that emerge over time after discrete interventions and refinements—stands apart from the view that dominated city planning and architecture in the second half of the twentieth century. The habit of drawing up simplified, ideal cities was centuries old. But with new enthusiasm for the future, the power of technology, and large-scale change, the postwar Modernists took grandiose utopianism to new levels (low and high). Some assumed the best city form would be the work of a single master. If the tangled and contradictory city of

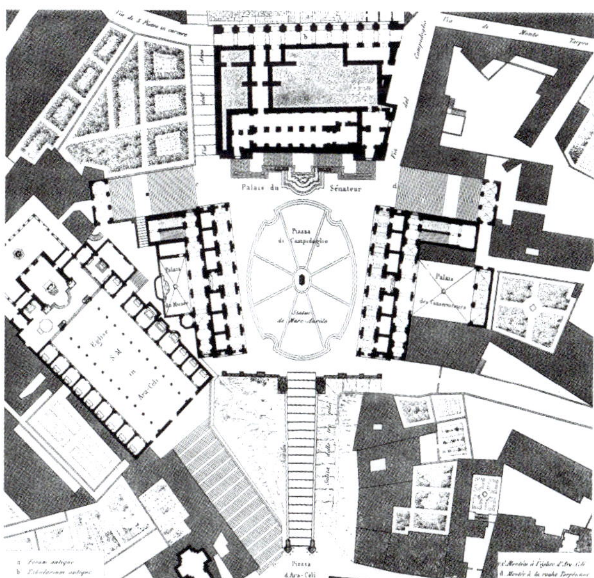

Figure 1.43: Piazza del Campidoglio, Rome, Italy. Michelangelo, 1536–1546. Plan from Paul Marie Letarouilly, *Édifices de Rome Moderne* (1840).

the past could be unified by one vision, some thought, we could make a brilliant new utopia. Predictably, the approach instantly racked up many failures, glaringly in Chandigarh and Brasilia.

Colin Rowe and Fred Koetter helped guide urbanism back to a happier path with their critical writing in *Collage City*, dismissing the idea of the authoritarian, single-authored, supersized, "perfect" utopian vision. They suggested instead that we could build more complex and interesting cities by juxtaposing bite-size utopias.[59]

That is a useful way for the urban designer to see the streets of the contemporary metropolis—as a series of interconnected spatial experiences stitched together and perfected independently and gradually (Figures 1.44–1.46). That makes the city simultaneously the beneficiary of both personal artistic visions and collaborative art. Working within the framework of the urban plan, builders, developers, planners, architects, and landscape architects work on small and large projects spread out over centuries without meeting.

> Rowe and Koetter dismissed the idea of the authoritarian, single-authored, supersized, 'perfect' utopian urbanism, suggesting instead that we can build more complex, interesting cities by juxtaposing bite-size utopias.

CONTEXT MATTERS

Street designers must respond to the context of their site. A small-town residential street should have a different character than a Parisian boulevard, and a winding road in the country should not have the same details as an urban street inside Chicago's Loop. Today, most New Urban designers and developers use the Urban to Rural Transect developed by Andrés Duany and Douglas Duany. The Transect has six "transect zones." At one end of the scale is T-1, wilderness like Yosemite National Park or the Amazon rainforest, and at the other is T-6, the metropolitan core. T-2 is rural land, like farmland. Zones T-3 through T-5 describe different densities and forms of building, adjusted to local conditions, so T-4 is different in a small country town than in Manhattan. Some modify the Transect by adding T-zones or subzones, especially to differentiate the area where high-rise development should be allowed (Figure 1.47).

People not familiar with the Transect sometimes have difficulty visualizing how that works. It can help to think of the Transect in relation to how we dress for different places. A banker making a presentation in a corporate boardroom on Wall Street (T-6) in Manhattan wears a dark suit and black leather shoes with laces (or at least that was still true when we wrote the first edition of this book). The farmer harvesting wheat on his farm (T-2) might wear blue jeans, a flannel shirt, and brown work boots. The third grader walking to school (T-3) has khaki

Figure 1.44: Carrer Avinyó, Barcelona, Spain. Streets can be thought of as segmented spatial experiences, perfected independently and stitched together.

Figure 1.45: High Street, Dumfries, Scotland. A beautifully informal series of public spaces. *Library of Congress / Detroit Publishing Company Photograph Collection*

Figure 1.46: Main Street, Galena, Illinois. The city is an assembly of public works of art that are never finished, produced by many hands collaborating over time. Galena is discussed in a case study in Chapter Two, beginning on page 152.

Figure 1.47: Urban to Rural Transect. Plan and section illustrating the six T-zones in the Urban to Rural Transect (also known as "the Transect"). Duany Plater-Zyberk & Company, 1996–2013. *Courtesy of Duany Plater-Zyberk & Company*

Figure 1.48: Urban to Rural Transect. Transects are locally calibrated. Planners can modify the Transect by adding T-zones or subzones. This rendering illustrates two T-5 zones. Dover, Kohl & Partners, 2006. *© 2006 Dover, Kohl & Partners*

pants and Jordan Aero Flight sneakers, while the writer in her home office in Park Slope, Brooklyn, might put on heels to go out to lunch in Manhattan.

It is important to emphasize that the Transect is about much more than just fashion or style. The size of a building, the massing of a building, the relationship of a building to the street, and indeed the way the street is made are all part of the Transect. An office building on Wall Street in Manhattan (T-6) is taller than a farmhouse in T-2 and has a different relationship to the street. The office building sits at the edge of a large sidewalk, with no setback, while the road in front of the farmhouse probably has no sidewalk, and the house may be far from the road. The rural road in the country (T-2) has no curbs, but a main street in T-4 might have granite curbs and certainly has sidewalks. The traffic lanes in T-4 and T-6 should be narrow, so cars will drive slowly in the space between the buildings they share with pedestrians.

The buildings lining the streets and sidewalks on Wall Street are stone, while the village stores might be in structures covered with wood clapboard. A split-rail fence or a stone wall is appropriate in T-1, but not on Wall Street, where an iron railing is more common. This range of materials and buildings is unlike the prevailing trends in contemporary architecture, where mirror-glass towers can be used in the Central Business District, in an exurban office park, and every T-zone in between.

The Transect figures prominently in the best new form-based codes. New Urbanists use these codes to regulate "the relationship between building facades and the public realm, the form and mass of buildings in relation to one another, and the scale and types of streets and blocks," rather than to separate activities and land uses.[60] Form-based codes emphasize creating urban fabric that shapes a public realm, a significant improvement over the use-based and automobile-centric zoning that contributed to the physical degradation of our towns and cities. To give one example, parking requirements in those auto-centric zoning regulations called for large numbers of parking spaces appropriate for sprawling suburbs—numbers often met by tearing down buildings on city streets, leaving unwelcome gaps in the scene. Form-based

Figure 1.49: Main Street, Cooperstown, New York. Looking west from River Street. The relationship of the buildings to the street shapes the public realm. On a good street like this, the sidewalk, tree, and porch are far more important to the street's character than the traffic lanes.

codes, on the other hand, typically emphasize the importance of continuous streetwalls in commercial areas and call for filling in the gaps on important streets—or at least arranging buildings so that the parking is behind them (see Figures 1.46, 1.47, and 1.48).

WALKABILITY

Restore human legs as a means of travel. Pedestrians rely on food for fuel and need no special parking facilities.

— Lewis Mumford

Figure 1.50: Jones Street, Savannah, Georgia. Savannah's variation on the rowhouse tradition reflects both climate and society. The porch makes a transition between the public street and the private interior. It gives a rhythm to the street and protects us from storms and glare as we enter the house.

This bears repeating: until recently, cities and towns were always walkable. Note that we use "walkability" as an indicator of a community's livability and completeness, not simply its friendliness towards pedestrians. Walkable streets tend to be environments where households, businesses, and institutions prosper. They tend to be the ones where cyclists are most comfortable, and the ones that make public transit most practical. They tend to be the ones where investments in infrastructure and property are rewarded with revenue. They tend to be the safest, most beautiful, and most sustainable. Perhaps most importantly—because these streets allow face-to-face interaction—they encourage the social bonds between neighbors and strangers that help solve problems and let democracy flourish.

Walkability is the baseline in street design for a sound city. From there, we add the other ingredients in the amounts necessary for the individual streets: cycling, driving, deliveries, garbage pickup, emergency response, utilities, parking, and other considerations can be incorporated in a balanced way once walking is established as the foundation of the design.[61] To return to the time-tested model of successful mixed-use cities and towns, the pedestrian's needs must come first: vehicles should be accommodated but not at the expense of the pedestrian.[62]

PRINCIPLES OF WALKABILITY

As we saw in our visit to East 70th Street, the basic principles of making streets where people want to walk are simple. These places are shaped, comfortable, safe, connected, interesting, and memorable.

Shaped

Walkable streets are well-defined spaces shaped by buildings and trees like an outdoor room. The "walls" of the room are the facades of the buildings and/or the column-like trunks of street trees. The proportional relationship between building height and street width and the continuity of the streetwalls produce what urban designers call "the degree of spatial enclosure," the main determinant of the sense of place and character. When the space between the buildings is too wide to give a sense of enclosure, or when there are gaps in the wall, street trees can be used to shape the space.

Comfortable

When the weather is hot, people walking want shade. In cold places, pedestrians like access to sunlight. In many climates, deciduous trees are an ideal solution, blocking the sun in the summer but not during the winter. In the subtropics, people seek protection from sudden storms. Awnings, marquees, arches, colonnades, galleries, and other architectural devices also work well. Awnings over storefronts make walking more comfortable and protect the interior from glare. Architectural features that address

the effects of weather can be a crucial ingredient in local distinctiveness. For example, in the Middle East or Central America, the close placement of buildings on the narrowest village streets keeps more of the street in shade during the hot part of the day. In Bologna and Turin, arcades over the sidewalk became signature features of the two cities.

Safe

When walking, we choose streets and routes where we feel reasonably free from danger. Without necessarily stopping to think, we size up the risks around us and move towards safe ways to move about. Streets that seem to be watched over by windows, doors, storefronts, or balconies feel safer than streets lined with blank walls; this impression of "natural surveillance" has been shown to reduce street crime. Having other people around going about their business is additionally reassuring. Safety also means having the confidence that if we fall ill or have an accident, a call for help will be heard.[63] Finally, pedestrians want to feel safe from the dangers posed by cars. The best defense against mayhem in pedestrian–auto collisions is to keep motoring speeds low by design. Any excessive width in the area devoted to motoring—whether in the number of lanes or their dimension—pushes speeds higher and discourages walking.

Figure 1.51: Grant Avenue, San Francisco, California. Walkable streets are like comfortable outdoor rooms, places where we want to get out of our cars and enjoy the public space.

Connected

When walking, we choose the paths that take us where we want to go. Walkable streets are almost always part of an integrated network—ideally one with small blocks—so many possible routes exist. The best street networks offer a variety of routes and experiences. When street vistas create a sequence of legible parts marked by landmarks that help with wayfinding and orientation, walkers have a sense of how the street space fits into the neighborhood's fabric. Not surprisingly, studies undertaken to analyze "space syntax" consistently show that city streets with the most connections to the rest of the network tend to be streets where stores get the most traffic and business.[64]

Interesting, Memorable

Public places entertain us. We are attracted to places that are beautiful and distinctive. The need can be satisfied with large design "moves," like a formal axis framing a ceremonial building, or with something as subtle as a canopy of leafy autumn color on a quieter street.

Places with richness, texture, and character attract us; we tend to return to the places where the three-dimensional geometry of urban design and architecture is used, in conspicuous or subtle ways, to create street scenes that unfold in some theater-like progression. We naturally want the backdrop of daily life to be agreeable, not drab. Beauty is not something extra to be added after all the other decisions about a street are already made; it is the one ingredient without which no street is ever "complete" (Figures 1.52 and 1.53).

When we build a street scene according to these principles, we get more than a mere transportation mechanism or even an address for pieces of real estate. The street becomes a *communications device*, sending messages about what is important to society. A memorable street is one with human creativity on display. That might come from architecture, or landscape, or signs, or art, or some or all combined. There is no better way to send a message of welcome and inclusiveness than to offer our neighbors and visitors a street made by humans for humans.[65]

Figure 1.52: Rue de Seine, Paris, France. We are drawn to places with richness, texture, and character, where street scenes unfold in a theater-like progression.

Figure 1.53: Nanjing Road, Shanghai, China. Memorable streets are places with human creativity on display, in ways subtle or bold: distinctive architecture, art, artful signs, color, and plantings can all contribute.

SLOW DOWN—SPEED KILLS

Drivers have fewer collisions and cause fewer traffic deaths when they slow down. That is true on limited-access highways and on city streets where motor vehicles, people walking, and cyclists are all in close contact. An important step in making better, safer streets for all, therefore, is to slow cars down when they enter walkable areas. An important benefit is that we can then replace a lot of the traffic engineer's detritus that gives streets an auto-based scale with human-scaled elements, making better places.

Drivers in speeding cars need lots of time and space to avoid hitting things: the faster they go the less they see (Figure 1.55), and the more ground their vehicle will traverse in the time it takes them to react, lift up on the accelerator, and move their foot to the brake pedal. Simple physics tells us that cars going faster will take more time and distance to come to a stop than cars moving slowly.

Figure 1.54: A Vision Zero image from the City of Seattle illustrating how rapidly pedestrian deaths rise when vehicle speeds increase. At 20 miles per hour, 90 percent of the victims will survive. At 40 miles per hour, 90 percent of the victims will die. *Seattle Vision Zero / Public Domain*

15 mph

20 mph

25 mph

30 mph

Figure 1.55: Cone of Vision Simulation. NACTO, 2010. The diminishing circles show that small increases in speed greatly decrease what a driver sees. The combination of reduced vision, the increased distances required for stopping, the greater amount of time needed for the driver to react to an obstacle, and the increased harm from impact as speeds go up explain why speed kills. *Courtesy of NACTO*

Then there is the question of what happens when a car hits a person walking. Experience shows that reducing the speed limit from 30 to 20 miles per hour brings an 80 percent reduction in the number of casualties. If a car going 20 mph starts to brake and hits a pedestrian while traveling 15 mph, most pedestrians will survive the crash, frequently with only minor injuries. At 30 mph, almost all collisions result in severe injuries, with half the crashes fatal: at 40 mph, the pedestrian will die 90 percent of the time.

People walking naturally feel more comfortable in a human-scaled space that has not been disfigured by bold striping, large reflective signs, and cheap, ugly elements like white plastic sticks. A fundamental principle in the design of shared space is to take away all the signs and markings that help the speeding driver feel comfortable while also making the pedestrian feel like an intruder in an unsafe space made for machines. Even when urban streets are *not* shared spaces, however (which is most of the time), pedestrians should feel comfortable crossing the street.

"Twenty miles per hour feels slow in a modern car," we wrote in the first edition of *Street Design*, "but the injury and fatality statistics for accidents at higher speeds are as dramatic as the numbers for drunk driving accidents." Since that time, a large number of towns and cities around the world have passed 20 mph (30 kph) speed limits on many streets.[66] On February 20, 2020 (an auspicious date), traffic safety advocates from 130 countries adopted the "Stockholm Declaration" at a road safety conference in Sweden. It requires 20 mph/30 kph limits where "vulnerable road users and vehicles mix"—for safety, air quality, and climate action.[67] We hope that groups like Families for Safe Streets, an advocacy group with 20 chapters across the United States, will be as successful as earlier groups like MADD (Mothers Against Drunk Driving) and the Stop de Kindermoord (Stop the Child Murder) movement that led to road safety reform sixty years ago in the Netherlands.[68] As we prepare to publish this second edition, Families for Safe Streets has successfully persuaded New York State to pass Sammy's Law, a road safety bill named for a 12-year-old boy killed by a speeding car in 2013.[69]

Figure 1.56: Edison Lane, Fort Myers, Florida. The painting by James Dougherty captures the sense of liberty that could come from the design of healthy, low-traffic neighborhoods that put people first. On this proposed street, parking and through traffic would be reduced to a minimum, impervious surfaces and curbs would be downsized and minimized, space for walking, biking, and planting trees would be generous, and unnecessary markings and signs would be eliminated. *Courtesy of Dover, Kohl & Partners*

COMPLETE STREETS, COMPLETER STREETS AND INCOMPLETE STREETS

The National Complete Streets Coalition has been phenomenally successful. Now a part of Smart Growth America, the Coalition has brilliantly and effectively written and promoted policies for the guidance of transportation planners and traffic engineers in the design and operation of "the entire roadway with all users in mind," including cyclists, public transit riders, and pedestrians of all ages and abilities. As of 2025, over 1,700 Complete Streets policies have been passed in the United States. Legislators and policymakers are not placemakers or urban designers, however, and many Complete Streets are just the old transportation corridors with painted bike lanes added.[70]

Figure 1.57 shows a complete street in suburban Florida. With the exception of the painted bike lanes, the street could have been built seventy-five years ago. There is little reason to walk there, and the painted bike lanes do not protect the cyclists from speeding traffic. Figure 1.58 shows a one-way, sub-urban arterial in the heart of America's most walkable and least car-dependent city. The sidewalks, narrowed by New York's Department of Traffic—yes, that was its name then—in the 1950s and 1960s, remain unchanged.[71]

Figure 1.57: A Complete Street in Fort Lauderdale, Florida. This unpleasant, pedestrian-repellent space should be a Completer Street, supporting more than multi-modal transportation.

Figure 1.58: First Avenue, New York, New York. View looking north from East 6th Street. NYC DOT, 2008. A multi-modal Complete Street in Manhattan, where three-quarters of the households are car-free. A picture-perfect example of the Monderman Rule (page 513), the one-way design emphasizes traffic flow and makes suburbanites feel at home, encouraging them to drive in and out of the city. *Courtesy of NACTO.*

BEYOND FUNCTIONAL CLASSIFICATION: A CATALOG OF TEN ESSENTIAL STREET TYPES

We close Chapter One with new drawings for ten essential street types for creating walkable and sustainable neighborhoods, towns, and cities. As we've seen, Functional Classification's meager menu of thoroughfares—the limited-access highway, the arterial, the collector road, and the local road—put traffic and Level of Service first and kick the pedestrian to the side of the road. In the twenty-first century, America and the world need to reduce dependence on motor vehicles by making places where people want to get out of their cars to enjoy the public realm.

People have built places like that for thousands of years. In Chapters Two, Three, and Four, we will look at some of the best examples, organized by street type. Before that, the catalog of essential street types introduces the different streets and the elements that define them. That begins with their physical form and how they are used.

The multiway boulevard moves the most traffic but also has places for people to walk, sit, and gather. Another type later in the catalog, pedestrian streets, puts people first. Sometimes, but not always, the pedestrian streets are car-free. Other streets in the catalog have different forms and functions. Combined, the ten essential street types are the building blocks for great cities and towns.

The drawings on the following pages illustrate, with parallel graphics and annotations, twenty-seven variations of the ten street types. These are just the tip of the iceberg. When the street types are used in real places, local conditions will bring specific problems. Design is a way to solve problems. Expanding the range of choices exponentially increases the number of possible designs; human ingenuity, once unleashed, will produce unique places.

Interactive, three-dimensional models of the ten essential street types are online at www.street.design/models. We hope urban designers, planners, engineers, elected representatives, and activists will find the types valuable and seek new ways to use them.

Figure 1.59: Canal Street (unrealized), Palm Beach County, Florida. This painting by David Oliveira captures the sense of liberty that could come from the design of healthy, low-traffic neighborhoods that put people first. This proposed canal street shows a way to adapt to sea level rise in coastal Florida. It is more green, more blue, and less gray; we can leverage compact designs to provide abundant housing while minimizing the town's environmental footprint. Space for walking, biking, and planting trees would once again be generous, because the minimum space would be devoted to moving and storing cars. *Courtesy of Dover, Kohl & Partners*

TEN ESSENTIAL STREET TYPES

Avenues and Boulevards
- Avenue
- Boulevard
- Multiway Avenues and Boulevards

Promenade Streets
- Promenade Street with planted island
- Promenade Street with paved island (*Rambla*)

Main Streets and High Streets
- Main Street in a Town
- Main Street in an Urban Core

Core Streets
- Market Street
- Work Street
- Village Street
- Rowhouse Street

Garden Streets
- English Garden Street with Shared Garden
- English Garden Street with Parking Grove
- American Garden Street
- American Garden Street with Trail

Slow Streets
- Yield Street
- Bicycle Street (*Fietsstraat*)
- Living Street (*Woonerf*, Home Street, Play Street)
- Shared Space

Pedestrian Streets and Passages
- Pedestrian Mall
- Midblock Arcade
- Pedestrian Passage (*Paseo*)
- Step Street
- Green Street

Canal Streets
- City Canal Street
- Village Canal Street

Drives
- City Drives
- Park Drives

Parkways
- Local & Regional Parkways
- Limited Access Parkways

**Figure 1.60:
Avenues and Boulevards**

Example 1: Avenue.

See Park Avenue, New York, page 115;
or Monument Avenue, Richmond, page 118.

Avenues and Boulevards are wide, tree-lined streets with formal geometry, most often with a center median. These are the grandest streets in the traditional city.

Avenues go from point to point and are visually terminated at one or both ends.

Boulevards are designed to go through the fabric of the city. By the traditional definition, they do not have terminated vistas.

Ⓐ Multiple Rows of Trees

Ⓑ Wide Sidewalks

Ⓒ Buildings Tall Enough to Contain the Space

Ⓓ Dedicated Lanes for Public Transit

Ⓔ Bikeways

Ⓕ On-Street Parking and/or Pickup-Dropoff, Loading Zones, Parklets, etc.

Ⓖ Terminated Vistas

**Figure 1.61:
Avenues and Boulevards**

Example 2: Boulevard.

See Boulevard St. Germain, Paris, page xv;
or Queens Road, Charlotte, page 123.

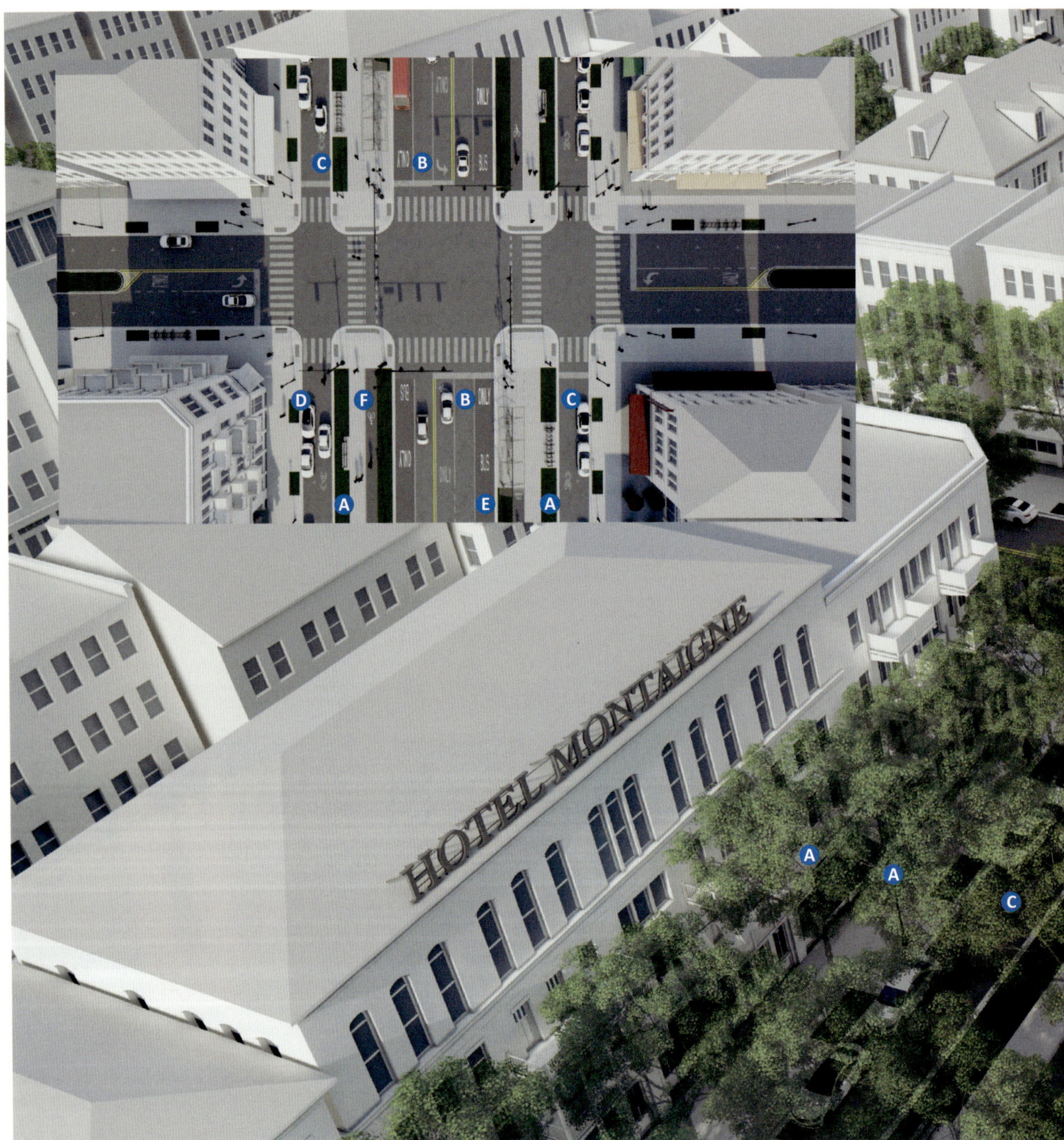

Figure 1.62:
Avenues and Boulevards

Example 3: Multiway Avenue or Boulevard.

See Avenue Montaigne, Paris, page 104; or Eastern Parkway, Brooklyn, page 112.

Multiway Boulevards have lanes in the center for transit and through traffic, slower side lanes for local traffic, and on-street parking. The through lanes are separated from the side lanes by planted medians. The medians, side lanes, parking, sidewalks, and perhaps bicycle lanes combine to form a slow-speed, pedestrian-dominated realm on each side of the multiway boulevard, making it reasonable to face a wide, busy street with commercial storefronts or residential stoops. The effect is an address that is both prominent and green.

A Multiple Allées of Trees

B Through Lanes

C Side Lanes

D Wide Sidewalks, On-Street Parking, and/or Pickup-Dropoff Zones

E Dedicated Lanes for Public Transit

F Bikeways/Promenades

**Figure 1.63:
Promenade Streets**

Example 1: Promenade Street with
planted island.

See Boulevard de Rochechouart, Paris,
page 134; or Commonwealth Avenue,
Boston, page 137.

Promenade Streets include a wide walkway for strolling and people-watching.
In northern Europe and North America, the promenade streets are usually planted
like parks to bring nature into the city. The walkway, or promenade, can be added
to avenues and boulevards, either in the center of the street or to one side. In Spain
and South America, promenade streets are called "*Ramblas*." Both types can vary in
length from a few blocks to many blocks.

- **A** Central Walkway with Rows of Trees
- **B** Benches
- **C** Café and Restaurant Seating
- **D** Bikeways
- **E** Traffic Lanes

Figure 1.64:
Promenade Streets

Example 2: Promenade Street
with paved island (*Rambla*).

See Paseo del Prado, Havana, page 133;
or Les Rambles, Barcelona, page 130.

Figure 1.65:
Main Streets (High Streets)

Example 1: Main Street in a Town.

See High Street, Oxford, page 17;
or Main Street, Galena, page 47.

Main Streets or High Streets are typically the armature of a town or city plan, with shops and offices shared by adjacent neighborhoods and the larger community. They are found along the natural paths between everyday destinations. The details vary, but one thing is constant: a successful main street is a walkable, mixed-use place where people want to get out of their cars and explore.

A Storefronts on Ground Level, with Apartments, Offices, or Lodging Above

B Streetwall

C Wide Sidewalks

D On-Street Parking and Pickup-Dropoff, Loading Zones, Parklets, etc.

E Expression Lines on Facade

**Figure 1.66:
Main Streets (High Streets)**

Example 2: Main Street in an Urban Core.

See Broadway, Portland;
or Ginza Dori, Tokyo, page 167.

**Figure 1.67:
Core Streets**

Example 1: Market Street.

See Broad Street, Oxford, page 169;
or High Street, Marlborough, page 173.

Core Streets do not necessarily have the consistent grouping of storefronts found on Main Streets, but they are a mainstay of city neighborhoods. Core streets connect with the larger street network. The buildings lining Core Streets can contain workplaces, stores, apartments, and houses.

Market Streets have a large central paved area for trucks, tents, and carts on market days.

Work Streets vary because office work, creative arts, and light industry all take place in myriad building types, sizes, and settings.

- (A) Connected to the Surrounding Street Network
- (B) Sidewalks
- (C) Storefronts
- (D) Wide Flexible Paved Space
- (E) On-Street Parking
- (F) Office or Factory Buildings

Figure 1.68:
Core Streets

Example 2: Work Street.

See Cary Street, Richmond; or Railroad Avenue, Great Barrington, page 179.

**Figure 1.69:
Core Streets**

Example 3: Village Street.

See Longmoor Street, Poundbury, Dorchester, page 474; or Swift Street, Buena Vista, page 515.

Village Streets are narrow streets lined with houses, gardens, workspaces, and the occasional shop. The streets follow local topography.

Rowhouse Streets are a staple in middle-density urban neighborhoods. Attached houses frame the street. They have small setbacks and may have stoops or porches, bay windows, shallow dooryards, and/or raised basements below the level of the sidewalk.

A Narrow Traffic Lanes

B South-Facing Gardens and Garden Walls

C Attached Houses

D Detached Houses

E Stoops and Porches

F Dooryards and English Basements

G Rear Gardens, Carriage Houses, and Accessory Dwelling Units

Figure 1.70:
Core Streets

Example 4: Rowhouse Street.

See Sydney Place, Brooklyn, page 187;
or Bartram Street, Glenwood Park, Atlanta, page 481.

Figure 1.71:
Garden Streets

Example 1: English Garden Street with Shared Garden.

See Lexham Gardens, London; or
Cornwall Gardens,
London, page 210.

English Garden Streets are a unique blend of streets, squares, and parks. British developers used this form to compensate for a shortage of private gardens in the compact lots around the park. The shared park has tall shade trees and may be defined by wrought iron fences, hedges, or low walls. An open variation has space for parking within the grove of trees (Figure 2.168).

A Shared Gardens
B Defined Edges (Fences, Hedges, Low Walls)
C Attached Building Types Enclose the Space
D Parking Grove

**Figure 1.72:
Garden Streets**

Example 2: English Garden Street with **Parking Grove.**

See Beaufort Gardens, London, page 210.

**Figure 1.73:
Garden Streets**

Example 3: American Garden Street.

See Pine Street, Boulder; or
Greenway Terrace, Forest Hills Gardens,
Queens, page 298.

American Garden Streets are the tree-lined streets with single-family houses prevalent in the streetcar-era garden suburbs and many recent Traditional Neighborhood Developments. These are usually **two-way** streets. Planting strips with shade trees separate the sidewalk from the automobile space on each side. Low garden walls, picket fences, and hedges can define the boundaries of the public right-of-way. Front yards and porches form an intermediate space between the private houses and the public street.

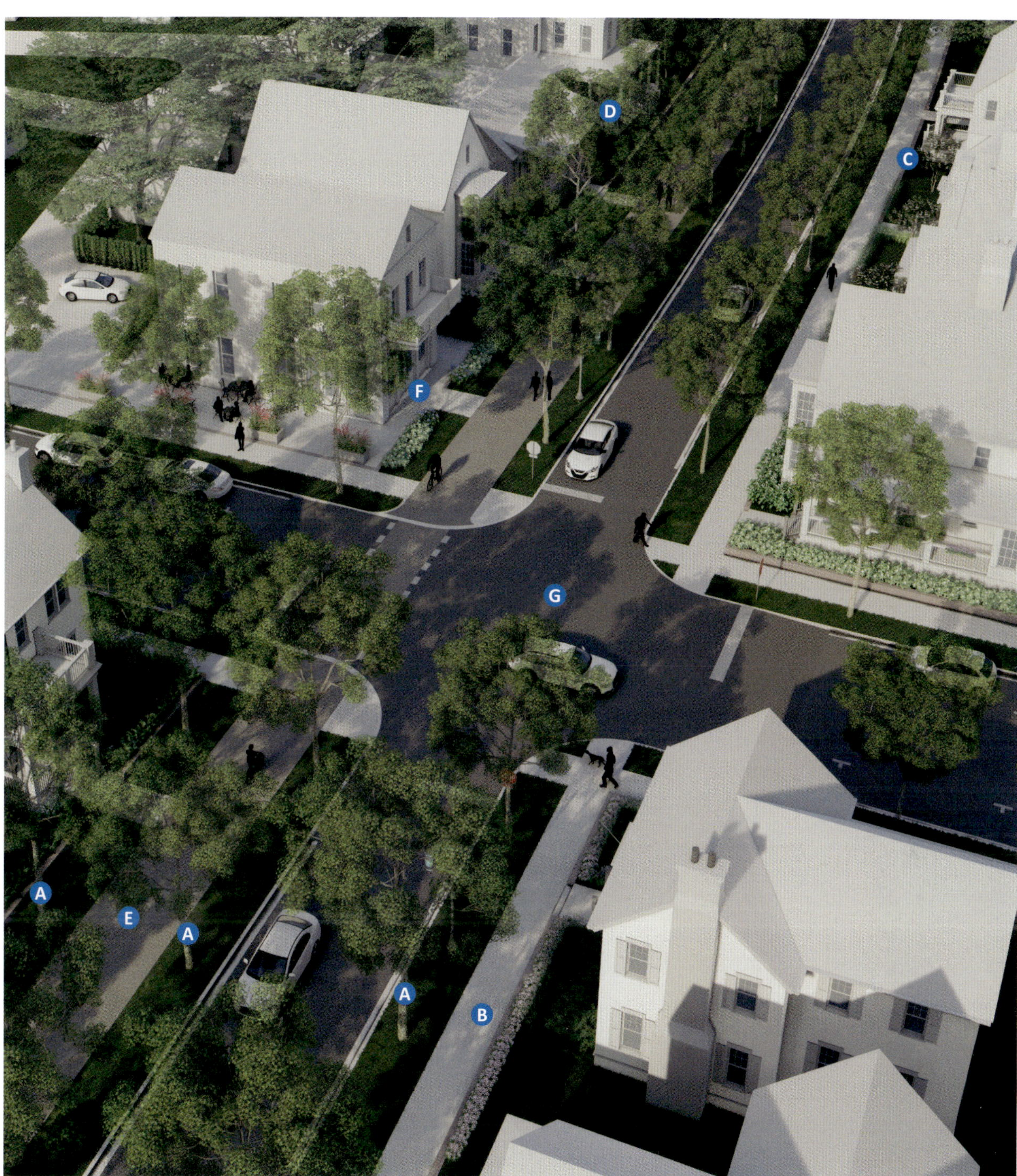

A Rows of Trees

B Sidewalks

C Garden Walls, Fences, or Hedges

D Front Porches

E Third Row of Trees and Connection to Regional Trail

F Occasional Corner Store

G Raised Intersection

**Figure 1.74:
Garden Streets**

Example 4: American Garden Street with Trail.

See West Drive, Warren Park, Rogers, page 494.

Figure 1.75:
Slow Streets

Example 1: Yield Street.

See Twain Avenue, Davidson, page 519;
or Balfour Place, London, page 209.

Slow Streets control driving speeds by giving priority to pedestrians and cyclists.

Yield Streets or "give-way" streets allow slow traffic on narrow streets. Only one car can go through at a time, forcing the car traveling in the opposite direction to "yield."

A Bicycle Street (Dutch "fietsstraat") gives bicycles priority over cars and trucks. In the Dutch version, signs say *"Auto te gast"* ("Your car is a guest"). The carriageway is divided into three narrow lanes instead of two. The middle lane has a rough texture that tells drivers to go slowly.

A Narrow "Give-Take" Carriageways
B Intermittent On-Street Parking and Trees
C Storefronts
D Smooth Pavement for Cycling
E Rough Texture to Slow Passing
F Limited On-Street Parking and Pickup and Dropoff Zones
G High-Visibility Crossings

Figure 1.76:
Slow Streets

Example 2: **Bicycle Street**
(*Fietsstraat*).

See Adriaan Pauwstraat, Delft; or **Dorpsstraat,**
Rotterdam, page 521.

**Figure 1.77:
Slow Streets**

Example 3: Living Street
(*Woonerf*, Home Street, Play Street).

See Tuinstraat, Amsterdam; or Hennepstraat,
Utrecht, page 522.

Living Streets (Dutch "woonerfs" or "woonerven") have a maximum speed of four miles per hour (walking speed). The speed limit applies to cyclists as well as cars. Living Streets are narrow, usually curbless shared spaces where children can safely play.

Pedestrian Malls are car-free shopping, entertainment, and main streets. Briefly fashionable in the 1960s, but then discredited as a failed experiment, they are now making a comeback. "Ped malls" require careful consideration.

A Mixed-Use Main Street Buildings

B Walkways

C Rough Paver Texture for Slow Passing

D Limited On-Street Parking and Pickup and Dropoff Zones

E Tight Geometry for Slow Vehicle Maneuvers

F High-Visibility Crossings

G Play Equipment, Plantings, and Seating

Figure 1.78:
Pedestrian Streets and Passages

Example 1: Pedestrian Mall.

See Pearl Street, Boulder;
or Lincoln Road, Miami Beach, page 523.

Figure 1.79:
Pedestrian Streets and Passages

Example 2: Midblock Arcade.

See Westminster Arcade, Providence, page 217;
or Burlington Arcade, London, page 213.

Passages are generally narrow, car-free streets in the middle of a block. They introduce routes for pedestrians that cut the length of long blocks.

Midblock Arcades are formal passages, often roofed and/or encased within buildings, that showcase storefront businesses by concentrating foot traffic.

Pedestrian Passages and Paseos lend themselves to leisurely walks and may be paired with parallel boulevards or avenues to offer an intimate alternative to those busier, wider streets.

- **A** Walkways
- **B** Skylights
- **C** Storefronts
- **D** Café and Restaurant Seating

Figure 1.80:
Pedestrian Streets and Passages

Example 3: Pedestrian Passage or Paseo.

See Cour du Commerce, Paris, page 216;
or Centre Place, Melbourne, page 305.

Figure 1.81:
Pedestrian Streets and Passages

Example 4: Step Street.

See The Vennel, Edinburgh, page 212; or
Ninety-Nine Steps, Charlotte Amalie,
St. Thomas, page 212.

Step Streets give pedestrians direct routes up and down slopes or hills that are
too steep for a straight street. When topography requires that roads going up the
hill must wind back and forth or that access roads are far apart, step streets provide
convenient access for pedestrians.

Green Streets have no paved roadways and are closed to cars. Houses face the
street as in the American Garden Street, but the space between the sidewalks
has gardens and parkland. All cars are in the alleys behind the houses or on the
cross streets.

A Houses
B Common Space
C Walkways
D Stepped Passage
E High-Visibility Crossings

Figure 1.82:
Pedestrian Streets and Passages

Example 5: Green Street.

See Radburn B Park Footpath, Fair Lawn;
or Hobbs-Hanley, Baldwin Park, Orlando, page 546.

Figure 1.83:
Canal Streets

Example 1: City Canal Street.

See Steenhouwersdijk, Bruges; or
Reguliersgracht, Amsterdam, page 506.

Canal Streets have a canal in the center and buildings on each side. City Canal Streets have streets on at least one side of the canal. Village Canal Streets may have only one walkway.

A Trees
B Sidewalks
C On-Street Parking
D Walkways

**Figure 1.84:
Canal Streets**

Example 2: Village Canal Street.

See LaFayette Canal, I'On, Mount Pleasant.

Figure 1.85:
Drives

See Ocean Drive, Miami Beach, page 583;
or Riverside Drive, New York, page 222.

Drives have a park or natural scenic view on one or both sides of the street. Historically, drives were created for relaxing carriage rides, before motorcars existed. They can be in a park or on the edge of the city. When buildings are on one side and a park or scenic view on the other, the walkway on the park side can be treated like a park walk, with park benches and room for strolling.

Parkways are limited-access roads designed to combine easy long-distance travel with the experience of driving in the countryside or a park. In the country, parkways can open to long vistas. In more crowded areas, trees on both sides of the parkway buffer the view. Originally, parkways stopped at the city's edge, where they met green boulevards.

A Forest, Park, or Other Open Space
B Walkways, Bikeways, and Trails
C The "Townless Highway"

**Figure 1.86:
Parkways**

See George Washington Parkway, Arlington County; or
Bronx River Parkway, Westchester County, page 236.

NOTES

1. Kunstler has a variation on that quote in his TED Talk. This version comes from correspondence with the authors. James Howard Kunstler, "The Ghastly Tragedy of the Suburbs," TED Talk, February, 2004, https://www.ted.com/talks/james_howard_kunstler_the_ghastly_tragedy_of_the_suburbs. Also see Kunstler's first two urban planning books *The Geography of Nowhere: The Rise and Decline of America's Man-Made Landscape* (Simon & Schuster, 1993) and *Home from Nowhere: Remaking Our Everyday World for the 21st Century* (Touchstone Press, 1998).

2. In *The Experience of Place* (Knopf, 1990), Tony Hiss refers to this as "simultaneous perception." Harvard brain scientist Jill Bolte Taylor has a related take on how we experience place, formulated after she experienced and observed her own stroke. See pages 607–608 in Chapter Five.

3. "Beauty is truth, truth beauty," the English Romantic poet John Keats wrote in his poem "Ode on a Grecian Urn." Poets, philosophers, artists, and architects have expressed similar sentiments at least since the time of Socrates and Plato, who said that beauty reflects ideal, unseen forms. "Beauty is a manifestation of secret natural laws, which otherwise would have been hidden from us forever," Johann Wolfgang von Goethe wrote over two millennia later. See J.W. von Goethe, translation by Bailey Saunders, *The Maxims and Reflections of Goethe* (Macmillan,1906): No. 183. For a concise but scholarly discussion of the ideal conception of beauty, see Crispin Sartwell, "Beauty," *The Stanford Encyclopedia of Philosophy* (Fall 2024 Edition), Edward N. Zalta & Uri Nodelman (eds.): https://plato.stanford.edu/archives/fall2024/entries/beauty/. For a selection of well-known quotes on the subject, see Brainy Quotes: https://www.brainyquote.com/topics/beauty-quotes.

Before Modernism and architectural education rejected the idea that humans respond to universal principles of beauty found in nature, architectural treatises and textbooks frequently discussed the principles and how to use them. John Belcher's *Essentials in Architecture, an analysis of the principles & qualities to be looked for in buildings* (Royal Institute of British Architects, 1907), is an example that can easily be downloaded: https://archive.org/details/070131-essentials-in-architecture-by-john-belcher-1907. Belcher's first chapter is about the principles of Truth and Beauty. In *The Architecture of the Classical Interior*, Steven W. Semes discusses Classical design principles and how to use them in interiors. Steven W. Semes, *The Architecture of the Classical Interior* (W.W. Norton & Company, 2004): 53–64.

Today, architects and neuroscientists look for scientifically measured, reproducible evidence that increases our understanding of beauty and how it affects human beings. That knowledge would help us get beyond the debate and also allow us to design places that make people feel good. So far, we have more conjecture and hypothesis than scientific proof, but new studies slowly but surely move the field forward. "Goodness, beauty, truth, dignity, and compassion all share the cognitive signature of neurological coherence, bound by synchrony. When the mind is pervaded by these qualities we feel deeper, more alive, and more whole." Jonathan F. P. Rose writes in *The Well-Tempered City: What Modern Science, Ancient Civilizations, and Human Nature Teach Us About the Future of Urban Life* (Harper, 2016): 382.

Two recent articles specific to how we experience place are Sheng Wang, Guilherme Sanches de Oliveira, Zakaria Djebbara, and Klaus Gramann, "The Embodiment of Architectural Experience: A Methodological Perspective on Neuro-Architecture," *Frontiers in human neuroscience*, May 9, 2022, https://pubmed.ncbi.nlm.nih.gov/35615743/; and Stephen E. Palmer, Karen B. Schloss, Jonathan Sammartino, "Visual aesthetics and human preference," *Annual review of psychology*, September 27, 2012: 64: 77–107, https://pubmed.ncbi.nlm.nih.gov/23020642/. Also see Semir Zeki, "Architecture: Not a Luxury—Only a Necessity," *Architectural Design*, September 5, 2019: 89, 5, 14–19, https://onlinelibrary.wiley.com/doi/10.1002/ad.2473. "The appreciation of beauty, no matter where found, is hardwired into our brains," Zeki writes.

Two valuable studies for street design and urban design are Tenna D.O. Tvedebrink, Lars B. Fich, Elisabetta Canepa, Zakaria Djebbara, Asbjørn C. Carstens, Dylan Chau Huynh, and Ole B. Jensen, "Motion and Emotion: Understanding Urban Architecture through Diverse Multisensorial Engagements," *The Journal of Somaesthetics*, (October, 2023), https://www.researchgate.net/publication/370211545_Motion_and_Emotion_Understanding_Urban_Architecture_through_Diverse_Multisensorial_Engagements; and Lara Gregorians, Zita Patai, Pablo Fernandez Velasco, Fiona E. Zisch, and Hugo J. Spiers, "Brain dynamics during architectural experience: prefrontal and hippocampal regions track aesthetics and spatial complexity," BioRxiv 2025.01.09.631831, 10.1101/2025.01.09.631831.

A good book on the subject is *Brain, Beauty, and Art, Essays Bringing Neuroaesthetics into Focus* (Oxford, 2021), edited by Anjan Chatterjee and Eileen Cardillo. Also see the sections on embodied cognition in Mark Hewitt, *Draw in Order to See: A Cognitive History of Architectural Design* (ORO, 2020) and John Massengale *et al.* "Neuroscience & Measuring the Experience of Place," Place Science, https://www.place-science.com/engelsberg/.

The question remains how to translate these ideas into practice. Daniele Quercia is the Director of Responsible AI at Nokia Bell Labs in England. He has worked for several years on two related projects based on Google Maps. One project is to teach Google Maps to make beautiful routes for cycling and walking, rather than simply choosing the most direct route. The other proposes using Artificial Intelligence to analyze Google Street Views and redesign the streets as beautiful places. See "The shortest path to happiness: Recommending beautiful, quiet, and happy routes in the city" at https://scholar.google.com/citations?user=nPyDLd0AAAAJ and Quercia's TED Talk, "Happy Maps" at https://www.ted.com/talks/daniele_quercia_happy_maps.

More traditionally, *Of Streets and Squares, Which public places do people want to be in and why?* (Create Streets,

2019) by Maddalena Iovene, Nicholas Boys Smith, and Chanuki Illushka Seresinhe summarizes public and professional understanding of the experience of place in historical and contemporary books and papers. Nicholas Boys Smith and Nikos A. Salingaros, "LLMs Judging Architecture: Generative AI Mirrors Public Polls" (Preprints. org, 2025). We will say more about this in Chapter Five. Also see endnotes 5 and 12 in this chapter and "The Experience of Place" on pages 605–609.

4. Gertrude Stein said about Oakland, California, that there is "no there there," in *Everybody's Autobiography* (Random House, 1937): 289. Gordon Cullen, the noted advocate of picturesque Modernist planning, thought the concept of "Here and There" (or here versus there) was an important one. See Cullen, *The Concise Townscape* (Architectural Press, 1961): 182.

5. Léon Krier, *The Architecture of Community,* ed. Dhiru A. Thadani and Peter J. Hetzel (Island Press, 2009): 28. Also see Richard Economakis, Léon Krier, *ANTA, Vol. 8, Fall 2025: Léon Krier* (University of Notre Dame School of Architecture, 2025). Krier (1946–2025) participated closely in the preparation of both books. Each contains material not seen in other publications.

6. Laurence Aurbach, the author of *A History of Street Networks: From Grids to Sprawl and Beyond* (PedShed Press, 2020), wrote a short online history of the grid with more information about early grids: https://pedshed.net/?p=12. In *The Well-Tempered City*, the planner and developer Jonathan Rose succinctly compares functional, symbolic, and philosophical differences in a history of urban grids in the Western and Eastern worlds. *Op. cit.*, pages 56–93.

7. Michel Pauls, "the american grid," Recivilization, https://recivilization.net/UrbanDesignPrimer/111theamerican grid.php. Quoted by Paul Knight, "The American Grid," The Great American Grid, July 25, 2012, https://web. archive.org/web/20131107184731/http://www.thegreata-mericangrid.com/archives/1922. An excellent and comprehensive book on city grids has come out since we wrote the first edition of *Street Design*: see Joan Busquets, Dingliang Yang, and Michael Keller, *Urban Grids: Handbook for Regular City Design* (Oro Editions, 2019). Another book that came out after the first edition of *Street Design*, the *Atlas zum Städtebau, Band 2, Straßen* (Hirmer, 2015) by Vittorio Magnago Lampugnani, Harald R. Stühlinger, and Markus Tubbesing, focuses on fewer streets but in amazing detail, with beautiful drawings of the urban context, and street details of the adjacent buildings. All the streets are European.

8. Charles Dickens, *American Notes for General Circulation* (Harper & Brothers, 1842): 39.

9. John W. Reps, *Monumental Washington: The Planning and Development of the Capital Center* (Princeton University Press, 1967): 97.

10. George R. Collins and Christiane Craseman Collins, *Camillo Sitte: The Birth of Modern City Planning* (Rizzoli, 1986): 195–196.

11. Sir Raymond Unwin, *Town Planning In Practice: An Introduction to the Art of Designing Cities and Suburbs* (Princeton Architectural Press, 1994): 115.
In the context of formal, picturesque, and hybrid design, the Corso Vittorio Emanuele II makes a good case study. After Rome became the capital of the new Kingdom of Italy in 1871, planners for the Comune di Roma (City of Rome) that replaced the Papal State's rule prepared Piani Regolatori (regulating or master plans) in 1873 and 1883 that included large new corsi (avenues and boulevards) that went from the center of Rome to the east and the west. The street to the west was named after the new king, Victor Emmanuel II. Construction of the Corso Vittorio began in 1886 and continued until 1900. It originated at an enlarged Piazza Venezia (practically and symbolically the crossroads of modern Rome) to the Ponte Vittorio Emanuele II that crosses the Tiber and links the center to the Vatican. The Corso Vittorio was one of the two new streets that began at the Piazza Venezia, the other being the Via Quattro Novembre. Together with the Via del Corso, an old Roman road that runs north as straight as an arrow from the Piazza Venezia to the Piazza del Popolo, the new streets "completed" the Trident of Pope Sixtus V, which connects pilgrimage churches and the Vatican. Steven W. Semes, author of *New Building in Old Cities: Writings by Gustavo Giovannoni on Architectural and Urban Conservation* (Getty Publications, 2024—Jeff Cody, Francesco Siravo, eds.), compares the street diagram to a compass overlaid on an anchor.
The Corso Vittorio was laid out by widening some existing streets and cutting new ones through the existing fabric. Generally, the designers widened the old streets by partially or completely demolishing buildings on the north side of the streets. The designers carefully preserved architectural monuments on both sides, however, skillfully bending the street to create a series of scenographic views of the historic buildings for people traveling on the Corso. Spiro Kostof called the picturesque combination of traffic flow, urban design, and historic preservation "a miracle of common sense." See Kostof, *The Third Rome 1870–1950: Traffic and Glory* (University Art Museum, 1973): 18. It "is a graceful curve from Piazza Venezia to the river that honors the course of the historic Via Papale and utilizes a string of great monuments in an eventful sequence of revealing prospects and pleasant stops," Kostof continued. Also see Thomas Michael Dietz, *The Road from Pope to King: Il Corso Vittorio Emanuele II* (MIT, 2005). For more information see "Missing Street Study: Corso Vittorio Emanuele II" on the Street Design blog at https://street.design/blog/corso-vittorio.
The design of the Corso was a reaction against the straight Hausmannian avenues proposed in the 1873 and 1883 master plans, which some condemned as sventramenti (guttings) of the historic city fabric. The city modernized more streets in the historic center of Rome in the twentieth century. Early in the century, different factions debated whether new designs should be formal (in the manner of the plans in *Der Städtebau*, Josef Stübben's publication),

picturesque (like Camillo Sitte's analyses of medieval streets), or a hybrid like Raymond Unwin's English Garden Towns. Examples of all three approaches were built.

For a free online English translation of Josef Stübben, *Der Städtebau* (Alfred Kröner Verlag, 1890), see https://urbanism.uchicago.edu/joseph-stubbens-city-building. Also see the record of a conference organized by the Royal Institute of British Architects to resolve the differences between formal picturesque, hybrid designs: *Town Planning Conference, London, 10–15 October 1910: Transactions* (Royal Institute of British Architects, 1911). (Facsimile reprint: Routledge, 2011; Studies in International Planning History.)

12. Yodan Rofè, "The Meaning and Usefulness of the 'Feeling Map' as a Tool in Urban Design and Architecture," in *The Oxford Conference: A Re-evaluation of Education in Architecture*, ed. Susan Roaf and Andrew Bairstow (WIT Press, 2008): 243–46. Also see endnote 3, which begins, "Beauty is truth, truth beauty."

13. Allen lives in one of the houses on the south side of the block: see Christopher Gray, "Streetscapes, East 70th Street, Along Millionaires' Row, at the Crest of Lenox Hill," *New York Times*, September 17, 2006, https://www.nytimes.com/2006/09/17/realestate/17scap.html. The article also says, "In 1939, *Fortune* magazine called it 'probably New York City's most beautiful residential block,' and Paul Goldberger, in his 1979 book, *New York: The City Observed* (Random House), described it as having 'a perfect balance between individuality and an overall order.'" Gray wrote several columns about East 70th Street, including "Streetscapes: 70th and 71st Street Between Madison and Park Aves.; How 7 Rear Yards Became a Secret Garden," *New York Times*, Real Estate, May 21, 2000, https://www.nytimes.com/2000/05/21/realestate/streetscapes-70th-71st-street-between-madison-park-aves-7-rear-yards-became.html (PDF: https://www.mottschmidt.com/uploads/add_resources/nyt-Streetscapes_70th_and_71st_Street_Between_Madison_and_Park_Aves.pdf); "Streetscapes: Readers' Questions; 70th Street Hospital, Hilltop House in Inwood," *New York Times*, Real Estate, December 3, 2000, https://www.nytimes.com/2000/12/03/realestate/streetscapes-readers-questions-70th-street-hospital-hilltop-house-inwood.html; "Streetscapes: 46 East 70th Street; Home of the Dakota's Owner and a Club for Explorers," *New York Times*, Real Estate, December 10, 2000, https://www.nytimes.com/2000/12/10/realestate/streetscapes-46-east-70th-street-home-dakota-s-owner-club-for-explorers.html; "Streetscapes: 40 East 70th Street; A Growth Plan for a Neo Georgian Garage," *New York Times*, Real Estate, April 26, 1992, https://www.nytimes.com/1992/04/26/realestate/streetscapes-40-east-70th-street-a-growth-plan-for-a-neo-georgian-garage.html; "Streetscapes: 110–118 East 70th Street; A Block of Mansions With a Shared Past," *New York Times*, Real Estate, May 2, 2010, https://www.nytimes.com/2010/05/02/realestate/02streetscapes.html; "Streetscapes: The Garden at the Frick, and How It Grew," *New York Times*, Real Estate, August 3, 2014, https://www.nytimes.com/2014/08/03/realestate/the-garden-at-the-frick-and-how-it-grew.html; and "A Museum's Former Life," *New York Times*, Real Estate (video), n.d., https://www.nytimes.com/video/realestate/1247467736944/a-museum-s-former-life.html.

In *The AIA Guide To New York City*, 4th ed (Crown, 2000), Norval White and Elliot Willensky wrote, "East 70th Street, between Park and Lexington avenues: A block as diverse, friendly, inviting, tactile, dappled, intricate, and surprising as anyone might wish. A masterpiece of the culture rather than of narrow architectural or planning decisions." Several of the individual houses are discussed in *The AIA Guide* and in Robert A.M. Stern, Gregory Gilmartin, and John Montague Massengale, *New York 1900, Metropolitan Architecture and Urbanism 1890–1915* (Rizzoli, 1983). Architect and urban designer Léon Krier prefers the Greenwich Village block where Barrow Street and Commerce Street come together to form a picturesque intersection. Krier would not like the way that the NYC DOT filled the space with bold striping, destroying some of the quiet charm of the irregular space.

14. See Bruce Appleyard, *Livable Streets 2.0* (Elsevier Ltd, 2019): 3–18.

15. See "Vision Zero (New York City)," Wikipedia, https://en.wikipedia.org/wiki/Vision_Zero_(New_York_City). For an early perspective on Vision Zero in the United States, see John Massengale, "Getting to Vision Zero In America's Most Walkable City," Street Design Blog, January 24, 2014, https://street.design/blog/vision-zero-in-americas-most-walkable-city; John Massengale, "Vision Zero Changes Everything," There Are Two Types of Architecture, January 24, 2014, https://blog.massengale.com/2014/01/24/vision-zero-changes-everything/; John Massengale, "Getting to Vision Zero," There Are Two Types of Architecture, July 8, 2014, https://blog.massengale.com/2014/07/08/2visionzero/; John Massengale and Victor Dover, "Vision Zero Changes Everything," Street Design Channel—YouTube, https://www.youtube.com/watch?v=k8Lx_asu3f0; and John Massengale, "WALKABILITY: A Street Is A Terrible Thing To Waste," *New York State Conference of Mayors Summer 2014 Municipal Bulletin*, see https://photos.massengale.com/nycom/. For later views, see John Massengale, "There Are Better Ways to Get Around Town," *New York Times*, May 15, 2018, https://www.nytimes.com/2018/05/15/opinion/there-are-better-ways-to-get-around-town.html; John Massengale, "Designing Streets for People," *Vision Zero Cities Journal*, October 23, 2018, https://medium.com/vision-zero-cities-journal/designing-streets-for-people-13b8078abd07; John Massengale, "Whose streets? Our streets, not cars': A post-COVID lesson for NYC," *New York Daily News*, November 11, 2021, https://www.nydailynews.com/2021/11/01/whose-streets-our-streets-not-cars-a-post-covid-lesson-for-nyc/; Gersh Kuntzman, "Vision Zero: Ten Years of Mixed, Inequitable Results, Report Shows," Streetsblog NYC, February 6, 2024, https://nyc.streetsblog.org/2024/02/06/vision-zero-ten-years-of-mixed-inequitable-results-report-shows; and Kea Wilson, "Five Things to Learn From NYC's Decade of Vision Zero Successes And Shortcomings," *Streetsblog USA*, February 12, 2024, https://usa.streetsblog.org/2024/02/12/five-things-to-learn-from-nycs-decade-of-vision-zero-successes-and-shortcomings.

16. Christopher Gray, "Streetscapes/Cars; When Streets Were Vehicles for Traffic, Not Parking," *New York Times*, March 17, 1996, https://www.nytimes.com/1996/03/17/realestate/streetscapes-cars-when-streets-were-vehicles-for-traffic-not-parking.html.

17. We can observe, measure, and analyze traditional proportions in old city and village streets, as urbanists frequently do to appreciate the local context in places where they work. In 2000, a group of New Urbanists assembled some of this information in a new section in *Architectural Graphic Standards* called "Elements of Urbanism" in the Ninth edition of *Architectural Graphic Standards*, written by Gary Greenan, Andrés Duany, Elizabeth Plater-Zyberk, Kamal Zaharin, and Iskander Shafie: see Charles G. Ramsey and Harold Reeve Sleeper, *Architectural Graphic Standards*, 9th Edition, John Ray Hoke, Jr., FAIA, Editor in Chief (John Wiley & Sons, Inc., 2000): 93–98. "Elements of Urbanism" added standards for urban design to the architectural standards in the book, but the section is unfortunately not included in later editions.

18. Francis Terry, "Proportions in Architecture and Music," Francis Terry & Associates, https://ftanda.co.uk/thoughts/proportions-in-architecture-and-music/.

19. *Ibid.*, "The same patterns can be seen in different-sized bells, different lengths of metal pipes in brass bands or church organs, and the different sizes of wood blocks for a xylophone. The maths throughout these various ways of making different pitches of sound is the same, and so they work with each other. If, for example, a trumpet player was to blow down a long brass tube they could make the notes of an arpeggio. This wind instrument's arpeggio would fit exactly into the previously mentioned octave of the stringed instrument, if they are tuned to the same pitch."

20. *Ibid.*, As Terry discusses, many buildings have proportions that are close to harmonic proportions without being precise, and the Classical orders do not have whole-number proportions. Scholars like Steve Bass and Keith Critchlow would argue the orders were created with proportioning systems other than numbers, however. Traditionally, many architects used dividers set for specific proportions, like the square root of two or the square root of five, also known since Euclid as the "golden ratio." Le Corbusier used the golden ratio for his Modulor system. See Steve Bass, *Beauty Memory Unity: A Theory of Proportion in Architecture* (Lindisfarne Books, 2019); Keith Critchlow, *Islamic Patterns: An Analytical and Cosmological Approach* (Inner Traditions, 1999); and Steven W. Semes, *The Architecture of the Classical Interior, op. cit.* Those are the pages for Chapter 7, "Proportion," which has diagrams based on the work of Richard Sammons. An architect, Sammons teaches proportion and is writing a long-awaited book. See "Practical Proportion," Fairfax & Sammons, https://www.fairfaxandsammons.com/practical-proportion-with-richard-sammons-at-glasgow-hall/.

21. Christopher Gray discusses the individual houses in his two "Streetscape" articles in the *New York Times*, "The Best House on the Best Block," October 6, 1996, https://www.nytimes.com/2009/07/12/realestate/12scapes.html, and "Along Millionaires' Row," *op. cit.*

22. For a broad view of this topic, see the first chapter in *City Rules, How Regulations Affect City Form* (Island Press, 2012) by Emily Talen. For New York City, see Richard Plunz, *Housing in New York City* (Columbia History of Urban Life, 2016), A. Jackson, *A Place Called Home: A History of Low-Cost Housing in Manhattan* (MIT Press, 1976), and Andrew Dolkart, *Biography of a Tenement House in New York City: An Architectural History of 97 Orchard Street* (The Center for American Places, 2006), and *The Row House Reborn: Architecture and Neighborhoods in New York City 1908–1929* (Johns Hopkins, 2009). In addition to *New York 1900, op. cit.*, see Robert A.M. Stern, Gregory Gilmartin, and Thomas Mellins, *New York 1930: Architecture and Urbanism between the Two World Wars* (Rizzoli, 1987), and Robert A.M. Stern, Thomas Mellins and David Fishman, *New York 1880: Architecture and Urbanism in the Gilded Age* (Monacelli Press, 1999). Before that series of books, Stern and Massengale wrote "With Rhetoric, The New York Apartment House," *VIA IV* (MIT Press, 1981): 77–107 and https://blog.massengale.com/wp-content/uploads/2024/12/Stern-Via-1980.pdf. A short history of the apartment building in New York City, "With Rhetoric" particularly focused on the courtyard apartment building and the multi-story apartment.

23. The first subway line in New York ran from City Hall to the new Grand Central Terminal at 42nd Street, then across to Times Square (on what are now the tracks for the Times Square Shuttle), and then underneath Broadway to West 145th Street. Part of the reason for the shift to the west was that the East Side already had train tracks on Fourth Avenue (now Park Avenue), with regular stops on the East Side. These included the railroad line now under Park Avenue, streetcar lines, two elevated train lines on Second and Third avenues, and the IRT Subway line under Lexington Avenue. The original line became the basis for the West Side IRT trains (1, 2, 3), the East Side IRT (4, 5, 6) and the Times Square Shuttle.

24. Until 1898, New York City was only the island of Manhattan. Between 1800 and 1900, the population of Manhattan grew from approximately 60,549 to 1,850,093. The population of consolidated New York in 1900 was 3,437,202.

25. The Painted Ladies are Victorian houses built in the Victorian and Edwardian eras. See "Painted Ladies," Iconic SF, https://www.sftravel.com/things-to-do/attractions/iconic-sf/painted-ladies. The first reference to the Painted Ladies is unknown, but the term was revived in the 1970s by Elizabeth Pomada and Michael Larsen in their book *Painted Ladies: San Francisco's Resplendent Victorians* (E.P. Dutton, 1978).

26. As we write this, America is in a housing crisis. Many aspects of that problem are beyond the scope of this book, but when the housing discussion involves urban form, street design is involved. For the bigger picture, we particularly recommend three books on the general topic: Patrick M. Condon, *Broken City: Land Speculation, Inequality, and Urban Crisis* (UBC Press, 2024); Lynn Ellsworth, *Wonder City: How To Reclaim Human-Scale Urban Life* (Fordham University Press, 2025); and Cameron Murray, *The Great Housing Hijack: The Hoaxes and Myths Keeping Prices High*

for Renters and Buyers in Australia. (Allen and Unwin, 2024). John Massengale has written a number of articles on the problem in New York City: "Big Real Estate's Continuing Stranglehold Over New York City," *Common Edge*, February 14, 2022, https://commonedge.org/big-real-estates-continuing-stranglehold-over-new-york-city/; and articles on the "City of Yes" at "The City of Yes: Unaffordable Housing in the Real Estate State," https://www.cnu.nyc/newurbanism/coyho/. The latter page also lists articles on the "City of Yes" from other authors.

27. The tallest residential building in New York City is 225 West 57th Street. See Wikipedia's list of the tallest buildings in New York: https://en.wikipedia.org/wiki/List_of_tallest_buildings_in_New_York_City. The north side of the tower sits on West 58th Street. It is over 25 times as tall as West 58th Street is wide.

28. Jane Jacobs, *The Death and Life of Great American Cities* (Random House, 1961): 14.

29. *Ibid.*, "Erosion of cities or attrition of automobiles," 348.

30. *Ibid.*, 187–190. At the time that Jacobs wrote *Death and Life*, New York City was suffering from migration to the suburbs. She could see, therefore, which neighborhoods stayed more popular than others. See page 30 and note 35.

31. Gray, "Along Millionaires' Row," *op. cit.*

32. Manhattan's high density usually means that stores run continuously along the avenues, rather than concentrated in discrete neighborhood centers or town centers, which is more normal in America. An interesting aspect of this is that because the north-south blocks are only two-hundred-feet across, there are rarely any long blocks on the avenues. Pedestrians do not like waiting for walk lights at the intersections (and frequently cross against the lights), but the short blocks are visually interesting for people walking. Traditionally divided into narrow storefronts, the blocks are good for window shopping. Today, however, online shopping and delivery services hurt Manhattan stores, and there are many vacancies. See Corey Kilgannon, "This Space Available," *New York Times*, September 6, 2018, https://www.nytimes.com/interactive/2018/09/06/nyregion/nyc-storefront-vacancy.html, and Steve Cuozzo, "Empty NYC retail space casts pall on key corridors," *New York Post*, May 30, 2022, https://nypost.com/2022/05/30/empty-nyc-retail-spaces-cast-pall-on-key-shopping-corridors/.

33. Christopher Gray, "Streetscapes, 131 East 70th Street, Architect's Own Brownstone Doesn't Fit the Mold," *New York Times*, April 23, 2006, https://www.nytimes.com/2006/04/23/realestate/23scap.html.

34. Jacobs, *op. cit.*, 150–151. Jacobs wrote, "To generate exuberant diversity in a city's streets and districts four conditions are indispensable:

1. The district, and indeed as many of its internal parts as possible, must serve more than one primary function; preferably more than two....

2. Most blocks must be short; that is, streets and opportunities to turn corners must be frequent.

3. The district must mingle buildings that vary in age and condition, including a good proportion of old ones so that they vary in the economic yield they must produce. This mingling must be fairly close-grained.

4. There must be a sufficiently dense concentration of people, for whatever purposes they may be there...."

For many years, the most expensive neighborhood in Manhattan was also a part of Manhattan with the smallest blocks: the Upper East Side between Central Park and Third Avenue, and between 59th Street and 110th Street (in other words, the small blocks east of Central Park). The Upper West Side blocks along Central Park were also expensive, but the blocks between Amsterdam Avenue (Tenth Avenue) and Central Park West (Eighth Avenue) are monotonously long. The most expensive streets in Manhattan today are in the areas outside of the Commissioners' Plan, like Greenwich Village, SoHo, and Tribeca, which have short blocks and low buildings (Figure 1.87). Also page 90 in the section 574 in Chapter Five, "My Manhattan."

35. See Michael R. Montgomery, "Keeping the Tenants Down: Height Restrictions and Manhattan's Tenement House System, 1885–1930," *Cato Journal*, vol. 22 (3): 502–3, https://ciaotest.cc.columbia.edu/olj/cato/v22n3/cato_v22n3mom01.pdf. Montgomery's article for the Libertarian Cato Institute takes the position that all zoning is restrictive, raising costs and lowering supply. He ignores that in a normal market, tall buildings increase land costs and construction costs, the two most important factors in the cost of any new building. Montgomery also states that more apartments would have lowered costs throughout the history of New York, but he provides no proof. In modern-day New York, the creation of luxury and superluxury apartments has not "trickled down" to more inexpensive apartments. See note 27, above.

36. James Ford, *Slums and Housing (with Special Reference to New York City): History, Conditions,* Policy (Harvard, 1936): 502–503.

37. Sarah B. Landau and Carl W. Condit, *Rise of the New York Skyscraper, 1865–1913* (Yale, 1996): 112.

38. For the New York State Multiple Dwelling Law, see https://babel.hathitrust.org/cgi/pt?id=nnc1.ar53650689&view=1up&seq=6.

39. In practice, there was room to modify the formula. The base of the San Remo is taller than fifteen stories, because terraces and setbacks crown the podium. The towers were 10 stories tall, but elaborate, uninhabited tops made the buildings taller than 300 feet. After the 1961 Zoning Resolution, the tops were converted to penthouse apartments.

40. New York Power Broker Robert Moses was in charge of constructing the housing projects built by the New York City Housing Authority (NYCHA). See the discussion of Washington Square Village, page 36. Also see "Washington Square Village," Wikipedia, https://en.wikipedia.org/wiki/Washington_Square_Village.

41. See Joseph B. Rose, "Reforming the New York City Zoning Resolution," a talk given on April 20, 1999, with a transcript on TenantNet, http://www.tenant.net/land/zoning/unifiedbulk/reforming.html. Also see Lynn Ellsworth, "For New York City, height limits are a radical step. But they are clearly needed—Joe Rose former Chair of City Planning," *Tribeca Trust*, https://tribecatrust.org/for-new-york-city-height-limits-are-a-radical-step-but-they-are-clearly-needed-joe-rose-former-chair-of-city-planning/.

42. In 2001, Rose described this in a talk he gave about his proposed zoning changes at the Ninth Congress of the Congress for the New Urbanism, which met in New York City. There is no transcript, but John Massengale and Victor Dover were in the audience.

43. Joseph B. Rose, *op. cit.*

44. In 2022, New York Governor Kathy Hochul asked the New York State Legislature to remove height limits on new building across the state. See "URGENT: Hochul Plan to Lift Residential Density Limit in NYC Advances to State Budget; Write Legislators in Opposition TODAY!," *Village Preservation*, February 7, 2022, https://www.villagepreservation.org/campaign-update/urgent-hochul-plan-to-lift-residential-density-limit-in-nyc-advances-to-state-budget-write-legislators-in-opposition-today/ and "Budget Victory Stops State From Lifting 12 FAR CAP," The New York Landmarks Committee, https://nylandmarks.org/news/budget-victory-stops-state-from-lifting-12-far-ca/.

45. The longer story can be seen at John Massengale, "Location, Location, Location: Affordable Housing and Urban Form in Manhattan," There Are Two Types of Architecture, October 18, 2024, https://blog.massengale.com/2024/10/18/qandq/; and John Massengale, "Capping the Heights," There Are Two Types of Architecture, October 17, 2024, https://blog.massengale.com/2023/10/17/cappingheight/.

46. Jacobs, *op. cit.* This relates to Jacobs' observation in the previous note that "exuberant diversity" requires a mix of "buildings that vary in age and condition, including a good proportion of old ones so that they vary in the economic yield they must produce. This mingling must be fairly close-grained." Also see note 30 in this chapter.

47. For more information, see Peter Norton's indispensable history of Organized Motordom, *Fighting Traffic, The Dawn of the Motor Age in the American City* (MIT Press, 2011).

48. Used as a noun, the word "pedestrian" dates back to 1763, but according to *Etymonline*, in the twentieth century the meaning changed from "a walker" to "mean especially [a] 'person walking on a road or pavement' as opposed to a person driving or riding in a motor vehicle:" https://www.etymonline.com/word/pedestrian. Google's Ngram Viewer shows that the use of the word sharply increased with the rise of the automobile: https://books.google.com/ngrams/graph?content=pedestrian&year_start=1800&year_end=2019&corpus=en-2019&smoothing=3.

49. In Chicago in 1926, as in most cities, "nothing" in the law "prohibits a pedestrian from using any part of the roadway of any street or highway, at any time or at any place as he may desire." Peter Norton, "When Cities Treated Cars as Dangerous Intruders," *The MIT Press Reader*, https://thereader.mitpress.mit.edu/when-cities-treated-cars-as-dangerous-intruders/.

50. The Secretary of Defense, Charles Erwin Wilson, was the former CEO of General Motors.

51. Le Corbusier, *Urbanisme* (Crès, 1925): 113. Translated by Frederick Etchells as *The City of Tomorrow* (J. Rodker, 1929). According to Stanislaus von Moos, *Le Corbusier: Elements of a Synthesis* (010 Publishers, 2009): 188.

52. Le Corbusier, *Oeuvre Complète*, 1910–1929 (Editions d'Architecture Erlenbach-Zurich, 1946), 129 ff. According to von Moos, *op. cit.*

53. Le Corbusier, *La Ville radieuse: Éléments d'une doctrine d'urbanisme pour l'équipement de la civilisation machiniste* (Editions de L'architecture d'aujourd'hui, 1935). Translated by Pamela Wright, Eleanor Levieux, and Derek Coltman as *The Radiant City: Elements of a Doctrine of Urbanism to Be Used as a Basis of Our Machine-Age Civilization* (Orion Press, 1967): 74.

54. Robert A. Caro, *The Power Broker: Robert Moses and the Fall of New York* (Knopf, 1974).

55. Laurence Aurbach documents the prewar history of the movement towards Functional Classification in *A History of Street Networks, op.cit.* "The term *functional classification* was used as early as 1911. In 1914, Raymond Unwin offered a scheme for street classification that was literally tree-like, from trunk to main branches to minor branches to twigs. Martin Wagner, future chief planner of Berlin, proposed in 1915 an ideal suburb with superblocks and rows of cul-de-sacs. Numerous planners and designers carried those ideas forward in the plans and officially adopted policies of the 1920s and later," Aurbach wrote in an email to the authors. "By 1934, the Bureau of Public Roads was using the classifications of primary, secondary/feeder, and land-service roads." The construction of millions of miles of federally funded arterials, collectors, and local roads did not happen until the 1950s and later, however.
For a succinct study of modern Functional Classification, see Aurbach's "Towards a Functional Classification Replacement (Part One)," Ped Shed, February 22, 2009, https://pedshed.net/?p=227.

56. Transportation Planners and Traffic engineers use "Average Daily Trips" or peak hour volume to model future Level of Service. ADT is different than VMT (Vehicle Miles Traveled) because it assumes that motor vehicle trips will always grow. The transportation consulting firm Fehr & Peers created an alternative "MMLOS" (multimodal level of service) to mathematically reward systems for achieving greater Mode Split between walking, biking, and transit. See *Traffic Data Computation Method*, US Department of Transportation, Federal Highway Administration, August 2018, https://www.fhwa.dot.gov/policyinformation/pubs/pl18027_traffic_data_pocket_guide.pdf.

57. Road width study: Peter Swift P.E., Dan Painter AICP, and Matthew Goldstein, "Residential Street Typology and Injury Accident Frequency," There Are Two Types of Architecture, https://massengale.typepad.com/venustas/files/SwiftSafetyStudy.pdf. Intersection study: Norman Garrick, "Traffic Safety, Travel Mode Choice and Emergency

Services," https://www.slideshare.net/CongressfortheNew Urbanism/norman-garrick-cnu-2009.
A more recent study finds that "roads with 10–12-foot lanes at 30–35 mph speed limits have a significantly higher number of crashes compared to those with 9-foot lanes. Narrowing lane widths at these speeds provides city leaders with an opportunity to improve safety for all roadway users." See "A National Investigation on the Impacts of Lane Width on Traffic Safety," Johns Hopkins Bloomberg School of Public Health, November 2023, https://narrowlanes.american-health.jhu.edu.

58. Charles L. Marohn Jr., *Confessions of a Recovering Engineer* (John Wiley & Sons, 2021): 4–8. Also see John Massengale's review of Marohn's book in Common Edge, October 18, 2021, https://commonedge.org/confessions-of-a-recovering-traffic-engineer/.

59. See Colin Rowe and Fred Koetter in *Collage City* (MIT Press, 1984). *Collage City* presents the theories and principles of urbanism taught by Colin Rowe and "the Texas Rangers" at Cornell University. See Smilja Milovanovic-Bertram, "In the Spirit of the Texas Rangers," Georgia Tech College of Design, http://hdl.handle.net/1853/29118. In a three-volume edition of his collected writings, Rowe talks about many aspects of Cornell's urban theories and histories, including the Texas Rangers. See *As I Was Saying: Recollections and Miscellaneous Essays, edited by Alexander Caragonne* (MIT Press, 1995). More recently, several of Rowe's students assembled their own writings about Rowe and Cornell. See Steve W. Hurtt and James T. Tice, eds., *The Urban Design Legacy Of Colin Rowe* (AR+D, 2025). Mark Hewitt reviewed the book in "The Urban Design Legacy of Colin Rowe," Common Edge, February 10, 2026, https://commonedge.org/the-urban-design-legacy-of-colin-rowe/. Cornell urban design graduate Michael Dennis has written two books that build on that legacy: *Court and Garden: From the French Hotel to the City of Modern Architecture* (MIT Press, 1986); and *Temples and Towns: A Study of the Classical Tradition in Urbanism* (MIT Press, 2020).

60. According to the Form-Based Codes Institute, "Form-based codes foster predictable built results and a high-quality public realm by using physical form (rather than separation of uses) as the organizing principle for the code. They are regulations, not mere guidelines, adopted into city or county law. Form-based codes offer a powerful alternative to conventional zoning." See "What Are Form Based Codes," Form Based Code Institute, November 2022, https://formbasedcodes.org/wp-content/uploads/2022/11/What-Are-FBCs.pdf.

61. The AASHTO Green Book states that transportation officials should first determine the function(s) of the street, then assign a design appropriate for the function(s). If walking is identified as a function of the facility from the beginning, the design will be different; walking ought to be one of the typical functions, but rarely is.

62. Also see Victor Dover's street design essay "Twenty-three" in Emily Talen, Ed., *Charter of the New Urbanism* (McGraw-Hill Professional, 2nd Ed., 2013): 211. "It's time to reunite architecture and the creation of public spaces into complementary and seamless tasks. The details of the right-of-way and the design of adjacent buildings should work together to comfort, satisfy, and stimulate pedestrians. People will walk through areas where they are provided with precise orientation, visual stimulation, protection against the elements, and a variety of activities. Moreover, they must feel safe—both from fear of crime and from fear of being hit by a vehicle."

63. See also CPTED (Crime Prevention through Environmental Design) document, or *Defensible Space* by Oscar Newman (Macmillan, 1973).

64. Some of the best studies on the subject are done by the company Space Syntax. Many are available at https://www.spacesyntax.com/downloads/.

65. Victor Dover says more about this in an interview with Carie Penabad in her *On Cities* podcast, "Learning from the Past to Design Tomorrow's Streets," recorded March 22, 2024, https://podcasts.apple.com/ca/podcast/learning-from-the-past-to-design-tomorrows-streets/id1667651571?i=1000650139237 and https://open.spotify.com/episode/0T2hsZlC2b6YGQasA0M086?si=f6873fd30be0450d. Also see, or listen to, John Massengale in the On Cities podcast, "Street Design, The Secret to Great Cities and Towns," recorded August 11, 2023, https://podcasts.apple.com/ca/podcast/street-design-the-secret-to-great-cities-and-towns/id1667651571?i=1000624185495 and https://open.spotify.com/episode/7hYd5encWQTu9Wdng3qobB?si=8a0bd5e701fc43dc.

66. Wales had one of the most ambitious plans: see "Introducing default 20mph speed limits," Welsh Government, July 8, 2021, updated September 25, 2024, https://www.gov.wales/introducing-default-20mph-speed-limits. In the United States, cities that adopt Vision Zero policies institute 20 mph limits on many streets, but American Departments of Transportation rarely support them with adequately slow design speeds for the streets. See the discussion of Streets for People and the "Triangle of Death" on page 580.

67. See the "Stockholm Declaration," *World Health Organization and Government Offices of Sweden*, https://www.roadsafetysweden.com/about-the-conference/stockholm-declaration/. Also see the Press Release from 20's Plenty for Us, https://www.20splenty.org/global_ministers_mandate_20mph.

68. See "Stop de Kindermoord Protests Led to NL Road Safety," *Dutch Reach Project*, https://www.dutchreach.org/car-child-murder-protests-safer-nl-roads/. Also see the Families for Safe Streets website, https://www.familiesforsafestreets.org/.

69. "Transportation Alternatives and Families for Safe Streets Applaud Governor Hochul on Signing Sammy's Law," Transportation Alternatives, May 9, 2024, https://transalt.org/press-releases/transportation-alternatives-and-families-for-safe-streets-applaudnbsp-governor-hochul-on-signing-sammys-lawnbsp.

70. See "Complete Streets Policies," Smart Growth America, https://smartgrowthamerica.org/program/national-complete-streets-coalition/policy-atlas/

71. An online search for Complete Street photos and renderings shows this problem. Their primary purpose is to move vehicles through the city, and little or no thought is given to being in the city, on a particular block. It is good engineering, in other words, but weak urban design and poor placemaking. In a borough where more than three-quarters of the households don't own cars and only 20 percent of the workers commute by car, what is the justification for making a suburban-style arterial for use primarily by suburbanites?

Street Directory. — 421

DISTANCES BETWEEN THE AVENUES.

SOUTH OF 23D STREET.

Aves. D and C...676 ft.	Aves. 4th and 5th...920 ft.
" C and B ...676 "	" 5th and 6th ...920 "
" B and A...666 "	" 6th and 7th...800 "
" A and 1st..613 "	" 7th and 8th...800 "
" 1st and 2d..650 "	" 8th and 9th...800 "
" 2d and 3d..610 "	" 9th and 10th..800 "
" 3d & Ir. pl..420 "	" 10th and 11th..800 "
" Ir.pl.&4th..425 "	" 11th and 12th..800 "

23D TO 34TH STREETS.

Aves. D and C...646 ft.	Aves. Mad. and 5th..420 ft.
" C and B...646 "	" 5th and 6th...920 "
" B and A...646 "	" 6th and 7th...800 "
" A and 1st..613 "	" 7th and 8th...800 "
" 1st and 2d..650 "	" 8th and 9th...800 "
" 2d and 3d..610 "	" 9th and 10th..800 "
" 3d and Lex..420 "	" 10th and 11th..800 "
" Lex. & 4th..425 "	" 11th and 12th..800 "
" 4th & Mad..425 "	

34TH TO 42D STREETS.

Aves. D and C...646 ft.	Aves. 3d and Lex...420 ft.
" C and B...646 "	" Lex. and 4th 405 "
" B and A...646 "	" 4th and Mad..405 "
" A and 1st..613 "	" Mad. and 5th..420 "
" 1st and 2d..650 "	" 5th and 6th...920 "
" 2d and 3d..610 "	" 6th and 7th...800 "

Aves. 6th and 7th..800 ft.	Aves. 9th and 10th..800 ft.
" 7th and 8th..800 "	" 10th and 11th..800 "
" 8th and 9th..800 "	" 11th and 12th..800 "

42D TO 110TH STREETS.

Aves. B and A ..646 ft.	Aves. 5th and 6th. ..920 ft.
" A and 1st..613 "	" 6th and 7th...800 "
" 1st and 2d..650 "	" 7th and 8th...800 "
" 2d and 3d 610 "	" 8th and 9th...800 "
" 3d and Lex..420 "	" 9th and 10th..800 "
" Lex. & 4th..405 "	" 10th and 11th..800 "
" 4th & Mad..400 "	" 11th and 12th..800 "
" Mad. & 5th..420 "	

NORTH OF 107TH STREET.

Aves. 10th & 11th 775 ft.	Aves. 11th & 12th..775 ft.

NORTH OF 110TH STREET.

Aves. B and A...646 ft.	Aves. 5th and 6th...895 ft.
" A and 1st..613 "	" 6th and 7th...750 "
" 1st and 2d..650 "	" 7th and 8th...775 "
" 2d and 3d 610 "	" 8th and 9th...800 "
" 3d & Lex. 420 "	" 9th and 10th..800 "
" Lex. & 4th..405 "	" 10th and 11th..775 "
" 4th & Mad..400 "	" 11th and 12th..775 "
" Mad.& 5th..420 "	

Madison Avenue to 4th, between 120th and 124th Streets, is 405 feet.

WIDTH OF THE AVENUES AND STREETS.

All the avenues are 100 feet wide, except the following :

Avenue A, south of 23d Street .... 80 ft.
" B, " " .... 60 "
" C, " " .... 80 "
" D, " " .... 60 "
Boulevard............................150 "
Lexington Avenue..................... 75 "
Madison Avenue, south of 42d Street...... 75 "
" north " " ...... 80 "
" bet. 120th & 124th Sts.. 100 "
4th Avenue, north of 34th Street........140 "
6th " " 110th " ....150 "
7th " " 110th " ....150 "
11th " " 107th " .... 150 "

All streets are 60 feet wide, except the following, which are 100 feet :

14th	72d	116th	165th
23d	79th	125th	175th
34th	86th	135th	195th
42d	96th	145th	205th
57th	106th	155th	215th

185th Street is 80 feet.
122d Street, west of 9th Avenue, 80 feet.
127th Street, west of 11th Avenue, 100 feet.
110th Street, west of 8th Avenue, 80 feet.

LENGTH OF BLOCKS NORTH OF HOUSTON STREET.

THE DISTANCES BETWEEN

1st and 3d Streets are 211 feet 11 inch.	16th and 21st Streets are 184 feet — inch.
3d " 5th " 192 " 1 "	21st " 42d " 197 " 6 "
5th " 6th " 194 " 1¼ "	42d " 71st " 200 " 10 "
6th " 7th " 181 " 9 "	71st " 86th " 204 " 4 "
7th " 8th " 195 " — "	86th " 96th " 201 " 5 "
8th " 9th " 187 " 10 "	96th " 125th " 201 " 10 "
9th " 10th " 184 " 6½ "	North of 125th " 199 " 10 "
10th " 11th " 189 " 7 "	121st and 122d, W. of 9th Ave., 191 " 10 "
11th " 16th " 206 " 6 "	122d and 123d, " " 191 " 10 "

The monuments on Avenues A, B, C, D—1st, 2d, 3d and 4th—stand in the angle of the northwesterly corners. On 5th, 6th, 7th, 8th, 9th, 10th, 11th and 12th Avenues the monuments stand in the angle of the northeasterly corners.
All the above distances are horizontal measures of medium temperature.
The above was prepared by the Bureau of Buildings.

Figure 1.87: A Street Directory from the New York Bureau of Buildings showing street and block dimensions in Manhattan, published by *The World Almanac* (Press Publishing Co., 1892): 421. See https://archive.org/details/worldalmanac1892newy/page/421/mode/1up.

CHAPTER TWO

HISTORIC STREETS

There is no new world that you make without the old world.

— Jane Jacobs, in an interview with James Howard Kunstler

Some streets are better than others: to be on, to do what you came to do.

—Allan B. Jacobs, *Great Streets*

OLD STREETS aren't what they used to be. Before the age of the automobile, the space between the buildings in towns and cities was public space, used for transportation, commerce, and public life.[1] Think of the scenes in the movie *The Godfather* with streets full of carts that immigrants used to gain a foothold in the new world. Their children invented games they played in the streets,

◄ **Figure 2.1:** Les Rambles, Barcelona, Spain. Looking north from the waterfront, where the promenade ends. Built on a dry riverbed, the street looks like it flows to the sea. Mature London Plane trees unify the great space. Also see Figure 2.47. © *2012 Oh-Barcelona.com / Flickr / CC BY 2.0*

like stickball, stoopball, Ringolevio, kick the can, Double Dutch, and many more. Not every American city and town was New York, but everywhere had public life that took place in the public realm.

In the last quarter of the nineteenth century, many countries built great public transportation systems. America, growing by leaps and bounds, had one of the best: a national network of railroads, subways, streetcars, and ships that connected walkable cities and towns across the country. In the cities, the space between the buildings was public space, used for transportation, commerce, and public life.[2]

Today, we have photos of old streets in cities and towns around the world (with many examples in the book such as Figures 1.1, 2.1, 2.5, 2.73, and 4.122),[3] and digitally restored films on the internet that show life on those streets a century ago.[4] We see that all streets were shared-space places. On busy streets, most people in the photos and videos walk or stand on wide sidewalks, but men and women cross the street whenever and wherever they want. They look comfortable standing in the street talking to each other. What we do not see in the photos is all the detritus of traffic engineering: stoplights, stop signs, traffic signs, no-parking signs, crosswalks, turn lanes, traffic lanes, bus lanes, or the multicolored paint traffic engineers use without considering how the paint carves up the street, disrupting the harmony of the space.

If there are cable cars in the streets, they are in the middle of the street. People wait there for the cable cars to arrive, while others walk in front of horse-drawn carriages that stop or slow while pedestrians pass. It reminds the authors of one of their European trips to look at streets in other countries. John and his wife arrived in Amsterdam first, on a beautiful spring day. The photo of the modern shared-space Singelgracht on the cover of the first edition of *Street Design* was taken directly in front of their hotel (Figure 5.1). "We've found an urban paradise," John wrote to Victor in a text.

We recognize that the photo of the Singelgracht is new because it shows modern cars parked on the street, as well as a few discreet traffic signs.[5] Bollards prevent drivers from parking next to the buildings and make a protected space for people who do not want to mix with cars and bikes. The protected walk is part of a network around the city for people who are sight-impaired or physically disabled.

One week later, we were in Paris. The year was 2011, and neither of us had been to Paris for several years. On our first morning, we walked to the famous intersection where generations of artists and writers sat in sidewalk

Figure 2.2: Mulberry Street, New York, New York. Looking north circa 1900. Our city streets were not made for cars. Historic photos of Mulberry Street and Broad Street, Figure 1.1, show two New York City streets shortly before the Ford Motor Co. introduced the Model T in 1908. *Detroit Publishing Co. / Library of Congress, Prints & Photographs Division*

*In 1981, the architect Donald Appleyard published a study of the effect of traffic on social relationships on residential streets. He found that streets with light traffic have stronger social connections and public life on individual blocks than streets with medium or heavy traffic. No one has made a similar study of commercial streets, but there is no question that most people walking or sitting prefer streets with little traffic to streets with more traffic.[6]

cafes at les Deux Magots, the Café des Flores, and the Brasserie Lipp, but the loud, smelly cars, trucks, and buses filled the beautiful and historic boulevard Saint-Germain and drove us away (so to speak).*

After we returned home we wrote about what we saw, and frequently what we saw was busy traffic damaging street life across France, England, and Spain. Car culture had a surprisingly strong grip on cities like Paris and London at the time. That has been much on our minds recently, because our increasing awareness of climate change and the urgency of dealing with it the last few years makes us all reexamine the role of the automobile in the growth of greenhouse gases.

The COVID-19 pandemic both initiated and accelerated important changes to the streets discussed in this book. *Street Design* is a book, not a work of journalism, but we sometimes write about current trends in street design because during the ten years between the two editions there was tremendous change in the ways we use our streets. Disappointingly, every European country has made far more progress than any American city or state.

As we say above, old streets no longer look or function the way they did when they were built. Consider the avenues and boulevards with which we begin this chapter. Miami planner and professor Ramon Triàs, a scholar and authority on nineteenth- and twentieth-century planning in Barcelona, says the city changed the grand streets like the Avinguda Diagonal (page 111) many times.[7] Most of the buildings along the avenue still look the same as they did one hundred years ago, but Barcelona changed the roadway and the sidewalks between the buildings several times.

In the 1970s, the city dug up the stone paving on the Diagonal, narrowed the symmetrical linear parks in the streets, widened the traffic lanes, and tore out an extensive system of tram lines—so that a greater number of cars could go faster within city limits. Now, as we write the second edition, the city is restoring the street to something like its earlier state. The city is not bringing back the old tram network, but it is extending the existing tram line to the north (see Figure 2.23) farther into the center. It is important to remember that many "historic" streets around the world were changed to improve traffic flow and that they can be better for other users, even in walkable cities like Barcelona and Paris.

In the end, of course, all the changes affect how we experience the street, for better or worse. That is another reason for visiting the streets in person.

AVENUES & BOULEVARDS

Avenues and boulevards are broad streets lined with regularly spaced trees. The earliest and many of the best examples are in Paris, where, by definition, an avenue is visually terminated at one or both ends, and a boulevard is a through street. *The Lexicon of the New Urbanism* continues that useful distinction, which we endorse, but it should be noted that a few streets in Paris called avenues are not terminated. An example is the avenue de New York, a short street that is part of a long continuous boulevard along the Seine that regularly changes name. In America, we frequently do not make this semantic distinction between avenues and boulevards. With few exceptions, the north–south streets called avenues in the New York Commissioners' Plan of 1811 are actually boulevards.[8]

"Boulevard" and "avenue" are both French words. Paris's boulevards and avenues do the jobs assigned to them well and are among the most beautiful in the world. The success of these streets stems in part from their adherence to principles that have been a part of French design for centuries. We think of them as having been built by Baron Georges-Eugène Haussmann for the Emperor Napoleon III in the nineteenth century—and many of them were—but some were older, built during the reigns of Louis XIII and Louis XIV, the Sun King.

The long French tradition of formal, tree-lined avenues began with the allées in grand French gardens like those at Vaux-le-Vicomte and Louis XIV's palace in Versailles, where they were Baroque symbols of power and absolutism.[9] That tradition also led to the design of rural allées that extended from the ramparts of Paris to chateaus and hunting lodges in the countryside.

The avenues "approached" the chateaus and lodges (from Old French *"avenir,"* which meant "approaching, or arriving").[10] These long axes were better for travel by horse than on foot. The avenues became drives where aristocrats paraded in carriages and on horseback to see and be seen. Over time, the avenues became armatures for the expansion of Paris.

Medieval cities like Paris expanded into the countryside when the city walls were torn down. The first step in Paris was planting trees on top of the old walls, in double rows that followed the path of the ramparts as they angled around the city. The original French meaning of boulevard is "top of the bulwark."

As one still sees in French parks, the trees were planted in regular geometric patterns, carefully proportioned.

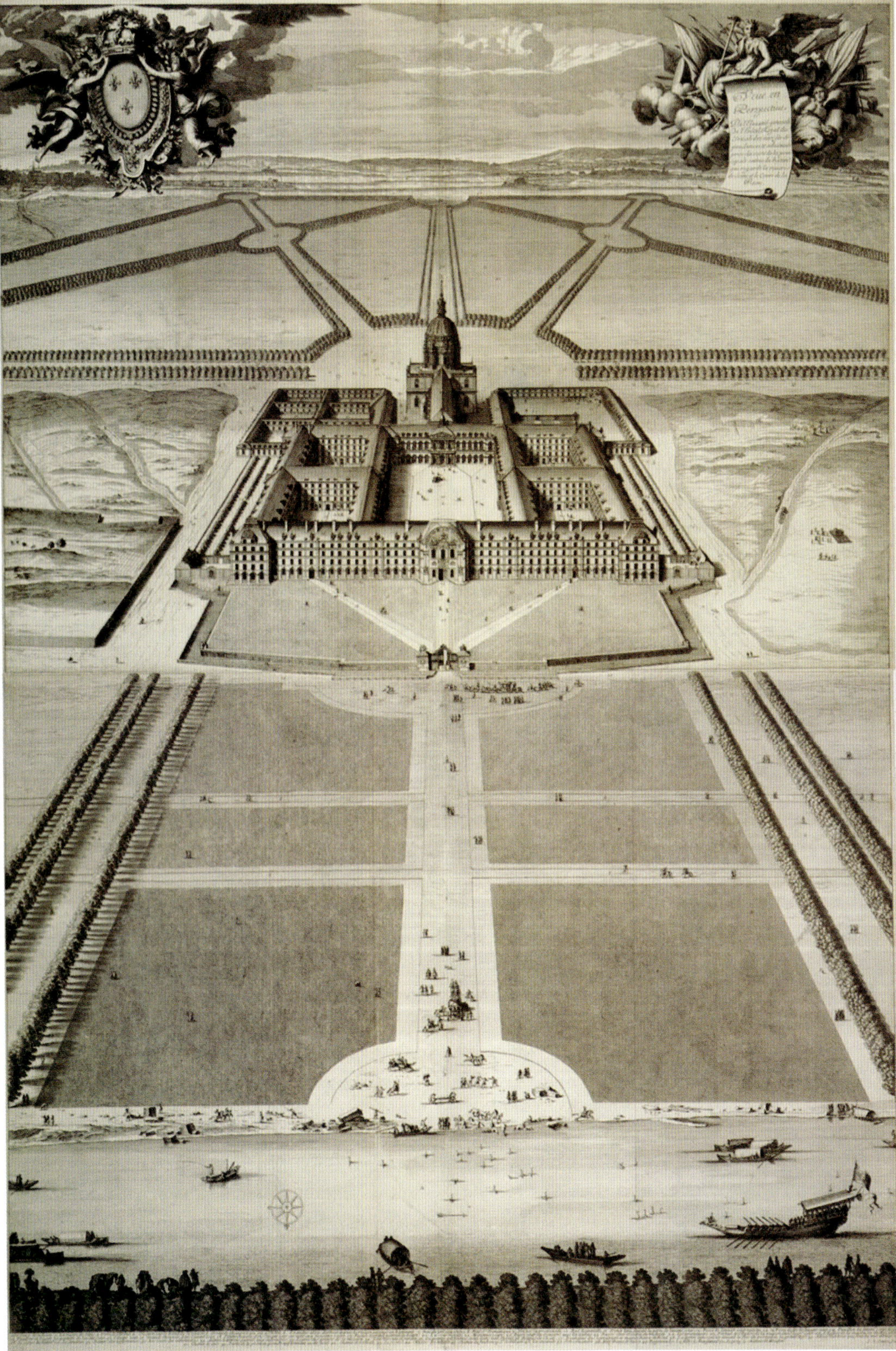

Figure 2.3: Libéral Bruant and Jules Hardouin-Mansart, Hôtel des Invalides, Paris, France, 1678. Birds-eye view looking south in 1679, etching and engraving by Jean Lepautre. Rural avenues approach the Invalides. *With permission of Royal Academy of Arts*

Figure 2.4: The Champs-Élysées, Paris, France, in 1740. A bird's-eye view from above the Champs-Élysées looking towards the palais du Louvre, painted by Charles Leopold Grevenbroeck. The rural avenues became modern streets like the avenue des Champs-Élysées, seen on the left, and the avenue Georges-V. The approach to the Invalides is across the Seine. *Courtesy of RMN-Grand Palais / Art Resource, NY*

Typically, four trees would form a perfect square (1-to-1), but sometimes the ratio of width to length would be 1-to-1½ or 1-to-2. All of these dimensions are pleasing. Few designers know about the old tradition, but comparing new, less rigorous designs to the old ones, we can see that the old patterns were more beautiful. Sadly, France cut down thousands of street trees in recent decades to widen roads. The French traffic engineers did not cut them *all* down, as their American counterparts frequently did (leaving Elm Street and Maple Street with no elms or maples), but by creating large gaps in the allées, they diminished the beauty of the streets—particularly for the pedestrians.

There are three primary types of boulevards and avenues. The first is simply a wide, tree-lined street, which can be either one-way or two-way. Examples include the boulevard Haussmann in Paris and Massachusetts Avenue in Boston. The second has a median in the center, tree-lined traffic lanes, and sidewalks on both sides: Park

Avenue in New York and Royal Palm Way in Palm Beach are exemplars of the type (Figures 2.32 and 2.5). The third is what Allan Jacobs, a co-author of *The Boulevard Book*, has dubbed "the multiway boulevard." The multiway boulevard has center lanes for through traffic and side lanes for local drivers. Often, but not always, tree-lined pedestrian malls or promenades sit in the middle of the street or to one side, but the medians may be as narrow as planting strips. Wide medians are sometimes also used for bicycle lanes and streetcar routes.

In the twentieth century, traffic engineers frequently eliminated multiway boulevards, for three reasons: they sacrificed medians and trees ("Fixed Hazardous Objects") to widen or add traffic lanes; multiway boulevards did not fit in the single-use functional classification system engineers adopted; and the engineers thought that the intersections where pedestrians, bicycles, cars, and streetcars came together on multiway avenues were too complicated and dangerous.

Figure 2.5: Royal Palm Way, Palm Beach, Florida. Avenues and boulevards frequently have a center median, rows of trees that reinforce formality, and sidewalks on each side, flanked by street-oriented buildings. © *2008 Joe Shlabotnik / Wikimedia Commons / CC BY 2.0*

Figure 2.6: Avenue des Champs-Élysées, Paris, France. Photograph looking northwest towards the Arc de Triomphe around 1900. When vehicles traveled more slowly, lampposts could be in the street, and pedestrians had a place to stand in the wide roadbed. *Library of Congress / Frank and Frances Carpenter Collection*

Figure 2.7: Rue de Castiglione, Paris, France. Perhaps by Percier and Fontaine, 1801. A hand-colored photograph from around 1900 looking towards the place Vendôme. The column erected in the center of the square by Napoleon to commemorate the victory of his troops in the Battle of Austerlitz illustrates how objects can activate a space, particularly when the object engages the axis. *Joseph Hawkes / Public Domain*

Subsequent studies have shown, however, that properly designed multiway boulevards are not dangerous, and recent interest in narrowing lanes to slow traffic and designing "multimodal Complete Streets" reminds us of the best qualities of multiway boulevards. As we will see, multiway boulevards come in a variety of forms, for different situations: some handle high traffic volumes well, some provide many parking spaces, while others have places for various forms of transportation. New boulevards are useful for taming wide, sub-urban transportation corridors and for stitching back together large tears in the urban fabric left by highway removal.

Paris has the world's highest concentration of multiway boulevards, but boulevards were imported to America in the nineteenth century, when a parks movement blossomed across the country. Multiway boulevards were used by Frederick Law Olmsted and later became popular with the City Beautiful movement, which included many designers educated in Paris. Frequently, multiway boulevards were part of developing streetcar networks and City Beautiful plans to promote urban expansion.

Typically, four trees would form a perfect square (1-to-1), but sometimes the ratio of width to length would be 1-to-1½ or 1-to-2.

THE AVENUE AND THE DEVELOPMENT OF THE MODERN CITY

By the early eighteenth century, writers described Paris as one of the most beautiful cities in the world, praised for its Classical buildings and the new avenues and boulevards.[11] In 1755, Abbott Marc-Antoine Laugier argued for city-improvement plans in his *Essai sur Architecture*.[12] Laugier, well-known to architects for ascribing the origins of Classical architecture back to what he called the Primitive Hut, described an ideal city plan that harmoniously combined Classical architecture and urbanism. The city Laugier had in mind was a new Paris, approached on broad, straight avenues. His ideal city had wide streets, arranged in a radiating network that connected commodious squares. Laugier cited the Piazza del Popolo in Rome as a precedent for his ideal city, but so was Henri IV's place de France (1608, never completed).[13]

Other writers and architects expanded on these ideas, calling for generously sized streets lined by Classical buildings. Properly proportioned, the buildings would spatially contain the wide streets, which would still have ample daylight, despite the taller buildings. The ideal modern city, they agreed, would be denser, airier, and more beautiful than the medieval city.

French kings and the Emperor Napoleon built avenues and boulevards during their reigns (pages 97 to 101). Under Napoleon III, Georges-Eugène Haussmann transformed Paris with the new streets he built. In the twentieth century, Modernist critics routinely dismissed Haussmann's work, saying it was all for the purpose of creating straight streets where the Emperor could deploy his troops to mow down crowds of rioters with musket volleys and cannonballs.

Napoleon III and Haussmann, however, believed they built a modern, healthier, more beautiful city. In seventeen years, they doubled the size of the city and increased its density at the same time. Haussmann built one-fifth of the streets in central Paris, a new

Figure 2.8: Avenue de l'Opéra, Paris, France. Bureau de la Voirie Parisienne, 1864. The opera building is known as the Palais Garnier, in honor of the architect, an important graduate of the École des Beaux-Arts. It is an archetypal French example of the axial terminated vista. *ikouneni / Blogger / Public Domain*

sewer system and a new water system that increased the freshwater supply five-fold. He routed the new avenues and boulevards through what he considered unhealthy slums—"It was the gutting of Paris," he proudly wrote in his *mémoires*[14]—and created large new parks. He connected Paris to the rest of France with new train stations. Near the end of his time, one in five Parisian workers worked in the building trade. Before Napoleon III hired Haussmann, however, the Emperor had the vision. "Thanks to his impetus," the Franco-American architect and historian Stephane Kirkland tells us, "Paris became the archetype of a modern, functional city, still perceived as the epitome of urban beauty."[15]

For two centuries, French monarchs and emperors built avenues and streets that opened vistas to monuments new and old. The grandest French avenue, the Champs-Élysées, approached the royal palais du Louvre. The architect Jacques-Ange Gabriel won a royal competition to design a new cross-axis at the place Louis XV that opened a vista to the church of the Madeleine (Figure 1.8). The pont Louis XVI extended the axis across the Seine, centered on the Palais Bourbon. The Emperor Napoleon rebuilt La Madeleine in a Neoclassical style and constructed the Arc de Triomphe in the Étoile at the top of the Champs-Élysées. Under Napoleon III, Haussmann put the grandest architectural statement of his era—the Palais Garnier, the national opera building—at the head of the avenue de l'Opéra. At the opera's opening, the Empress Eugenie asked the architect Charles Garnier what style he called the building. "Napoléon III," Garnier astutely replied.

THE FLÂNEUR VERSUS HORSEPOWER

The avenues and boulevards of Haussmann's Paris initiated the rise of the *flâneur*, a stroller on the city's streets. Charles Baudelaire described this metropolitan character as a "passionate spectator" who "enters into the crowd as though it were an immense reservoir of electrical energy." Walter Benjamin, a philosopher and cultural critic who wrote about Baudelaire, called the *flâneur* a pedestrian with "a detective's nose." The painter Édouard Manet used the city's streets, gardens, and cafes as muses, like many artists and writers of the time.[16]

They found the life they observed on the *Grands Boulevards* and in the cafes and shops along them endlessly fascinating. The boulevards and avenues are shaped by confidently scaled buildings with ground floors that are either beautiful—and therefore interesting to look at and walk along—or by storefronts, cafes, and public institutions that entertain and educate us.

The car has the power to damage that, however. The noise and smell from the traffic jam on the boulevard Saint-Germain when we visited it in 2011 exhausted us, literally and figuratively: the exhausts on the cars, buses, trucks, motorcycles, and *mobylettes* pumped out bad air and a cacophony of irritating noise (page 97). It was a hot day, and when we got close to the cars, heat radiated from the engines and the tailpipes. The "electrical energy" from the people on the street and in the cafes that drew Baudelaire to the street was overwhelmed by the bad energy from the road.

Figure 2.9: Avenue des Champs-Élysées, Paris, France. André Le Nôtre, 1667. Originally a hunting road in the country, the avenue des Champs-Élysées was improved and urbanized during the reigns of Louis XVI, Napoleon I, and Napoleon III. View from the top of the Arc de Triomphe, looking southeast towards the place de la Concorde and the Louvre Museum.

Another downside: having started as long, wide streets with a scale best suited for horses and horse-drawn carriages, avenues like Pennsylvania Avenue in Washington, DC can seem endless to anyone walking on them on a hot summer day. In addition to the problems cars can add, many American avenues have large, dull buildings with blank ground floors. Parking lots for the large buildings break the streetwall, and over the years, many Public Works Offices and Departments of Traffic have cut down the majestic trees that once lined the streets of practically every American city. Sometimes they replace them with a variety of small trees that will never form a canopy, but sometimes they prefer to simply get rid of the Fixed Hazardous Objects (or FHO, see page 39).

The Franco-American engineer Pierre Charles L'Enfant designed the master plan for Washington.

Born in Paris, he traveled to America to serve in the Continental Army, first on the staff of the Marquis de Lafayette, and later on General Washington's staff. After the war, he stayed in America, working as an urban planner and engineer. André Le Nôtre's designs for the avenues at Vaux-le-Vicomte and Versailles influenced L'Enfant's plan for Washington: a simple grid with diagonal avenues overlaid. However, the city did not have a beautiful old medieval core to build on, nor powerful autocrats like the Sun King, Baron Haussmann, and the Emperor Napoleon III to execute the plan. Washington, in other words, is not Paris, and Pennsylvania Avenue is not the equal of the Avenue Montaigne or the Champs-Élysées. Like many American avenues, it is better for horse-powered travel (or high-horsepower cars) than for the *flâneur*.

MULTIWAY AVENUES & BOULEVARDS

Avenue Montaigne, Paris, France
Bureau de la Voirie Parisienne, 1850
Multiway Boulevard

The avenue Montaigne extends from the unusual traffic circle or "rond point" at the western end of the jardin des Champs-Élysées to the place de l'Alma at the edge of the Seine. Wide sidewalks and formal Parisian architecture share a simple palette of materials and colors along the avenue. The combination of a mix of uses, proportional street space, and effective street trees creates a balanced relationship between walkers, cyclists, and drivers.

The beautiful street could be the poster child for Parisian avenues—a free-moving thoroughfare with a finite length, it is visually terminated at both ends.[17] The grand avenue is 126 feet wide, which is compact for a Parisian multiway avenue or boulevard: Montaigne efficiently moves traffic but remains a comfortable place

for pedestrians. Three travel lanes and a parking lane make up the central roadway. One of the travel lanes today is a counterflow lane for buses, taxis, and cyclists. The side-access lanes have one lane of traffic and parking on both sides. At several points along the street, one of the side-access lanes becomes slightly wider, expanding to approximately twenty-four feet and accommodating an additional row of parking.

Traffic moves well on the avenue Montaigne, even though it is narrower than many Parisian avenues and boulevards. According to studies published by Allan B. Jacobs, Elizabeth MacDonald, and Yodan Rofè, 850 vehicles per hour use the three central travel lanes and 42 vehicles per hour use the side-access lanes. At the same time, a remarkable number of pedestrians are strolling along the street—more than 1330 people can be counted on the sidewalk within an hour.[18] Roughly half the width of Barcelona's Passeig de Gràcia (Figure 2.20), the avenue Montaigne is just as efficient. It embodies the characteristics of a great street by effectively transporting local patrons without excessively wide travel lanes. The story of this grand avenue begins and ends with its proportion, variety, and beauty (Figures 2.11–2.13).

Figure 2.10: Multiway Boulevard. © *2022 Dover, Kohl & Partners*

Figure 2.11: Avenue Montaigne, Paris, France. Bureau de la Voirie Parisienne, 1850. Looking towards the avenue Montaigne from the rue François 1er. Note the harmonious, muted color palette, with soft gray asphalt sidewalks (rather than stone) on this fashionable, expensive street.

Figure 2.12: Avenue Montaigne, Paris, France. Bureau de la Voirie Parisienne, 1850. What later became the avenue Montaigne was originally a country path or lane that was given an allée of trees in 1770. The aligned rows of trees are key elements in multiway boulevards and avenues. A plan would show that in France their placement is carefully considered, following the traditions started by landscape architects like André Le Nôtre in royal gardens and rural allées.

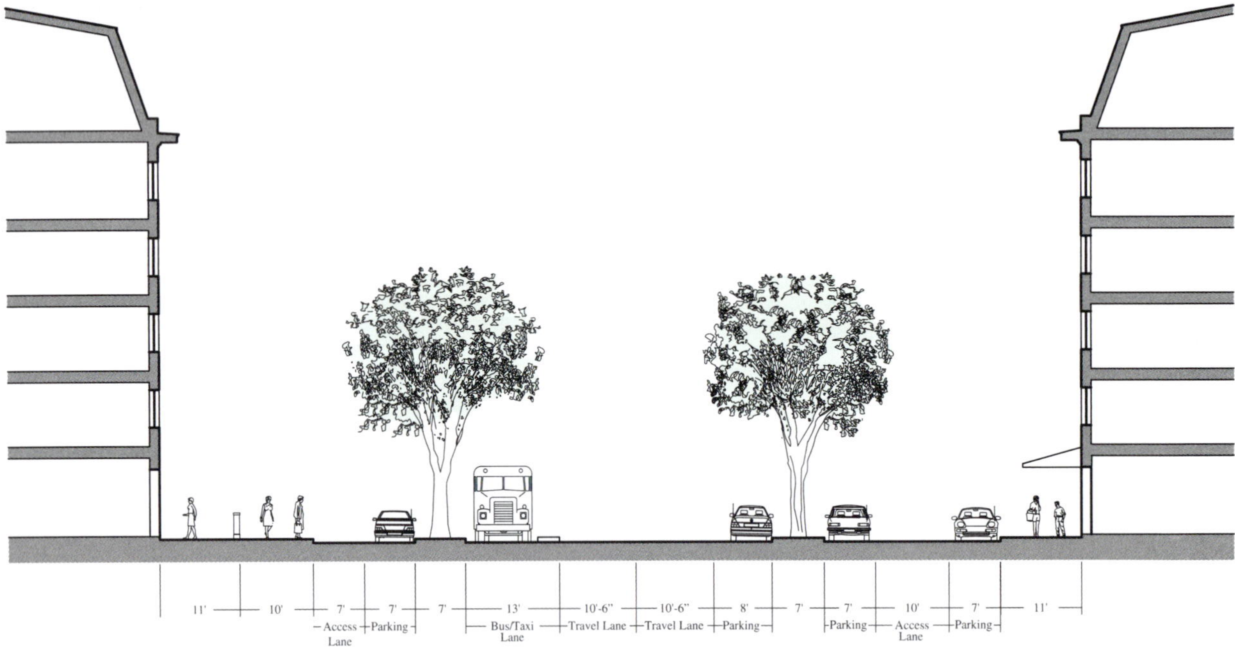

Figure 2.13: Avenue Montaigne, Paris, France. Bureau de la Voirie Parisienne, 1850. Section. © 2013 Dover, Kohl & Partners

THE MOST BEAUTIFUL PARKING LOT ISN'T A PARKING LOT… IT'S A STREET

Avenue d'Iéna, Paris, France
Bureau de la Voirie Parisienne, 1858
Multiway Avenue

The avenue d'Iéna extends south from the Arc de Triomphe and then deflects southwest towards the Trocadéro gardens (Figure 2.14). Like many other grand avenues in Paris,[19] the avenue d'Iéna has parallel rows of plane trees symmetrically dividing the center lanes from the side lanes, the ensemble flanked by graceful buildings. But the avenue's chief distinction may lie in the number of automobiles parked on it without harming the public realm. This could not be more different from the American parking lots where we find generous amounts of surface parking, even

Figure 2.14: Avenue d'Iéna, Paris, France. Bureau de la Voirie Parisienne, 1858. Looking south from the top of the Arc de Triomphe. The most beautiful parking lot in the world.

Figure 2.15: Strip Shopping Center, Miami-Dade County, Florida. The avenue d'Iéna makes this typical approach to parking look ugly and wasteful.

on the day after Thanksgiving at the mall (Figure 2.15). But on the avenue d'Iéna, six rows of parking along one block provide storage space for an astonishing 208 cars and more than 20 motorcycles (Figure 2.16).

This significant feat was accomplished simply by coupling two rows of parking in the side-access lanes with two more rows alongside the through lanes (Figure 2.17). The example shows, among other lessons, how important the regular alignment of the trees can be in the overall design of the street space. Standing on the street, a pedestrian is acutely aware of the tree canopies high overhead, the trunks planted like columns marching down the avenue, and the

beautiful, harmonious architecture on either side—but not very aware of the parked cars. The limited palette of materials and colors also unifies the space three-dimensionally. The roadway is paved with gray Belgian blocks, complementary to the colors of the tree trunks, the curbs, the building facades, and the Arc de Triomphe in the distance.

The avenue d'Iéna was constructed in 1858, so its designers did not envision this as a street for the movement of cars, much less for their storage. The fundamental elegance of the urban form—of the multilane, multiway avenue—allowed it to be adapted for new needs. The avenue accommodates parking efficiently and stealthily.

Figure 2.16: Avenue d'Iéna, Paris, France. Bureau de la Voirie Parisienne, 1858. Looking north towards the Arc de Triomphe. A stealthy parking lot: pedestrians appreciate the trees and the architecture on either side—but hardly notice the parked cars.

Figure 2.17: Avenue d'Iéna, Paris, France. Bureau de la Voirie Parisienne, 1858. Section. © 2013 *Dover, Kohl & Partners*

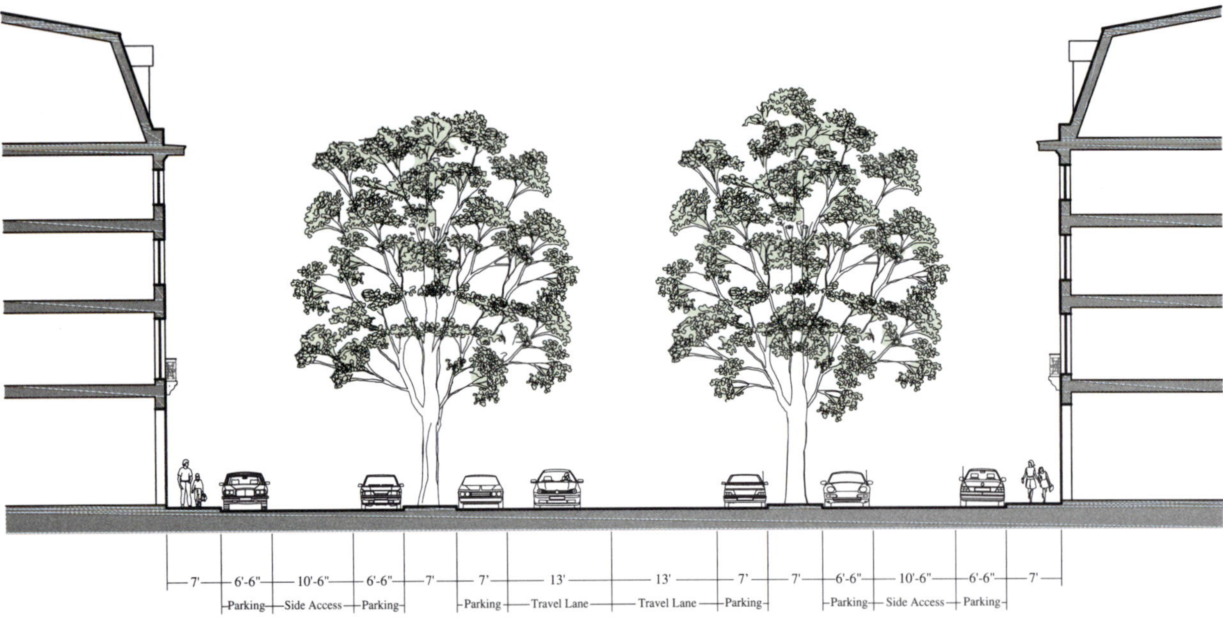

Gran Via de les Corts Catalanes, Barcelona, Spain

Ildefons Cerdà, 1859

Multiway Avenue

There are many spectacular streets in Barcelona, but none are more spectacular than the Gran Via de les Corts Catalanes, the city's spine.[20] Designed in 1859 as a crucial part of Ildefons Cerdà's plan for the Eixample (a large, gridded expansion of the city), Gran Via intersects the Passeig de Gràcia and the Rambla Catalunya one block north of Plaça de Catalunya. The street connects the great addresses and famous streets of Barcelona, tying them together and extending them into an accessible system. Gran Via is a place where people want to be—on all 13.1 kilometers (8.14 miles). It is the route that pedestrians, cyclists, public transit patrons, and drivers use to get to school, to go to work, or to go out for lunch. The ground level of the boulevard is lined with shops, offices, and cafeterias, while hotels and residences occupy the levels above the commercial spaces. One part street for locals, and another part traditional boulevard, Gran Via moves vast numbers of people in multiple modes of travel (Figures 2.18 and 2.19).

The space dedicated to the pedestrian on Gran Via—including the sidewalks and the large medians—is approximately eighty-four feet across. The combined paved surfaces dedicated to buses and vehicles measure approximately eighty-two feet, giving motor vehicles and pedestrians roughly the same amount of space. This proportion makes crossing Gran Via more comfortable than crossing streets with less generous ratios of roadbed to pedestrian space but similar traffic volumes and overall widths. Other essential characteristics of the boulevard include:

- The approximate dimensions of the multiway boulevard include a fifty-two-foot-wide central carriageway with travel lanes, bookended by medians that are thirty-two feet in width. Side lanes are located on each flank. Ten-foot-wide sidewalks on both sides of the boulevard complete the right-of-way.

- A buffer of hedges lines the side of the median adjacent to the five travel lanes. The hedges provide additional protection from the moving traffic in the center.

- The medians have parking for bikes and mopeds.

- Bici, a Spanish bike-share service, has locations at metro stops along Gran Via.

- Mature trees symmetrically line each side of the median, spaced at approximately fifty feet on center, creating a canopy that partially shades the boulevard.[21]

- The historical side lanes were changed in places from a slow local lane and a parking lane to two traffic lanes; these double lanes act as express lanes next to the pedestrian on the sidewalk. This is good for the driver, but bad for the people walking. In better segments, the side lanes now consist of a protected bike lane and one slow traffic lane.

Figure 2.18: Barcelona, Spain. Satellite view. *Courtesy of Google Earth*

Figure 2.19: Gran Via de les Corts Catalanes, Barcelona, Spain. Ildefons Cerdà, 1859. The Gran Via is the spine of the city. This photograph shows the rows of mature trees, the altered side lanes discussed in the text, and the chamfered corners characteristic of blocks in the Eixample, the extension to Barcelona designed by Cerdà.

Figure 2.20: Passeig de Gràcia, Barcelona, Spain. Looking southeast from an apartment in the Casa Milà, designed by Antonio Gaudi in 1906. The four rows of aligned trees, the wide sidewalks, and the distinctive lighting fixtures (foreground) are hallmarks of the street design. *Courtesy of Gianni Longo*

Passeig de Gràcia, Barcelona, Spain
Multiway Boulevard

Passeig de Gràcia begins at Plaça de Catalunya, crosses the Gran Via, and ends after intersecting with Avinguda Diagonal (Figures 2.20 and 2.21). It connects the old Gothic Quarter and the nineteenth-century Eixample extension. Several notable components contribute to this old-world complete street:

- Mature street trees form a canopy for both pedestrians and motorists, offering shelter from the weather and creating visual consistency along the multiway boulevard.

- The sidewalks are approximately thirty-six feet wide and paved with a patterned hardscape similar in color to the architecture surrounding it. Typically, they are full of people dining and shopping, making the multiway boulevard both a destination and a travel route.

- Unique lampposts made with elaborate ironwork cantilever above the street. Installed under municipal architect Pere Falqués in 1906 (Figure 5.65), they satisfy the practical need for lighting with a delightful, artful solution.

- Medians range from six feet to twenty-two feet wide: frequently used by pedestrians crossing the street, they allow people to navigate the wide road easily and comfortably.

Avinguda Diagonal, Barcelona, Spain

Ildefons Cerdà, 1859

Multiway Avenue

Ildefons Cerdà first proposed the Avinguda Diagonal in 1859 as part of his plan for the expansion of the city. Fifty meters wide, the avenue begins at the Ronda de Dalt and runs to the Mediterranean, traveling diagonally across the city and crossing Gran Via de les Corts Catalanes and Avinguda Meridiana at Plaça de les Glòries Catalanes (Figures 2.22 and 2.23). The diagonal is a boulevard with a heavy volume of traffic that can be exhausting. It nevertheless handles high levels of traffic better than most roads of similar volume and dimension.

Figure 2.22: Avinguda Diagonal, Barcelona, Spain. Ildefons Cerdà, 1859. Looking east towards the Passeig de Gràcia. The multiway boulevard stretches across the Eixample askew to the rest of the city pattern, uniting the city, shortening trips, and relieving the monotony of the grid in a monumental way.

Figure 2.23: Avinguda Diagonal, Barcelona, Spain. Ildefons Cerdà, 1859. Looking west from the Plaça de Francesc Macià. The streetcar line runs in a broad grassy median, a welcome axis of green amid the crowds and concrete of the busy street. This treatment is known as "green track." According to Charles Mulford Robinson, Frederick Law Olmsted designed the first green track for Beacon Street in Brookline, Massachusetts, in 1868.[22]

◄ **Figure 2.21:** Passeig de Gràcia, Barcelona, Spain. Throngs of pedestrians use the Passeig de Gràcia's wide sidewalks, passing some of Barcelona's most popular addresses. Note that the Spanish engineers used much less paint in the crosswalk than American engineers would.

Eastern Parkway, Brooklyn, New York

Frederick Law Olmsted and Calvert Vaux, 1870–1874

Multiway Boulevard

The influential European examples of multiway boulevards inspired a generation of North American experimentation with the form, with memorable results. Brooklyn's Eastern Parkway, designed by Frederick Law Olmsted and Calvert Vaux during the early 1870s (see Figure 2.214), was the first multiway boulevard in the United States.[23] More than 210 feet wide, the boulevard has six central travel lanes, redesigned in the past half-century to include turn lanes. Wide medians separate the central traffic lanes from narrow local access lanes on each side lined with rows of parallel parking. Each median accommodates walkways, bicycle paths, and park benches (Figure 2.24).

Shaded by mature trees, the medians create a park-like atmosphere along the high-traffic throughway (Figure 2.25). Rowhouses and apartment buildings make block-long facades typically three to five stories tall (Figure 2.26).

The results of a livability study conducted by Peter Bosselmann and Elizabeth MacDonald demonstrate that residents of streets like Eastern Parkway live comfortably despite the high volume of traffic. Bosselmann and MacDonald also concluded that air quality and noise levels along the boulevard are either the same or better than on typical residential streets.[24]

Eastern Parkway is refreshing proof that streets with extraordinary levels of automobile traffic do not have to be placeless spasms of asphalt. According to Bosselmann and MacDonald's study, at least 42,000 vehicles travel through the Brooklyn neighborhood every day. (In contrast, Tamiami Trail in Miami-Dade County, Florida, transports a similar amount of daily car traffic without providing much dedicated space for anything other than cars and buses: it is therefore not a place where people typically like to be.) A remarkable reality on Eastern Parkway is that a resident can comfortably sit on a bench in the median while commuters by the thousands simultaneously navigate their way home.

A sense of place is an essential part of the street design puzzle that contributes to the design of a city as a whole. When a street scene fails to incorporate distinct cultural references, it fails to be attractive—even to the motorists who have to use it. On Eastern Parkway, the traditional rowhouses not only create a solid urban wall; their shape, size, and placement give the street its distinctive, local feel (Figure 2.27).

Figure 2.24: Eastern Parkway, Brooklyn, New York. Frederick Law Olmsted and Calvert Vaux, 1870–1874. The paired trees on the medians are part of a design where pedestrians are comfortable despite the extraordinary volume of automobile traffic.

Other features include,

- The median on one side of the road is now part of the Brooklyn-Queens Greenway, a 40-mile route for walking and cycling that links major parks, cultural institutions, and neighborhoods across Brooklyn and Queens. It stretches from Coney Island in the south to Long Island Sound in the north.

- Mature elm trees protect more than 50 percent of the street space from bad weather. The organization and spacing of the trees establish visual order and define the public realm within the large medians (Figure 2.28).

- Front yards as large as twenty-five feet by thirty feet make it possible for families to socialize without interruption as pedestrians pass by on the sidewalk.

- Wide sidewalks next to the yards also provide enough room for interaction when desired.

Eastern Parkway is a unique American street with an architectural style and urban character that evokes a sense of place. Although cities like San Francisco also have streets lined with rowhouses, when you arrive on Eastern Parkway, you know you are in Brooklyn.

Figure 2.25: Eastern Parkway, Brooklyn, New York. Frederick Law Olmsted and Calvert Vaux, 1870–1874. Wide medians with paths for cycling and walking separate through-going drivers from the slow, narrow access lanes with on-street parking.

Ocean Parkway, Brooklyn, New York

Frederick Law Olmsted and Calvert Vaux, 1874–1880

Multiway Boulevard

Frederick Law Olmsted and Calvert Vaux proposed both Eastern Parkway and Ocean Parkway in 1868. Eastern Parkway was built between 1870 and 1874, while Ocean Parkway was constructed from 1874 to 1880. The road is nearly five and a half miles long and was modeled on Unter den Linden in Berlin and the avenue Foch in Paris (Figure 1.15).[25] Ocean Parkway has seven central travel lanes, including a median in the center that alternates as a left turn lane. Unlike Eastern Parkway, the street facade is not entirely continuous—detached housing and apartment buildings define the edges rather than rowhouses (Figures 2.29 and 2.30). Pedestrians going to work or using the medians as parks benefit from wide spaces between the roads. Other notable features include:

- Apartment buildings and houses with small setbacks from the street create a stable streetwall.

- Bike paths in the medians, which cyclists use to travel from Prospect Park (also designed by Olmsted) to Coney Island.

A continuous tree canopy graces the mall. It is enjoyable to walk or ride in a place sheltered from sun, rain, and snow. Both Ocean Parkway and Eastern Parkway are successful boulevards that easily accommodated their neighborhoods' changing needs over the past century. Ocean Parkway, however, is slightly larger, and that makes a remarkable difference in its quality. The six central travel lanes on Eastern Parkway have an average total width of sixty-five feet, whereas the seven travel lanes on Ocean Parkway add up to an average width of seventy feet. The five-foot differential makes Eastern Parkway noticeably more compact, easier to navigate as a pedestrian, and visually more agreeable.

Figure 2.26: Eastern Parkway, Brooklyn, New York. Frederick Law Olmsted and Calvert Vaux, 1870–1874. Looking west from Washington Avenue. Recent improvements make Eastern Parkway an important link in New York's bicycle network. *Courtesy of Kenneth García*

Figure 2.27: Eastern Parkway, Brooklyn, New York. Frederick Law Olmsted and Calvert Vaux, 1870–1874. Satellite view. *Courtesy of Google Earth*

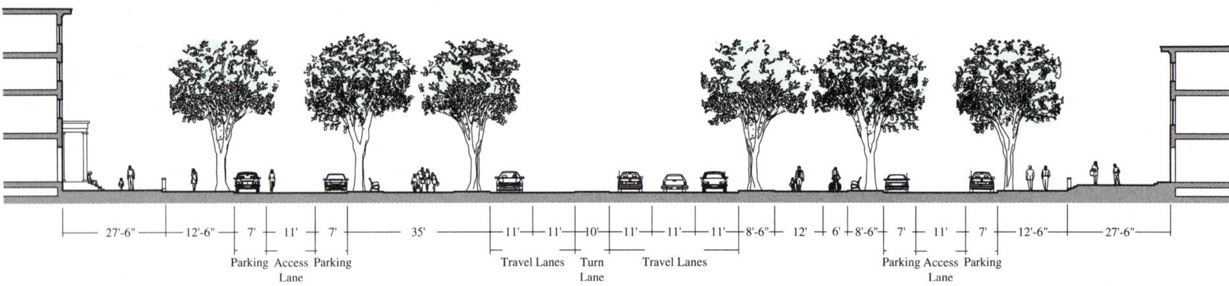

Figure 2.28: Eastern Parkway, Brooklyn, New York. Frederick Law Olmsted and Calvert Vaux, 1870–1874. Section. The traditional rowhouses and apartment buildings give Eastern Parkway its distinctly New York feel; the six rows of trees tame the sizable boulevard, making it a comfortable place for people. © 2013 Dover, Kohl & Partners

Figure 2.29: Ocean Parkway, Brooklyn, New York. Frederick Law Olmsted and Calvert Vaux, 1874–1880. Looking north from Cortelyou Road. The Eastern and Ocean Parkways grew from an 1867 proposal by Olmsted to connect Manhattan's Central Park to Brooklyn's Prospect Park and Coney Island with parkways. *Courtesy of Kenneth García*

Figure 2.30: Ocean Parkway, Brooklyn, New York. Frederick Law Olmsted and Calvert Vaux, 1874–1880. Satellite view. *Courtesy of Google Earth*

AVENUES & BOULEVARDS

Park Avenue, New York, New York
Avenue

In the Commissioners' Plan of 1811 that laid a street grid over most of Manhattan, the north-south streets were called avenues. What later became Park Avenue was called Fourth Avenue. Like all the original avenues, it was one hundred feet wide (see Figure 1.87). Despite being on the east side of the island, where most of the development took place after the Commissioners' Plan, Fourth Avenue remained unfashionable because of the railroad tracks that ran up its center—first from a terminal at 23rd Street and later from 42nd Street. The moving railroad cars and the noise and smoke from the locomotives made the avenue unpopular for development during most of the nineteenth century. Eventually, however, the tracks south of the new Grand Central Station on 42nd Street were removed, and the tracks north of Grand Central were buried until they came above ground north of Carnegie Hill at 96th Street. In 1903, New York State required that all trains to Grand Central have electric locomotives, and the city widened the street to 140 feet and renamed Park Avenue.

That same year, a Park Avenue property owner persuaded the society figure Senator Elihu Root to build a large house on the avenue, designed by the architects Carrère & Hastings, and the wide street became attractive to the rich. Within a decade, the once sparsely inhabited Park Avenue had become a long street of grand mansions. The west side of Park between East 68th and East 69th streets illustrates the scale of the avenue at the time. The four Georgian-style houses on the block were built between 1909 and 1926—two designed by McKim, Mead & White, one by Delano & Aldrich, and one by Walker & Gillette.

In 1911, however, Senator and Mrs. Root moved to 998 Fifth Avenue, an apartment house thirteen blocks north on Fifth Avenue that was the first apartment building to break the wall of mansions on Fifth Avenue above 59th Street opposite Central Park. Designed by McKim, Mead & White, 998 Fifth Avenue was managed by the real estate broker Douglas Elliman, who convinced Root to move there in return for a 50 percent reduction on his rent. Events like that can determine the character of streets: while the New York middle class had lived in large apartment buildings since the 1870s, people with the social aspirations of the Roots had previously lived only in houses. Soon many of the mansions on Park Avenue were torn down and replaced by the apartment buildings that still set the character of the broad avenue today (Figures 2.32 and 2.33).

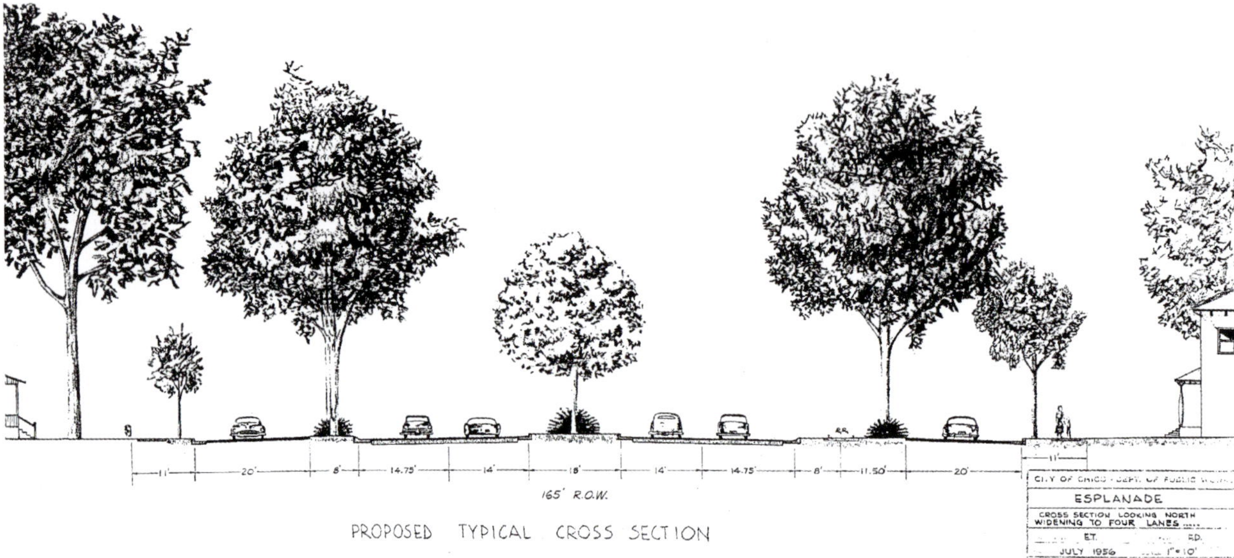

Figure 2.31: The Esplanade, Chico, California. Fred Davis, 1956. Section. The multiway boulevard tradition, so closely associated with Paris, Olmsted, and the City Beautiful movement, occasionally resurfaced during later decades. With the Esplanade, the form was adapted to Northern California and predominantly single-family neighborhoods. Although the Esplanade is one of Chico's widest streets, it is also the perennial local favorite.

Figure 2.33: Park Avenue between 85th and 86th Streets, New York, New York. A view of the Reginald De Koven House (John Russell Pope, 1911), sandwiched between two 1920s apartment houses. Until relatively recently, the pedestal-mount traffic signal was the only traffic signal used on Park Avenue, but the city has now added redundant mast arm traffic signals. Washington, DC, still uses mostly pedestal mount signals on downtown streets.

Three elements in the new apartment houses on Park Avenue combined to make a streetwall that defined the space well: the buildings had stone or brick facades, boxy massing, and a common twelve-story cornice line imperceptibly taller than the avenue was wide. In the boom years of the Roaring Twenties, some of the buildings—like 720 Park Avenue, designed by the noted apartment house architect Rosario Candela (Figure 2.34)—took advantage of the Multiple Dwelling Law of 1929 to rise

◄ **Figure 2.32:** Park Avenue, New York, New York. Looking south from 87th Street in 1929. The median in 1929 was almost twice as wide as it is today, and the sidewalks were wider too, so that each roadbed is now almost twenty feet wider. *Courtesy of NYC Vintage Images*

above the cornice with setbacks that evoked fantasies of romantic European architecture, simultaneously holding the cornice line and enriching the view down the avenue. On the ground, broad sidewalks, planted medians above the railroad tracks, and a prohibition against trucks and buses gave the sunny street a generous feel.

One of the most appealing aspects of Park Avenue is the way it goes uphill and down. That breaks up the views along the long, straight street so that the unterminated vista—the Achilles heel of the relentless American grid—is avoided. The broad medians reinforce the space of the street and make crossing it more comfortable for pedestrians, as bumpouts rarely do. Until recently, Park Avenue's stoplights were mounted on poles, but the engineers at the New York City DOT added the hanging traffic

signals that are nearly ubiquitous across America (see *Making Ugly Things Compulsory Details in Cities and Towns*, page 591). The traffic lanes on the street are wide enough that some drivers exceed fifty miles per hour, to go ten blocks (half a mile) or more between light changes. The speeders reflect the changes that the New York City DOT made to the street in the 1950s, when they narrowed the median and the sidewalks to make wider traffic lanes. In theory, the speed limit on Park Avenue is twenty-five miles per hour, but there is little or no enforcement, even after speeding cars collide at the cross streets.

Monument Avenue, Richmond, Virginia
Begun by CPE Burgwyn, 1887
Avenue

Richmond's Monument Avenue is an axial, tree-lined street, defined along its edges by distinguished houses and apartment buildings. Wide sidewalks that measure between ten and twenty feet are next to the front yards and porches, establishing a quiet pedestrian realm within the neighborhood. Despite its overall width, the street space dedicated to plantings or pedestrian use has nearly a 1-to-1 proportion with the area dedicated to auto use.

Figure 2.34: Park Avenue, New York, New York. Looking north from 68th Street in 1930. The large apartment house on the west side of Park Avenue at 70th Street is 720 Park Avenue, designed by famous New York apartment architect Rosario Candela in 1929. Older buildings on Park Avenue were limited to a 150-foot height until 1929 by the City's 1929 Multiple Dwelling Law which allowed taller buildings, if they set back above that height (see *A Short Discussion of Residential Building Heights in New York City* on page 31). Several wonderful examples of what could be done quickly rose on Park Avenue around 70th Street (some still under construction in this photo). Compare this to Figure 2.32, which shows a slightly less wealthy section of Park Avenue, where development took longer to recover. *Courtesy of the Museum of the City of New York*

When compared to an iconic Parisian street like the avenue Montaigne, the difference in street character is immediately apparent. Monument Avenue is slightly larger, but it is also primarily residential (Figures 2.35–2.39). Detached houses and apartment buildings of various sizes and styles create a less enclosed atmosphere that contrasts with the homogenous facades and mix of uses found on the avenues of Paris and Barcelona. All have some similar attributes and terminated vistas, but the American adaptation of the European type was a unique evolution of the type, widely built in cities across the country (and since adopted in other parts of the world).

The individual elements of the new American type built in Richmond are simple. Wide sidewalks made for strolling combined with broad medians and street-oriented, freestanding buildings define the avenue. Trees are planted approximately forty feet apart. Houses and apartment houses of various sizes account for a range of incomes along the avenue. Modest details—like asphalt paving blocks and hardscape in neutral tones—make Monument Avenue seem grand and approachable at the same time.

The economical and durable asphalt blocks also give a texture to the street that would improve the look of many wide streets. Street arrangements like the cross-axis at North Davis Avenue—where the Metropolitan Community Church visually terminates the axis of the small green perpendicular to the avenue—give variety and richness.

▶ **Figure 2.36:** Monument Avenue, Richmond, Virginia. Begun by CPE Burgwyn, 1887. The generously sized median has paired rows of trees creating a linear park that joins the city together. Compare this with the meager concrete "traffic separator" that runs down the middle of the average suburban arterial. *Courtesy of Christopher Podstawski*

Figure 2.35: Monument Avenue, Richmond, Virginia. Begun by CPE Burgwyn, 1887. The grand avenue was gradually adapted to fit the American urban form. *VCU Libraries / Wikimedia Commons / Public Domain*

Figure 2.37: Monument Avenue, Richmond, Virginia. Begun by CPE Burgwyn, 1887. The front porches facing Monument Avenue are slightly elevated above the sidewalk, establishing privacy and dignity.

Figure 2.38: Monument Avenue, Richmond, Virginia. Begun by CPE Burgwyn, 1887. Equestrian statue of Robert E. Lee by Antonin Mercié, 1890 (now removed). At key intersections, monuments commemorated historical figures, ranging from Confederate generals to, more recently, African American tennis star Arthur Ashe. *Courtesy of James Dougherty*

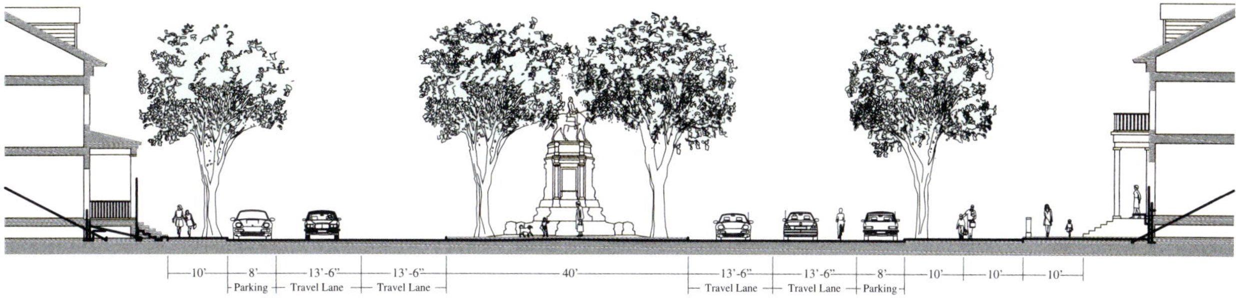

Figure 2.39: Monument Avenue, Richmond, Virginia. Begun by CPE Burgwyn, 1887. Section. A close inspection reveals that the avenue is not entirely symmetrical, that its lanes are inadvisably wide, and that the building heights and widths vary substantially. However, the broad median and the parallel rows of majestic trees have a powerful unifying effect. © *2013 Dover, Kohl & Partners*

Figure 2.40: Monument Avenue, Richmond, Virginia. Protesters surround the Robert E. Lee Statue. On the statue, graffiti reflects the emotions of the protesters. *Courtesy of Richmond Times Dispatch*

Streets have always been much more than mere addresses or movement corridors. Like other public spaces, streets are also communications devices, sending messages about what a community considers important. From time to time, those messages need to change, as society evolves and its values mature. On those occasions, a street can be both the medium for delivering the new message and the scene where those moments of historical importance unfold. That's exactly what happened in Richmond in 2020 and 2021.

Monument Avenue was already famous, but the message it communicated was heard differently by different groups. On one hand, its elegance and geometry had long sent a message about pride of place. But, the statues of Confederate generals sent a more sinister message, directly celebrating slavery and glamorizing an Old South built on racial injustice. For generations of African Americans, Monument Avenue felt like a slap in the face. It is clear that the statues were part of a campaign to romanticize cruelty and reinforce institutions of oppression; they were installed in the run-up to Jim Crow laws, voter suppression trickery, redlining, and white supremacy promoted via popular culture in films like *Birth of a Nation*. The author of *The Clansman*, Thomas Dixon Jr., was a major defender of the Richmond monuments, which by contrast drew opposition in the Northern press from the beginning. (On May 29, 1890, the day before the dedication of the Robert E. Lee monument, the *New York Mail and Express* carried the headline "Treason Glorified!") According to historian Bruce Stephenson, "Robert E. Lee personified the Lost Cause. Cast as a devout Christian who abhorred slavery and labored tirelessly after the war to unify the nation, Lee was [in fact] a slave-owning aristocrat who fought to keep slavery and his privileged position."[26]

The murder of George Floyd by police in Minneapolis on May 25, 2020, prompted the Black Lives Matter movement to lead months of protests demanding justice. The protesters creatively used streets to transmit this message, painting immense BLACK LIVES MATTER murals directly on streets in at least eighty-one American streets as well as in Australia, Canada, and the United Kingdom.[27] Photos of these murals grabbed the world's attention in press reports and the symbols went viral on social media. Seizing the moment, citizens renewed calls upon their mayors and governors to remove symbols of inequality like Confederate flags and statues of rebel soldiers and brutal colonizers. In Richmond, nightly protests centered on the Robert E. Lee statue on Monument Avenue. Protesters covered the pedestal of the statue in graffiti and projected images on it. A lengthy court battle followed, and on September 8, 2021, crane operators lifted the statue away as crowds gathered to watch.

Great public spaces often include landmarks like the Robert E. Lee statue. But the meaning of these landmarks, and our understanding of their messages, can and should change with time. In the harsh light shone on it after the brutality of Floyd's murder, the full meaning of the Lee statue on Monument Avenue became clear to many more people.

On ordinary days, a public space might merely be a good spot to meet a neighbor or enjoy the scenery, but there come days when that same space needs to be used for large gatherings, and for standing up for what's right. Monument Avenue became the canvas on which to broadcast the injustices, to change the message, and to mark an evolution in American values. On that day, Monument Avenue transcended its role as a path for cars going from here to there and became a path for people moving on to the future.

Figure 2.41: Queens Road West, Charlotte, North Carolina. John Nolen, 1911. Today, the trees planted by Nolen feel like a majestic forest—but one that easily accommodates houses and yards, and that we comfortably drive through.

Queens Road West, Charlotte, North Carolina

John Nolen, 1911

Boulevard

Queens Road was designed in 1911 as part of John Nolen's plan for the leafy streetcar suburb of Myers Park, where he sought to blend town and country. The street is a key through route with substantial daily traffic; at the same time, it showcases the grandest houses on some of the largest lots in Myers Park. Queens Road proves that it is possible for addresses that are part of a continuous, connected street network to retain—and even acquire—prestige, postwar prejudices notwithstanding. Houses in Myers Park have steadily appreciated in value and are among the most sought-after in the Charlotte region. In large part, this is a result of Nolen's brilliant yet simple tree plan for Queens Road. Being on Queens Road feels like being in a mature forest under a high canopy of native oak, elm, and tulip poplar trees (Figures 2.41 and 2.42).

Nolen experimented by transplanting one hundred mature trees in Myers Park in the first year of development; after a year, only one of the trees had died.[28] Concluding that the high survival rate of the transplanted trees was due to the use of native species, he thereafter enthusiastically encouraged the planting of native trees. He directed developer George Stephens to plant large numbers of trees throughout Myers Park, including the street trees following the alignment of Queens Road, and he provided design advice to individual lot owners. He also demanded a specific number of trees in front of and behind each house at a time when this was not common practice.

The streets in Myers Park range in width, depending on their type and their proximity to the large public park that is integrated into the neighborhood layout. Some of the streets are as narrow as forty feet. Queens Road West is the widest of the designs, with a right-of-way of 110 feet. A central median thirty-two feet wide is a source of open space along the arterial, while wide planting strips frame the sidewalks on each side. Nolen was a landscape architect

Figure 2.42: Queens Road West, Charlotte, North Carolina. John Nolen, 1911. The Queens Road houses have substantial front yards.

by training and felt free to depart from orthogonal grids, instead applying gently rounded forms more like those observed in nature. Queens Road has a curving, picturesque design, fitted to the rolling topography, and Nolen's fine-tuned plan allowed for differentiating the sizes (and costs) of houses, lot depths, and front yard setbacks.

> To plan a residential area in a shape that respects the lay of the land obviously necessitates blocks of irregular shapes and sizes, and consequently varied building lots.
>
> — John Nolen, *New Towns for Old*

If we look only at the cross section of the street between the curbs, Queens Road shares some characteristics with Commonwealth Avenue in Boston and St. Charles Avenue in New Orleans: all three have wide central medians and tall trees planted in rows (Figure 2.43). However, the effect on Queens Road is radically different than on the other two. To begin with, Queens Road lacks the central promenade, and the attached buildings lining Commonwealth Avenue knit it together and provide urbanity. The immersive environment of consistent architecture on Commonwealth—the result of a strict design code—gives a different feel than the eclectic mix of Colonial Revival, Arts and Crafts, and Tudor Revival houses on Queens Road. The Queens Road median was originally intended to provide space for a streetcar, which might have given the street a character reminiscent of St. Charles Avenue. In the end, however, very little of the streetcar line was built. In the absence of tracks, the broad lawn on today's Queens Road lends the corridor a more suburban feel. The great depth of the front yards accentuates this impression; whereas the fronts of many buildings on St. Charles Avenue are within conversational distance of the sidewalks, the fronts of houses on Queens Road can be as far as eighty feet away. Finally, both Commonwealth Avenue and St. Charles Avenue are flat and have long segments of arrow-straight formality. Queens Road winds and dips and rises in a streamform pattern; the effect is more picturesque but also more highway-like than Nolen probably intended.

On the other hand, one of the many attributes of a well-designed street is that it gracefully accepts revision, adapting to changing times. Nolen's foresight in reserving the land for the broad median means that someday, when Charlotte fully restores its once-robust streetcar system, Queens Road will be ready for it.

PROMENADE STREETS

A *promenade* is a place for strolling, usually tree-lined. A number of street types and park designs incorporate a form of the promenade. On some of these, like the famous *ramblas* in Spain, the promenade is the signature element. These examples belong in a category of their own: Promenade Streets.

Some streets in this book contain promenades as part of a larger ensemble. Commonwealth Avenue in Boston has a central promenade that offers a pleasant alternative to the sidewalks shaded by buildings on a cold but sunny winter's morning. The boulevard de Rochechouart in Paris has a bustling central promenade, too. But comparing these to Barcelona's ramblas, we see that the Spanish roads have a far smaller proportion of their cross section devoted to moving and parking vehicles and a wide median devoted to places for walking, dining, and vendors rather than planted beds or lawn.

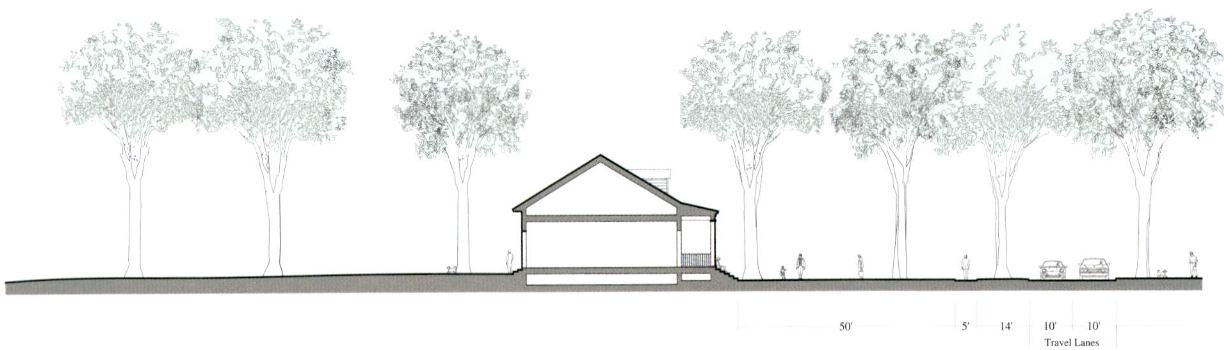

Figure 2.43: Queens Road West, Charlotte, North Carolina. John Nolen, 1911. Section. © 2013 Dover, Kohl & Partners

SUBURBS: THEY DON'T MAKE 'EM THE WAY THEY USED TO

Figure 2.44: Elmwood Village, Buffalo, New York. The classic American streetcar suburb was designed to create a walkable, dignified public realm and to support civic pride.

Figure 2.45: Doral, Florida. In the sprawl era, garage doors and wide driveways supplanted front porches and street trees, leaving the streets unwalkable and unloved.

The Promenade Street is a sibling of the boulevard and the avenue. All three are important street types in the urban scene, and they are usually arranged in a grand manner—the Promenade Street might be visually terminated like an avenue or open at its end like a boulevard, for instance. Like them, it might be a favored location for stores and restaurants, because it serves as an organizer and attractor of pedestrian activity. A Promenade Street is not typically the fastest route across town for motorists, though. Its travel lanes are few and narrow, and the flow of auto traffic in many instances tends to be impeded by an almost continuous crossing of pedestrians from the flanking buildings and side streets.

The Promenade Street is like a park, too, bringing a wide swath of public space and tree canopy into the heart of the city. But its linear form makes the street more of a spatial *connector* rather than just a green exception to the built-up blocks that surround it. The Promenade Street is like a square or plaza, as well, in that it assumes a role as the neighbors' shared public space, apart from their homes, yards, and ordinary streets. Like a square, the Promenade Street is a space where people are drawn

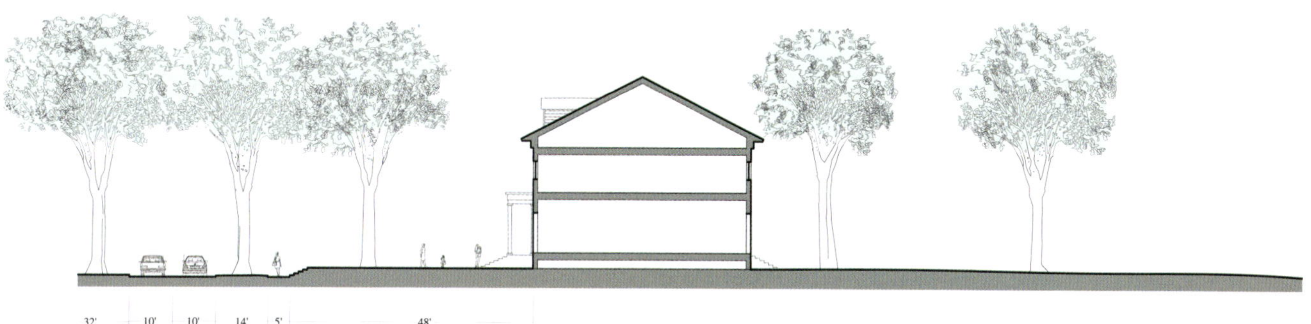

Figure 2.46: Promenade Street with planted island. © 2022 Dover, Kohl & Partners

to gather and socialize and eye one another—but unlike a square, which condenses this activity, the Promenade Street stretches it out and sets it in motion. Our favorite examples offer a daily parade of people of all sorts.

The case studies that follow clearly show there was a golden era of building Promenade Streets, perhaps culminating in the Paseo del Prado in Havana. As we talk about reducing auto use to lower our carbon footprint, the promenade is a highly relevant street type for today. Not only are the built examples beloved—more popular now than ever—but they offer solutions to modern problems. Today's parks departments are under budgetary siege, struggling to maintain their existing manicured acreage, whereas a promenade is a kind of public space that generates money, with phenomenal economic activity flowing out of its cafes, flower markets, and shops. A rambla address has a magnetic appeal for tourists, who favor hotel rooms nearby, and for the leaders in creative industries, who know that their designers, thinkers, and visitors want to step out into that scene at the day's end. A Promenade Street is also a practical way to introduce a great open space into the city without acquiring vast tracts of land; its linear form makes it a natural replacement for an

obsolete elevated freeway or formerly oversized arterial road, for example. What a trade!

The Promenade Street is one of the street types that quietly dropped off the professional street-design menu in the twentieth century, but it should be back on our menu for the twenty-first.

THE RAMBLA TYPE / STEFANOS POLYZOIDES

Dedicated to Gabriel Garcia Aragon

The rambla is a multimodal street type common to Catalonia and the regions of Valencia and the Balearic Islands that share the Catalan language and culture, from whence it spread to other parts of Spain and Spanish colonies. Distinguished by a large and central pedestrian island, limited narrow carriageways, and a monumental streetscape, the rambla is the Catalonian and Spanish equivalent of the promenade, a design solution for the need to sanitize cities in response to the pollution and congestion resulting from the early Industrial Revolution.

Figure 2.47: Promenade Street with paved island. © 2022 Dover, Kohl & Partners

As a design opportunity, it was presented by the demolition of city walls, which began in this part of the world at the end of the eighteenth century and lasted until the 1860s (Figure 2.47).

The kinds of residual land outside these urban fortifications were of two possible topographic patterns. The first kind was made up of irregular, abandoned flat fields left open for defensive purposes and often used for open-air commerce. The second kind was made up of dry riverbeds, residual parts of the natural landscape, which channeled the surface water of surrounding agricultural lands into the sea. Close to dense cities, these areas had become unsightly and unhealthy places. Yet, they offered all the necessary ingredients for potentially generating a new kind of constructed urban landscape.

The rambla type at its origin was introduced as a large-scale linear park, a promenade providing badly needed recreational and civic space for urban dwellers. At the same time, it was a grand element of hydrological infrastructure that solved the twin civil engineering challenges of conducting water runoff and channeling vehicular traffic through cities.

After two hundred years, such ramblas have become the single most prominent, iconic, and beautiful public places in Palma de Mallorca, Barcelona, Tarragona, Vilanova i la Geltrú, Figueres, Sant Feliu de Guixols, Arenys de Mar, Sabadell, Tarrassa, Lleida, Girona, and Igualada—the Catalan cities where the prime examples of the type are located.

The most prominent design element of the rambla type is its central and dimensionally imposing median island. Originally finished in decomposed granite, the islands are now paved. Pedestrian circulation is channeled there, to mark the dominating presence of walkers over vehicles, and to provide a place for a variety of temporary commercial activities and ritual annual events. A single, narrow traffic lane, parking lane, and sidewalk are placed symmetrically on either side of the central island to complete the typical rambla plan.

The cross section of a rambla is also a distinctive ingredient. *Alamedas* of large, mature plane trees are typically laid out rhythmically and symmetrically along the central median.[29] When in full foliage in the spring and summer, the trees fill the entire right-of-way and completely shade the median. In this part of the world, plane trees are native to creek and river beds. Their planting pattern produces a spatial effect that recalls the memory of a natural watercourse in the middle of a busy metropolitan street.

LES RAMBLES, BARCELONA, SPAIN / STEFANOS POLYZOIDES

Promenade Street

Les Rambles of Barcelona is located on the central north–south axis of the city. It is one of the most formally complex and vital streets anywhere. The name *les Rambles*, plural, is the Catalan name, generally used in the region, for the six interconnected sections that together make the main promenade of Barcelona. These have been identified throughout history with adjacent institutions or activities: Rambla de Canaletes, Rambla dels Estudis, Rambla de Sant Josep, Rambla dels Caputxins, Rambla de Santa Monica, and Rambla del Mar.

But in fact, the form of les Rambles is more easily understood and experienced as a three-part composition, each part enriched by some of the greatest historic buildings and places in the city, such as the Mercat de la Boqueria, the Plaça Reial, and the Dressanes. A long, straight space dominates the middle part of the street and establishes its dominant character. The alameda planted along the entire length produces the effect of a cathedral nave. To the north and south, two funnel-shaped residual spaces, more open and less planted, transition to the rest of the city, leading to the Eixample and the harbor, respectively.

The collective architectural character of les Rambles was established in the late eighteenth century. The juxtaposition of buildings previously attached to the demolished western medieval wall with the irregularly plotted suburban buildings in the Raval across from them initially produced a heterogeneous architectural fabric. In the last two centuries, this pattern has become dominant. The addition of buildings diverse in type, style, scale, height, profile, materials, and decorative details has resulted in les Rambles being defined by an extremely varied built edge (Figure 2.48).

Pedestrian traffic along the street is constant and very high in volume, the result of its privileged location along the central movement axis of Barcelona. This high pedestrian volume generates a powerful retail economy. Stores occupy the ground floors of almost all buildings, and a variety of picturesque, temporary retail activities

are accommodated in pavilions on the central island. All kinds of public spectacles also take place along its length. The experience of continuous waves of energy and vitality washing over the place produce a sense of great joy and conviviality (Figure 2.49).

Yet, the single most important formal ingredient of les Rambles is in the geometric definition of its plan. The place is a virtual symphony of diverging dimensions, an ode to asymmetry and irregularity. The variation of the combined right-of-way, sidewalk, carriageway, and pedestrian island dimensions is extreme. The right-of-way varies between 75 feet at its narrowest to 190 feet at its widest. In places, sidewalks can be as narrow as 4 feet or as wide as 41 feet. The carriageway measures between 15 and 33 feet. The pedestrian median squeezes down to 32 feet and opens up to 101 feet. Nowhere in the kilometer length of les Rambles are the dimensions of the four elements described above ever repeated singly or in combination (Figure 2.50).

Figure 2.48: Les Rambles, Barcelona, Spain. The buildings along the edge of les Rambles are diverse and frequently highly articulated. Building heights, widths, facades, and ornamentation change from one end to the other.

Figure 2.49: Les Rambles, Barcelona, Spain. We walk down the center of the street, which is a rare but comfortable vantage point for pedestrians. The constant flow of strolling people under the great canopy is marvelous.

The visual effect is as extraordinary as it is unique. Moving on foot from the unusual vantage point of the center of the right-of-way, surrounded by a vertically and horizontally varied architectural enclosure, following in the vector of a lightly sloping ground, through a constant dimensional variation in plan and section (Figure 2.51) and the shifting light of the sun, produces a rare sensation: the ground, buildings, trees, vehicles, pavilions, people, animals, and inanimate objects seem to be engaged all together in gentle motion. This is the magic that is experienced on les Rambles, and the foundation of its reputation as one of the most famous streets in the world.

After two hundred years, ramblas are the most prominent, iconic, and beautiful public places in Catalan cities like Barcelona, Palma de Mallorca, Vilanova i la Geltrú, and Figueres.

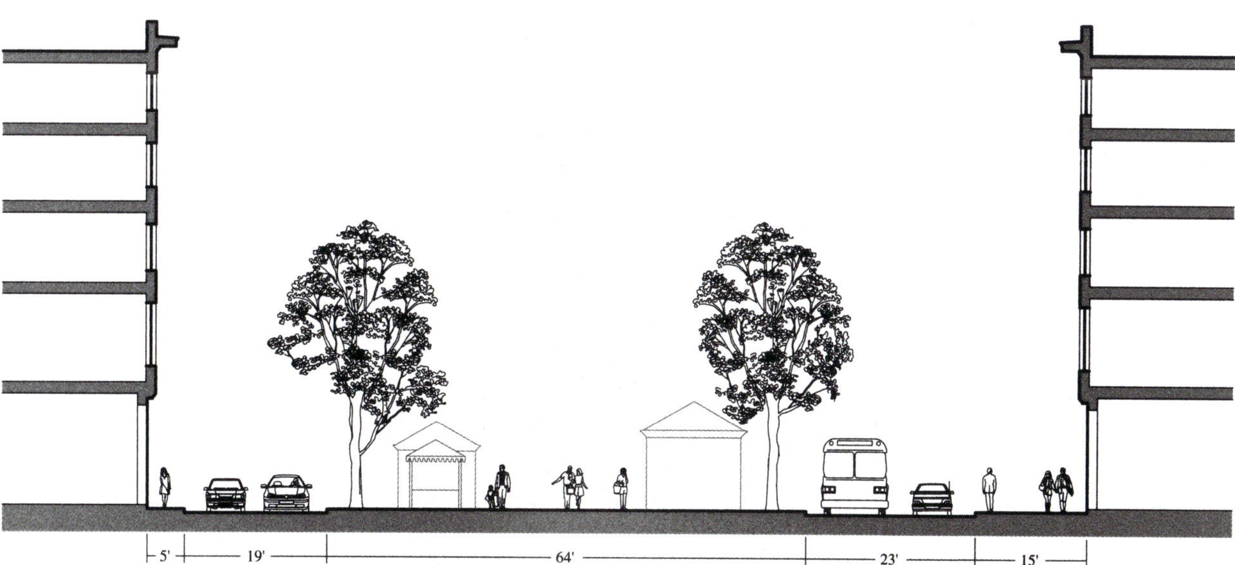

Figure 2.50: Les Rambles, Barcelona, Spain. Section. © 2013 Dover, Kohl & Partners

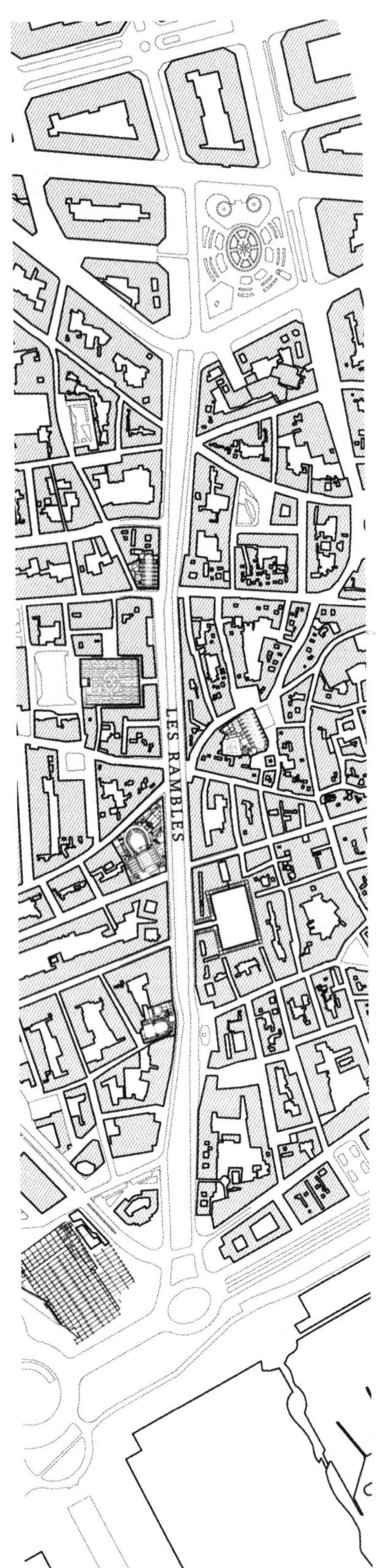

RAMBLA CATALUNYA, BARCELONA, SPAIN / STEFANOS POLYZOIDES

Ildefons Cerdà, 1859

Promenade Street

The Rambla Catalunya is one of Barcelona's most distinctive and elegant urban streets. Together with the Passeig de Gràcia, it occupies the geographic center for the Eixample, the heroic nineteenth-century regional enlargement of the city designed by Ildefons Cerdà. This rambla was built along ten new, uniform, repetitive city blocks. Conceived as an extension of the historic les Rambles, it connected the medieval core of the city with the village of Gràcia.

As with all the other street types that Cerdà designed throughout the Eixample, the Rambla Catalunya has a dimensionally stable plan along its entire length. The right-of-way measures 98 feet in width; the pedestrian central island, 42 feet; both sidewalks, 11 feet each; and both the single, one-way traffic and parking lanes, 16 feet each. The small sidewalks are particularly noteworthy as they force most pedestrians to circulate on the central island. There are elegant retail stores lining the ground floors of the buildings along most of the rambla's length. Many of these ground-floor businesses are cafes and restaurants. Their tables are often located under umbrellas in the shade of the alameda of the central island and are served by waiters who cross the moving traffic to reach their customers.

The dimensions of the two carriageways are tight for traffic movement and parking, and as a result, they appear diminutive compared to the width of the pedestrian island (Figure 2.52). Pedestrian traffic is favored by the continuous slope of the ground plane and various physical and functional impediments to the speedy flow of traffic. These include continuous lines of tree trunks at both edges of the pedestrian islands, the constant and random street crossing by pedestrians, the vibrant store graphics, the merchandise in the storefront windows, and the distracting decorative detail of most of the buildings.

Figure 2.51: Les Rambles, Barcelona, Spain. Figure-ground drawing. Les Rambles is actually several segments laid end-to-end, rising from the funnel-shaped end at the harbor, through the semi-straight central section, and widening again as it nears the Eixample. © *Moule & Polyzoides, Architects & Urbanists*

Figure 2.52: Rambla Catalunya, Barcelona, Spain. Ildefons Cerdà, 1859. Narrow traffic lanes allow deliveries, neighborhood traffic, and parking without interfering with the coming and going of waiters for the café and restaurant tables in the center.

The sectional configuration of the Rambla Catalunya is also very distinctive, just under 1-to-1 in proportion of street width to height (Figure 2.53). The visual effect it conveys is one of intense enclosure and memorable urbanity. The continuous and rhythmic sequence of street trees planted as an alameda, the light poles, and the street furniture further contribute to the stable visual character of the place. But the formal signature of this rambla is the quality of the buildings defining its edges.

The immense Eixample project, extending over three hundred blocks, was built out over approximately a century through architectural designs operating on two lot and building types only: stacked flats in rectangular lots and stacked flats in triangular corner lots. How could such a rational, repetitive, relentless degree of typological repetition ever produce such a unique and distinguished built environment?

The answer is that mixed-use buildings of spectacular individual design relieve each block face. There are six fifty-foot-wide lots on the orthogonal portions of every block facing the Rambla Catalunya. Each of these lots is occupied by a separate building designed by the hand of an accomplished architect. Some of the best examples of Catalan *Modernismo* (Art Nouveau) are to be found there. The stone facades, rich architectural details, and deeply projecting frontages of these buildings form an architectural fabric that animates and enriches the public realm of the Rambla, providing a theatrical stage for the common and the ritual events that unfold day after day along this spectacular urban promenade.

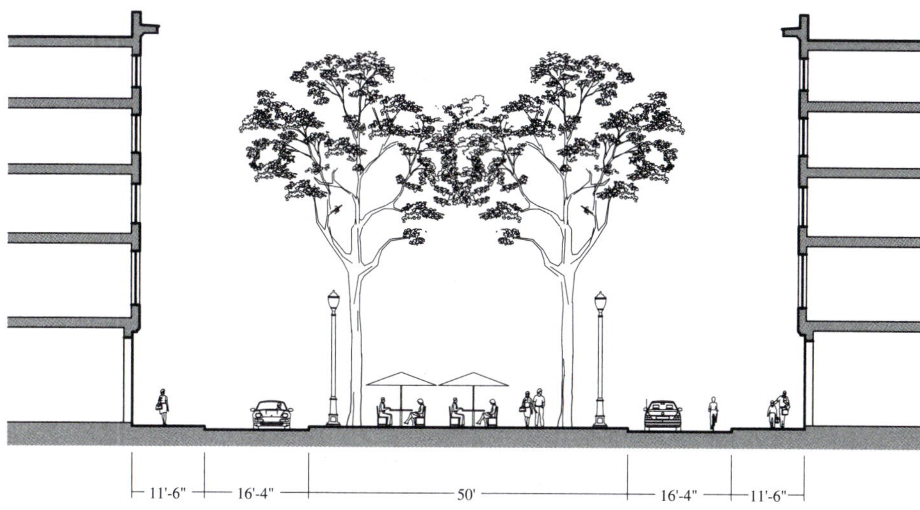

Figure 2.53: Rambla Catalunya, Barcelona, Spain. Ildefons Cerdà, 1859. Section. © *2013 Dover, Kohl & Partners*

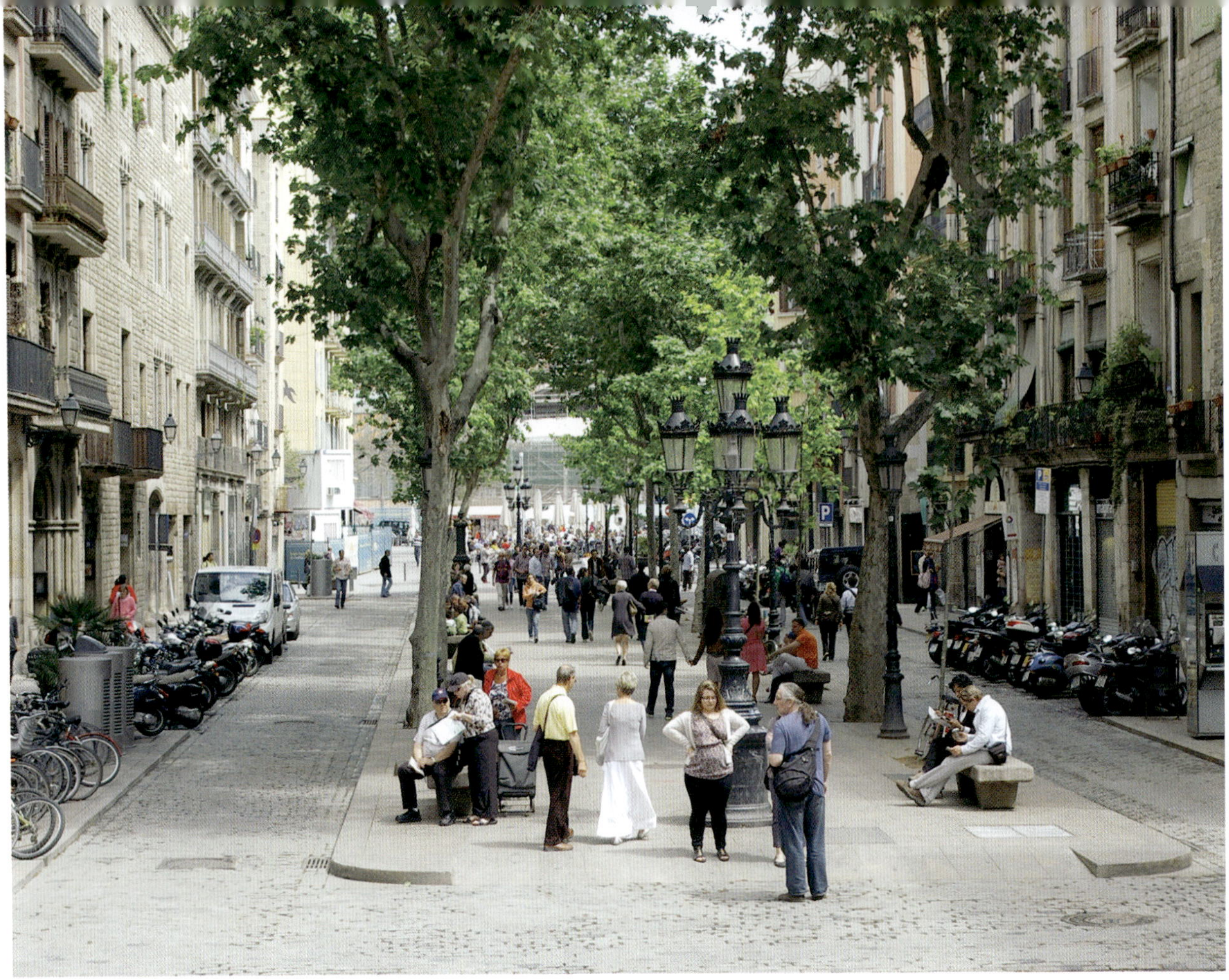

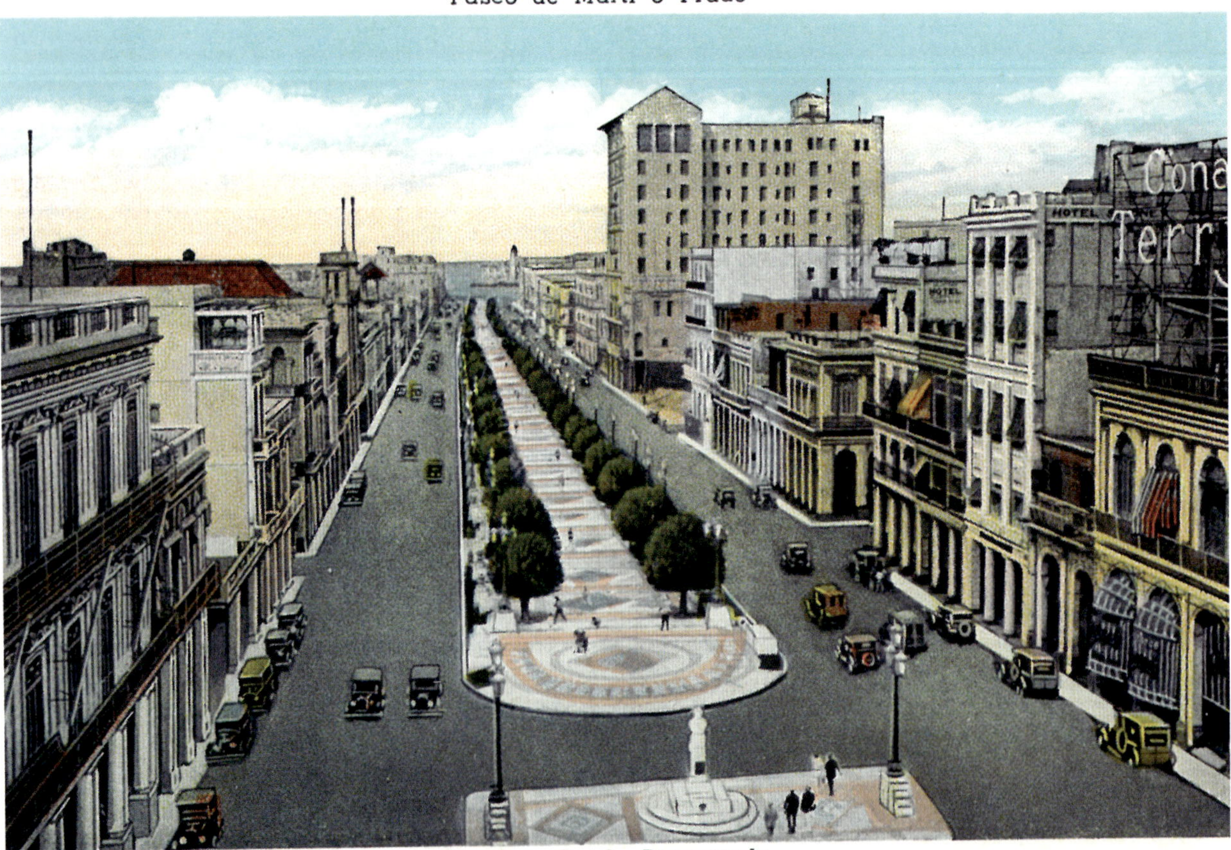

Figure 2.54: Passeig del Born, Barcelona, Spain. View from the steps of the Church of Santa Maria del Mar, looking northeast. The rambla form appears in shorter examples around the region like the delightful Passeig del Born.

Paseo de Marti o Prado

Marti or Prado Promenade

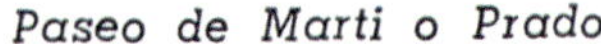

PASEO DEL PRADO, HAVANA, CUBA / ANDRÉS DUANY

Begun, 1772; rebuilt, 1834; redesigned by Jean-Claude Nicolas Forestier, 1928

Promenade Street

Paseo del Prado (El Prado) in Havana is the last of those great streets, like the Ringstrasse in Vienna and the ramblas of Barcelona, that replaced demolished fortifications. El Prado was redesigned by Jean-Claude Nicolas Forestier, the French Beaux-Arts-trained landscape architect and planner of the 1920s, as part of his overall plan for Havana. This included the waterfront Malecon, as well as two other avenues—one that leads to the university and one dedicated to the presidents. These are excellent, but neither, and perhaps no promenade in the world, is as accomplished a design as El Prado (Figure 2.56).

The Prado leads from the sea to the new Capitolio, which is from the same period. The Capitolio together with that other great Beaux-Arts ensemble, the Universidad de La Habana, are superb. The Prado itself is special not only because of its section but also for its detailing. The floor is interior-quality terrazzo, and there are masterfully integrated banisters, steps, benches, and lamps (Figure 2.57). El Prado deploys a unique section: the promenade portion is elevated so that it lifts the pedestrian above the traffic on the side streets. The integral seating faces not only inward to the tree-lined promenade as would be expected but also outward—so that the edge of the elevated area is not a socially impervious "hull." Beyond the promenade, the flanking buildings are required by code to have a continuous arcade. They have a uniform height, but the buildings are by various designers. This produces just enough contrast between the private realm and the uniform civic architecture of the central promenade.

One termination of the Prado is the slim dome of the Capitolio, perfectly framed. The walkways and entourage of the Capitolio (and an intermediary square) are designed in an integral way with the Prado, forming a unified ensemble. While this is brilliantly handled, the

Figure 2.56: Paseo del Prado, Havana, Cuba. Jean-Claude Nicolas Forestier, 1928. Satellite view. The Prado stretches from the sea to the Capitolio. © 2012 DigitalGlobe / Google Earth

◀ **Figure 2.55:** Paseo del Prado, Havana, Cuba. Seen here soon after its twentieth-century reconstruction and initial replanting, "el Prado" was envisioned to have a similar character along its length. In reality, the results varied from one end to the other, from shady to sunstruck, reflecting the failure of the trees closest to the sea breezes. *Courtesy of Maricé Chael*

other termination is the great flaw of El Prado. Instead of a building (the Palacio Presidencial could have easily been located there instead of nearby), it is open to the sea. Winds have always swept up the length of El Prado. As a result, about a third of the trees that were integral to the Prado's design are dead and another third are damaged. Only the trees farthest from the sea are sufficiently protected to show the original intention (Figure 2.58).

The ocean can be destructive to urbanism. This lesson of what not to do is visible in El Prado, as it is along Havana's great drive by the sea, the Malecon, which is constantly besieged by the ocean, weather, and neglect. Havana has lessons on the good and the bad on offer, not to be ignored.

Figure 2.57: Paseo del Prado, Havana, Cuba. Jean-Claude Nicolas Forestier, 1928. The promenade is elevated above street level, with a terrazzo finish worthy of any luxury interior, and benches and lampposts fully integrated into the design. © 2013 Stephen A. Mouzon

Boulevard de Rochechouart, Paris, France
Service des Promenades et Plantations, 1864
Promenade Street

Located at the base of the Montmartre hill, the boulevard de Rochechouart has a compelling street section with a wide median that accommodates walking, cycling, and entrances to the *Métro*. An intricately constructed boulevard, Rochechouart varies slightly in total width as it crosses the Parisian city fabric. The large, central median with four rows of trees gives the street its unique visual signature.

The boulevard is named after a seventeenth-century abbess of Montmartre. It was fashioned into its current form in the nineteenth century, following the alignment of a section of the notorious Wall of the Farmers-General, where tax collectors infuriated the populace with their levies just prior to the French Revolution.[30] Today, it anchors a notably happier neighborhood, just far enough from haunts more overrun by tourists to remain relaxed and popular with ordinary Parisians.

Rochechouart is a mixed-use *tour de force* that puts to shame our contemporary "multimodal" designs (Figure 2.59). Two vehicle lanes travel in each direction, four in total, with two of the lanes dedicated to buses and taxis. Thus, an ample proportion of the right-of-way

Figure 2.58: Paseo del Prado, Havana, Cuba. Jean-Claude Nicolas Forestier, 1928. The dramatic tree canopy intended for the Prado is seen in the foreground, but the damaging effects of the sea winds on the trees closest to the waterfront are clearly visible in the background. © 2013 Stephen A. Mouzon

Figure 2.59: Boulevard de Rochechouart, Paris, France. Service des Promenades et Plantations, 1864. The famed Métro entrances (H. Guimard, 1900) are within the median, as seen here at the Anvers station.

is given over to public transportation. Beyond the travel lanes, on each of the outer sides, is a row of on-street parking and then a twelve-foot sidewalk typical of major streets in Paris. The storefronts along the sidewalks are virtually continuous and define the outer edges of the street space. Offices and apartments overlook the scene from the upper floors.

The Paris Métro runs below the boulevard. Within the median, the famous Art Nouveau Métro entrances designed in 1900 by Hector Guimard instantly identify the stations.[31] The median is slightly raised above the level of the travel lanes—not as dramatically at the Paseo del Prado in Havana, but just enough to allow pedestrians on the central promenade to sense their command of the street. Hedges, benches, and plantings define the median, subtly demarcating the space for pedestrians and cyclists without an excessive amount of striping or signage. Cycle tracks within the median travel in each direction, providing a consistent and protected space for cyclists on the boulevard (see Figure 2.60).

Figure 2.60: Boulevard de Rochechouart, Paris, France. Service des Promenades et Plantations, 1864. The cycle track is slightly elevated above the roadway and is framed by rows of bushes and shade trees. As cycling increases in the new 15-minute city, the track looks increasingly narrow.

Typically, a curb and hedges form the edge of the median; at specific intervals, however, the curb is replaced with two or three steps. As on the recently retrofitted Cours Mirabeau in Aix-en-Provence, the stepped edge naturally slows the flow of auto traffic. The detail changes the psychology of the space: the steps give a visual impression that pedestrians would and should feel welcome to step out into the travel lane or cross to the center median at many points along the block, and they reinforce the feeling that the buses and cars are moving through

Figure 2.61: Boulevard de Rochechouart, Paris, France. Service des Promenades et Plantations, 1864. View of the broad, planted promenade.

Figure 2.62: Boulevard de Rochechouart, Paris, France. Service des Promenades et Plantations, 1864. Section. The stepped-curb detail changes the psychology of the space, suggesting that the whole street is pedestrian space through which cars are invited to carefully travel, not vice versa. © 2013 Dover, Kohl & Partners

pedestrian-dominated space, rather than vice versa (Figures 2.61 and 2.62). Crossing the street as a pedestrian is easy—in fact, all modes of traffic seem to operate with ease.

One block south of the Anvers Métro station, the avenue Trudaine parallels the boulevard de Rochechouart. Only four blocks long, the quiet street has a narrow median in the center and trees planted on both sidewalks. In between the two avenues is the green place d'Anvers, which is attached to the block on the western side, with a narrow road on its eastern side. The place and the avenues create a set of green gathering spaces, making prestigious addresses along their edges for local institutions.

The boulevard de Rochechouart extends seamlessly at each end to connect with the boulevard de Clichy in the west and the boulevard de la Chapelle in the east. Together these form a major east–west route across the city, yet Rochechouart is not bloated with motoring lanes. This multimodal balance offers a lesson for many other cities.

Figure 2.63: Commonwealth Avenue, Boston, Massachusetts. Arthur D. Gilman, 1856. Satellite view. *Courtesy of Google Earth*

Commonwealth Avenue, Boston and Newton, Massachusetts

Arthur D. Gilman, 1856

Promenade Street

Commonwealth Avenue is a formal boulevard that adjusts to the surrounding fabric of the city as it travels through different neighborhoods. The most pleasing section of "Comm Ave" today begins at Boston Common and then crosses through the flatlands of Back Bay, where the city filled in a bay between 1857 and 1882 to create a new neighborhood (Figure 2.63). That part of the boulevard is straight and unusually wide. Sitting in the rectilinear nineteenth-century grid, it brings variety to the street plan with a promenade down the middle. Called the Commonwealth Avenue Mall, the promenade is approximately one hundred feet wide, with two parallel rows of trees on each side of a central walk (Figure 2.64). The trees sit in wide lawns, and the walk is periodically interrupted by monuments along the central axis. Most of the time, Commonwealth Avenue has less traffic than the surrounding streets, and a walk down the center can be peaceful and relaxing.

The Mall falls apart when half the lanes dive under Massachusetts Avenue while the other lanes rejoin the perpendicular avenue at grade. After another multilevel crossing at Kenmore Street, the greenway drops off altogether, and the center of Commonwealth is taken over by streetcars. This section is the spine of Boston University, but the trolleys and cars dominate the street, and the heavy hand of an engineer is evident. Comm Ave then runs for approximately a mile and a half before turning up the hill and becoming a wide, twisting road that goes uphill and down through the town of Newton. Like the flat, straight portion of the boulevard at Boston University,

Figure 2.64: Commonwealth Avenue, Boston, Massachusetts. Arthur D. Gilman, 1856. Wide-angle view showing the promenade flanked by rows of trees. *Courtesy of Christopher Podstawski*

Figure 2.65: The Paseo, Kansas City, Missouri. George Kessler, 1893–1899. Postcard view. Commonwealth Avenue has many siblings; the Promenade Street type was deployed widely as the City Beautiful movement evolved and spread, as seen in this Midwestern example.[32] *Courtesy of Kansas City Public Library*

this section has suffered at the hands of car-focused transportation officials over the years.

Ill-considered improvements for the benefit of the car on and around Commonwealth Avenue not only damaged the boulevard but also the city as a whole. Originally Commonwealth Avenue was part of Frederick Law Olmsted's Emerald Necklace, a seven-mile-long greenway that ran from the Common and through the Fens to Brookline.[33] That connection was severed at Charlesgate by the construction of the Massachusetts Turnpike, and to a lesser degree by entry and exit ramps for Storrow Drive. The Turnpike and Storrow Drive also cut the Emerald Necklace's meeting with the Charles River Esplanade.

MAIN STREETS

Main Streets are both a street type and a metaphor. They have played prominent roles in Hollywood movies like *It's A Wonderful Life,* where they symbolize the All-American

Main Street we all have in our mind's eye. Physically, they have shops on both sides of the street with offices or apartments above. The buildings might be two or three stories tall and only go for a block or two, or they might be eight to twelve stories high and be part of a large downtown. "Wall Street versus Main Street" is a current political trope to describe the interests of the super-rich and global capitalism versus Middle America, but until auto-based planning and development killed our downtowns, every city of a certain size had what was known as its Main-Main Corner, where the two most important streets intersected and where the most profitable stores, like the local department store, wanted to be.

Main Street is usually a spine, or what Douglas Duany calls the armature, of the town or city plan (Figures 2.66–2.70). They aren't all the same. In our smaller towns and cities, a single street in a downtown neighborhood typically functions as *the* main street, like the ones we write about in the Massachusetts towns of Nantucket and Great Barrington and in Galena, Illinois. Main streets are the mixed-use heart of town, and locals living nearby meet many daily needs there. In the largest and most complex cities, on the other hand, certain main streets take on a regional role; for instance, Manhattan has continuous retail along its avenues yet rarely a location one can identify as *the* neighborhood center. London has many neighborhood High Streets (the British equivalent of Main Streets), but in the center of London, certain regional shopping streets like Oxford Street or regional entertainment streets like Shaftesbury Avenue take on the dominant role.

While main streets vary, one thing is constant: a successful main street in a sustainable, walkable town or city is a place where people want to get out of their cars and explore. There are many factors that contribute to that, including things to do (like shopping or going to a concert), interesting sights to look at (including people—we like to people-watch), and a level of physical and psychic comfort. As we walk along, we like to feel safe and secure. Beauty is also important to making a place where we want to be. No one likes walking in an ugly place that feels unsafe.

Figure 2.66: King Street, Alexandria, Virginia. A classic American main street: the mixed-use heart of its community.

Figure 2.67: High Street, Chelmsford, Essex, England. Note the cyclists—originators of the "good roads" movement—in this historic photograph. *The Story of Some English Shires by Mandell Creighton, 1897 via fromoldbooks.com, with permission*

Figure 2.68: Main Street, Rhinebeck, New York. A classic downtown in the Hudson River Valley, with small storefronts in two- and three-story brick and granite buildings. *Juliancolton / Wikimedia Commons/ Public Domain*

Figure 2.69: Main Street, Georgetown, Colorado. At two stories, the mining town's main street was modest in scale but designed with human delight in mind: deep corbelled cornices, high ceilings, and storefronts face the space.

Figure 2.70: Main Street, Springfield, Massachusetts. A view from around 1905 of a small New England city rich with manufacturing and rich with architectural detail. *Library of Congress / Detroit Publishing Company Photograph Collection*

MAIN STREET

Figure 2.71: Main Street. © 2023 Dover, Kohl & Partners

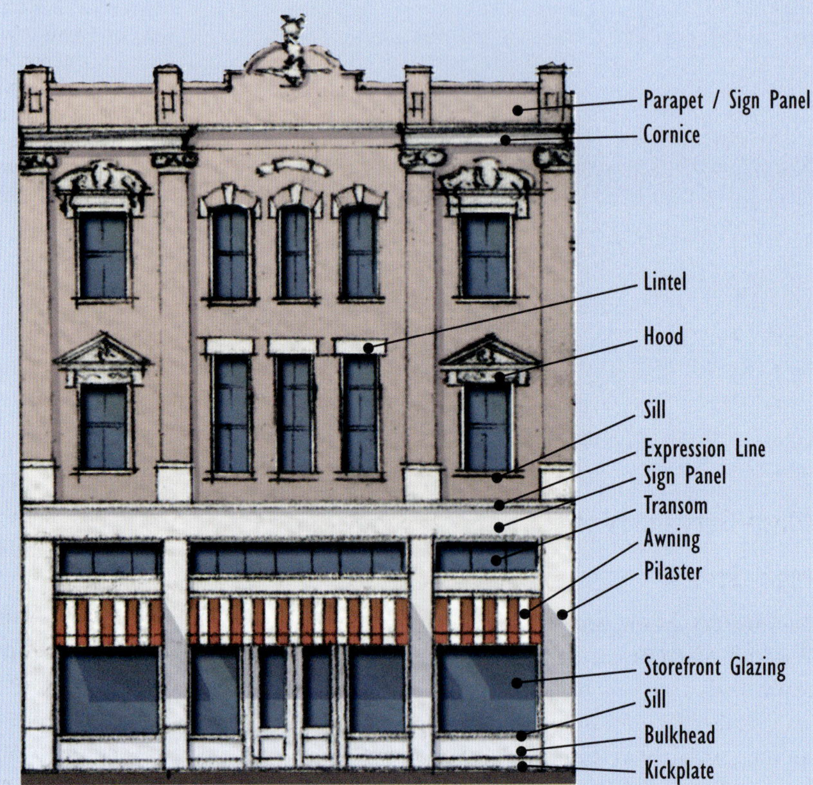

Figure 2.72: Anatomy of a Main Street Building.

Main Street, Nantucket, Massachusetts

Main Street

Nantucket's Main Street is an archetypal coastal New England Main Street. It has a well-proportioned public realm with a canopy of mature trees and a strong street-wall made of quintessential New England storefronts in quintessential New England buildings (Figure 2.73). It has good materials: a cobblestone roadbed bordered by brick and bluestone sidewalks with granite curbs. In the middle of the street is a green metal fountain that could only be historic, because the Massachusetts Department of Transportation (massDOT) would not allow it today—and yet people love the fountain, and against all conventional DOT wisdom manage to avoid running over it.

Every commercial building on the street is more than one hundred years old. Nantucket, an island thirty miles off the coast of Cape Cod, was once the whaling capital of the world, but these durable and well-loved buildings were mainly built after both the whaling trade and the island's economy went into a severe decline. From before 1900 to the present, tourism has been the main economic driver on this small island. Today Nantucket is one of the favorite places for the rich and the super-rich to buy vacation homes: Nantucket has the highest average house price of all American counties by a significant margin. The effect of all that on the form and uses of Main Street is a story of many parts, but it is a story worth telling—one that touches on main street retailing, street

Figure 2.74: Main Street, Nantucket, Massachusetts. Looking west on Lower Main Street in December 2010. In the center of the street, bothering no one and pleasing many, sits a horse fountain commemorating the only Nantucket resident to die in the Spanish-American War. In 2013, the Nantucket Preservation Trust honored the Nantucket Garden Club for maintaining and beautifying the fountain with flowers and wreaths. See Figure 2.80 for a view of another Fixed Hazardous Object on Main Street.

design, historic preservation, placemaking, its place in development today, and the effect of supply and demand on real estate prices.

Lower Main Street

Main Street is the spine of a small downtown that is the commercial and residential center of the island. Stores are on Lower Main Street (Figure 2.74), a three-block wide stretch that begins near the harbor at South Water Street and runs uphill to Center Street. Because Nantucket is an island, the downtown is the heart of the community in a way that few American places are today. No one can drive to the next town for shopping or a movie, so the islanders and the summer residents value what they have. There is some modern retail outside the town, but the citizenry has limited its expansion: in 2006, the town not only restricted all future commercial development to areas on the island that already had it but also cut back the areas zoned for commercial uses in 1972.

Nantucket's love for its Main Street, and its debate about what that Main Street should be, goes back to the nineteenth century. After losing 50 percent of its population between 1850 and 1870, the townspeople resolved to make Nantucket one of the most popular resorts in America. They succeeded well enough that organized resistance from the summer residents stopped the town fathers when they wanted to pave over the cobblestones on Main Street in 1919.

In 1937, a summer resident named Everett U. Crosby wrote a book called *"95% Perfect": The Older Residences at Nantucket* that documented Main Street buildings, Nantucket houses, and their architectural details; Crosby's book led, in turn, to an informal consensus among elected leaders and developers about how to build on Nantucket.[34] A second study by Crosby two years later suggested restricting the modernizing of the business district. Passage by the

◄ Figure 2.73: Main Street, Nantucket, Massachusetts. Looking west from Union Street. A view of Lower Main Street around 1920, before Dutch elm disease arrived in America. Upper Main Street, the residential section of the street, still has many of its old American Elms. The elements of the street are simple: this streetscape can easily be created today. *Courtesy of the Nantucket Historical Association*

town of Historic District Legislation in 1955 is sometimes credited to Crosby's efforts.

In the late 1950s, S&H Green Stamp heir Walter Beinecke Jr., another summer resident, saw the potential of Nantucket's rundown waterfront at the foot of Main Street and started investing in it, buying waterfront properties, fixing them up, and filling in the holes with traditional-*ish* Nantucket buildings. Before long, he owned a lot of the downtown, and he formed the Sherburne Trust for his holdings. Because of the size of his investment, he became king of all he surveyed, with a strong influence over Nantucket's planning.

Beinecke was famous for an elitist yet profitable view that favored those with money to spend over those with less money. "Instead of selling six postcards and two hot dogs," he told *Time*, "You have to sell a hotel room and a couple of sports coats." In the end, Beinecke sold tourists t-shirts and trinkets too, which wasn't always popular with the locals.

"Not everyone saw eye to eye with Mr. Beinecke's plan at first," *Time* reported in 1968, after Beinecke had built a new grocery store and a marina. Buttons circulated on Nantucket that said, "No man is an island" and "Ban the B" because of the amount of power he held over the town. As he hoped, however, his efforts raised the value of his buildings, many of which he had bought for as little as $30,000. In 1987, he sold the entire portfolio to a Boston Real Estate Investment Trust for $55 million.

Since 1990, the United States has seen an enormous increase in the wealth of the super-rich in the country. One result is that many of our most beautiful places like Nantucket, Aspen, Santa Fe, and Charleston sell for prices that only a few can afford. Aspen was just one place that had to institute affordable housing programs for residents with $150,000 or even more in income, because skyrocketing real estate prices led not only to a local shortage of low-income workers but doctors and lawyers, as well. Nantucket had a problem attracting veterinarians, among other professions. From the point of view of development and placemaking, unbalanced markets like these tell us, among other things, that there is a shortage of great places like Nantucket and Santa Fe.

The Success of Lower Main Street

Many great individual buildings were built around the world in the last half-century, but few if any new streets built during the period rival Nantucket's organically produced Main Street. That is because in most other places architects, builders, developers, and planners emphasized auto-centric development, isolated buildings and a separation of uses, "unprecedented invention," and an "architecture of our time," viewing traditional architecture as kitsch. But making great streets is about making places where people want to be, not about ideological architectural expression. Current theories that limit architecture to the "autonomous" self-referential creation of buildings that do not play well with others can harm old streets and prevent the construction of good new streets. Contemporary development patterns and formulas that produce throwaway architecture and unwalkable suburban Power Centers are bad placemaking. This architecture of time and invention does not add up to good streets. That requires an architecture of *place*.

The successful construction of a place like Nantucket's Lower Main begins with the making of the street. The buildings lining Lower Main Street reinforce one's sense of the street itself—the space between the buildings—by making a solid streetwall. None of the buildings are individual masterpieces, but they all have the urban virtues of traditional materials, composition, rhythms, and beauty (Figure 2.75). "Traditional composition" means that they have distinct bottoms, middles, and tops; that their design involves the use of principles like centering, symmetry, the appearance of obeying the laws of gravity (their walls have substance, with light and shadow, vertical openings, and openings spanned by lintels); and that they have architectural scale and ornament that relates to human scale, human culture, and even the human body. "Traditional composition" does *not* mean that they are in any particular style, or that they cannot appear modern.

These traditional architectural motifs are urban virtues that contribute to the making of a place where people want to gather (Figure 2.76). Lower Main Street is a comfortable place to be or even park, lined with interesting things to look at. It sits on a slope, which tilts the square up in a way that puts it on display for those coming up from the harbor. This highlights the beauty of the cobblestones, which also slow traffic. So does the horse fountain in the middle of the street as you enter the square (Figures 2.74 and 2.75).

The end result is one of the most beautiful places in America. The space, the streetwall, the trees, the quality of the materials, and the island light all contribute. So do the consistently Classical proportions of the quiet background buildings, which sing to us in harmony. The best building visible from the square is the Methodist Church's Greek temple on Centre Street, which angles towards Main Street, but which urban designers might put in place of the residential-looking Pacific Bank at the head of the square.

At night, the storefronts light the sidewalks and create a warm glow behind the trees. The pity is that so few of the stores still provide the daily necessities of life. Studies in Copenhagen by Jan Gehl show that buildings with active storefronts have more people in front of them than immediately adjacent buildings with blank walls, even though many of the people in front of the shop ignore the store and just hang out there. Social beings, we need other people to socialize. When Main Street was a normal town center, people would go there daily to buy what they needed and socialize with their neighbors. Today, most of the stores are oriented towards tourists, but islanders still visit Main Street without a hardware store or a barbershop to draw them. Main Street remains the social center, even in a modern, expensive Nantucket that caters to tourists.

Nantucket caters to tourists well. Lower Main Street is the same width as the contemporary "Market Square Center" model that the urban retail consultant Robert J. Gibbs calls one of the best layouts for retail development. The author of the book *Principles of Urban Retail Planning and Development,* Gibbs recommends a maximum width of 85 feet and an optimum length-to-width ratio of 2.5 to 1, or approximately 212 feet by 85 feet.[35] Lower Main, once called Market Square, is 85 feet wide but 495 feet long. However, the beauty of the space combines with the interesting storefronts to draw the pedestrian in. Retail studies show that Americans do a lot of their shopping while on vacation, even though other studies show that shopping is a stressful activity. Gibbs advises his retail clients that one of the best antidotes to this stress is a beautiful shopping street.

Moreover, the gentle tilt of the street showcases the beautiful space, making it seem inviting rather than long. From the bottom, you can easily survey the square, while from the top you see the harbor and, during the winter, the surrounding church steeples. The street is wide enough to feel like a square, but narrow enough for pedestrians on one side of the street to see the stores on the other side.

Figure 2.75: Main Street, Nantucket, Massachusetts. Commercial buildings on the north side of Main Street between Union Street and Coal Alley, with the horse fountain in the foreground.

Figure 2.76: Main Street, Nantucket, Massachusetts. Looking west on Upper Main Street from the corner of Federal Street. Many of the expensive stores cater to tourists, but residents look for reasons to come to Main Street.

It has enough space for easy, head-in parking convenient to the stores, and even accommodates without problem the large SUVs that dominate Nantucket today. When the drivers of the parked cars back out, other drivers pay attention and slow down.

Gibbs advocates the use of street trees, citing studies that show trees like those on Nantucket have an economic value for merchants because they create a calming sense of place and, like the solid brick buildings on Lower Main, give a feeling of quality. (Also see *The Seven Roles of the Urban Street Tree* in Chapter One.) Cars park in the shade under the trees, which softens their presence on the street.

A photograph of Lower Main Street from the 1920s shows a magnificent allée of elm trees that arc over the street in a full canopy (Figure 2.73). They majestically cover the space, but at the same time, their limbs rise above the storefronts, showing off the stores and the buildings. Some of the elms survive, mixed with sweet gums and oaks. The sweet gums and oaks are beautiful, but they will never form a canopy quite like the old one (Figure 2.74). See the discussion of Great Barrington's Main Street in Chapter Four for a description of why it would be better to slowly phase in other, disease-resistant elms.

The buildings facing the square support modern principles of urban retail. One of the most important of the principles is that at least 60 percent of a store's street frontage should be glass. It is important to show potential shoppers what the store is selling, because of what retailers call the Eight Second Rule: the storeowner typically has only eight seconds to interest the passerby enough so that he or she will walk in, which doesn't allow much time for distractions.

Gibbs tells us that the division of the traditional facade helps focus the walker's attention on the lower part of the building, aided by the sign band between the storefront and the upper facade that most traditional buildings have. Merchants want pedestrians to look at their storefronts rather than at fancy new sidewalks or exciting, "unprecedented" architecture.

In the first edition of *Street Design*, we were critical of the formulaic, unthinking application of brick sidewalks with granite curbs in the standard Main Street renovation. Gibbs's comment emphasizes one of the reasons why: streets should have a simple harmony that emphasizes the street as a whole. "'Streetscape' is a dirty word," says the nonprofit Project for Public Spaces. What they mean is that formulaic streetscape designs—fancy benches, new street lights, and colored pavements—can draw attention away from the experience of the street, encouraging us to focus instead on its individual parts. But Nantucket's brick and stone buildings harmonize with Nantucket's brick and stone sidewalks. The red bricks of the sidewalk are muted versions of the red bricks in the buildings, and there are no multicolor bikeways, busways, or pedestrian crossings.

These retail rules are not a matter of personal taste. They come from the experience and experimentation of national retailers—first in shopping malls and later on city streets. The retailers can tell you precisely how changing the variables can affect sales price. They know that if one of their stores has a particular light bulb with a cold light, sales per square foot will go up X percent if they change to a particular warm light bulb, and they know that an empty lot next to their building on Main Street will cost them Y percent in sales per square foot. The *Secrets of Successful Retail* break-out box in this chapter has a short list of some of the main rules for urban retail.

Building before the time of earth-moving equipment, the road follows the path of least resistance along the contours of the island, sometimes curving with the contours of the hill and changing directions slightly every time the grade of the hill changes.

Beinecke's Sherburne Trust owned approximately half the commercial buildings on Main Street. Between Main Street and the water are several blocks of shingle-covered shacks, many of which Beinecke built. They were probably intended to be modest, like fisherman's shacks. They all have minimal trim and minimal trim details, no evident composition, and are all painted gray. Despite what Beinecke said about his target audience, they look cheap and attract stores that sell inexpensive items like t-shirts. With the exception of an attractive movie theater and a supermarket with a well-landscaped parking lot that Beinecke built for the locals to use, the stores in these shacks mainly appeal to tourists.

Chain stores, also popular with tourists, are another matter, however. There will be no Starbucks café on Nantucket's Main Street, for example, even though the company's name was taken from the name of the first mate on the Pequod in *Moby Dick* (inspired by a Nantucket whaler of the same name). In 2006, Nantucket passed a bylaw that limits stores and restaurants downtown to companies with fewer than fourteen outlets and fewer than three standard features among items like trademarks, menus, and employee uniforms.

Residential Main Street

At the top of the square, the Pacific National Bank, built in 1818, sits on axis, visually terminating Lower Main Street. But instead of stopping, Main Street narrows there and passes to one side of the bank building, continuing up and then down the hill, and becoming one of the most beautiful residential streets in America. Many of the houses were built by Quakers and are in a modest New England Classical Vernacular (Figure 2.77). The highest points on the street have many of the grandest houses, built by whaling captains and merchants. That includes the famous Three Bricks, three superb Greek Revival houses built in 1837 by the Quaker whaling merchant Joseph Starbuck (Figure 2.78). By today's McMansion standards, these large and grand houses are modest. They do not go for the architectural composition that New Urbanist Andrés Duany calls "the House of Twenty-Seven Gables," which evokes "an entire village in a house." Instead, all the houses have simple, box-like massing and consistent Classical Vernacular details and proportions. By the standards of modern suburbia, the houses also have small setbacks and narrow side yards. Many of the houses come right to the sidewalk, so there is no room for a fence or wall along the street. But a number of the houses are set back three to five feet and have white wooden fences, while the Three Bricks have black wrought-iron fences.

The elements of Upper Main Street are simple, similar to those on Lower Main: a cobblestone roadway; brick sidewalks with granite curbs; simple Classical or Classical Vernacular buildings close to the street; and a regular rhythm of majestic, mature trees (many of them American elms) that form a canopy over most of the street. Built before the time of earth-moving equipment, the road follows the path of least resistance along the contours of the island, sometimes curving with the contours of the hill and changing directions slightly every time the grade of the hill changes. At the top of the hill above

Figure 2.77: Main Street, Nantucket, Massachusetts. Looking west on Upper Main Street. The beautiful cobblestone roadway follows the contours of the land.

Figure 2.78: Main Street, Nantucket, Massachusetts. Looking east, back towards Lower Main. The Three Bricks are on the left. The cobblestone street follows the gentle contours of the island.

Figure 2.79: Main Street, Nantucket, Massachusetts. A pair of Greek Revival mansions sits across the street from the Three Bricks.

the square, Main Street angles towards the south. The street then goes downhill and turns again at the lowest point, where a street crosses. The Three Bricks and two large Greek Revival houses stand at a leveling off where another road comes in and the street angles slightly once again (Figure 2.79). The highest point on Main Street has an intersection where Main, Milk, and Gardiner streets cross. Here, there is a beloved object in the middle of the road, a Civil War memorial with a small but heavy granite obelisk (Figure 2.80). Very effective for traffic calming, it is but one of many similar monuments and fountains in the middle of New England streets. Any town that built one today would risk being sued, however, and if they *were* sued, engineers and lawyers would testify against the town. Sometimes it seems as though it would be better to take away the licenses of drivers who are incapable of avoiding the monument but, as we shall see in the final chapter (and Figure 1.55), slowing the cars down makes it easy for drivers to avoid hitting things.

The Future of Main Street

Nantucket and its downtown are beloved places, and people want them to remain that way. Those who live in Nantucket have chosen either to stay there (often a difficult choice, in light of local incomes and expenses) or to move there, despite the inherent problems. Those

who vacation in Nantucket have intentionally chosen a beautiful New England island, and they want to preserve its character. As we've seen, resorts are where Americans do much of their shopping these days, and national retailers make more money than local shops, so in most locations, the chains can drive out the locals. In Nantucket, year-round and summer residents want to keep a sense of where they are and not lose it to Starbucks, The Gap, Lady Gap, Baby Gap, and Junior Gap.

In 1972, Nantucket worked to continue the preservation of the buildings that had begun with "*95% Perfect,*" the creation of the Historic District, and more controversially, Beinecke's Sherburne Trust, by bringing out *Building with Nantucket in Mind: Guidelines for Protecting the Historic Architecture and Landscape of Nantucket Island.*[36] A set of guidelines for renovation and new construction, written by J. Christopher Lang and Kate Stout Lang, did not address retail issues, but, like the architectural pattern books recommended by Gibbs, it supported the creation of a place where retailers wanted to be.

The book was expanded in the 1990s, and in 2003 the town created a special downtown commercial zoning district tailored to its existing character, with smaller lots and setbacks than in the previous zoning. That was followed in 2009 by the adoption of a new master plan that expanded the area covered by the downtown commercial district and described implementation plans

Figure 2.80: Civil War Monument, Main Street, Nantucket, Massachusetts. Looking east on Main Street from the intersection with Milk Street. Erected in 1875, the obelisk commemorates the 69 islanders who died in the Civil War. Most Departments of Transportation would call the heavy granite memorial a Fixed Hazardous Object, but the rough cobblestone street makes drivers go slowly enough that they avoid hitting it without the need for reflectors and colored striping. The residents of Nantucket love their beautiful streets and work to keep what they have. The result is a great public realm and one of America's most beautiful streets, rather than a place for vehicular throughput.

for the development of key downtown sites. The result was an amalgamation of the pattern books and the form-based codes that Gibbs recommends. A single form-based code that combined all the regulations and guidelines in one document would be even stronger. An advantage of a good form-based code over a pattern book is that much of the judgment involved in interpreting pattern book guidelines becomes simple enforcement of the clear rules in a good form-based code.

An important point in coding Main Streets is that while uses come and go, our much-loved places can keep their physical character over time if we build and code well. Nantucket has been rich and poor and rich again, but the buildings on Main Street are the same buildings that were there 100 years ago, and that is something the community values. Nantucket may have been built when it was the whaling capital of the world, but it has been a resort for almost 150 years. There are now ten thousand residents on the island in the winter and fifty thousand people in the summer, and the majority want to keep Main Street a real street where they can live their daily life, whether they are there to work or vacation. A problem, of course, is that Nantucket's great success as a resort has threatened the economics of Main Street and full-time residence.

The newest chapter in the story of Nantucket's Main Street is that another summer resident, Wendy Schmidt, stepped into the ongoing effort, and with the help of Google money formed ReMain Nantucket. Focused on Main Street and the downtown, it uses philanthropic funds and private investment to keep the old buildings and local uses of Main Street thriving. ReMain Nantucket bought buildings, saved a local bookstore, and founded a new downtown bakery. It emphasizes adapting buildings to deal with problems of climate change while supporting businesses that keep the downtown strong and lively.

The American Planning Association named Main Street one of the Top Ten Great Streets of 2011. We cannot disagree, but we can point out that if we made more great places, Nantucket would not be so expensive. From the time in 1919 when the summer residents of Nantucket preserved the cobblestones on Main Street until now, the population of the United States has grown from 104 million to over 300 million people, more than tripling in size. By some estimates, more than 90 percent of the buildings in America have been built during that period. But a billionaire is willing to pay $20 million for a 150-year-old house on Main Street because of the rarity of the house, the street, and the network of streets that make the downtown.

SECRETS OF SUCCESSFUL RETAIL

Retail design expert Robert Gibbs knows how to help traditional Main Streets compete with shopping malls and strip shopping centers (Figure 2.76). Some of Gibbs's advice is about management, like promoting common hours of operation. However, much of Gibbs's advice is about design. Here are some of the highlights regarding streetscapes and street design, all based on studies that show these produce greater sales per square foot.[37]

- Storefront design

  - Retailers want predictability and quality on the street. A form-based code or pattern book can give that.

  - Storefronts should appear open; at least 60 percent of the storefront should be glass.

  - In a mixed-use building, the ground floor should be distinguished from the rest of the building facade.

  - A top-to-bottom curtain wall in a tall building causes the upper levels to dominate the building's appearance and minimize the curb appeal of ground-floor retail.

- Signs

  - Sign bands above the storefronts are essential to hold the pedestrian's view.

  - Sign bands make it easier to change storefronts over time.

  - Nothing contributes to strong retail sales and an attractive downtown as much as well-designed and properly scaled signs.

  - A single background color on all signs is bad, because it eliminates the sense of unique stores and goods.

  - Rule of thumb for size: One square foot of sign for each linear foot of street frontage.

- Maximum letter height should be eight to ten inches.

- Lighting should be external only. Backlighting solid letters or signs should be allowed, but not internal illumination.

- Some regions have their own sign patterns: New Englanders like painted wooden signs with gold-leaf lettering, but these may be inappropriate elsewhere.

- The base of the sign must be at least eight feet above the sidewalk and should extend no more than three feet over the sidewalk.

- There should be at least fifteen feet of separation between signs for different stores.

- Some businesses, such as cinemas, require bigger signs.

- Awnings

  - Awnings define the storefront and the brand.

  - Awnings should be made of canvas, cloth, metal, or glass.

  - Cloth in an awning should be or look like natural fabric and be limited to two colors (no plastic).

  - Awnings should not have internal illumination.

  - Logos and letters should be limited to eight inches tall and should only be on the front flap, not on the slope of the awning.

  - Shed-type awnings without side panels appear lighter, which is generally beneficial.

  - Awnings should complement the building facade.

  - Awnings should not hide architectural elements.

- Awnings should have no more than a 25-degree pitch.

- When every storefront has an awning, the effect is dreary or monotonous.

- Sidewalks

  - Sidewalks provide the first and last impression the shopper sees.

  - Sidewalk materials and scale should harmonize with their location.

  - Sidewalks should be wide enough to allow shoppers to pass each other.

  - Major urban centers like Michigan Avenue in Chicago call for sidewalk widths of twenty to twenty-five feet.

  - Small hamlets or villages need sidewalks at least five to eight feet wide.

  - In hot climates, shady sidewalks are the most popular; in cold climates, sunny sidewalks are sought.

- Street furniture

  - Trendy, "cutting-edge" furnishings will go out of date; buy medium-cost items that will wear well until they are replaced every five to seven years.

  - Planters or merchandise along the sidewalk at the street's edge can distract shoppers away from the stores.

- Lighting

  - Illuminate sidewalks with light from the store windows until 11 p.m.; supplement with streetlights where necessary.

  - Use color-corrected light sources for warmth.

In this book, we show how simple it is to make streets where people want to be, if one has the will and the permission. If we judge success by market demand, then New Urbanism has had many successes. New Urban developments always sell for higher prices than neighboring projects, because they are selling something that the market has not supplied. The average house price in the New Urban resorts on the Florida panhandle is almost as high as the average house price on Nantucket. That's not because the resorts are elitist, as is sometimes said, but because they sell beautiful communities where people want to be, much like Nantucket, Aspen, Charleston, and Santa Fe. The challenge for America is to build again in our neighborhoods, towns, and cities in a way that produces great places we all can afford. Then we can concentrate less on preservation and more on making the future better.

The challenge for America is to build again in our neighborhoods, towns, and cities in a way that produces great places we all can afford. Then we can concentrate less on preservation and more on making the future better.

Main Street, Galena, Illinois

Main Street

Galena, Illinois, is a classic American town. Located in the rolling landscape of the Mississippi River Valley, it has brick-clad buildings along Main Street, with quaint store-fronts on the ground floor and offices and apartments on the two or three stories above. The town's topography defines its streets—some with very steep slopes and winding corners—making it an interesting place to walk, bike, or drive. Visibly the densest development in the city, Main Street is clearly *the* center of Galena, and the heart of local economic and social life (Figures 1.46, 2.81, 2.82, and 2.83).

Galena's topography is a result of a glacial shift that occurred in the last ice age that flattened most of the surrounding land but missed the hills and valleys that define the city itself. By the early 1800s, the land was known to be rich in lead, and the community that grew up there quickly became an active part of the Upper Mississippi Lead Mine District. Named Galena after the technical term for the sulfide of lead, the settlement became a town in 1826 and was chartered as a city in 1841.[38] Until the middle of the nineteenth century, the buildings on Main Street were constructed of wood. After several fires in the 1850s, most of the buildings were rebuilt in brick or stone.

Like many of the best streets, Galena's Main Street changes shape and size as it stretches across the town. At one of its narrowest points, the street is approximately fifty feet wide with a one-way travel lane, flanked by a row of parallel parking and nine-foot-wide sidewalks on each side (Figure 2.84). The cross section changes as the street alignment cranks, maintaining traffic and accommodating diagonal parking spaces (Figure 2.85). The overall width becomes slightly larger at this point but mostly remains the narrow, historic dimension that it had during the town's industrial boom in the early 1800s.[39] On the original plat for the City of Galena, a fifty-foot-wide curb-to-curb dimension is shown at both the beginning and the end of Main Street (Figure 2.86).

The slight curve or "crank" on Main Street is an excellent urban feature in central Galena. The crank provides a sequence of changing vistas for the user of the space. Although the curve matches the topography, the result of a natural event, the effect is often purposely replicated by urban designers. The deflected vista makes the street feel as though it has an end rather than appearing to extend far into the distance. This is good for business: the storefronts down the street tip into view, showcasing the variety and quality of goods and services. Main Street has an understated color palette established by the brick facades and concrete hardscape.

Figure 2.81: Main Street, Galena, Illinois. The relatively flat street is aligned with a natural "bench" in the topography; behind it, the town rises on a steep slope.

Figure 2.82: Main Street, Galena, Illinois. The height of the buildings and the width of the street combine in a nearly perfect proportional relationship.

Figure 2.83: Main Street, Galena, Illinois. The little curve or "crank" produces a gently deflected vista, tipping storefronts into view, and making the space feel finite and intimate. Note that the white building is an intentional focal point; the building itself is bent to match the crank, making for upstairs rooms with unique views in both directions from the bay windows.

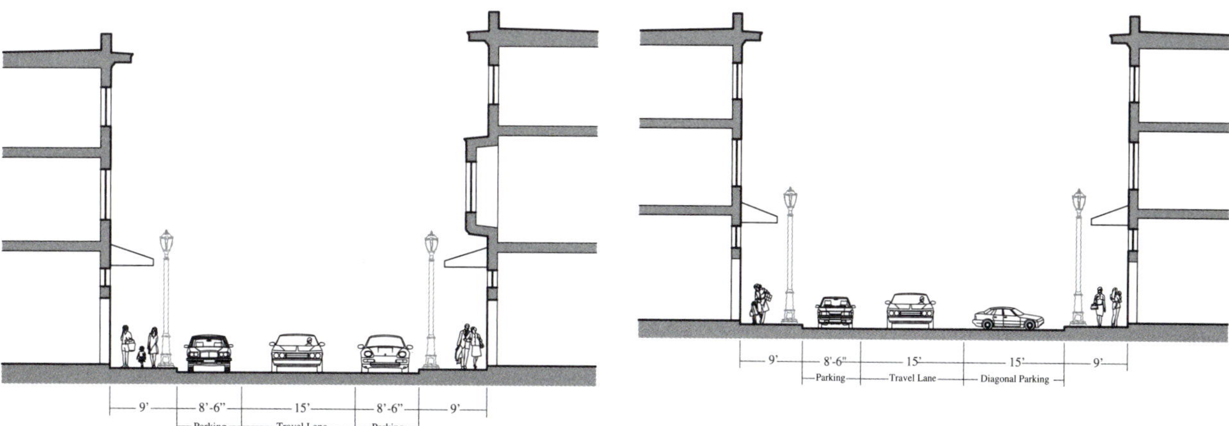

Figure 2.84 and 2.85: Main Street, Galena, Illinois. Sections. The design changes segment by segment, like several streets laid end-to-end. Some segments have parallel parking, some have diagonal parking, and some have two-way travel. © 2013 Dover, Kohl & Partners

Figure 2.86: Galena, Illinois. Figure-ground drawing. Main Street (center) curves to match a natural "bench" in the topography, with most of the town behind it (upper left) sloping up away from the riverfront (lower right). © 2013 Dover, Kohl & Partners

Figure 2.87: Trinity Street, Cambridge, England. Looking from St. John's Street towards Trinity Street and the entrance to Trinity College. Cyclists and pedestrians use Trinity Street throughout all seasons. © 2012 Kim Fyson / CC BY 2.0

TRINITY STREET / EMILY GLAVEY

Cambridge, England

Main Street

Trinity Street is a shared space that leads to the center of the old, medieval city of Cambridge and the heart of the University of Cambridge, founded in 1209. On most days, it is full of students, professors, and tourists walking or cycling to their destinations. It was once part of a longer High Street, but this section between St. John's College and Kings College is now named after Trinity College, one of the original colleges at the university.

Trinity Street begins at a small plaza in front of the iconic entry to Trinity College (Figure 2.87). A hydraulic bollard on St. John's Street blocks access to Trinity Street for anyone without the required electronic pass. Therefore, people on foot or on bikes dominate, and cars slow down to their pace. The travel lane is only ten feet wide or so throughout its length, with simple gray or beige pavers and a low curb about half an inch high and a sidewalk on each side that varies in width. The whole street measures between twenty-four and thirty feet wide. While the subtle curb defines a constant-width travel lane, the street is experienced as a meandering shared space.

Along most of the street, the facades of the university define one edge of the space, while the other side includes two- to four-story buildings with local shops on the ground floor and apartments above. Beyond the university facades are large, beautiful, and generous courtyards that have defined life in the colleges for centuries. There are also intersecting and interesting pedestrian passages to wander along—including Green Street, Rose Crescent, and Trinity Lane. These are narrower than Trinity Street, but just as beautiful, each lined by typical medieval buildings, full of historic and well-designed storefronts, university departments, and places that are part of university life, including pubs and restaurants.

Trinity Street dramatically ends as the narrow space empties into Senate House Hill and King's Parade at the site of the Senate House, one of the most important buildings in town. It is place where all Cambridge students receive their diploma in a traditional ceremony.

Trinity Street demonstrates that a place can simultaneously be a beautiful city and a university campus. It feels best as a pedestrian or cyclist, whether wandering along on a less crowded winter day watching the snow accumulate on the historic buildings, cycling to class in the fall, or when it is packed full during the summer as people make their way to the open markets, cafes, university activities, and local events.

Figure 2.88: Trinity Street, Cambridge, England. The author measuring Trinity Street's dimensions. *Courtesy of Iulia Colescu*

Figure 2.89: Trinity Street, Cambridge, England. The entire space is pedestrian-dominated, but note the low curbs and the textures that shape the part of the street that is shared with slow-moving vehicles. © *2012 John Sutton / Flickr / CC BY 2.0*

Figure 2.90: Trinity Street, Cambridge, England. Looking south, towards the intersection where the street opens up to King's Parade. © *2014 John Sutton / Creative Commons Attribution-ShareAlike 2.0 Generic*

TRINITY STREET IN CONTEXT

If you are lucky enough to become familiar with Trinity Street and its surroundings, you realize what an extraordinary place it is. On all four sides, the short street has world-famous buildings, institutions, and sights, each quite different. Writing about the setting, it is difficult to avoid running out of superlatives.

When the street opens out to Senate House Hill and King's Parade, it arrives at some of the most beautiful and important buildings at the university. Students receive their diplomas in the Senate House, a dignified design by James Gibbs. Built in the 1720s in Portland stone, the Classical building abuts Kings College Chapel, one of the most famous churches in England. An outstanding example of the High Perpendicular Gothic style, the chapel is home to England's preeminent Choir of the college, known around the world for daily Evensong and the annual Christmas Festival of Lessons and Carols. The chapel has a Rubens altarpiece, outstanding carved choir stalls, and an impressive rood screen. Next to the Gothic chapel, Gibbs designed simpler, large Classical buildings for King's College that set back from the Parade behind broad lawns that open expansive, iconic views. The River Cam runs along the back of the college, next to a rural-looking park where sheep slowly roam in an idyllic English landscape. In his book *England's 100 Best Views*, Simon Jenkins called the vista from the Backs the fifth most beautiful in the country (Figure 2.91).[40]

Figure 2.91: Kings College, Cambridge, England. A view of the Chapel and the College from the Backs, along the River Cam. © *2004 Andrew Dunn / Wikimedia Commons / CC BY SA 2.0*

Trinity College and Gonville and Caius College also sit between the Backs and Trinity Street, with Tudor and Gothic courtyards that epitomize Oxbridge college architecture. Trinity has a wonderful Classical library designed by Sir Christopher Wren, as well as the setting for the famous courtyard race in the movie *Chariots of Fire*. On the other side of Trinity Street are picture-perfect medieval passages. At the northern end of the street, St. John's Street curves to the east, framing the entrance to the ancient college. Judging by the great number of "Trinity Street" photographs on the internet, that view of St. John's Street is also picture perfect (while many of the photos are taken from Trinity Street, they actually show St. John's Street).

Figure 2.92: Cambridge University, Cambridge, England. Satellite view. The experience of Trinity Street is shaped by the knowledge that Trinity College and Gonville & Caius sit between the medieval city on one side and the River Cam and the rural Backs on the other. *Courtesy of Google Earth*

LA PASSEGGIATA, ORVIETO, ITALY / DOUGLAS DUANY

Main Street

Every evening in Orvieto, people gather along a long and narrow street—talking, clustering and dispersing, and only slowly moving. They do so as a community, and they do it daily.

All over the world, this quintessential urban behavior occurs spontaneously. It becomes a custom, a ritual, a pleasure. This greeting and gathering help to unite the community. It is not an abstract notion; it is how the town takes possession of its public space, and how individuals take possession of themselves at the end of the day.

The Corso in Orvieto

The two main streets of Orvieto are the Corso Cavour and the Via del Duomo. The Corso Cavour runs uphill from the Piazza della Repubblica, intersecting with the Via del Duomo at a three-way crossroad that leads from the lower town. The Via del Duomo then leads to the Piazza del Duomo, where one encounters the city's magnificent cathedral. Together, these streets and their anchoring end-squares constitute the pedestrian armature of the town's public realm (see Figure 2.106 at the end of this case study). I will call both the Corso, a name redolent of its most important function, which is to serve as the social gathering place of the town in the early evening.

The Piazza della Repubblica is an interesting square with a town hall and an ancient church (Figure 2.93). The best feature of the space is a beautiful Renaissance-era loggia—modest in appearance but one of my favorites—that closes the entrance to the piazza. Its placement is as elegant as its architecture (Figure 2.94).

As one leaves the civic piazza behind to stroll along Corso Cavour, one finds oneself in the heart of the old city's armature, the shops interspersed with restaurants, cafes, craft and food shops, as well as stalwarts like hardware stores.

The Furlong as a Unit of Pedestrian Life

The Corso is a miracle of balance, an ancient market street anchored at the edges by the cathedral, a municipal theater, and the city hall. The distances between them measure about 660 feet, and this is a critical dimension: the furlong. Why a furlong? Because it seems to be a distance that measures excellent short walks, measuring both pedestrian engagement and retail activity. It is the distance between centers of interest, one that grounds our hyperactive minds and propels our bodies to the next one, irrespective of whether we are engaging in a good country walk in the south of England or rushing between avenues in Manhattan. Shorter than the traditional half-mile diameter that defines a neighborhood, furlongs remind designers of the smaller or overlapping scales of pedestrian life.

> Why a furlong? Because it seems to be a distance that measures excellent short walks, measuring both pedestrian engagement and retail activity.

Figure 2.93: Piazza della Repubblica, Orvieto, Italy. Looking east towards the Piazza della Repubblica, approaching the loggia on axis.

Figure 2.94: Orvieto, Italy. In the Piazza della Repubblica, looking east towards the loggia and the Corso. The tower appears on the Corso.

Street as Place

When we see a wide street on a map in the center of town, we assume it is important. While that often holds true, the narrowness of the Corso contradicts this. People simply enjoy the closure and intimacy of narrow streets. Across the world, streets that are sixteen to twenty-four feet wide are considered "sweet" for the way they contain space and aid shopping. You can see the goods on both sides very easily. Further exploration in cities will reveal that this narrowness marks the best market concourse in many cultures, from shambles to bazaars to malls, and this is something to remember.

Both the curves and the range of widths of the Corso add visual closure to the corridor space of the street. The curves and alternating counter-curves (the Classical cyma) are marked with small building jut-outs that emphasize the shift from one to another. The unfolding vista has a rhythm that reminds us that the experience of walking along the street can pleasantly involve time. People respond by moving more slowly, and time slows down with slowing movement. At the end of the day, they meet and form groups to talk, and the enclosed space now becomes a public drawing room.

Organic Geometry

It is unorthodox to say this in America, but parallel curves are hard to lay out without modern surveying equipment! Urban planners have reduced geometry to just a few options, and streets are all parallel, all either curved or straight, and all intersections are four-way. How can there be any other way, if we've lost the ability to imagine it?

In organic streets, geometry is open. Straight parts alternate with curves, and you can mix straight and curves or have different curves on either side. One beauty is to have a straight segment on one side, while still curving on the other!

But organic geometry is not simply about variety; it has an objective, a general rule that was once stated as: "open before closing, close before opening." So a series of visual spaces are created and streets become a *series of places*.

Orvieto's narrow meander is ideal in the way it responds visually to the slow pedestrian movement. One architectural element after another dominates the visual field and then "disappears" as another comes in to hold the visual attention (Figures 2.95–2.97).

Figure 2.95: Orvieto, Italy. Looking east at the central part of the Corso Cavour, with the orienting central tower.

Figure 2.96: Orvieto, Italy. Looking east. The central tower begins to disappear as one approaches from the Piazza della Repubblica.

Figure 2.97: Orvieto, Italy. Looking east. Approaching the intersection of the Corso and the Via del Duomo, the tower has disappeared, replaced by the palazzo in the center of the photograph.

We associate the sequencing of visual events with "medieval streets." This is a rubric that covers everything that is not "planned" or "designed." A standard explanation is that this sequencing is the result of organic processes, improved paths, and architecture responding to and exploiting the visual space. What is extraordinary in Orvieto is how quickly the visual events follow one another, and how they differ. Good medieval streets best respond to the fourth dimension, which is time. In comparison with other superb examples of streets that unfold events, Orvieto's Corso makes sequences like the Governo Vecchio in Rome, or the central armature of Montepulciano, seem a bit slow.

The Magic Tower

Old Italian cities were full of towers. They formed a clustered skyline from afar and when one walked the streets, the towers appeared and followed one another sequentially. We know that from the only remaining complete example, the small *città* of San Gimignano.

In San Gimignano, the towers do a perfect dance of appearing and disappearing magically in response to the pedestrian movement in the winding streets. Was this "planned"? We can assume that the medieval aristocracy wished to show them off as centers of power and prestige, but there is a more functional reason that is overlooked. The towers had to be defended, and that required that shooters in the tower could survey the streets as completely as possible, both near and far. The towers sited defensively work well aesthetically and make perfect landmarks. What this means is that in urbanism there are two ways of "seeing" a building: from within the building itself and from the street.

The Corso in Orvieto has that same visual responsiveness to movement as San Gimignano, but it has only

Figure 2.98: Orvieto, Italy. Looking south. Going uphill from the central tower towards the Duomo, passing the café in the Largo Luigi Barzini, a piazzetta on the Via del Duomo.

Figure 2.99: Orvieto, Italy. Looking south. At the end of the Largo Luigi Barzini, the facade of the Duomo first appears.

Figure 2.100: Orvieto, Italy. Looking south, near the end of the Via del Duomo. The first bracketing of the Duomo.

one tower marking its central intersection, the Torre del Moro. When you visit Orvieto, this tower is eminently worth tracking as you walk the *passeggiata*. It will open up the world of dynamic street-making, a lesson that you can then apply to the other elements that unfold.

The Torre del Moro is simple and not very high, but it is visually impressive. It is seen in tripartite: in a level approach from the Piazza della Repubblica (Figures 2.95 and 2.96), during the descent from the Piazza del Duomo (Figure 2.103), and in ascent from the lower town. It is stable through much of the ascent from the lower city, pulling people into the passeggiata, then engaging the rooflines of the winding street beautifully before disappearing behind an irregular palazzo, to be replaced immediately by a small campanile (Figures 2.104 and 2.105). From the Via del Duomo, it appears centered on the road at the beginning, and then disappears, only to reappear with a different face and sink slowly, while from the passeggiata in the Corso, it descends to the left before disappearing.

The effect is magical. The moment it disappears from the cone of vision, another object—a palazzo or a nearby simple campanile—catches your eye, and then, just as

that is passed, the tower reappears, always marking the central intersection.

So what lesson can one learn from medieval towers? Well, there is always the old one suggested by the great historian, architect, and urban designer Camillo Sitte: When you blow the budget on a single building, frame it in as many ways as possible to exploit its potential.

But there is also a second, more practical lesson for this secular age. A city of high-rises looks great from the interstate, but in the city their tops can disappear, leaving us with a less impressive base along the street. Not bad, you say, but no longer the magical city one saw from the distance.

In an era of austere budgets, one can learn how to place buildings so that they are worth observing from every angle, in interaction with the surrounding fabric. In plan, one places a halo at some distance from the tower and then one deflects segments of streets towards the tower within the "doughnut." Steeples, with churches at their base, are perfect for this, which is why good urban designers have a motto, "If churches didn't exist, we would have to invent them." Very subtle deflections in the streets suffice.

Figure 2.101: Orvieto, Italy. Looking north from the Piazza del Duomo, back to the beginning of the Via del Duomo.

Figure 2.102: Orvieto, Italy. Returning to the Largo Luigi Barzini, walking north. Small trucks navigate the streets in the morning.

Figure 2.103: Orvieto, Italy. Walking north after passing the Largo, the central tower reappears on the Via del Duomo.

Figure 2.104: Orvieto, Italy. Looking west. The tower dominates the uphill approach from the lower town, beautifully framed by the serpentine street.

Figure 2.105: Orvieto, Italy. Looking west. When the tower disappears, it is immediately replaced visually by a small campanile.

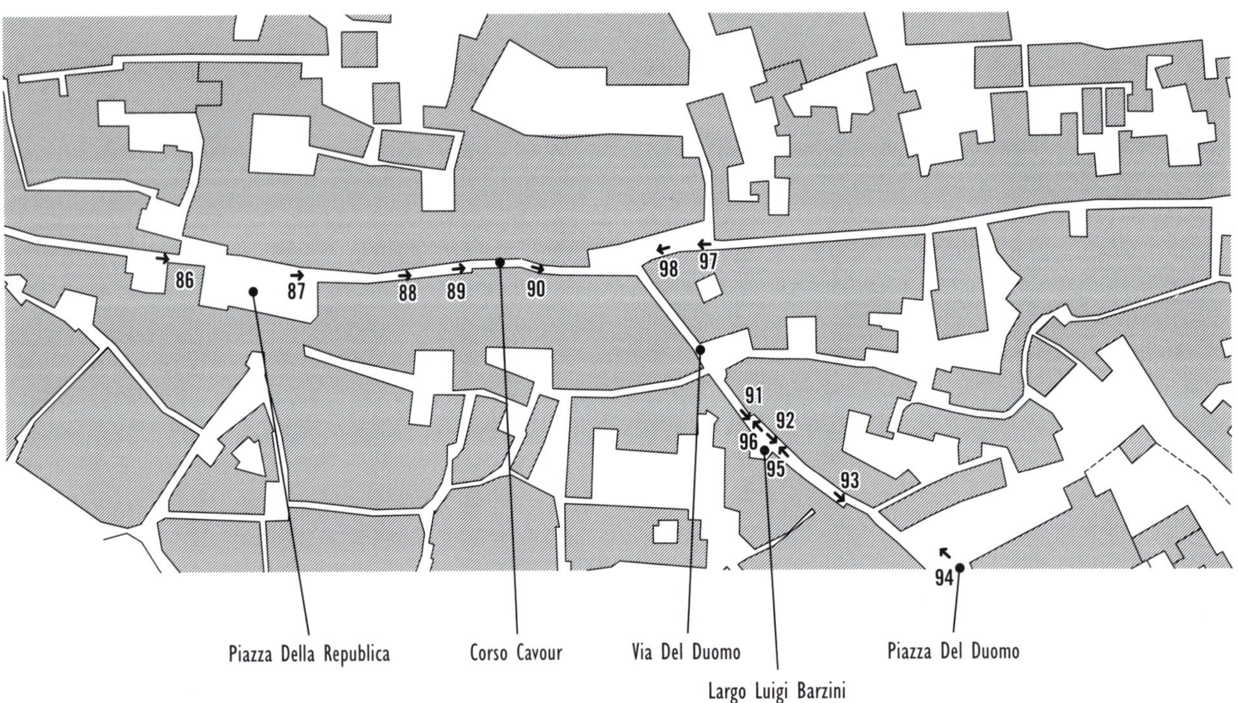

Figure 2.106: Orvieto, Italy. Figure-ground drawing. © 2013 Dover, Kohl & Partners

Why not place towers simply, at the end of an axis? Again, not bad—especially over a short distance if the tower has an engaging base. But a perceptually stable object over a long distance gets boring quickly and the mind wanders to the next thing.

Why not make the "next thing" another piece of a beautiful world, instead of another (usually repetitive) thought? We need interesting urbanism to ground us when we shut down our computers and walk outside, and in an increasingly abstract age, we need a way back to the concrete. Interesting urbanism turns our walk into a passeggiata, and our surroundings into a promenade.

The Visual Angles of Hans Maertens

It is worth decoding the underlying mechanism of this magical dynamic effect. The key is the natural cone of vision of the human eye, which changes what we see when we turn our head or, when we are moving, replaces one thing with another.

The urban effects of the cone of vision were worked out by a German, Hans Maertens, but his theory doesn't have to be read or explained. One learns it simply by approaching a good building and noting how our perception of it changes with our distance from it.

Note that, as you get closer, everything becomes more finely textured: instead of seeing the outline, for example, you can read the composition of a building, and how its elements hold together. As you get even closer, an impressive building becomes even more impressive, until from close up (45 degrees from your eye to the top) the cornice and the edges of the building lose importance. But close up, the whole doesn't hold together; the elements take precedence, and if you're standing next to the base, it is the door that you see.

This is all a function of the cone of vision. The active cone of vision is less than 33 degrees, and this explains why buildings appear and disappear in the Corso even before you pass them, as another building, an element, or a café engages your attention.

The cone of vision is universal, a key tool in design, the one that makes the relation between building and space intimate. Imagine designing a square: you cannot simply take a bird's eye view. Instead, you must imagine the perspective of the eye from both the square's edges and from its center; the long view and crosswise. It is the cone of vision that reveals how we will experience the closure of the space.

A Great Masterpiece

The approach to the cathedral is the climax of the visual effects of the Corso. The Duomo is itself a masterpiece—reason enough by itself to go to Orvieto—its many-hued facade a brilliant stage set. Approached from the end of the passeggiata along the Via del Duomo, this facade and its two towers are bracketed and re-bracketed in an astonishing manner by the street (Figure 2.100). The street frames and reframes different parts of it in a short distance, each a perfect picture before it shifts again. The square of the Duomo is at a high point on the edge of the great plateau of Orvieto. Some of the edges of the square are low. After the narrow enclosure of the street, the piazza releases us into the sky and open space.

It is still part of the passeggiata. Baby carriages, grandfathers, mothers, conversations. And, on the other side of the Duomo, teenagers. But the cathedral is also a meaningful counterpoint to the market and the society of the Corso. It calls us into a larger order than society: nature, art, religion, and this, the nonmarket. That which is not everyday or mundane is a meaningful component of life in every *città*—or of every town or village that is balanced, that meets the needs of human habitation.

Steeples, with churches at their base, are perfect for this, which is why good urban designers have a motto, "If churches didn't exist, we would have to invent them."

Quinta Avenida, La Antigua, Guatemala
Main Street

Antigua was established as a splendid Spanish capital in the New World, but devastating earthquakes led to the relocation of the capital in 1776. The result of its near abandonment was that the original Spanish Colonial form of the town remained intact and many of its impressive ruins were left undisturbed, enabling a gradual restoration beginning one hundred and fifty years later. Today, Quinta Avenida's yellow arch framing the Volcán de Agua is the city's iconic view. In the other direction, the view also ends dramatically, culminating with the side facade of La Iglesia de la Merced, a massive, brightly colored church with baroque details (Figure 2.107). The arch itself is Antigua's most recognizable landmark; it was built in the 1600s so that nuns

Figure 2.107: Quinta Avenida, La Antigua, Guatemala. The view looking north culminates at the facade of La Iglesia de la Merced, with green hills beyond. *Courtesy of Kenneth García*

from the Santa Catalina convent could cross to a school without using the street.

Other attributes include:

- The cobblestone street operates like a shared space—automobiles drive on the road at slow speeds, but cyclists and pedestrians dominate.

- Colorful building facades form a continuous streetwall (no more than two or three stories tall), establishing a balance between the height of the buildings and the width of the street.

- Sidewalks are narrow—usually only five or six feet wide—and pedestrians often spill out into the space beyond the sidewalks.

Antigua's oversized grid is emblematic of the pattern followed by many Latin American cities, per the *Law of the Indies*. The resulting streets have nearly identical dimensions throughout each city block.

Figure 2.108: Quinta Avenida, La Antigua, Guatemala. The view looking south, in which the arch frames the distant volcano, is the city's iconic vista.

Figure 2.109: Quinta Avenida, Antigua, Guatemala. View looking south under the Santa Catalina arch. The stone street operates like a shared space; automobiles drive on the road at slow speeds, but motorcyclists and pedestrians dominate. Typical of Spanish Colonial streets, colorful building facades form a continuous streetwall, balancing the height of the buildings with the width of the space. *Courtesy of Kenneth García*

GINZA DORI, TOKYO, JAPAN / GIANNI LONGO

Main Street

Ginza Dori, in Tokyo, is the attractive "main street" of a region of thirty-two million people: an eminently walkable and well-scaled street at the center of a shopping, residential, and business district. Following a major fire, the Ginza district was deliberately rebuilt in 1872 as the showcase of Japan's modernization efforts. At its core stood Ginza, a large street lined with two- and three-story-high Georgian brick buildings: a novelty in a city of narrow winding streets and wood buildings.

Ginza (Figures 2.110 and 2.111) is 3500 feet long and 90 feet wide with 22-foot sidewalks. Building heights range from six to eight stories, giving the street ample yet well-contained proportions. Harumi Dori, a shorter east–west street, divides Ginza into two equal segments. Their intersection is Tokyo's preeminent 100 percent corner, a distinctive meeting and gathering place. Together, they anchor a compact district of 0.2 square miles that is home to an estimated 10,000 shops.

One of the unique characteristics of Ginza's 400-foot-long blocks is that they contain a larger number of buildings than are typically found on major streets. Some of these skinny buildings have frontages as narrow as

fifteen feet. They are for the most part anonymous, thirty to forty years old, not particularly beautiful or unique, and typical of most of Tokyo's postwar architecture. They establish the street-level character of Ginza with shop entrances and elegant window displays that change rapidly from building to building. High land values have pushed shops, restaurants, cafes, hotels, and nightclubs to the upper floors of buildings, adding an exuberant mix of uses and a vertical dimension to the street's diversity. Multiple-story retailing is a condition that has rarely worked in other parts of the world; in Tokyo, where it is a necessity more than a choice, upper-story uses succeed without detracting from the already rich street vitality.

Ginza is a complete and complex street that supports and protects a great variety of street uses. Pedestrians and bikes share the sidewalks, as in the rest of Tokyo. Deliveries, however, take place using the extensive network of service alleys, eliminating all trucks from the street. Speed limits are low and scrupulously respected, leading to what appears to be relaxed and unhurried driving. The most extensive network of public rail transportation in the world drastically reduces the number of private automobiles on the road. Three major subway lines converge under Ginza, and a major underground concourse that is clean, well-lit, and elegantly appointed efficiently distributes passengers throughout the district using twenty-one clearly marked exits. These three lines connect Ginza to a system of subways and commuter rails that carry over

Figure 2.110: Ginza Dori, Tokyo, Japan. The evergreen trees creates a screen separating the sidewalks from the roadway. *Courtesy of Gianni Longo*

Figure 2.111: Ginza Dori, Tokyo, Japan. The vertical signs along Ginza have a function similar to the flag poles along New York's Fifth Avenue. They bring order and enrich the visual environment of the street. *Courtesy of Gianni Longo*

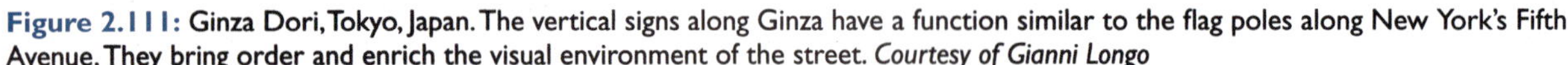

forty million passengers daily and make Ginza fully accessible to the region. This system is also what liberates the street from vehicular congestion, creating a dedicated pedestrian environment that is both vibrant and safe.

Ginza has rubberized pavement, which reduces noise. Pedestrians can converse at normal speaking levels, and the effect is reminiscent of some of Europe's most successful pedestrian precincts. The street is also impeccably maintained and clean, in spite of its high usage and the complete absence of public trashcans. Its maintenance is the result of efficient public sanitation, the support of organized volunteer groups, and the continuous efforts of individual shop owners. The fact that many people in Japan consider it bad manners to eat while walking in public also reduces a major source of public waste.

Signs, tree planting, and transparent storefronts add to the pleasure of walking along Ginza. Vertical signs are better suited to the orientation of Japanese calligraphy; they also better serve those narrow buildings advertising shops and restaurants on their upper stories.

Well-designed graphics, illuminated at night, give rhythm to the street, enhance the anonymous architecture, and keep the walker's eye interested. Trees planted at the edge of the sidewalk alternate with flowerbeds planted flush with the pavement. The trees are evergreens and are kept at a height of seven to eight feet, an unusual arrangement. They are designed to provide a distinct visual screen to separate the sidewalks from the roadway. Storefronts are often fully open to the street in the Japanese tradition of the merchant townhouses. This merchandising technique gives pedestrians the ability to experience and react to the stores' offerings and to transition from the public to the private realm of the stores.

Ironically, Ginza's extraordinary success has, in recent years, begun to erode its idiosyncratic charms. The combination of anonymous buildings and vertical retailing along Ginza has begun to change as larger buildings by signature architects replace multiple narrow ones, and international brands exclusively occupy space previously shared by many.

BROAD STREETS, NOT WIDE STREETS

Since the end of World War II, Americans have made a staggering number of streets that function solely as auto sewers or isolated cul-de-sacs. The biggest problem with many new and renovated streets is that they are too wide. Countless examples of the latter were widened to add more lanes or create wider lanes, or both. Sidewalks were narrowed and trees were chopped down on the old roads. Many of the new streets never had trees or sidewalks.

Before modern planning and zoning, however, some of our best streets were made by simply widening sections of streets where important activities took place. These are places made less formally than squares and plazas. Broad Street in Oxford is an example that comes from a British market street tradition: streets on the edge of town that were widened for markets hundreds of years ago (Figures 2.112 and 2.113—also see Figure 2.118). As the towns grew, the broad streets were at the center.

American towns have their own variations on the tradition (Figures 2.114 and 2.115). Broad Street in lower Manhattan is wide because there was a canal running

Figure 2.112: Broad Street, Oxford, England. English broad streets commonly have parking in the center, which is removed for farmer's markets and street fairs.

Figure 2.113: Broad Street and its environs, Oxford, England. Figure-ground drawing, showing the marvelous variety of broad and narrow spaces in central Oxford. Narrow Turl Street runs between Broad Street and High Street. The green symbols on Turl Street and the High Street mark the two famous lean-in trees seen in Figure 2.136. © 2013 Dover, Kohl & Partners

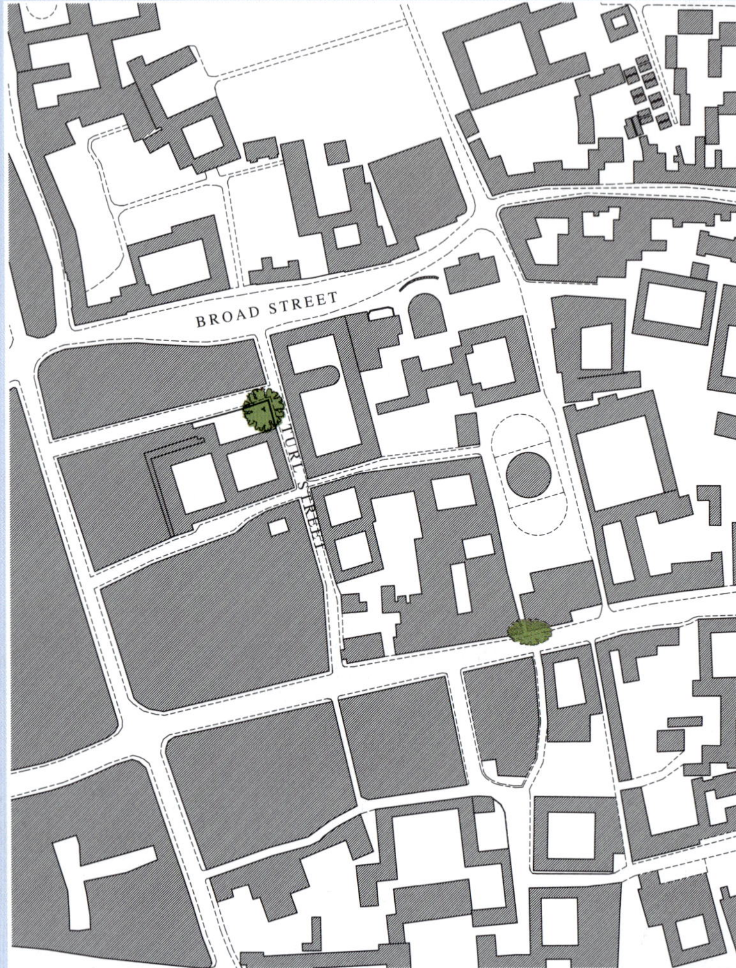

Figure 2.114: Pitt Street, Mt. Pleasant, South Carolina. "That's not a real town," goes an old Southern saying, "that's just a wide spot in the road." Pitt Street splays slightly to form an extraordinary funnel-shaped space in the Old Village of Mount Pleasant, faced by housing, lodging, the old post office (now a restaurant), and the Pitt Street Pharmacy, arguably the social center of the town. *Courtesy of James Dougherty*

Figure 2.115: Pitt Street, Mt. Pleasant, South Carolina. Satellite view. Most of the Old Village is comprised of single-family detached houses on spacious lots, but for the briefest of moments, where Pitt Street broadens, more urban buildings are grouped closely together and tightly frame the street. *Courtesy of Google Earth*

down the center when New York was Nieuw Amsterdam, but when New York started building tall buildings in its financial district, its width made Broad Street a favored location for the new skyscrapers (Figure 1.1).

Marlborough High Street in England is an unusually long broad street, presumably because it was built in an important Wiltshire market town. Each end of the street has a civic building that "plugs" the street and visually contains the space (Figure 2.116). The interesting broad street in another Wiltshire town, Lacock, is not the wide High Street but the end of Church Street near the church, where the street widens, after being very narrow (Figure 2.117) This follows the medieval street design principle "narrow before widening, widen before narrowing."

Main Street in Northampton, Massachusetts, breaks most of the rules of placemaking and yet is a popular gathering place. The Project for Public Spaces put it on a list of favorite places, and the American Planning Association named it a Great American Street.[41] Its width comes from the annexation of an old New England common that was paved over long ago. Sitting at the top of a hill, the street broadens to the sky as we approach.

The garden streets of London are another type of broad street, some drawing on the city's tradition of private squares (Figures 2.166 and 2.167).

Figure 2.117: Church Street, Lacock, England. Lacock is listed in the Domesday Book that recorded the physical state of England in 1086 for the Norman conquerors. With a variety of old street types and spaces, it is used for the filming of English period dramas.

◄ **Figure 2.116:** High Street, Marlborough, England. Looking northeast from the tower of St. Peter's Church. When medieval broad streets were on the edge of town, they commonly had a large entrance at the end near the town and split at the far end, with exits to more than one destination. Marlborough's unusually long High Street splits at both ends. *Courtesy of Neil Goodwin*

Figure 2.118: Broad Street, Oxford, England. View looking west. Less formal than squares and plazas, market streets were formed by widening a segment of a traditional street, pulling open the space to make room for vendor stalls. Once at the edge of villages, some became centers of the towns that grew up around them.

Figure 2.119: This version of a market street, which can be compared to High Street, Marlborough (See Figure 2.116), is an example of how a broad street can be closed to vehicles for market day. Note how buildings and curbs subtly depart from parallel to both open and enclose the space. © 2023 Dover, Kohl & Partners

Figure 2.120: Maximilianstrasse, Augsburg, Germany. Aerial view, 2023. *Courtesy of Google Earth*

Figure 2.121: Maximilianstrasse, Augsburg, Germany. View looking towards the Perlachturm in the early 1900s. *Library of Congress / National Photo Company Collection / Public Domain*

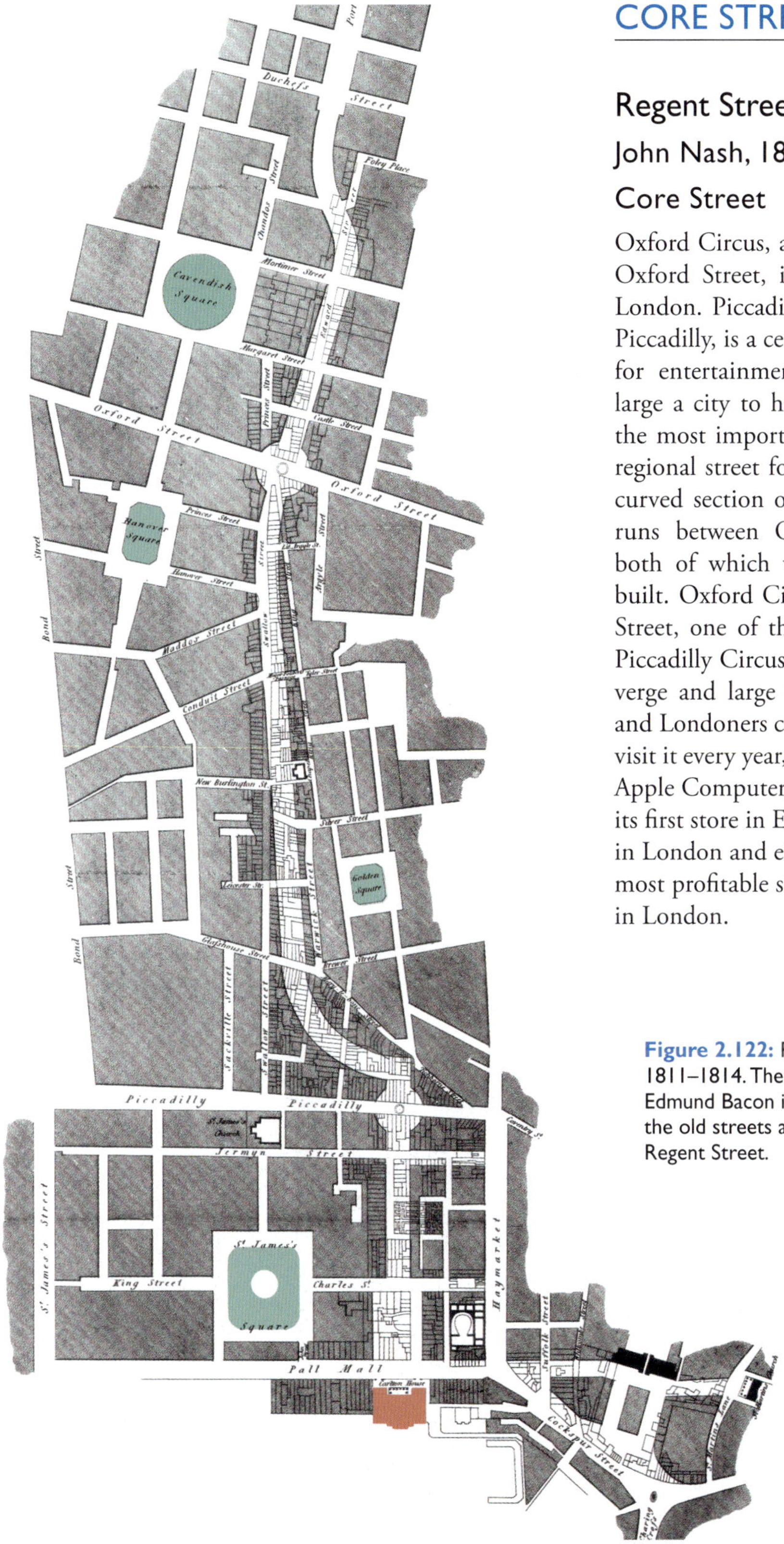

CORE STREETS

Regent Street, London, England
John Nash, 1811–1814
Core Street

Oxford Circus, at the intersection of Regent Street and Oxford Street, is the Main-Main shopping corner of London. Piccadilly Circus, where Regent Street crosses Piccadilly, is a center for tourists and Londoners looking for entertainment in the metropolis. London is too large a city to have a single Main Street that is clearly the most important, but Regent Street is an important regional street for Greater London and its visitors. The curved section of the street known as "the Quadrant," runs between Oxford Circus and Piccadilly Circus, both of which were created when Regent Street was built. Oxford Circus connects Regent Street to Oxford Street, one of the busiest shopping streets in London. Piccadilly Circus is a public space where tube lines converge and large billboards are mounted. Both tourists and Londoners converge there. Over 7.5 million tourists visit it every year, and the world's most profitable retailer, Apple Computer, chose Regent Street as the location for its first store in Europe. Until Apple opened larger stores in London and elsewhere, the Regent Street store was its most profitable store, as well as the most profitable store in London.

Figure 2.122: Regent Street, London, England. John Nash, 1811–1814. The classic Regent Street plan, discussed by Edmund Bacon in *The Design of Cities*. The drawing shows the old streets and buildings along the path of the new Regent Street.

Figure 2.123: The Quadrant, Regent Street, London, England. Sir Richard Norman Shaw, 1908, Sir Reginald Blomfield and others, 1928. Commerce and business demanded larger buildings than the original designs by Nash, and Shaw gave the Crown Estate an Imperial Edwardian Classicism. © *2009 Jon Curnow / Wikimedia Commons / CC BY 2.0*

Regent Street's prominence came about by design in 1811 after an act of Parliament gave the Prince Regent full authority to rule for his father, "Mad" King George III. The fashionable son admired Napoleon's urban interventions in Paris, so one of his first decisions was to hire Sir John Nash to push a wide, north–south axis through a maze of narrow streets to create a formal, processional street from Pall Mall to a new park called Regent's Park. Many Londoners wanted a reordering of medieval London at least since the Great Fire of 1666, but until the Prince Regent and Nash, no one had both the vision and the authority to make it happen.

The Crown-owned land to the north was cut off from fashionable and wealthy London by the narrow medieval streets. Nash laid out Regent's Park on the royal land, surrounded by what looked like Classical palaces that were actually English "terraces" with luxurious apartments where members of the growing upper-middle class could live like kings. The southern end of the street was a broad axis terminated by Carlton House, the Prince Regent's residence on Pall Mall, which is the royal ceremonial route from Trafalgar Square and the Admiralty Arch to Buckingham Palace.

In his book *Design of Cities*, the planner Edmund Bacon makes the point that Nash masterfully adapted the form of Regent Street to the functional requirements of the city.[42] Unlike the planners who worked for Napoleon and Napoleon III, Nash did not simply drive great avenues and boulevards through jumbled streets. He used some existing streets by widening them, frequently respected the backs of properties on surrounding streets, and primarily kept within the street grids that floated at

Figure 2.124: Park Crescent, London, England. John Nash, 1806. View looking south from Park Square West. A crescent marks the beginning of Regent Street at Regent's Park (although the northern end of Regent Street is called Portland Place, and there is a small green called Park Square connecting the Crescent to Regent's Park). The Prince Regent said Regent Street would rival Napoleon's Paris, but the Regency architecture of the Crescent is much quieter than the Imperial Classicism built on Regent Street less than hundred years later.

different angles between the new park and Pall Mall. That meant that the street could not go in a straight line from Regent's Park to Carlton House but had to shift to the east before it arrived at Pall Mall.

The original design for Regent Street has a smooth curve between Portland Place and the place a little north of where Regent Street crossed Oxford Street (Figure 2.122). Nash changed the plan to preserve the backs of mansions on Cavendish Square. He then persuaded the authorities to hire him to design a church at the knuckle of this armature and used the design for All Souls Church to turn a potentially awkward juncture along the street into a beautiful point of focus and transition. Going south, Regent Street flows around the circular entrance to the church. From the south, Regent Street is visually terminated by the church and its spire.

At the crossing of Oxford Street, to counter the "fashionable objection" of living farther north, Nash blurred

the boundary by introducing the circular, omnidirectional Oxford Circus. The size of the circus was determined once again by the backs of neighboring properties. Early plans had a large square south of the Circus, entered at the corners, to allow easy eastward movement. That plan gave too much land to open space, so Nash replaced the square with the beautiful curve of lower Regent Street, an unusual urban space sometimes called the Quadrant (Figure 2.123). The curve ends at Piccadilly Circus, another circle designed by Nash. It terminates the axis to the Mall, redirects the street, and marks the beginning of Shaftesbury Avenue, Coventry Street, and Piccadilly (originally called Portuguese Road).

A little over seventy-five years after Nash's plan was made, the son of Charles Dickens wrote, "Piccadilly, the great thoroughfare leading from the Haymarket and Regent Street westward to Hyde Park-corner, is the nearest approach to the Parisian boulevard of which London

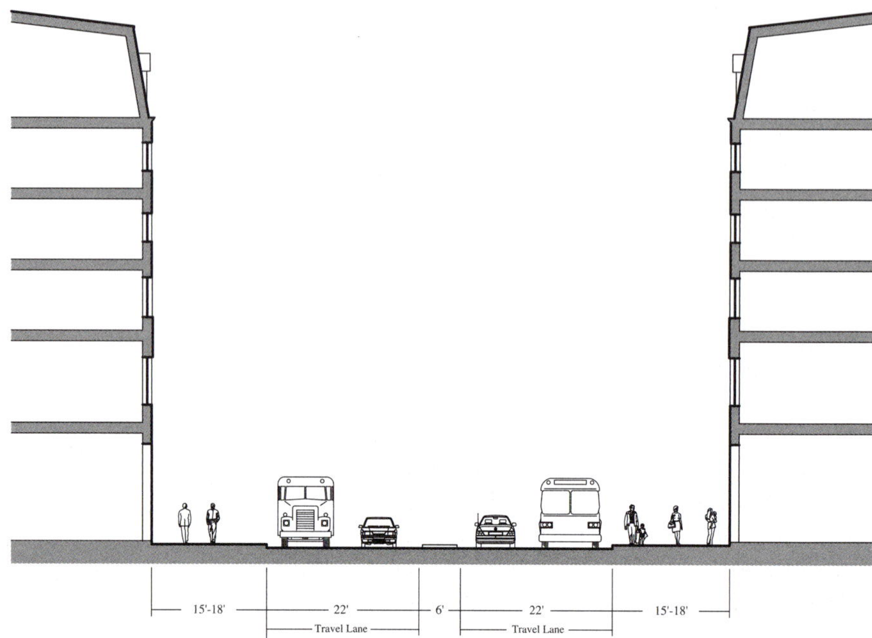

Figure 2.125: Regent Street, London, England. Section at the Quadrant. *© 2013 Dover, Kohl & Partners*

can boast."[43] Nash cleared a large axis south of Piccadilly Circus, which runs downhill. He further widened the street at the lower end, where Carlton House sat on axis. Flanking buildings designed by Nash on the north side of the square framed the view back to Piccadilly, where another building designed by Nash terminated the view. Between Piccadilly Circus and Carlton House, Nash cleared a cross-axis from St. James Square to Haymarket. He (or perhaps the Prince Regent) persuaded the owner of the Haymarket Theatre to move the theater to the end of the axis and to let Nash design the new theater with a Classical portico and pediment to terminate the vista.

The Prince Regent left Carlton House in 1820, when he succeeded his father as king and moved his residence to Buckingham Palace. That provided the opportunity to open a grand vista from Piccadilly Circus to the Mall and St. James Park. Carlton House was torn down in 1825, and the Duke of York Monument was placed on axis, at the top of stairs leading down to the Mall. Standing 137 feet, nine inches tall, the monument was a Tuscan column topped by a statue of the Duke, who was the brother of the new king and the commander of the British forces during

Figure 2.126: Air Street, London, England. View from Piccadilly, looking north on Air Street. Air Street is one of the old streets that crosses Regent Street. Sir Richard Norman Shaw incorporated it into his grand Edward Classicism with bold arches and columnar screens in buildings along the Quadrant. *Courtesy of the Crown Estate*

the French Revolutionary Wars. At the same time, the lower end of Regent Street was renamed Waterloo Place.

Waterloo Place and the nearby Trafalgar Square commemorated British battles on land and sea. Regent Street was a real estate deal for a Prince Regent who was always in debt, but it was also an expression of Britain's modern imperial power and the Prince Regent's position as a leader of architecture and fashion. The imposing curve of Regent Street and the new axes around Waterloo Place tell us that London was a world capital, designed by the leading architects of the day in what came to be known as the Regency Style (Figure 2.124).[44] When commerce demanded more functional spaces for new businesses at the end of the nineteenth century, the Crown Estate built even larger and more imperial buildings. Sir Richard Norman Shaw invented a grand British Classicism so popular that it became characteristic of the Edwardian age, and the Crown Estate acquired neighboring properties so that it could make the new buildings larger. Shaw designed the Piccadilly Hotel, completed at the intersection of Regent Street and Piccadilly Circus in 1908, and Sir Reginald Blomfield was brought in two years later to oversee the completion of the Quadrant.[45] Each block had harmonious Classical facades built in Portland Stone, with a uniform cornice sixty-six feet above the sidewalk (dormers and turrets decorate the mansard roofs above the cornice—Figure 2.125). Impressive rusticated arches and majestic columnar screens pierce the facades but continue the buildings where small lanes cross Regent Street (Figure 2.126).

Regent Street is a British variant of the Grands Boulevards of Paris. Paris would not be the Paris we love if it had only the medieval streets of old Paris, without the grand urban interventions of Louis XVI, Napoleon, and Napoleon III with Baron Haussmann. London would be a very different place without Piccadilly Circus, Regent Street, and Regent's Park.

Large-scale urban interventions got a bad name in the twentieth century, when American planners inspired by Le Corbusier's urban design principles built disastrous urban renewal projects across the country. Corbusier advocated the demolition of the street and the construction of towers in parks, but as Jane Jacobs pointed out, we actually built towers in parking lots. She also labeled urban renewal "urban removal," and hand in hand with that went a program of building highways that ran through cities and tore apart neighborhoods. Today we have begun to pull down those highways, leaving us

with large swaths of cleared land and badly ripped urban fabric that needs to be reknit in creative ways. Ambitious designs like the ones for Regent Street and the Grands Boulevards are once again relevant, useful precedents for modern designs.

Railroad Street, Great Barrington, Massachusetts

Work Street

Railroad Street, a short, mixed-use street in Great Barrington, Massachusetts, shows how easy it can be to make a good street, because the elements and details of the street are so simple and easy to replicate. It has one medium-length block that goes uphill from Main Street before the street makes a right turn and becomes a service street for a parking lot next to the railroad tracks (Figures 2.127 and 2.130). The sidewalks are poured concrete slabs with granite curbs, and the streetlights are old cobra heads. The buildings along the street are two- and three-story mixed-use buildings, with storefronts on the ground floor and workplaces or apartments above— all typical of Massachusetts in the nineteenth century. The Western Massachusetts town has enough tourists to support a high occupancy rate, but the rents are not expensive, and no one is getting rich owning the stores or buildings.

Let's catalog why it is good. The space is a comfortable "outdoor room" because:

- It is well proportioned; the tallest buildings are approximately three-quarters as tall as the street is wide.

- From wall to wall, the space is 50 feet wide and 350 feet long (1-to-7).

- The slight upward slope gives the building that terminates the vista more prominence and increases the sense of enclosure.

- The streetwalls that define the room are almost continuous.

- The streetwalls have great "firmness," a solidity that comes from their weighty materials and the light and shadow that show the depth of the facades.

- The materials that make the streetwall, primarily brick and stone, are pleasing to the senses.

- The vertical window openings above the ground floor and the number of individual buildings on the block give a visual counterbalance to the horizontal space.

- The weathered asphalt "floor" is a pleasant gray. The asphalt harmonizes with the building colors and unifies the space.

- The parked cars and the painted stripe alongside the parked cars visually narrow the space.

- The cars, the stripe, and the sidewalks are all parallel to the streetwalls.

- The street plan works well with the block.

The buildings are not great works of architecture, but they are all beautiful in a simple way that reflects the visual principles described by Christopher Alexander in *A Pattern Language* and *The Nature of Order*. The simple compositions of their facades and their natural materials make excellent background buildings, background buildings that can withstand scrutiny and that give a pleasant feel to the street (Figure 2.128).

Streets that do not have good architecture need trees to screen the buildings while providing order and beauty. The buildings on Railroad Street are good enough that the street feels good without trees. The street used to have one tree, in the bumpout on the north side of the street where the Railroad meets the town's Main Street, a situation that is worth discussing (Figure 2.129). In Chapter Four, we will see that there has been a lot of discussion about Great Barrington's street trees, related to the fact that Bradford pear trees have weak "crotches," and this tree came down in a freak Halloween blizzard (see *Street Trees* in Chapter One).

When it was still standing, the location of the tree on the bumpout was subtle. Trees on both sides of Main Street usually line up along the street and on opposite sides of the street. One of the Main Street trees is directly across from the end of Railroad Street, on axis with its center. The single tree on Railroad Street partially

Figure 2.127: Railroad Street, Great Barrington, Massachusetts. Looking west from Main Street, with the Bradford pear trees in bloom. The tree in the bumpout at the end of Railroad Street is gone.

balanced the tree on the other side of Main, while also softening the width of Railroad Street without blocking it, functionally or visually. The bumpout provided a place for a pair of mailboxes on the sidewalk without blocking the narrow sidewalk on Railroad or the wider but busier sidewalk on Main. And it did all this without calling attention to itself as a bumpout. Part of the reason for that is that it does not stick out as far as the parked cars, but partially shields the parked cars on one side of the street in a way that feels like relaxed placemaking, rather than the imposition of a rigid, one-size-fits-all streetscape formula. Unfortunately, after we wrote the first edition, the Massachusetts Department of Transportation's formulaic makeover of Great Barrington's Main Street significantly damaged one of the best Main Streets in New England (see page 411 *Retrofitting a Main Street That's Also a State Highway: A Cautionary Tale* in Chapter Four).

The final bullet point above says, "The street plan works well with the block." To understand what that means, it helps to look at both sides of Main Street (Figure 2.130). Railroad Street is to the west of Main Street, with nineteenth- and early-twentieth-century mixed-use buildings on both sides. Some of these are flexible New England loft-buildings that over the years have accommodated offices, manufacturing space, apartments, hotel rooms, and stores on the ground floor. One building has a large theater called The Mahaiwe behind and beneath the offices and apartments.

The other side of Main Street has similar buildings, but behind those buildings, to the east, are parking lots that were at some point cleared in a typical "urban removal" plan. The lots come out to the sidewalks on both side streets, creating an unpleasant pedestrian experience along the streets as well as a view of the backs of buildings that were never meant to be seen by passersby.

There is parking to the west of Main Street, too, but it is screened by buildings along the side streets. Railroad Street does that well and screens the railroad tracks at the top of the slight hill. When you get to the top of the hill, however, where a building with a restaurant and an

▼ **Figure 2.130:** Railroad Street, Great Barrington, Massachusetts. Figure-ground drawing. © 2013 Dover, Kohl & Partners

▲ **Figure 2.129:** Railroad Street, Great Barrington, Massachusetts. View of the bumpout at the intersection of Railroad Street and Main Street, showing the Bradford pear tree mentioned in the text. Unlike the ubiquitous, one-size-fits-all bumpouts commonly built today, this bumpout is a civic amenity.

outdoor terrace blocks the view of the tracks just behind the building, Railroad Street takes a sharp right-hand turn and enters a parking lot along the railroad track. Facing the parking on the right are some inexpensive utilitarian buildings that screen another parking lot behind the buildings on Main Street. Many of the Main Street buildings have rear doors that open onto the parking lot, and just enough work has been done on the lot (there is a mural on a windowless wall, there are trees, and a small terrace for a coffee shop) that it functions as both parking and public space. One pedestrian alleyway lets you walk back to Railroad Street and another pedestrian walk, next to the café, goes out to Main Street. In the center, between the two parking lots, sits a new three-screen movie theater.

The parking to the east of Main is useful but a visual blight. Partly thanks to Railroad Street, however, the parking to the west is a civic amenity.

Figure 2.131: Chartres Street, New Orleans, Louisiana. View looking north between Iberville and Bienville streets. The French Quarter is the oldest neighborhood in New Orleans. At its heart is a small grid of similarly sized streets and blocks that are a tourist mecca. One would be hard pressed to say which one is the main street. They all have bars, restaurants, stores, hotels, residences, and offices. They are urban core streets.

Aviles Street, St. Augustine, Florida

Village Street

One of the oldest American streets, St. Augustine's Aviles Street is generally full of people—dining outdoors, stopping to peer into the windows of storefronts, casually walking and biking through the right-of-way. Although predominantly a street for pedestrians, motorists and cyclists use Aviles Street too, and the sidewalk cafes spill out into the right-of-way. The total width of the road varies: at its narrowest, it measures only a little more than ten feet from curb to curb. The dimension widens at intersections, but only slightly, which accommodates parallel parking at its southern end. The width does not deter cars from using the street, but it does make them move slowly, and this is reinforced by the textured pavers installed in a recent renovation. The hardscape unites the space (Figure 2.132).

Because it is such a narrow road, Aviles Street functions similarly to many of the shared spaces of Amsterdam, such as Tweede Tuindwarsstraat in the Jordaan District. Shared space of that sort is prevalent in many Dutch cities, but it is still quite rare in America—although, ironically, the pattern still works so well on one of our oldest streets.

Figure 2.132: Aviles Street, St. Augustine, Florida. Small is beautiful: a single narrow lane, tightly spaced buildings, and the slight cant combine to make one of the most comfortable "shared-space" streets in America.

Formosa Street, London, Great Britain

Village Street

Urban designers debate where neighborhood centers should be located, particularly neighborhood retail centers. Those who mainly work in places where everyone drives want to put the retail centers on high-traffic roads. Designers working in more urban settings where most people walk, or at least park and walk, want some models for getting away from roads dominated by cars. Formosa Street is an interesting model for the second group.

Formosa Street is a small neighborhood center in the Little Venice section of Maida Vale in west London. The local Underground stop, Warwick Avenue, is two short blocks to the south. From the Underground station, one walks to Formosa Street via Warrington Crescent, which begins at the Underground stop and curves off to the northeast.

There is a debate among urban designers about the best locations for neighborhood centers, particularly neighborhood retail centers.

"The Crescent" is a fancy street, with large "terrace houses" (English for identical or mirror-image row-houses) that have high ceilings and large, shared private gardens in the center of their large blocks. To the west of Warrington Crescent is Castellain Road, which has smaller houses, with smaller gardens. Formosa Street begins at Warrington Crescent and runs west, crossing Castellain. In between Castellain and Warrington, Formosa bends in the middle so that it is perpendicular to both cross streets. Only the block of Formosa Street between the two has stores (Figure 2.133).

The short block has a travel lane in each direction, with a parking lane on one side. But people park on both

Figure 2.133: Formosa Street, London, England. Looking east on Formosa Street, from Castellain Road. The Prince Alfred pub is on the corner.

Figure 2.134: Formosa Street, London, England. Looking north on Warrington Crescent. The eastern end of Formosa Street is seen on the left, flanked by decorative arches.

sides, sometimes leaving a yield lane in the center that is less than 9 feet wide. Even in the best of conditions, the sidewalks are 9 feet wide on both sides and the parking lane is 6½ feet, leaving 15 feet for the travel lanes, or 7½ feet per lane.

Most of the buildings on the street are two-bay, three-story, mixed-use buildings with small stores on the ground floor. A Victorian pub at one end also has a dining room in the middle of the block. At the other end, on the Crescent, the small mixed-use buildings nod to the formality of the large terrace houses with applied triumphal arches on mainly blank walls (Figure 2.134). But small, understated doors in each facade, as well as a small side window in one elevation, give a quirky informality to the monumental elevations.

Warrington Crescent is wide and lightly trafficked. The few cars there move quickly through the quiet residential neighborhood, on their way to somewhere else.

Step off the Crescent onto Formosa, and you are in a quieter, smaller, less formal place. The bend in the road focuses the space on the center of the block. Very few cars enter the block unless that is their destination, and most people on the street arrive by foot. Since there is so little traffic, pedestrians frequently walk in the street rather than on the sidewalks, automatically creating a shared space where cars go slowly. If the locals didn't walk in the street, the narrow sidewalks would be crowded, because several establishments on the street have outdoor tables or stands. Shops include a convenience store, a café, a launderette, a dry cleaner, a small restaurant that serves breakfast and lunch, a bakery and small restaurant, and a large restaurant attached to the corner pub. Some larger neighborhood centers with more stores and variety are within a quarter of a mile walk. When coming from the center of London on the Underground, one can choose to go to the next stop, where there are more stores.

Figure 2.135: Tweede Tuindwarsstraat, Amsterdam, the Netherlands. A film strip showing the constantly changing activity in Tweede Tuindwarsstraat. Even a passing truck driver engages in the social life of the street. *Courtesy of Ethan Kent, Project for Public Spaces*

TWEEDE TUINDWARSSTRAAT, AMSTERDAM, NETHERLANDS / ETHAN KENT

Shared Space

A Street as a Place

The patterns of activity that occur in one place on this street offer the outcomes for which all streets should be planned.

When wandering through a network of streets we are naturally drawn to its center. On each street we seek out its slowest point—often an intersection—which usually brings us to progressively more engaging and slower streets and spaces.

Amsterdam's Jordaan District offers such a network of streets, where the external thoroughfares feed into progressively more pedestrian-oriented and comfortable spaces. Moving into the district, shops get smaller and increasingly concerned with the street; walking shifts to strolling; conversations grow longer; and people feel justified, and often compelled, to just stop and linger, taking in the scene.

The desire to pause and experience the street peaks in a section of Tuindwarsstraat that happens to be near the center of the district. The nexus of a flower shop, a bar, and a café across the street creates what is perhaps the slowest place in the neighborhood. Each passerby moves through the space with attention engaged, making eye contact and frequently stopping to read a menu, greet a friend, or purchase some flowers.

These patterns of use all beget themselves.

While the pace is slow, the space supports dynamic experiences, exchanges, and interactions for its participants. The vibrancy of these interactions is buttressed through the broad diversity of people, all allowed to feel equal and respected in this space while being brought together by the strong cultural identity of the setting. The greatest indicator of its comfort—and perhaps its greatest achievement—is its sociability, reflected in consistent displays of engaged conversation, picture-taking, and affectionate greetings.

While not technically defined as a "shared space" or "Complete Street," this spot succeeds as both (Figures 2.135 and 2.136). All types of traffic can easily and comfortably access this place and benefit from its qualities. Even a passing truck driver can engage in

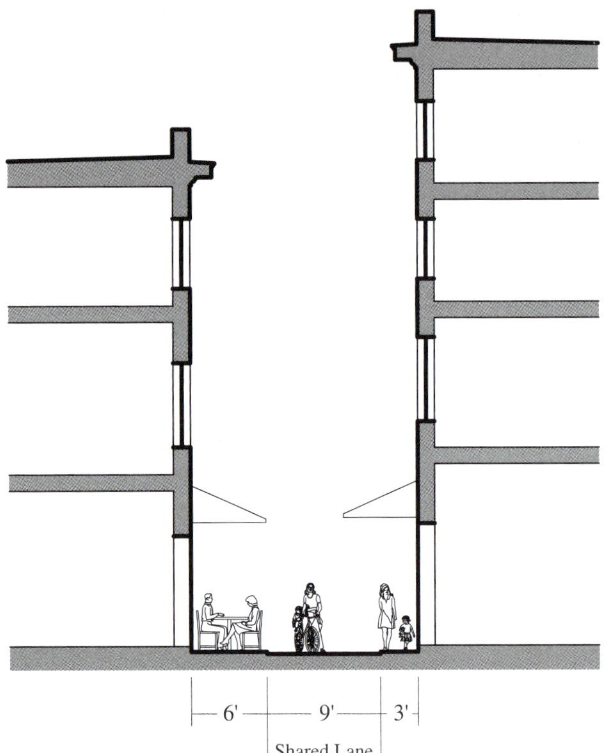

Figure 2.136: Tweede Tuindwarsstraat, Amsterdam, the Netherlands. Section. © 2013 Dover, Kohl & Partners

the social life of the street, stopping briefly to talk with café patrons. Because of the strong hierarchy of streets surrounding the Jordaan District, the transit and bike orientation of the city, and the slowness of this street, it does not carry significant traffic.

There are many features of the street to which one can attribute its dynamism, from the affluent neighbors and tourists to the Dutch culture and related scale, density, and style of urban design. Indeed, it is through these contexts that we usually evaluate, explain, and experience such spaces. Yet the experience of these fundamental patterns of social life—of a street performing as a place—is something infinitely possible and prevalent in human settlement.

Across contexts and cultures, when allowed, people naturally gravitate to participating in and creating street life. And it is this life that is the building block of strong communities, cultures, and local economies. This dynamic street hums along because of vibrant Dutch culture and context, yet that strength emerged because of a society nurtured around shared places like this

It may well be the presence of the flower vendor, who spends time in front of his shop adjusting his display and talking with passersby, that truly anchors the place and embodies what this street makes possible. Retail that competes to contribute to the public experience, not just benefit from it, is usually the greatest contributor to successful streets and is what allows public streets to compete with malls and chains. The life and vitality of cities are defined at this human scale of the street, yet no one but pedestrians and those serving them really pay this scale any attention.

Despite our deep affection for scenes like these, few plan a street for the potential of what it can become as a place. Tweede Tuindwarsstraat centers on the daily patterns of social life, business activity, personal mobility, and people being allowed to feel comfortable in the city. While most streets and transportation systems are planning for mobility and attempting to mitigate its negative impacts, what if we planned streets and transportation systems with the goal of supporting settings like this, of planning first for places where people want to be?

As Chuck Marohn of Strong Towns says, "We did not build places like this because we were rich; we became rich because we built places like this."

ROWHOUSE STREETS

Sidney Place, Brooklyn, New York
Rowhouse Street

There are two reasons for including Sidney Place in this book. First, like Railroad Street in Great Barrington (Figure 2.127), it shows how easy it can be to make a good street. Sidney Place is one of the most beautiful streets in New York, and yet it was made almost entirely without the help of architects or urban designers. Second, Sidney Place demonstrates that a shift in a simple grid can contribute a lot to the act of placemaking.

Like Railroad Street in Great Barrington, Sidney Place shows how easy it can be to make a good street. Sidney Place is one of the most beautiful streets in New York, and yet it was made almost entirely without the help of architects or urban designers.

Sidney Place (Figure 2.138) is in Brooklyn Heights—one of the oldest parts of Brooklyn, which sits on a bluff across the East River from lower Manhattan. Development of Brooklyn Heights began soon after the start of ferry service between Brooklyn and Manhattan in 1814. In 1818, the Brooklyn trustees wanted to extend two long north–south streets, Columbia and Willow streets, across the estate of Hezekiah B. Pierrepont.[46] The Pierrepont family allowed east–west streets to cross the property but blocked the construction of the north–south streets, which soon leapfrogged over it. When the estate was later sold and then subdivided in 1842, this resulted in the construction of short, discontinuous north–south streets, including Columbia Place, Willow Place, and Sidney Place. Each runs between State Street and Joralemon Street. All are tree-lined streets with rowhouses, visually terminated by houses on the cross streets. These small changes in the Brooklyn Heights grid makes the short street spaces enough like plazas that someone appropriately named them "places" rather than "streets."

State Street and Joralemon are not quite parallel to each other, so the blocks between them get longer as the streets go east; Sidney Place is the farthest east of the discontinuous blocks. About seven hundred feet long, Sidney Place has an interruption midblock on one side, where a new east–west street, Livingston Street, starts. This

THE LEAN-IN TREE: TURL STREET, OXFORD, ENGLAND

Oxford is a place where the city and the university combine in a uniquely British neighborhood fabric. The Gothic, Neo-Gothic, and Classical buildings create a cityscape with a distinguished sense of place. Turl Street is a skinny street that connects Broad Street to the High Street (Figure 2.137). One of its most notable features is the large tree that overhangs from an adjacent courtyard. The "lean-in tree" naturally unifies the street scene. In many American cities and towns, ordinances require trees that lean over the right-of-way to be trimmed or cut back. This Oxford Street is an example of how the lean-in tree can instead add comfort and character to an urban environment. An additional element on Turl Street is the hydraulic bollard—which can be adjusted to let cars travel through Turl Street when desired. When car traffic is not wanted, the bollards remain up.

Figure 2.137: Turl Street, Oxford, England. View looking north towards Broad Street. The "lean-in tree" gently sculpts the narrow, intimate space.

Figure 2.138: Sidney Place, Brooklyn, New York. Looking north from State Street. A house on Joralemon Street is visible at the opposite end of Sidney Place.

Figure 2.139: Sidney Place, Brooklyn, New York. Looking south towards State Street from midblock. Everything lines up on the straight block: the buildings, the stoops, the sidewalks, and the curbs. Even the windows and the cornices frequently line up.

interruption visually and psychologically breaks down the scale of Sidney Place: even though it is not a through street, Livingston Street prevents Sidney from feeling like a long block. A Catholic church sits on the northeast corner of Livingston and Sidney, and its parochial school is on the other side of Livingston. Livingston Street is now one-way from Sidney Place, which doesn't have much traffic, so that intersection becomes something of an assembly ground for the church and the school.

Lined with rowhouses on standard lots, Sidney Place is typical of many blocks in Brooklyn (Figure 2.139). When the Brooklyn Heights grid was platted by Brooklyn's engineers in 1839, before Brooklyn was part of New York City, they used the 200-foot by 400- to 900-foot blocks that were already becoming the standard across the river in New York. All over the two cities, rowhouses were commonly built on 100-foot by 16- to 32-foot lots that went back-to-back on the 200-foot blocks.

Today, great Brooklyn neighborhoods like Park Slope, Fort Greene, Cobble Hill, and Brooklyn Heights have block after block that remain largely intact from the nineteenth century. In 2006, the editors of *Time Out* magazine picked one of the blocks of South Portland Avenue in Fort Greene as the most beautiful block in the city, although there are many other blocks that are

almost identical. Their common elements are mature trees and rowhouses with stoops. The rowhouses on both sides of the street will often be in just one or two styles, either repeated or mirrored. The houses facing each other and on the same side of the block will typically be all brownstone, or primarily brick, and the trees will all be the same genus, planted equidistant apart and all in a row. There is lots of repetition, in other words, but only contemporary architects and landscape architects say the street needs more variety and "creativity." Most New Yorkers just wish they could live there.

Sidney Place has more architectural variety than South Portland Avenue. The rowhouses are a combination of redbrick Greek Revival houses (some have been painted) and Connecticut-brownstone Neoclassical houses of the sort built all over New York during the late-nineteenth-century period that Lewis Mumford called "the Brown Decades." Many have their original stoops; some have been converted to "the English plan" (which means their stoops were removed and the first two floors remodeled - see page 25); and some are apartment house conversions with no stoop. The buildings are unified by their similar heights and widths, Classical details and proportions, masonry fronts, vertical double-hung windows, and the trees that form a canopy over the street.

Alta Vista Terrace, Chicago, Illinois

J. C. Brompton, 1904

Rowhouse Street

Alta Vista Terrace is a straight, block-long, narrow residential street developed by the prolific builder Samuel Gross in 1904 (Figures 2.140 and 2.141). Gross was fascinated with European urbanism and with the terrace houses of London in particular, so he directed architect Joseph C. Brompton to adapt the type to American conditions in Chicago's Lake View neighborhood. On Alta Vista Terrace, the compact design of both the forty lots and the street allowed substantial development to fit in an unusually shallow parcel while keeping the street framed by matching house types on either side—like facing like—to create a unified street scene. Brompton and Gross alternated the facade designs so that the houses across from each other were never the same; that introduced variety and improved privacy. The lots are twenty-four feet wide and forty feet deep.

This skinny street space is open at both ends—knitted into the regular grid of Chicago blocks—and yet intimate and distinctively its own place. Visitors sense they are entering the residents' territory. With stoops and bay windows overlooking the street and friendly neighbors coming and going, the scene is very clearly watched over. Alta Vista Terrace feels private and self-policing, but there is no gate or guardhouse here.

This combination of intimacy, sociability, and security is partly accomplished by its narrow width: the street includes just a one-way travel lane, a row of parallel parking, and sidewalks. The sidewalks are approximately ten feet wide and meet the edges of small dooryards in front of each rowhouse.

Unique in Chicago, Gross's block is now a National Historic Landmark, designated the Alta Vista Terrace Historic District.

Figure 2.140: Alta Vista Terrace, Chicago, Illinois. J.C. Brompton, 1904. Satellite view. The shallow lots, compact building type, and narrow street (center) permitted the development of distinctive addresses despite the extremely tight parcel. Compare the shallow houses to the full-size rowhouses nearby (right and bottom). *Courtesy of Google Earth*

Figure 2.141: Alta Vista Terrace, Chicago, Illinois. J.C. Brompton, 1904. Like facing like, with a twist: Identical house facades are alternated diagonally, imparting visual variety within the group of rapidly produced, similar houses.

CHURCH STREET, CHARLESTON, SOUTH CAROLINA

Village Street

Some of the most memorable and delightful streets unfold in sequences, like a play or novel, and vary dramatically from one scene to the next. Church Street is one such episodic street, changing in width and character as it travels north along the peninsula of Charleston's historic district. The experience of the street breaks down into five short, distinct segments just in the section between White Point Gardens and St. Philip's Church. The spatial sequence from south to north changes in formality and scale to match the kinds of addresses rooted along each segment. This sequence accumulated slowly over time, with no single designer or grand, overarching plan. The effect is as if each segment had been treated as its own unique compositional assignment, but the idiosyncrasies are endearing and make Church Street an example of what François Spoerry called "gentle urbanism."[47]

The traditional buildings framing the street easily adapted to the twenty-first century. Meanwhile, although there are many beautiful works of architecture on Church Street, nothing about the road details is particularly fancy or elaborate. In many ways, the street is an ode to simplicity. A visitor will notice the near total lack of roadway striping and signs. Church Street proves that parallel parking does not need excessive amounts of white paint to work, and that speed can be controlled without brightly painted objects that detract from the aesthetic quality of a town. Unpretentious local plants and flowers add value and beauty.

From the south, the sequence begins at White Point Gardens, Charleston's formal waterfront square (Figure 2.142). This is not just the edge of the neighborhood; it is the end of the peninsula. Wide vistas across the harbor extend to distant barrier islands and the horizon. The contrast is dramatic once we cross Battery Street to begin the walk along Church Street.

Here, Church Street is at its tightest and most enclosed, and the views are short. On this segment between Battery Street and Water Street, the carriageway is paved with red bricks and the curb-to-curb dimension is approximately seventeen feet. Vehicles travel at slow speeds through this single lane, leaving ample space for parallel parking (Figure 2.143). Contrary to what many traffic engineers might expect, motorists and emergency vehicles have no problem navigating the narrow street. This portion of the Church has few street trees in the conventional sense. Instead, the pattern is that of the satisfactory lean-in tree, reminiscent of Turl Street in Oxford (Figure 2.137).[48] There are sidewalks on both sides, but most people walk in the street.

The experience changes again when Church Street cranks—it doesn't curve, bend, or arc, it cranks—in a dogleg at Water Street (Figure 2.144). On the west side, the property lines on the Church Street lots continue straight to the corner lot, and the result is a shady triangular space on one side, in which the walls of gardens and houses are no longer parallel to the curbs. We are told Water Street is appropriately named; it used to be a creek, in the early days of the settlement, and—like a lot of special places in Charleston—its urban form emerged gradually as earthquakes and hurricanes knocked down buildings that became the fill material for much of the town's upland. The dogleg, the crank, and the triangular space at the intersection of Water and Church streets are all artifacts of that evolutionary process. The long, straight vista is softly concluded; the space feels enclosed but not claustrophobic, and street trees are introduced. The variation from a regular grid removes the sensation that the street might go on forever. But there is another effect—on traffic. The cranked street slows down any motorist without being annoying.

It is like a giant, invisible speed bump, only better, because it is a place rather than an anti-place.

Between Water Street and Tradd Street, Church Street is still one-way, but here the lane widens and there is parking on both sides. Narrow one-way streets are okay in this kind of residential fabric, where there is a rich network of streets, blocks are fairly small, and there are plenty of surrounding streets to support circulation (Figure 2.145). This neighborhood is one of Charleston's most beloved and livable, and in part, this is because it has a generous menu of street types and widths, including the charming Stoll's Alley (Figure 2.146) and—narrowest of all—Longitude Lane (Figure 2.147).

Between Tradd Street and Broad Street, Church widens again and there is a brief, two-way segment. By this point, Church Street is a different sort of street from the intimate place seen earlier (Figure 3.5). It is still beautiful, and the oak canopy

Figure 2.142: White Point Gardens, Charleston, South Carolina. Looking south towards the Ashley River. Church Street begins at Charleston's formal waterfront square.

Figure 2.143: Church Street, Charleston, South Carolina. Looking north to White Point Gardens. Here, Church Street is narrow, and trees "lean in" from adjacent gardens.

Figure 2.144: Church Street, Charleston, South Carolina. Looking north towards Water Street. The dogleg at this intersection provides a series of delightful effects, including the shady triangular space and tamer traffic.

Figure 2.145: Church Street, Charleston, South Carolina. Looking north towards Stoll's Alley, which is on the right.

arches overhead in a spectacular way, but pedestrians keep mainly to their sidewalks. The scene is becoming more citified (Figure 2.148).

The climactic moment in the walk comes just before Broad Street, where the steeple of St. Philip's, still nearly three blocks away, becomes visible over the tree canopy (Figure 2.149). Today this is one of the most celebrated examples of the "terminated vista" in American urbanism, but the church buildings came first and, happily, the street was compromised to match. The present St. Philip's church building was completed in 1838 and substantially encroaches on the alignment of the Church Street centerline, as had its predecessors. The steeple, added in 1848, is about two hundred feet tall.[49]

Between Broad Street and St. Philip's, the street reverts to one-way traffic, and the giant oaks give way to smaller trees and palmettos to avoid obscuring the steeple. Once the street reaches the church, the roadway skirts around the steeple and porticoes. The pavement on Church Street is perennially cracked, patched, and broken, but hardly anyone notices. It is the view of that steeple that counts (Figure 2.150). As Douglas Duany has pointed out, until modern engineering came along, curving roads were usually edged with straight sections of curb, and that is the case at St. Philip's.[50] The whole thing is the opposite of smooth, friction-free, streamform auto geometry (Figure 2.151). Indeed, Church Street makes demands of the drivers who travel it, and that makes it safer. Almost everything about its design would be labeled substandard by most contemporary engineering manuals, but there it sits, and it works.

Figure 2.146: Stoll's Alley, Charleston, South Carolina. The neighborhoods around Church Street have a wide range of street types, from broad to tiny.

Figure 2.147: Longitude Lane, Charleston, South Carolina. The narrow passage is both a conversation piece and a convenient shortcut.

Figure 2.148: Church Street, Charleston, South Carolina. In this short, straight segment, Church Street briefly opens up visually and has two-way travel, under the arched canopy of live oaks.

Figure 2.149: Church Street, Charleston, South Carolina. Looking north from Broad Street. The climactic moment comes when the steeple of St. Philip's Church, still nearly three blocks away, becomes visible over the tree canopy.

Figure 2.150: Church Street, Charleston, South Carolina. St. Philip's Church (J. Hyde, 1838; steeple by E.B. White, 1848) instantly produced a quintessential terminated vista (and the iconic Charleston street scene). The street goes around the steeple in a faceted jog. At more than 200 feet tall, the steeple has long been the skyline's most recognizable feature, and thus it was used as a target during the bombardment of Charleston during the Civil War. Although the sanctuary was damaged by shelling on several occasions, thankfully the steeple survived.

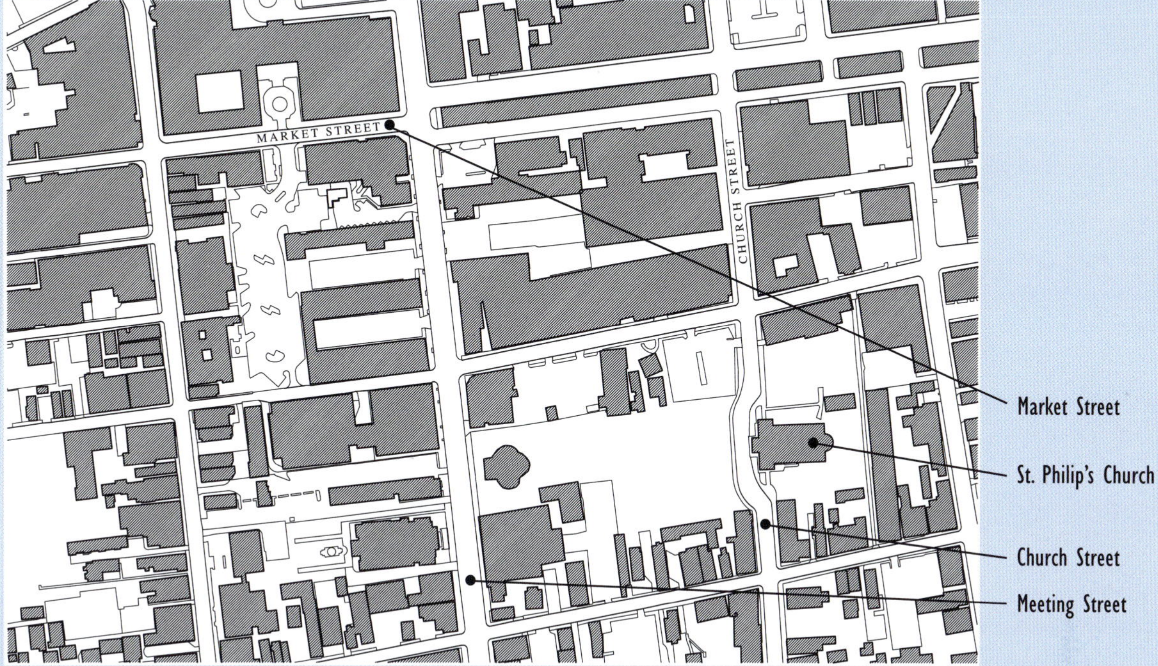

Figure 2.151: Church Street, Charleston, South Carolina. Figure-ground drawing. The street alignment takes a sudden westward jog, allowing the church steeple to step forward into view. *Courtesy of Dover, Kohl & Partners with data provided by the City of Charleston*

LEGARE STREET, CHARLESTON, SOUTH CAROLINA

Neighborhood Street

Many great streets do not comply with the rigid public works standards in force today. Frequently, the narrowest streets with the most idiosyncratic details are the best places to be. Venerable Legare Street (pronounced "Luh-GREE") is one of Charleston's most popular addresses, but its physical form would be tough to replicate within the current rules. Legare Street (Figures 2.152 and 2.153) provides plenty of evidence to suggest the rules need to change.

Legare Street—roughly parallel to Church and Meeting streets—is a key organizing element of the historic district. The oak trees provide a beautiful, continuous canopy along the street. There are both "lean-in" trees and street trees within the right-of-way on Legare. Some are quite old trees, with large roots coming out of the sidewalk, and several enter the travel lanes. These trees may not comply with typical regulations, but that does not mean that they are not safe or effective. In fact, their massive presence calms traffic without looking like a "traffic calming device." (Also see Turl Street, Figure 2.137.)

> If typical suburban residential streets were designed like Legare Street, there would be no need for "traffic calming devices."

Legare Street is not quite as narrow as the southernmost segment of Church Street, but it is close. At its smallest, the measured curb-to-curb dimension is just under twenty feet wide—comparable to Twain Avenue in Davidson—expanding somewhat as the road travels north from Battery Street. The narrow dimension accommodates informal parking and effectively slows traffic. Another characteristic that Legare shares with Church Street: the perspectival effect of a deflected vista, rather than a seemingly endless vista. Subtle bends in the street (shown in the plan, Figure 2.154) limit what can be seen by the walker, cyclist, or motorist looking down it.

Figure 2.152: Legare Street, Charleston, South Carolina. Legare Street breaks some rules; trees emerge straight from the pavement, for example. The alignment of the street makes a very slight crescent, bent plenty (in two places) to close off the view and create a delightful intimacy, but barely enough to be noticeable as a designer's maneuver unless one is looking for it.

Figure 2.153: Legare Street, Charleston, South Carolina. View looking north by the Simmons-Edwards House, built around 1800. The live oaks make the scene and yet break all the rules. Some of the most memorable, prized streets defy the criteria enforced by modern specialists; a traffic engineer, arborist, zoning official, stormwater expert, or utility clerk could all give reasons why a street like Legare is unprofessional—yet there it sits, beautiful, safe, practical, and perennially valuable.

Figure 2.154: Legare Street and Church Street Charleston, South Carolina. Figure-ground drawing. *Courtesy of Dover, Kohl & Partners with data provided by the City of Charleston*

BENEFIT STREET, PROVIDENCE, RHODE ISLAND / DAVID BRUSSAT

Neighborhood Street

The most beloved street in Providence goes by the name of Benefit. It sits midway up College Hill, one of the most beautiful and well-preserved neighborhoods in America, looking over the city founded in 1636 by Roger Williams. From between the houses of Benefit, you look west across the narrow Providence River at the best-preserved mid-sized downtown in America.

Some might say that it looks as though time stood still on Benefit. No. In fact, its stretch of simple clapboard houses and grand merchant mansions runs an architectural "Mile of History," from the late Colonial and Federal periods through the Gilded Age, the Classical Revival, and the City Beautiful—even including a bare smidgeon of the Modern era, which then steps back from Benefit with a rare and becoming modesty.

If this sounds paradisiacal, the blessing is magnified by the physical, streetly character of Benefit and its neighboring thoroughfares, such as North and South Main Street, College Street, and Thomas Street—not to mention the recently built Memorial Park along a short stretch of the Providence River. Benefit Street was laid in 1756–1758. Originally called Back Street, it is up the hill from and parallel to the town's original main street—first known as the Towne Street and since called Main Street. The buildings on the west side of Main Street ran along the Providence River, which was paved over in the nineteenth century but (as we shall see) uncovered again in the 1990s.

Benefit, which lies generally flat but curves gently along the steep topography, was originally intended for houses displaced by the growing bustle of the Towne Street (its commercial district was called Cheapside), before the construction of the Providence Arcade in 1828 in what is now downtown led commerce to the other side of the river. The great houses are mostly on the south end of the street and on its uphill side, from which mercantile princes might gaze from their rooftop window's walks downriver towards Narragansett Bay.

The narrower, northern end of Benefit features a large collection of clapboard houses erected between the Revolution and the Civil War, fronting right up to the sidewalk, with facades of soft pastel coloration. The streetwall is largely unbroken except by steep cross streets, many with views of the Rhode Island State House designed by McKim, Mead & White and completed in 1901. Street trees meet overhead, contributing to the feeling of intimacy. The sidewalks are of brick, and period lampposts add to the sense of history. Cars pass gingerly.

This stretch boasts the first Rhode Island State House (1762, add. 1851). Among many buildings of note are the Sullivan Dorr House (1809), where Thomas Dorr, leader of the Dorr Rebellion of 1842, was raised; the John Reynolds House (1785), where poet Sarah Helen Whitman dallied with Edgar Allan Poe; the John Mawney House (1764), the street's oldest and the setting for one of H.P. Lovecraft's short stories, "The Shunned House"; and the Benefit Street Arsenal (1840), with its twin crenellated towers.

To the south Benefit widens, as does the ambition of its architecture. The John Brown House (1786) was described by John Quincy Adams as "the most magnificent and elegant private mansion I have ever seen on this continent." On the very next block is the Col. Joseph Nightingale House (1791). The First Unitarian Church (1816) is one of several buildings on Benefit by John Holden Greene, a prolific practitioner among early Providence architects.

Between the north and south portions of Benefit stand two blocks of grand civic and institutional architecture. The Providence Athenaeum (1838), a Doric temple by the Philadelphia architect William Strickland, still operates as a private library and is where Poe made love, literarily, to Miss Whitman. The Rhode Island School of Design (founded in 1877), with its string of elegant institutional structures, gives Benefit Street an urbane, classical ambience; the Providence County Courthouse (1924–1933), by Jackson, Robertson & Adams, steps up in three gabled Georgian stages from South Main. Athenaeum Row's five splendid townhouses sit just south of the Athenaeum itself.

Figure 2.156: Benefit Street, Providence, Rhode Island. Looking east on Power Street, from Benefit Street. A portion of the street that has Brown family mansions uphill from Benefit Street. John Quincy Adams described the John Brown House, just out of sight on the left, as "the most magnificent and elegant private mansion I have ever seen on this continent." *Courtesy of Diana S. Faria, creator of the blog Long Island Daily Photo*

Benefit's fine looks today belie its middle years of dilapidation. The well-to-do left Benefit partly because the center of urban gravity had moved across the river in the nineteenth century and because early motor vehicles lacked the horses to climb College Hill. At the depth of Benefit's decline, urban renewal called for the demolition of much of the street. This was thwarted by a study, in 1961, urging a tepid preservationism. It would only have saved some of the historic houses while embracing, without any apparent reluctance, new multifamily housing, including towers, insensitive to the street's historic appearance.

Thankfully, locals led by Beatrice "Happy" Chace, advised by early preservationist Antoinette Downing, purchased historic houses, restored their exteriors, and sold them for next to nothing to families willing to finance their own interior restorations or renovations. With these private activities in full swing, talk of demolition evaporated. In the 1970s, the sidewalks were bricked, and "faux" gas lamps replaced cobra-head lampposts.

"This is not preservation, this is Las Vegas!" said architecture critic Ada Louise Huxtable of these embellishments during a visit around that time. She was wrong. Whether the period lampposts are preservation or beautification, they are equally authentic. A street is for people, for their use and their pleasure. It is not the head of a pin on which experts are invited to dance with theory.

◄ **Figure 2.155:** Benefit Street, Providence, Rhode Island. Looking north from South Court Street.

Which brings us downhill via College Street, Waterman Street, and Thomas Street. All of the fifteen streets climbing down from Benefit to North or South Main have a different sort of charm, small-grained or high caliber. Many are lined by wood-frame houses of the nineteenth century, modest but of considerable variety in style. Perhaps the most extraordinary of these streets are the major connectors, College, Waterman, and Thomas, which run from Benefit down to where North Main becomes South Main at Market Square, next to the Providence River.

College, a relatively wide avenue of two lanes carrying traffic east and west, opens at Brown University's ornate campus gate a block above Benefit. It widens as it crosses Benefit by the Athenaeum, then dips between the Providence County Courthouse and the RISD College Edifice (1936). Both are by the same firm and of Georgian Revival style, large buildings that seem to bend with the gentle curve of the street (though only the RISD building does). They form a gateway of monumental brick bookends between College Hill and downtown. The new brick vehicular bridge that spans this newly uncovered stretch of the Providence River is flanked by two softly arched pedestrian bridges, also of brick, an ensemble that echoes the gateway.

One block north, Waterman takes traffic headed uphill out of downtown, and Thomas, the next block north, takes traffic downhill into downtown. In between, on the slope, sits the First Baptist Church (1775), America's first of the denomination, by Joseph Brown. Adjacent to Waterman, opposite the church, is the entrance to a tunnel built in 1914 to ease the grade for traffic heading up College Hill. It emerges on Thayer Street, near Brown.

Angell Street, College Hill's main westbound avenue, becomes Thomas Street for one block between Benefit and North Main, then becomes Steeple Street before hopping another elegant new bridge into downtown. Stepping down Thomas are the Providence Art Club's four magnificent townhouses, spanning 1786–1885, forming a lovely backdrop for the church. However, vehicles hurtling downhill and around the curve at

Benefit have turned Thomas into a danger zone for pedestrians that trumps the block's beauty for many who travel its lovely, slender, unprotected, treeless brick sidewalk. The church railing often bears mangled evidence of Thomas's hazard.

South and North Main both head north by such splendid buildings as the courthouse—with an impressive colonnade on this side—and the church, but many more as well, including the Old Stone Bank (1898) and the ogee-gabled Joseph Brown House (1774). Modernism has been bolder on Main than on Benefit: Old Stone Square (1985), by Edward Larrabee Barnes, and RISD's recent museum addition, by Rafael Moneo, diminish the waterfront's beauty without adding much, if anything, to its interest.

Between South Main and the river lies Memorial Park (1996), dominated by Paul Cret's World War I memorial. The park is the southern terminus of a new riverfront created in the 1990s when the state "daylighted" the river, which had been covered by streets since the nineteenth century. The widest bridge in the world (1,147 feet, as noted in the Guinness Book of World Records), has been replaced by twelve elegant bridges and a river walk terminating at Waterplace Park. The new riverfront's traditional style is almost as much a miracle as its passage through local, state, and federal bureaucracies (pushed by its designer, Rhode Island architect and planner William D. Warner) as an adjunct to an earlier major redevelopment/transportation project.

Miraculous well describes the survival of Providence's urbane beauty during a period when so many other cities mutilated some of their loveliest features. The territory between Benefit and the Providence River may show off the blessings of the T3–T4 transect at its best—please excuse the New Urban nomenclature—but it is only the beginning of this small city's large *civitas*.

A street is for people, for their use and their pleasure. It is not the head of a pin on which experts are invited to dance with theory.

LOCAL NEIGHBORHOOD STREETS, REDISCOVERED

Figure 2.157: Main Street, Southport, Connecticut (also see Figure 1.13).

Southport has a classic New England Main Street. It begins at a small public square facing the Southport Harbor. Up the hill are houses of various sizes and styles, sometimes close to the street and each other, sometimes not. Beautiful, mature trees line the street. Figure 1.13 shows a nearby street in the neighborhood.

Bronxville is a small railroad suburb north of New York City. Developer William Van Duzer Lawrence laid out the streets and built most of the housing (page 303). It has some of the highest real estate prices in the New York region, reflecting market demand for the housing. It's worth noting, therefore, that in addition to large, single-family houses, Lawrence built conventional urban apartment houses, courtyard apartment buildings, pedestrian courts, rowhouses, and small attached houses like the ones just visible in Figure 2.158 that were grouped in buildings of multiple sizes. He usually arranged those around greens facing the streets. He even put trees in the middle of the street.

Figure 2.158: Willow Street, Bronxville, New York (also see Figure 3.61).

CANYON ROAD, SANTA FE, NEW MEXICO / STEFANOS POLYZOIDES AND JOHN MASSENGALE

Neighborhood Street

The people of Santa Fe are justly proud of their rich and heterogeneous cultural heritage. Native Americans lived a thousand years ago where the city now stands; Spain and then Mexico later occupied the area for several centuries before the Americans took the territory in 1846. The history produced several architectural influences: Pueblo, Spanish Colonial, and Territorial. And yet so much has either been built or significantly altered since New Mexico began preparations to become a state in 1912 that almost everything we see in downtown Santa Fe can be described as new since 1910. That's why we have called it elsewhere "the most beautiful city of the twentieth century."[51]

> That's why we have called Santa Fe "the most beautiful city of the twentieth century."

Even the Palace of the Governors—the "oldest continuously inhabited building in the United States"—was dramatically renovated by the United States Army in the nineteenth century while Santa Fe was still part of the New Mexico Territory. In 1910, architects remodeled the Victorian fantasy and created a new "historical" version never seen before, giving us the building we have now.

In 1912, the city fathers prepared a master plan for the city that mandated the use of the regional architectural style we now associate with Santa Fe. When the Palace of the Governors was rebuilt in 1913, it served as one of the style's first examples. Based on both Native pueblo and Spanish Colonial architecture, this combination eventually evolved into two official Santa Fe styles: Territorial and Spanish Pueblo. The first was based on the character of native pueblo architecture, but with Spanish Colonial overtones. Thick walls, rounded adobe forms, and *portales* come from the local tradition. A thin Classicism mainly expressed by white trim around doors and windows adds the Colonial influence. Spurred by the expectations of tourists and fueled by boosters and promoters, a second style emerged a little later, based more literally on Pueblo architecture precedents. These included volumetric compositions, earth-hugging profiles, deepwalls,

portales, and spare openings. A city preservation ordinance in 1957 mandated the use of either the Territorial or Spanish Pueblo style for all construction in more than half the rapidly expanding city.

Far away from quarries, in an arid climate next to the high desert, Santa Fe was originally built with adobe bricks, covered by mud troweled on wet. Dried mud slowly washes away in the rain and must be regularly repaired (which accounts for the rounded forms of the architecture of the native pueblos). Because of the high maintenance required on adobe buildings, once freight trains arrived, most projects in the twentieth century were finished with stucco plaster rather than with local adobe. Stucco is just as malleable as mud, but doesn't dissolve in the rain. Tinted with the colors of the local earth, it is magical in the famous desert light of Santa Fe.

Initially, the stucco was always applied over an adobe brick structure, later over a concrete frame with tile or concrete block infill, and, lately, increasingly over highly insulated wood-frame and plywood structures. This is Santa Fe's default building technology. It provides good thermal mass and thermal gain or loss protection, in a climate where the nights can be cold and the days hot. It is a simple and durable technology, and the resulting architectural forms are well understood and loved in Santa Fe. Good, new interpretations of the Santa Fe Style architecture are characterized by simplicity, honesty of construction, and beauty of materials and form.

Canyon Road is a half-mile-long old country road that comes into Santa Fe's downtown from the east. Built long before earth-moving equipment was available, it gently follows the natural topography, sometimes flat and sometimes sloped, taking the path of least resistance. Its right-of-way roughly parallels the *Acequia Madre* (Mother Ditch) that flows into the city from the Santa Fe Mountains. The overall urban form of Canyon Road is that of a traditional New Mexico *Cordillera* village. Similar linear villages, often one building deep, edge many New Mexico rural roads.

The early houses on Canyon Road are eighteenth-century farmhouses that were later folded into family compounds encouraged by local patterns of inheritance established under Spanish rule. When New Mexico became a state, a number of American artists moved to Santa Fe, and many settled on Canyon Road. Galleries followed them, and a second generation of artists

Figure 2.159: Sketches cataloging frontages on Canyon Road, Santa Fe, New Mexico. Stefanos Polyzoides. © 2012 Moule & Polyzoides, Architects & Urbanists

and galleries arrived in the 1940s, attracted by what was then known as the Santa Fe Colony. In 1962, the city designated the road a "residential arts and crafts zone."

The most extraordinary characteristic of the urbanism of Canyon Road—and the subject matter of the sketches in Figure 2.159—is the way the frontages of the buildings function. A frontage is a threshold device that mediates between the private realm of each house and the public realm of the street it fronts. On a short two-hour visit, Stefanos Polyzoides documented no less than twenty frontages on Canyon Road. This catalog has a rare and exceptionally rich array of frontage types that are essential ingredients in Santa Fe's distinctive building design and memorable urban placemaking.

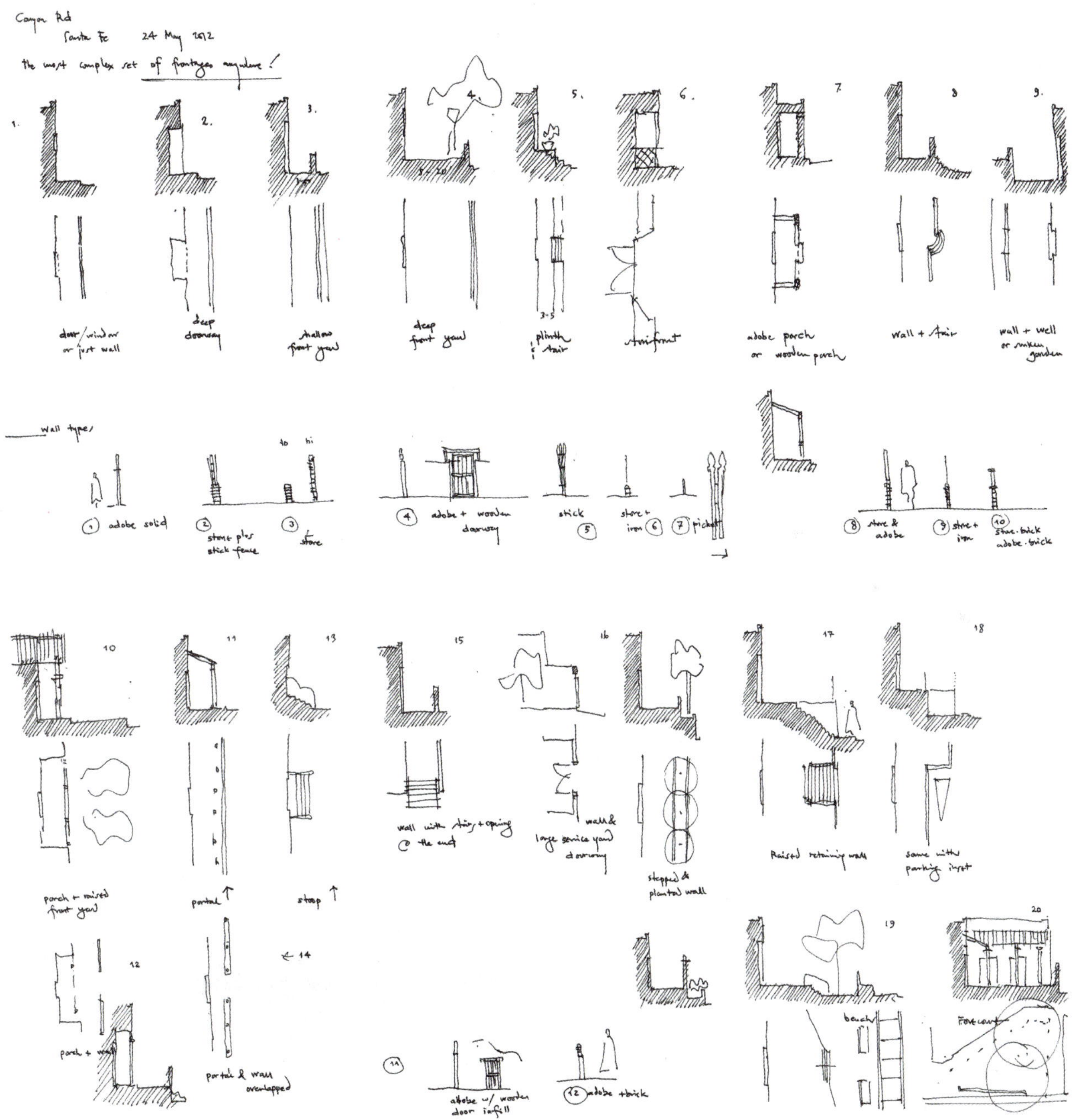

Looking at why Canyon Road has so many types of frontages can be instructive:

- Canyon Road was sensitively sited on hilly topography. The variety of the frontages respond to flat and hilly lots.

- Hispanic New Mexican society traditionally valued both family privacy and community form—creating a pattern of a walled residential compounds organized around internal courts and balanced by an active, public realm.

- Northern New Mexico's extremes of climate are notable. Many of the frontages serve to moderate the local climate through the seasons, providing cover from rain and shade from the sun.

- Many Canyon Road owners supervised the construction of their houses. The variety and spontaneity of the designs are the result of the freedom these individuals enjoyed in choosing between making thresholds for their particular house and its particular place on the street— whether on a flat spot, a terminated vista, an uphill or a downhill site, a large lot, or a tiny one.

- Until the advent of statehood, there was no regulation of use or form in this part of Santa Fe. Frontages were, therefore, initially designed to accommodate residential, local commercial, local craft, agricultural, and other uses. Over time, and as building uses changed, owners modified the forms based on their evolving needs. The visual complexity, the picturesque composition, and the sheer beauty of Canyon Road give us an essay on successional urbanism, based on the small contributions of hundreds of people.

- The width and geometry of the street reflect its origins as a thoroughfare built to accommodate animal-drawn carts, people on horse, mule, or donkey-back, and pedestrians. The charm of Canyon Road's insistent irregularity reflects not only that the design of its right-of-way evolved over time, based on the ever-changing needs of its residents, but also that the frontages mediate between the form of the street and the fabric of its buildings.

When the city's economy was poor, Canyon Road was one of the hippie capitals of the West, with inexpensive rentals. Today, Canyon Road is thought of as one of the most beautiful places in America and Santa Fe has a thriving high-end real estate market and a treasured architectural heritage. But Santa Fe's buildings are not valued because they have been preserved in amber. In fact, as we said, most of what we see dates from the twentieth or twenty-first centuries. The buildings are valued because the city's inhabitants have developed a disciplined yet organic architectural approach to the region's extraordinary cultural legacy.

Figure 2.160: Santa Maria Street, Coral Gables, Florida. F. M. Button and Pierson & Skinner, 1925. Originally conceived by Coral Gables founder George Merrick as a Neoclassical Village, Santa Maria Street is the archetypal American "T3" street.

Santa Maria Street, Coral Gables, Florida

F. M. Button and Pierson & Skinner, 1925

American Garden Street

Santa Maria Street is an appropriately scaled street in a suburban neighborhood (Figures 2.160 and 2.161). The street is not too wide; it measures approximately twenty-one feet across, which is just narrow enough to moderate traffic speed, but it could be narrower. Slow speeds encourage pedestrian and cyclist activity.

From the perspective of a cyclist, a neighborhood street like Santa Maria that is well shaded by large canopy trees and often has "eyes on the street" (or residents in their front yards, casually monitoring road activity) is ideal for biking. Noticeably absent from this cycling-friendly street: separate zones or lanes marked with brightly colored paint, designated for bike use only. On an adequately sized suburban street where beauty is seamlessly integrated with function, specialized lanes are not needed as long as cars drive slowly (Figure 2.162). Noticeably present on Santa Maria: sidewalks with an informal appeal. The curbless edges are safer and create a pleasant aesthetic. The sidewalks measure five feet—a typically narrow dimension—but the lack of curbs at the edge of the pavement *and* at the edge of the sidewalks enables pedestrians to casually overflow into the adjacent swale.

Figure 2.161: Santa Maria Street, Coral Gables, Florida. F. M. Button and Pierson & Skinner, 1925. Wide swales have permitted the trees to grow massive by South Florida standards, giving year-round shade. The front yard fences define the private space with elegance, as a natural extension of the architecture of the houses.

Figure 2.162: Santa Maria Street, Coral Gables, Florida. F. M. Button and Pierson & Skinner, 1925. In the single-family residential sections of traditional North American neighborhoods, prestigious and valuable streets almost always follow a time-tested recipe: street-oriented architecture, spatial definition, agreeable proportions, slow speed, narrow width, and trees lined up to make a canopy over the street.

La Alameda de Santa Rosa, Antigua, Guatemala

American Garden Street

Antigua Guatemala has three *alamedas*, the most exceptional of which is the Alameda de Santa Rosa (Figure 2.163). Like the other two, Santa Rosa is paved with cobblestones. Vehicles and pedestrians co-exist within the space; motorists drive slowly because the surface of the street is not smooth and flat. The street has a steep inverted crown, draining towards the center rather than the sides. While a bit disconcerting, this undoubtedly also slows down drivers.

While Quinta Avenida (Figures 2.107 to 2.109) is the civic emblem of Antigua, the streets in the city with alamedas are immediately noticed by visitors because of their rarity. Most streets in the historic center do not have trees, so Santa Rosa stands out. Existing within the pattern of the large, regular grid, La Alameda de Santa Rosa has wide swales at each edge of the street. Narrow sidewalks are located next to the swales, and many buildings and garden walls have been constructed immediately adjacent to the sidewalks, interrupted occasionally by *zaguans* or tall, wide gates, as is customary in Spanish Colonial towns in the region. Antigua has become famous for its growing number of Spanish-language immersion schools, and at the beginning and end of each day the sidewalks of La Alameda de Santa Rosa are filled with foreign students finding their way to and from all-day conversation sessions.

The mature trees that line the street do not simply add "columns" to the public room with their trunks. The canopy they create provides a sense of "ceiling" enclosure overhead as well, making the wide street feel even more contained.

Figure 2.163: La Alameda de Santa Rosa, Antigua, Guatemala. A rare exception in the city's streets, its formal rows of trees grow in wide swales on both sides of the inverted-crown roadway; high garden walls with *zaguans* line the sidewalks (seen at right).

YIELD STREETS AND GARDEN STREETS

Yield Streets

Yield streets are two-way streets so narrow that one car must pull over to let a car going the other way pass by. England has many examples, on narrow country lanes, in old villages, and in cities. (Figure 2.186). Pictured here is Balfour Place, in Mayfair, London (Figure 2.164). Out in the countryside, many roads between hedgerows are so small that they are barely wide enough for one car, let alone two (Figures 2.165 and 2.187). Pullovers are made by the side of the road every mile or two, or when visibility is particularly bad.

American developers dealing with truculent bureaucracies demanding wide roads have been known to build one-way streets with parking on both sides, taking down the one-way signs after all permits are issued. A more legal approach, when building privately deeded roads, is to use local standards for parking lots and alleys: a yield street can frequently be built by labeling it "parking" or "alley," because the regulations for those tend to be less restrictive.

Figure 2.165: Deep Lane, South Hams, England. A yield street in the country just outside the city of Plymouth. The hedgerows occasionally leave space to one side or the other to pull over and wait for a car to pass.

Figure 2.164: Balfour Place, London, England. View looking north towards Mount Street. The 'yield street' is a time-honored way to handle light two-way traffic without an unduly wide roadway.

English Garden Streets

▲ **Figure 2.166:** Knightsbridge, London, England. Satellite view. The developers of Georgian London perfected variations on a unique and useful form, in which the streets are stretched and planted with tall trees as lush, elongated parks, faced by terrace houses. The maneuver simultaneously made the city greener and healthier, added shared outdoor open space to offset the scarcity of private gardens on the compact house lots, aided with storm-water management, and created special addresses of enormous value. Beaufort Gardens is an example of these English Garden Streets. The wide street at the end of Beaufort Gardens is Brompton Road. *Courtesy of Google Earth, © 2012 Bluesky*

◀ **Figure 2.167:** Cornwall Gardens, London, England. Satellite view. Cornwall Gardens is a variation on the private squares of London, showing how flexible the type can be; other examples are bent, mashed, squeezed, and adapted to great effect. Despite some blurring of the usual spatial distinctions between street, square, and park, they remain in the end streets, perhaps owing to the fact that the architecture shaping them is consistently street-oriented and resolutely urbane. In architectural terms, the way the large houses at the western end of the street are "plugged into" the park creates a valuable variety in the character of the gardens. Two houses have private yards on the private park (which in most cities would be a public park). *Courtesy of Google Earth, © 2012 Bluesky*

◀◀ **Figure 2.168:** Beaufort Gardens, London, England. George Adam Burn, architect, Jeremiah and Henry Little, builders, 1861–1870. Looking north towards Brompton Road. Beaufort Gardens is just one block long, functioning like a Close. A single row of immense trees hovers over the space, which, instead of a green, has a beautiful place for parking and walking.

▶**Figure 2.169:** Beaufort Gardens, London, England. George Adam Burn, architect, Jeremiah and Henry Little, builders, 1861–1870. Looking north towards Brompton Road. At four stories plus attics, the street has a height-to-width ratio of approximately 1:1.5. The graceful houses prove that repetitive designs can indeed be agreeable.

PEDESTRIAN PASSAGES AND STEP STREETS

Much of the focus of urban design is on the big, important streets, but there is wonder, promise, and richness of skinny and special streets too. Small laneways, pedestrian passages, canal streets, and step streets can be useful and enriching additions to the mix of streets in any neighborhood. These exceptional streets can be a way to solve problems, like breaking down oversized blocks into walkable ones, taming topography too steep for vehicles, or simply squeezing in a direct route to make things convenient in circumstances where a full-size street just won't fit. A standout little street is at times just the right recipe for an address of distinction; for example, the tony Burlington Arcade, a calmly elegant shopping street in London, is wholly unlike its bustling surroundings, which is precisely the point.

These streets benefit a city in many ways, besides the practical business of providing a direct route. The skinny streets offer up alternative addresses, expanding the choices in local real estate. They can be used to convey prestige (by setting up especially tranquil home sites) or to enable affordability (by using a low-cost type of infrastructure while getting a few more buildable lots into the block). The urban designer might also employ exceptional little streets to showcase special views, or to overcome a topographic feature that threatens the urban framework's continuity.

But the biggest benefit is variety. The little streets are inherently human-scaled, and they introduce contrasts into what might otherwise be a relentless or boring set of similar spatial experiences.

The skinniest charming streets make a great argument for a fine-grained grid and tightly wound intersections. Most of these exceptional streets are either pedestrian-only or allow just one-way traffic; neither pattern is recommended as the norm for a whole neighborhood (Figure 2.170). When dealing with a well-connected grid, however, one can afford to demote a few links to a less-trafficked status. The result can elevate the quality of the whole urban ensemble.

 Quince Street, Philadelphia, Pennsylvania. One of Philadelphia's Trinity streets, named for the tiny, three-story (Father, Son, and Holy Ghost) houses that line them. Each floor has one room: the three floors combined amount to 400–700 square feet. Built as mid-block worker housing, the oldest may date from 1720. *Courtesy of Paul Murrain*

The Ninety-Nine Steps, Charlotte Amalie, St. Thomas, United States Virgin Islands

Step Street

Charlotte Amalie has one of the largest collections of stepped streets in the world, forty-five in total. Part of the original Danish layout of the town, the staircases extend the ordinary streets up slopes too steep for vehicles. (Some historians think the pattern originated with designers back in Denmark who specified the street grid without realizing the site was so hilly.) Hardy pedestrians have more options than motorists on the hillsides above the harbor: long views gradually unfold as they climb the stairs. Over the past fifty years, many of the stepped streets fell into disrepair and disuse. The Ninety-Nine Steps—now a major landmark and tourist destination in the U.S. Virgin Islands—was the first of the stepped streets to be restored and is one of the most prominent examples of this unique street type (Figure 2.171).

In 1998, developer Michael Ball and designer Felipe Ayala restored the Ninety-Nine Steps as part of an upgrading of the Blackbeard's Castle hotel, welcoming visitors deeper into the town and showcasing the views from the hillside.[52] They cleared the path, reconstructing the steps where necessary. (Note, there are one-hundred-and-three of them, not ninety-nine.)[53] Next, they strategically placed plantings that frame the space with color and texture. They renovated houses on both sides of the street and filled vacant lots with buildings that open directly onto the staircase. This process continued, house by house, right up to the present. Today, the Ninety-Nine Steps form not just a path up the hill but a place in its own right.

Figure 2.171: Ninety-Nine Steps, Charlotte Amalie, St. Thomas, U.S. Virgin Islands. The step street is an alternative to full-sized streets where topography is too demanding.

Figure 2.173: Longitude Lane, Charleston, South Carolina. The passage cuts one of Charleston's oversized blocks down to size, making the neighborhood more permeable and more interesting.

Longitude Lane, Charleston, South Carolina

Passage

Longitude Lane averages ten feet wide (Figure 2.173). The narrow passage begins at Bay Street and goes west. Lined with houses and residential compounds, it has a brick and stone hardscape that unifies the street.

One of the skinniest streets in America, Longitude Lane is an example of a type that is an integral part of the urban fabric in many cities of the world. Like the picturesque "laneways" in Melbourne, small streets in the Gothic Quarter of Barcelona, or ancient passages in Rome, Longitude Lane complements the larger network of roads in its city. The narrow passage adds a surprise in the walk around town, a welcome contrast with the full-size streets.

Figure 2.172: Vennel Street, Edinburgh, Scotland. View looking north towards Edinburgh Castle. Also called "The Vennel," the street is more than a shortcut. The step street is a destination, with a view that makes it popular with wedding photographers and visitors recording their trip on social media. *Courtesy of Sophie Pearce – Third Eye Traveller*

MIDBLOCK PASSAGES

Figure 2.174: Burlington Arcade, London W1, England. Samuel Ware, 1819. Between Piccadilly and Burlington Gardens, parallel to Old Bond Street. The Burlington Arcade cuts through a large block in Piccadilly that was part of the London estate of a wealthy aristocratic family with many titles and names, including Burlington (the Royal Academy of Arts next door is their former house). The Arcade is a convenient passage that also has a lavish architectural expression, making it a natural home for jewelry stores and other luxury shops. © 2005 Andrew Dunn / Wikimedia Commons / CC BY SA 2.0

Midblock passages come in many shapes and sizes, from the informal shortcuts like Longitude Lane (Figure 2.173) to the grand Galleria Vittoria Emanuele (Figure 2.181). Walter Benjamin wrote about the in-between arcades like the Galerie Vivienne (Figure 2.179), seen in London variants like the Burlington Arcade, above): these were part of the life of the *flâneur*, page 103. In the twentieth century, his unfinished but famous work, *Das Passagen Werk (The Arcades Project)* about the *passages couverts* of Paris, became an iconic text of cultural criticism, in which

he outlined the birth of modern consumerism and its "dreamworld." You can see some of his notes at street.design/benjamin. The full English translation of *Das Passagen Werk* is online at monoskop.org.

The passages and arcades break up the monotonous experience of long blocks for pedestrians, adding quiet shortcuts and oases that can also be destinations. In London, Paris, Milan, and Providence, they are some of our favorite destinations. New York's 6 1/2 Avenue is a modern, expedient variation (page 537).

Figure 2.175: Sicilian Avenue, London WC1, England. R.J. Worley, 1910. Looking north from Southampton Row towards Bloomsbury Square Gardens. A popular outdoor space in Bloomsbury, the Edwardian interpretation of an Italian street is a diagonal cut between two buildings designed by Worley.

Figure 2.176: Lansdowne Row, London W1, England. Between Berkeley Street and Fitzmaurice Place. A handy shortcut near Berkeley Square, lined with convenience stores and inexpensive food shops, most of which would be unable to pay the rents on the neighboring streets of Mayfair near Berkeley Square.

Figure 2.177: Crown Passage, London SW1, England. Between Pall Mall and King Street. A convenient shortcut in a long block, with less expensive stores than the surrounding upscale streets.

Figure 2.178: Stratford Studios, London W8, England. A quiet, pedestrian cul-de-sac. The cozy space gives variety to the West London neighborhood.

Figure 2.179: Galerie Vivienne, Paris, France. Between the place de la Bourse and the rue des Petits Champs. A popular Parisian *passage couvert*. © 2011 Groume / Wikimedia Commons / CC BY SA 2.0

Figure 2.180: Cour du Commerce Saint-André, Paris, France. Between the rue Saint-André des Arts and the boulevard Saint-Germain, looking towards the boulevard Saint-Germain. The Cour is an old street, closed to traffic. The large wooden gates at each end lend it the air of a "secret" place—even when it is crowded.

Figure 2.181: Galleria Vittorio Emanuele II, Milan, Italy. Between the Piazza del Duomo and Via T. Marino. One of the grand shopping places of Italy: a place for the Milanese to see and be seen. © 2008 Alterboy / Wikimedia Commons / CC BY SA 3.0

Figure 2.182: Westminster Arcade, Providence, Rhode Island. Between Westminster and Weybosset streets, known locally as The Arcade. In 2013, a developer converted the shops and offices on the top two floors of this distinctive passage into small apartments.

Figure 2.183: Latta Arcade, Charlotte, North Carolina. William H. Peeps, 1914. Between South Tryon and South Church streets. On the National Register of Historic Places, the Latta Arcade is a popular destination in a city working hard to revive the walkability of Uptown Charlotte (its downtown). The arcade brings variety to the pedestrian experience in the regular grid of Charlotte's center. *Courtesy of Steve Minor*

Figure 2.184: Warren Place, Brooklyn, New York. Between Warren and Baltic streets. A semi-private passage through model housing in Cobble Hill, built in 1878 by the developer Alfred Tredway White, whose motto was "philanthropy plus five percent." The cottages for "working men" originally rented for $18 per month. Today the 14-foot-wide cottages sell for $2.5 million and up.

PRINCIPLES FOR DESIGNING BEAUTIFUL RURAL ROADS

- Follow the contours of the land, as though you have no access to modern machinery for grading. Avoid long, single-radius curves.

- Alternate straight lines with natural curves.

- Make edges on one or both sides of the road with rows of trees, hedgerows, stone walls, or rural fences. Walls and rows of trees can be combined.

- When there is a vista, a single edge or a single edge with a low wall on the opposite side can be good.

- When there is no vista, both sides of the road should have a well-defined edge. An allée with a tree canopy over the road is good.

- Coming into a clearing with a vista is good.

- When the road passes through a clearing, put the road along the edge of the clearing or have a good reason for not doing so.

- Alternating clearings with allées in woods or forests is good.

- Alternating clearings with sequences in allées or forests to create a sequence of vistas is good.

- When there are ridges and valleys, put the road on the ridge or at the bottom of the valley, or pass from one to the other.

- Clearing the hillside so that the line of the top of the hill is visible, with or without trees along the crest, can be good.

- Make the roads drain without curbs.

- Minimize the use of signs and interstate-style striping.

- Do not use suburban arterial elements such as Jersey barriers, center turn lanes, large signs, and modern galvanized light fixtures.

- The pure geometry of modern roundabouts looks out of place in a rural setting.

- When rural roads come into hamlets or villages, the buildings should be close to the road.

- When rural roads have very few buildings, individual buildings may be very close to the road. In that case, they should be parallel to the road.

- When rural roads have frequent buildings, the buildings should either be close to the road or distant from the road. Avoid frequent, repetitive suburban-style setbacks (twenty-five to fifty feet), with parking or poorly defined front yards along the road.

Figure 2.185: Streever Farm Road, Pine Plains, New York. The trees date from the era of the Depression program in Dutchess County that paid farmers to plant allées along country roads.

Figure 2.186: Lane at Littlebredy, Dorset, United Kingdom. The green wall made by the hedgerow highlights the view on the other side of the road. It is precisely because the road has not been widened, straightened, or flattened for the convenience of drivers that speeds are mild and the rural character undisturbed. *Courtesy of Ben Pentreath*

Figure 2.187: Lane, Littlebredy, England. A one-lane yield road in Dorset. *Courtesy of Ben Pentreath*

Figure 2.188: Strada della Stazione, Orvieto, Italy. The road leaving the Umbrian hill town.

Figure 2.189: Via Appia Antica, Rome, Italy. Appius Claudius, 312 BC. "The Queen of the Long Roads." Street design has always defined advanced civilization. Today, visitors who fly into Rome's small original airport can drive into the city along the old Appian Way and then via the Roman Forum. One of the most scenic entries to any large city.

DRIVES

RIVERSIDE DRIVE, NEW YORK, NEW YORK / JOHN MASSENGALE

New York City Parks Commission, 1873, West Side Improvement Plan, 1934, New York City Department of Traffic, 1950.

Frederick Law Olmsted, 1873, Calvert Vaux, 1881, Clarke & Rapuano, 1934.

Drive

Riverside Drive is a beautiful, unique street designed by Frederick Law Olmsted, the father of landscape architecture in America.[54] In the 1930s, Robert Moses created and oversaw a West Side Improvement Plan that made many changes to Riverside Drive and Riverside Park, some good and some bad. Two years after the publication of *Street Design*, my wife and I had the good fortune to move to an apartment on Riverside Drive overlooking Riverside Park and the Hudson River. The experience has taught me that Riverside Drive is one of the world's great streets, full of lessons for designing streets.

Our apartment on the eighth floor is a little above a forest of trees that stretch from our building to the river—some of them planted by Olmsted, we tell ourselves. From our living room, we watch the change of seasons and never tire of the sunsets (Figure 2.190).[55] When we leave the building, we face one of the "islands" designed by Olmsted. These are small parks between the narrow "Side Road"[56] that runs along the front of some Riverside buildings and the wide boulevard to the west.

At the street corner near our front door, a path leads into the island. If we walk west and then across the wide boulevard, we can go down the Carrère & Hastings steps into Riverside Park. Thomas Hastings designed the stairway in remembrance of his architecture partner John Merven Carrère. The partnership designed many great buildings, including the New York Public Library on Fifth Avenue at 42nd Street, one of the most impressive buildings in America (and one of my personal favorites).

Halfway across the island, there is an intersection where the path splits in two. The bushes here are favorite places for House Sparrows, who happily chirp most of the day.[57] Go north at the fork, and you are on a path that continues uphill to a beautiful monument for New York's bravest, called the Firemen's Memorial.[58] I didn't know the memorial before we moved, but it is now one of my favorite monuments in the world.

The combination of the islands, the wide boulevard, the narrow Side Road, the City Beautiful monuments, and the park create a unique urban experience recognized as a National Scenic District on the National Register of Historic Places. Today, the park is a little run down, and the streets are clearly under the control of traffic engineers,

Figure 2.190: Riverside Drive and Riverside Park, New York, New York. Frederick Law Olmsted, 1873; Calvert Vaux, 1881; Clarke & Rapuano, 1934. Looking north from West 91st Street, up the Hudson River to the George Washington Bridge. The street here curves around the Bowl, as seen in the section (Figure 2.193). *Courtesy of Peter Whitehead*

but the national designation recognizes the quality of the design, starting with the original plan by Olmsted. He built into the design a variety and richness that has disappeared from street design since engineers took control of our streets and the roadway became a space devoted to the movement of machines.

History

A brief discussion of the Manhattan grid and the platting of the Upper West Side gives some insight into Olmsted's design for Riverside Drive. The New York Commissioners' Plan of 1811 (Chapter One) laid a grid of streets across Manhattan that ignored the ragged edges of the island: east–west streets and north–south avenues ended abruptly whenever they reached the shorelines.[59] The plan showed Twelfth Avenue on what is now the Upper West Side, but the Hudson River sometimes interrupted it. Today, Twelfth Avenue only exists for fourteen blocks between St. Clair Place and West 138th Street. Several blocks of Riverside Drive sit underneath the Riverside Viaduct (making an interesting urban space).

The Commissioners' Plan called for small parks like Washington, Madison, and Union squares. But between 1820 and 1855, the population of New York quadrupled, the city expanded northward at a rapid rate, and the city and state agreed that New York needed a larger park. In 1853, the New York State Legislature passed the Central Park Statute, authorizing the purchase of 843 acres in the center of the island between 59th and 110th Streets. Four years later, Olmsted and the British architect Calvert Vaux won the competition to design the park. The creation of Central Park spurred development around the park on the sparsely settled Upper East and Upper West Sides. Development happened more rapidly on the East Side than the West Side, then called "the West End."[60]

In 1865, New York Parks Commissioner William Martin proposed a park along the Hudson River to attract development to the far side of the West End. The land there was a combination of farms, country estates, and woods west of Eleventh Avenue (now West End Avenue). Glacial furrows, ridges, and ledges dominated the site, which ran from 72nd Street to 125th Street.

Commissioner Martin wanted a scenic carriage drive above the river, but one year after Martin's proposal, a new Parks Commissioner called for simply extending the grid from the Commissioners' Plan of 1811 over the hilly site and creating a straight avenue that would have required extensive cutting and filling. The projected expense delayed the plans, and in 1873 the Parks Commission hired Olmsted to prepare a new plan for a street he called "Riverside Avenue" (known by that name until 1908). The Commission proposed that the avenue would be high on the hill and the land to the west would be the site of a new "Riverside Park." Tracks for the New York Central Railroad lay along the river at the bottom of the hill, sometimes on the land where the Commissioners' Plan put Twelfth Avenue.

Olmsted changed the future of both the avenue and the park when he made a design that combined the land purchased for the avenue with land to the west designated for use as a park. With a larger site, he could design a street that followed the contours of the hills, saving on earth-moving costs and creating a picturesque route along the Hudson. Construction began in 1877 and proceeded in stages between West 72nd Street and the Manhattan Valley at West 125th Street. No part of the broad, rolling boulevard had a slope greater than 1 in 27, the preferred gradient for horse-drawn carriages.

The Changing Cross Section

One of the beauties of Riverside Drive is the way the cross section through the boulevard, the sidewalks, and into the park regularly changes as it moves from the relatively flat land at its southern end, up and down hills, and around natural plateaus and valleys. Because most development at the time was in the lower third of the island, where the land is relatively flat and often low, the rolling avenue rapidly became a favorite place for drives in the country.

Between 72nd Street and 95th Street today, the boulevard has two traffic lanes and two parking lanes. North of 95th Street, the road sometimes has four travel lanes and two parking lanes. Along the way, the walks on both sides of the Drive come and go and become wider and narrower, responding to the local conditions. This is the opposite of the formulaic section promoted by traffic manuals and the popular Streetmix website, a good site that can easily convince users that an abstract, flattened diagram produces a real street design (see the discussion of "the endless section" on page 40).[61]

In the 1930s, the designers of the West Side Improvement Plan reworked elements of Olmsted's plan, significantly changing it above 95th Street. The New York City Department of Transportation later made changes too, although the records of those changes are buried deep in the city archives, where they are hard to find.

Between 79th Street and the northern end of the original park at Grant's Tomb, a retaining wall and a broad walk—Olmsted called it a "promenade"—separate

Riverside Drive from Riverside Park most of the time. Olmsted thought the steep hills that covered most of the site made the park unusable for recreation, and the proximity to the loud, smoke-belching trains at the bottom of the hills could be unpleasant. But the first seven blocks of the Drive between West 72nd Street and West 79th Street sat on a broad and relatively level bluff above the river, and a sharp drop at the western edge of the plateau put the railroad tracks out of sight. Olmsted took advantage of the conditions by pushing the Drive all the way to the east and then opening much of the flat parkland to the road by making the sidewalk along the western side of the boulevard simultaneously part of Riverside Drive and Riverside Park.

For the first four blocks, Riverside Drive is straight as an arrow as it goes up a slight slope. At West 74th Street, the sidewalk leaves the Drive and eases to the west while the park goes gently downhill. A greensward and then a retaining wall made of cyclopean brownstone blocks separates the walk from the Drive as the path falls away until it comes out again at West 79th Street.

At 76th Street, Riverside Drive turns slightly to the east and goes straight downhill. A small plaza on the western side of the Drive marks the transition from uphill to down. Warren & Wetmore, the great architects of Grand Central Terminal and the New York Yacht Club, designed a surprisingly ill-proportioned horse fountain for the plaza. Because there is no sidewalk on the western side of the street between 74th and 79th Streets, steps lead from the plaza down to the park.

At 79th Street, Riverside Drive and the park sit at the same level. But at that point, the Drive begins a long, gentle rise, while the park goes down into a glacial furrow. From there on, with only a few small exceptions, a brownstone retaining wall and the promenade mark the western edge of Riverside Drive. A row of trees a few feet from the wall makes a sheltered place for benches that sometimes face the park and sometimes face the promenade. Another row of trees next to the street creates a canopy over the walk, which is usually around 20 feet wide. When the distance between the walk and the street is wider, the trees sit on rectangular lawns. When that distance is narrower, rectangles of stone blocks surround the trees. Both zones where the trees sit have various combinations of stone blocks, lawns, and the asphalt pavers used on the walkway.

By the time the promenade reaches 81st Street, it is more than 20 feet above the park on the other side of the retaining wall. The wall stops for a few hundred feet at 83rd Street, where a large stone outcropping known as Mount Tom rises above the street and the park. North of Mount Tom the retaining wall again rises above a steep hill. The name dates from before the creation of the Esplanade, which buried a lot of the outcropping. Edgar Allan Poe lived nearby and liked to sit on the rock.

The Drive continues uphill to a plateau at 88th Street. A paved plaza on the west side of the street subsumes the promenade, creating a foreground for the Soldiers and Sailors Monument built in 1902 at the northern end of the plaza. On the other side of the monument is a

Figure 2.191: Frederick Law Olmsted, Riverside District, New York, New York, 1875. Signed drawing showing an early "outline" plan, and a section and "expanded" plan near West 106th Street. Olmsted's drawing for the Riverside District included early ideas for two carriage drives, a walk, an island (labeled "turf"), and a Side Road. Broadway was still called The Boulevard, and Twelfth Avenue was imagined as a street along the Hudson River with a pier at the end of every city street. The bridle path was not part of the design yet, and the railroad was apparently wished away. *Courtesy of Municipal Archives, City of New York*

Figure 2.192: Riverside Drive, New York, New York. Frederick Law Olmsted, 1873; Clarke & Rapuano, 1934. Looking north from the plaza at the Soldiers' and Sailors' Monument (Stoughton & Stoughton, 1902) at the first of the islands. The Drive circles around the Bowl here before heading down the hill. Sitting on top of hills on the western edge of Manhattan, Riverside Drive frequently has spectacular sunsets.

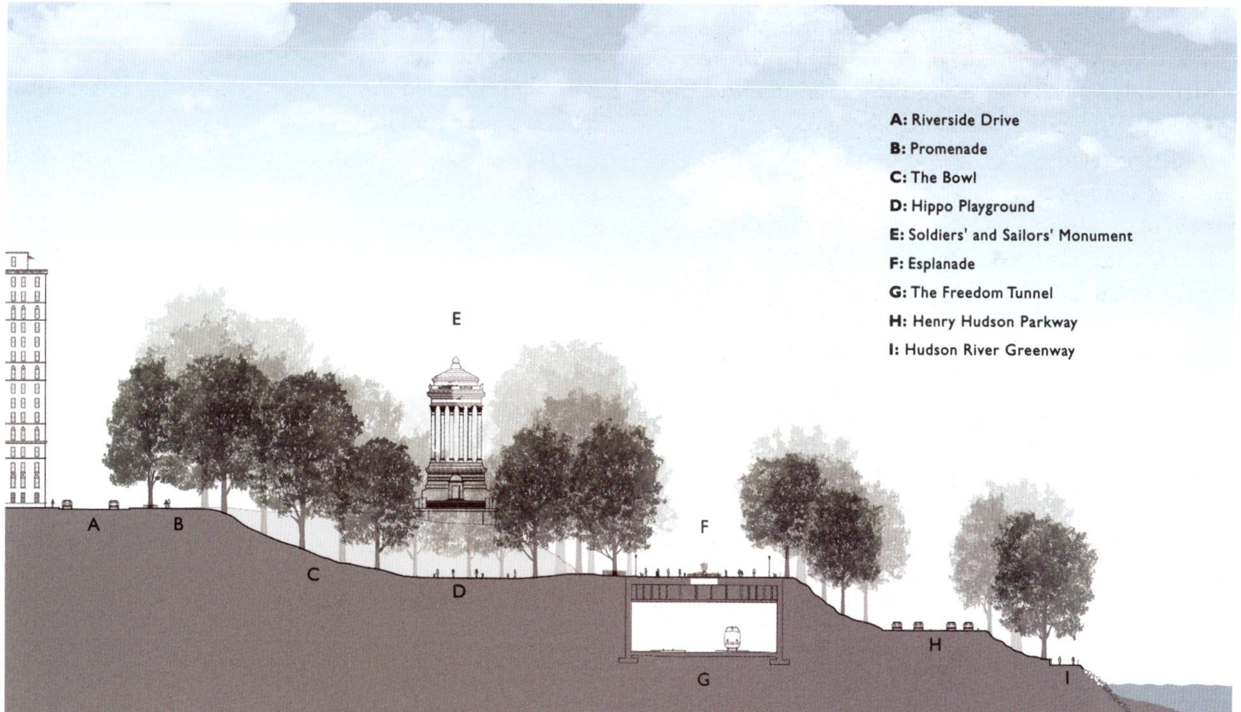

Figure 2.193: Riverside Drive and Riverside Park, New York, New York. Frederick Law Olmsted, 1873; Calvert Vaux, 1881; Clarke & Rapuano, 1934. Section, looking south from West 91st Street. The Soldiers' and Sailors' Monument sits above the Bowl. The photograph in Figure 2.190 was taken from the building of 173–175 Riverside Drive (J.E.R. Carpenter, 1926), seen in the section at the top of the hill and on the right in Figure 2.193. A second section at West 100th Street can be seen at https://street.design/riverside.
© 2024 Massengale & Co LLC and Dover, Kohl & Partners; section drawn by Robin Crowder

natural feature known as "the Bowl," presided over by the Classical temple. Riverside Drive curves around the bowl and then swings west again to head down to 96th Street. The plaza and the curving promenade next to the Bowl are two of the best places for sitting on Riverside Drive.

The Drive edges towards the river as it goes down the hill. On the east side of the street, between 91st and 95th streets, Olmsted created the first island (Figure 2.192). It takes up the slope between the boulevard and the Side Road east of the island. Uphill, at the western end of 93rd Street, a statue of Joan of Arc sits on axis in a plaza in the island, making a formal entrance into the park. New York City placed the memorial in the island in 1915, when the United States was France's ally in World War I.

In the early nineteenth century, a natural valley at West 96th Street ended in an inlet called Strycker's Bay. A bridge over the low ground avoided expensive grading. Late in the century, the city filled the bay, and built docks in the river. In 1900, they built a steel and stone bridge, perhaps designed by Carrère & Hastings (the Carrère Memorial is nearby).[62] Traffic going to and from the docks went under the bridge. The separation of traffic was similar to the separation of traffic on the transverse roads Olmsted designed for Central Park, including one a half-mile east on 96th Street.

Between 95th and 97th streets, Riverside Drive is a wide roadway that feels cut off from the park, and a lack of street trees or greenery of any kind outside the park makes the Drive an uncomfortable place to walk. Historic photographs show a few trees on the bridge one hundred years ago, but they are long gone. When Moses built the Henry Hudson Parkway as part of the West Side Improvement Plan, he put beautifully landscaped and detailed exit and entry ramps at the foot of 96th Street, and a similar exit ramp at 95th Street. But the ramp dumps drivers onto the bridge, making it feel like an auto sewer. Sometimes it is clear that the drivers have had a hard trip, and they bring suburban road rage onto city streets.[63]

Figure 2.194: Riverside Drive, New York, New York, circa 1920 (before the West Side Improvement Plan). Frederick Law Olmsted, 1873, Calvert Vaux, 1881. Looking north from 96th Street. This photo and others from the period show a bridle path beginning at West 97th Street, rather than 5½ blocks farther north, but later photographs show that the lower section of the bridle path was removed by the mid-1920s. Several places discussed and illustrated later are also visible in this view. In Figure 2.202, we show before a redesign for the wide bridge that spans 96th Street, built in 1900. The Side Road that goes up the hill at 97th Street is the location for the before and after views on page 234. The Firemen's Memorial seen in Figure 2.196 is at the top of the hill on the right. The young trees in the island show how differently the islands were treated in the past. *Courtesy of New York Public Library*

The Islands and Their Monuments

At 97th Street, Riverside Drive goes uphill and to the west. For the next seventeen blocks, five more islands sit between the boulevard and the Side Road. Initially, the islands had few trees, bushes, or plants. For years the islands served as public front lawns for the houses and apartment buildings on the Drive.

In the 1960s or 1970s, the New York City Department of Traffic (as it was called then) widened the Side Road that goes behind the island at 97th Street by cutting into the park all the way from 97th Street to 110th Street and building a concrete retaining wall. Either the DOT or the Parks Department put a chain-link fence on top of the wall. Tall bushes behind the ugly fence frequently create a visual barrier between the park and the Side Road (Figure 2.200).

Today, the five islands have four monuments. Two of the islands are small—one of the islands has two monuments, and one of the monuments sits between two islands. New York City and private organizations built the monuments in the years after 1900 when the City Beautiful movement was strong.[65] Our apartment faces the southernmost and biggest of the islands. The smallest and northernmost of them can sometimes be seen in the Amazon Prime Video series "The Marvelous Mrs. Maisel." The fictional Midge Maisel lived in the real Strathmore apartment house on the corner of 113th Street (Schwartz & Gross, 404 Riverside Drive, 1913). A long scene in which Mrs. Maisel argues with a hung-over Lenny Bruce shows both the building and the island in front of it.

"My" island contains my favorite monument, the Firemen's Memorial, designed by the architect Harold Van Buren Magonigle (Figure 2.196). The beautifully proportioned monument evokes Classical *stele* and *sarcophagi*. First intended for Union Square downtown, it was built on Riverside Drive in 1913. The monument fits the site so well that one would never guess Magonigle originally designed it for another site.

Figure 2.195: Equestrian Statue of General Franz Sigel, Riverside Drive at West 106th Street, New York, New York. Karl Bitter, Sculptor; William Welles Bosworth, Architect, 1907. View looking across Riverside Drive towards West 106th Street. General Sigel was a German liberal and patriot who fled Germany after the Revolutions of 1848–1849. The statue sits in the narrow island and visually terminates West 106th Street, one of Manhattan's wide cross-streets. Pedestrians can go down the steps around the statue and cross the Drive, but cars must go to West 104th Street or West 108th Street.[64]

Figure 2.196: Firemen's Memorial, Riverside Drive and West 100th Street, New York, New York. Harold Van Buren Magonigle, architect; Attilio Piccirilli, sculptor, 1913. Looking southeast towards the Side Road from the island between the Side Road and the main boulevard. Only a block north of the author's apartment, it has become his favorite monument.

The east side of the monument abuts the Side Road and visually terminates West 100th Street. Symmetrical steps on both sides lead down to a fountain on the west side of the monument. A large bronze bas-relief above the fountains depicts firemen in horse-drawn carriages rushing to a fire. Larger-than-life allegorical sculptures on each end of the memorial represent "Duty" and "Sacrifice." A bronze plaque embedded in the plaza in front of the fountain reads, "May 1927 / This Tablet Is Dedicated / To The Horses That Shared / In Valor And Devotion / And With Mighty Speed / Bore The Rescue."

For years, the New York City Fire Department held a moving ceremony at the memorial honoring their comrades who died on duty. Hundreds of firefighters in dress uniforms fill Riverside Drive and the broad steps leading up to the monument during the commemoration. Since 9/11, when 343 firefighters from 75 firehouses died, the FDNY has also held an annual ceremony every September 11. An officer reads the name of each firefighter killed and rings a bell before a moment of silence.

On other days, the monument is a happier place. The stele and curving walls of Knoxville granite shape a brick plaza with stone benches that is a popular place for people who live nearby. My wife and I frequently take our morning coffee there when the weather is nice and watch the neighborhood dogs and joggers pass through. The fountain helps drown out noise from the highway in Riverside Park.

Riverside Park and the West Side Improvement Plan

An intelligent discussion of Riverside Drive must include a little discussion of Riverside Park. As mentioned, Olmsted solved the grading problems on the hilly site by annexing some of the parkland to the west of Riverside Drive and creating the islands to the east. Riverside Drive and Riverside Park, which includes the islands, are now part of a single National Scenic Landmark District.

Olmsted thought most of the hills in the park were too steep for park use. That was only one of the problems in the park, which ended at the railroad tracks the Hudson River Railroad built along the river in 1846. In some places, the shoreline west of the tracks had a tangled mess of docks, warehouses, and sheds. The trains themselves were pulled by coal-burning, smoke-belching engines. The prevailing winds blew the smoke up the hill.

When the New York State Legislature expanded the boundaries of the park in 1894 by including all the land to the west of the tracks, the New York Central Railroad (the new owners of the right-of-way) asserted their control by increasing the number of tracks from two to six. West End residents complained loudly. Multiple groups proposed covering the tracks.

By 1913, trains entering Manhattan were increasingly electrified. New York's Board of Estimate and the New York Central Railroad began discussing a plan for roofing over the tracks between 72nd Street and the Manhattan Valley at 125th Street.

But it wasn't until 1924 that all trains were electrified. By that time, the city was building an elevated highway from lower Manhattan to 72nd Street. Advocates for a West Side Improvement Plan pushed to bury both the railroad tracks and the highway in a tunnel north of 72nd Street. The architects McKim, Mead & White designed a train tunnel that looked like a Roman aqueduct, with a Classically detailed highway on the roof.

Robert Moses became Commissioner of Parks in 1934, taking charge of the West Side Improvement Plan. Quickly dismissing the Classical style used by McKim, Mead & White, Moses proposed putting the trains in a tunnel while giving the highway the place of honor along the Hudson. The automobile drivers had some of the best views in the park and beautiful entry and exit ramps.

Moses hired two of his favorite landscape architects to design the park: Gilmore Clarke and Michael Rapuano. They used landfill to extend the park into the river, expanding it by more than 132 acres. The park included playgrounds, playing fields, a marina, and a variety of public spaces like the Bowl, two tree-lined esplanades above the train tunnel, and a riverwalk. The esplanades created flat land where there had been steep slopes or tracks. Today, the tunnel is called "the Freedom Tunnel," because of the graffiti there created by Chris "Freedom" Pape.

Beyond connecting the highway to Riverside Drive with extensive ramps, Moses's team made a few changes to the Drive. The most notable are between West 103rd Street and West 120th Street, where Olmsted had included a bridle path. Clarke & Rapuano widened the promenade there and turned the bridle path into a wide, green buffer between the street and the promenade. Playgrounds and lovely square flower beds at the end of 104th, 106th, and 108th streets interrupt flat lawns edged with low fences that make comfortably proportioned spaces. Rows of trees between the promenade and the lawns shape the walk.

The former bridle path is wider than the broad walkway in the promenade. Another parallel row of trees shapes it too. Altogether, there are five rows of trees on this stretch of the Drive: they define the promenade, the bridle path, the southbound side of Riverside Drive, and the northbound side along the islands. Olmsted designed a diamond pattern for the trees planted in the median between the traffic lanes. The DOT hasn't maintained that pattern: their median is narrower than Olmsted's, and the trees are sporadic. But the combined effect of the trees in the park, along the road, and in the islands is still beautiful.

The trees simultaneously shape spaces, like the promenade, and unify the Drive under a continuous canopy. Walking on the promenade, the leaves and branches of the mature trees in the park below are at eye level. The color of the leaves and the light shining through them changes with the seasons, bringing nature into the city. The rough brownstone wall feels appropriately rural. The broad green space where the bridle path was and an almost-continuous row of parked cars effectively separate people walking on the promenade from the traffic on the street. Today, the islands on the other side of the Drive are all planted with trees, bushes, flowers, and lawns. The result is one of the most pleasant places for walking on the Drive.

The islands to the east of the boulevard vary in width and length, but they make this part of the Drive feel like a parkway, a favorite road type for Olmsted (see page 246). The walks on the islands are interior walks. Except for a couple of carefully chosen exceptions, the islands have no sidewalks along the Drive or on the Side Road. Pedestrians on the Side Road walk on the broad sidewalk along the buildings or in the quiet street.

The section on the Side Road, therefore, has parkland on one side and a preponderance of 10- to 12-story masonry apartment buildings that rise straight up from the sidewalk. Here and there, old mansions and opulent townhouses remain from the early days of Riverside Drive. Many of the apartment houses, like the one we live in, have curving facades that follow the line of the street. Most of the walls in our apartment align with the Manhattan grid, but our living room turns seventeen degrees towards the southwest, because of the curving facade.

For almost twenty blocks above West 103rd Street, the Drive and the promenade alternate between straight stretches with small changes in slope or direction that keep the straight vistas from being long enough to become boring. The land between Riverside Park and the islands is frequently flat and level from the east to the west, but gently tilts up or down in the north–south direction. When the slopes change, the direction of the street changes.

Figure 2.197: Riverside Drive and Riverside Park, New York, New York, date late 19th century. Frederick Law Olmsted, 1873; Calvert Vaux, 1881. Looking north from West 101st Street towards Grant's Tomb, one mile to the north, in the early twentieth century. The bridle path and promenade are on the right, beginning between 102nd and 103rd streets. There are more historic photos of Riverside Drive photos at https://street.design/riverside. *Courtesy of Museum of the City of New York*

Figure 2.199: Riverside Promenade, New York, New York. Frederick Law Olmsted, 1873; Clarke & Rapuano, 1934. View looking north around West 114th Street, in the early spring, in the morning. Riverside Drive and the Promenade curve as the Drive starts up hill towards Grant's Tomb.

At West 114th Street, the drive turns towards the east and briefly goes downhill. At West 116th Street, which runs uphill to the main entrance of Columbia University, there is a small valley. Then the Drive turns slightly west and goes straight uphill.

Between 103rd and 114th streets, the islands are long and narrow, because the rise from the boulevard to the Side Road is less than in the larger islands. When West 104th Street breaks through the islands at a level spot between a curve and a straight uphill slope, the land between the park and the Side Road is strikingly flat, almost forming a plaza. Even the camber of the streets seems inadequate for draining, but appearances are deceiving.

The green median and the trees in the middle of the road stop at West 120th Street, where the Drive loops around a large park. Grant's Tomb sits in the center of the park. High above the Hudson, carriages and riders could loop around the island to go back downtown. Olmsted's Drive ended there, at the edge of a ravine at West 125th Street. Today, Riverside Drive continues over a viaduct

constructed in 1900 and goes all the way to West 181st Street (although technically the street on top is the Riverside Viaduct, and only the street directly underneath is called Riverside Drive).

Altogether, there are eleven monuments between 72nd Street and 125th Street. A statue of Eleanor Roosevelt at 72nd Street is the newest. It was dedicated in 1996 by First Lady Hillary Rodham Clinton. Grant's Tomb is the oldest and most northern of the monuments.

When Ulysses S. Grant died in 1885, more than 90,000 people around the world donated over $600,000 towards construction of a tomb for the former President and his wife Julia Dent Grant (connecting three First Ladies to the Riverside monuments). The architect John H. Duncan designed the largest mausoleum in North America, modeled after the Mausoleum of Halicarnassus. Twin sarcophagi based on Napoleon's tomb at Les Invalides give an "unmistakably military character." Over 1 million people attended the dedication in 1897, presided over by President William McKinley.[67]

Grant's Tomb and the Soldiers' and Sailors' Monument (dedicated in a grand ceremony by President Teddy Roosevelt) are impressive civic monuments. Today, we look at civic statements like these differently than we did one hundred years ago (see pages 14 and 118), but in our time of political division and placeless sprawl, it is worth looking at our examples of "civic adornment" and the City Beautiful movement. Elsewhere in the book, we say that good urbanism and good street design require a combination of richness and order. The Manhattan grid has plenty of order. Riverside Drive is a welcome addition of richness to the grid. Riverside Park and the monuments add richness to the Drive.

Our cities could use more design competitions like the one held in 1909 for a gateway on the Hudson River to Morningside Heights and Columbia University. The winning design by H. Van Buren Magonigle, the designer of the Firemen's Memorial, was for a monumental Classical structure between West 114th and West 116th

◄ **Figure 2.198:** Riverside Drive Promenade, New York, New York. Frederick Law Olmsted, 1873; Clarke & Rapuano, 1934. View looking north on the promenade towards West 104th Street, in the early fall. Riverside Park is on the left. The wide greensward on the right that replaced Olmsted's bridle path buffers the promenade from the green, leafy Drive, making a beautiful place for strolling or walking. Every piece works together. Is there another street like this in the world? In any case, it is a wonderful example of "the work of many hands" and Olmsted's principle of Subordination. There are more Riverside Drive photos at https://street.design/riverside.[66]

streets in Riverside Park, on the river downhill from the main entrance to the new Columbia campus designed by McKim, Mead & White. It was never built, but you can look it up, as the great New Yorker Casey Stengel used to say.[68]

Live from the Drive

Living on Riverside Drive, I've discovered that it has some of the best places in Manhattan for walking and sitting outdoors. Riverside Park is a great place for bike rides. The East Coast Greenway, a walking and cycling route that goes from Maine to Florida, runs along the edge of the park, next to the Hudson River. I can bike seven miles to my office by City Hall and only pass through five stoplights. That's unusual in Manhattan.[69]

Sometimes traffic on the Henry Hudson so-called Parkway comes to a stop, and cars overflow onto the Drive. Normally, however, traffic on Riverside Drive is light, which makes the M5 bus that goes up and down the Drive one of the most useful buses in New York (because it doesn't get stuck in traffic). Instead of walking 9 minutes to the express subway stop at Broadway and West 96th Street, I can ride the bus to the next express stop at West 72nd Street and trade getting to my destination a few minutes later for a calm and beautiful ride along the park. The bus also takes me directly to Lincoln Center, as well as several stores and restaurants I like on Broadway. And in addition to the subway lines beneath Broadway and Central Park West, I can take buses from the Upper West Side to the Upper East Side that move quickly through Central Park, avoiding what would otherwise be three long blocks between Eighth and Fifth Avenues. I think our neighborhood has the best bus service in Manhattan.

In Chapter Five, I talk about traffic problems in New York (page 537). Here on Riverside Drive, we suffer from the congestion, noise, and pollution created by Robert Moses's decision to put a highway along the river.[70] It is a prime example of the way we allowed planners and engineers to give far too much importance to the automobile in America and even New York City. More than three-quarters of the households in my neighborhood do not own a car, and we live at the center of the best transportation system in North America. But the island we live on is ringed by highways and filled with suburban-style, one-way transportation corridors that encourage nonresidents to drive in and out and damage

the quality of life in the city. And not only the *quality* of life, those drivers are a big part of why *New York* magazine architecture critic Justin Davidson, who lives nearby, called our neighborhood "a Bermuda Triangle of pedestrian death."[71]

Climate change and decades of experience with automobiles poisoning the air make us see the car differently than Moses did. We live with his decision to promote cars and driving in the city. People walking in the park listen to the din of traffic noise from a six-lane highway less than a hundred feet away. The prevailing west winds blow across the highway, bringing noise and polluted air into the park.

The Hudson River is west of the park and the highway. Many people have worked for decades to improve the water quality, and CBS News filmed nearly a dozen whales in front of our apartment recently.[72] The park could benefit from the natural smells from the tidal river. Instead, we get unhealthy auto emissions.

Above 100th Street, playing fields run along the highway for a quarter mile. Beautiful mature plane trees shade the playing fields, holding the polluted air under the leaves. The train tracks and the esplanade above make a tall barrier along the east side of the playfield that holds the air in too. In California, the law prohibits building playgrounds or playing fields within 500 feet of a highway.

From late fall through early spring, when the leaves are down, we see from our window a relentless tide of cars coming and going on a highway a little over a thousand feet away. The noise from the highway never stops, even at three in the morning. We have an eighth-floor apartment with three exposures and cross-ventilation (and there can be a lot of wind above the trees coming across the continent), but we're forced to use HEPA-filter screens when we open the windows facing the park and the river. If we don't, we notice the grime on the floors and the furniture.

I love Riverside Drive and Riverside Park. I love seeing nature outside our windows and door. I love walking into the island, seeing the Firemen's Memorial every day, walking on the broad path along the retaining wall, biking along the river, and walking or sitting on the Esplanade.

One might argue that all this nature outside my window isn't very New York. But what is New York except variety and life? And the view out my window is practically the famous Saul Steinberg cover that shows the Midwest and Los Angeles just beyond New Jersey (see note 55). I love traveling and working across America, but I am also happy here on Riverside Drive.

RIVERSIDE SIDE ROAD

After the leaves come down in the fall, one can look south from Riverside's Side Road on the hill between 98th and 99th Streets and see the stoplight on Riverside Drive at 95th Street change color. Fourteen seconds later, a wave of traffic arrives at 97th Street from the Henry Hudson Highway. That's where the Side Road starts again, after Riverside Drive crosses the bridge over 96th Street.

The cars accelerate up the hill behind the island, with only a few going more slowly than New York's citywide speed limit of 25 miles per hour.[73] Drivers from the suburbs, who have almost finished their trip into the city, drive into my neighborhood to find free parking spaces. Many still carry road rage from the highway, because the highways they came in on were overcrowded and unpleasant. And since Covid began, drivers on the highway are more aggressive. From our apartment, we hear the sounds of cars and motorcycles racing on the highway every night. Videos online from dash-cams chronicle young men's attempts to lap the perimeter of Manhattan faster than anyone else.

In front of our building, people walking to the park for peace and quiet or to walk their dogs are frequently honked at by speeding drivers who think the people crossing the street need to get out of their f*****g way, right now! Sometimes the drivers stop to say what they think about people in the street blocking their way.

So here is a modest proposal for the Riverside Side Road that,

- Closes the entrance to the Side Road at West 97th Street by widening the sidewalk on Riverside Drive beyond the width of the Side Road. Cars cannot turn into 97th Street, a one-way street going towards the Drive.

- Turns the Side Road into a two-way shared space between 97th Street and 110th Street. A continuous sidewalk on the east side of the Side Road between 97th Street and 106th Street alerts drivers to the fact that they are entering a shared space.

- Has no crosswalks, stop signs, or traffic signs other than some small historic district signs. Traditionally styled bishop's crook lampposts replace highway-style street lights. This is a shared-space street for people.

- Follows the New York City Parks Department guidelines for Parks Without Borders[74] to make "parks more open, welcoming, and beautiful by improving entrances, edges, and park-adjacent spaces." The guidelines say that all New York City parks should have frequent entrances and barriers between the park and adjacent sidewalks and streets.

- Removes the galvanized chain-link fence separating the street from the park and covers the concrete retaining wall the NYC DOT built when they widened the Side Road with a stone wall.

- Improves the experience in the no-man's land between 95th and 97th Streets by widening sidewalks, creating rain gardens in former street space, and planting trees.

Originally, when the islands had few trees or bushes, they functioned as front yards for the buildings on Riverside (Figures 2.194 and 2.197). The ugly wall and cheap fence the DOT built restricted access points into the park, and the Parks Department planted a green barrier that further separated the island and the Side Road. North of 100th Street, the breaks in the wall are farther apart, and the plantings larger and more difficult to see through: a few blocks feel more like a tunnel than a narrow street next to a beautiful park. When crime rose in the 1970s and 1980s, that separation made the islands and the Side Road more dangerous. Not surprisingly, the Parks Department now refers to the Side Road as "the service street," like the access roads next to highways.

Riverside Drive and Riverside Park are on the National Register of Historic Places and are New York City Scenic Landmarks. The changes the Department of Traffic has made over the years are out of character with the place Frederick Law Olmsted designed. Some of the decisions by the Riverside Park Conservancy aren't much better. Their landscape architect doesn't want the islands open to the Side Road. "People will just walk wherever they want," she says—as they did when Olmsted was alive.

Making a human-scaled, shared-space street next to the islands will bring the Side Road into the park, unifying the public space between the promenade and the buildings facing Riverside Drive.

Figure 2.200: Riverside Drive Side Road, New York, New York. View looking north towards 99th and 100th streets. As discussed on page 233, this is what the street looks like today (although north of the Firemen's Memorial at 100th Street, the distance between the entrances into the park increases). Figure 2.201 reimagines it as a shared-space street. There is an animated GIF and more information at https://street.design/riverside. © 2024 Massengale & Co LLC, rendering by Dover, Kohl & Partners

Figure 2.201: Riverside Drive Side Road, New York, New York. View looking north towards 99th and 100th streets. As discussed on page 233, this image shows the street reimagined as a two-way shared space alongside the Riverside Park island. Changes include: a stone wall that sits on the curb camouflages the concrete retaining wall; human-scaled lamps replace highway-scale lighting; stop signs, traffic signs, and crosswalks are all gone; a continuous sidewalk announces the slow zone to drivers from the side streets; and new entrances grace the park. There is a GIF and more information at https://street.design/riverside. © 2024 Massengale & Co LLC, rendering by Dover, Kohl & Partners

Figure 2.202: Riverside Drive, New York, New York. Satellite view showing proposed changes to Riverside Drive between 95th and 97th Streets. The design adds trees and rain gardens to the wide roadway, as well as a rain garden on the western side of the bridge over 96th Street. See page 226 for more information. Figure 2.194 shows the condition of the bridge in the 1920s. Today, the DOT has lots of white and yellow paint that clearly marks the vast space as a place for machines. (Larger before and after views can be seen at https://street.design/riverside.) © 2024 Massengale & Co LLC, rendering by Dover, Kohl & Partners

PARKWAYS

Of course I litter the public highway. Every chance I get. After all, it's not the beer cans that are ugly; it's the highway that is ugly.

— Edward Abbey, *The Second Rape of the West*

America's long-distance, high-speed roads were not always ugly. Before the interstate highway, we had parkways, a collaboration between engineers, architects, and landscape architects that were beautiful, by design. A new type of road invented in the early twentieth century (different than the "parkways" Frederick Law Olmsted and others built), they allowed city dwellers to escape to the countryside while offering a driving experience that was pleasurable in itself. Parkways were places, not just "transportation."

Bronx River Parkway

The first of these parkways, the Bronx River Parkway, was also the first limited-access highway in the world (Figure 2.203). The parkway runs alongside the Bronx River in northern New York City and Westchester County, New York, the county to the north of the Bronx. Planning began in 1906, and construction in 1913. Originally planned to be 13.2 miles long (the parkway was later expanded to both the north and the south), the Bronx River Parkway opened to traffic in 1922 and was completed in 1925 (Figure 2.204). The beautiful road was set in parkland along a badly polluted river (Figure 2.205). One of the purposes of building the parkway was to remove the sources of pollution along the river and create parks along both sides of it (Figure 2.206).

Particularly important for the design were a German-trained forester named Hermann Merkel and the landscape architect Gilmore Clarke, who went on to

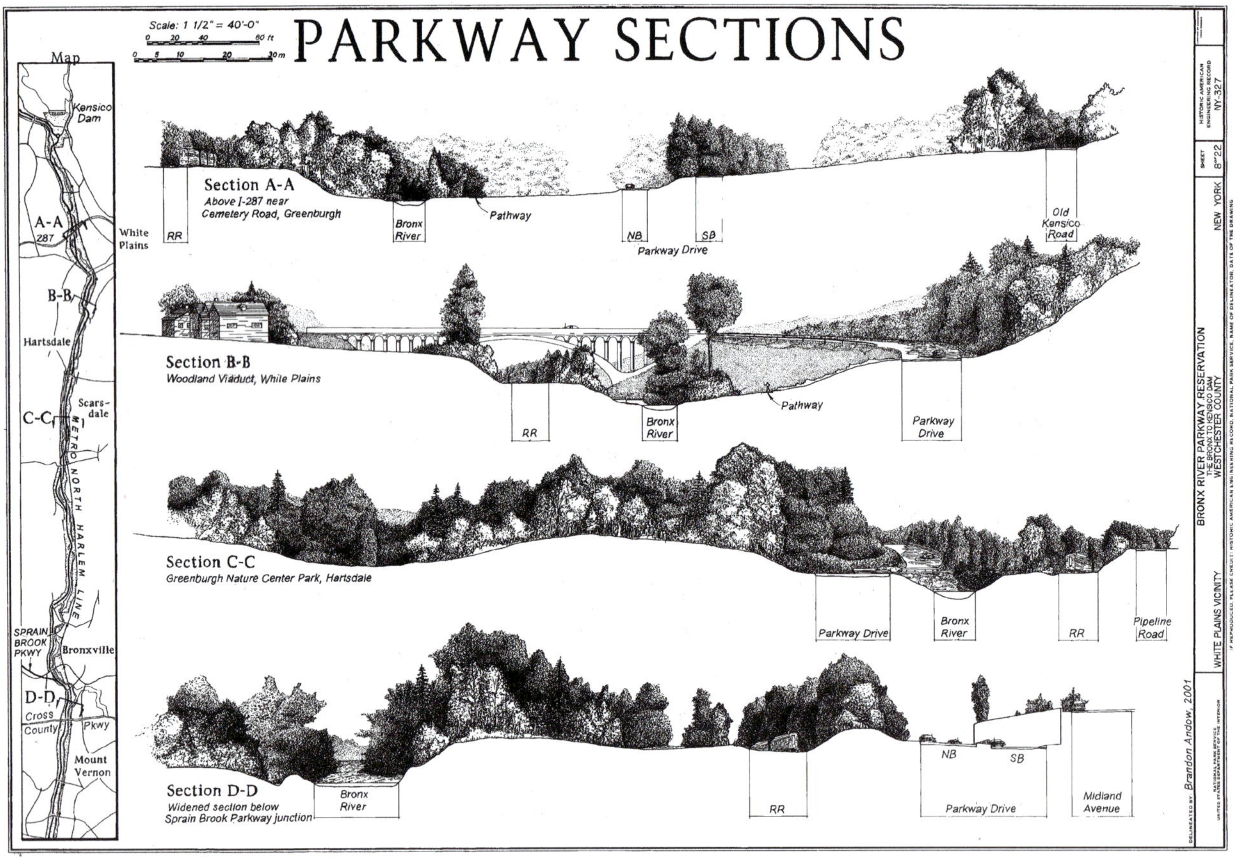

Figure 2.203: Bronx River Parkway, The Bronx, New York, and Westchester County, New York. Gilmore Clarke, Landscape Architect and Civil Engineer; Arthur Hayden, Designing Engineer; and Hermann Merkel, Forester, 1907–1925. Sections of the parkway in the vicinity of White Plains, New York. *National Park Service, Brandon Andow Delineator, 2001 / Public Domain*

Figure 2.204: Bronx River Parkway, The Bronx, New York, and Westchester County, New York. Gilmore Clarke, Landscape Architect and Civil Engineer; Arthur Hayden, Designing Engineer; and Hermann Merkel, Forester; 1907–1925. A 1922 photograph showing an early section of the parkway in Yonkers, New York. Median dividers soon followed. *Library of Congress / Historic American Buildings Survey / Historic American Engineering Record / Historic American Landscapes Survey Collection*

design many other early parkways (and parts of Riverside Drive, page 222). Clarke also praised the Designing Engineer, Arthur Hayden, saying that his "sensitivity to aesthetic matters" was exceptional among engineers.

Working for the Bronx Parkway Commission, the team set the standards for America's early parkways, capturing the public's imagination with this new type of road that combined efficiency and beauty (Figure 2.207). Carefully coordinating landscape design with highway engineering, they created a safe, comfortable roadway that gave its users the feeling of a ride through the countryside, even when town centers were nearby. The parkway helped the environmentalist and planner Benton MacKaye formulate his idea of "the townless highway" (and the "highwayless town"),[75] which was influential before engineers took sole control of the highway building process after World War II (Figure 2.208).[76] The primary concepts were:

- The parkway was part of linear park that buffered the roadway from anything other than the road and the park.

- The roadway and the park looked as though they fit seamlessly and naturally into the existing landscape (including in places where the park landscape was a new creation).

- The roadway was a limited-access highway separated from its surroundings (although early parkways had exceptions).

- Bridges separated local traffic from parkway traffic (with exceptions in early parkways).

- Medians separated traffic going in opposite directions (with exceptions in early parkways). Wide grassy or tree-filled medians became the preferred choice.

The Bronx River Parkway was not strictly a limited-access road. It had seven grade crossings where drivers on the parkway could stop and wait to turn left, before crossing against oncoming traffic. Today, the Bronx River Parkway still has level crossings with stoplights, narrow underpasses, and short entry and exit ramps.

Design precedents for the parkway included English parks, country estates, and New York City's nearby Central Park, designed by Frederick Law Olmsted and Calvert Vaux. Central Park had man-made landscapes that looked rural and natural, limited-access drives, and innovative grade- and use-separation for the vehicles, cyclists, and pedestrians in the urban park. Three

Figure 2.205: Bronx River Parkway, The Bronx, New York, and Westchester County, New York. Gilmore Clarke, Landscape Architect and Civil Engineer; Arthur Hayden, Designing Engineer; and Hermann Merkel, Forester; 1907–1925. A 1912 photograph documenting the conditions along the Bronx River before the construction of the parkway. This "before" view from White Plains, New York, shows privies draining into the river. The "after" view in 1915 is shown in Figure 2.206. *Library of Congress / Historic American Buildings Survey / Historic American Engineering Record / Historic American Landscapes Survey Collection*

transverse roads connect the Upper West and Upper East sides of New York without crossing the paths of cars, cyclists, and pedestrians inside the park.

The Parkway Commission identified two other models for study. For grade separation and access to the parkway, they studied the East and West River drives along the Schuylkill River in Philadelphia's Fairmount Park, also designed by Olmsted, Vaux & Co. Between the river and steep slopes in the park, a mix of roads, pathways, trolley lines, and railways came together, crossed by bridges over the Schuykill (Figure 2.209).

Figure 2.206: Bronx River Parkway, The Bronx, New York, and Westchester County, New York. Gilmore Clarke, Landscape Architect and Civil Engineer; Arthur Hayden, Designing Engineer; and Hermann Merkel, Forester; 1907–1925. A 1915 photograph of the Bronx River Parkway Reservation in White Plains, New York, showing the "After" view of the site in Figure 2.205. *Library of Congress / Historic American Buildings Survey / Historic American Engineering Record / Historic American Landscapes Survey Collection*

The second model was Boston's Emerald Necklace, where Olmsted created a linear network of parks and streets that helped drain a fetid marshy area unsuitable for urban development.[77] The Bronx River Parkway was in the Bronx River Reservation, a newly-created park with environmental problems.

As the design progressed, the Bronx Parkway Commission identified key principles for the Parkway:[78]

- The roadway should conveniently accommodate the large amount of traffic expected and display the principal interesting features.

- Bridges should be carefully designed and built for permanence, with architectural treatment in harmony with their natural surroundings.

- Exposed surfaces of bridges, retaining walls, etc., should be of native stone, avoiding formal, cut-stone effects. Only long-span viaducts would have outside surfaces of concrete.

Figure 2.207: Bronx River Parkway, The Bronx, New York, and Westchester County, New York. Gilmore Clarke, Landscape Architect and Civil Engineer; Arthur Hayden, Designing Engineer; and Hermann Merkel, Forester; 1907–1925. A view of the stone bridge built for the Elm Street access road in Tuckahoe, New York. *Library of Congress / Historic American Buildings Survey / Historic American Engineering Record / Historic American Landscapes Survey Collection*

- Natural features should be preserved, which included adequately protecting existing trees.

- Reforestation and landscaping treatment should be carried out along natural lines, avoiding exotic plantings.

- "All objects foreign to, or distracting from, the naturalness of the valley must be hidden by natural objects wherever possible."

- "In planning the planting, therefore, as in the rest of the design, a humanized naturalness should be aimed at, sufficiently diversified to create woodland groups and vistas of all of the types that belong; broad enough that he who runs (or rides) may see; with intimate bits for those who wish to pause; with material prevailingly indigenous, but always suitable to the situation and its requirements."

Regional Parkways

In 1922, Westchester County opened the Bronx River Parkway and announced its plan to build a parkway system (Figure 2.210). "A chain of parks, connected by parkways… would form, as planned, one of the most extensive and elaborate pleasure grounds in the world," the *New York Times* wrote.[79] Three of the four parkways run alongside rivers that flow between the county's glacially formed ridges running north and south: the Saw Mill River Parkway (planned 1923, construction began

Figure 2.208: Bronx River Parkway, The Bronx, New York, and Westchester County, New York. Gilmore Clarke, Landscape Architect and Civil Engineer; Arthur Hayden, Designing Engineer; and Hermann Merkel, Forester; 1907–1925. A 1927 photograph of the Palmer Road bridge over the Bronx River Parkway in Bronxville, New York, designed by Cecil Stoughton. When the trees matured, the buildings seen here disappeared from view. *Library of Congress / Historic American Buildings Survey / Historic American Engineering Record / Historic American Landscapes Survey Collection*

Figure 2.209: Fairmount Park, Philadelphia, Pennsylvania. Robert Morris Copeland, 1855; Olmsted & Vaux, 1867. A racing car in Fairmount Park in 1909. The separation of traffic in Fairmount Park was one of the models for the Westchester County Parkways. *Courtesy of Motor Age / Public Domain*

1925, opened 1926),[80] the Sprain Brook Parkway (planned 1923, delayed until the 1950s),[81] and the Hutchinson River Parkway (planned 1923, construction began 1927, opened 1928).[82] In 1925, the Westchester County Park Commission planned the Cross County Parkway to connect the parkways and cities and towns in the southern part of the county. A later extension turned to the north to parallel the Boston Post Road (U.S. Route 1) and eventually connect to Connecticut's Merritt Parkway (see below).

Following Westchester County's example, New York State, New Jersey, and Connecticut all built parkways in the greater New York region. The first was the Taconic State Parkway, which began at the northern end of the Bronx River Parkway.[83] Planned with the idea that city dwellers could use the road to drive upstate to a series of state parks in the Hudson Valley, the Taconic goes north through Putnam and Dutchess counties before ending in Columbia County. Northern Westchester, Putnam and Dutchess counties have steep hills that sometimes open scenic vistas. Park lands protected the long views.

Franklin Delano Roosevelt, who owned houses in Manhattan and the Hudson Valley, was a long-term advocate for rural parkways. He first argued for a road like it when he ran for State Senate in 1910, but when he later lost the full use of his legs to polio in 1920, he relaxed by touring the countryside in his car and became an even stronger advocate. Two years later, he became Chairman of the Taconic State Park Commission, which supervised the construction of the Parkway. (This became the source of long-running friction between Roosevelt and New York highway czar Robert Moses.[84]) Noted critic Lewis Mumford was another fan of the parkways. Although he frequently wrote about the negative effect highways had on cities, Mumford advised friends and professionals visiting him in Dutchess County, New York, to drive up on the Taconic,[85] which he described as "a consummate work of art, fit to stand on a par with our loftiest creations"[86] (Figures 2.211 and 2.212).

Moses, the infamous New York power broker,[87] built parkways from the 1920s in Manhattan, the Bronx, Brooklyn, and Queens, as well as on Long Island and the western side of the Hudson River in New York and New Jersey.[88] The Henry Hudson Parkway connected Manhattan to the Bronx and the mainland. Built between 1934 and 1937, it was part of the West Side Improvement Plan that rebuilt Riverside Park and Riverside Drive (page 228).[89] North of the George Washington Bridge, the design took full advantage of the topography and two parks in the West Side Improvement Plan: Inwood Hill Park and Fort Tryon Park (where the Met Cloisters museum stands[90]), making the drive down the Hudson River a beautiful way to enter Manhattan. Well-designed exits and entrances all along the route add to the experience. South of the George Washington Bridge today, however, much of the roadway feels like the state-owned, substandard highway it is.

Moses proposed what is perhaps the most scenic parkway in New York, the Bear Mountain Parkway. Built by the Westchester Parks Commission in northern Westchester, the parkway winds up, down, and around steep hills and ravines along the Hudson River on its way to the Bear Mountain Bridge. At the time it opened in 1932, the Bear Mountain Bridge was the first bridge to cross the Hudson north of the George Washington Bridge.[91]

Two other parkways in the New York region deserve mention. The Palisades Interstate Parkway in Orange and Rockland counties in New York and Bergen County, New Jersey is a late parkway.[92] Proposed in the early 1930s by an executive at the Palisades Interstate Park Commission, it was built by Robert Moses between 1947 and 1958. Before that, however, John D. Rockefeller bought and preserved 700 acres of the Palisades cliffs on the western side of the Hudson River. The parkway begins at the George Washington Bridge and runs through the Palisade preserve to the Bear Mountain Bridge, making it a major commuter route to New York City. The Palisades Parkway also connects to smaller rural parkways in the Harriman and Bear Mountain state parks.

The Merritt Parkway connects to the Cross County Parkway at the Connecticut–New York border and continues to the Housatonic River and Route 8 in Stratford, Connecticut.[93] For most of its route, the parkway runs parallel to the Connecticut shoreline, a few miles south. A scenic and much-loved parkway, the Merritt is the subject of several books.[94] It shares two qualities with some of the other parkways mentioned here: it is on the National Register of Historic Places and it does not allow trucks or commercial vehicles. Many SUV, commercial van, and pickup truck drivers ignore the ban, but unlike some of the parkways, the Merritt is relatively well preserved. It is worth a visit today, like all the parkways mentioned here.[95]

From the Townless Highway to the City Beautiful Parkway

It is difficult to draw a tidy conclusion to a succinct discussion about parkways. First, only two cities in the world built large networks of limited-access parkways, New York and Washington (see below), and their

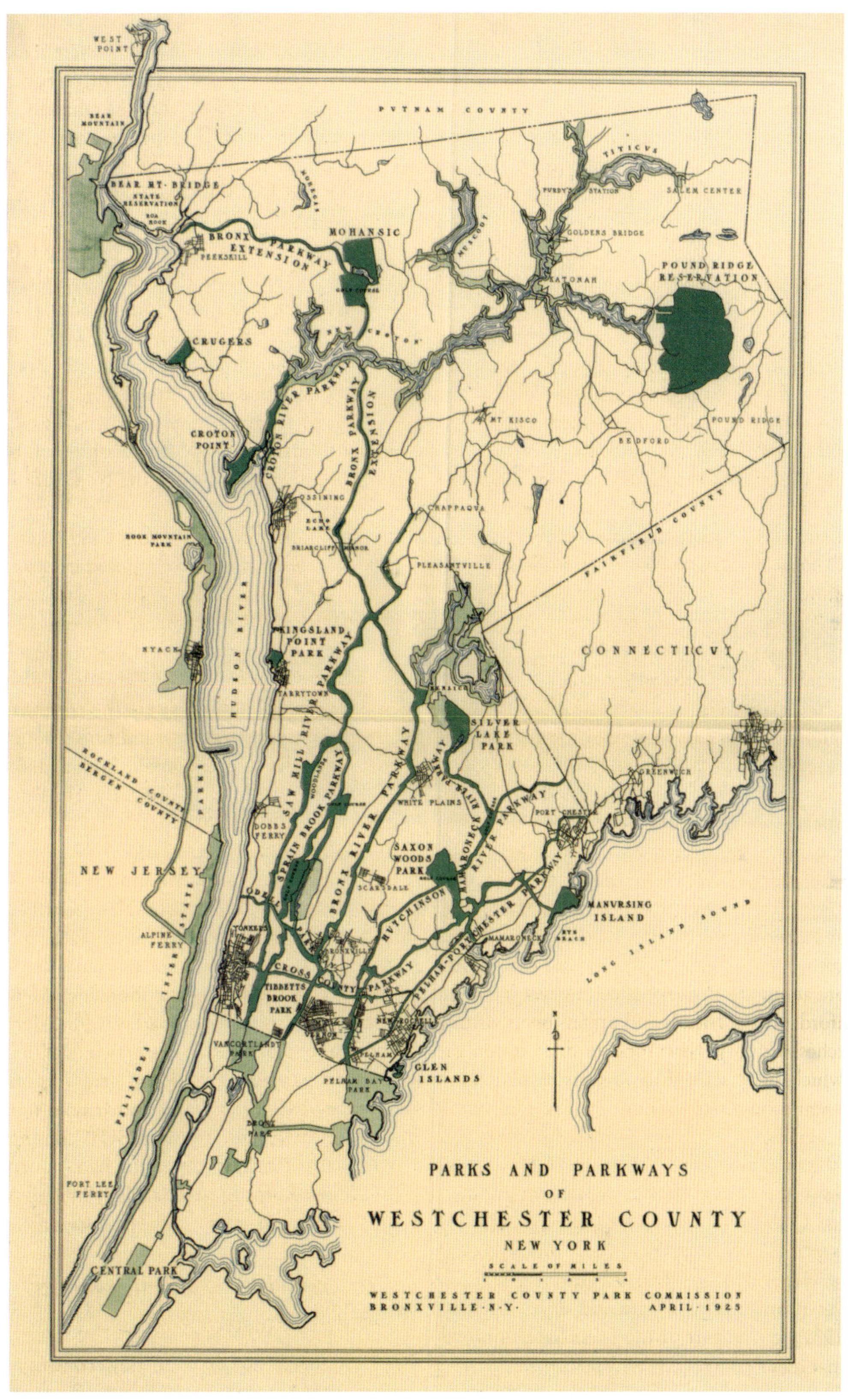

Figure 2.210: Westchester County Parkway System, New York. Jay Downer, Chief Engineer, Gilmore Clarke, Landscape Architect, 1923. 1928 Map. Most of the parkways ran alongside rivers, like the Bronx River Parkway. *Courtesy of Westchester County Archives*

Figure 2.211: Taconic Parkway, Hudson River Valley, New York. Gilmore Clarke, 1925–1949. Looking north on the Taconic Parkway in Dutchess County, New York. Gilmore designed the Taconic Parkway to provide a variety of views and vistas. Sometimes there are large open fields between the north and south roadways, and at other times the roadways go their separate ways in the woods.

Figure 2.212: Taconic Parkway, Hudson River Valley, New York. Gilmore Clarke, 1925–1949. Looking north on the Taconic Parkway in Dutchess County, New York. The roadways are narrow enough that the trees can form canopies over the parkway.

histories differed. Second, the discussion overlaps with the history of parkways of another type in Boston, Buffalo, Chicago, Denver, and other American cities (see below). And third, the New York discussion closes with Robert Moses, a complicated and contradictory character. On the one hand, Moses built more parkways than anyone else, including some of the best. He built the last, the Palisades Parkway, officially named the Palisades Interstate Parkway. But no one did more than Moses to make the

parkways seem old-fashioned and inadequate in a new regional network of high-speed freeways and toll roads.

Notably, the Henry Hudson Parkway promoted a City Beautiful vision for cars entering and traveling through the city, leaving behind the philosophy of the highwayless town. The parkway began at the Bronx border in Van Cortlandt Park, the city's third largest park, where it connected to the Saw Mill River Parkway coming from the north. The road then ran a little more than two miles through the Riverdale section of the Bronx, in a treeless highway with little buffering from the surrounding, semi-urban neighborhood. But when the highway reached the Spuyten Duyvil Creek—which flows to the Hudson River through a wide and deep ravine between the Bronx and Manhattan—Moses built a beautiful steel arch bridge with a panoramic view of the Hudson and the Palisades. For the next five miles, the parkway runs along the river, sometimes high above it, sometimes next to it, through the Inwood, Fort Tryon, and Riverside parks. Despite some unfortunate changes over the years, the Henry Hudson is still one of the most impressive ways to enter a large city by car.[96]

Even in 1937, however, induced demand was a problem. Within a year of the opening of the Henry Hudson Parkway, Moses discovered that the four-lane Henry Hudson Bridge was too small, forcing him to add a second four-lane deck.[97] From the beginning, the section of the Henry Hudson between Van Cortlandt Park and the bridge into Manhattan was a problem. There wasn't enough land available to build a leafy parkway, so Moses tore down the old-style Spuyten Duyvil Parkway, a tree-lined boulevard with no buffer zones, and built a conventional and wider highway. In Manhattan, the parkway was originally a narrow four-lane, two-way road with no traffic separation, but Moses long ago turned it into a six-lane highway with twelve-foot traffic lanes that separated Manhattan's residents from the river. Moses buried the railroad tracks under his parks but gave the cars the place of honor along the Hudson.

Always a pragmatist in search of power, Moses learned better than anyone how to secure Federal funding for interstate highways, and how to use that funding to control decisions important for the future of New York. Just as he changed how he built housing in response to Title I funding, when post-war regulations and funding called for interstate freeways, Moses pivoted from the old to the new. With the exception of the hybrid Palisades Interstate Parkway, he built the rest of the regional network with standard-issue interstate highways that dominate local transportation (and today, Google Maps and Waze favor freeways over parkways.) As recounted in Robert Caro's great work *The Power Broker*, Moses built one of the most anti-urban, destructive highways when he rammed the Cross Bronx Expressway through crowded neighborhoods in the Bronx.[98] The Cross Bronx was his most brutal highway, but he built miles of contenders for that title, like the Brooklyn-Queens Expressway and the Long Island Expressway.

During his career, Moses built 627 miles of roads in and around the city, inducing regional demand. Greater New York has the best public transportation system in North America, but Moses was at center of the creation of an enormously expensive alternative transportation system that encouraged people to move out of the city and then drive back and forth for work. After he began planning the Hudson Parkway in 1934, the number of personal and commercial motor vehicles in the United States has increased by more than 11,000 percent (with 21,544 cars and 3,717 commercial vehicles then, compared to more than 285,000,000 today). Forgetting all the problems that cars bring, there is the simple problem that so many people want to drive on the region's highways that building enough highways, bridges, and tunnels for them is impossible. Today, the Henry Hudson Parkway is an under-engineered, over-used highway that pumps cars, pollution, and impatient drivers into Manhattan twenty-four hours a day, seven days a week.

> Since the time he began planning the Hudson Parkway in 1934, the number of personal and commercial motor vehicles in the United States has increased by more than 11,000 percent (with 21,544 cars and 3,717 commercial vehicles then, compared to more than 285,000,000 today).

Washington and the City Beautiful

The development of Washington's parkway system had two phases. The first came in the aftermath of Washington's Centennial and the World's Columbian Exposition in Chicago in 1893, when the City Beautiful movement was symbolically born. In 1901, the City Beautiful philosophy led to the creation of the Senate's McMillan Plan, which brought America's best architects and landscape architects to Washington to study the capital's urban planning problems (Figure 1.10).[99] That in turn slowly but directly led to the National Park Service's construction of parkways in the region in the 1920s and 1930s.[100]

Modernist architects and historians have been strongly critical of the City Beautiful movement. Consequently, it was never well chronicled. Therefore, it is worth mentioning that the City Beautiful movement was a product of the Progressive Era that called for social, economic, and political reform. Allied with the anti-corruption and anti-Robber Baron impulses of Progressive Era leaders like Teddy Roosevelt, the City Beautiful movement embodied

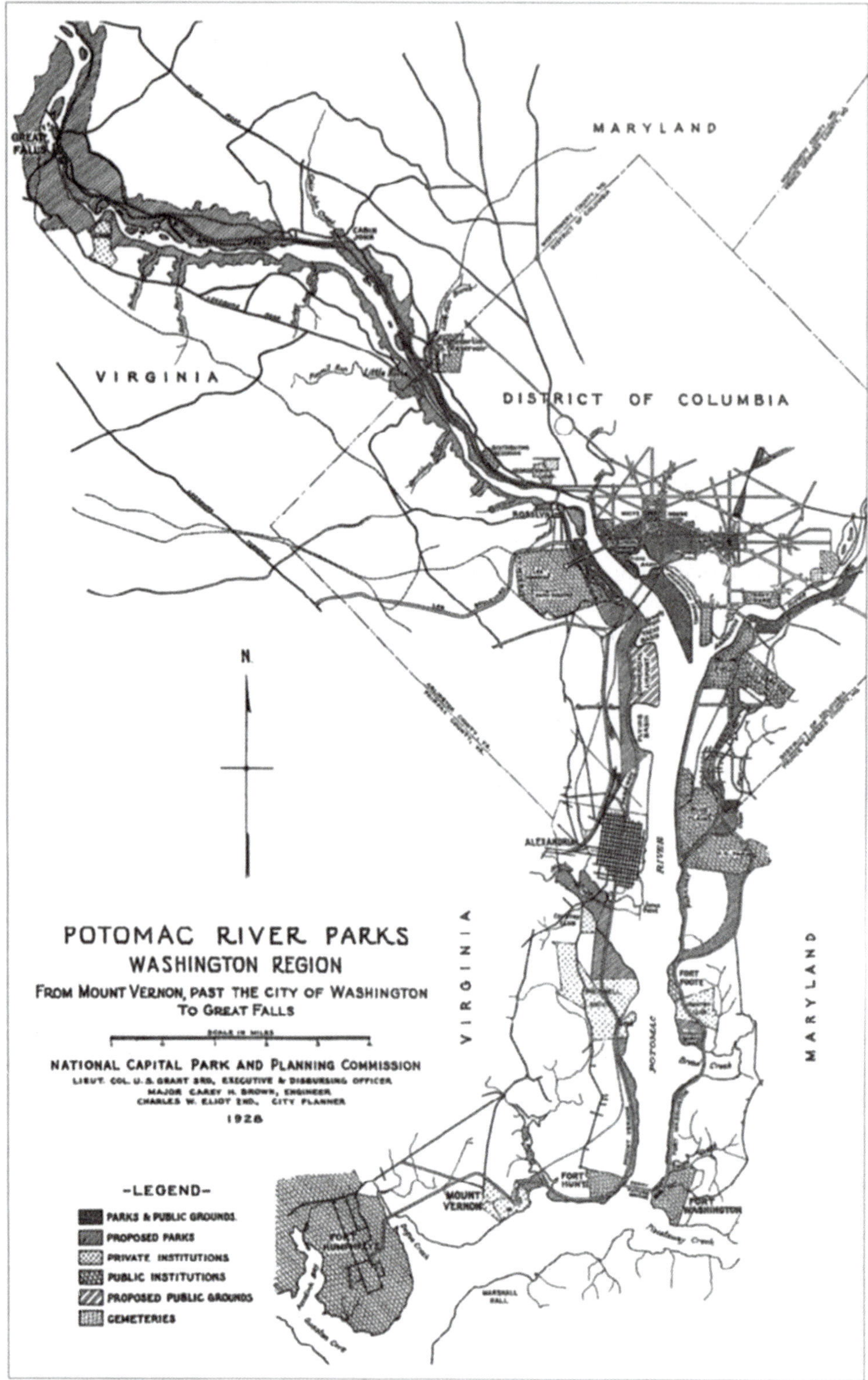

Figure 2.213: Potomac River Parks, Washington Region. National Capital Park and Planning Commission, 1928. A map of Washington area parks used to plan the parkway system built in the 1920s and 1930s. *Library of Congress / National Capital Park and Planning Commission*

the idea that a more equitable society should have good cities for all. The movement was not simply about beauty, in other words. The City Beautiful also embraced the creation of new forms for adapting to modern technologies like the motorcar and steel frame construction of buildings and promoted building beautiful civic buildings like libraries and schools for all.

An obvious question for the McMillan Plan authors was how drivers would enter the monumental capital, planned with axially sited buildings, memorials, and streets. When Charles McKim, Frederick Law Olmsted, and Daniel Burnham extended the plan of Washington towards the Potomac, they reshaped the river basin and created sites for the Jefferson and Lincoln memorials. McKim, Mead & White designed a new bridge across the Potomac that axially connected the Arlington National Cemetery and the Lincoln Memorial. The cemetery is a military cemetery, with strong connections to the Civil War. Annual ceremonies at the cemetery on Memorial Day and Veterans Day symbolically take the president and other U.S. leaders to and from the Lincoln Memorial. At other times, Washingtonians going about their everyday lives, cross the Potomac from the George Washington Parkway focused on the Lincoln Memorial.

In 1902, the McMillan Plan called for the creation of what became the George Washington Memorial Parkway along the Virginia side of the Potomac River.[101] The plan drew on an 1880s proposal for a national road from George Washington's home at Mount Vernon to Washington, DC, and expanded the scope. Congress authorized payment for the road in 1928, and the parkway was built in stages between 1929 and 1970, with a significant amount of construction in the 1950s and 1960s. Today, the parkway begins at Mount Vernon and runs all the way to the Capital Beltway (I-495). It sits in more than 7,000 acres of land owned by the National Park Service that protect a beautiful natural landscape across from the capital. Unlike the Henry Hudson Parkway, it never feels like a rundown, under-engineered highway.

Early sections of the Washington Parkway, built by the Federal Bureau of Public Roads, are some of the most interesting. We say that because when the Parkway goes through Old Town Alexandria on its way from Mount Vernon to Washington, the road stops being a parkway and turns into the city's main streets (called Washington Street). On the other side of town, the road turns back into a parkway. The newer parts of the parkway share the aesthetic of the Palisades Interstate Parkway (but sit

below a palisade rather than on top of one). The curves are longer and conform less to the topography, but the parkway is a pleasant place to drive, with beautiful scenery around it. There are some spectacularly scenic entries and exits, like the Spout Run exit in Arlington. The National Park Service maintains the shorter but similar Clara Barton Parkway on the Maryland side of the Potomac.[102]

Two more Washington parkways deserve mention. The first is the Rock Creek and Potomac Parkway, in Washington's largest and oldest park.[103] Since the beginning of Rock Creek Park, many roads have been built in it, with long debates about the appropriate balance between preserving the park and creating travel routes through it. In the 1930s, the National Park Service extended the park into Maryland and used New Deal funding to upgrade park roads. When the last piece of the Rock Creek and Potomac Parkway opened in 1936 after decades of construction, the die was cast. Today, there are beautiful sections of the aptly-named parkway, and other sections that look like a modern freeway in an inappropriate setting. And while the park is large, traffic generated by the parkway disturbs the peace and quiet of the park in many areas.[104]

The Baltimore–Washington Parkway is the other parkway worth mentioning.[105] The Bureau of Public Roads began the process for the design and construction of the parkway in 1942. It sits in a national forest, bought by the National Park Service to protect its sylvan character. Primarily built between 1947 and 1954, the road is a hybrid between a parkway and a freeway, with wide traffic lanes, gentler, more regular curves, grade-separated interchanges, and long, modern merge lanes at entrances and exits. At 65 miles per hour, the six- to eight-lane parkway is less ugly and unpleasant than the usual Federally-regulated freeway. As a memorable driving experience, it can't compare to the Taconic or George Washington parkways.

All the Washington parkways were subject to comment or more active participation by the National Park Service, the Senate Park Commission, the National Capital Park and Planning Commission, the Fine Arts Commission, and, of course, Congressional funding. All the groups looked at Washington as a young city that needed improvements that were both modern and worthy of the nation's capital. That is undoubtedly why the Federal Bureau for Public Roads, which was to become the Federal Highway Administration, promoted standard traffic engineering in the rest of the country but brought in important members of the parkway design teams from New York's and Boston's parkway systems to design parkways in Washington.

That is not to say that Washington didn't get more than its share of bad highways after 1945. Washington became a city that increasingly relied on private automobiles to move workers in and out of the city every day, and the Bureau of Public Roads / Federal Highway Administration was at the heart of that change.[106]

Parkways, "Parkways," Drives, and Boulevards

The history of America's parkways, drives, and boulevards remains to be written. Frederick Law Olmsted, Frederick Law Olmsted Jr., Benton MacKaye, Charles Eliot, Charles Eliot II, Gilmore Clarke, George Kessler, and other designers and planners like Arthur Shurcliff and John Nolen worked across America at a time when the country was rich and rapidly growing to develop new models for tree-lined boulevards, drives, and parkways.[107] The work began in the second half of the nineteenth century with Olmsted Senior and ended in the first half of the twentieth century when Modernism supplanted the City Beautiful movement. The purpose was to create a system of beautiful streets appropriate for their context, the context being what we now call the Urban to Rural Transect (page 46).[108] The designers succeeded, leaving us a legacy of unusual and exceptional public roads. Cities from Los Angeles, to Kansas City, to Boston have wonderful examples of their work. Some of the streets were not built as intended, and many of the boulevards and parkways have been inappropriately "upgraded," but one can also see portions in their original state.

Frederick Law Olmsted, the father of American landscape architecture, is also the father of these green, tree-lined streets. Most credit Olmsted for popularizing the word "parkway" when he proposed the Eastern Parkway (page 112) and Ocean Parkway (page 113) in his 1868 Parkway Plan for Brooklyn. At the time, Olmsted and Vaux were designing Central Park in Manhattan and Prospect Park in Brooklyn, which was still a separate city. But Olmsted thought the two cites should be connected by a network that would also link to a great park on the Atlantic Ocean. Unlike the parkways we have been looking at in the previous section, the Eastern and Ocean parkways were linear parks in urban grids, with sidewalks and building lots on both sides rather than green buffers (Figure 2.214). "Eastern and Ocean Parkways are major

presences" [in Brooklyn], Elizabeth MacDonald says in her groundbreaking book *Pleasure Drives and Promenades, A History of Frederick Law Olmsted's Brooklyn Parkways*. "No other streets are quite like them. They feel like parks, but are simultaneously residential streets and major traffic corridors."[109]

In 1880, Olmsted began the design of Boston's famous Emerald Necklace (pages 254 and 324) a series of connected tree-lined streets and parks planned by Olmsted. In the 1890s, Boston's Metropolitan District Commission began expanding the leafy, curving road network with a series of streets like Memorial Drive ("Mem Drive," originally called Charles River Road), which runs along the northern side of the Charles River Basin. Mem Drive connected to other River Basin roads,[110] as well as to the new Fresh Pond and Alewife Brook parkways that led out to Boston's western suburbs. The Metropolitan District Commission soon created a Metropolitan Boston Parkway system encircling Boston.[111] The parkways were not limited-access parkways, but tree-lined boulevards within the local roads, planned as a regional transportation network. They frequently connected parks and "reservations" created in and around Greater Boston.[112] After World War II, Federal funding and Functional Classification standards moved Massachusetts to the construction of conventional, Federally-funded arterial roads.

Chicago, Illinois, Louisville, Kentucky, and Buffalo, New York have networks of parks and connecting parkways designed by Olmsted Sr., sometimes with his sons.[113] Buffalo built a network of parks and tree-lined avenues and boulevards distinguished by their size. Olmsted designed three large parks in Buffalo and connected them with a web of "parkways" and avenues known as "the Park Approaches." The four parkways are two-hundred feet wide, and three avenues are one-hundred feet wide. Their seven-mile, 125-acre route includes four traffic circles designed by Olmsted. In 1902, Buffalo's Park Board rebuilt one of the circles, Chapin Place, with a City Beautiful design. When the City of Buffalo later assumed control of the parks and avenues, it "improved" (for traffic flow) two of the circles. In 1960, one of the parkways was demolished and replaced with the Kensington and Scajaquada freeways.[114]

America has one more type of parkway, the scenic parkways built by the National Park Service in rural settings around the country: the Colonial Parkway that runs from Jamestown to Williamsburg, Virginia; the Blue Ridge, Shenandoah Skyline, and Great Smoky Mountains parkway system in Virginia, North Carolina, and

Tennessee; and the Natchez Trace in Tennessee and Mississippi.[115] The Colonial Parkway, the Blue Ridge Parkway, and the Natchez Trace have different characters and provide different experiences, but they are all beloved places. The interstates and arterials we build are not.

Olmsted and Vaux proposed the Eastern and Ocean Parkways forty years before the Ford Model T went into production. They built their multiway boulevards for horses, horse-drawn wagons and carriages, and people on foot. Two decades later, cycling became an important vehicle for transportation, exercise, and personal freedom. Male cycling clubs became popular, and women and suffragettes in particular adopted the bicycle as a way to move around the city on their own. The first cycling lanes in America became part of the parkways. Twenty years later, Organized Motordom controlled the roadway. The cars are still with us, and so is the carbon they spewed into the atmosphere. But the multiway parkways, the drives like Riverside and Mem Drive, and the limited-access parkways provide inspiration for building a future with fewer cars, in a greener world, on streets that are places suitable for a variety of users and types of travel.

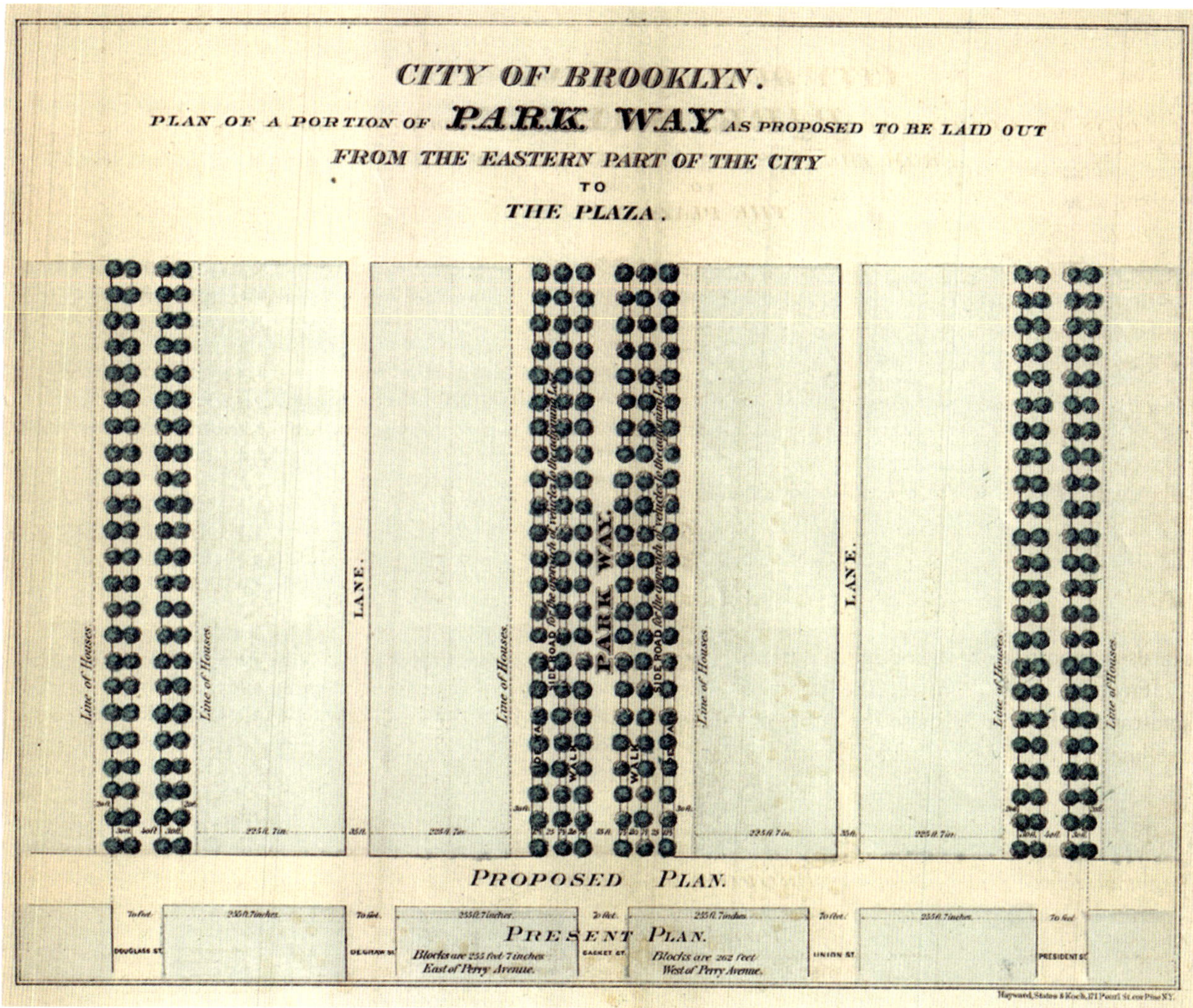

Figure 2.214: Eastern Parkway, Brooklyn, New York. Olmsted & Vaux, 1868. Proposed plan. Unlike the limited-access parkway that came later, Olmsted & Vaux's multiway "parkway" was part of the city grid and in the city fabric, with houses on both sides of the street. *Geographicus Rare Antique Maps / Wikimedia Commons / Public Domain*

NOTES

1. The Ford Model T went into mass production in 1908. That was a turning point in the history of American streets. For the next 20 years, Organized Motordom successfully fought to kick pedestrians to the side of the road. Peter Norton chronicles the rise of Organized Motordom in his book *Fighting Traffic: The Dawn of the Motor Age in the American City* (MIT Press, 2008). Here are two quotes from the 1920s quotes by people actively working to advance the interests of Organized Motordom: "Americans are a race of independent people, even though they submit at times to a good deal of regulation and officialdom. Their ancestors came to this country for the sake of freedom and adventure. The automobile satisfies these instincts."—Roy Chapin, "The Motor's Part in Transportation," *Annals* 116 (November 1924): 1–8 (4–5). "We shall no doubt see automobile traffic . . . more and more collected in great arterial ways, from which the pedestrian will be excluded, being overpassed or underpassed, and being kept from playing or wandering in the roadway."—William Cox, paper presented at a conference of the American Society of Civil Engineers, New York, May 1927; reprinted as "Population Density as a Factor in Traffic-Accident Rates," *American City* 37 (August 1927): 207–209 (209). In "Car Harms Monday: 'Gridlock Sam' Says We Have Lost Our Lives to the Automobile," Gridlock Sam Schwartz talks about New York children still playing in the street when he was young. Schwartz grew up in Brooklyn and was the former NYC Traffic Commissioner prior to the Department of Transportation changing to its current name. "Before the car population boom, the 'street,' aka the 'gutter,' was our playground," Schwartz says. "And it was lost to the car in the name of progress and improvement. Let's look back." See the article in StreetsblogNYC, June 2, 2025, https://nyc.streetsblog.org/2025/06/02/car-harms-we-have-lost-our-lives-to-the-automobile. For more on Gridlock Sam, see Aaron M. Renn, "When New York City tried to ban cars – the extraordinary story of 'Gridlock Sam,'" *The Guardian*, June 1, 2016, https://www.theguardian.com/cities/2016/jun/01/new-york-city-ban-cars-gridlock-sam-schwartz-manhattan.

2. Between 1875 and 1900, the population of the United States grew from 45.8 million to 78.8 million. Between 1900 and 1930, the population of New York City grew from 3.4 million to 6.9 million.

3. Robert A.M. Stern's series of New York books only used period photographs. See Robert A.M. Stern, Thomas Mellins, and David Fishman, *New York 1880: Architecture and Urbanism in the Gilded Age* (New York: Monacelli Press, 1999); Robert A.M. Stern, Gregory Gilmartin, and John Montague Massengale, *New York 1900, Metropolitan Architecture and Urbanism 1890–1915* (Rizzoli, 1983); and Robert A.M. Stern, Gregory Gilmartin, Thomas Mellins, *New York 1930: Architecture and Urbanism Between the Two World Wars* (Rizzoli, 1987).

4. The quantity and quality of digitally remastered movies available online that show old cities grows so rapidly that a printed list would quickly become outdated. John Massengale's talk for the New Haven Preservation Trust has some New York clips: "We Made This City To Walk And Stroll," September 17, 2019, https://www.youtube.com/watch?v=9pJ7jhgvSe0.

5. The Singelgracht has a pair of one-way streets bordering the Singelgracht (or Single Canal). In Figure 5.1 you can see two small blue and white traffic signs. One has a simple white arrow on a blue background that shows the direction of the street. The other has a white "P" on a blue background. That shows parking is allowed there. Neither sign is designed to be read at high speed, and neither sign has its own lighting. Cars are welcome on the street, but the comfort of the people walking and biking comes first.

6. The name of the book was *Livable Streets* (MIT Press, 1981). Donald Appleyard's son Bruce greatly expanded the book and published it as *Livable Streets 2.0* (Elsevier, 2019). *Livable Streets 2.0* has an excellent bibliography on the topic. "Who Gets Peace and Quiet? Urban Noise in the COVID-19 Pandemic," describes work by an Assistant Professor of Epidemiology at Brown University named Erica Walker (*Vision Zero Cities Journal*, October 4, 2023, https://medium.com/vision-zero-cities-journal/who-gets-peace-and-quiet-urban-noise-in-the-covid-19-pandemic-c821e0a08887). She wrote, "The health impacts associated with higher noise levels are brutal. They can limit the quantity and quality of our restorative sleep and disrupt our mood so profoundly that our body activates a flight-or-fight response. In other words, each and every fiber of your being is coordinating to keep you alive during a situation it believes can potentially kill you. Your heart rate and blood pressure increase, you start to sweat, you become tense, and you may start trembling. Over time, this consistent stimulation in response to noise can lead to the manifestation of serious health outcomes ranging from hypertension to cardiovascular-related mortality."

7. In conversation with the authors. Triàs is the former planning director of the City of Coral Gables, Florida.

8. Today, most of Manhattan's avenues are one-way, sub-urban, tree-less stroads. For a discussion of stroads and changing terminology, see pages 504 and 505.

9. There is a story about allées and Peter the Great, the Tsar of All Russia in the late seventeenth and early eighteenth centuries. In 1703, Peter founded the city of St. Petersburg on the Baltic Sea to open a "window to the West." Requiring the members of his court to move to St. Petersburg, Peter gave them large grants of land in the forests around the city. The recipients had to build impressive estates, which Peter sometimes visited. On an early visit to one, Peter remarked that he was not impressed. "Where are the French allées?" he asked. A year later, he revisited the estate. The landowner and Peter stood behind the country house and looked out over the landscape. The landowner raised his handkerchief, giving the signal for thousands of serfs in the woods to pull down thousands of trees already prepared for clearing. Three great avenues appeared in the forest, fanning out from the terrace where Peter stood.

10. From etymonline.com: *avenue* (n.) c. 1600, "a way of approach" (originally a military word), from French *avenue* "way of access" (16c.), from Old French *avenue* "act of

approaching, arrival," noun use of fem. of *avenu*, past participle of avenir "to come to, arrive," from Latin *advenire* "to come to, reach, arrive at," from *ad* "to" (see ad-) + *venire* "to come" (from a suffixed form of PIE root *gwa- "to go, come"). The meaning was extended to "a way of approach to a country-house," usually a straight path bordered by trees, hence, "a broad, tree-lined roadway" (1650s), then to "wide main street" (by 1846, especially in the United States). By the late nineteenth century in U.S. cities, it was used to form the names of streets without reference to character.

11. Anthony Sutcliffe, *Paris: An Architectural History* (Yale University Press, 1993): 50.

12. M.-A. Laugier, *Essai sur l'architecture* (Paris, 1755). We are indebted to Sutcliffe's description of the development of planning in eighteenth century Paris: *op. cit.* See "The Architectural and Urban Codes of Paris," page 277.

13. Sutcliffe, *op. cit.,* 52.

14. Quoted by Jonathan Glancey, "The Man Who Created Paris," BBC Culture, January 26, 2016, https://www.bbc.com/culture/article/20160126-how-a-modern-city-was-born. "No other major city, before or since, has been transformed so radically during peacetime," Glancey wrote.

15. Stephane Kirkland, *Paris Reborn: Napoléon III, Baron Haussmann, and the Quest to Build a Modern City* (St. Martin's, 2013): 1. Kirkland begins the Prologue saying, "From the outset, Louis-Napoléon Bonaparte had a great ambition for Paris. He wanted to transform it into the most modern and functional city in the world, a city where wide, convenient boulevards suitable for modern transportation would replace narrow streets, where elegant ladies could walk without treading in filth and decay, where new neighborhoods would rise to house the swelling population; he wanted a city that would represent the principles of order and modernity of his presidency. A pragmatist and an idealist, Louis-Napoléon was committed to personally overseeing the realization of this vision." For drawings, photographs, and analysis of many European streets, see Vittorio Magnago Lampugnani, Harald R. Stühlinger, and Markus Tubbesing, *Atlas zum Städtebau, Band 2: Straßen* (Hirmer Verlag, 2018). The boulevard Haussman is on page 102, and the Passeig de Gràcia is on page 30.

16. From Stephanie Rosenbloom, "The Art of Being a Flâneur," *New York Times, Travel,* June 19, 2023, https://www.nytimes.com/2023/06/19/travel/walking-travel-cities.html.

17. Duany, Plater-Zyberk & Co., *Lexicon of the New Urbanism,* https://www.dpz.com/wp-content/uploads/2017/06/Lexicon-2014.pdf.

18. Allan B. Jacobs, Elizabeth MacDonald, and Yodan Rofè, *The Boulevard Book* (MIT Press, 2002): 20–22.

19. Alistair Horne, *Seven Ages of Paris* (Alfred A. Knopf, 2002): 232–240.

20. Joan Busquets, *Barcelona: The Urban Evolution of a Compact City* (Nicolodi editore, 2006).

21. Jacobs, MacDonald, and Rofè, *op. cit.,* 43, 83–84, 101–102, 108–109.

22. L.J. Aurbach writes, "The turf improved the railbed's appearance and reduced noise and dust, and was widely adopted elsewhere." Laurance Aurbach, *A History of Street Networks, from Grids to Sprawl and Beyond* (PedShed Press, 2020): 118.

23. "Eastern Parkway," *New York City Department of Parks and Recreation,* http://www.nycgovparks.org/parks/B029/history.

24. Peter Bosselmann and Elizabeth MacDonald, "Boulevard Livability Study," *Places, A Forum of Design for the Public Realm* 11, 2 (1997): 66–69.

25. "Ocean Parkway," *New York City Department of Parks and Recreation,* http://www.nycgovparks.org/about/history/historical-signs/listings?id=10787 and Jacobs, MacDonald, and Rofè, *op. cit.,* 44–53. Too new to be included in our first edition text was an excellent study of the Eastern and Ocean Parkways by Elizabeth MacDonald: *Pleasure Drives and Promenades, History of Frederick Law Olmsted's Brooklyn Parkways* (Center for American Places at Columbia College, 2012). For Eastern Parkway, also see Jacobs, MacDonald, and Rofè, *op. cit.,* 29.

26. Stephenson wrote, "Thomas Dixon was Janus faced, but he was not a conflicted soul. A foe of Tammany Hall and a favorite of Republican reformers, the Baptist minister remained a committed Democrat and passionate defender of the Lost Cause. Early in his pastorate, he broadcasted this bare conviction in a highly publicized sermon, 'Sectional Journalism.'" See Bruce Stephenson, "The Clansman and the City Plan: Thomas Dixon's Vision of a Racial Utopia," 1990 [Unpublished manuscript].

Eliot Shepard, the partisan Republican owner of the *New York Mail and Express*, was the target of Dixon's wrath. On May 29, 1890, the day before the unveiling of the now-infamous monument honoring Robert E. Lee in Richmond, Virginia, the *Mail and Express* ran a doubled-leaded headline: "Treason Glorified! The Solid South Pays Tribute to the Memory of Benedict Arnold Lee. Traitors are on the March to Richmond." Besides labeling Lee a traitor, Shepard claimed that the massive statue erected in the former Confederate capital signified a "new rebellion." A Union Army veteran and president of the New York Bar Association, he proposed that Congress prohibit building monuments memorializing Confederates and displaying rebel flags at occasions such as the Richmond ensemble, which attracted over 100,000 people. "The New York Letter," *Shepherdstown Register*, June 13, 1890: 1; "A Letter to Mr. Shepard," *Henderson Gold Leaf,* June 12, 1890: 1.

Dixon held nothing back in chastising Shepard, claiming that he and Satan were brethren. "In the name of God and of truth, of honesty and of integrity, I, for one, repudiate this so-called newspaper as in any sense representative of Christianity," Dixon thundered. "The God that presides over the editorial office of the *Mail and Express* and such papers is not my God. About as near as I can make it, his God is the devil." A man who "steals the livery of heaven in which to serve the devil," Shepard "has ridden to death" the terms "traitor and rebel." Dixon considered such language superfluous in a world where Ulysses S. Grant and Robert E. Lee were no longer opposing generals, but heavenly brothers consecrated in holy union. A generation removed from the Lee's surrender at Appomattox, it was time to "listen to the voices of the heroic dead. They all speak for peace and harmony." With the plan for Grant's tomb complete and the unveiling of the monument honoring Lee, "we must take each other by the hand and crush these influences that seek to perpetuate strife for a base ignoble purpose," Dixon concluded. "God help us that we may have, in deed and in truth, one glorious, unified nation." "We are Brethren," *Raleigh State Chronicle*, June 5, 1890: 2.

Robert E. Lee personified the saliency of the Lost Cause that empowered Dixon's Radical theology. Cast as a devout Christian who abhorred slavery and labored tirelessly after the war to unify the nation, Lee was a slave-owning aristocrat who fought to keep slavery and his privileged position. Federal prosecutors attempted to try him for war crimes that were hanging offenses, the most egregious being the action of his troops the first time they encountered large numbers of black soldiers at the Battle of the Crater. When these Union troops tried to surrender, no quarter was shown, and they were brutally executed. When Lee surrendered, Grant offered gracious terms, and Lee would later be pardoned by President Andrew Johnson for his treasonous acts. A year after Appomattox, Grant was less than enamored with his former rival, writing that he was "setting an example of forced acquiescence so grudging and pernicious in its effects as to be hardly realized." Adam Serwer, "The Myth of the Kindly General Lee," *Atlantic Monthly*, June 4, 2017, https://www.theatlantic.com/politics/archive/2017/06/the-myth-of-the-kindly-general-lee/529038/.

An opponent of both enfranchising African Americans and securing racial equality, Lee's concept of reconciliation was to unify whites on the condition that blacks were denied civil rights. This belief also imbued Reverend Thomas Dixon, who considered "Robert E. Lee, a Christian and a gentleman," and "one of the greatest men who ever lived." See "We are Brethren," *op. cit.*

Also see Mariah Williams, "Designing Cities for Grief and Remembrance," Next City, March 10, 2026. https://nextcity.org/urbanist-news/designing-cities-for-grief-and-remembrance.

27. "List of Black Lives Matter Street Murals," Wikipedia, https://en.wikipedia.org/wiki/List_of_Black_Lives_Matter_street_murals.

28. John Nolen, *New Towns for Old, Achievements in Civic Improvement in Some American Small Towns and Neighborhoods* (Marshall Jones Co., 1927): 104–10.

29. An *alameda* is a formal alignment of paired trees with a public walkway or promenade, found in Spain and former Spanish colonies.

30. Jacques Hillairet, *Dictionnaire historique des rues de Paris* (Éditions de Minuit, 1963); and "Mapping Paris," France in the Age of Les Misérables, https://rmschwartz.wordpress.com/paris/the-new-streets-of-paris-an-investigation-into-baron-haussmanns-controversial-urban-planning-project/two-maps-a-side-by-side-examination-of-paris-in-1836-and-1864/.

31. Maurice Rheims and Felipe Ferre, *Hector Guimard* trans. Robert Erich Wolf (Harry N. Abrams, 1988): *passim*.

32. "The Paseo (Kansas City, Missouri)," Wikipedia, https://en.wikipedia.org/wiki/The_Paseo_(Kansas_City,_Missouri).

33. Charles Beveridge, *Frederick Law Olmsted: Designing the American Landscape*, rev. ed. (Universe, 2005): 96–98.

34. Everett U. Crosby, *"95% Perfect": The Older Residences at Nantucket* (Tetaukimmo Press, 1937).

35. Robert J. Gibbs, *Principles of Urban Retail Planning and Development* (John Wiley & Sons, 2012): 67–71.

36. J. Christopher Lang and Kate Stout, *Building with Nantucket in Mind, Guidelines for Protecting the Historic Architecture and Landscape of Nantucket Island* (Nantucket Historic District Commission, 1996), https://www.nantucket-ma.gov/DocumentCenter/View/12329/Building-With-Nantucket-In-Mind-1992-PDF.

37. Gibbs, *op. cit.*, adapted from pages 81–95.

38. Diann Marsh, *Galena, Illinois: A Brief History* (The History Press, 2010): 13–14.

39. "Galena History," City of Galena, Illinois, https://www.cityofgalena.org/our-community/galena-history.

40. Simon Jenkins, *England's 100 Best Views* (Profile Books, 2013): 85.

41. The Project for Public Places named Northampton's Main Street one of their Great Public Spaces. See https://www.pps.org/places/main-street-northampton-ma. In 2007, the American Planning Association honored the street with one of their annual Great American Places awards: http://www.planning.org/greatplaces/streets/2007/mainstreetnorthampton.htm.

42. Edmund N. Bacon, *Design of Cities* (Penguin, 1976): 201.

43. Charles Dickens Jr., *Dickens's Dictionary of London,* http://www.victorianlondon.org/districts/piccadilly.htm.

44. In addition to Nash, two of the most prominent architects were Sir John Soane and Charles Robert Cockerell.

45. World War I delayed construction, and Blomfield continued until 1928, with architects Sir John James Burnet, Arthur Joseph Davis, and Henry Tanner also involved.

46. Clay Lancaster and Edmund V. Gillon Jr., *Old Brooklyn Heights: New York's First Suburb* (Dover Publications, 1979): 74. Also see Max Carr, "Ones and Twos: The State of Joralemon West," boxed out, http://max-carr.blogspot.com/2012/08/ones-and-twos-state-of-joralemon-east.html.

47. François Spoerry, *A Gentle Urbanism* (John Wiley & Sons, 1991).

48. Sir Raymond Unwin, *Town Planning In Practice: An Introduction to the Art of Designing Cities and Suburbs* (Princeton Architectural Press, 1994).

49. National Park Service, U.S. Department of the Interior, "Religious Architecture of Charleston," https://www.nps.gov/articles/religious-architecture-of-charleston.htm.

50. Prominent exceptions may have been the designs of the seventeenth and eighteenth centuries that celebrated pure geometry in places like the Circus in Bath. Since those curbs haven't survived centuries of repaving, we don't know.

51. John Massengale, "The Most Beautiful City of the Twentieth Century," There are Two Types of Architecture, https://blog.massengale.com/2015/09/08/vv-redux-santa-fe/.

52. St. Thomas Source staff, "Signs of the Renaissance, Haagensen House." *The St. Thomas Source*, March 23, 2000, http://stthomassource.com/content/business/st-thomas-business/2000/03/23/signs-renaissance-haagensen-house-more.

53. See "Charlotte Amalie, St. Thomas," *Virgin Islands Now*, https://www.vinow.com/stthomas/attractions_stt/charlotte-amalie/.

54. Despite all the words that have been written about Frederick Law Olmsted, there is surprisingly little written on Riverside Drive, Olmsted's best and most elaborate street design. The best sources of information are the New York City Landmarks Preservation Committee's "Riverside Park and Riverside Drive Landmark Designation Report," http://s-media.nyc.gov/agencies/lpc/lp/2002.pdf, and the New York City Parks Department Master Plan, https://issuu.com/nycparksplanning/docs/10.19.17_riverside_park_master_plan. The Wikipedia article on Riverside Drive has one of the best lists of sources: see "Riverside Drive (Manhattan)," Wikipedia, https://en.wikipedia.org/wiki/Riverside_Drive_(Manhattan). Stephanie Azzarone's popular history of Riverside Park, *Heaven on Hudson: Mansions, Monuments, and Marvels of Riverside Park* (Fordham University Press, 2022) also is good source of information. Also see *New York 1900*: 360–364, *passim.*, and *New York 1930*: 450–451.

For Olmsted as the father of American landscape design, see Witold Rybczynski, *A Clearing in the Distance: Frederick Law Olmsted and America in the Nineteenth Century* (Scribner, 1999) and George W.S. Trow, "The Harvard Black Rock Forest," *The New Yorker*, June 3, 1984, https://www.newyorker.com/magazine/1984/06/11/the-harvard-black-rock-forest.

55. The view from our apartment always reminds visitors of the famous Saul Steinberg cover for *The New Yorker*, "View of the World from 9th Avenue." See "View of the World from 9th Avenue," Wikipedia, https://en.wikipedia.org/wiki/View_of_the_World_from_9th_Avenue. For the actual view, see the Riverside Drive photos on the *Street Design* website at https://street.design/photos/riverside.

56. The small road to the east of the islands is part of Riverside Drive: buildings there have Riverside Drive addresses. Olmsted's early drawings for Riverside Drive show that he also called it the "Local Road," but he later changed that to the "Side Road."

57. In addition to the omnipresent House Sparrows and Pigeons, we regularly see or hear Robins, Cardinals, Blue Jays, Chickadees, House Starlings, Mourning Doves, Sea Gulls, Ducks, Geese, and Red Tailed Hawks. And not irregularly, Carolina Wrens, Northern Mockingbirds, and Tufted Titmice.

58. "Firemen's Memorial," *New York City Department of Parks & Recreation*, https://www.nycgovparks.org/parks/riverside-park/monuments/482.

59. The Commissioners' Plan called the new north–south streets "avenues." As discussed earlier in the chapter, avenues should have a terminated vista. Typologically, most New York avenues are boulevards, with unterminated vistas.

60. Nevertheless, in the 1880s, the West End was still sparsely settled. Writing about the famous Dakota apartment house, built in the 1880s on Central Park West at 72nd Street, Christopher Gray said, "Many popular guidebooks retell the well-known story that the huge apartment building was ridiculed for being so far out of town that it might as well have been in the Dakotas, and that the name stuck." See Gray's "Streetscapes" column, "The Dakota; The Elusive Mystery of Its Name," *New York Times*, August 15, 1993, https://www.nytimes.com/1993/08/15/realestate/streetscapes-the-dakota-the-elusive-mystery-of-its-name.html. But Gray also tells us that the developer of the Dakota, Edward Clark (President of Singer Sewing Machine Company) thought names from the Western United States would be appropriate street names for the West End. "'The names of the newest states and territories have been chosen with excellent taste' he said, and suggested Montana Place for Eighth Avenue, Wyoming Place for Ninth Avenue, Arizona Place for 10th Avenue and Idaho Place for 11th Avenue."

61. As children, we all heard about the "infinite monkey theorem." It states that a monkey hitting keys on a typewriter or keyboard at random for an infinite amount of time will almost certainly type any given text, such as the complete works of William Shakespeare. However, the probability that even an infinite number of monkeys would type Shakespeare's *Hamlet* in a finite period of time is so tiny that we can say it is almost certainly zero (and absolutely zero to anyone but a mathematician). But an infinite number of monkeys using Streetmix would never design Riverside Drive, because Streetmix eliminates elements like topography, varying width along the third dimension, varying use, adjacent land use, and so on and so forth. The Streetmix website is https://streetmix.net/.

62. See "The Much Discussed Riverside Drive Viaduct, at 96th Street, is Now Approaching Completion," *New-York Tribune*, June 27, 1902, 30, online at Ancestors.com: https://www.newspapers.com/article/new-york-tribune-the-much-discussed-rive/130176420/. A notice in the *Tribune* on March 3, 1901 on page 19 reported that construction began in the fall of 1900: https://www.newspapers.com/article/new-york-tribune-the-viaduct-at-ninety-s/130172634/. A photo at the Museum of the City of New York shows the intersection of Riverside

Drive and 96th Street before 1900. The photo is online at Sergey Kadinsky, "Strycker's Bay, Manhattan," Hidden Waters Blog, June 14, 2019, https://hiddenwatersblog. wordpress.com/2019/06/14/stryckers/. A *New York Times* article attributes the design to Carrère & Hastings: Frank Prial, "Riverside Bridge Stripped of Bronze, Part of a Wave of Thefts," *New York Times*, June 27, 1974, https://www.nytimes. com/1974/06/27/archives/riverside-bridge-stripped-of-bronze-part-of-a-wave-of-thefts-police.html.

63. Justin Davidson, "The Upper West Side's Zone of Pedestrian Death," Curbed New York, January 24, 2023, https://www. curbed.com/2023/01/upper-west-side-zone-pedestrian-death.html.

64. For more information on General Sigel and the Sigel Monument see the New York City Parks Department list of Riverside Park Monuments at https://www.nycgovparks. org/parks/riverside-park/monuments/page/1 or the page on General Franz Sigel at https://www.nycgovparks.org/parks/ riverside-park/monuments/1447.

65. See Stern, Gilmartin, and Massengale, *op. cit.*, *passim*.

66. See Charles Beveridge, "Frederick Law Olmsted: The Seven S's of Olmsted Design," *The Olmsted Network,* June, 20, 2023, https://olmsted.org/frederick-law-olmsted-the-seven-ss-of-olmsted-design/ and Charles Beveridge, "Frederick Law Olmsted: His Essential Theory," *The Olmsted Network*, June 20, 2023, https://olmsted.org/frederick-law-olmsted-his-essential-theory/. Beveridge is the series editor of *The Papers of Frederick Law Olmsted* vol. 1–9 (Johns Hopkins, 1977–2015), which is online at https://rotunda.upress.virginia. edu/founders/default.xqy?keys=OLMS-print&mode=TOC. In *The Experience of Place*, Tony Hiss describes the experience of visiting Olmsted's Prospect Park in Brooklyn, designed to "trigger something in people who walk down it into the park, something that makes many of them begin almost automatically to experience everything around them, whether they've come there with that intention or not." Hiss, *The Experience of Place* (Alfred A. Knopf, 1990): 28.

67. Old funerary joke: "Who's buried in Grant's Tomb?" is a question Groucho Marx frequently asked guests on his quiz show *You Bet Your Life*. The correct answer is "no one," because Grant and his wife lie in above-ground sarcophagi on the ground floor of the memorial. "However, Marx often accepted the answer 'Grant' and awarded a consolation prize to those who gave it. He used the question, among several other easy ones, to ensure that everyone won a prize on the show." See Azzarone, *op. cit.*, 199.

68. Stern, Gilmartin, and Massengale, *op. cit.*, 414.

69. "About the East Coast Greenway," East Coast Greenway, https://greenway.org/about/the-east-coast-greenway.

70. Many people complained about the noise and smell from the railroad along the river before Moses buried it. In 1909, Riverside Drive resident Julia Rice founded the Society for the Suppression of Unnecessary Noises, according to Justin Davidson in his book *Magnetic City: A Walking Companion to New York* (Random House, 2017): 147–148. But, car noise from the highway is more constant now, and the pollution worse, even though it is less visible.

71. "The Upper West Side's Zone of Pedestrian Death," *op. cit.*, https://www.curbed.com/2023/01/upper-west-side-zone-pedestrian-death.html. Davidson is one of the more astute observers of driving in New York City (and he writes well). For more on both traffic and Davidson's articles about traffic, see Justin Davidson, Article Archive, https://nymag. com/author/justin-davidson/.

72. "Nearly a dozen humpback whales spotted near New York City," CBS News, August 17, 2022, https://www.cbsnews. com/newyork/news/new-york-city-humpback-whales/.

73. The New York City Police Department doesn't ticket anyone going less than 10 miles per hour above the speed limit. On Riverside Drive, 35 miles per hour is 40 percent faster than the posted limit and in turn more deadly. And as we mentioned in the first edition, Streetsblog looked at the data for speeding tickets in New York City and found that a driver speeding in New York City can expect to speed for thirty-five years before getting a ticket. See Brad Aaron, "TA Report: Reckless Driving Casualties Rising as NYPD Enforcement Lags," Streetsblog New York City, July 14, 2009, https://old.nyc.streetsblog.org/2009/07/14/ta-report-reckless-driving-casualties-rising-as-nypd-enforcement-lags/.

74. See "Parks Without Borders Making our parks more open and welcoming," New York City of Parks and Recreation, https://www.nycgovparks.org/planning-and-building/ planning/parks-without-borders.

75. Benton MacKaye, "The Townless Highway," *The New Republic* 62, 792 (March, 1930): 93. MacKaye's research for the Townless Highway is archived at Dartmouth University: "The Townless Highway And The Highwayless Town", 1929, 21, Box: 178, Folder: 13. MacKaye Family papers, ML-5. Rauner Library Archives and Manuscripts, https://archives-manuscripts.dartmouth.edu/repositories/2/ archival_objects/425746.

76. However, ideas developed for the Bronx River Parkway influenced the design of interstate highways built after the war. These included limited access and traffic separation.

77. Aurbach, *op. cit.*, 206–207.

78. New York (State) Bronx River Parkway Commission, *Report of the Bronx River Parkway Commission: Appointed under Chapter 669 of the Laws of 1906* (Trow Press, 1907). Also see "Historic American Engineering Record Bronx River Parkway Reservation," http://westchesterarchives.com/ BRPR/Report_fr.html.

79. Aurbach, *op. cit.*, 210.

80. "Saw Mill River Parkway," Wikipedia, https://en.wikipedia. org/wiki/Saw_Mill_River_Parkway.

81. "Sprain Brook Parkway," Wikipedia, https://en.wikipedia. org/wiki/Sprain_Brook_Parkway.

82. "Hutchinson River Parkway," Wikipedia, https:// en.wikipedia.org/wiki/Hutchinson_River_Parkway.

83. "Taconic State Parkway," Wikipedia, https://en.wikipedia. org/wiki/Taconic_State_Parkway.

84. In 1929, Roosevelt was elected Governor of New York. Robert Moses was his Secretary of State. Moses, expert at obtaining and controlling funding, diverted state funds away

from construction of the Taconic to the parkways Moses built on Long Island, beginning with the Northern State Parkway. See "Taconic State Parkway," Wikipedia, http://en.wikipedia.org/wiki/Taconic_Parkway.

85. Donald L. Miller, *Lewis Mumford: A Life* (Grove Press, 2002): 480. "[A]nd whenever he was in Leedsville Mumford would tell friends from New York City who were coming by car to visit him to take [the] Taconic State Parkway, a winding ribbon of road through the Hudson River Valley."

86. Harold Faber, "Metropolitan Baedeker; Savoring the Scenic Delights Along the Taconic State Parkway," *The New York Times*, August 14, 1987, http://www.nytimes.com/1987/08/14/arts/metropolitan-baedeker-savoring-the-scenic-delights-along-the-taconic-parkway.html.

87. Robert A. Caro, *The Power Broker: Robert Moses and the Fall of New York* (Knopf, 1974).

88. Wikipedia has a list of all of Moses's projects: Robert Moses projects, Wikipedia, https://en.wikipedia.org/wiki/Category:Robert_Moses_projects.

89. "Henry Hudson Parkway," Wikipedia, https://en.wikipedia.org/wiki/Henry_Hudson_Parkway.

90. See The Met Cloisters location, Met Museum, at https://www.metmuseum.org/visit/plan-your-visit/met-cloisters and "The Cloisters," Wikipedia, https://en.wikipedia.org/wiki/The_Cloisters.

91. "Bear Mountain State Parkway," Wikipedia, https://en.wikipedia.org/wiki/Bear_Mountain_State_Parkway.

92. "Palisades Interstate Parkway," Wikipedia, https://en.wikipedia.org/wiki/Palisades_Interstate_Parkway.

93. "Merritt Parkway," Wikipedia, https://en.wikipedia.org/wiki/Merritt_Parkway. The Merritt Parkway Conservancy has a list of resources for learning more about the Merritt: see https://www.merrittparkway.org/parkway-resources. A recent article from the *New England Historical Society* referenced recent news about the Merritt: "7 Fun Facts About Merritt Parkway," https://newenglandhistoricalsociety.com/7-fun-facts-about-the-merritt-parkway/.

94. In particular, see Bruce Radde, *The Merritt Parkway* (Yale, 1996) and Laurie Heiss, *The Merritt Parkway: The Road That Shaped A Region* (History Press, 2014).

95. Parkways in New York City, The Roads of Metro NYC, http://www.nycroads.com/roads/pkwy_NYC/.

96. The Bronx River Parkway and the Taconic Parkway illustrate the model of the townless highway. Later parkways in Westchester County sometimes showed the influence of City Beautiful ideas. Farragut Parkway is the entrance to Hastings, New York, from the Saw Mill River Parkway. Named after a favorite son of the town, the wide tree-lined street (not a limited-access parkway) begins in a small park before passing the town's schools on its way to Main Street. Playland Parkway in Rye, New York, was first planned as an extension of the Cross County Parkway. It is a wide tree-lined street leading to Rye Playland, an amusement park and beach on Long Island Sound designed by Walker & Gillette

with Gilmore Clarke and built by the county in 1927. See "Playland (New York)," Wikipedia, https://en.wikipedia.org/wiki/Playland_(New_York).

97. "Henry Hudson Parkway Historic Overview," The Roads of Metro NYC, http://www.nycroads.com/roads/henry-hudson/.

98. Caro, *op. cit.,* 880–885.

99. The McMillan Plan was overseen by the Senate Park Committee, led by US Senator James McMillan and his Chief of Staff Charles Moore. See John W. Reps, *Monumental Washington; the Planning and Development of the Capital Center* (Princeton, 1967): 93–108.

100. See "Parkways of the National Capital Region, 1913–1965," United States Department of the Interior, National Park Service, https://dcpreservation.org/wp-content/uploads/2021/11/Parkways-of-the-National-Capital-Region-1913-1965.pdf.

101. "George Washington Memorial Parkway," Wikipedia, https://en.wikipedia.org/wiki/George_Washington_Memorial_Parkway. In the nation's capital, even interstate highways sometimes act like City Beautiful streets. I-395 approaches the Jefferson Memorial on axis as it crosses the Potomac.

102. "Clara Barton Parkway," Wikipedia, https://en.wikipedia.org/wiki/Clara_Barton_Parkway. Congress approved construction as part of the Washington Parkway, but construction of the two- to four-lane, 6.8-mile parkway only began in 1961. It was renamed the Clara Barton Parkway in 1989.

103. "Rock Creek and Potomac Parkway," Wikipedia, https://en.wikipedia.org/wiki/Rock_Creek_and_Potomac_Parkway.

104. The Henry Hudson Parkway has the same effect on Riverside Park. And it starts in the largest park in the Bronx, which Moses cut into six pieces with the Henry Hudson, the Mosholu Parkway, and I-287, one of the busiest highways in the New York area.

105. "Baltimore-Washington Parkway," Wikipedia, https://en.wikipedia.org/wiki/Baltimore-Washington_Parkway.

106. Also important to the simultaneous movements away from parkways and towards more reliance on cars was the Chair of the National Capital Planning Commission (NCPC), Harland Bartholomew. Bartholomew was one of the first important planners (as opposed to urban designer) who promoted the early twentieth-century ideal of the "City Efficient," prioritizing car travel over mass transit. The emphasis of the City Efficient on wider roads, utilitarian infrastructure, and modern traffic engineering was a conscious replacement for the City Beautiful. See Nathan Jackson, "Harland Bartholomew: Destroyer of the Urban Fabric of St. Louis," Nextstl, April 10, 2021, https://nextstl.com/2021/04/harland-bartholomew-destroyer-of-the-urban-fabric-of-st-louis/ and Harry Kollatz, Jr., "A Man With a Plan," Richmond, April 14, 2017, https://richmondmagazine.com/news/richmond-history/city-planner-harland-bartholomew/. For Bartholomew's connection to the City Efficient, see Joseph Heathcott,

"The Whole City Is Our Laboratory: Harland Bartholomew and the Production of Urban Knowledge," Journal of Planning History, vol. 4, no. 4, November, 2005, https://doi.org/10.1177/1538513205282131; and Benjamin Clarke Marsh, *An Introduction to City Planning: Democracy's Challenge to the American City* (Benjamin Clarke Marsh, 1909): Chapter VII. Marsh originally published the book himself. Legare Street Press republished the book in 2022.

Bartholomew believed that mass transit would fail in Washington, as the city's extensive streetcar system did by the mid-1950s (helped along by Bartholomew's policies), and that the future lay in developing a network of highways in and around Washington. The architects, urban designers, and landscape architects who argued for beautiful parkways proved to be no match for the combination of modern "scientific" planners, the authority of the highway engineering profession, and the almost-free money that flowed from the Federal agencies. Accepting the money meant accepting AASHTO's standards for post-war interstate highways.

The National Capital Park and Planning Commission called for a highway around the city in 1950, included in the first Federal-Aid Highway Act of 1956. Construction of the Capital Beltway began in 1957, and congestion soon followed. Initially built with three lanes in each direction, the Beltway now has sixteen lanes in some portions and still has infamous traffic jams.

The Beltway is now I-495, and I-66, I-295, I-395, and I-695 connect it to central Washington. The result is not "efficient." In 2025, the local NBC News station reported that Washington has the worst traffic in the country. See Joseph Olmo, "DC beats out Los Angeles for worst traffic in the country, study says," 4 Washington, July 23, 2025, https://www.nbcwashington.com/news/local/dc-beats-out-los-angeles-for-worst-traffic-in-the-country-study-says/3963031/.

Two points worth mentioning are that "Inside the Beltway" is a pejorative phrase used to describe Washington's political culture, and that President John F. Kennedy responded to criticism of Washington's auto dependency by appointing planners to the NCPC who promoted mass transit. In 1960, the Federal government created the National Capital Transportation Agency to develop "Metrorail," but sixteen years passed before the first portion of the network opened. Since that time, there have been many complaints about its operation. In 2016, the Washington Post ran "A complete guide to the major problems facing Metrorail." Written by Chiqui Esteban and John Muyskens, the article ran on April 24, 2016: see https://www.washingtonpost.com/graphics/local/metro-history-failures/.

107. You can find many of their names in the notes for this chapter. Several are described in Laurance Aurbach's book, *op. cit*. One of the most prolific designers was the landscape architect, engineer, and city planner George Kessler. Based in Kansas City, Missouri, Kessler worked in over 100 cities around the world, but mainly practiced in the Midwestern United States and the Great Plains. For his work in Kansas City, see George E. Kessler, "Kansas City Parks & Boulevard System," georgekessler.org, http://www.georgekessler.org/index.php?option=com_content&view=section&id=8&Itemid=77; Board of Park Commissioners. *The Park and Boulevard System of Kansas City, Missouri: Souvenir* (Board of Park Commissioners, 1920); and Suzanne Hogan. "The History Behind The Paseo, One of Kansas City's First Boulevards," *Dallas Morning News* and KCUR, May 23, 2017, https://www.kcur.org/post/history-behind-paseo-one-kansas-citys-first-boulevards. Also see "George Kessler," Wikipedia, https://en.wikipedia.org/wiki/George_Kessler. The Kansas City Public Library has an online video of a talk by Kurt Culbertson about Kessler's work in Kansas City: see "Kessler's Historic Parks and Boulevards: A Green Framework for a Great Kansas City," Kansas City Public Library, https://kclibrary.org/events/kesslers-historic-parks-and-boulevards-green-framework-great-kansas-city.

108. The Transect was briefly discussed in Chapter One on Page 46. For more on the Urban to Rural Transect, see the Center for Applied Transect Studies, https://transect.org/.

109. MacDonald, *op. cit.,* 1.

110. Soldier's Field Road, the Charles River Greenway, and Greenough Boulevard were some of those roads. Storrow Drive, named for James J. Storrow and built 1950–1951, was always a low-grade, ugly highway that despoiled the Charles River Esplanade. James Storrow supported preservation of the Charles River Basin and opposed construction of Storrow Drive.

111. A number of parkways in Essex, Middlesex, Norfolk, Plymouth, and Suffolk Counties that are part of the Metropolitan Park System are also listed in the National Register of Historic Places. Internal and border roads were an integral part of the Metropolitan Park Commission plans for natural and scenic, river, and ocean reservations from the beginning. These roads connect the various park sites in an expansive network, providing convenient, publicly accessible passage from one site to another along historic and scenic routes, and all within 10 miles of the Massachusetts State House. As the system evolved in the late nineteenth and twentieth centuries, parkways were in almost all of the metropolitan parks and became an integral part of Boston's regional transportation system during the early years of the streetcar and automobile, and beyond. Many of the National Register listed parkways, including the Blue Hills Parkway, Lynn Fells Parkway, and Charles River Reservation Parkways, were designed and landscaped by famous landscape architects such as Arthur A. Shurcliff, landscape architect for Colonial Williamsburg (1928–41), and the firm of Olmsted, Olmsted & Eliot. Arthur Asahel Shurcliff was born Arthur Asahel Shurtleff on September 12, 1870. Wikipedia explains why he changed his name: https://en.wikipedia.org/wiki/Arthur_Asahel_Shurcliff. Best known for his work in Williamsburg, Virginia, Shurcliff began his career

working for Charles Eliot on the Metropolitan Park parkway system. For more information, see "Arthur Asahel Shurcliff," Frederick Law Olmsted National Historic Site, https://www.nps.gov/people/arthur-asahel-shurcliff.htm; and works by or about Shurcliff at the Internet Archive: https://bit.ly/aashurcliff. The Old Harbor Reservation Parkway, which was originally known as the Strandway, was in fact the final component of Olmsted's "Emerald Necklace" in Boston. The Neponset Valley Parkway connects the Stony Brook Reservation in Boston and Dedham to the Blue Hills Reservation in Milton and Quincy. Many also have rotaries that were typically landscaped as miniparks with memorials or monuments in the center, for example, the Horace James Memorial Circle in the Hammond Pond Parkway. River Parkways follow the contours of the existing rivers, like the Alewife Brook Parkway, which, naturally, follows along the Alewife Brook in Cambridge and Somerville. Similarly, the ocean parkways follow the contours of the shoreline and are usually close to the shore. Winthrop Shore Drive is one of the eight ocean parkways in the system. Like all of the ocean parkways, its primary reason for existence is its connection to the adjacent beach and the ocean views. All of the parkways, no matter what their subtype, share the design characteristics of the system—the natural and scenic views and historic landscapes—but each has its own specific characteristics derived from its function and from the existing topography of the environments. In addition to their natural and scenic views, many historic roadside structures can also be seen along some of the parkways. The 1933 Metropolitan District Commission bathhouse was built along the VFW Parkway at Havey Beach. The Mystic Valley Parkway has adjacent elements that were added to the National Register of Historic Places through the Water Supply System of Metropolitan Boston Thematic Resource Area, including the Medford Pipe Bridge (1897–1898), Mystic Dam (1864), Mystic Pumping Station (1862–1864), and Mystic Gatehouse (1862–1868). The Furnace Brook Parkway features a view of the Quincy Armory (1924), as well as a partial view of the seventeenth to eighteenth century Quincy Homestead. Quincy Homestead, a National Historic Landmark added to the National Register in 1971, is a 1.8-acre site that has been owned and interpreted by the National Society of the Colonial Dames of America in the Commonwealth of Massachusetts since 1904. The Quincy Shore Drive Parkway features views of the historic Squantum and Wollaston Yacht Clubs, both of which extend out on piers to the water. Both two-story wood-frame buildings date to 1903 and were part of the original design for the Quincy Shore Reservation. By the final decades of the nineteenth century, the urban parks movement had already begun with many cities throughout the country creating large country parks or park systems. While Boston's Emerald Necklace park system neared completion, the loss of open space outside of Boston troubled many. Landscape architect Charles Eliot and journalist Sylvester Baxter proposed an extremely ambitious new idea for a regional park system. They hoped to generate public support to acquire several thousand acres of parkland in over two dozen communities within a 10-mile radius of Boston. See: "Metropolitan Park System of Greater Boston," National Park Service, https://www.nps.gov/places/metropolitan-park-system-of-greater-boston.htm;" Charles River Reservation Parkways," Wikipedia, https://en.wikipedia.org/wiki/Charles_River_Reservation_Parkways; "Blue Hills Parkway," Wikipedia, https://en.wikipedia.org/wiki/Blue_Hills_Parkway; "Blue Hills Reservation Parkways," Wikipedia, https://en.wikipedia.org/wiki/Blue_Hills_Reservation_Parkways; "Lynn Fells Parkway," Wikipedia, https://en.wikipedia.org/wiki/Lynn_Fells_Parkway; "Fells Connector Parkways," Wikipedia, https://en.wikipedia.org/wiki/Fells_Connector_Parkways; "Middlesex Fells Reservation Parkways," Wikipedia, https://en.wikipedia.org/wiki/Middlesex_Fells_Reservation_Parkways; "Revere Beach Parkway," Wikipedia, https://en.wikipedia.org/wiki/Revere_Beach_Parkway.

112. The landscape architect Charles Eliot was in charge of their planning. Eliot was briefly a partner at Olmsted, Olmsted and Eliot. He died in 1898, when he was only 37 years old, and the firm became Olmsted Brothers. His son Charles Eliot II and Frederick Law Olmsted Jr. were important for the design of Washington's parks and parkways.

113. Charles Beveridge, "Parks, Parkways, Recreation Areas and Scenic Reservations," The Olmsted Network, July 5, 2023, https://olmsted.org/parks-parkways-recreation-areas-and-scenic-reservations/; and "Louisville Park System," The Olmsted Network, https://olmsted.org/sites/louisville-park-system/. Also see Julia S. Bachrach, "The Olmsted Legacy in Chicago," Julia Bachrach Consulting, https://www.jbachrach.com/blog/2021/9/28/the-olmsted-legacy-in-chicago.

114. See "Buffalo Olmsted Park System Map | 6 Parks, 7 Parkways, 8 Circles," Buffalo Olmsted Parks Conservancy, https://www.bfloparks.org/mapandguide/; and "Buffalo's Olmsted Parks and Parkways System," Olmsted in Buffalo, https://www.olmstedinbuffalo.com/. Also see "Parks and recreation in Buffalo, New York", Wikipedia, https://en.wikipedia.org/wiki/Parks_and_recreation_in_Buffalo,_New_York.

115. See "Lying Lightly on the Land: Building America's National Parks Roads and Parkways," National Park Service, http://www.nps.gov/history/hdp/exhibits/lll/overview.htm. See also "Highways in Harmony," National Park Service, https://www.nps.gov/parkhistory/online_books/hih/generals/index.htm.

CHAPTER THREE

STREET SYSTEMS AND NETWORKS

STREETS DON'T FLOAT IN ISOLATION. Good streets are inseparable from the city or town where they sit, integrated into a larger system of public and private spaces. We cannot separate our understanding of a street or block from the particular details of the place where

◄ **Figure 3.1:** Rome, Italy. Historic Center. See Figures 3.89 and 3.90 and the section "Networks and Connectivity: Analysis Method" (pages 320-323) for a discussion of the street network in Rome's *Centro Storico. Courtesy of Google Earth*

we experience them. The context might be the immediate one of the adjacent streets, or it might be a larger one, such as the history of the city where the street is found or the form of the city itself.*

An urban designer must simultaneously think of the design of the individual street and how that street will operate in its larger setting. In this chapter, we explore useful streets, both historic and new, in pairs and in sets, and in grids, patterns, and networks. Here, policy and design intersect. In this holistic way of thinking, removing an anti-urban inner-city highway to make a new boulevard and reconnect the grid, inserting a new street to create more routes for walking, rewriting the development rules for groups of corridors, or measuring a neighborhood's walkability to improve it are all part of the same work.

---

*Think of the experience of getting to know a new city, town, or neighborhood. The first time you walk to a new place without consulting a map—or zooming out to look at the surrounding context on a phone—streets can feel isolated. But after walking the neighborhood and perhaps the route several times, your mind pieces together the streets, and the experience on the street changes. Places where walking is comfortable and inviting offer a variety of routes and experiences.

Figure 3.2: Charleston, South Carolina, and Savannah, Georgia. Historic maps drawn at the same scale. Both cities have beautiful streets, but they became beautiful in different ways.

THE STREETS OF CHARLESTON AND SAVANNAH

Some of our favorite streets are found in Charleston, South Carolina, and Savannah, Georgia, two southern American cities only one hundred miles apart. Two of the most beautiful cities in North America, Charleston and Savannah have different histories, reflected in their streets (Figure 3.2).

Charleston and Savannah are beautiful because their historic streets are beautiful, but the cities and their streets became beautiful in different ways.[1] Savannah, planned by General James Oglethorpe in 1733, has the most intelligently varied pure grid in America, perhaps the world. In contrast to that orderly plan, the old streets of Charleston are a ragged agglomeration of more conventional American grids built over time, first by the Lords Proprietor who owned the Carolina colony and later by the developers who came after them. As a result of that piecemeal process, a map of Charleston can look in places like a patchwork quilt (Figure 3.3). With some notable exceptions, the beauty of the individual streets of Charleston depends more on the buildings that line the streets than the artfulness of the streets. Savannah, on the other hand, has good and sometimes great buildings, but the planning of the streets is an essential part of the city's greatness and beauty.

Savannah, planned by General James Oglethorpe in 1733, has the most intelligently varied, pure grid in America, perhaps the world.

Figure 3.3: Charleston, South Carolina. Figure-ground drawing. *Courtesy of Dover, Kohl & Partners*

In his book *Architecture, Men, Women and Money in America, 1600–1860,* the banker and historian Roger Kennedy explains that different political and economic histories in the two cities affected the evolution of their physical form. Before the American Revolution, Charleston was the fourth-largest port in the American colonies, the wealthiest and largest city south of Philadelphia. The port continued to grow after the war, but Charleston's population was half the size of New York's, and for a long time its planters were content to let the British control the trading and shipping from Charleston to England and its colonies. The city became known for the luxury items its wealthy residents imported from the home country. A number of its inhabitants were English Loyalists who sat out the Revolution in London.[2]

In the early nineteenth century, Charleston's economy and port grew more slowly than those in Boston, New York, Philadelphia, and Savannah. According to Kennedy, Charleston was led by an oligarchy dominated by a large contingent of wealthy planters who stayed in the city and lived an increasingly patrician lifestyle, with more emphasis on conserving money than making it. The importing of luxury items from Europe continued, and in the War of 1812 many Charlestonians were reluctant to oppose British sanctions.

In contrast, Savannah's growth came during and after the War of 1812. Thanks to new inventions of the Industrial Revolution like the cotton gin and the steamboat, the market boomed and the city's share of the shipping of cotton surged. Cotton and other goods flowed down the Savannah River from land previously closed off from easy access to trade. In the first quarter of the century, Savannah raced past Charleston as a center of trading and shipping.

Architecture in Charleston and Savannah

The architecture built in the booming city of Savannah in the early nineteenth century was innovative and adventurous, particularly in the hands of transplanted English architect William Jay. The architecture of Charleston at the same time was a conservative continuation of Charleston's unique and beautiful building type, the Single House (Figure 3.4). This reflected its patrician society, a contrast to the more entrepreneurial culture of Savannah.

The image of a row of refined and restrained Single Houses on shaded, tree-lined streets became the dominant architectural and urban image of Charleston. The Single

House type was built in the city for over one hundred years, continuing a tradition that never lost its stylistic connection to eighteenth-century Classicism. The tradition included a long-lasting Classical Vernacular related to eighteenth-century Georgian architecture but simultaneously particular to the region. The simple Classical aesthetic also dominated the design of other building types in the city, giving Charleston a character strongly connected to its roots in the 1700s. No other American city has such a consistent collection of outstanding Classical buildings shaping its streets (Figure 3.5).

When we walk around a city we know well, we keep an unconscious connection to other parts of the city. Walking the old parts of Charleston's main commercial street, King Street, surrounded by mixed-use, party-wall buildings with ground-floor stores, we consciously or unconsciously draw connections between the architecture of King Street and the Classical Vernacular we know from the residential streets. The fact that the commercial buildings are rooted in the same cultural past and architectural sensibility as the freestanding Single Houses supports those connections.

Not coincidentally, the buildings on both the residential streets and the commercial streets tend to be on narrow building lots. There is an urban legend that Charleston's lots are narrow because the Lords Proprietor taxed the lot frontage or the number of windows along the street. That is supposed to explain why the Single House has its narrow side, just one room deep, along the street. Architectural historian Robert Russell, former Director of the College of Charleston's Program in Historic Preservation and Community Planning, says there is no historical record to support the legend, however, which was not published until the 1980s.[3]

To explain the narrow lots, Russell notes that as landowners subdivided the large original lots over time, long, narrow lots worked best on the deep blocks, because early business owners wanted to be on Broad Street, where there were only a few blocks to subdivide. In other old cities, large blocks were frequently divided over time with alleys and new streets that gave smaller lots and a separation of houses and businesses. But the Lords Proprietor discouraged subdivision of the blocks, so businesses and residences were frequently combined on the deep, narrow lots, with stores in the front and houses behind. Supply and demand therefore naturally led to narrow storefronts on Broad Street.

Figure 3.4: Maiden Lane, Charleston, South Carolina. This 1937 photograph of Maiden Lane shows the characteristic streetscape of Charleston, created by the Single House form. "Single" refers to the one-room width of the houses: they also have a porch facing a side yard, perpendicular to the street. The regular pattern of yard-porch-house, yard-porch-house along the street establishes a spatial rhythm in Charleston streets. Modern development replaced the Single Houses seen here with larger new buildings. *Library of Congress / Carnegie Survey of the Architecture of the South*

Figure 3.5: 90–94 Church Street, Charleston, South Carolina. Two Single Houses from the 1760s flanking an 1830s replacement show the continuity over time in Charleston's home-grown building type. *© 2013 Max L. Hill, III*

In time, the residents of the growing city built more and more houses away from the commercial streets, while keeping the distinctively narrow and deep lots, and the form of the Single House evolved: a simple building mass, one-room wide for cross ventilation, was turned perpendicular to the street. Each floor (with a minimum of two floors) had two rooms, separated by a stair hall in the center of the block, removed from the street. A porch, called a "piazza" by Charlestonians (who put no T-sound in the word), ran along the side of the house—usually on the south or the west side, where it would block the sun. At the street, a Classical doorway enclosed the piazza, making it a semipublic space; this was unlike the typical Southern porch, where there was more interaction with the street. The kitchen, the outbuildings, and the slaves and servants were at the back of the lots, so the residents of the houses socialized along the side garden, while office space gravitated towards the back of the house, at the end

of the piazza. Built from the 1780s until the 1890s, these narrow houses—made of brick, stucco, stone, or wood, with Classical columns and ornament—gave beauty, rhythm, and continuity to the simple streets and blocks of Charleston.

These streets still exist because Charlestonians value their past. Since 1783, the motto of the city has been *Aedes Mores Juraque Curat,* which Charlestonians translate as "She guards her buildings, her manners, and her laws." They built new buildings in their own traditions, which include the Single House and its "northside manners." While not required by law, the builders religiously followed the concept of northside manners: the convention dictated that the northern (or eastern) sides of the Single Houses have small windows, because one shouldn't watch one's neighbors sitting on their piazzas. (You *could* watch, but it was bad manners to talk about what you saw.)

Charlestonians also preserved many buildings constructed before the advent of the large building types of the late nineteenth and twentieth centuries. In 1931, the city passed the first preservation ordinance in America, to protect its "Old and Historic District." The area was declared a National Historic District in 1960, just as the urban renewal that plagued so many American cities was gaining sway. That is another important reason why the city has so many beautiful streets.

Civic Art and Street Design in Charleston

The design of the individual streets in the variety of grids in downtown Charleston is rarely particularly artful, but because the designs embody the basic principles of placemaking, they provide a strong public realm and a good setting for Charleston's buildings. Most of the streets are not wide, and the beautiful buildings and trees lining them visually define their edges. (Meeting Street can be very wide, but when it has a canopy of live oaks, as it has south of Broad Street, it feels contained and comfortable.)

Figure 3.7: Meeting Street, Charleston, South Carolina. Looking north towards the South Carolina Society and St. Michael's Church in the distance. The church stands on the corner of Meeting and Broad streets, at the intersection known as the Four Corners of Law. Just to the south of the vantage point from which this photograph was taken is the Branford-Horry House with its piazza covering the sidewalk (Figure 3.6), similar to the way the porticos on St. Michael's and the Society building span the public sidewalk. This view from 1865 also shows that before Charleston's streets were paved, trees were planted in the roadbed. When the city paved the streets, they moved the trees to the sidewalks, giving pedestrians very little space. One way to narrow streets without the expense of moving the drains that run alongside the sidewalks is to create permeable tree-planting strips with bicycle lanes between the trees and the side-walks. This would also create better conditions for the trees to grow. *Library of Congress / Selected Civil War photographs*

The basic elements that make up the streets in the simple grids are usually parallel or perpendicular: the streets on most blocks are straight, the sidewalks and the buildings align with the street, the building setbacks are usually consistent, and there is often a regular rhythm established by the buildings.

Charleston traditionally planted street trees in rows, close enough to each other that the mature trees form beautiful canopies over the street. Like the buildings, the trees give a rhythm to the street. Palm trees, like the new palmettos on Market Street, function as architectural elements, providing a sense of enclosure. Old photographs show that when Charleston had dirt streets, the utility poles and trees were frequently in the roadway, and the sidewalks ran between the trees and the edge of the right-of-way (Figures 3.7–3.8). Unfortunately, Charleston today frequently has sidewalks that are too narrow for more than one person at the points where the trees and utility poles were moved out of the street and into the pedestrian space.

The original plan for Charleston had a continuous, orthogonal grid. As the city grew, individual developers platted private tracts without government oversight. Creeks and swamps on the low peninsula sometimes separated the extensions of Charleston from the old grid, so that when the new neighborhoods were built, there might be no direct connection to many of the older streets. One of the earliest extensions was Ansonborough, laid out by the British naval hero, Lord George Anson,

who named two of the new streets for himself (George and Anson streets), and three for his ships (the *Centurion*, the *Scarborough*, and the *Squirrel*—these three names were later changed). His land extended to what is now Market Street, which was then a creek bed. When the market was built over the creek, the original Charleston streets Church and State were connected to Ansonborough simply where they fell by chance.

Three important buildings have prominent sites, terminating the views on significant city streets. The City Market building sits on the Market Street axis at the edge of Ansonborough, with a Classical temple raised on columns above a covered market space, providing a picture-perfect

Figure 3.8: Meeting Street south of Broad Street, Charleston, South Carolina. A beautiful mature canopy of live oak trees, only slightly south of the blocks shown in Figures 3.6 and 3.7.

example of a terminated vista (Figure 3.9). The Old Exchange and Dungeon is the building from which the Lords Proprietor ruled Charleston. A Classical monument terminating one of the widest and most important streets in Charleston, it said to the city, "We're large and in charge." The third example, also a beautiful Classical building, is St. Philip's Church in a bend on Church Street, brought about by felicitous circumstances described in Chapter Two (page 196 and Figure 2.149).

St. Philip's is unusual among Charleston's churches. Once called "the Holy City" because of its early tolerance towards different religions and the steeples that dominated its skyline, Charleston nevertheless has a tradition of midblock churches with no setbacks; the mundane siting diminishes their civic importance. Like their residential neighbors, their primary contribution to the streetscape is usually a monumental scale with a high level of Classical beauty. A number of the churches are white Greek Revival temples in a sea of red-brick Single Houses.

Like many of the handsome churches and temples, Charleston's civic buildings have sites with little of the civic importance that urban designers would normally give them. The Four Corners of Law at the center of the city has a unique and interesting plan, but it is stronger in conception than in execution. Four civic buildings stand proudly on the four corners of Meeting and Broad streets, the most important intersection in the city.

Gradually assembled from 1752 to 1896, the four corner sites hold the City Hall, the County Courthouse, the Federal Courthouse, and St. Michael's Episcopal Church (representing God's law). The original idea was that the civic buildings sit in a park surrounding the intersection of Meeting and Broad, with all four buildings slightly set back from the streets. In reality, the park spaces that remain are small and the sense of a larger park is weak. Nevertheless, the combined architectural presence of the buildings is strong, and the long history of commitment to this address on the part of the four institutions makes the Four Corners of Law an anchor in the downtown (Figure 3.3).

Civic Art in the Plan of Savannah

Savannah's more eclectic architecture does not consistently rise to the level of beauty seen in Charleston's buildings, and yet Savannah too is one of the most beautiful cities in America. Oglethorpe's plan for the city combines a variety of well-proportioned streets and spaces in an intelligent layout that shows how rich a grid can be: its regular rhythm of squares and streets of varying widths give a hierarchy, order, and majesty to the grid (Figure 3.2).

The original design for Savannah was an expression of Enlightenment ideals in which the rational grid symbolized a correlation between the underlying natural order of the plan and divine principles of social equality and civic beauty. In his book *The Oglethorpe Plan: Enlightenment Design in Savannah and Beyond*, the planner and historian Thomas D. Wilson documents that these included the belief that a balanced and just government depended on the fair distribution of land and the perceived benefits of land ownership: Oglethorpe had a plan for a slave-free society of small farmers owning land in what he called "agrarian equality."[4]

Likely precedents for the plan were numerous: colonial towns planned by the ancient Greeks and Romans, ideal cities of the Italian Renaissance, English gardens, and sources for Freemasonry in the Old Testament. Oglethorpe was a Mason, and he designed the plan of the city in cubits, a measurement favored by Freemasons, because the Old Testament described the use of the cubit for the planning of ancient cities and Solomon's Temple. Also found in the Classical text *Ten Books of Architecture* by Vitruvius, the cubit was equal to 1½ feet.

Oglethorpe organized his master plan with modular units he called "Wards." The wards were square (450 cubits by 450 cubits, or 675 feet square), designed to be repeated as the city grew—although Wilson argues convincingly that Oglethorpe was an agrarian who never intended Savannah would grow as large as it has. He made

all land in a square mile around Savannah common land owned by the city. Originally, Savannah residents farmed on the common land. Over time, when the city needed to expand or raise revenue, the city fathers could plat and sell land for the private development of new wards.

Oglethorpe designed six wards and initially built four. He organized each ward around a central square; the first four built were Johnson Square, Wright Square, Ellis Square, and Telfair Square. North–south and east–west streets ran through the centers of the wards, on axis with the centers of the squares. The broad streets had 50-cubit (75-foot) rights-of-way until they reached the squares, where the streets split into roadways just half as wide that ran around the perimeters of the squares. Bull Street (Figures 3.10 and 3.11) was the first north–south broad street, and President Street (originally named King Street) is an example of the early broad streets on the cross-axis.

Each square has only one north–south street going through, but the streets along the top and bottom of each square continue as 25-cubit (37½-foot) streets going east and west. Oglethorpe designated the four blocks flanking the square "Trust Lots," which were reserved for civic buildings. Those are small lots for civic buildings, 40 cubits by 120 cubits, or 60 feet by 180 feet.

Figure 3.10: Bull Street, Savannah, Georgia. Looking south towards Johnson Square and Wright Square, in a photograph taken in 1901 from the Savannah City Hall. The view clearly illustrates how close the squares are to each other. *Library of Congress / Detroit Publishing Company Photograph Collection*

Figure 3.11: Bull Street, Savannah, Georgia. Looking north towards Wright Square and the Savannah City Hall around 1910. Today the squares are full of mature trees that block all long vistas except at street level. *Library of Congress / Detroit Publishing Company Photograph Collection*

The other blocks in the ward are four Tything (residential) Blocks, two to the north of the square and two to the south. These are bisected by narrow east–west lanes that are only 15 cubits, or 22½ feet, wide. The larger blocks are 135 cubits by 200 cubits (202½ feet by 300 feet). Originally, each block had ten residential lots that together shared one square mile of farmland outside the city. Over time, landowners sometimes divided or combined lots, and there are a few places where the lanes were closed to combine the 90-foot lots with the 22½-foot alleys to make a lot 202½ feet long—a superblock in the context of Savannah.

By 1851, Savannah had twenty-four wards based on Oglethorpe's model. Some street and block dimensions were modified slightly in the later wards. Oglethorpe himself made Johnson Square larger as construction began, to accommodate government buildings and make it the most important square. The next important change came in the building of the roads around the wards. The original plan had a 50-cubit (75-foot) east–west through street between the wards (Broughton Street was the first), and a 50-cubit east–west through street, Bay Street, along the river. A north–south through street between the original wards, Whitaker Street, had a narrower 30-cubit (45-foot) right-of-way. But as Savannah grew, two of the 50-cubit perimeter streets were built as wider boulevards with green medians. The first was Oglethorpe Avenue, which was 110 cubits wide, or 165 feet (Figure 3.12), and the second was Liberty Street, 90 cubits, or 135 feet wide.

Figure 3.12: Oglethorpe Avenue, Savannah, Georgia. A nineteenth-century view of one of the broad boulevards between Savannah's wards before Organized Motordom turned the central space into a Vegetative Containment Zone where Moving Hazardous Objects (the picnickers) were prohibited.

This change from Oglethorpe's plan was easy to make because the wide streets were built in the open land surrounding Savannah that the city held for expansion. The boulevards enriched the variety of the plan with a different type of park space: old photographs show people picnicking in the linear parks. Eventually, streetcars ran down the medians, as part of the system that enabled Savannah to develop streetcar suburbs. Today, unfortunately, the streetcars are gone but the public works department has planted the medians with bushes that prevent people from crossing the boulevards anywhere except at the intersections, because the east–west avenues carry a lot of the through traffic in the city. The medians still enrich our mental image of the plan, but more simplistically than they once did.

Details in the plan highlight its clarity, like the sidewalks that continue straight through the squares and the regular repetition of the squares themselves, which always puts the pedestrian close to a square (Figure 3.13). On the broad streets that bisect the squares, the sidewalks on the long straight streets are aligned with the sidewalks that go through the squares, so the walker understands that this walk literally goes on for miles. The long vista is softened by the trees in the square, and the walker feels a longer axis than he or she actually sees. The axiality is powerful (Figure 3.14).

The cool, shady squares have a commanding presence. In the hot, humid climate of Savannah's summer, it is noticeable that the checkerboard plan reverses the pattern of medieval cities. There, narrow streets are frequently in shade, and walkers see sunlight in the distance where the squares open up. Savannah's squares are heavily planted with live oaks that provide restful shade, while the trees on the wide streets are usually too small to provide a canopy overhead. There are noticeable exceptions, like Jones Street—a wide, canopied street where some of the short blocks feel almost like piazzas (Figure 3.15).

With South Carolina to the north, blocking municipal purchase of common land on the far side of the Savannah River, the city grew to the south along the

Figure 3.13: Bull Street at Madison Square, Savannah, Georgia. As the wide north–south streets come into the squares, they narrow to half the width and become part of the squares, functioning almost like shared spaces. Without the design elements in modern roundabouts that prioritize traffic flow over pedestrian safety, such as splitter islands and gentle curves, drivers slow down, knowing there might be people crossing the street around the corner, in the middle of the block. *Courtesy of Kevin Klinkenberg*

Figure 3.14: Madison Square, Savannah, Georgia. Looking south along one of the walks that continues on axis through the squares. In person, the walker sees the next square ahead, and the stores along the street in between.

Figure 3.15: Jones Street, Savannah, Georgia. Under the state tree of Georgia (the live oak), the wide-but-short street feels almost like a piazza. The gentle undulation of the red brick road surface slows cars down and contributes to the comfortable feel of the space. See "The Evolution of Street and Sidewalk Pavements in North America" later in this chapter.

axis of Forsyth Park. The one-way, north–south streets like Whitaker Street, which have narrow sidewalks and few trees or stop signs, frequently became raceways. But the traffic on the broad streets that goes through the rectangular squares—like Savannah's central axis, Bull Street—naturally calmed by the conditions around the squares. The narrow streets there are almost shared spaces, and pedestrians feel safe slowly crossing the streets in the Southern heat.

The architect and urbanist Christian Sottile, former Dean of the School of Building Arts at the Savannah College of Art and Design and the designer of an extension of the city based on the ward system, says the residents of Savannah are well aware of the elements of the grid and how they interrelate. The squares and the various street types are the dominant image of the city for them. The legibility and clarity of what he calls its "radically clear rationality" are a comfortable part of their daily lives as they move around the city.[5]

Sottile also points out that the building lots created by the small blocks influence the way we experience the city. The modestly sized Trust Lots flanking the squares seem smaller with modern building types on them than they did in the eighteenth century. Now each lot is frequently filled from lot line to lot line by a single church, institutional building, or government building, often raised on a Classical plinth. The effect can be monumental and dignified, giving civic importance to the squares that have these buildings. For the largest or most complex buildings on the civic lots, the streets at the top and bottom of the square become service streets with loading docks, while the larger streets on axis with the squares remain prime pedestrian streets. Away from the center, city residents turned many of the Trust Lots into mansion sites, because as the number of wards grew, there were more Trust blocks than were needed. Some of the mansions were later converted to civic uses, like the Telfair Museum, which occupies one of William Jay's best designs (Figure 3.16).

The 90-foot lots in the Tything Blocks encourage modest urbanism and build-out. Most of the buildings on these sites extend to the lot lines and go straight up from the street without additional setbacks. When the city allowed some developers to close the narrow lanes, the discipline was relaxed. One of the most egregious examples is the Hilton Savannah DeSoto hotel, which inappropriately pulls back from the street to make a drop-off space for guests. Low terraces along the sidewalk have clumsy details and feel out of place in Savannah. Equally out of place is the looming mass of the hotel's minimally detailed concrete tower.

Bull Street and Broughton Street
Main Street

The Hilton Savannah DeSoto sits on Bull Street, Savannah's central north–south spine. City Hall sits on axis at the northern end of the street, creating the only vista terminated by a building in the old part of Savannah. To the south, Forsyth Park interrupts the Bull Street axis with a broad, green swath that makes a good juxtaposition to the smaller squares. Bull stops at the park and resumes at its southern end.

Bull Street is not Savannah's main retail street. There are many stores along Bull, but the strongest shopping street is Broughton Street, a wide street perpendicular to Bull. That is because long-distance traffic in the center of the city, heading south to the suburbs, moves to north–south streets that are not interrupted by squares. These become raceways, where fast traffic makes the streets unpleasant for pedestrians. They deliver traffic to the east-west streets, Broughton and Bay. Broughton Street is a shopping destination, and Bay Street has many bars and restaurants, showing that even in auto-oriented cities, commercial streets don't have to be on the main traffic routes, as long as they are convenient to them.

Forty or fifty years ago, Broughton Street and Savannah itself fell on hard times. Shopping centers, shopping malls, suburban office parks, and white flight moved much of the money in the city to its periphery. But in 1978, four individuals founded the private Savannah College of Art and Design (SCAD), which has grown from 71 students in the first year to more than 10,000 today. The school and its students brought money and investment to Savannah. The old department store on Broughton Street houses the school's library, and the movie theater is one of SCAD's theaters. The college has followed a program of buying and restoring old buildings around the city that has greatly benefited Savannah and shown the durability of good architecture and great urbanism.

Figure 3.16: Alexander Telfair House/Telfair Academy of Arts and Sciences, 121 Barnard Street, Savannah, Georgia. William Jay, 1818–1819. Jay was an English architect who brought an inventive combination of the latest Regency Style architecture and the Greek Revival to the booming city of Savannah. At the same time, Charlestonians one hundred miles to the north were continuing to build the traditional Single Houses they had built for decades. Since 1886, the mansion has housed the Telfair Museum, which also owns another Savannah house designed by William Jay. *Library of Congress / Carnegie Survey of the Architecture of the South*

THE EVOLUTION OF STREET AND SIDEWALK PAVEMENTS IN NORTH AMERICA / ROBIN B. WILLIAMS

Introduction

Given the push to convert conventional, automobile-focused roadways into more "complete streets" and the growing concern for urban environments to offer a "sense of place," urbanists would benefit from a clearer understanding of how modern streets evolved. The transformation of streets from dirt roadways prevalent before 1850 to the corridors of asphalt that dominate our cities today responded to a range of public health, safety, social, economic, and technological needs and concerns. As cities grew with industrialization, so too did the need to address public health issues posed by dirt streets and to combat the growing number of urban fires. Separately, the emergence of bicycles and automobiles as new modes of transportation put pressure on civic officials to improve streets by replacing dirt with some form of pavement—a challenge that became known as "the pavement problem"—with each city devising a distinctively local set of solutions. Cities experimented with a wide variety of paving materials, with private citizens playing an active role in the selection and funding of such improvements. The outward spread of cities into rapidly growing suburbs, facilitated by the installation of street railways during the second half of the nineteenth century, led to a gradual change in social uses of streets and an increased focus on transportation. The dramatic rise of automobile usage, coupled with the development of synthetic sheet asphalt in the 1920s, allowed for the faster and more efficient flow of cars, which in turn prompted the adoption of jaywalking laws that defined the car-centric nature of streets.

Cities of Dirt

The installation of sidewalks and street pavements represented important steps in the modernization of urban environments in the nineteenth century. Before being paved, dirt streets posed significant problems to citizens, especially in wet weather when they became

Figure 3.17: Horses and delivery wagon stuck in mud. Toronto, Canada. 1914. *Unknown author / Wikimedia Commons / Public Domain*

boggy quagmires that not only harbored diseases but also could impede the movement of people and vehicles (Figure 3.17). Pavement became an important factor in fighting urban fires (to facilitate the movement of fire engines), so much so that Sanborn Fire Insurance Atlases often indicated whether a street was "paved" or "unpaved." Such atlases helped insurance brokers assess risk by looking at various fire-fighting-related variables.

Privileging Pedestrians in the Early Nineteenth Century

In most North American cities, the first pavements installed benefited pedestrians, with a focus on sidewalks and crosswalks. Wood plank sidewalks typically came first and were gradually replaced by more durable materials, such as brick or flagstone. Although sidewalks generally exist on public property as part of the street corridor (defined as the space between opposing property lines), the installation of sidewalk pavement was initially the responsibility of the abutting property owner, resulting in a patchwork of differing sidewalk materials that can still be seen in older sections of cities—or by the random absence of a sidewalk in later suburban neighborhoods. Crosswalks, or "crossings," provided pedestrians with a means of traversing dirt streets without soiling their clothes and afforded better footing when streets were muddy. While permanent materials were desirable, such as parallel strips of elongated rectangular flagstones or slabs of granite, often combined with smaller blocks of stone in between, a more expedient method involved compacting dirt into a raised berm that resembled a modern speed hump.

The Street Pavement Problem

Paving streets proved challenging and costly. To solve the pavement problem, cities experimented with diverse materials in an attempt to balance durability, traction, economics, availability, and noise considerations. These street pavement experiments fell into two general groups: natural materials (stone, wood, or shell) simply cut or shaped for use and introduced mostly before 1850 (though they would see continued use after 1850); and natural materials altered through some process, such as heating (asphalt, vitrified brick, and scoria), impregnation (wood treated with creosote) or forming a mixture (concrete), that were introduced after 1850. It was not uncommon by the beginning of the twentieth century for a city to employ as many as twelve different paving materials at the same time, as documented in specialized paving maps created by municipal engineers or streets departments.

Cobblestones

The earliest street pavements in many cities, especially port cities, were cobblestones, naturally rounded stones that arrived as discarded ship ballast (Figure 3.18). Evidently first used in America by the Dutch in New Amsterdam (now New York) in 1658 on what became known as Stone Street [see the Stone Street essay in Chapter Four, page 527], cobblestones were later introduced in Boston and Philadelphia during the eighteenth century and became common in many cities by the mid-nineteenth century. They were cheap and durable, but bumpy, crude, and noisy. William Gillespie, a professor of civil engineering at Union College in Schenectady, New York, criticized them in his 1847 manual on road-making as a "common but very inferior pavement which disgraces the streets of nearly all our cities."[6] Today, by contrast, for the few cities retaining cobblestones, they offer a charming reminder of a distant past and, for the astute, a glimpse into the complexity of global trade, as the individual stones represent astonishing geological diversity. In Savannah, one cobblestone displays Chinese characters that date to 1798.

Figure 3.18: Cobblestone pavement, Prince Street, Alexandria, Virginia, 2016. *Courtesy of Robin B. Williams*

Macadam and Oyster Shells

Most early pavements involved some form of shaping, usually done on site. As developed by John Loudon McAdam in the 1780s, macadam pavement involved layers of differently sized gravel cut by hand, with the size of rocks regulated by passing them through a ring. Popularized with McAdam's publication *Remarks on the Present System of Road-Making* in 1816 (with eight subsequent editions through 1827) and *Practical Essay on the Scientific Repair and Preservation of Roads* (1819), "macadamized" pavement technology spread to the United States by 1823. But it wasn't until in the 1850s, when the use of blasting powder to break stones to the proper sizes made macadam pavement financially feasible, leading to its widespread and extensive use in the United States. As of 1904, macadam remained the most common type of pavement in some major cities, including Boston, New York, and St. Louis.[7] A cheaper regional variant, popular in the American South, was made with oyster shells, which were laid like modern gravel and cost a fraction of any other paving type. Both macadam and oyster shells suffered from issues of durability and complaints about dust.

Wood Pavements

Various forms of wood pavements offered the promise of a smooth, noiseless, dust-free, and reasonably affordable road surface. While plank roads gained widespread use of wood as a pavement outside of cities, a more urban use of wood as a paving material solution involved various forms of blocks—cubes, bricks, and cylinders—employed beginning in the 1840s. Samuel Nicholson popularized wood block paving through publications and aggressive marketing. Initially untreated, he experimented with different block sizes, foundations of sand and coal tar, and eventually treating the blocks with creosote. Despite these efforts, wood blocks suffered from poor durability, concerns about hygiene as they absorbed horse urine, and being slippery when wet. Yet, their smoothness and especially their noiselessness, what some called the "silent pavement," kept wood blocks popular through the early twentieth century, especially near financial exchanges, city halls, and courthouses (Figure 3.19). By 1904, Chicago had 738 miles of wood block streets (representing almost 18 percent of the city's pavement), and Detroit had 290 miles, constituting almost 77 percent of that city's pavement.[8] The species of wood used varied by region, with cedar most common in the north, mesquite in Texas, and yellow pine in the South.

Belgian Blocks and Iron Slag (Scoria) Blocks

In areas of a city where durability and load resistance were most important, such as in industrial areas and port facilities, Belgian blocks, rectangular cut blocks usually of granite but sometimes of sandstone or other

Figure 3.19: Wood block pavement, Hessler Court, Cleveland, Ohio, 2017. *Courtesy of Robin B. Williams*

material, became the dominant material from the 1850s onward. Their hard and somewhat rough surfaces could withstand the pounding of heavy loads in areas where few people cared about the noise created by horses' hooves and metal-rimmed wheels. Extensive expanses of Belgian blocks can still be found in the Soho and Meatpacking districts of Manhattan, DUMBO in Brooklyn, the Society Hill area of Philadelphia, and the Fells Point area of Baltimore (Figure 3.20). Another hard and durable, though rare, pavement involved casting blocks from iron slag (or scoria), a waste product from iron smelting. Expensive to produce, they found limited use despite an attractive blue glaze appearance. A few streets of scoria blocks survive in Philadelphia and numerous streets in Old San Juan, Puerto Rico.

Natural Sheet Asphalt

The challenge of a smooth, durable, and relatively quiet pavement appeared to have its solution in the development of natural sheet asphalt. The first method, developed in 1849 by Swiss engineer André Merian, involved grinding, heating, and compressing rock asphalt with a roller and was first used in the rue

Bergère in Paris in 1854. More revolutionary was the method first employed in the United States in 1870 by Belgian-born chemist Edmund J. DeSmedt, involving natural asphalt harvested by hand from Pitch Lake in Trinidad, then part of the British West Indies, and processed in the United States. Compared to the rock asphalt used mainly in Europe, natural asphalt had a higher percentage of bitumen, which made the pavement more durable. Despite being the most expensive form of pavement in the late nineteenth century, natural sheet asphalt quickly gained appeal for its smoothness and superior durability over wood. But like wood blocks, it absorbed horse urine and presented difficulty in cleaning by machine. The issue of absorption was resolved by a contemporaneous pavement innovation.

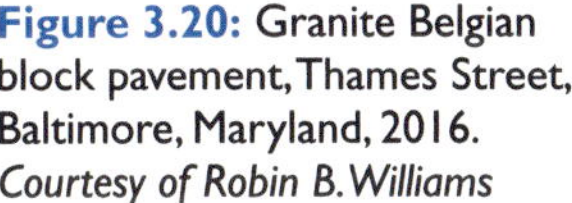

Figure 3.20: Granite Belgian block pavement, Thames Street, Baltimore, Maryland, 2016. *Courtesy of Robin B. Williams*

Vitrified Bricks, Asphalt Blocks and Concrete

Discovered by accident in Charleston, West Virginia, in 1870, vitrified bricks quickly spread to become the most popular paving solution for streets—strong, waterproof (no bad odors from horse waste), smooth, and moderate in cost. More than any other historic paving material, vitrified bricks survive in many cities and towns across North America, ranging in color from yellow or buff to red, dark purple, and black. Although some city dwellers initially opposed its use in residential areas due to the shrill sound created by metal horse hooves clattering on its surface, vitrified bricks nonetheless enjoyed a long period of use through the 1930s, perhaps since automobiles and their rubber tires generated less noise. The extensive networks of brick streets that characterized so many cities can still be found in St. Petersburg, Tampa, and Orlando, Florida, and in the German Village section of Columbus, Ohio (Figure 3.21). Asphalt blocks and concrete with irregular and large aggregate, both introduced near the end of the nineteenth century, joined vitrified bricks as smooth pavements, the former quieter but less durable and the latter stronger and more capable of carrying heavy loads, but subject to cracking and challenging to conduct below-grade repairs.

How cities selected which pavement to use and where to install it depended on a variety of variables—cost, availability, local needs, the performance of the material, and local or personal preferences, depending on the degree of input citizens were allowed. As a major municipal improvement, analogous to the onerous cost of installing public transit today, city leaders had to be selective as to which streets would be paved and with which material. Each city devised its own complex solution to the challenge of paving streets, creating a uniquely local paving "fingerprint" that was vividly documented in a 1904 study of street pavements in American cities (Figure 3.22).[9] The varying degrees to which historic pavements survive in cities further enhance this sense of a local pavement identity.

Innovations in Sidewalk Pavements

Sidewalk pavement and curbing exhibit even greater variety. While granite is the most common historic curbing material, slate, sandstone, schist, marble, and, in a few larger cities, metal of varying widths and sometimes featuring stamped patterns can often be found. More than any other part of the historic built environment, though, sidewalks possess and retain the greatest material and design variety. In New York and Philadelphia, one can find massive monolithic slabs of granite spanning the full depth of the sidewalk and sporting a variety of hand-chiseled textures. Pavement types exclusive to sidewalks include decorative glazed bricks with embossed geometric designs, hexagonal and octagonal concrete pavers produced in different colors, and, most extravagant, terrazzo, a composite material, either precast or poured in place, consisting of small chips, usually of marble, set within a

Figure 3.21: Vitrified brick pavement, Mohawk Street, Columbus, Ohio, 2016.
Courtesy of Robin B. Williams

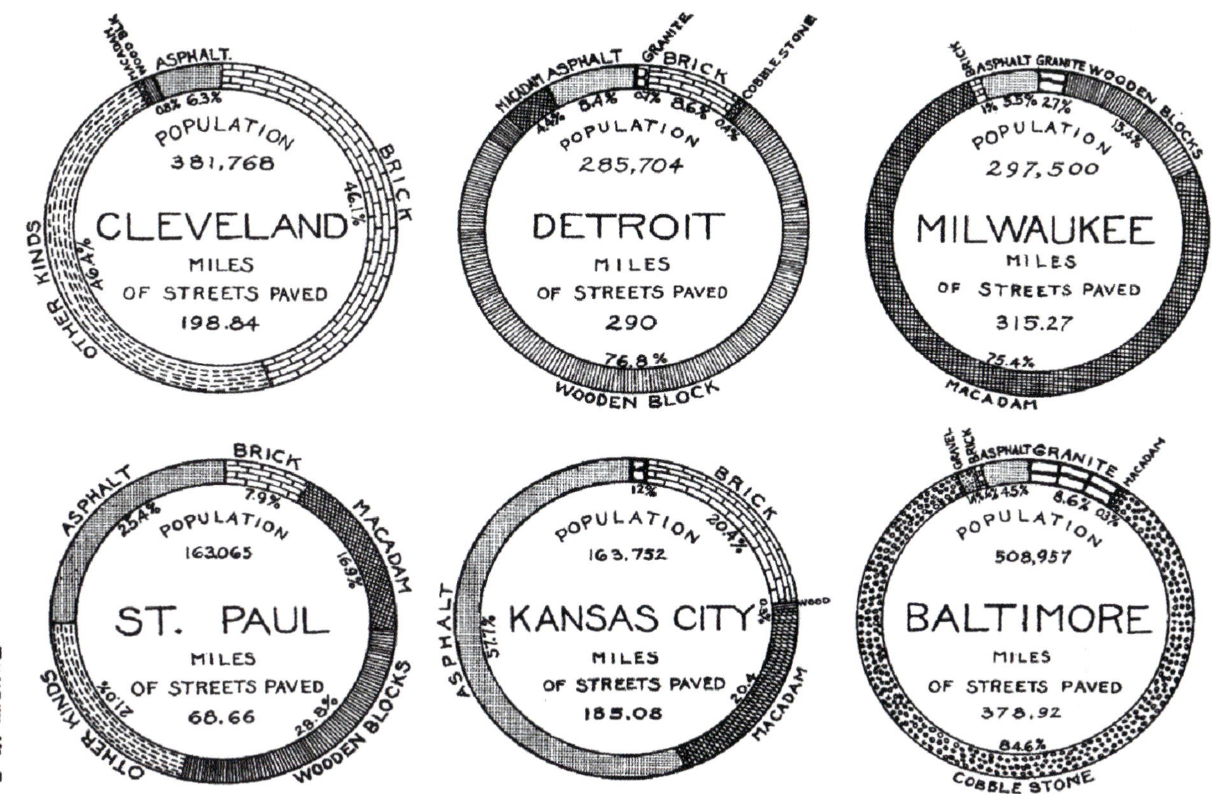

Figure 3.22: Street pavement diagrams. *John W. Alvord, The Street Paving Problem of Chicago, 1904 / Public Domain*

cementitious or polymeric binder. Introduced to the United States in the late 1890s, it gained widespread popularity in the 1920s with the introduction of thin metal divider strips, which helped control cracking and facilitated bold design patterns and the inclusion of words—both ideal for commercial applications. Downtown Los Angeles showcases the greatest variety and extent of terrazzo sidewalks (Figure 3.23).

Figure 3.23: Terrazzo sidewalk pavement, Clifton's Cafeteria, Broadway, Los Angeles, California, 2017. *Courtesy of Robin B. Williams*

Embedded Features

Other elements of the modern city—street signs, address numbers, commercial signage, manhole covers, prismatic glass vault lights (which permitted sunlight to filter into basements that extended under sidewalks), among others—were embedded into the surfaces of sidewalks and curbs. Such elements varied greatly in how they appeared in cities and could be sources of great local pride. Though encaustic six-inch-tall tiles used for spelling street names exist in other cities, they are most famously associated with New Orleans, which appears to have been the first to use them, beginning in the 1880s. A later tile technology, the one-inch square tile mostly used for bathrooms, found an alternative application in Houston beginning in the 1920s, where they formed thousands of blue and white address signs affixed to curb faces across the city.

The rise of automobile ownership corresponded with the introduction of cheaper, mass-produced synthetic asphalt by the 1920s. Together, they dramatically altered the modern street. More cars on the road and their faster movement, facilitated by the spread of asphalt, led to jaywalking laws that banished people out of roadways. Smooth, sufficiently durable, cheap, and no longer plagued by horse excrement, locally produced synthetic sheet asphalt allowed cities to cover over or replace older pavements, even in cities such as Charleston, South Carolina, and New Orleans, which were simultaneously spearheading a nascent historic preservation movement focused on buildings.

Civic Engagement and Pavement Preservation Battles

Recognition of the significance of historic pavement emerged with the expanding historic preservation movement in the mid-twentieth century. The relaying of a brick street in Wilmette, Illinois, as a 1930s W.P.A. project and a 1941 monument in Bellefontaine, Ohio, commemorating an early concrete street, anticipated later efforts to save, protect, and restore historic pavement that mainly emerged in the 1970s. Calls for preservation typically came from local residents, whose sense of place was defined by more than their private houses. Residents on Carlton Street in Toronto successfully persuaded civic officials in 1970 to leave their beloved red brick street alone. In Wilmington, North Carolina, more aggressive citizen action was required in 1973, when nearby residents, armed with

rakes, shovels, and hoses, quickly removed a layer of asphalt slurry seal mistakenly laid on a brick street (Figure 3.24). A decade later, in Columbus, Ohio, U.S. Congressman (and future governor) John Kasich assisted in the restoration of the city's brick streets, a campaign promoted with the slogan "Let them eat bricks." Recognition on the National Register of Historic Places arrived in 1975 when Hessler Court, Cleveland's only surviving wood block street, was the first street listed for its pavement. The nation's first legal protection for street pavement began in Racine, Wisconsin, in 1983, while St. Petersburg, Florida, was the first to protect sidewalks with its preservation ordinance of 2003.

Pavement and a Sense of Place

St. Petersburg's prevalence of hexagonal concrete pavers, with many sidewalks exploiting the variety of colors available to create alluring patterns, inspired the neighborhood of Old Northeast to adopt the hex pavers as a local identifier, gracing flags flown by private residents from their homes advocating to "Preserve Historic Old Northeast." More recent preservation

Figure 3.24: Residents remove slurry seal, Wilmington, North Carolina, 1973. *Wilmington Star*

initiatives highlight the broad social appeal of distinctive local signage. In Houston, the Blue Tile Project, established by Joey Sanchez in 2015, has successfully documented over 4000 of the city's curb-mounted tile signs through crowdsourcing, using a social media app he created. The project has spawned local branding with commercial signage, T-shirts, and even a local beer celebrating the signs. Sanchez explained how, "Through all channels—preservation, art/images and local business—we want to bring the blue tiles back to the city. If the blue tile font becomes associated to all things Houston on a national scale, we will have created the civic image of our dreams."[10] Meanwhile, the broad popularity of the New Orleans street name signs prompted local businesses to reproduce the distinctive tiles for souvenir merchandise and other products, further attesting to the power of street name signs as signifiers of local identity.

Pavement Restoration and the Practical Benefits of Traditional Pavements

Some cities have demonstrated the feasibility of restoring historic pavements in a manner that allows streets to meet modern traffic and accessibility needs. Notable restoration efforts have taken place in New York City, with numerous granite Belgian block streets in the Meatpacking, Soho, TriBeCa, and Dumbo sections relaid with smooth granite crosswalks to better accommodate wheelchairs, walkers, and strollers. In Orlando, many miles of asphalt have been removed by means of melting and scraping to reveal vitrified brick streets below. And in Toronto, a reimagining of the city's waterfront boulevard, Queen's Quay, as a complete street included in 2015 the creation of a broad sidewalk with the pink and gray granite paving cubes laid in large maple leaf patterns that celebrated that city's long local connection to the national symbol. There are multiple reasons to preserve and protect, let alone restore, historic street and sidewalk pavements—heritage value, economics, environmental benefits, and public safety among them.

Note: Visitors to Savannah, Georgia, can contact Professor Williams at the Savannah College of Art and Design (SCAD) to ask about his Historic Pavement tour of Savannah.[11]

THE ARCHITECTURAL AND URBAN CODES OF PARIS

The streets of Paris are among the most regulated and ordered in the world. The standardization of Parisian architecture and streets began with King Henri IV. He brought this urban standardization about through both regulation and example, setting a pattern that lasted for hundreds of years. In 1600, Henri IV made the first of two royal orders that set maximum dimensions for building projections and required urban buildings to follow build-to lines (*lignes d'alignement*) along streets. He also built two developments as king that established the classicizing and regularization of Paris architecture: the place Dauphine and the place Royale (now the place des Vosges). A third unbuilt square, the place de France, would have shared with the other two a common architectural language of repeated, often identical Classical facades. Between the place Dauphine and the place de France, Henri IV planned a new street called the rue de Turenne that was to run straight from one square to the other with rows of repeating facades. This street was not built until the time of Louis XIII, but south of the place Dauphine, Henri IV ordered the private owners of sites by the Pont Neuf to build identical Classical facades for the "beautiful ornament" of the city.

During the reign of Louis XIV (1661–1715), building types were regularized, and the dominant new type—private *hôtels* built by aristocrats—had a Classical architectural style that flowed from the Sun King and the Royal Academy of Architecture he founded. Palaces built by Louis XIV were influential and imitated, while the Royal Academy of Architecture disseminated didactic knowledge, with standards for the design and composition of harmonious Classical facades, as well as for proportions, materials, and details. These trends influenced all buildings in Paris, in what became known at the time as the *architecture d'accompagnement* (accompanying or supporting architecture). Pierre le Muet's *Maniere de bien bâtir pour toutes sortes de personnes* (*Good Building for All Kinds of People,* published in 1647), documented and supported the *architecture d'accompagnement* and was dedicated to the king. Le Muet presented a series of urban residential buildings that went from small to large while maintaining a consistent Classical taste, with supporting details.[12]

In 1666, influenced by the Great Fire of London, the French Treasury regulated building projections, and the following year, the Royal Bureau of Finance set a maximum height (8 *toises*, or 15½ meters) for the front walls of new houses in Paris and required that exposed timbers of existing houses be covered with plaster. In 1669, the rue de la Ferronnerie—a narrow Parisian street where Henri IV had been assassinated when his carriage was blocked—was widened as part of a speculative real estate venture by the church that owned the buildings along the street. As king, Henri's grandson Louis XIV imposed the first street ordinance in Paris. A fifty-two-bay row of houses was given Classical symmetry, rhythm, and details, with many of the classic elements of later Parisian architecture, such as uniform floor heights, moldings that ran across multiple buildings, an arched *entresol*, repeated elevations, and mandatory balconies. "By 1715," Anthony Sutcliffe writes in *Paris: An Architectural History*, "the triumph of Classicism had launched the belief that Paris was one of the world's most beautiful cities."[13] Contemporary authors identified symmetry, uniform heights, and architectural regularization as critical elements and pointed to the rue de la Ferronnerie and the royal squares built by Henri IV as examples of beautiful places. The trend grew, and in 1755 the Abbé Laugier published the important *Essai sur l'architecture* (*Essay on Architecture*) that promoted urban interventions with new streets lined with buildings made harmonious by regulation. Later authors like Pierre Patte endorsed the creation of urban beauty by code.

By the early eighteenth century, according to Sutcliffe, the Royal Academy of Architecture had firmly established the idea that there was a French national manner of architecture, based on proportion, a high standard of materials and execution, and the correct use of Classical design. Architecture was a frequent subject of publication, and the tone was usually didactic, with an almost exclusive focus on French examples. Jacques-François Blondel (who was unrelated to François Blondel, the founder of the Royal Academy a century earlier) was the most prolific and influential author. His four-volume *Architecture française* (*French Architecture*) and his nine-volume *Cours d'architecture* (*Architecture Course*) marked a move away from the principles taught by François Blondel and towards the emulation of examples. An important part of his work was an emphasis on the design of houses for the middle class.

It became common for all buildings on new streets to be designed by a single architect. Individual lots could come with the drawings for the facade, or even with the facade already built. In the eyes of the Crown, one of the most important urban improvements of the eighteenth century was the creation of the place Louis XV, now the place de la Concorde, where the axis of the Champs-Élysées meets the royal palace. At the time, the Champs-Élysées was a royal promenade on the western edge of the city, designed by André Le Nôtre for Louis XIV. Louis XV created the place Louis XV—now the place de la Concorde—and placed the statue of himself there that marked the edge of the city for expansion. A design competition for the site was won by Jacques-Ange Gabriel, who proposed two grand *hôtels* on the north side of the large square, bordered on three sides by formal plantings of trees. Instead of houses on the north side, Gabriel proposed that hôtels frame a new street to be called the rue Royale. Louis XVI expanded the square, and in the early nineteenth century, Napoleon ordered the great vista from the place de la Concorde to be completed with the construction of the Church of the Madeleine at the end of the axis (Figure 1.42).

The two hôtels and all the buildings on the rue Royale were designed by Gabriel, with royal approval. There was no use for the hôtels at first, and only the facades were built until uses could be found. For the rue Royale, the Crown imposed an architectural ordinance that precisely coded Gabriel's design. Royal letters of the patent required that Paris ensure that any surrounding streets be suitably harmonious if the streets were widened or realigned.

Louis XVI had a great interest in the improvement of Paris. In 1783, the royal authorities issued under his name a new set of building regulations for the city. At the same time, Louis XVI ordered an accurate survey of the city as the basis for the revision of the build-to lines established in the seventeenth century. The codes, revised a year later, were the first Parisian codes to relate building height to street width: the maximum height for new buildings was set at 60 feet (*pieds*, or 18 meters), but lower heights were set for streets less than 30 feet wide. The preamble to the code stated that this was in response to the building of apartment houses that, if too tall, were unhealthy for the populace.

> All of the codes operate under the Parisian belief that harmony and good taste are more important for a streetscape than complete artistic freedom, the antithesis of most architectural theory today. And yet how many new streets approach the quality of typical Parisian streets, famous around the world for their beauty?

The primary tactical strategy was a focus on natural light. Maximum heights were determined by a line drawn at an angle of 67.5 degrees from the bottom of a building properly aligned to the top of the opposite building. The maximum cornice height of 54 feet was set for 30-foot streets, including the cornices below attics and mansard roofs built in the place of attics. Formulas for roof heights gave an overall height of 64–69 feet, depending on the depth of the building.

Little building took place after the French Revolution in 1789 until Napoleon came to power in the early nineteenth century, but the royal regulations for the city remained in place. Napoleon's imperial ambitions came out in his work on the great axes of Paris, including the Champs-Élysées and the rue Royale, where he authorized the construction of the Arc de Triomphe on the former and the Church of the Madeleine on the latter. In 1801, Napoleon authorized the construction of the first stage of the rue de Rivoli, which was the most obvious way to extend the east–west axis of the Champs-Élysées blocked by the Louvre. The government built the arches and the vaults of the arcade designed by Percier and Fontaine but

had difficulty attracting private investment for the completion of the buildings. What little private construction there was elsewhere in Paris continued under the code of 1783–1784, with the exception of a revision in 1823 that changed the regulations for building projections and a height-limit change in 1848.

Apartment construction in Paris revived in the 1830s, but the next significant transformation of Paris didn't take place until the Second Empire. In 1852, the ambitions of Napoleon III, the ideas of Baron Haussmann, and the fruits of the Industrial Revolution combined to form much of the physical city we know today. Important for a book on street design are the *Grands Boulevards* and the Parisian building code of 1859. The roots of the new code lay in the regulations of 1783–1784, but the code was adapted to the new scale and scope of construction in the 1850s when the *Grands Boulevards* were plowed through the medieval neighborhoods of Paris. Buildings on the wider new streets could be 2½ meters taller than before (roughly one story), and the new code emphasized a horizontality desired by Haussmann to reinforce the dramatic perspectives of his long avenues and boulevards.

Figure 3.25: Rue de Rivoli, Paris, France. Begun by Percier and Fontaine, 1801, extended 1847. A nineteenth-century view looking west. See Figure 5.61 for a modern view of the rue de Rivoli, once again dominated by bicycles. *Public Domain*

Code revisions in 1882, 1884, and 1902 were evolutionary. An expert in the history of the codes can walk the streets of Paris and identify which codes were in place when the buildings were built, but for most of us, the changes are difficult to detect. The entresol comes and goes, balconies move up and down or in and out, and so the buildings at the east end of the rue de Rivoli are slightly different than the buildings at the western end, because they were built under a different code. The 1902 code was a response to a call for more stylistic freedom, enabling the design of Art Nouveau buildings. At the same time, all of the codes operate under the Parisian belief that harmony and good taste are more important for a street-scape than complete artistic freedom, the antithesis of most architectural theory today. And yet, how many new streets approach the quality of the average Parisian street, famous around the world for its beauty?

RUE DE RIVOLI AND RUE DE CASTIGLIONE, PARIS, FRANCE / JAVIER CENICACELAYA

Rue de Rivoli

Begun by Percier and Fontaine, 1801–1830; extended 1847–1854

Avenue

The rue de Rivoli in Paris is located on the north side of the Louvre Museum and runs for more than two kilometers (about a mile and a quarter), going from the district of the Marais on the east (near the place des Vosges), to the place de la Concorde on the west (Figure 3.25). The street was named in honor of the victory of Napoleon's army in the Battle of Rivoli against Austrian troops in 1797.

Conceived by Napoleon I, the street was designed by Charles Percier and Léonard Fontaine, the *Architectes de l'Empereur* (the imperial architects). Their architecture, interior design, and decorative arts, known as *le Style Empire*, were influential across Europe. They practiced a radical Neoclassicism that added references from Egypt to the common ones from Rome and Greece to make an eclectic and sophisticated Classical style.

When Napoleon moved the imperial residence from the palace at Versailles to the Tuileries Palace at the Louvre, he commissioned Percier and Fontaine to create a suitable street on the north side of the Louvre. As laid out by Percier and Fontaine, the rue de Rivoli was one of the longest straight streets in Paris. Later rulers extended the street

until it reached its present length. The part of the street designed by Percier and Fontaine is the most interesting.

Their portion of the rue de Rivoli is inexorably straight. To achieve such an alignment on the north side of the Louvre, it was necessary to literally chop off the *Cours d'architecture,* the "running course" of the facades of the existing *hôtels*; the absence of a continuous alignment of the facades of those hôtels was therefore replaced with a long, arcaded facade. The repetitive facade and the long roofline increase the grandeur of the street. The result is a kind of megastructure capable of competing with—or at least standing up to—the long north facade of the Louvre. It is interesting to note the way the roof is treated with a barrel vault that softens the roofline and recalls other significant Classical buildings, such as the Basilica of Vicenza.

Inside the arcades are shops that give the street some of its vitality. Percier and Fontaine opted for arches in the porticos, a choice in tune with their eclecticism. This became popular in later stylistic revivals, both Classical and medieval, as in the *Rundbogenstil* in Germany. If we compare those to the design by Percier and Fontaine, though, the rue de Rivoli appears more restrained—most likely due to the scale of the intervention. In smaller projects like the nearby rue des Colonnes, Percier and Fontaine showed a higher level of sophistication and invention. There we find an exquisite portico with arches, built in the Empire Style but at a different scale.

The arcade of the rue de Rivoli is one of the longest in Europe to have a single architectural treatment, and the street itself served as a precedent for the boulevards Baron Haussmann later cut through the medieval fabric of Paris during the reign of Napoleon III.

Rue de Castiglione

Perhaps by Percier and Fontaine, 1801

Avenue

The impressive effect of the rue de Rivoli can turn into a surplus of repetition over time. After a certain amount of walking, one feels like turning around the next corner and escaping into a street with more variety. One of those streets is the rue de Castiglione, which runs from the rue de Rivoli to the place Vendôme. Wider than the other streets that start from rue de Rivoli, Castiglione is also arcaded (Figure 2.7). It was conceived as a continuation of the longer street; the strong axial connection to the place Vendôme was the reason for keeping the same treatment of the facades.

The architecture around the magnificent place Vendôme is the work of Jules Hardouin Mansart, conceived around the end of the seventeenth century. It is one of the baroque *Places Royales* built during the reign of Louis XIV. The center of this type of square was meant to be occupied by a statue or monument to the reigning king. Such a landmark should be visible from as far away as possible. The effect of perspective was very important, therefore, and this explains the special treatment given to rue de Castiglione, as the anteroom or "entrance" to place Vendôme. The statue of Louis XIV was torn down during the French Revolution and later replaced by Napoleon with an obelisk modeled after Trajan's Column to celebrate the French victory at Austerlitz.

The rue de Rivoli and the rue de Castiglione are two of the best examples of Neoclassical urbanism. Following French and Baroque precedents, Neoclassicists laid out long perspectives, or allées, with the clear intention of making elegant promenades of some grandeur. The urban designers of the Neoclassical period were also interested in straight alignments to get more rational arrangements of all types of infrastructure. In addition, they wanted to improve urban hygienic conditions and create places where people could socialize.

STREETS TO DREAM OF IN TOULOUSE / NICHOLAS BOYS SMITH

Over the last decade French cities, towns, and villages have been getting better: greener, safer, more attractive, and far more pleasant to visit. The French are falling back in love with their streets, reimagining them emotionally, not just technocratically, and reinventing them as public places, not just roads through which to speed. None are doing it better than Toulouse, the ancient city of Romans, Visigoths, and troubadours, and *préfecture* of the Haute-Garonne in France's far southwest.

Figure 3.26: The Capitole de Toulouse (1750–1760). Surely one of France's most sumptuous city halls inside and out. *Courtesy of Nicholas Boys Smith*

Have you ever heard of Toulouse as a city center whose improved street design is revolutionizing local prosperity, benefiting from and in turn further catalyzing the region's booming economy? Probably not. It is not normally lauded as a case study. But the quality of Toulouse's city center street design is staggering. The results are profound. Walking round the city in October 2022, I could literally not find a single empty shop in the city center. (There must be some.) How have they done it? Well, it has to be admitted, they have a trick up their sleeve: the city center's buildings are almost uniformly beautiful, largely unscarred by war or traffic-modernism. None are finer than the Capitole in the city central's square, the epitome of Gallic civic pride.

Many of the city streets are built from the region's traditional "*foraine*" brick, which is large and flat (like a Roman brick) and comes in a lovely range of ochres, reds, pinks, and oranges. You can see why the French call it "*La Ville Rose.*"

They do have some white stone historically brought down the Garonne River from the Pyrenees. They also paint some bricks. This gives a gloriously polychromatic effect as stone and brick alternate from bay to bay, particularly in the wider, longer shopping streets such as the *rue d'Alsace Lorraine.*

Figure 3.27: *Foraine* bricks in modern Toulouse. *Courtesy of Nicholas Boys Smith*

Figure 3.28 and Figure 3.29: White stone and painted brick, alongside deep lime mortar, create a joyful red and white polychromy. *Courtesy of Nicholas Boys Smith*

In 2012, the city hired the Spanish architect and urban planner Joan Busquets to improve the city center.[14] What did he do? The overall strategy is one of putting people first. Cars are permitted in much (not all) of the city center, but they are second-class citizens compared to the teeming hordes on foot, bicycle, or scooter. In some streets, cars are contained by elegant bollards in narrow carriageways (Figures 3.30 and 3.31).

This is the approach taken in most of the central streets but there are very few cars. It's hard to drive through the town, and it is just easier to walk or cycle. The bollards are elegant, and the pavements and carriageways are "at grade," i.e., are the same level.

Figure 3.30 and Figure 3.31: Permitting cars, but as guests. *Courtesy of Nicholas Boys Smith*

Figure 3.32: Narrow street, wide pavement. *Courtesy of Nicholas Boys Smith*

Figure 3.33: Investing in stone for all time, not tarmac for tomorrow. *Courtesy of Nicholas Boys Smith*

Globally, many street improvement programs in historic cities seem embarrassed by their finely hewn beauty, littering the streets with lurid and unsustainable plastic bollards, smearing their cobbles and setts with tar and tarmac. But Toulouse has not just reveled in its physical beauty but has enhanced it. The materials are excellent. In Figures 3.33 and 3.34, you can see the pavement, edge of carriageway, and carriageway. There is no garish paint or ugly signs cluttering up the city. Details matter.

There is a very modest 20 kmph speed limit in place nearly everywhere. But, in Figure 3.30, all cars and mopeds (there are lots of mopeds) drive carefully because they have to. Look how close the moped in Figure 3.35 is to the pram. It would give most British or American traffic engineers a heart attack. However, this is, in fact, a very safe street, because people come first.

The rue de Taur is another street that does permit cars but uses design to slow them down. If you look carefully, you can see a car in the distance. In front, young children walk safely right beside the carriageway. They are not in danger. It is a place for *flâneurs,* not fast driving, and the shops are busier for it. Nearly everything modernist town planning and highway engineers did undermines urban prosperity and quality of life.

There is very little on-street parking in the city center other than for disabled permit holders (Figure 3.37). There is some parking below ground and more parking further out. Above all, there is no incentive or need to drive round and round "looking for a place." This is certainly not undermining shops' vitality though.

In contrast, there are lots of places to park bikes, including the city's shared bikes. The bikeshare system started in 2007.

Figure 3.34: A complex street with no need for signs. *Courtesy of Nicholas Boys Smith*

Other easy and pleasant alternatives include the city's tramway line (opened in 2010 with some gorgeous "green track" along boulevards) and the metro system created over the last twenty years with driverless, rubber-tired trains.

Did I mention the trees? Wherever the city opens up, for example at place Saint-Sernin, the city is planting gorgeous street trees with benches beneath. The combination of trees and those ochre bricks is magical, even against an autumnal slate sky on the day I visited. Street trees are proven to improve places, clean air, and slow traffic. Here, they perform their task magnificently.

There is some so-called "filtered permeability" that keeps cars away for at least part of the day. This is done in some town squares, between cafe tables or underneath those ubiquitous trees.

However, everywhere the same principle is core: people come first. That's often (very practically) expressed through continuous crossings (sometimes called

Figure 3.35: Mopeds and prams within a few feet reflect a safe, slow street, not a dangerous one. *Courtesy of Nicholas Boys Smith*

Figure 3.36: A street with cars and children. *Courtesy of Nicholas Boys Smith*

Figure 3.37: Some rare on-street parking. For disabled drivers. *Courtesy of Nicholas Boys Smith*

Copenhagen crossing or Copenhagen sidewalks) where the pavement continues across the carriageway of the secondary street. You can see this everywhere.

The results are glorious. City streets teeming with life, prosperity, and movement, fulfilling the role that cities have played since the dawn of history as efficient places to come to buy, to sell, to work, to meet, and to have fun in the process.

Figure 3.38: It's easy to park your bike though. *Courtesy of Nicholas Boys Smith*

Figure 3.39: Toulouse trees. Pink and white: gray and green. *Courtesy of Nicholas Boys Smith*

All aspects of a place's economic and human prosperity are interconnected. Of course, Toulouse's prosperity is not *just* due to superlative transport planning and city center street design. But what Toulouse's well-judged improvements have done is strengthen the local "economics of place," encouraging people to wish to live and work, invest and set up companies, start children, and raise families locally. Nice places attract people. People create jobs and demand. And this in turn attracts more people and more funds to improve the place. This is the virtuous circle of place that Toulouse is supporting, spinning it faster, more confidently, and more joyously.

Toulouse's streets are rich in people: by day and long, long into the evening. Busy cafes, at-grade streets, and full bicycle racks tell a powerful story that too many town councils in too many countries ignore.

It is important to add that Toulouse gives back to children the liberty of movement lost over much of the last century. It is a hard thing to capture in photos, but the city's children are constantly moving about unaccompanied on bikes, feet, or scooters.

Every corner or niche that could possibly have a shop squeezed into it does. I did not spy a single boarded-up shop front in the city center in several hours of exploration. Global brands and chain stores are content

Figure 3.40: Walking, working, talking, and living; the virtuous circle of place. *Courtesy of Nicholas Boys Smith*

Figure 3.41: A so-called Copenhagen crossing in Toulouse. *Courtesy of Nicholas Boys Smith*

Figure 3.42: A child cycling in the center. The child is accompanied, but many were not. *Courtesy of Nicholas Boys Smith*

to comply with clearly strict local codes on acceptable design.

I have not tried to check the data but the well-maintained state of upper stories and overall buildings clearly showed that many are keen to live or work in the beautiful, busy city. There are rental incomes coursing through buildings. This is sumptuous and very Gallic "Gentle Density".

It is hard to summarize all of Toulouse's reforms and wise decisions over the last fifteen years, but here is an effort.

- Do create or preserve beautiful buildings;
- Use beautiful and resilient materials that will last;
- Do put people (and bikes and scooters) first;
- Do provide copious bike hire and parking;
- Do create trams (it is time for trams and light trains);
- Do plant street trees;
- Limit car parking;
- Prevent cars from zooming through (the city is not a motorway); and
- Ban garish paint and signs.

Underpinning these are interwoven concepts of beauty, of respect for the precious and particular, of the fifteen-minute city, and urban regreening. Some ideas might seem "left wing." Some might see "right wing." But who cares? The evidence of their benefits for human prosperity and sustainable living is clear and brought into radiant perspective by *La Ville Rose*. The renaissance of Toulouse's city center streets over the last generation points the way to a form of living that is happier and better connected, one in which we can breathe cleaner air, know more of our neighbors, permit our children to dare to wander more freely and in which we can all tread more lightly upon the planet. That seems worth doing.[15]

Figure 3.43: Gentle density that dignifies and does not demean. *Courtesy of Nicholas Boys Smith*

Figure 3.44: La Ville Rose. *Courtesy of Nicholas Boys Smith*

THE ARCADES OF BOLOGNA / GABRIELE TAGLIAVENTI

Bologna is famous for having the most arcades in Europe, built in four distinct networks. Arcades provide multiple benefits, including protecting pedestrians from the elements and adding spatial definition to a street. Arcades also shade the storefronts. In Bologna, the arcades link together a series of commercial streets, forming a web of comfortable and coherent public spaces (Figures 3.45 and 3.46).

Bologna's arcades come from a unique urban code dating to the fourteenth century, when the city began to expand beyond its existing walls to accommodate a rapidly growing population. New walls—that would

Figure 3.45: Via Santo Stefano, Bologna, Italy. Bologna's network of arcades gives the city a unique feel.

Figure 3.46: Piazza Santo Stefano, Bologna, Italy. The arcades of Bologna, built over the course of centuries, come in many different styles and forms. *Courtesy of Gabriele Tagliaventi / Photo by Bernard Durand-Rival*

define the city limits until 1889—encompassed nearly 456 hectares (about 1,125 acres), making Bologna one of the largest European cities of the time. The growing city, reflecting an urban renaissance that was embracing nearly all of Italy in the 1300s, was planned according to a law that required all new buildings to have arcades at the ground floor if they were located along a main street. Today, Bologna has four distinct networks of arcades (Figure 3.47).

1. The first network of arcades, laid out between 1561 and 1563, occurs at the monumental core of the city, along the Pavaglione and connecting to the two plazas around the Basilica of San Petronio.

2. A second network occurs along the main avenues, including Strada Maggiore, Via Zamboni, Via Santo Stefano, and Via San Vitale. These Renaissance streets were built using the palazzo building type as a reference.

3. A third system of arcades—the vernacular—corresponds to the secondary streets.

4. Lastly, the network of *extra moenia* arcades—the arcades built outside the fourteenth-century walls—are based on the 3.5-kilometer-long arcade that connects Porta Saragozza to the Sanctuary of San Luca on top of the hills surrounding Bologna. The term "extra moenia" also refers to the arcades that connect the Sanctuary with the main Carthusian cemetery and to the Alemanni's arcades, which connect the Porta Mazzini with the hospitals of the 1500s.

Figure 3.47: Via Castiglione, Bologna, Italy. Beginning in the fourteenth century, the city coded Bologna's famous arcades. The city now has the most extensive network of arcaded streets in Europe, protecting pedestrians from the elements, linking the streets into a system, and narrowing the streets. © 2011 Stephen A. Mouzon

The construction of the arcades at the core of Bologna was the largest urban operation during the Renaissance, involving architects such as Vignola and Antonio Morandi (called *Terribilia*). A monumental axis and a series of important civic institutions—such as the Ospedale della Vita and the Biblioteca comunale dell'Archiginnasio—define the main commercial venues of the city. This institutional grouping was the first permanent establishment of the University of Bologna. Pope Pius IV initiated the design and implementation of the development to provide the city with a proper public plaza in accordance with new Classical ideals. (The pope also intended to stop the construction of the Basilica of San Petronio to prevent Bologna from having the largest church in Christendom.)

The extra moenia arcades of Bologna are part of one of the most successful examples of an intentionally designed peripheral urban neighborhood in Europe. The Masterplan for the Extension of the City in 1889 further reinforced the design with a series of new urban neighborhoods, each built around an arcaded square. The most famous of those, called *Bolognina*, was a new neighborhood built north of the historic center on a typical 1800s grid along a main commercial axis (Via Matteotti). Since Bolognina was conceived as a "little Bologna," it had its own main arcaded avenue and public monuments, including a new cathedral and a new town hall.

THE STREETS OF TEL AVIV—FROM GARDEN SUBURB TO GLOBAL CITY / YODAN ROFÉ

Tel Aviv is a young city. It was born in 1909 as a suburb of Jaffa, a city with over three thousand years of history. Separated from Jaffa in 1921 during the British mandate in Palestine, Tel Aviv annexed the area after the 1948 war, when many of Jaffa's Arab residents fled. The city grew rapidly in its short history and is now the center of a metropolitan area that extends throughout Israel's central coastal plain.

Tel Aviv's origins and three stages of its growth are described below. Each period has an emblematic street. These streets, through their original design and changes over the years, embody the history of this fast-moving city.

The Birth of Tel Aviv as a Garden Suburb

Tel Aviv was founded in 1909 as Ahuzat Bait (Home Estate), a garden suburb neighborhood of Jaffa. Ahuzat Bait was not the first Jewish neighborhood created outside the old city walls, but it was the first neighborhood built with the express intention of forming the nucleus for a new "Hebrew City." That's why the creation of Ahuzat Bait is considered the beginning of Tel Aviv. The neighborhood plan included a major North–South main street (Herzl Street) connecting to the main road leading from Jaffa. At the north end of the new neighborhood, fronting the main street, was the high school Herzliya Hebrew Gymnasium, built in an eclectic neoclassical style and terminating the vista along Herzl Street (Figure 3.48). A short boulevard named Rothschild Boulevard was built one block south of the school.

Before and after World War I, Tel Aviv continued to grow haphazardly, without a general city plan. New

Figure 3.48: Herzliya Hebrew Gymnasium seen from the corner of Herzl Street and Rothschild Boulevard circa 1920. One can see the garden suburb idea in the setback of the houses from the street, and their setting within a private garden. *Israel Internet Association / Wikimedia Commons / Public domain*

Figure 3.49: Sir Patrick Geddes, Master Plan of Tel Aviv. Preexisting neighborhoods are in darker gray. © 2023 Dover, Kohl & Partners

neighborhoods were usually formed by associations of individuals who banded together to raise the funds to buy property, receive authorization to plot the land, lay out the streets, and build their own homes.

The Geddes Plan and its Legacy

In 1924, the Zionist Organization, working to promote Jewish settlement in Palestine, bought a large tract of land stretching from the seafront to where Ibn Gvirol Street is today, and from the existing built-up area of Tel Aviv to the banks of the Yarkon River (see map in Figure 3.49). The organization invited Sir Patrick Geddes, one of the pioneers of modern city planning, to prepare a plan for the area. In 1926, Geddes stopped in Palestine on his way back to the United Kingdom from India and spent a week surveying the land and the growing city. Before leaving Palestine, he produced a report and a drawing. The original drawing was lost, but Tel Aviv's master plan, approved in 1929, probably followed it faithfully.

The Geddes plan (Figure 3.49) was remarkable in several aspects: one was the way it established a connection to the existing fabric of the city; a second was an open-endedness of the street system that laid the groundwork for future growth towards the east; and a third was its variety of streets. These had varying degrees of movement and patterns of use that allowed a unique interface of public and private life to develop.

Designed as an organic non-repetitive grid of streets running both parallel and perpendicular to the sea, the three north–south arteries and the six east–west major streets (the "main-ways") defined the basic cellular elements of the plan (the "home-blocks" roughly 150×250 meters in size). Arguably the most original and humane aspect of the plan, the home-blocks have a single or double enclosure of buildings surrounding a protected internal space. A web of minor streets ("home-ways") and pedestrian lanes lead to a public garden at the core of the block. Nearby, Geddes planned local institutions (primary schools, kindergartens, synagogues, clinics).

Commercial activities were concentrated along the north–south "main-ways." Primary among these is Dizengoff Street (Figure 3.49 [6]), which winds its way across the plan from the southeast corner, connecting it towards the east, and then changing direction to north–south and ending in the northern boundary of the plan on the Yarkon River.

Geddes envisaged a city at a human scale. The main-ways are twenty-five to thirty meters wide, and the buildings lining them were meant to be a maximum of three stories tall. The narrower home-ways are ten to fifteen meters wide, lined with buildings up to two stories high. The plot frontages are short, fifteen to twenty-five meters, with frequent entrances. The front gardens were intended to provide shade and greenery to the streets, giving them a pleasant garden-like aspect. Geddes gave little thought to the presence of private automobiles and made no particular allowance for them. The plan is similar in many ways to late nineteenth- and early twentieth-century extension plans of European cities, except for its more organic irregular grid, rich variety of streets and public spaces, and a low density found in European garden suburbs planned in the early twentieth century.

Greater realities intervened, however. With the rise of Nazism in Germany, antisemitic and Fascist regimes in Eastern Europe, and the Revolution in Russia, many Jews fled or looked for a haven for their savings. This caused rapid growth and severe housing shortages in Tel Aviv in the 1920s and 1930s. Lacking available space and capital,

the city allowed higher buildings, up to four stories, to be built where only one- and two-story buildings had been planned. Some of the internal public spaces in the home-blocks were converted to buildable land. This intensified life in the main commercial streets and made Tel Aviv a more vibrant city. The intellectual and cultural fervor created by the immigration of highly educated and creative people from major urban centers in Europe transformed Tel Aviv into the economic and cultural capital of Israel.

At the same time, many architects emigrating from Europe brought early modern style to Israel. They created a local architecture based on "functional" considerations. Lacking historical roots, it represented modern positivist thinking and a renewal of values. With the fast growth in the city, it became the dominant style in new buildings and the extension of older structures.

From City to Metropolis—From "Garden" to "Global" City

Fortunately, the costs of destroying the old center of the city proved too high, and the national government had other priorities. In the 1960s and 1970s, the professional climate changed, and a conservation movement began. Rather than seeing the buildings and urban fabric created in the early years of the city as outdated and decrepit, the value of the buildings and the urban structure, as well as the economic potential of the historic city, were beginning to be understood. The city developed a conservation plan to save and renew buildings. The plan included the provision of public services and rental and mortgage assistance to attract young families back to the city center. It promoted Tel Aviv's contribution to world culture, citing the Geddes Plan, and the prevalence and visibility of early modernistic buildings. The plan received UNESCO recognition as a World Heritage site in 2003.

Parallel to this, the expansion and integration of the metropolitan area continued. The redevelopment of the rail network, practically abandoned for forty years, strengthened a new business district that began to form around the Ayalon Freeway. Now, at the beginning of the twenty-first century, a light rail system connecting Tel Aviv with the suburban cities surrounding it is being built. The first line (opened in August 2023 after many delays) connects the suburbs of Petah Tikva and Bat Yam through the Central Business District, which now stretches from the area where Tel Aviv began, towards and along the Ayalon River and along Begin Road. This road is being transformed from one surrounded by light industrial areas to a major boulevard, surrounded by office and residential high-rise buildings and shopping centers. It is the heart of the financial and high-tech industry that has made Tel Aviv the generator of the "Start-up Nation."

Three Emblematic Streets

Rothschild Boulevard

Rothschild Boulevard was originally not much more than an elongated public garden, set in a natural depression that was filled and leveled but could not be built on. Later neighborhoods extending the city to the northeast continued the pattern, but gradually changed direction from an east–west direction to a north–south one. The boulevard has a central median of generous dimensions, with rather narrow sidewalks and two lanes for cars on each side. Throughout most of its length, one lane is used for parking, so crossing to the median on foot is easy. After deteriorating over the years, Rothschild Boulevard was renovated at the turn of the twenty-first century and now has a dedicated bicycle lane and a pedestrian lane in the median. Originally, like all the boulevards in Tel Aviv, Rothschild Boulevard was designed to be a residential boulevard: commercial activities were banned (Figure 3.50). For the most part, it remains, except at the western end, around the intersections with Allenby and Herzl Street, where the center of the old Central Business District was. Here, it becomes more commercial and is surrounded by a mix of residential and office towers.

Dizengoff Street

Dizengoff Street (see Figure 3.49) was the main street of Tel Aviv in the 1960s and 1970s. The street begins on the eastern boundary of the tract planned by Geddes, where it connects to the main street of Sarona German Colony (Kaplan Street), and then snakes its way first to the west, to pass along the Philharmonic Hall and the National Theater, before gradually moving northward to Dizengoff Square and beyond to the Yarkon River's mouth, where the Port of Tel Aviv and a fairground were built. Dizengoff Street changes in character along its length. Its most commercial part lies between King George Street (see Figure 3.49) and Ben-Gurion Boulevard. Roughly in the center of this section is Dizengoff Square, which is really a circle, designed in international style by Genia Averbuch. One of the first female architects in British Mandate Palestine, she won a competition for its design.

The street is about twenty-five meters wide in its commercial section, with wide tree-lined sidewalks, and two to three lanes of traffic. Recently it has been redesigned as a public transit mall, with a two-way bicycle path running along it.

Figure 3.50: Rothschild Boulevard approaching Habima Square showing the residential character of the street. *Courtesy of Yodan Rofé*

Figure 3.51: The reconstructed Dizengoff Square became again a center of activity and life. *Courtesy of Yodan Rofé*

Ibn Gvirol Street

This long street, separating the Geddes Plan from the New North, is considered an "urban design miracle." Stretching two and a half kilometers, it was designed with a colonnade throughout its length. The buildings facing it are typically five stories tall, with stores on the ground floor. Unlike most buildings in Tel Aviv, the structures on Ibn Gvirol Street lack setbacks and side yards and present a continuous façade to the street. Around the main square of the city (now Rabin Square), the buildings have six stories, creating stronger edges for the large square. Facing the square from its north side is the modernistic slab of the municipality building. This is the main political square of Israel, where major demonstrations have always been held. On the west side of the square, a boulevard connects it to the Philharmonic Hall and National Theatre Square (where Rothschild Boulevard ends). Ben-Gurion Boulevard connects Rabin Square westward to the beachfront. Ibn Gvirol continues northward, connecting to all the east–west main streets that run between the Geddes Plan and the New North, before ending at a bridge crossing the river. On the other side of the river, it will be connected to a new boulevard, which will form the spine of a new neighborhood to be built on the last remaining open land in the city. The light rail system now being built will run underneath its whole length, connecting the suburbs of Herzliya and Rishon Letzion, each a large city, through the center of Tel Aviv.

The street probably emerged as a result of an unofficial policy implemented gradually by many detailed plans on the plots surrounding the street, rather than from a unified statutory urban design plan.

The street is about thirty meters wide between building fronts. The sidewalks are rather narrow (four to five meters) and include a bike lane as well. This is rather constrained; however, a five-meter-wide colonnade which is legally part of the private plot, where the public is given a right-of-way, adds significantly to the pedestrian realm. There are three lanes of vehicular traffic each way, separated by a small median, and one of the lanes is dedicated to public transit (Figure 3.52).

Figure 3.52: Ibn Gvirol Street. *Courtesy of Tel-Aviv – Jaffa Municipal Archive*

Conclusions

Tel Aviv is a modern city. It was founded and grew in the first half of the twentieth century. However, its planning up until the 1950s was still "traditional," influenced by the ideas and plans of the garden city suburban neighborhoods prevalent in the same decades in Europe and North America. Patrick Geddes's plan was unique in its insistence on connection with the surrounding development and the local topography, as well as its small block structure, which allowed the city to become vibrant economically, as well as extremely walkable. The two emblematic streets of this initial period, Rothschild Boulevard and Dizengoff Street, demonstrate its two qualities: residential calm and green juxtaposed with commercial and cultural vibrancy, all within a rich and walkable "organic" grid.

This vision was lost in the decades following World War II, to create an automobile-dependent city, eroding some of its initial qualities. But even in those years, miracles happened such as the colonnaded Ibn Gvirol Street, with the main city square in its center. The architecture along it is functionalism at its ugliest, but the urbanism is sound, and it is a vibrant, two-and-a-half-kilometer-long urban street that is both the main street of the city today, and a local commercial street.

Fortunately, the continued growth and densification of the city at the center of an expanding metropolitan area are producing an understanding that Tel Aviv's transportation needs cannot be resolved by the automobile. The orientation is changing towards promoting public transportation, cycling, and personal mobility—improving the pedestrian environment. This is exemplified in the recent transformation of Begin Road into a boulevard that is the spine of the expanding Central Business District as it grows along the Ayalon River and extending into the neighboring city of Ramat Gan.

These three streets exemplify the transformation of the city from a garden suburb into the center of a metropolitan, global city. They also demonstrate the ability of major streets to adapt and find new meaning and function within a rapidly changing context. They have been the theater of Israel's public and political life, throughout its history, as great streets in a great city should be.

THE TRANSECT OBSERVED: FOREST HILLS GARDENS

Forest Hills Gardens is a railroad suburb in Queens, New York, built by the Russell Sage Foundation in the early twentieth century as a model development (Figure 3.53). The Olmsted Brothers, landscape architects, were the planners, and the primary architect was Grosvenor Atterbury. The streets, which are privately owned but open to the public, connect to the Queens grid around the development. The streets illustrate the Urban-to-Rural transect, even though the Olmsted Brothers and Atterbury planned them long before urban designers talked about the transect. Their arrangement is most formal at Station Square, at the center of the transit-oriented development (Figure 3.54), and most informal and winding at the periphery of the 142-acre site.[16]

As one walks along the Greenway from Station Square, the transect changes quickly. The Olmsted Brothers and Atterbury subtly and appropriately transformed the landscape, the streetscape, and the architecture to give

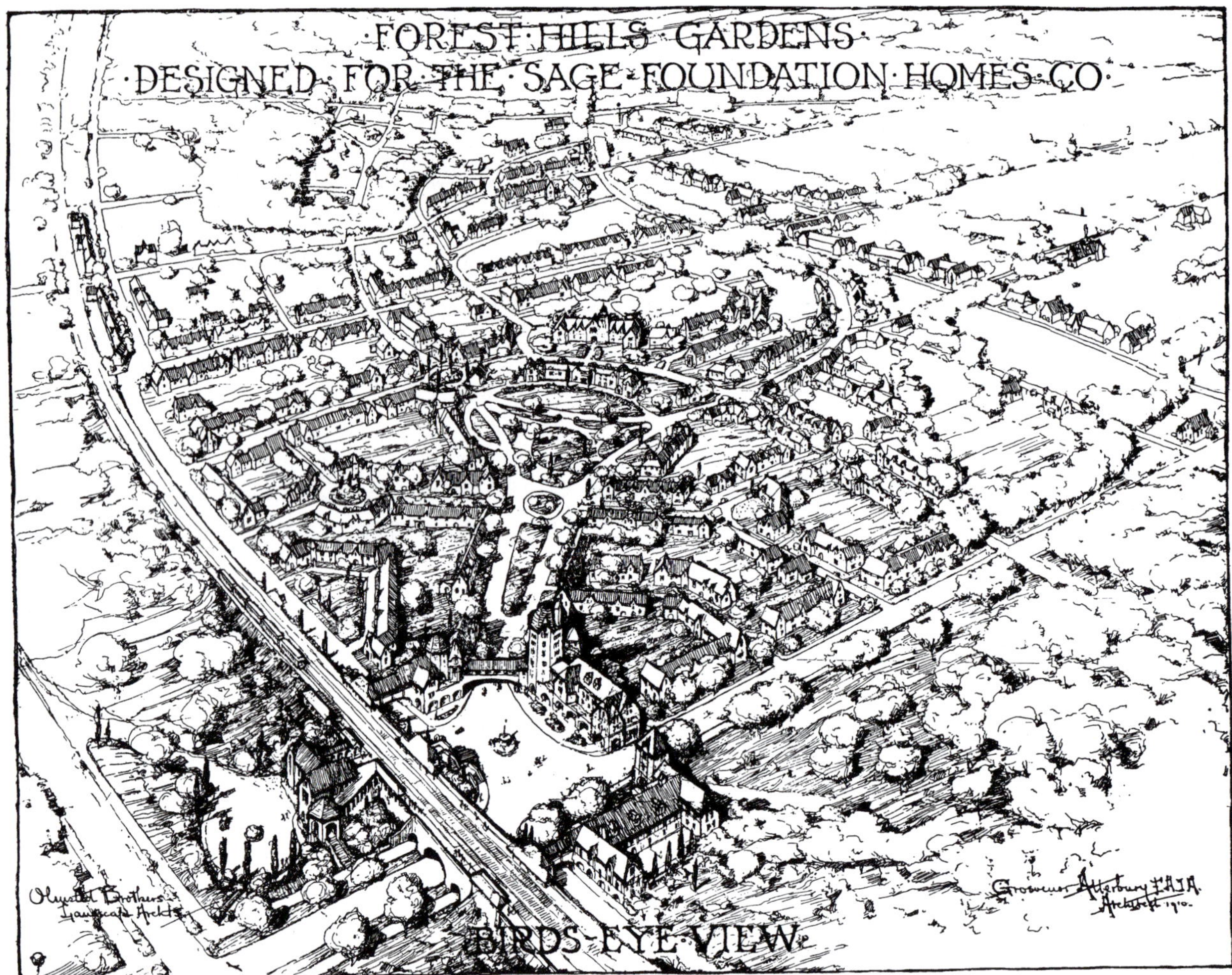

Figure 3.53: Forest Hills Gardens, Queens, New York. Olmsted Brothers and Grosvenor Atterbury, 1908. Aerial rendering, drawn by Grosvenor Atterbury, 1910. The drawing shows how the development goes from Transect zone T-5 to Transect zone T-3 as one walks along and away from the axis of the Greenway that extends from Station Square in the foreground. The railroad tracks in the foreground belong to the Long Island Railroad, connecting to and from Pennsylvania Station in Manhattan. The wide, diagonal street at the corner of Station Square is the Greenway. *Public Domain*

Figure 3.54: Station Square, Forest Hills Gardens, Queens, New York. Olmsted Brothers and Grosvenor Atterbury, 1908. Looking southeast towards the Greenway. Station Square has apartments above stores, restaurants, and offices.

Figure 3.55: Greenway Terrace, Forest Hills Gardens, Queens, New York. Olmsted Brothers and Grosvenor Atterbury, 1908. Looking back towards Station Square from the intersection of Greenway Terrace and Archway Place. Rowhouses behind stone walls face a sidewalk with brick strips between the concrete walk and the roadway.

people on the Greenway a sense of where they were in Forest Hills. Starting in the square, they used brick paving and three-and-a-half-story buildings to establish the urban center at the train station. The buildings on Station Square have the only stores, restaurants, and office space in Forest Hills Gardens, and once had a hotel as well. Bridges that span the streets leading out of the square connect the buildings, creating a strong sense of enclosure.

As the roads leave Station Square, the brick paving changes to asphalt. On the Greenway, the sidewalks enter arcades where the walks have small-scale patterns in the concrete. When the arcades end, the buildings are lower and slightly set back from the street. The small-scale pattern in the sidewalk stops, replaced by larger concrete panels. A little farther along, there are brick strips between the sidewalk and the road (Figure 3.55).

Close to the square, the park in the center of the street is only a few feet wide, but at the far end it broadens to a few hundred yards across (Figure 3.53). The roads on each side of the park have gentle curves that follow the topography. A lamppost at the beginning of the square has a family resemblance to the lights in

Figure 3.56: Forest Hills Gardens, Queens, New York. Olmsted Brothers and Grosvenor Atterbury, 1908. A street light in Station Square, with the silhouette of Dashing Dan, the symbol of the Long Island Railroad. Compare to the lamppost in Figure 3.55.

Station Square, but those were mounted on buildings rather than on posts (Figure 3.56). Across from the park, close to the square on both sides, are two-and-a-half story rowhouses. On the west side of the park, the rowhouses continue up and over an arch that allows a road to go through to Archway Place, recalling the covered roads a few hundred feet away at Station Square.

As the pedestrian moves south along the Greenway and the wider park, the sidewalks change from concrete with brick strips along the road to concrete with planting strips along the road (Figure 3.57). The blocks of rowhouses are shorter, and the setbacks in front of the rowhouses and group houses are larger. The walls separating the front yards of the houses from the sidewalk become more rustic, changing from stone and concrete to stone and brick, but—like the lights—they maintain a family resemblance. The houses themselves have more variety in the materials, including more details in wood. Within a few hundred yards of Station Square, the group houses seamlessly morph into single-family houses, and at the end of the park, the houses become freestanding in suburban-style grass lawns (Figure 3.58). The sidewalks become narrower with wider grass strips than fifty feet before, and the stone and brick houses turn into stone and stucco houses, or stone and wood houses. Throughout the development, places where roads intersect frequently have group houses that harmonize with the size and massing of the single-family houses just around the corner.

Figure 3.57: Forest Hills Gardens, Queens, New York. Olmsted Brothers and Grosvenor Atterbury, 1908. A photograph of the west side of Greenway Terrace, looking north, taken just fifty feet south of the vantage point in Figure 3.55. The strips between the sidewalk and the roadbed are wider, as well as planted rather than paved with bricks.

Figure 3.58: Forest Hills Gardens, Queens, New York. Olmsted Brothers and Grosvenor Atterbury, 1908. Looking south towards the intersection of Greenway Terrace and Slocum Crescent. The near side of the street has the block of rowhouses seen in Figure 3.57, while the other side of the street has freestanding, single-family houses. "Like faces like" along Slocum Crescent (see the discussion of East 70th Street in Chapter 1 for a discussion of like faces like).

CRANFORD STREET, FOREST HILLS GARDENS, QUEENS, NEW YORK

Olmsted Brothers and Grosvenor Atterbury, 1908

American Garden Street

We include Cranford Street in *Street Design* for one reason: to show the virtues of narrow streets, narrower than almost every city or town in America will allow today, even in a private development with private streets. If you visit Forest Hills Gardens, experience Cranford Street for yourself. The details are good, and the houses are pleasant, but the street is not extraordinary in any way except in the context of modern planning and street design. It feels good, however, to be on a quiet, two-way street with on-street parking that is only seventeen feet wide.

Cranford Street is approximately two-thirds of the way from Station Square to the edge of Forest Hills Gardens. The single-family houses are freestanding, unlike the rowhouses, group houses, and attached houses located closer to the Square. But the Cranford Street houses are smaller than the single-family houses located farther out on the transect.

Our unscientific survey of the residents on Cranford Street suggests that they like the variety of experiences and housing choices in Forest Hills Gardens. Although the Russell Sage Foundation built Forest Hills Gardens as a model development for many income levels, buyers today pay a premium to live there. They pay for well-designed, well-built houses, but also for well-designed public spaces. Many of the smaller houses have alleys with parking behind them. The gardens and lawns are in front of the houses, along the street.

Figure 3.59: Cranford Street, Forest Hills Gardens, Queens, New York. Looking southwest towards the Greenway. *Courtesy of Damon Hemmerdinger*

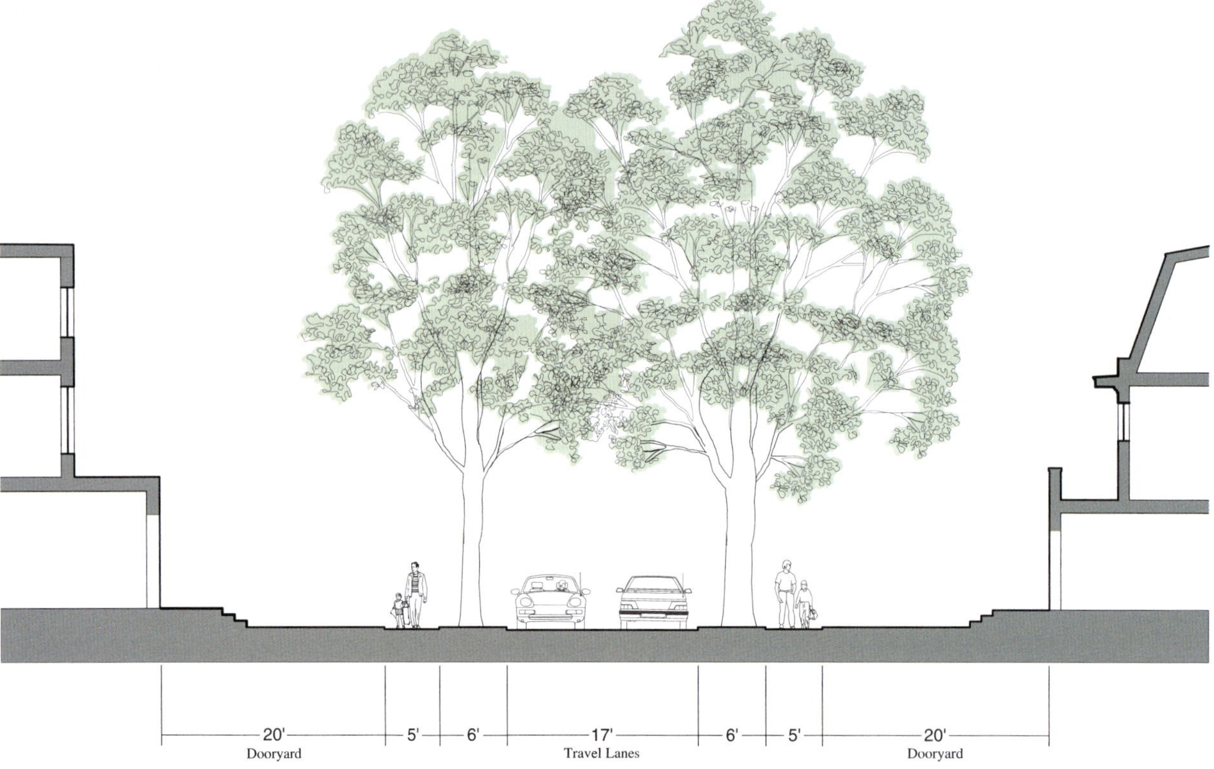

Figure 3.60: Cranford Street, Forest Hills Gardens, Queens, New York. Section looking southwest. © 2023 Dover, Kohl & Partners

CHESTNUT AVENUE, LAWRENCE PARK, BRONXVILLE, NEW YORK

William Van Duzer Lawrence and William Augustus Bates, 1889

Village Street

Today, Lawrence Park would be a PUD (Planned Unit Development) or perhaps part of a form-based code overlay. In 1889, it was the first part of mixed-use, mixed-income private development that became the Village of Bronxville, New York. The developer was a patent medicine magnate named William Van Duzer Lawrence, who later founded a college named after his wife, Sarah Lawrence.[17] He bought a hundred-acre farm on the top of a hill near a new railroad station forty minutes from Grand Central Terminal,[18] advertising that Lawrence Park would be an artistic community for "long-haired men and short-haired women."[19]

Wanting to respect the topography of the site, Lawrence and the architect William Bates walked the land and laid out the picturesque streets with a team of oxen. With a few exceptions, the streets are 14 feet wide—or three feet narrower than the "narrow" Cranford Street in Forest Hills Gardens. Many of the Master of the Universe bankers and hedge fund managers in Lawrence Park own the largest SUVs money can buy, but they don't seem to mind the narrow streets. In fact, they pay a lot of money to live on those streets.

Most of the streets are two-way yield streets. A few of the streets have sidewalks, but most do not. There are stop signs at selected intersections and some do-not-enter signs, but the two-way streets lack painted center lines. One street has a telephone pole in the street, while another has a tree in the middle.[20] In other words, everyone has to drive slowly, and newcomers have to go extra slowly because so many of the usual engineering clues are missing. They are nineteenth-century, American shared-space streets. Along them sit some of the most expensive houses in the world. We say that to make the point that these are not undesirable places.

> They are nineteenth-century, American shared-space streets. Along them sit some of the most expensive houses in the world. We say that to make the point that these are not undesirable places.

Chestnut Avenue is one of the wider streets in Lawrence Park. We include it in *Street Design* because one short hillside stretch feels like a country road, but it's only seventeen miles from Times Square in a dense suburban neighborhood that includes nearby apartment houses and a busy Main Street. The surrounding houses

Figure 3.61: Chestnut Avenue, Lawrence Park, Bronxville, New York. We intentionally include a view that shows a fire hydrant and a telephone pole, to emphasize that Chestnut Avenue is not in the country, despite its rural appearance.

disappear from view, and the split-rail fence and stone paving feel like rural details. Before you walk much further, however, a county fire hydrant and a telephone pole appear. They confirm what you already know: you are not in the country. But the illusion is pleasant, whether you're walking up the hill from Bronxville or down from Lawrence Park.

BIG BOULEVARDS AND TINY LANEWAYS IN MELBOURNE, AUSTRALIA / CHIP KAUFMAN

Begun by Robert Hoddle, 1837

Multiway Boulevards and Pedestrian Passages

The characteristic big boulevards and tiny laneways of Melbourne, Australia, are wonderful examples of how streets at very different scales can enliven the grid and expedite multimodal movement (Figures 3.62, 3.63, and 3.65). Robert Hoddle, Melbourne's original city surveyor, began the layout of its street network in 1837, making use of both narrow lanes and very broad multi-lane boulevards that went out to the suburbs.

Melbourne's orthogonal Central Business District is admired in part for its tiny, north–south pedestrian streets (both open-air and with glazed roofs), which complement and contribute to its "plaid" street network of north–south streets 30 meters wide and spaced 200 meters apart. The east–west streets are 100 meters apart and alternate between 10 and 30 meters wide. Many additional open-air, 6-meter-wide, north–south service lanes give access for deliveries and basement parking and have helped keep most single properties small—another key characteristic of Melbourne's highly diverse and attractive Central Business District.

Heavy rail, most of which is below ground, runs the perimeter of the District before radiating outward in several directions. New developments are gradually capping over the remaining at-grade heavy rail lines adjoining the District. On-street light rail lines lace

Figure 3.62: St. Kilda Road, Melbourne, Australia. View looking north towards the Shrine of Remembrance. At sixty meters wide, the major avenue was readily adapted to multimodal travel, including the city's famous trams. *Courtesy of Chip Kaufman*

Figure 3.63: Degraves Street, Melbourne, Australia. The city's downtown remains dominant, thanks in part to its pedestrian oriented, effective street network, devised before the automobile. *© 2011 Rob Deutscher / Wikimedia Commons / CC BY SA 2.0*

through the District and then continue out to the suburbs, giving Melbourne the fourth-largest light rail network in the world.

The pedestrian ways, ranging in width from 10 meters to 4 meters and directly fronted by mid-rise heritage buildings, have many amenities that attract small local boutique shops (Figures 3.63 and 3.64). The street-level stores are generally no more than 5 meters wide, and have offices and apartments above. Some of these arcades rival the covered passages (*passages couverts*) of Paris (Figure 2.179).

At the opposite end of the scale, large multimodal, multilane boulevards fan out from Melbourne's Central Business District. Several of the boulevards have new "Copenhagen" cycle lanes. The pedestrian experience is heightened by medians planted with trees spaced no more than two lanes apart, which usually means four or six rows of mature trees for each boulevard, and relative ease of crossing for pedestrians.

St. Kilda Road is Melbourne's highest-capacity boulevard. It supports a popular tram line, a footpath, and a bicycle path, as well as traffic lanes sufficient for more than 50,000 vehicles per day. It begins as Swanston Street, the central north–south 30-meter-wide spine of the CBD, and then extends many kilometers southward to the suburb of St. Kilda on Port Phillip Bay as an approximately 60-meter-wide multilane boulevard. Some of Melbourne's most prestigious businesses and apartment buildings are on St. Kilda Road.

Thanks in large part to the government's initiative to establish a well-organized and effective street network long before the advent of the car, Melbourne has matured into a vibrant, attractive, and transit-oriented mixed-use inner city, centered around a still-dominant Central Business District and close to many popular neighborhoods. It has Robert Hoddle, who first stretched the spectrum of street types and scales, to thank for its renowned urban character.

Figure 3.64: Centre Place, Melbourne, Australia. Melbourne's slender laneways are vital scenes, bustling with art, cafes, and pedestrians. Centre Place is open to vehicular traffic only during limited hours each day. *© 2013 Joseph Ip*

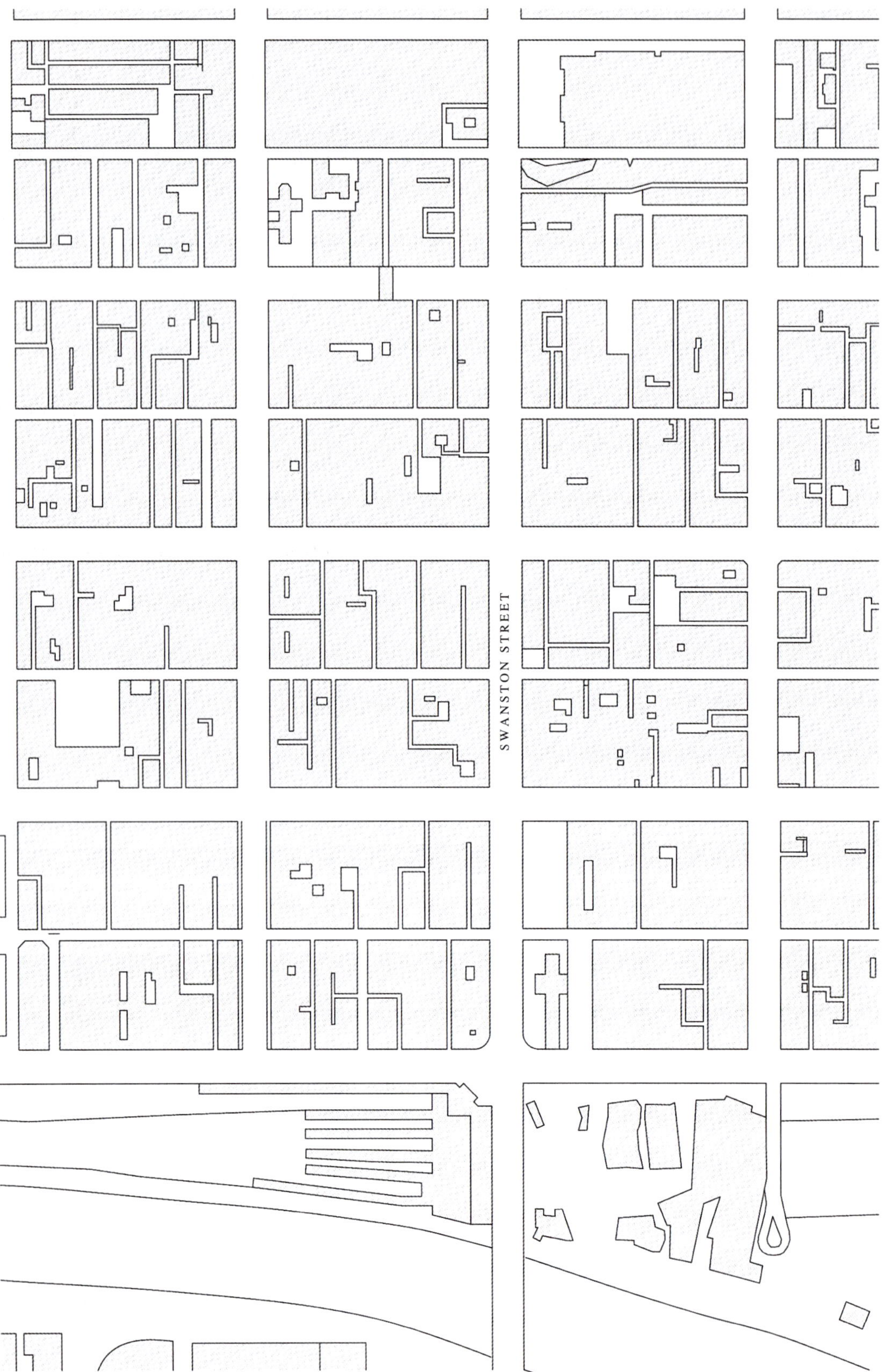

Figure 3.65: Central Business District, Melbourne, Australia. Figure-ground drawing showing the result of "the Hoddle grid," devised by surveyor Robert Hoddle in the 1830s. Note the range of widths in the streets, service lanes, and tiny north–south "laneways." © 2013 Dover, Kohl & Partners

SEVEN DIALS, LONDON, ENGLAND / HANK DITTMAR

Thomas Neale, circa 1690

Shared Space

But what involutions can compare with those of Seven Dials? Where is there such another maze of streets, courts, lanes, and alleys? ⋯ The stranger who finds himself in "The Dials" for the first time, and stands Belzoni-like, at the entrance of seven obscure passages, uncertain which to take, will see enough around him to keep his curiosity and attention awake for no inconsiderable time. From the irregular square into which he has plunged, the streets and courts dart in all directions, until they are lost in the unwholesome vapor which hangs over the housetops.

—Charles Dickens, *Sketches*, 1843

My favorite London street is actually the confluence of seven streets near London's Covent Garden into a circular space called the Seven Dials (Figures 3.66 and 3.67). At the center of the Dials stands an obelisk with six sundials (not seven) donated by the Worshipful Company of Mercers—one of London's guilds, and the owner of the area—in the 1690s.

Designed by Thomas Neale, a speculator who took the area on a lease from the Mercers Company, the area has certainly seen its ups and downs. In Dickens's time, it was part of the infamous St. Giles Rookery, an overcrowded slum famous for violence, prostitution, and brawls between the Irish and the English. With the redevelopment of the Covent Garden Market as an attraction in the 1970s, Seven Dials has grown in popularity, and today is loved by residents and tourists alike.

Over the past two years, the clutter has been removed from nearby streets, granite pavers have been restored in the pavement, and the monument has been cleaned and restored. In traffic-engineering terms, the junction operates as a shared space, with the monument serving as a

Figure 3.66: Seven Dials, London, England. Shared space in London. *Courtesy of Emily Glavey*

Figure 3.67: Seven Dials, London, England. Satellite view. The wide street with trees is Shaftesbury Avenue. *Courtesy of Google Earth, © 2012 Bluesky*

pedestrian haven and the junction itself used equally by cars, cyclists, and pedestrians. Each of the seven streets is similarly a shared space, and the surrounding area boasts one of London's best coffee roasters; a backcourt called Neal's Yard with London's best *fromagerie*; and many delightfully individual cafes and shops, with wares ranging from rare books to secondhand designer clothing. Each street has its own character (one with market stalls), and no matter where one sets out to go, one ends up in an interesting place.

Incidentally, not two blocks away is Central St. Giles—a celebrated new building covering a city block and designed by Renzo Piano—that is the opposite of Seven Dials. The buildings are made of steel and brightly colored glass, and the public space is internal, not exposed to either cars or pedestrians. As a result, the retail is struggling—less than a year after Central St. Giles opened to acclaim from some design critics for its contribution to the civic realm.

Comparing the two areas is instructive, for it reveals what makes Seven Dials special. It is part of a densely connected network of small streets, bounded by three-to-six-story buildings, with continuous retail, mostly on the small side. While there is a traditional delineation between street and pavement, it is made with paving materials and curbs—and without either the bright colors or railings that have become the default positions of the shared spacers and the engineers, respectively. The monument

Figure 3.68: Barri Gòtic, Barcelona, Spain. Long lost from the urban design palette, there is wonder and promise in skinny, exceptional streets and tightly wound intersections. They introduce contrasts into what otherwise might be a relentless or monotonous set of similar spatial experiences.

at the center has an inviting shelf upon which to sit and people-watch, and to get there pedestrians have to walk across the intersection. Finally, the area is managed for a mix of small unique shops and larger brands, with residential space above and office space nearby.

Traffic engineering is a small but essential part of what makes this a great street. The junction of seven streets could have easily become a no-go area for pedestrians, but the intimate scale of the streets, the quality of the pavement, the calm and understated detailing and materials, and the fine monument at the center of the junction all contribute to making this a space for people where cars are tolerated, and one of my favorite places to be.

FREEWAY TEARDOWNS / JOHN NORQUIST

The utopian dream of traveling through cities without interruption dates back at least to 1922, when the revolutionary architect Le Corbusier drew what he called the *Ville Contemporaine*. Before that, the movement of railroads on their own rights-of-way must have brought to mind the notion that streets could be designed to speed traffic on its own right-of-way, thus avoiding the hustle and bustle of the city. It may even have occurred to the ancient Romans that roads could be designed like aqueducts, to attain free-flowing cart and chariot movement. Regardless, the new technology of the motorcar created a demand for paved streets, all-weather

roads, and a taste for higher speeds. Within a decade of Corbusier's drawing, the first autobahn was constructed in Germany and work had started on New York City's West Side Highway. The freeway era had begun.

In 1941, Norman Bel Geddes, the designer often credited with conceiving of America's Interstates, warned that the revolutionary highway system he envisioned could harm cities. After contemplating his freeway-centric 1939 World's Fair exhibit, he wrote a book, *Magic Motorways*, in which he expressed his doubts. "If the purpose of the motorway as now conceived is that of being a high-speed, non-stop thoroughfare, the motorway would only bungle the job if it got caught up with the city. A great motorway has no business cutting a wide swath right through a town or city and destroying the values there; its place is in the country."

Like urban activist Jane Jacobs, Bel Geddes understood that the value of a city derives from its complexity, proximity, and diversity of economic and social activity. Its grid of streets, avenues, boulevards, transit, and sidewalks organized around blocks creates a setting for social interaction, business growth, and real estate development. The grid facilitates access and distribution. The hierarchy of road types (boulevards, avenues, streets, and alleys) is fairly flat. Whereas with expressways the hierarchy is steep (cul-de-sac, 120-foot-wide feeder arterials and above all in importance, grade-separated highways). Highways are intended to move traffic at high speeds. This goal is easily achieved in low-density rural settings. But in cities and populous suburbs, an expressway attracts traffic that congests at peak travel times and then pushes congestion into the city grid from its interchanges. By contrast, the street grid distributes vehicle traffic, bikes, transit riders, and pedestrians throughout its complex network. Perhaps the greatest difference is that expressways take up land creating dead space under and around them. The city grid provides a setting for commerce and development that produces value, particularly a tax base from which municipalities derive revenue to pay for librarians, police, and firefighters. This is especially important in Wisconsin, where towns, villages, and cities depend on property taxes as their main source of revenue.

Freeways proved to have several significant side effects in the urban context. First of all, the roads were expensive to build in cities with higher land values and existing buildings in the way. Some cities tried to build freeways with their own money. In 1949, Milwaukee started the Stadium North Freeway. The city built it through Washington Park, which had been designed by Frederick Law Olmsted. It was the site of the zoo, which had to be relocated at great expense. The Stadium North would probably have been Milwaukee's last freeway if the Interstate Highway Act of 1956 hadn't come along with its high-octane, 90 percent Federal and 10 percent state funding mixture. With free roads on the menu, few cities could resist.

In Western Europe, freeways exist between cities and around the perimeter but rarely intrude on city centers. Our Canadian neighbors, after flirting with urban freeways in the 1960s, have largely confined them to intercity travel. One reason for this may be that Canada has no equivalent of our interstate highway program. The cities and provinces pretty much pay for their own infrastructure—and that means that infrastructure had better add value to the place where it is built. Perhaps, as a result, all major Canadian cities have good transit and only have the roads they are willing to pay for. Vancouver, for example, has no freeways within its borders, but has an excellent system of boulevards, avenues, and streets, along with efficient transit lines. It should be noted that despite the lack of a national highway-financing program, Canadian roads do successfully connect across provincial and even international borders.

Another unanticipated side effect of freeways: the effect they have had on settlement patterns and the flow of commerce. Street networks coupled with streetcars, subways, and commuter trains attracted people to the center of cities. An auto-centric system of expressways has the opposite effect, pushing the distances people regularly travel—"the drive-sheds"—deep into the hinterlands and altering shopping and commuting patterns.

Corbusier's dream of unimpeded traffic influenced the world, especially the Americas. The well-known New York City planner and expressway enthusiast Robert Moses was paid by the Rockefeller family to visit São Paulo, Brazil, in 1949 to draft a new plan for the fast-growing city. In fact, São Paulo already had

existing plans, which Moses was seeking to replace; in 1930, the Brazilian architect, engineer, and planner Francisco Prestes Maia had put forward the *Plano das Avenidas*, which had significant support among designers and civic leaders. This plan envisioned a European-type system of streets with a hierarchy scaling up to boulevards (without the grade separations recommended by Corbusier and Moses). When Maia was the mayor of São Paulo, between 1938 and 1945, he began implementing the plan.

Maia's boulevards accommodated large traffic volumes, with three moving lanes in each direction, but the streets retained their connection to the city fabric.

Throughout the history of urban civilization, major thoroughfares had three important functions: to move people and goods, to provide a place for commerce, and to serve as a social gathering place for the community. Corbusier and Moses sought to restrict the urban thoroughfare's function to movement alone. Moses viewed the Maia Plan's complexity and detailed fabric as an obstacle to fast vehicular movement; in São Paulo, his vision of a city of expressways won over Maia's urban vision. A vast system of expressways was constructed.

In 1961, Maia was elected to a new term as mayor that lasted until 1965. He tried to change the direction established by Moses. If he had lived longer (he died in 1965), he might have enjoyed the commencement of the construction of a metro train system that now exceeds 380 kilometers. Maia would also enjoy São Paulo's recent decision to begin removing some of its freeway infrastructure. First to go will be what is called "the Worm," a giant roadway that travels through the city center.

New York City, San Francisco, Portland, Milwaukee, Toronto, Montreal, Chattanooga, and Rochester have all replaced freeway segments with avenues or boulevards. These moves proved popular, and now several removals are progressing in other North American cities. In Europe, several cities, including Stockholm and Paris, replaced urban expressways with surface thoroughfares integrated into their street networks. Perhaps the most spectacular example of highway removal is in Seoul, South Korea (Figure 3.69). An expressway built atop a stream called Cheonggyecheon that carried more than 150,000 vehicles per day was removed and replaced with two surface lanes on each side of the restored waterway. Since 2002, Seoul has removed 14 more freeway segments.[21] This is just the beginning. With property values skyrocketing

Figure 3.69: Cheonggyecheon, Seoul, South Korea. Remnants of the highway after its demolition. © *2007 Kyle Nishioka / Flickr / CC BY 2.0*

near demolished freeways, urban expressway deconstruction could be one of the biggest public works projects of the twenty-first century.

> Americans: we make huge mistakes, but we also correct them.

Rethinking Freeways

Tearing down expressways after fifty years of the greatest road-building binge in world history may seem strange, but we've been through this rethinking process before. Remember the U.S. Army Corps of Engineers? In the Florida Everglades watershed, the Corps once drained marshes and forced streams into concrete-lined channels in an attempt to tame waterways and make more land available for agriculture and development. Now we know that draining wetlands and "channelizing" streams not only damaged the environment but increased the likelihood of pollution and flooding downstream. Marshes, meadows, swamps, grasslands, and bogs slow down and filter the water that flows through them.

In the twenty-first century, the Corps has begun the process of undoing the damage that many of its twentieth-century projects caused. Having learned that forcing water into concrete-lined channels was foolish and counterproductive, it is now ripping out concrete and restoring marshes. In a similar way, traffic engineers are learning that urban street grids can distribute urban traffic more efficiently than superhighways.

Figure 3.70: Oakland, California. Aerial of Oakland I-980. The tight downtown street grid is interrupted by the underutilized interstate, impairing the connection from historic black communities to downtown. © *2015 Google Earth Pro*

Figure 3.71: Oakland, California. Aerial from Downtown Oakland Specific Plan. The boulevard, first proposed by ConnectOakland, will reknit the West Oakland neighborhood with downtown, replacing the interstate with a multiway boulevard. © *2015 Dover, Kohl & Partners*

The Cap at Union Station, Columbus, Ohio

Meleca Architecture, 2004

Downtown Street Retrofit: Hiding a Highway

When transportation officials and engineers pushed Interstate 670 through Columbus, Ohio in the 1970s, they severed an old neighborhood around North High Street from the downtown, leaving behind what the *New York Times* architectural writer Herbert Muschamp called an "engineered gash."[22] Like most postwar, inner-city highway construction projects, this one brought alienation and disruption to the people forced to live next to it (Figure 3.72).

In spite of the high value of land in this location—the highway divides the downtown area from the Short North arts and entertainment district—local officials and the Ohio Department of Transportation (ODOT) proposed in the early 1990s to further expand the highway, widening it from four lanes to eight. To appease angry neighbors, ODOT offered to construct a hardscaped "park" on top of the bridge across the interstate to offset the unattractive highway expansion.

The business owners and residents of the neighborhoods located just north of I-670, including Short North, opposed the highway widening and the perfunctory hardscaped bridge park, pointing out that the existing four-lane freeway and long bridge already discouraged patrons from making their way across. They realized that the widening would only make things worse.

In a political breakthrough, the City of Columbus, the Department of Transportation, developers from Continental Real Estate Companies, and local business owners compromised by agreeing on a "Cap" alternative. Along with the highway expansion, they converted the bridge above it into a walkable street, hiding the freeway and sewing the two parts of the city back together (Figure 3.73). The parties agreed to a complex arrangement in which both the costs and the profits of the development would be shared. Continental owned neighboring properties and did not want to see its land devalued; the Cap offered them a way to use their own work to improve those values and to link Columbus's successful arts district with the downtown convention center.

Completed in 2004, the Cap at Union Station is now a place to go rather than simply one to drive through. David Meleca of Meleca Architecture designed the buildings and public spaces. Arcades and sidewalks in front of the shops are more than twenty-five feet wide and constantly full of people, thanks to the Cap's varied mix of retail and restaurants (Figure 3.75). Lightweight materials were used in the construction of the Cap to address the structural challenge of constructing full-size buildings above the wide Interstate (Figures 3.75 and 3.76). The west side of the "cap" is nearly 73½ feet deep and the east side is 53½ feet deep; both span the total distance of the bridge.

Figure 3.72: The Cap at Union Station, N High Street over I-670, Columbus, Ohio. Meleca Architecture, 2004. View looking west, showing the High Street bridge before the construction of The Cap. *Courtesy of David Meleca*

Figure 3.73: The Cap at Union Station, N High Street over I-670, Columbus, Ohio. Meleca Architecture, 2004. View looking south on N High Street, showing The Cap crossing I-670. *Courtesy of David Meleca*

Figure 3.74: The Cap at Union Station, N High Street over I-670, Columbus, Ohio. Meleca Architecture, 2004. View showing an arcade at The Cap. *Courtesy of David Meleca*

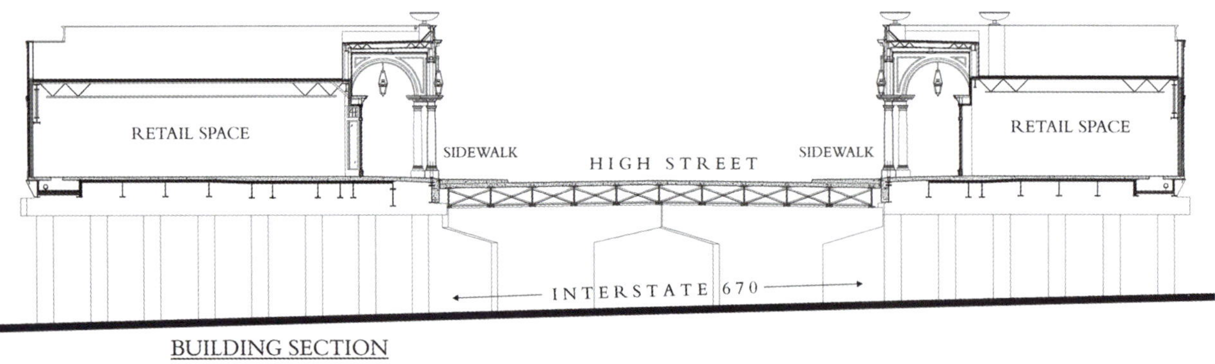

Figure 3.75: The Cap at Union Station, N High Street over I-670, Columbus, Ohio. Meleca Architecture, 2004. Section. *Courtesy of David Meleca*

Figure 3.76: The Cap at Union Station, N High Street over I-670, Columbus, Ohio. Meleca Architecture, 2004. Aerial view, looking north. © 2013 Pictometry International Corp.

Figure 3.77: Pulteney Bridge, Bath, England. Robert Adam, 1774, later expanded and altered. Spanning the River Avon, the Pulteney Bridge is promoted in Bath as one of four bridges in the world crossing a river that has shops on both sides. The Cap at Union Station shows that the type has tremendous promise for neighborhoods torn apart by half-buried highways.

Figure 3.78: Figure-ground map illustrating the potential for repair of the urban fabric previously torn out to build Interstate 81. Restoring the "community grid" will reconnect downtown, the 15th Ward, the campuses of local hospitals, and Syracuse University.
© 2023 Dover, Kohl & Partners

Figure 3.79: Syracuse, New York. Before: Ground-level view of the existing Almond Street corridor. Under the highway where Almond Street runs, paint can be seen chipping from the structure. Whole chunks are said to fall regularly. Hovering over the street, I-81 creates a dark, unfavorable scene. © 2022 Dover, Kohl & Partners

Syracuse Community Grid / Almond Street Boulevard

Dover, Kohl & Partners, 2022

Before the construction of the Interstate I-81 viaduct, the street system between present-day downtown Syracuse and Syracuse University generally followed a grid pattern. The blocks were well-sized for accessibility to the necessities of daily life, and the simple grid provided convenient locations for stores and made walking easy.

During the Great Migration in the first half of the twentieth century, the working-class neighborhoods adjacent to Almond Street grew. A predominantly African American community emerged on the land just south of a now-filled branch of the Erie Canal. As the population increased, racial discrimination and segregation intensified, leading to a concentration of African Americans directly east and south of downtown, in an area known as the 15th Ward. As in many U.S. cities, this part of Syracuse was then a target of federally-funded urban renewal

schemes; Syracuse's episode led to the displacement and relocation of people and the demise of many small businesses. When the Federal Highway Act authorized a new interstate system, maps called for Interstate 81 to cut directly through the heart of the city, demolishing dozens of properties, including remaining businesses and homes. The erection of I-81 increased the trauma and worsened impacts on health, and injustice already underway within the Southside community.

Undoing the Damage

With a joint Record of Decision announced on May 31, 2022, the New York State Department of Transportation (NYSDOT) in Albany and the Federal Highway Administration (FHWA) in Washington, DC officially began the process of removing Interstate 81 through the heart of Syracuse and replacing it with the "Community Grid alternative." This announcement came at the end of an exceedingly long and thorough Environmental Impact Statement (EIS) process that saw years of

Figure 3.80: Syracuse, New York. After: Ground-level view of the proposed Almond Street corridor. The revised plan will create a more lively and greener street that encourages walking, biking, and better redevelopment. © 2022 Dover, Kohl & Partners

community input and thousands of comments, questions, and answers.

Under the proposed Dover-Kohl redesign, the centerpiece of I-81's removal—a redesigned, reconfigured, and reconnected Almond Street—should become Syracuse's finest tree-lined street, complete with a central promenade inspired by Boston's Commonwealth Avenue (Figure 2.64). The new Almond Street will replace the four lanes of elevated, limited-access freeway. This is the flagship project in a larger effort implementing safer and more walkable street designs for Syracuse, intended to complement the first citywide overhaul of zoning since the mid-1960s.

In the redesign, the typical 166 feet total right-of-way includes 128 feet allocated to people and landscaping (Figures 3.81 and 3.82). This unusually generous public realm will help transform the corridor into a memorable address. Combined with the donation of NYSDOT properties to a community land trust, the reestablishing of Almond Street should provide a preeminent example of restorative justice within the built environment.

A redesigned, reconfigured, and reconnected Almond Street should become Syracuse's finest tree-lined street.

Figure 3.81: Syracuse, New York. Aerial view of I-81. Existing: The site of the interstate in Syracuse's southside neighborhood was once home to many black-owned businesses and entrepreneurs. The spaghetti-like interstate sliced the neighborhood in two. © *2022 Dover, Kohl & Partners*

Figure 3.82: Syracuse, New York. Proposed: Restored Almond Street corridor. The proposal displays how the removal of Interstate 81 could transform central Syracuse. The proposed corridor will reconnect the street network and facilitate replacing excessive parking lots with needed infill development. © *2022 Dover, Kohl & Partners*

NETWORKS AND CONNECTIVITY: ANALYSIS METHOD

Streets exist as part of a larger pattern or network. The level of connectivity in a network is a potent indicator of how people move around a neighborhood and the network's environmental and economic strengths and weaknesses. The US Green Building Council (USGBC), the Natural Resources Defense Council (NRDC), and the Congress for the New Urbanism (CNU) developed a rating system called LEED-ND (which stands for Leadership in Energy and Environmental Design for Neighborhood Development) that measures the pluses and minuses of network pattern, location, and green building technology in a neighborhood. A high score means a project can be "LEED-certified" as a sustainable development.

During the creation of the LEED-ND standards, Dover, Kohl & Partners devised a method for calculating the connectivity of the street plan. First, count the number of intersections in a network. Intersections with alleys count, but intersections that lead only to cul-de-sacs do not. Only a percentage of intersections with bikeways and trails count. Converting the count to a number per square mile produces a connectivity score. The higher the score, the better connected the network is, and the better connected the network is, the lower the vehicle-miles-traveled (VMT) per person is likely to be. Increasing choices for travel routes increases the options for walking or biking and lowers congestion at any one intersection. Not surprisingly, older neighborhoods built with grid plans score better than auto-centric sprawl patterns.

▲ **Figure 3.83:** The Woodlands, Texas. Satellite view. *Courtesy of Google Earth*

▶ **Figure 3.84:** The Woodlands, Texas. Diagrammatic plan highlighting street intersections. The suburban cul-de-sac pattern scores poorly on connectivity, forcing longer driving trips and harming walkability. © *2013 Dover, Kohl & Partners*

Figure 3.85: Celebration, Florida. Cooper, Robertson & Partners and Robert A.M. Stern Architects, 1997. Satellite view. *Courtesy of Google Earth*

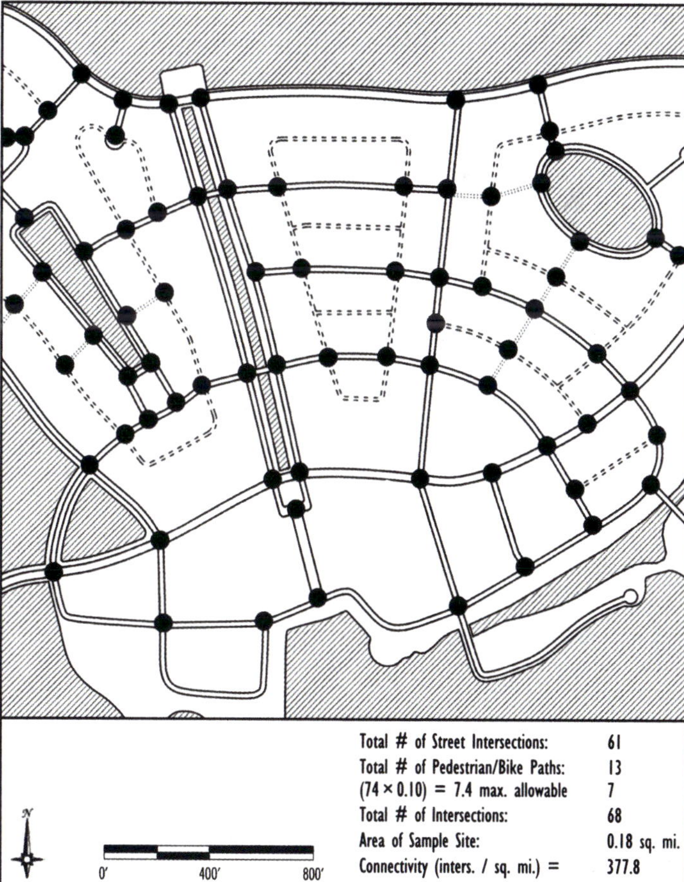

Figure 3.86: Celebration, Florida. Cooper, Robertson & Partners and Robert A.M. Stern Architects, 1997. Diagrammatic plan highlighting street intersections. Block sizes vary, improving the connectivity score. *© 2013 Dover, Kohl & Partners*

Figure 3.87: Miami Lakes, Florida. Lester Collins, 1962. Satellite view. *Courtesy of Google Earth*

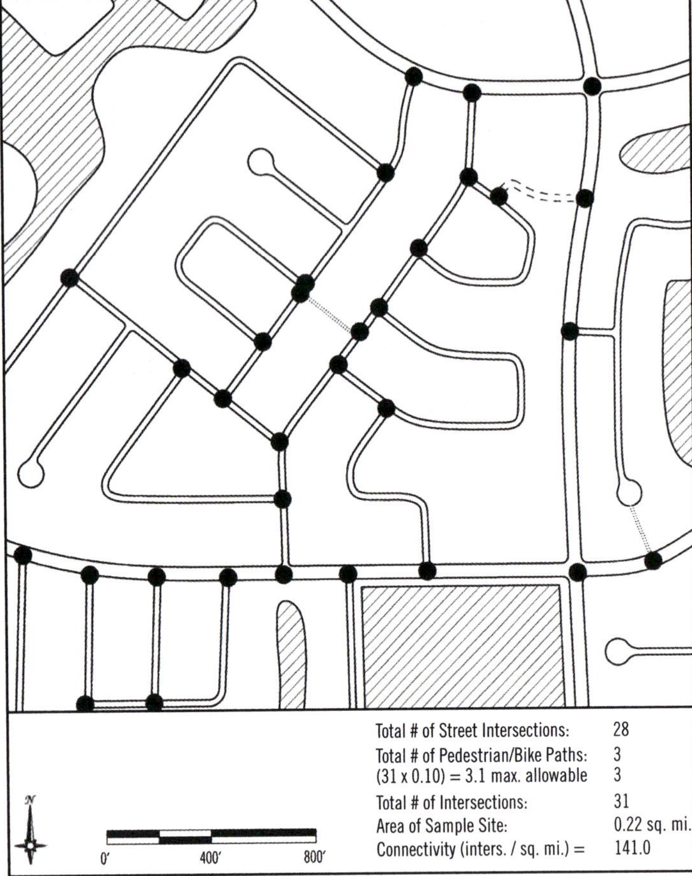

Figure 3.88: Miami Lakes, Florida. Lester Collins, 1962. Diagrammatic plan highlighting street intersections. Oversized blocks result in a poor connectivity score. *© 2013 Dover, Kohl & Partners*

Figure 3.89: Historic Center, Rome, Italy. Satellite view. The Pantheon is in the center of the photo, on the right, and the Piazza Navona is on the left. In the nineteenth and twentieth centuries, broad streets were cut through the fabric: the Corso Vittorio Emanuele II goes east–west, and the Corso del Rinascimento runs parallel to the Piazza Navona. *Courtesy of Google Earth*

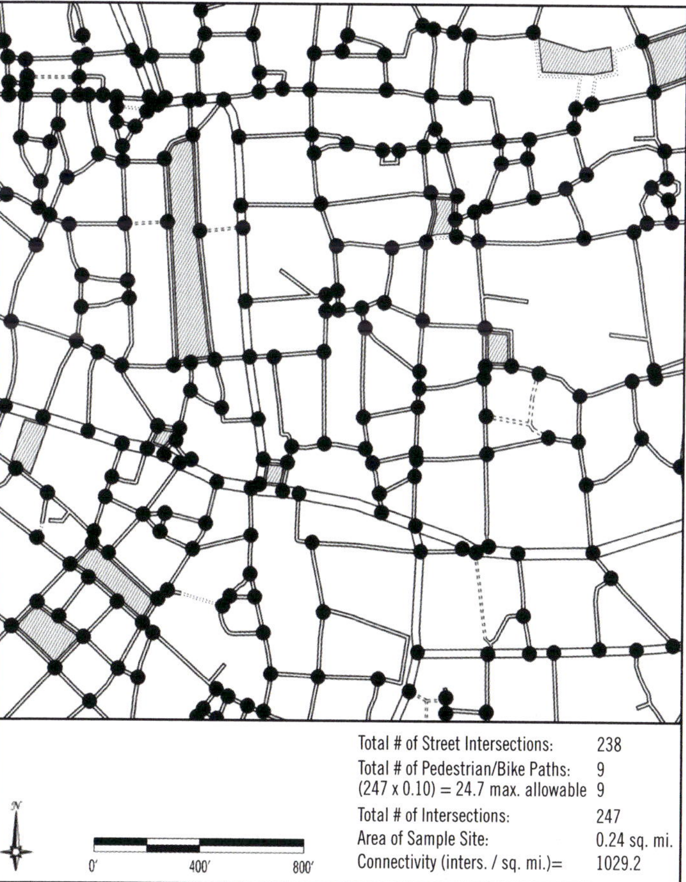

Figure 3.90: Historic Center, Rome, Italy. Diagrammatic plan highlighting street intersections. The full menu of street types, from wider main streets to tiny lanes, is combined with small blocks to result in extraordinarily high connectivity. © 2013 Dover, Kohl & Partners

NETWORKS: FOSTERING A CULTURE OF ACTIVITY

The larger lattice or grid of streets in a city, town, or neighborhood affects how useful and delightful our individual great streets can be. Although most of this book focuses on the detailed design of individual streets, the city map is designed, too. Urbanism is concerned with how we give shape to the city's networks, not just their parts, and with how the bigger forms of neighborhoods and corridors combine.

Because the form of our built environment has a profound effect on whether we choose to walk, bike, use transit, or drive, the design of the street network directly impacts our physical and mental health and well-being, just as the design of an individual street does. The city form determines our access to "greenness," whether we move our bodies regularly, how often we have casual social interactions, and the quality of our air and water.

Planning for Public Health

The city planning profession as we know it today sprang from public health reform and the parks movement. This was famously embodied by Frederick Law Olmsted's "Emerald Necklace" of interconnected parks and greenways for Boston, laid out over a two-decade period starting in the late 1870s. Olmsted addressed the urgent health problems of the day by considering and consciously assembling an ensemble of parts greater than the pieces. In the case of the Emerald Necklace, he saw connectedness as the key to simultaneously improving sanitation and flood control, resolving pollution in the Charles River, improving the drinking water, and relieving the stress of life in the industrial city. In Buffalo, New York, Olmsted and Vaux oversaw an even more ambitious (and more street-dependent) program to link parks and neighborhoods across the city via leafy boulevards, uniting these green parts of the urban form into a networked *system*. In the early twentieth century, the Olmsted Brothers firm incorporated these ideas into plans for cities large and small across the continent.[23]

Today's public health problems call for a similar whole-systems approach. Heart disease, childhood obesity, early-onset diabetes, and epidemic loneliness are national problems; better outcomes on all these are dependent on the next generation of green, networked streets linking the public spaces of our cities to make daily walking or biking feel natural again, encourage encounters among neighbors, and make driving optional for many more people.

The Lake Wales Way

Dover-Kohl's plans for the small town of Lake Wales, Florida, may serve as a model. The *Lake Wales Connected* downtown plan and the subsequent *Lake Wales Envisioned* big-picture plan revive city planning concepts devised for Lake Wales by Frederick Law Olmsted, Jr. between 1925 and 1931.

The younger Olmsted, himself a senior statesman for the landscape architecture and city planning professions by the 1920s, devised a plan for Lake Wales as a "city in a garden." His proposals for creating new neighborhoods, extending the city street network, systematically planting street trees, and establishing parks and preserves were only partially followed, but to significant effect.[24]

Lake Wales Connected brought Olmsted's street tree program back into the spotlight as part of a broader downtown revitalization effort. Local preservationists, including the Lake Wales Heritage group, have since planted hundreds of native shade trees along streets and trails and plan to plant thousands more.

In 2023, a coalition between the City and supporting organizations undertook a second planning effort that expanded upon the downtown plan and widened the lens to cover the whole city, plus its surrounding 55,000-acre utility service area. This *Lake Wales Envisioned* plan put forward a series of overlaid networks: the future street network, a discernible pattern of neighborhoods and corridors, a "neighborhood green network" of greenways and parks linked by trails that will eventually stretch across the region, and a "big green network" of farmlands and conservation areas intended to form greenbelts surrounding the town.[25]

The presence of such a combined green network—with large parks and conservation areas as well as also small squares and, just as importantly, tree-lined blocks in everyday neighborhoods—strongly affects public health. University of Miami professor Joanna Lombard, who holds dual appointments in both the School of Architecture and the School of Medicine, reports, "We did a study of about 250,000 Medicare

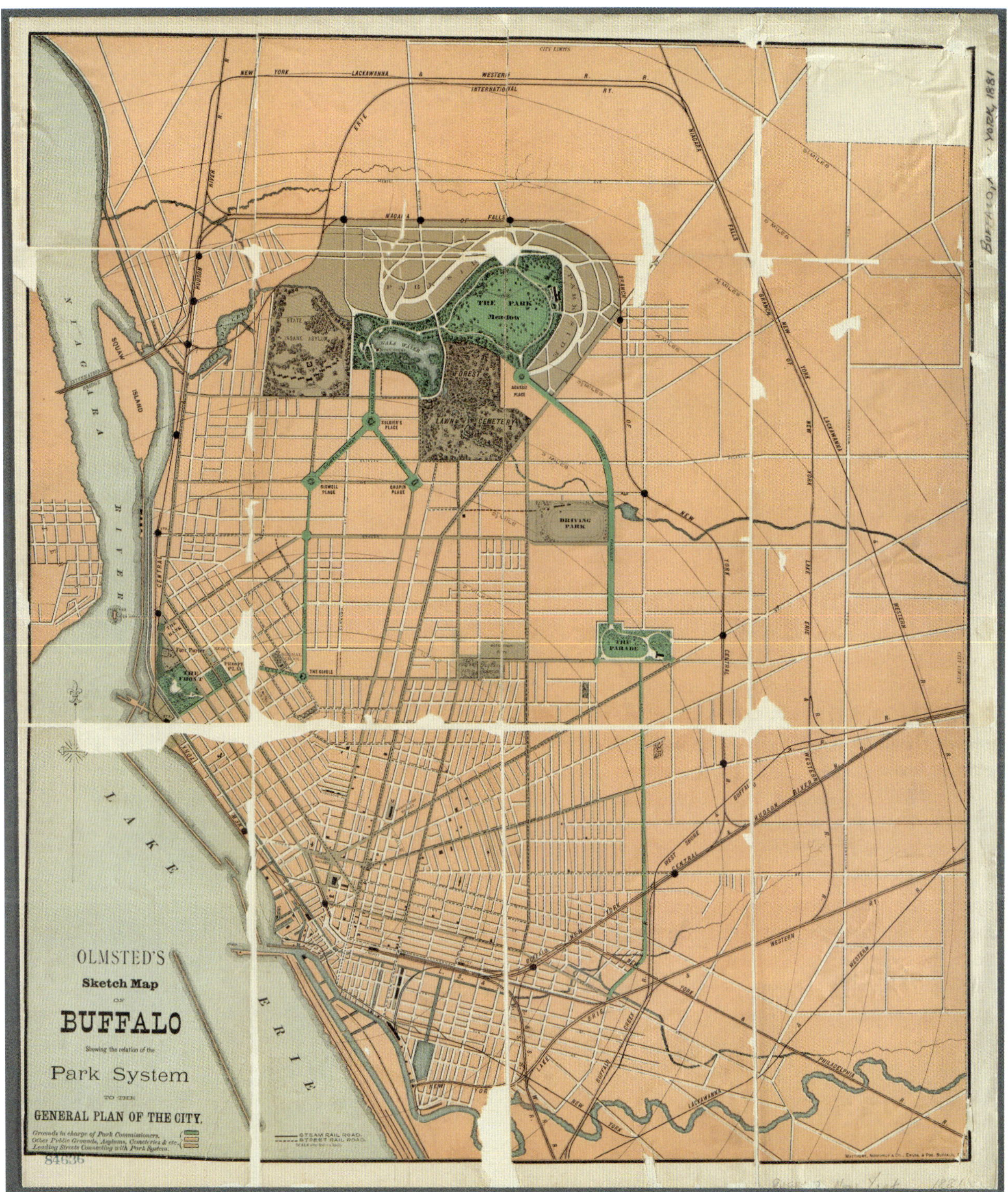

Figure 3.91: Buffalo, New York. Olmsted and Vaux, 1881. Olmsted's sketch map, showing the relation of the park system to the general plan of the city. *Courtesy of New York Public Library*

Figure 3.92: Lake Wales, Florida. Computer model. Before: Existing conditions, looking north from downtown Lake Wales (foreground) towards Bok Tower Gardens (in the distance at the upper right). © *2018 Dover, Kohl & Partners*

Figure 3.93: Lake Wales, Florida. Computer model. After: Future conditions, with new street trees and strategic infill development according to the *Lake Wales Connected* plan. © *2018 Dover, Kohl & Partners*

Figure 3.94: Lake Wales, Florida. Aerial view looking southeast. Concept for balancing development, agriculture, and conservation in the southern portion of Lake Wales. © 2023 Dover, Kohl & Partners

beneficiaries, and we found that the beneficiaries who lived on blocks with higher levels of **greenness** had lower levels of heart disease, lower levels of depression, lower levels of obesity, lower levels of dementia, and even lower levels of Alzheimer's."[26]

Not Just Healthier. Better for Transportation, Too

The same network design choices that improve health and happiness help with the problems of costly car dependence, traffic congestion, and violent collisions, too.

"Why am I stuck in traffic all the time?" asks Wade Walker, lead transportation engineer for *Lake Wales Envisioned*. "Well, it's really about networked versus isolated streets. In typical sprawl you've isolated your land uses by creating these pods that have very limited connectivity, most of those trips end up having to come right out on that one big road, and everyone is relying on that one spine road to get around. That approach is more set up for longer trips, so the streets end up being designed only for cars."[27]

In traditional neighborhood design, however, we mix the land uses, put commercial close to houses, put the schools closer to houses, tie the big green network and connect it all with a network of walkable streets that are linked to each other. Suddenly residents don't have to rely on just one road. Now there are multiple ways of getting around: because things are set up for short trips, it is easier to walk, bike, or even drive. These things allow us to design streets for people, not just cars.

Traffic engineers sometimes refer to walking, biking, and transit as "alternate modes" or "alternative modes" of transportation. We couldn't agree less. Walking is the *original* mode! Human-powered transportation, including walking and biking, is now often called *active mobility*.

> Walking is not an "alternate mode" of transportation; walking is the *original* mode.

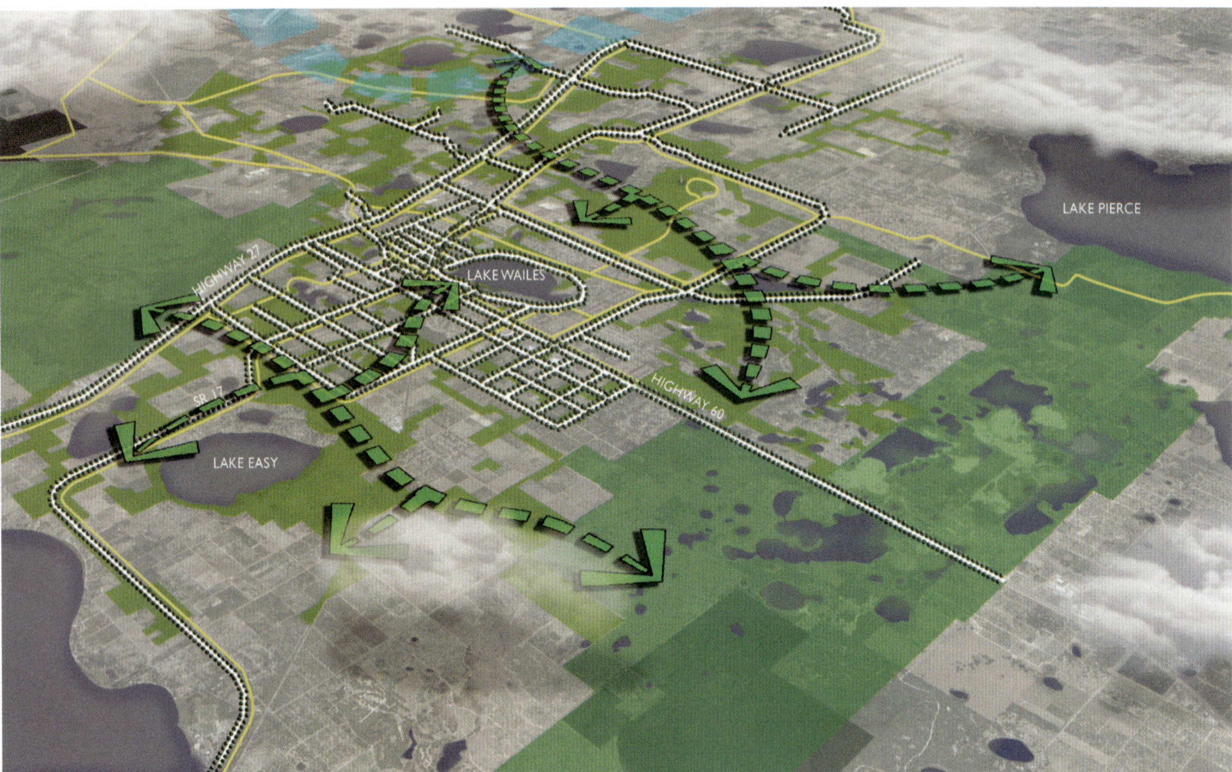

Figure 3.95: Lake Wales, Florida. Aerial diagram looking northwest. Concept for the "neighborhood green network" for the southern portion of Lake Wales. © *2023 Dover, Kohl & Partners*

Figure 3.96: Lake Wales, Florida. Aerial perspective: Hypothetical plan for expanding Lake Wales in a pattern of traditional neighborhoods and interconnected green corridors. © *2023 Dover, Kohl & Partners*

Figure 3.97: Lake Wales, Florida. Aerial perspective: Typical blocks in the future traditional neighborhoods in Lake Wales, showing streets, squares, housing variety, and parks. © 2023 Dover, Kohl & Partners

Figure 3.98: Lake Wales, Florida. Aerial perspective: A hypothetical future neighborhood square in Lake Wales. The tree-lined streets and greenways connect the neighborhoods to the larger scenic and ecological backdrop in the "big green network" beyond. © 2023 Dover, Kohl & Partners

John Simmerman, founder of the Active Towns Initiative, makes the connection between active mobility and street network design this way:

> Physical activity is a natural part of daily living for humans. But so is being sedentary. One of the big challenges in our current realm is that much of our environment is designed around using the automobile and taking the easy way. Part of what we are trying to do in communities [instead] is, making the active choice an easy choice.

> The network matters when creating or attempting to create a culture of activity because the person really needs to have the confidence that they can walk out their door and get to their meaningful destinations and not only be able to do it, but really feel comfortable in that environment and have it be a pleasurable experience.[28]

"The *Lake Wales Envisioned* plan does this by [proposing] a built environment that promotes mobility through walking, cycling, e-biking, or other means of transportation beside an automobile," Robert Steuteville wrote on the CNU website. "One way to do that is through proximity—to a park, multiuse pathway, protective bike lane, or walkable destination. Another is to ensure that the environment is pleasant for walking or using a bicycle."[29]

Simmerman usually prefers to say he promotes a "culture of activity" rather than "*physical* activity" because the latter just sounds too much like mere exercise, too much like Phys Ed class. He asks: "How comfortable is it to walk out your door and get to meaningful destinations?" If it feels natural to walk or bike, then many people will start moving and get daily "physical activity" without even thinking about it.[30]

> Jumping on the bike should be a most natural thing to do. If I have an interconnected, comprehensive, connected and comfortable network, an active mobility network, then, well, I have choice. I can still choose to drive if I want to, but if I decide to, I can walk, or I can decide to bike—and if it's a pleasurable, meaningful, utilitarian experience, I can make a habit of it.[31]

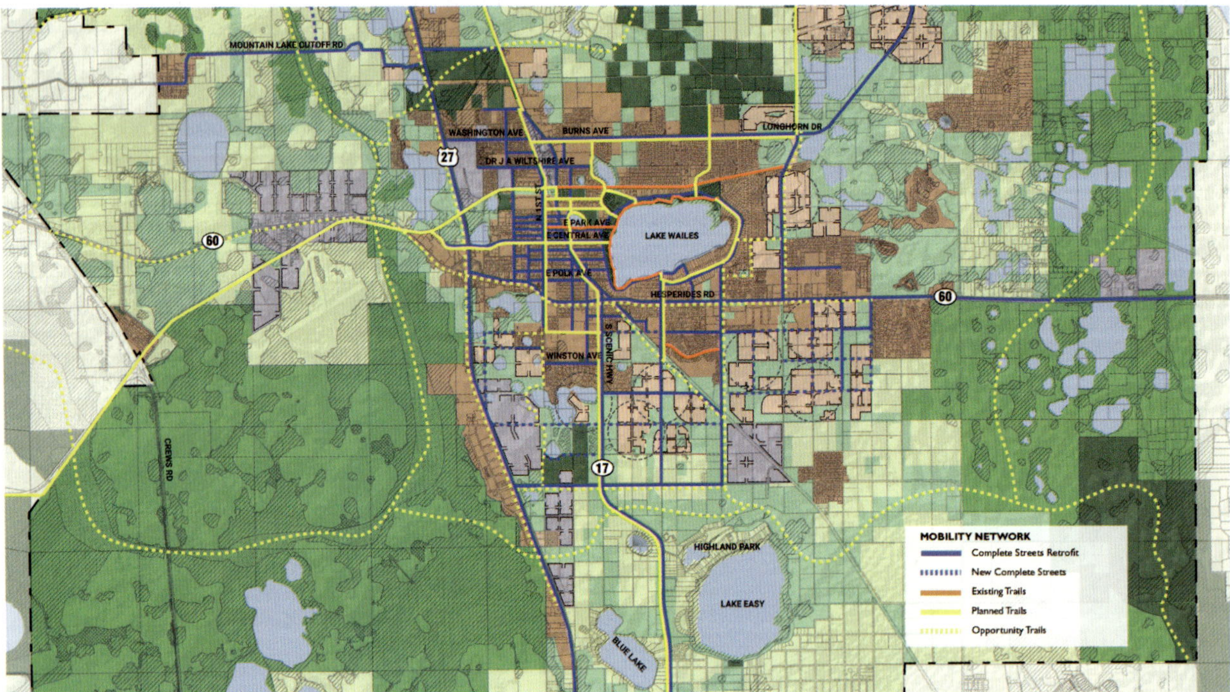

Figure 3.99: Lake Wales, Florida. Map: Combination of the fine-grained pattern of shaded, walkable, bike-friendly streets and the coarser, "unbraided" network of longer routes for active mobility in Lake Wales, overlaid on the proposed green networks. © 2023 Dover, Kohl & Partners

The Dutch Networks, Plural

Cities in the Netherlands are famous for dedicated cycling infrastructure intended to promote the outcomes Simmerman describes, but he points out that only about 30 percent of the Dutch street network has separated or protected bike lanes. The other 70 percent is either shared space—"incredibly comfortable, low speed, low volume, shared streets" within neighborhoods—or so-called *unbraided* corridors for cross-town travel.

Within the fine-grained neighborhood street network focused on short trips, the Dutch apply what Jason Slaughter of *Not Just Bikes* calls "filtered permeability" to prioritize certain routes for walking and biking and to discourage driving, resulting in low car traffic volumes.[32]

The Dutch also use a concept they call *ontvlechten* (which roughly translates as unbundling, untangling, or unbraiding). "For the cross-town trips, the Dutch look to creatively untangle the large-scale networks for driving, transit, or walking/biking longer distances so that they're not creating conflicts for each other," Simmerman says.

In the Dutch unbraiding approach, there are multiple networks overlaid—certain corridors are prioritized for driving and freight, other corridors for public transport, and yet another group of corridors where biking is prioritized— and at this level, whenever possible, they're trying to untangle them, to reduce conflicts. Where they *do* intersect, the goal is to put the top priority on safety, comfort and continuity for those walking and biking.[33]

MEASURING WALKABILITY: WIN VERSUS LOS

Hall Planning & Engineering, a New Urban firm in Tallahassee, Florida with work across the country, created a Walkability Index that uses a fixed set of criteria to measure pedestrian-friendliness (Figure 3.100).[34] Unlike the system at the popular website WalkScore®,[35] Hall's Walkability Index is based on physical attributes such as the frequency of intersections and the qualities of the block frontage. Blending transportation planning and engineering with urban design, the Walkability Index uses measures of context and density to determine the pedestrian's sense of freedom, comfort, and safety, block by block.

The Index uses ten criteria to determine the overall walkability:

1. Vehicle speed: Vehicle speed is measured outside of peak traffic times when traffic is moving freely. Taking at least ten samples with a radar gun is recommended. (Lower speeds are better for pedestrians.)

2. Thoroughfare width: The street width at each pedestrian crossing, measured from curb to curb. (Wider streets are more daunting for those on foot.)

3. Street parking: The presence of on-street parking (and its variations, including loading zones, ridehailing spots, "streateries," and bikeshare docks) and the percentage of block front dedicated to parking where it is permitted and in use. (On-street parking improves walkability.)

4. Sidewalk width: The width of the paved sidewalk. (The ranges offered vary, according to the urban-to-rural transect or "context zones," but wider sidewalks are still preferred within each category.)

5. Pedestrian connectivity: The distance between street intersections or midblock crossings. This measures network density. (The more options the pedestrian has, the more walkable the street or block is.)

6. Pedestrian features: The presence and quality of pedestrian amenities. (Shade trees, sidewalks in good condition, and the like raise the score.)

7. Street enclosure: The ratio of building height to street width. (Except on the grandest avenues and boulevards, simple ratios with low numbers like 1-to-1 and 2-to-1 are good for pedestrians.)

8. Land-use mix: The presence of different land-use types—retail stores, restaurants, private homes, for example. (Here, also, the Index is sensitive to the context—a country road is very different from a bustling town—but variety is valued by the pedestrian in all settings.)

9. Facade design: The number of doors and/or windows and the visual quality of the facades on the block face. (A long, blank factory wall would receive a poor walkability rating. A row of townhouses or small shops would receive a much better rating.)

10. Transit/bicycle features: The presence of bus shelters, bus stops, protected bike lanes or cycle tracks, bicycle lockers, and bicycle racks.

The Walkability Index Number Data Sheet, available on the Internet, allows one to evaluate a street, segment by segment, according to these criteria. In the plain version of the Index, each criterion has a maximum value of 10, making 100 points the highest possible score. However, the user is encouraged to adjust the weight given to each criterion according to its importance in their situation and to refine the criteria. The following table shows total scores, graded per street segment.

90–100: High walkability (A)

70–89: Very walkable (B)

50–69: Moderately walkable (C)

30–49: Basic walkability (D)

20–29: Minimal walkability (E)

19 points or less: Uncomfortable/Hazardous for walking (F)

Easy to understand and easy to apply, the Walkability Index Number (WIN) offers both professionals and concerned citizens an easy way of determining what works and what doesn't for pedestrians in their communities. An alternative to the vehicle-based Level of Service (LOS), its criteria are being used to produce designs for walkable places in a growing number of community workshops.

Figure 3.100: New Row, London, England. View looking east from Bedford Street. Exceptionally narrow streets might be either pedestrian-only or allow just one-way traffic. Neither pattern is recommended for a whole neighborhood, but when there are many connected streets, some links can be demoted to a less-trafficked status, to great effect.

STREET NETWORKS AND THE ENVIRONMENT

Building better streets is one tool for addressing climate change. On the one hand, an agreeable street design encourages people to replace driving with walking, cycling, or mass transit, reducing greenhouse gas emissions and the consumption of fossil fuels. But the auto sewers we've built all over America—even in our most walkable cities—encourage us to stay in our cars and drive everywhere.

According to Reid Ewing and the researchers behind *Growing Cooler: The Evidence on Urban Development and Climate Change*, the potential impact of this "mode shift" is vast.[36] Their analysis of VMT shows that a well-connected street network with good public transport can decrease auto use and have a positive effect on the environment. They show that increased density results in lower levels of VMT and that compact urban development promotes freedom from dependence on the car.[37] And they discuss the work of Peter Newman and Jeff Kenworthy, who conclude that higher-density neighborhoods lower VMT numbers and reduce fuel consumption far better than any other method of cutting fuel consumption.[38] Those who drive less also get more exercise.

Good streets help the environment on another, more subtle level. Walkable streets with bike lanes and mass transit reinforce a variety of broad environmental goals such as controlling sprawl, reducing regional energy consumption, protecting watersheds, and decreasing the loss of farmland and wilderness. They do this by encouraging growth in the right places.

On the other hand, no one wants to walk on the arterials and collectors built under the Functional Classification system. Combined with the vast interstate highway system, these roads advanced the sprawl process they were part of by inducing the construction of more and bigger roads. As generations of Americans used the roads to move farther and farther out, the increased driving had the perverse effect of making the traffic still worse—and, yes, inflating demand for still more and wider roads.

A good street where people want to be is naturally more marketable, so it sets up its adjacent places for perpetual use and reuse. This is the ultimate recycling process: reoccupying previously settled land more densely and giving historic buildings new life also accommodates population growth and economic change. Without this, population growth and economic evolution lead inexorably to the development of raw land in far-flung locations, skipping over the previously built-on land. To discourage sprawl, we have to make real cities and towns attractive to the large number of homebuyers and businesses who have a choice about where to locate. They'll choose the locations that balance beauty, comfort, and convenience with privacy and safety. Better design is the tool that allows compact neighborhoods to fulfill their ecological promise. This is why the great Michael Busha, former director of the Treasure Coast Regional Planning Council says, "New Urbanism is the operating system of smart growth."[39]

Figure 3.101: Koningsplein, Amsterdam, the Netherlands. Walkable, bike-friendly streets with mass transit can be great, "green" places. Amsterdammers use very little energy for cars because they prefer to walk, bike, or ride the tram.

NOTES

1. Since we wrote the first edition of *Street Design*, Thomas D. Wilson, the author of the Savannah history cited below in note 4 has written *Charleston and Savannah: The Rise, Fall, and Reinvention of Two Rival Cities* (University of Georgia Press, 2023). Readers eligible for a Project MUSE membership can download the book at https://muse.jhu.edu/book/110022.

The economies, infrastructure, and buildings of Savannah and Charleston relied on enslaved labor until the end of the Civil War. Two articles that discuss that history in Charleston are Phillip Smith, "African-American Classicism of Charleston: Talents, traditions of enslaved Africans built revered local architecture," *Preservation Progress,* Fall / Winter 2022, page 15; and Nathaniel Robert Walker, "Classicisms of Color: Transatlantic Exchanges in African and American Traditional Architecture," *Journal of Traditional Building, Architecture and Urbanism,* no. 2, November 2021, page 437. Walker's article is available online at https://doi.org/10.51303/jtbau.vi2.531. PDFs of Smith's article and an expanded version are available for download at https://blog.massengale.com/2022/11/17/african-american-classicism/. In addition, Bernard E. Powers Jr., *Black Charlestonians: A Social History, 1822–1885* (University of Arkansas Press, 1994) is a foundational study of Black labor, including enslaved artisans and builders; Walter J. Fraser Jr., *Savannah in the Old South* (University of Georgia Press, 2003) is the definitive economic and social history of Savannah, with extensive treatment of enslaved labor in the city's development; Clifton Ellis and Rebecca Ginsburg, eds., *Slavery in the City* (University of Virginia Press, 2017) is a collection of essays on urban slavery, including Charleston specific chapters; and Nathaniel Robert Walker and Rachel Ama Asaa Engmann, eds., *Architectures of Slavery: Ruins and Reconstructions* (University of Virginia Press, 2025), which includes the essay by Robin B. Williams, "Henry Ford, the Hermitage, and the Role of Slave-made Savannah Grey Bricks in the Construction of a Plantation Ideal in Twentieth-Century White Suburbia."

2. Roger G. Kennedy, *Architecture, Men, Women and Money in America, 1600–1860* (Random House, 1985): *passim.*

3. Robert Russell, "The Architecture of Politeness: Form and Meaning of the Charleston Single House," a talk delivered at the Center for the Study of the American South, UNC-Chapel Hill in February, 2006. In *Southern Architecture: 350 Years of Distinctive American Buildings* (Dutton Adult, 1981), Kenneth Severens proposes the theory that the Single House originated in Barbados, from where several Charleston planters came. Russell visited Barbados to look into this theory and found it unconvincing.

4. Thomas D. Wilson, *The Oglethorpe Plan: Enlightenment Design in Savannah and Beyond* (University of Virginia Press, 2012), *passim.*

5. In conversation with the author.

6. W. M. Gillespie, *A Manual of the Principles and Practice of Road-Making: Comprising the Location, Construction, and Improvement of Roads (Common, Macadam, Paved, Plank, etc.); and Railroads*; (A.S. Barnes & Co., 1847): 216–217.

7. John W. Alvord, *A Report to the Street Paving Committee of the Commercial Club on The Street Paving Problem of Chicago* (R.R. Donnelley & Sons, 1904): exhibit no. 2. Macadam represented 50.1, 30.1, and 28.35 percent of all street surfaces in Boston, New York, and St. Louis, respectively, where unpaved was still the largest percentage.

8. *Ibid.*

9. *Ibid.*, Exhibits nos. 9–14.

10. Kirsten Hower, "The Photogenic History Right Under Houstonians' Feet," *Saving Places,* November 4, 2016, https://savingplaces.org/stories/joey-sanchez-houston-blue-tile-project.

11. See the Contact page for the Historic Pavement website: https://www.historicpavement.com/contact-and-consulting.

12. Anthony Sutcliffe, *Paris: An Architectural History* (Yale University Press, 1993): 48–50. We are indebted to Sutcliffe for his concise and insightful history of the urban codes, which we have borrowed from liberally. An excellent book that has come out since we wrote the first edition of *Street Design* is Benoît Jallon, Umberto Napolitano and Franck Boutté, *Paris Haussmann* (Park Books, 2017).

13. *Ibid,* 57.

14. Joan Busquets, Dingliang Yang, and Michael Keller are the authors of the excellent, encyclopedic *Urban Grids: Handbook for Regular City Design* (Oro Editions, 2019).

15. For more photos taken by Nicholas Boys Smith in his walk around Toulouse see his post "Streets to Dream of in Toulouse…" on the Create Streets blog: https://www.createstreets.com/streets-to-dream-of-in-toulouse/.

16. See Sage Foundation Homes Company, *Forest Hills Gardens* (The Company, 1913); Susan L. Klaus, *A Modern Arcadia: Frederick Law Olmsted Jr. and the Plan for Forest Hills Gardens* (University of Massachusetts Press, 2004); Peter Pennoyer and Anne Walker, *Grosvenor Atterbury: Architect of the American Country House* (Monacelli Press, 2024): pages 148–187; Robert A.M. Stern with John Massengale, *The Anglo-American Suburb* (Architectural Design Profile, 1981): 33; Robert A.M. Stern, Gregory Gilmartin and John Massengale, *New York 1900, Metropolitan Architecture and Urbanism 1890–1915* (Rizzoli, 1983): 427–429; and Robert A.M. Stern, David Fishman, and Jacob Tilove, *Paradise Planned: The Garden Suburb and the Modern City* (Monacelli Press, 2013): 133–135 and 140–144 (for additional references, see the index and bibliography in *Paradise Planned*). Also see Paul Goldberger, "Design Notebook; An Honorable U.S. Tradition of Suburban Planning," *New York Times*, November 12, 1981, https://www.nytimes.com/1981/11/12/garden/design-notebook-an-honorable-us-tradition-of-suburban-planning.html.

17. William Van Duzer Lawrence (1842–1927) made a fortune marketing a patent medicine he named "Pain Killer." It could be taken either internally or externally, and its active ingredients were probably opium and alcohol. In *The Adventures of Tom Sawyer*, Tom feeds Aunt Polly's cat a dose of Pain Killer, with noticeably dramatic results.

18. The Village of Bronxville was incorporated in 1898, by which time Lawrence owned most of the land in the one-square-mile Village.

19. See *The Anglo-American Suburb*, op. cit.: 31. Also see: Wilbur Cross III and Anita Inman Comstock, editors, *Bronxville, Views and Vignettes, 1898–1973* (The Bronxville Diamond Jubilee Committee, 1974); Loretta Hoagland, *Lawrence Park: Bronxville's Turn-of-the-Century Art Colony* (Lawrence Park Hilltop Association, 1992); Eloise L. Morgan, editor, *Building a Suburban Village, Bronxville, New York, 1898–1998* (Bronxville Centennial Celebration, 1998); and *Paradise Planned*, op. cit.: 310–315.

20. One of the Bronxville streets Lawrence later built downhill from Lawrence Park has a tree in the middle of the street.

21. Philippe Mesmer, "Seoul demolishes its urban expressways as city planners opt for greener schemes," *The Guardian*, March 13, 2014, https://www.theguardian.com/world/2014/mar/13/seoul-south-korea-expressway-demolished.

22. Herbert Muschamp, "This Time, Eisenman Goes Conventional," *New York Times,* May 2, 1993, https://www.nytimes.com/1993/05/02/arts/architecture-view-this-time-eisenman-goes-conventional.html. For a case study from the Urban Land Institute, see "The Cap at Union Station," https://casestudies.uli.org/wp-content/uploads/2015/12/C035010.pdf.

23. Lucy Lawliss, ed., *The Master List of Design Projects of the Olmsted Firm 1857–1979* (National Association for Olmsted Parks, 2008).

24. In one case, the idea was revived forty years later: Tiger Creek Preserve, a 5000-acre wilderness area managed by the Nature Conservancy, was established in 1971.

25. For more information about the Lake Wales Envisioned Plan, see https://lakewalesenvisioned.com/.

26. Dover, Kohl & Partners, *Design & Health: Better Neighborhoods = Healthier Lives* (video interview with Joanna Lombard), June 27, 2023, https://youtube/watch?v=LoLfrU3O7Gk.

27. Dover, Kohl & Partners, *Deep Dive: Complete Streets/Mobility* (webinar with Wade Walker, PE), July 10, 2023, https://youtu.be/rD6oJXNWKik?si=nKEcFbSDNN13RtL9.

28. Robert Steuteville, "Planning a City That Gets People Moving," CNU Public Square, December 21, 2023, https://www.cnu.org/publicsquare/2023/12/21/planning-city-gets-people-moving.

29. *Ibid.*

30. *Ibid.*

31. Telephone interview with John Simmerman, December 27, 2023.

32. John Simmerman, "Secrets of Dutch Cycle Network Success" (Active Towns podcast interview with Jason Slaughter), December 22, 2021, https://youtu.be/sBRS4Mqg7To?si=bClgI_R1xYz3xOgr. Discussion about this concept begins at the 25-minute mark.

33. Telephone interview with John Simmerman, *op.cit.*

34. For more information about the Walkability Index, see https://hpe-inc.com/hpes-walkability-index-quantifying-the-pedestrian-experience/.

35. Walk Score˚ also makes apps for smartphones and tablets. See https://www.walkscore.com/.

36. Reid Ewing, Keith Bartholomew, Steve Winkelman, Jerry Walters, and Don Chen, *Growing Cooler: The Evidence on Urban Development and Climate Change* (Urban Land Institute, 2008).

37. *Ibid.*, 54–55. "Compact development has the potential to reduce VMT [Vehicle Miles Traveled] by anywhere from 20 to 40 percent relative to sprawl." Reducing VMT reduces the carbon footprint.

38. See also Peter Newman and Jeff Kenworthy, "The Transport Energy Trade-Off: Fuel-Efficient Traffic versus Fuel-Efficient Cities," Transportation Research Part A: General 22, no. 3, 1988, 163–174. https://doi.org/10.1016/0191-2607(88)90034-9.

39. Jean Scott, "An Overview of New Urbanism in South Florida," *Public Square – A CNU Journal*, https://www.cnu.org/publicsquare/guidebook-new-urbanism-florida-2005.

CHAPTER FOUR

NEW & RETROFITTED STREETS

◄ **Figure 4.1:** "Northgate Street," El Paso, Texas. Dover, Kohl & Partners, 2010. Watercolor rendering. Concept drawing for redeveloping the Northgate Mall. New and retrofitted streets are an opportunity to make welcoming places for all. © *2010 Dover, Kohl & Partners*

AROUND THE WORLD, cities and towns are changing their streets. That was true when we wrote the first edition of *Street Design*, but the pace of change is faster now: some of the streets we wrote about in Chapters Two and Three are significantly different than when we visited and wrote about them. Barcelona is rebuilding the Avinguda Diagonal as we write, undoing many of the changes made between 1950 and 1970 that gave traffic flow a higher priority on the great street. The layout and use of the rue de Rivoli in Paris are dramatically different than they were before Covid-19. In the new 15-minute city Paris is building, the rue de Rivoli is one of the main new bicycle routes in the city (Figures 3.25 and 5.61). Changes in the layout of the roadway have made it one of the world's most photographed and filmed streets.[1]

---

*Wikipedia says Gibson "is reported to have first said this in an interview on *Fresh Air*," the talk show on National Public Radio on August 31, 1993. He later repeated it, Wikipedia says, on the NPR show *Talk of the Nation*, prefacing the idea with "As I've said many times…" ("The Science in Science Fiction," *Talk of the Nation* on November 30, 1999.)

There are three types of places where change can happen most quickly. The first is in cities, towns, and neighborhoods built before 1920. These were once 15-minute cities with a public realm with shared-space streets. In the twentieth century, cities and towns redesigned and rebuilt their streets to prioritize traffic flow. But in the twenty-first century, many places are once again making them welcoming to everyone.

The second type of place ripe for transformation is in the parts of suburbia and exurbia where residents want change: this is particularly true in American sprawl. Americans spend too much time sitting in traffic jams, the cost of buying and operating our cars goes up and up,[2] and in the words of Robert Putnam, we are increasingly tired of "bowling alone."[3] Street transformations like Columbia Pike in Arlington, Virginia create new centers where people can get out of their cars, enjoy public life, and perhaps even live and work. Many also want to reduce their carbon footprint.

The third type of place is a new development, where design can showcase better ways to build. New streets in all-new neighborhoods or towns present opportunities to apply the lessons in this book holistically, without compromises or workarounds. Sometimes, developers build and market these neighborhoods to buyers and renters with particular interests. For example, Culdesac Tempe is for renters who want to live a car-free life.[4] The Bend in Chattanooga is designed to be a car-optional neighborhood (page 383). Seaside, Florida, is a resort built in the form of a walkable town, with an emphasis on place

rather than transportation, while Battery Park City was planned to be a normal, walkable New York neighborhood designed the old-fashioned way, beginning with the streets. These are all welcome changes.

Seaside and Battery Park City

Two very different developments from the early 1980s are important landmarks in the recent history of urban design and street design. Battery Park City, a ninety-two-acre extension of Manhattan in the Hudson River that was built on landfill from the construction site of the World Trade Center, has office towers, mid-rise and high-rise apartment buildings, and stores. Seaside, Florida, an eighty-acre development on the Florida Panhandle, is a resort built in the form of a town. What the two places have in common is that their streets were designed with many of the placemaking principles outlined in this book. Both projects were radical departures from conventional practice at the time. The histories of both demonstrate how auto-centric regulations across the country hinder the making of good streets.

It was not that people did not understand the principles; by the early 1980s, these ideas had been discussed and praised for at least two decades. Jane Jacobs wrote the enormously popular *The Death and Life of Great American Cities* in 1961, while Bernard Rudofsky published *Streets for People* (also very popular) in 1969.[5] In the 1980s, Holly Whyte published his influential studies of how people use urban space in two books, *The Social Life of*

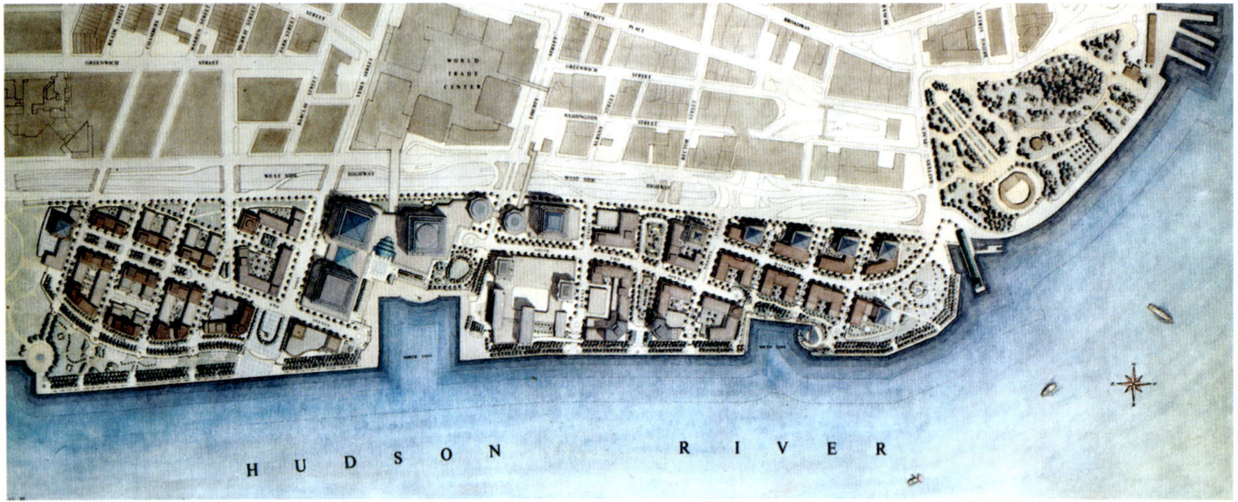

Figure 4.2: Battery Park City, New York, New York. Cooper Eckstut Associates, 1980. Hand-drawn master plan. Streets in downtown Manhattan continue through the site, maintaining views to the river. *Courtesy of Perkins Eastman*

Small Urban Spaces[6] and *City: Rediscovering the Center.*[7] Despite professional acceptance of the theories of Jacobs and others, most of the sprawl in America was built after the publication of *Death and Life*. Planners endorsed the works, but the American Planning Association and its members continued to promote regulations based on an auto-centric separation of uses, with road standards established by the engineering profession's anti-urban Functional Classification system. "The pseudoscience of planning," Jacobs wrote, "seems almost neurotic in its determination to imitate empiric failure and to ignore empiric success."[8]

In the design world, many architects and urban designers quoted Jacobs as though she were a biblical prophet, but the manifestation of her ideas in built works was slow in coming. A landmark event was New York Mayor John Lindsay's creation in 1966 of a municipal

> Despite professional acceptance of Jane Jacobs' theories, most of the sprawl in America was built after the publication of *The Death and Life of Great American Cities.*

Urban Design Group, staffed with young architects who later went on to become important urban designers, including Jaquelin Robertson, Jonathan Barnett, Alex Cooper, and Robert A.M. Stern. Working within the context of America's largest and densest city, they chipped away at New York's 1961 Zoning Resolution, which had institutionalized Le Corbusier's paradigm of making the building more important than the street.

Nevertheless, city regulations affecting street design continued almost unchanged, favoring cars over pedestrians. When Cooper and his partner Stan Eckstut

Figure 4.3: Battery Park City, New York, New York. Cooper Eckstut Associates, 1980. Aerial photo looking east. Battery Park City was built west of the Financial District on fill from the construction of the World Trade Center. © 2021 *Stephen Amiaga*

designed Battery Park City, New York's auto-centric regulations still required overscaled streets that were wider than the most common streets in the city's grid. The designers at Cooper Eckstut Associates knew how to design good streets and make good urbanism, but their hands were tied when it came to the size of the streets.

While they could not ignore city law, their design was often innovative, paying little attention to many conventional planning standards of the time. It was customary then for a single architect and one developer to plan and build the entire project, usually by beginning with the design of the buildings and then putting in streets as necessary—a method of planning sometimes called Big Architecture. In fact, Battery Park City had an earlier tower-in-the-park plan designed that way, but Cooper-Eckstut instead began with the street plan rather than the buildings, as urban designers should.

They intelligently extended adjacent city streets through the site, giving long views to the Hudson River and the open sky over it. Using those through streets as a framework, they made a pattern of streets that produced normal New York City blocks, the antithesis of the superblocks that were still in vogue with most architects and planners at the time. On top of that, they laid a simple form-based code over the plan, so that the Battery Park City Authority could sell building lots to different developers with some confidence about what would be built.

The Seaside Community Development Corp. began construction of the resort on the Panhandle several months after the Battery Park City Authority started building. The development benefited from a trend in architectural education that was perhaps the biggest change in the twenty years between the publication of *Death and Life* and the parallel achievements of Battery Park City and Seaside. During the late 1970s and much of the 1980s—a brief period when Postmodernism and Modernism peacefully coexisted in the world of architecture (particularly in architecture schools)—a

Figure 4.4: Seaside, Santa Rosa Beach, Florida. Duany Plater-Zyberk & Company, 1981. Watercolor master plan. A resort on the Florida Panhandle planned like a town. © *1981 DPZ CoDESIGN*

renewed appreciation for the design of traditional cities and streets was somehow floating in the air. Practitioners like Stern, Cooper, Robertson, and Eckstut were teaching as well as practicing. From 1973 onward, students in schools around the country seemed to find their way to books like *Civic Art* that had literally been gathering dust on library shelves.[9]

Occasionally, a new book made a strong impression. The bilingual *Rational Architecture Rationelle*[10] from Maurice Culot's Archives d'Architecture Moderne was particularly treasured for its glimpses of the work of the architect Léon Krier. Krier had no built work at that time, but his entry in a French competition to design a new neighborhood called La Villette came in second, behind a plan by Bernard Tschumi. For students poring over texts like *Civic Art*, Tschumi's scheme seemed like conventional planning of the time, dressed up with French intellectual conceits, but Krier's design was eye-opening. It had some of the most beautiful drawings in the recent history of architecture and urbanism, but the ideas they illustrated were even better. Krier's entry was a fresh design for a normal European neighborhood, with streets and squares and a public realm with civic monuments. The concept was simple, but executed with an astonishing richness and invention unlike any other work being published at the time. It gave encouragement and inspiration to students who went on to become New Urbanists, Classical architects, traditional designers, or all of the above.

At least part of the reason these students were open to studying old models was that they were among the first generations to grow up in a world of cul-de-sacs and suburban arterials. They had personally experienced both the old and the new, and they frequently found the latter lacking. Moreover, there were cases where the new models were proving to be just plain bad: the highly acclaimed and award-winning Pruitt-Igoe, a textbook tower-in-the-park housing project in St. Louis, was such a disastrous social experiment that the city demolished it in 1972.

It was the students of that generation who designed Seaside a few years later in 1981, under the leadership

Figure 4.5: Seaside, Santa Rosa Beach, Florida. Duany Plater-Zyberk & Company, 1981. Aerial photograph looking north, 2007. Robert Davis grew up spending summers with his grandparents in nearby Grayton Beach. Robert and Daryl Davis built a resort their grandchildren would enjoy. © *1981 DPZ CoDESIGN*

of Miami developer Robert Davis, his wife and partner Daryl Davis, and architects Andrés Duany and Elizabeth Plater-Zyberk. Davis was slightly older than the others, but the entire team had three advantages that helped them design Seaside: no one on the team had designed a new town; the county on the Florida Panhandle, where they built Seaside, had virtually no planning or building regulations; and although the market was in a recession when Davis started planning Seaside, he had inherited the land debt-free, so there was time to ponder what to do. Fate gave the team a *tabula rasa* to work on, and they designed a place on the Gulf Coast where they would enjoy spending time. Davis wanted a place where generations of families would come for the summer year after year, as he had with his grandparents. And there was talk of building Seaside so simply that even architects would be able to afford a second house there.

Davis had spent childhood summers on the Gulf Coast's "Redneck Riviera" with his grandparents. When he revisited the area to look at the site where he would build Seaside, he was appalled to discover that the simple paradise he remembered was overrun with Fort-Lauderdale-style condominium towers behind large parking lots. So Robert and Daryl Davis took Duany and Plater-Zyberk on a tour of southern Alabama and the Panhandle to look at the local small towns and building types the Davises admired (Figure 4.5). Eventually, they came up with a concept that was essentially the program for a large resort hotel broken down into small parts. Instead of hotel rooms, they planned small houses arranged in the pattern of a small town, along normal streets. The spa was placed at the back of the town and called The Country Club. The public meeting rooms and shops formed the downtown, and so on.

A model for the streets of Seaside came from a nearby beach town called Grayton Beach, where Davis had spent many happy summers as a boy. Grayton had unpaved sandy roads and occasional boardwalks. Most of the time, people on foot or bicycles shared the narrow roads with the cars. Davis was also influenced by *Streets for People*, calling author Bernard Rudofsky one of his heroes.[11] Duany, who had recently met Krier, brought him on board as a consultant.

The Seaside streets they designed had a variety of widths. Outside of the downtown and a few important spots, the streets are narrow, with no sidewalks: paved roadways eighteen feet wide are bordered by parking spaces covered in oyster shells. Davis experimented with various ideas before settling on the permanent details. He paved the first road with clay, but clay had wash problems in a hard rain and could be messy in a light rain. In the second year, Davis switched to using compacted oyster shells. The oyster shells also had wash problems when it rained, and they were dusty when the strong Florida sun dried them out. So Davis ultimately settled on concrete bricks with a red tint and oyster shells on each side for the parking spaces.

Just as in Grayton Beach, walkers, cyclists, and cars shared the road, creating some of the first purpose-built "shared spaces" in America in many decades. As mentioned in the next chapter, long before Hans Monderman popularized new shared-space streets in the Netherlands, Italian municipalities and urban designers created shared-space streets in *Centri Storici* around the country in the 1960s. Davis knew those streets well from his travels in Italy (Figure 5.53).

The team took great care to make the streets comfortable spaces. In place of minimum setbacks for buildings, there were build-to lines, placed so that the buildings would shape the street. In other words, rather than saying that a house could be no closer to the street than twenty feet, the Seaside code specified that all houses on a particular street must be exactly ten feet from the street's edge, and no more. The build-to line for the downtown buildings was on the front lot line, and no downtown building could be set farther back.

As much as possible, the plan followed traces on the land of things like paths through the dune grass to the beach. On the highest point in Seaside, the plan had a small traffic circle with a gazebo in the center (Figures 4.4, 4.216, and 4.218). When all was said and done, Seaside's famous water tower ended up on axis with Forest, Grove, and Natchez streets. "That's what happens when you design Classical plans," Andrés Duany said.

After construction started in 1981, Seaside developed slowly. The Davises set up a stand where they sold shrimp, sangria, and lots, and in the beginning, they sold more sangria than land. But, as the market proves time and again, because Americans in recent years have not made enough good *places*—places where people want to be—the law of supply and demand drives prices up for those smart enough to strike out on their own to meet an unfilled demand. Within a few years, the price of lots had risen from $11,000 to $100,000, and the last few of the four hundred or so lots sold for almost $4,000,000 each. Davis wanted to limit the number of houses simultaneously under construction, and every year there were too many buyers. Once he sold the number he wanted, he raised the prices in the fall or the winter to a point of sales resistance—but in the spring the buyers always came back, willing to pay more. It quickly became clear that Seaside would not be a place for impoverished architects. (Even the charrette members who received lots in lieu of cash or checks did not foresee Seaside's phenomenal success. Most sold too early in the process, thinking they were making a good deal.)

> The market shows that since 1945 Americans have not made enough good places—places where people want to be. The law of supply and demand always creates high prices for new places like Seaside.

With success, came problems. Impressed by Seaside, officials from the local county (Walton County) talked to Davis about how to spread the good fortune around. Davis sponsored some charrettes for critical pieces of land along the coast, and Walton County decided that planning was the secret to Seaside's success. After contacting the American Planning Association, the county received names of planning firms in and around Walton. To make a long story short, the Alabama firm hired by the county came up with a plan that immediately made many of the standards at Seaside illegal—including its distinctive street widths. Making matters worse, Davis was only partway through building Seaside when the rules were changed. To finish the rest as originally designed, he had to get a variance for each new section as he went along.

By the standards of the new countywide code, the streets were not only too narrow—the way they "lay lightly on the land" did not meet construction standards. Being designed to let water pass through the road and into the ground, the roads did not have the required gutters and drains. Instead, the bricks were laid with pervious joints, and the oyster-shell parking spaces were also pervious. The wider streets near downtown had a pipeless drainage system that directed the water to a large, grassy bowl in the center of town that fills with water in hard rains. The rest of the time, it serves as an open-air amphitheater.

The first "New Urban" design to be built, Seaside garnered a lot of praise. In *TIME*, Kurt Andersen wrote that it was one of the best designs of the 1980s, calling Seaside "one of the most influential projects of the decade, and, hopefully, decades to come."[12] Around the country, similar projects began to appear—not because they were copying Seaside, but because the time was right for mixed-use, walkable developments. The designers and builders of these projects quickly discovered that despite demand, planning and zoning rules made it difficult or even impossible to build such places. Most of America by that time had regulations that prohibited anything other than arterial roads, collector roads, or cul-de-sacs, usually with engineering standards that required wide roads, clear-cut areas on both sides of the roads, and one-size-fits-all requirements for drainage, on-site water retention, location of electrical supply boxes, and the like.

In 1993, Duany, Plater-Zyberk, Davis, Krier, and approximately two hundred others (including the authors) gathered at the Athenaeum in Alexandria, Virginia, for the first annual meeting of a new organization called the Congress for the New Urbanism. The mission of the CNU is the advancement of walkable, sustainable cities, towns, and neighborhoods. Perhaps half the original membership was made up of architects who had rejected most of the planning principles they had been taught in school. One requirement for membership was a pledge not to contribute to sprawl. Many of the participants were passionate about their desire to create great streets, and there were many war stories about how difficult it was to do that.

So when you see a new street or retrofit that is not quite right, remember that what you are looking at may not be what the designer wanted. The trailblazing new streets required the painful upending of two generations of entrenched bad practices, and many arguments about them were lost. Slowly, however, examples have emerged that prove Americans can again make civilized streets. The work is improving, and the evidence is mounting.

Working on the second edition, we have to add a few comments. The evolution of what seemed like a revolution ten years ago has been very slow, particularly in America. And in the last few years, particularly since the advent of Covid, Europe has moved much more quickly than America. We will look at all that in the final two chapters, including this one.

Figure 4.6: Dexter Avenue, Montgomery, Alabama. Mural by an unknown artist, circa 1938. Dexter Avenue has seen many events in American history, from the Civil War to the civil rights movement. *Image courtesy of City of Montgomery, Alabama*

Figure 4.7: Dexter Avenue, Montgomery, Alabama. Dover, Kohl & Partners, 2007. Before: Existing conditions, circa 2006. *© 2006 Dover, Kohl & Partners / UrbanAdvantage*

AVENUES AND BOULEVARDS

Dexter Avenue, Montgomery, Alabama

Dover, Kohl & Partners and Hall Planning & Engineering, 2007

Retrofit: Repairing an Historic Place

Avenue

Dexter Avenue, the main street of Montgomery, has seen momentous events in American history (Figure 4.6). It was where the first electric streetcars in the United States ran—as part of the city's "Lightning Route." It is the place where Confederate leaders sent off their fateful telegram to Charleston, giving the order to fire on Fort Sumter. Nearly a century later, Dexter Avenue was where Rosa Parks refused to sit in the back of a municipal bus, leading to the Montgomery transit boycott. Thus, the Court Square intersection is in a sense the birthplace of both the Civil War and the Civil Rights movement. Dexter Avenue was also the concluding leg of the marches from Selma to the capital and the site of some of the Reverend Dr. Martin Luther King, Jr.'s most famous speeches, delivered at the Dexter Avenue Baptist Church or in front of the Alabama statehouse at the eastern end of the avenue.

The dome on the statehouse is the focal point of one end of the avenue. A nineteenth-century fountain in a triangular "square" terminates the western end (Figure 4.121). During the long national decline of downtowns and main streets, Dexter Avenue became quiet, then quieter, then moribund. The city briefly converted Court Square and Court Street into a bland pedestrian mall. Many stores on Dexter Avenue closed over time. The avenue finally began to improve when Montgomery restored Court Square as a proper plaza, following a design by Rick Hall of Hall Planning & Engineering.

Paved with Belgian cobblestones, Court Square has a fountain at the center of the triangle. Cars, buses, pedestrians, and parades circle the fountain, safely sharing the street, which has minimal traffic markings. Montgomery completed reconstruction in 2007, making Court Square an early example of reclaiming shared space in the U.S. After fifty years of putting cars first, American cities began giving pedestrians equal rights. (Figure 4.121).

The transformation marked the beginning of the downtown revitalization, in accordance with the 2007 Master Plan prepared by Dover, Kohl & Partners. Today, historic buildings are being restored and reoccupied, and vacant lots filled in according to the new SmartCode. The plan's visualizations and recommendations are the result of public input from over 850 local residents, business owners, and community leaders. The SmartCode will ensure that all

Figure 4.8: Dexter Avenue, Montgomery, Alabama. Dover, Kohl & Partners, 2007. After: Computer simulation of proposed improvements and revitalization. © 2007 Dover, Kohl & Partners / UrbanAdvantage

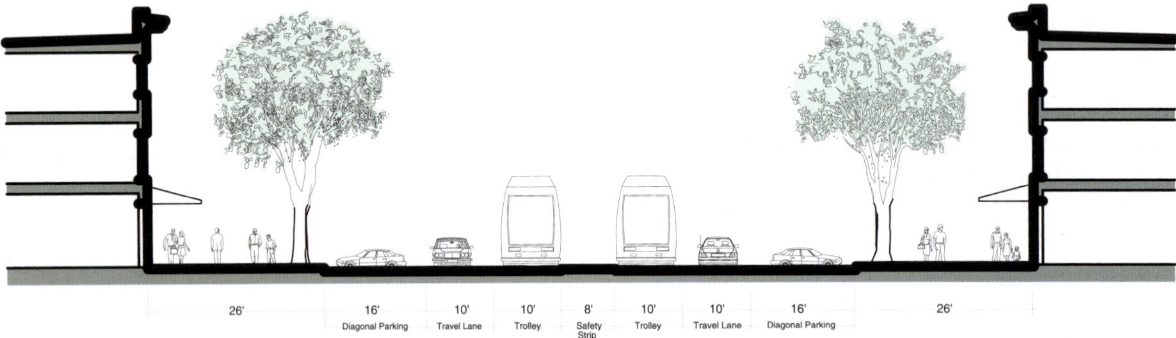

Figure 4.9: Dexter Avenue, Montgomery, Alabama. Dover, Kohl & Partners, 2007. Section showing proposed improvements.
© 2013 Dover, Kohl & Partners

future development promotes architecture, civic space, and street design appropriate to the city.

Fundamentally, this meant shifting from an emphasis on land use and parking requirements to an emphasis on design, focusing on the relationships between the buildings and the street. The supersized new buildings developed in downtown Montgomery in the 1970s and 1980s tended to face sidewalks with the blank walls and cold facades one might expect from fortifications, not from office buildings or stores on a main street. Now the rules promote normally-sized buildings that face the public spaces with doors, storefronts, outdoor cafés, balconies, and the like. The plan proposed overhauling the avenue's right-of-way, correcting the dimensions of the sidewalks, lanes, and parking spaces.

LANCASTER BOULEVARD, LANCASTER, CALIFORNIA / KAID BENFIELD

Moule & Polyzoides, 2010

Retrofit: Arterial Strip to Boulevard

Boulevard

A terrific street redesign is assisting economic recovery in a Southern California community that has suffered from deteriorating economic conditions but is nevertheless seeing significant population growth. This is a story of municipal foresight, excellent recent planning, and green ambition.

Lancaster experienced a period of very rapid growth, and now has a population of a little over 170,000 in far northern Los Angeles County, about seventy miles from downtown L.A. Its population has more than tripled since 1980; it increased by nearly a third from 2000 to 2010. It is racially mixed (45 percent Latino, 26 percent white non-Hispanic, 21 percent African American) and, like so many fast-growing western cities, decidedly sprawling. The satellite view on Google Earth reveals a patchwork pattern of leapfrog development, carved out of the desert. It is a city with a very suburban character.

Lancaster's economic condition in 2010 was not among the country's very worst, but it had certainly been better and has improved in the years following the Lancaster Boulevard transformation. According to City-Data. com, the median price of home sales in the city plummeted by almost two-thirds between 2007 and 2009, from $350,000 to about $125,000—which is more or less where it stood for years. [As of early 2024, the median home price had recovered to just over $400,000, still about half the median price in California as a whole.] Because the city is not far from Edwards Air Force Base and related industry, its fortunes have long been associated with aerospace engineering and defense contractors. In the years leading up to the Lancaster Boulevard transformation, however, some major employers, including Lockheed-Martin, had begun moving their investments elsewhere. As of August 2012, unemployment stood at 15.7 percent, way above the state average of 10.4 percent. [Notably, by 2023, the unemployment rate in Lancaster had fallen to 6.6 percent, roughly on par with the statewide long-term unemployment rate.]

Sprawl and disinvestment in Lancaster have left their scars. Greg Konar writes in the *San Diego Planning Journal*:

By the late 1980s the City's historic downtown was in serious decline. Most retailers and commercial services had long since migrated to commercial centers and strip malls in other parts of the city. For years big box retailers and regional malls had captured nearly all new commercial growth. Much of it was concentrated along the Antelope Valley Freeway (I-14). Meanwhile the historic downtown deteriorated rapidly. Crime became an increasing problem and the surrounding older neighborhoods were suffering.[13]

That is a pattern all too typical of America in the late twentieth century, but Lancaster moved to do something about it, including the adoption in 2008 of a form-based zoning code for Lancaster Boulevard, a downtown corridor. (Form-based codes encourage walkability by promoting mixed uses and a pedestrian-friendly streetscape.) The city also hired the well-known architecture and planning firm Moule & Polyzoides to capitalize on the opportunities created by the code by redesigning the boulevard to attract businesses and people.

The results—a reinvigorated section of downtown now known simply as The Boulevard—have been spectacular (Figures 4.10–4.13). The project has won multiple awards, including EPA's top national award for smart growth achievement. Moule & Polyzoides describe the design features: "Among the plan's key elements are wide, pedestrian-friendly sidewalks, awnings and arcades, outdoor dining, single travel lanes, enhanced crosswalks, abundant street trees and shading, and added lighting, gateways, and public art. Lancaster Boulevard has been transformed into an attractive shopping destination, a magnet for pedestrian activity, and a venue for civic gatherings."

Greg Konar's article, cited above, provides an excellent review of what makes the design features of the project work so well.

Justly proud of their work, the architects recount some of what happened in the area in the first several years after the project was completed:

- Forty-nine new businesses along the boulevard and an almost doubling of revenue generated, compared with just before the work began.

- An almost 10 percent rise in downtown property values.

- Eight hundred new permanent jobs, 1,100 temporary construction jobs, and an estimated $273 million in economic output.

- Eight hundred new and rehabbed homes.

- Dramatically increased roadway safety, with traffic collisions cut in half and collisions with personal injury cut by 85 percent.

This is a great example of how the right legal framework and the right design at the right time can help make a difference. It is also a great example of how our suburban communities can be improved. Is Lancaster Boulevard the best or most walkable district in America? Not by a long shot. But the change is tangible. This enterprising redesign gives the city something to build upon and sets an example for similarly situated communities—while at the same time reusing infrastructure and reducing emissions from car travel by taking advantage of a central location that shortens driving distances and encourages walking.

Figure 4.10: Lancaster Boulevard, Lancaster, California. Before: Existing conditions, circa 2008. © 2010 Moule & Polyzoides, Architects & Urbanists

Figure 4.11: Lancaster Boulevard, Lancaster, California. Moule & Polyzoides, 2010. Illustrative plan. The new promenade is multi-functional, able to accommodate both parking and civic events. © *2010 Moule & Polyzoides, Architects & Urbanists*

Figure 4.12: Lancaster Boulevard, Lancaster, California. Moule & Polyzoides, 2010. Moule & Polyzoides transformed the street by removing traffic lanes and installing a central promenade, with aligned trees and lampposts. *Courtesy of the City of Lancaster*

Figure 4.13:
Lancaster Boulevard, Lancaster, California. Moule & Polyzoides, 2010. Lancaster reclaimed its public realm. *Courtesy of the City of Lancaster / Photo by Curt Gideon Photography*

Fairfax Boulevard, Fairfax, Virginia

Dover, Kohl & Partners, 2008

Retrofit: Arterial Strip to Multiway Boulevard

Multiway Boulevard

Like most aging traffic corridors with shopping centers and low-density commercial uses, Fairfax Boulevard is a wide arterial dominated by cars that invites high speeds and chases away pedestrians. Built in 1934, it was once the "new bypass" connecting the eastern portion of Lee Highway with the part of Fairfax known as Kamp Washington. It became the desired location for the emerging retail model in the middle years of the twentieth century: fast-food chains, strip centers, discount tire stores. The boulevard is still an active commercial street, but it has lost much of its economic value, in part because the old stores and strip centers cannot compete with the newer malls, "lifestyle centers," and revitalized downtowns in the area, where customers would prefer to be. Retail is always changing, and Fairfax Boulevard feels out of date (Figure 4.14).

A comparison with other major roads in mature metropolitan areas is useful. The boulevard de Rochechouart in Paris and Eastern Parkway in Brooklyn are examples of boulevards that are similar in width to Fairfax Boulevard, but Rochechouart and Eastern Parkway are busy and full of life, well connected to their surrounding neighborhoods. Unlike Eastern Parkway, Rochechouart is a local commercial center, but both boulevards have medians for walking and stops for important subway lines below. In contrast, Fairfax Boulevard serves little but the automobile. To a great extent, the boulevard's commercial activity still reflects the car-happy era that produced it: car-oriented discount retailers, car dealerships, muffler shops, and garages dominate. Its separated land uses and haphazardly scattered commercial buildings only work well for motor vehicles, and residential buildings are mostly located away from the boulevard.

In 2008, Fairfax hired Dover, Kohl & Partners to design a new master plan for Fairfax Boulevard and its surroundings. The city has had an uneasy relationship with growth in recent decades, warily eyeing the real estate boom all around it in Northern Virginia. The tranquil suburban scenes of single-family houses flanking the corridor are in high demand, but the boulevard itself languished. Fairfax missed out on much-needed jobs, businesses, and housing variety.

Residential uses had not previously been a part of the mix on the corridor, but by 2008 a clear opportunity existed for Fairfax to retrofit the existing fabric, embrace

Figure 4.14: Fairfax Circle, Fairfax, Virginia. Before: Photo taken from within the circle in 2008. The circle is a local landmark as a traffic solution only, not as a *place*.

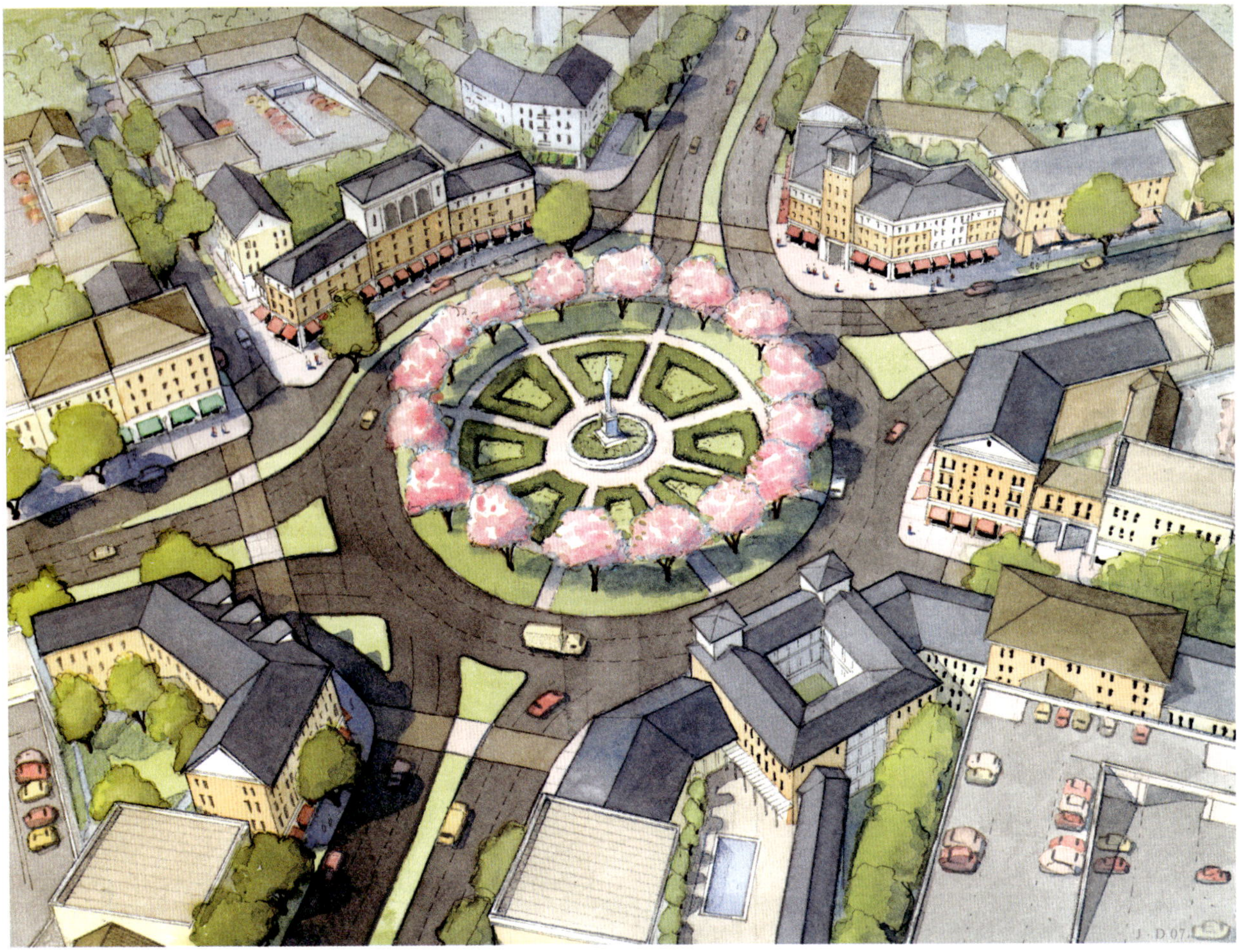

Figure 4.15: Fairfax Circle, Fairfax, Virginia. Dover, Kohl & Partners, 2008. Watercolor rendering, aerial view. The redesign recommended correcting the traffic flow for slower speeds, planting a ring of trees within the circle to make its form more spatially evident, and amending development regulations to require space-shaping, street-oriented buildings. © *2008 Dover, Kohl & Partners*

growth, and build its way out of its problems. The market had changed, and Fairfax Boulevard became a prime location for mixed-use, walkable development. It became critical to set appropriate limits for building size, adopt standards for architecture and landscaping, and require thoughtful transitions from the newly citified corridor to the suburban houses lying just beyond it (Figure 4.15).

Fairfax Boulevard was still a bypass that drivers used to get from one destination to another. However, creating a new boulevard that balanced traffic capacity, safety, placemaking, and local character could have turned it into a great street—a destination rather than a bypass (Figures 4.16–4.19).

Proposed Retrofit(s)

- Multistory, mixed-use buildings along the boulevard step down to smaller, residential development closer to existing houses.

- Side access lanes with parallel parking provide a comfortable pedestrian environment. Street trees in the medians next to the side access lanes were a key part of the strategy.

- Two lanes of on-street parking in the side lanes would have buffered the sidewalks from the traffic and provided convenient access to the stores.

Figure 4.16: Fairfax Boulevard, Fairfax, Virginia. Dover, Kohl & Partners, 2008. Before: Existing conditions in 2008. © *2008 Dover, Kohl & Partners / UrbanAdvantage*

Figure 4.17: Fairfax Boulevard, Fairfax, Virginia. Dover, Kohl & Partners, 2008. Computer Simulation. After: The corridor reimagined as a multilane, multiway boulevard. © *2008 Dover, Kohl & Partners / UrbanAdvantage*

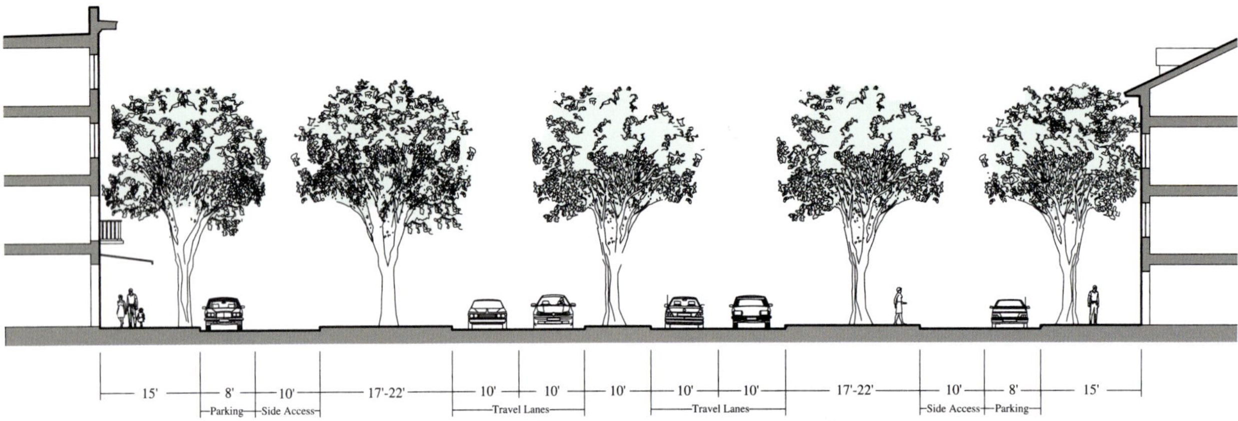

Figure 4.18: Fairfax Boulevard, Fairfax, Virginia. Dover, Kohl & Partners, 2008. Proposed section. © 2013 Dover, Kohl & Partners

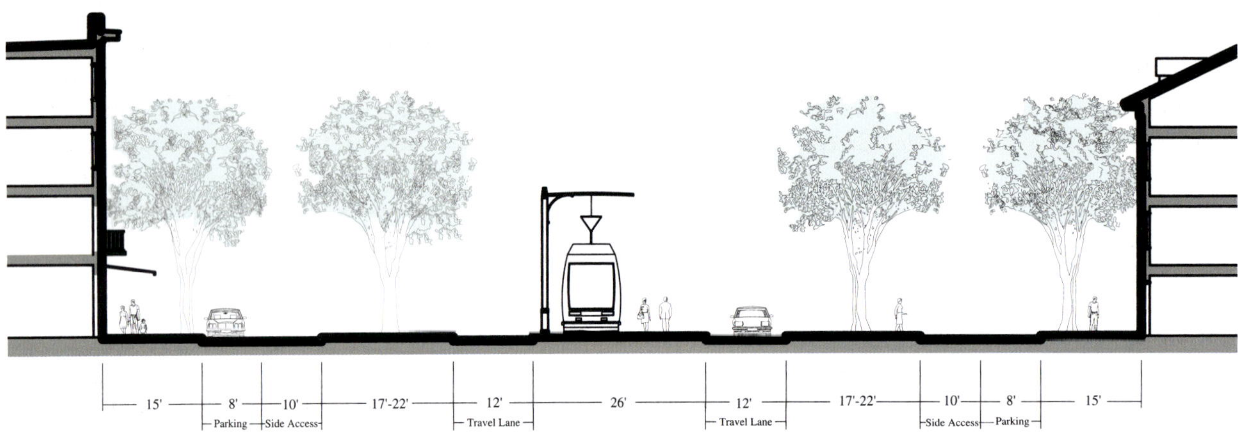

Figure 4.19: Fairfax Boulevard, Fairfax, Virginia. Dover, Kohl & Partners, 2008. Proposed section showing streetcar concept. © 2013 Dover, Kohl & Partners

New public spaces, mixed-use development, wide sidewalks, multiple transit options, and consistent tree lines would have transformed Fairfax Boulevard into a more productive, more pleasant place (Figure 4.20)— but that did not happen. Though the plan received widespread acclaim, it stalled and was eventually shelved. The City of Fairfax had commissioned the plan, but the city's longtime mayor steadfastly opposed mixed-use development in the corridor and was unwilling to consider any housing in the mix. Although the mayor was shortly thereafter convicted on various charges and served a prison term, he had successfully paralyzed the project, leaving little hope for the future of Fairfax Boulevard.

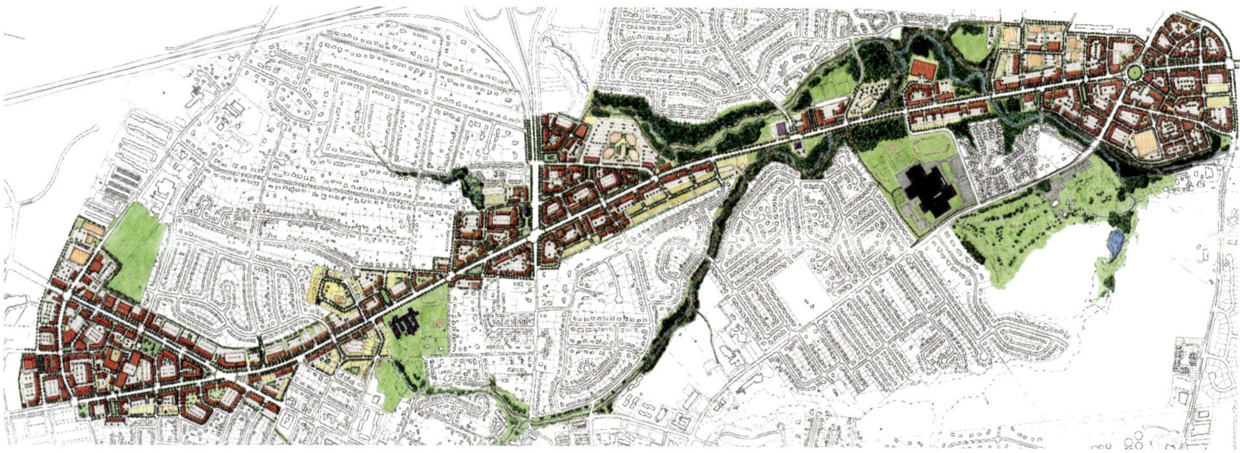

Figure 4.20: Fairfax Boulevard, Fairfax, Virginia. Dover, Kohl & Partners, 2008. Illustrative Plan. The plan proposed that redevelopment efforts should be focused in three distinct nodes: around Kamp Washington (left), Chain Bridge Road (center), and Fairfax Circle (right). © 2008 Dover, Kohl & Partners

Queens Boulevard, Queens, New York

Massengale & Co LLC, 2014

Retrofit: Fixing the Boulevard of Death

Multiway Boulevard

In 2014, Queens Boulevard was one of the most dangerous streets in New York City. Sixty percent of the traffic fatalities in New York happened on 10 percent of the streets, and Queens Boulevard was one of the two or three most dangerous. But it is also the central artery in New York's Borough of Queens, with five subway lines that run underneath it and a number of bus lines that run along it or across it. In some stretches the Long Island Railroad is nearby. When New York Mayor Bill de Blasio and NYC DOT Commissioner introduced a new 25-mile-per-hour, city-wide speed limit and new Vision Zero policies in 2014, many immediately saw that Queens Boulevard could become dramatically different. The influential civic group Transportation Alternatives hired Massengale & Co to create concept sketches.[14] John Massengale easily convinced them that taming the boulevard could create great housing sites and that they should illustrate that too. In theory, affordable housing was a high priority for the De Blasio administration (although they didn't manage to produce much for New Yorkers with income in the bottom 50 percent of the city residents).

The "Before" photo (Figure 4.21) shows the market clearly said over the years that housing away from the artery (even just half a block away) was more desirable than apartments on the busy boulevard. Even the main shopping areas were one block away from the boulevard or perpendicular to it in some locations. The most obvious reasons for that are the twelve lanes of parking and traffic, with lots of pavement and few trees. In Europe, as we have seen, multiway boulevards, are some of the best streets in the grandest cities, like the Champs-Elysées and the Avenue Montaigne in Paris, or the Gran Via and the Avinguda Diagonal in Barcelona.

Unlike those streets, Queens Boulevard did not have grand allées of majestic street trees or side lanes designed to be comfortable for pedestrians. Queens Boulevard is barren, and the cars on the wide side lanes sometimes go faster than the cars in the center. Plus, even though there are usually not many pedestrians on Queens Boulevard, the sidewalks can be so narrow that they become congested when people come up out of the subways.

The two "After" images (Figures 4.22 and 4.24) began with visions of tree-lined streets and wider sidewalks. The wider sidewalks make the side lanes narrower: they become slow lanes where cars share the space with cyclists and pedestrians. Next to the traffic lanes are protected bicycle lanes, planted with permeable pavers that make the trees part of a stormwater management system that also gives the roots room to grow.

Figure 4.21: Queens Boulevard, Queens, New York. Aerial view from 67th Road, looking southeast. Before: "The Boulevard of Death," an auto-sewer where few people want to live, despite the convenient subway service to Manhattan. *Courtesy of Transportation Alternatives*

Figure 4.22: Queens Boulevard, Queens, New York. Massengale & Co LLC, 2014. Photo-realistic rendering. After: Aerial view from 67th Road, looking southeast. Compare with Figure 4.21. Rendering by UrbanAdvantage. *© 2014 Massengale & Co LLC*

Figure 4.23: Queens Boulevard, Queens, New York. Photo looking southeast from between 67th Drive and 68th Avenue. Before: Traffic on the wide outer traffic lanes frequently goes faster than traffic in the center.

Figure 4.24: Queens Boulevard, Queens, New York. Massengale & Co LLC, 2014. Photo-realistic rendering made from the photo in Figure 4.23. Looking south from between 67th Drive and 68th Avenue. Rendering by UrbanAdvantage. © 2014 Massengale & Co LLC

The result would be that each side of the boulevard would become a comfortable place for pedestrians. When they come out of the subways or get out of their cars, the wide sidewalks, new stores, and space-enclosing buildings would make Queens Boulevard a place to be. That would create foot traffic that could help the stores, restaurants, and cafés flourish. Above the stores and restaurants could

be new apartments, in tall buildings appropriately sized for the unusually wide street.

Crossing the street would be easier on the redesigned street, because the heavy traffic volume would be limited to the six center lanes. Those will have a new speed limit of 25 miles per hour. They would also have a wider, planted median, much like the one on Park Avenue in

Manhattan. Building sites that have been undesirable for more than three-quarters of a century would suddenly be attractive and valuable on a beautiful and lively boulevard with convenient access to the subway.

Since 2014, the NYC DOT has made Queens Boulevard slower and safer.[15] But that did not happen without resistance from the old guard, i.e., the career bureaucrats at the top of the five thousand engineers in the DOT. There were suggestions at public meetings that it would be a mistake to lower the traffic speed to 25 miles per hour, even though that was one of the core principles of the city's new Vision Zero policy. The traffic lights on the Boulevard of Death were set to keep traffic flowing at 45 miles per hour: "If we slow it down too much, we'll have more rear-enders at the intersections," one top engineer said.[16]

Community pressure in public meetings and guidance at the top from DOT Commissioner Polly Trottenberg and her staff, like Deputy Commissioner Michael Replogle, gradually led to slower speeds, narrower traffic lanes, new bike lanes, and tricks like fake speed cameras supplementing the rare actual cameras. But no one will confuse the new Queens Boulevard with the *grands boulevards* of Paris. Cars still come first, and the current design does not produce zero annual deaths or make a place where people want to live, shop, and stroll.[17]

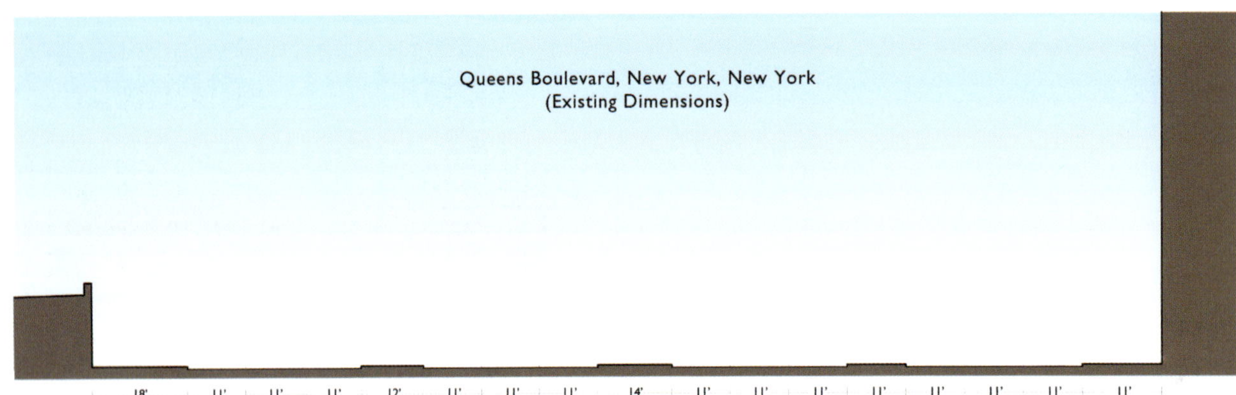

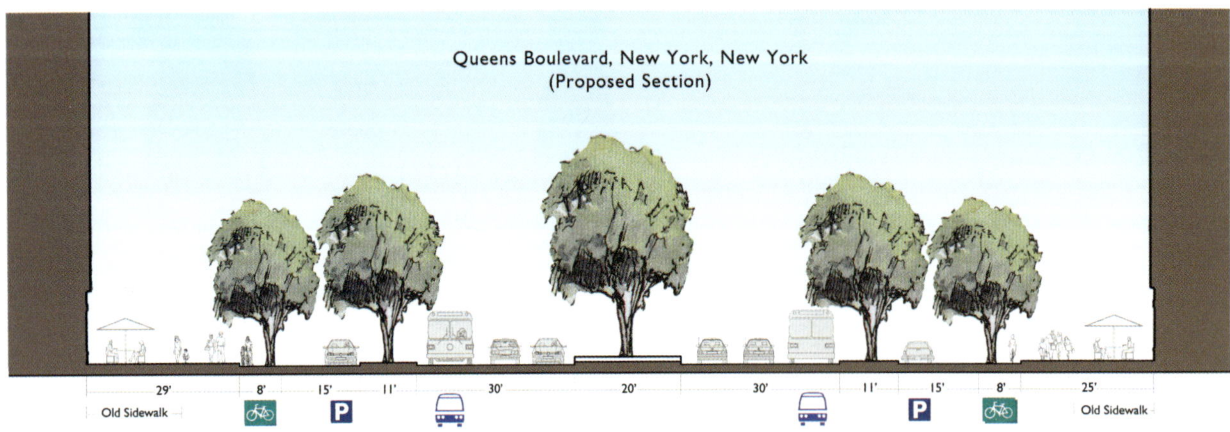

Figure 4.25: Queens Boulevard, Queens, New York. Massengale & Co LLC, 2014. Section Before & After. © 2014 *Massengale & Co LLC*

Corridor Transformations in Jeddah, Saudi Arabia

Dover, Kohl & Partners with Hall Planning & Engineering, 2008

Multiway Boulevards and Main Streets

In 2008, the municipality of Jeddah in the Kingdom of Saudi Arabia undertook the Jeddah Street Improvement Project, hiring urban designers and transportation engineers to produce a heavily illustrated street improvement manual (Figure 4.27). The municipality had the experts study a number of corridors, addressing the form and regulation of the vehicles, the pedestrians, and the private realm on the corridors. Over time, the standards in this manual helped the Kingdom produce a coherent and context-responsive street network. The manual contains detailed *how-to* instructions for architects, engineers, and developers, as well as extensive background material on the reasons why the standards specify what they do (Figures 4.26 and 4.28–4.30). To make the manual easy to use, it included a series of simple question-by-question checklists for confirming compliance.

The story of a legible city is told in its streets. In Jeddah, the needs are great. The significance of this project is that the most crucial corridors in the city were all considered at once, as an ensemble. With a standardized approach to the analysis and a highly customized, context-specific approach to the individual designs, the city can proceed with the improvements in an efficient way.

Figure 4.26: Corridor Transformations, Jeddah, Saudi Arabia. Dover, Kohl & Partners with Hall Planning & Engineering, 2008. Rendering of Al-Malik corridor. © *2008 Dover, Kohl & Partners*

Buildings on MAJOR COMMERCIAL STREETS

The following building configurations were created for structures along "major commercial streets" (as defined by the Jeddah Municipal Zoning Ordinance). They were designed specifically with **Al Amir Sultan Street** and **Al Nuzha Road** in mind, however, they may be applied to all other corridors with the "major commercial streets" designation. These design parameters govern the dimensional and configuration requirements of building fronts. These standards are intended to guide the redevelopment of existing structures, as well as assist with the development of new structures.

Primary entrances
- Primary entrances to all buildings must face the street.
- Side entrances shall not be primary entrances.

Parking
- All parking shall be located behind buildings, or in garages behind habitable space or free standing liner buildings, or to the building's side but behind a garden wall.
- Parking is not permitted within front setback areas.

Habitable space
- Habitable Space is required to avoid large blank walls facing streets, such as from parking garages, places of assembly, and retailers with large multi-story spaces.
- Floors facing streets shall have usable habitable space.
- The depth of habitable space shall be a minimum of 6 meters.

Building elements
- All buildings within this zone are required to have either an arcade or awnings along the front façade at the ground level for the purpose of shading the sidewalk. This arcade may be multiple stories in height.
- Where used, arcades shall be a minimum of five (5) meters in width and may extend twenty five (25) to seventy five (75) percent of building façades.
- Portions of buildings with arcades may be constructed forward of the setback line. See diagrams to the left that correspond to this street type.
- Where used, awnings must extend out over the sidewalk a minimum of two (2) meters.
- Awnings shall extend along ground floor exteriors that are not covered by arcades. However they are permitted to break for structural columns therefore shading the window bays between them.
- Corner buildings may have arcades or awnings on the sides that face streets.

Shopfronts / storefronts / openings
- A minimum of ninety (90) percent of all building frontages shall have shopfronts or storefronts at the ground floor.
- Upper stories shall have a minimum of forty (40) percent window and/or balcony openings.

JEDDAH MUNICIPAL ZONING ORDINANCE DIMENSIONAL REQUIREMENTS

First story height	5 meters maximum
Additional story heights	4 meters maximum
Front setback	5 meters
Side & rear setback	2 meters minimum for buildings that are 1-3 floors high 3 meters minimum for buildings that are 4-8 floors high 4 meters minimum for buildings that are 9-12 floors high 5 meters minimum for buildings that are 12 and more floors high
Side setback for corner buildings	5 meters maximum
Floor area ratio (FAR)	Varies between 2.4 and 4.2, based on property size. Consult Municipal Zoning Code
Maximum height	Varies between 4 stories and 7 stories with a 60% lot coverage based on property size. Additional height is allowed up to 12 stories with smaller lot coverage. Consult Municipal Zoning Code

Mezzanines are allowed within the first floor, up to a maximum of 50% of the floor area.
Note: Existing zoning is subject to change independently of this document.

ADDITIONAL REQUIREMENTS

Maximum front setback	8 meters, for the first 8 stories of height
Minimum and maximum setbacks for arcades*	See diagram at left
Recommended side setback*	0 meters/ none
Maximum side setback for corner buildings	3 meters
Required building elements	Arcade** or awning
Building frontage (with 1 or more buildings)	50% to 100% of lot frontage is to be occupied by building(s)
First finished floor elevation	Same as sidewalk level within the public ROW

** additional approval may be required.*
*** arcades are further described in Façade Design, found in page 4.28.*

Table 4.6: Existing and additional dimensional requirements for buildings on MAJOR COMMERCIAL STREETS

Towers
- Floors of buildings above the eighth floor shall set back to match that of current zoning. No maximum front setback applies. Side setbacks above the eighth floor shall be the same as existing zoning allows.
- Floor plates of towers should not exceed two thousand three hundred (2,300) square meters.

Figure 4.7: Building configurations on MAJOR COMMERCIAL STREETS. Use this for Al Amir Sultan Street and Al Nuzha Road. *Note: These diagrams show the maximum height allowed for this street type. When proposed buildings are less than the maximum height, they shall follow the configuration requirements for the numbers of floors shown. All heights are subject to limitations set by existing zoning, please review the Municipal Zoning Code.*

Figure 4.27: Corridor Transformations, Jeddah, Saudi Arabia. Dover, Kohl & Partners with Hall Planning & Engineering, 2008. Sample pages from the Jeddah Streetscape and Urban Design Manual. © *2008 Dover, Kohl & Partners*

Figure 4.28: Corridor Transformations, Jeddah, Saudi Arabia. Dover, Kohl & Partners with Hall Planning & Engineering, 2008. Rendering of Old Mekkah corridor. © *2008 Dover, Kohl & Partners*

Figure 4.29: Corridor Transformations, Jeddah, Saudi Arabia. Dover, Kohl & Partners with Hall Planning & Engineering, 2008. Rendering of a park/plaza along a street corridor. © 2008 Dover, Kohl & Partners

Figure 4.30: Corridor Transformations, Jeddah, Saudi Arabia. Dover, Kohl & Partners with Hall Planning & Engineering, 2008. Rendering of a corridor interrupted by a neighborhood square. © 2008 Dover, Kohl & Partners

Figure 4.31: Corridor Transformations, Jeddah, Saudi Arabia. Dover, Kohl & Partners with Hall Planning & Engineering, 2008. Rendering of a regional park and adjacent corridor. © 2008 Dover, Kohl & Partners

Boundary Street, Beaufort, South Carolina

Dover, Kohl & Partners, 2006

Retrofit: Arterial Strip to Multiway Boulevard

Multiway Boulevard

In 2006, the City of Beaufort hired Dover, Kohl & Partners to prepare a Master Plan for Boundary Street, a major transportation corridor that runs from the city limits to downtown Beaufort. A key street in Beaufort itself, Highway 21 also traverses the city's marshlands to points further east and is used for both short local errands and longer regional trips. Victor Dover, Amy Groves and Margaret Flippen led a charrette for Beaufort that produced a community vision for transforming the corridor from a typical suburban strip (Figure 4.32) into an

urban street suitable for more economically productive uses than fast-food chains and tire shops. With a design strategy that maintains traffic flow while improving safety and character, Boundary Street will evolve from a route that is only used to get to another destination. It will become a destination itself, with its own memorable character.

Dramatic vistas across marshlands are a hallmark of South Carolina's Lowcountry. During the design and planning process for the Master Plan of Boundary Street and its surroundings, therefore, the team gave a lot of attention to the Battery Creek marshes next to the southern side of a long stretch of the corridor. The design team proposed a continuous marsh-front park and trail to showcase the views and reconnect natural places that had been divided by the construction of the highway. More marshlands lay just to the north, narrowing the landmass

Figure 4.32: Boundary Street, Beaufort, South Carolina. Dover, Kohl & Partners, 2006. Watercolor rendering, aerial view. Before: Existing conditions, circa 2006. At Jean Ribaut Square, one-story, single-use retail buildings are set behind large parking lots in a typical strip shopping center configuration. © *2006 Dover, Kohl & Partners*

Figure 4.33: Boundary Street, Beaufort, South Carolina. Dover, Kohl & Partners, 2006. Watercolor rendering, aerial view. After: Redevelopment scenario for Jean Ribaut Square. A pattern of streets and blocks is to be established, creating addresses for new multistory, mixed-use buildings, public spaces, and civic buildings; on-street parking will be combined with parking within the block. © *2006 Dover, Kohl & Partners*

Figure 4.34: Boundary Street, Beaufort, South Carolina. Dover, Kohl & Partners, 2006. Before: Existing conditions, circa 2006. © 2006 Dover, Kohl & Partners / UrbanAdvantage

Figure 4.35: Boundary Street, Beaufort, South Carolina. Dover, Kohl & Partners, 2006. Computer simulation. After: The north side of the corridor will eventually have the side access lane of a classic multilane, multiway boulevard. © 2006 Dover, Kohl & Partners / UrbanAdvantage

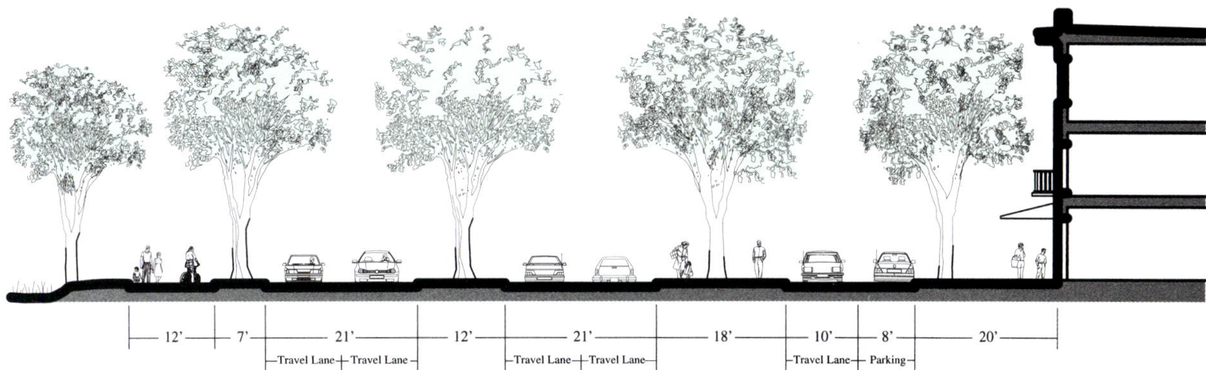

Figure 4.36: Boundary Street, Beaufort, South Carolina. Dover, Kohl & Partners, 2006. Proposed section. © *2013 Dover, Kohl & Partners*

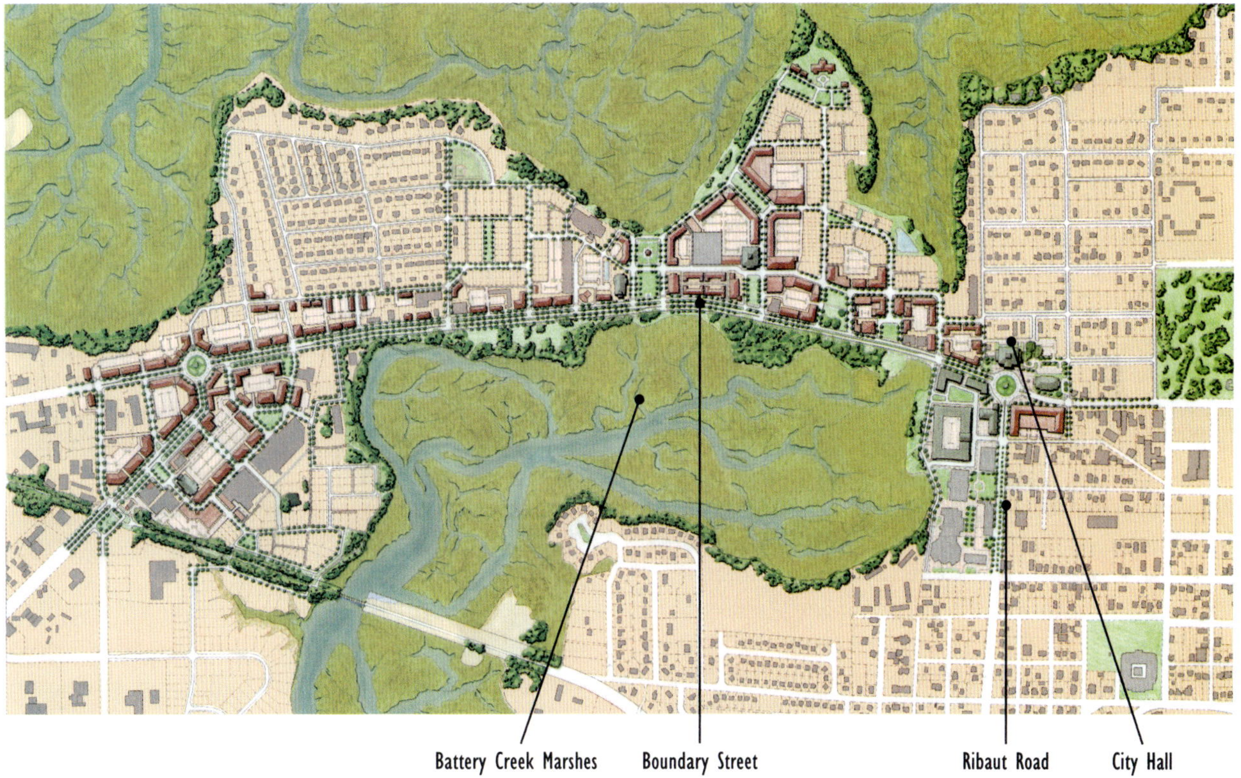

Figure 4.37: Boundary Street, Beaufort, South Carolina. Dover, Kohl & Partners, 2006. Illustrative Plan. In the central segment, the new buildings on the north side will face views of the Battery Creek marsh across the street and parallel trail. At the far right, facing the roundabout originally recommended at the Ribaut Road intersection, is the new City Hall, axially terminating the perpendicular street vista. © *2006 Dover, Kohl & Partners*

along the corridor. The master plan shows a new town square or village green at the narrowest point to reunite the divided landscape (Figure 4.33).

The overall concept for the master plan was to create a heightened sense of connectedness in this fragmented part of Beaufort. Accordingly, in addition to unifying the long views and natural landscapes, the plan establishes a coherent network of streets. As each parcel is redeveloped, the Boundary Street plan calls for gradual assembly of a "slow lane," with adjacent on-street parallel parking (Figures 4.34 and 4.35). This is meant to establish a pedestrian- and business-friendly environment without unduly sacrificing capacity for vehicles going through. Converting commercial shopping centers with large, ugly parking lots into town blocks will encourage local activity. The combination of new public space and the retrofit of the existing shopping centers on the strip will set up a new generation of economic opportunity.

Walkable streets have travel lanes with narrow dimensions, large sidewalks, parking that is appropriately located, and buildings adjacent to sidewalks—qualities that encourage interaction and accessibility. A few simple changes in the placement of buildings, parking, and landscape can improve the character of any outdated and placeless corridor (Figures 4.36 and 4.37).

A few defining elements can help change a good road into a great street. The Boundary Street plan prescribed shade trees to add comfort and protection to the street space; awnings to make the street consistently comfortable for pedestrians; small curb radii at the intersections to slow traffic; and frequent crosswalks.

Architecture creates and reinforces the sense of place, often reflecting the weather as much as the culture. The drawings for the Boundary Street Master Plan reference a regional vernacular—one derived not just from tradition, but from the famously sticky Lowcountry weather, in which porches, verandas, awnings, overhangs, and breezeways make life more comfortable and help conserve energy. The hope is that Boundary Street will look like it belongs in Beaufort but will also work with the local climate. The City of Beaufort adopted a detailed form-based code for the corridor, and private developers have since been gradually filling in vacant lots and redeveloping others, with various degrees of adherence to the concept and the code. The City also implemented the Master Plan's concept for a new City Hall terminating the vista along Ribaut Road (seen at far right in the illustrative plan, Figure 4.33).

The Master Plan took on new importance in the last decade when it was used to secure a grant for reconstruction under the federal Transportation Infrastructure Generating Economic Recovery (TIGER) stimulus program. The City of Beaufort, Beaufort County, and South Carolina Department of Transportation collaborated to rebuild 1.4 miles of the corridor, installing the primary tree lines, adding new, wider sidewalks and pedestrian-scaled lighting, adjusting lane widths, and building segments of a new marsh-front trail according to the plan.

Columbia Pike, Arlington, Virginia

Dover, Kohl & Partners and Ferrell Madden Associates, 2002–2013

Retrofit: Transit-Oriented Development with a Form-Based Code

Boulevard

Columbia Pike is one retrofitted corridor where all the pieces are coming together. Among similar projects underway in the United States, it is also one of the farthest along in its transformation. Columbia Pike has appeared on maps for more than two centuries; it starts at the Pentagon and runs west. The latter half of the twentieth century was hard on the Pike. Arlington County adopted a revitalization plan and form-based code for portions of the corridor in 2002, after thirty years of disinvestment and decline. The plan and code covered a series of mixed-use nodes along the corridor where redevelopment is now gradually producing higher densities and walkable areas, stimulated at first by simplified regulations and anticipation of improved public transit. (Figures 4.38 and 4.39). Two years later, the county finalized public-space standards that now control the rebuilding of the Pike, and in 2010 it took over the road from the Virginia Department of Transportation and implemented the first modest portions of the ambitious street design, adding street trees. (An updated street design was adopted in 2012 and nears completion in 2024). Seventeen major private redevelopments have since been completed, replacing low-slung stores, parking lots, and car dealerships with multistory, street-oriented, mixed-use buildings containing housing, office space, and stores, including a full-size grocery store tucked within an urban block. The character of the corridor has changed rapidly from despair to optimism. As of 2019, 270,000 square feet of new retail space had been added since 2003, and an estimated 38 percent of retail square footage "is in new development built in an urban, walkable format. The remaining 62% of retail is in legacy storefronts constructed prior to implementation of the form-based code."[18]

All of the progress on Columbia Pike can be attributed to elected officials and neighborhood leaders who got impatient and doggedly persisted until somebody listened. In 2002, the primary objective was to get investment—*any* investment—going again along the Pike. By 2010, rapid development following the form-based code and anticipation of the future streetcar had reversed the situation so completely that rents had begun rising faster than incomes, and a new look at the corridor was necessary. Local leaders sought to figure out how to maintain a mixed-income community over the long term. Arlington County reconvened the planning team in 2012 and adopted their companion volume to the original revitalization plan, after an extensive new round of research and public consultation. The 2012 *Columbia Pike Neighborhoods Plan* addresses a much larger context—including the surrounding neighborhoods and the largely residential segments of the Pike between the original plan's nodes—and a wider range of topics, including mixed-income housing and sustained funding for the revitalization initiative (Figures 4.40 and 4.41). The county board then adopted a second form-based code, which interlocked with the original code and gave the *Neighborhoods Plan* the force of law, this time adding extensive provisions for the inclusion of permanently affordable housing. This has proven effective, but at the same time the mixed-income strategy has not diminished the appeal to wealthier households choosing to live along the Pike; in 2017, an HR&A economic study showed that in the preceding seven years household incomes had risen along Columbia Pike *eight times* faster than in the Washington, DC, region as a whole.[19]

Figure 4.38: Columbia Pike, Arlington, Virginia. Before: Existing conditions in 2002 after decades of blight and disinvestment. The auto-only phase on the two-hundred-year-old corridor had run its course. © 2002 Dover, Kohl & Partners / UrbanAdvantage

Figure 4.39: Columbia Pike, Arlington, Virginia. Dover, Kohl & Partners and Ferrell Madden Associates, 2002. After: The form-based code facilitates redevelopment in a series of mixed-use nodes distributed along the corridor. © 2002 Dover, Kohl & Partners / UrbanAdvantage

Figure 4.40: Columbia Pike, Arlington, Virginia. Before: Existing conditions between the "nodes" in 2011. © *2011 Dover, Kohl & Partners / UrbanAdvantage*

Figure 4.41: Columbia Pike, Arlington, Virginia. Dover, Kohl & Partners and Ferrell Madden Associates, 2011. After: The followup *Columbia Pike Neighborhoods Plan* directs the redevelopment of areas between the mixed-use nodes, increasing residential density, converting the development pattern to street-oriented, transit-worthy urbanism. The Columbia Pike Streetcar has since been canceled. © *2011 Dover, Kohl & Partners / UrbanAdvantage*

Figure 4.42: Lessard Design, Westmont Apartments, Columbia Pike, Arlington, Virginia, 2023. Photograph looking northeast. With the code in place, mixed-use developments like six-story Westmont are rising along the corridor. *Courtesy of Westmont Arlington & EarthCam*

In a setback, the Columbia Pike Streetcar project was abruptly canceled following the political pendulum swing of 2016. Until the County revives the project, it has improved the frequency and quality of the bus service along the corridor and widened the sidewalks, buried powerlines, and replaced aging underground utilities.

In 2023, the retail vacancy rate along the Pike stood at 2 percent, less than half the vacancy rate countywide. Office space vacancy ballooned nationally during COVID, but the office vacancy rate on Columbia Pike was just 7 percent, compared to 25 percent for Arlington County as a whole.[20]

Mesa Street, El Paso, Texas

Dover, Kohl & Partners, 2010

Retrofit: Right-Sizing the Roadway

Main Street and Multiway Boulevard

As cities mature and grow, they can spread up, out, or fill in overlooked and poorly-used spaces within their boundaries. The latter, usually called "infill development," is the most promising option for many American cities. Mesa Street in El Paso, Texas, a wide thoroughfare flanked by half-developed urban land, cries out for infill

Figure 4.43: Mesa Street, El Paso, Texas. Before: Existing conditions, circa 2010.

Figure 4.44: Mesa Street, El Paso, Texas. Dover, Kohl & Partners, 2010. Computer simulation. After: In this segment, right-sizing the street will allow for wide sidewalks, street trees, on-street parking, and a cycle track. © *2010 Dover, Kohl & Partners / UrbanAdvantage*

Figure 4.45: Mesa Street, El Paso, Texas. Before: Existing conditions, circa 2010.

Figure 4.46: Mesa Street, El Paso, Texas. Dover, Kohl & Partners, 2010. Computer simulation. After: In this segment, the street can be reconfigured into a multilane, multiway boulevard with a Texas twist, using drought-tolerant native trees. © *2010 Dover, Kohl & Partners / UrbanAdvantage*

development. The six-lane route is the primary connection between northwest El Paso and downtown. As it approaches central El Paso, the road narrows to a four-lane street. This corridor invites large volumes of high-speed traffic but the road is difficult to navigate as a pedestrian, cyclist, or transit rider.

Mesa Street's first generation of one-story commercial developments, surrounded by large expanses of surface parking, failed to define the street's public realm. This condition prevails in many American cities and suburbs, resulting in an auto-centric atmosphere that misses out on the synergy of real towns. Ironically, a "sell-scape" meant to synchronize with motoring and draw customers in becomes swamped with traffic and pushes customers away. The traffic isolates businesses from potential customers on foot, bikes, or public transit. A great commercial street is a *place*, full of people—browsing, shopping, dining—and accommodates all modes of traffic. Retrofitting the right-of-way and implementing infill development is the key to transforming transportation corridors.

In 2010, the city initiated *Plan El Paso*, a comprehensive plan that targets the revitalization of specific neighborhoods and streets.[21] In an effort to grow the community in a smarter, healthier, and more sustainable

Figure 4.47: North Oregon Street, El Paso, Texas. Photograph looking northwest, circa 2015. Before: Historic buildings along the Mesa/Oregon corridor fell into a cycle of disrepair (and, in some cases demolition by neglect), when enthusiasm waned for historic preservation. © *Google Earth*

Figure 4.48: North Oregon Street, El Paso, Texas. Photograph looking northwest. After: New interest for historic preservation and adaptive reuse followed *Plan El Paso* policy changes and the return of the streetcar. *Courtesy of Carlos Gallinar*

way, the plan identifies areas for infill and redevelopment. Mesa Street's strategic location can become a dynamic public space under *Plan El Paso*, a destination for local patrons, as well as a through-going route.

The future design converts some of the stroad into a main street (Figures 4.43 and 4.44). Another stretch becomes a multiway boulevard (Figures 4.45 and 4.46). These street types—newly reintroduced to El Paso—accommodate large volumes of cars but also dedicate large parts of the right-of-way to walking, biking, and mass transit. Installed in a series of planned phases, this urban evolution alters the character of the street,

reactivates pedestrian life, and turns Mesa into a vibrant public space.

Creating a lively multiway boulevard for the majority of Mesa Street involves establishing a right-of-way that is approximately 150 feet wide and accommodates multiple lanes of through traffic; a lane for express buses and two lanes lined with on-street parking complete the boulevard cross-section. There are median spaces located in the center of the road and on either side of the access lanes. These spaces, shaded by mature trees, help make the wide street comfortable for the pedestrian.[22] The climate-conscious retrofit of Mesa Street provides comfort, and gives shelter from the sometimes harsh local weather. Awnings, arcades, and shade trees are placemaking tools that make the street comfortable for pedestrians.

Plan El Paso designates key spots in the city as walkable places, and connects Mesa Street to other neighborhood centers with transit along primary corridors. The city eventually arrived at the idea of restoring the historic streetcar lines with rails on Oregon Street and Stanton Street, paralleling Mesa (one block away in each direction). Six streetcars connect downtown to the University of Texas at El Paso, linking the international bridge and several bus transit lines. *Plan El Paso* is

heavily illustrated with images showing how Mesa Street and the other corridors should be transformed along their length.

El Paso has a fine inventory of buildings dating from before the Great Depression. Fundamental to *Plan El Paso* is the idea that historic preservation can reinvigorate the Borderland economically. The implementation of the plan has given the North Oregon Street portion of the plan a wave of success stories. Derelict buildings have been renovated and occupied to take advantage of their addresses along the restored transit line, bringing pedestrians and their money to the businesses on the street.

Bedford Road, Katonah, New York

B.S. and G.S. Olmstead, 1897

Boulevard with Planted Median

Katonah is one of three villages in the Town of Bedford, New York, in northern Westchester County. In the early 1890s, New York City told the residents of Katonah that the city's new reservoir system would eventually put the village underwater. At a meeting of the Katonah Village Improvement Society (KVIS), the residents decided to move the hamlet "buildings and all." In 1894, the town arranged to purchase land

Figure 4.49: Bedford Road, Katonah, New York. B.S. and G.S. Olmstead, 1897. Photograph looking north. The local landscape architects B.S. and G.S. Olmstead were unrelated to Frederick Law Olmsted. *Courtesy of Collection of the Bedford Historic Society*

for a new town to the south of the old location[23] and hired two landscape architects named B.S. and G.S. Olmstead.[24] They designed a new town with two main streets (Figure 4.49). Facing a station and tracks owned and operated by the New York Central and Hudson River Railroad was a one-sided commercial street called Katonah Avenue. Train service today is run by New York's Metro-North Railroad.[25]

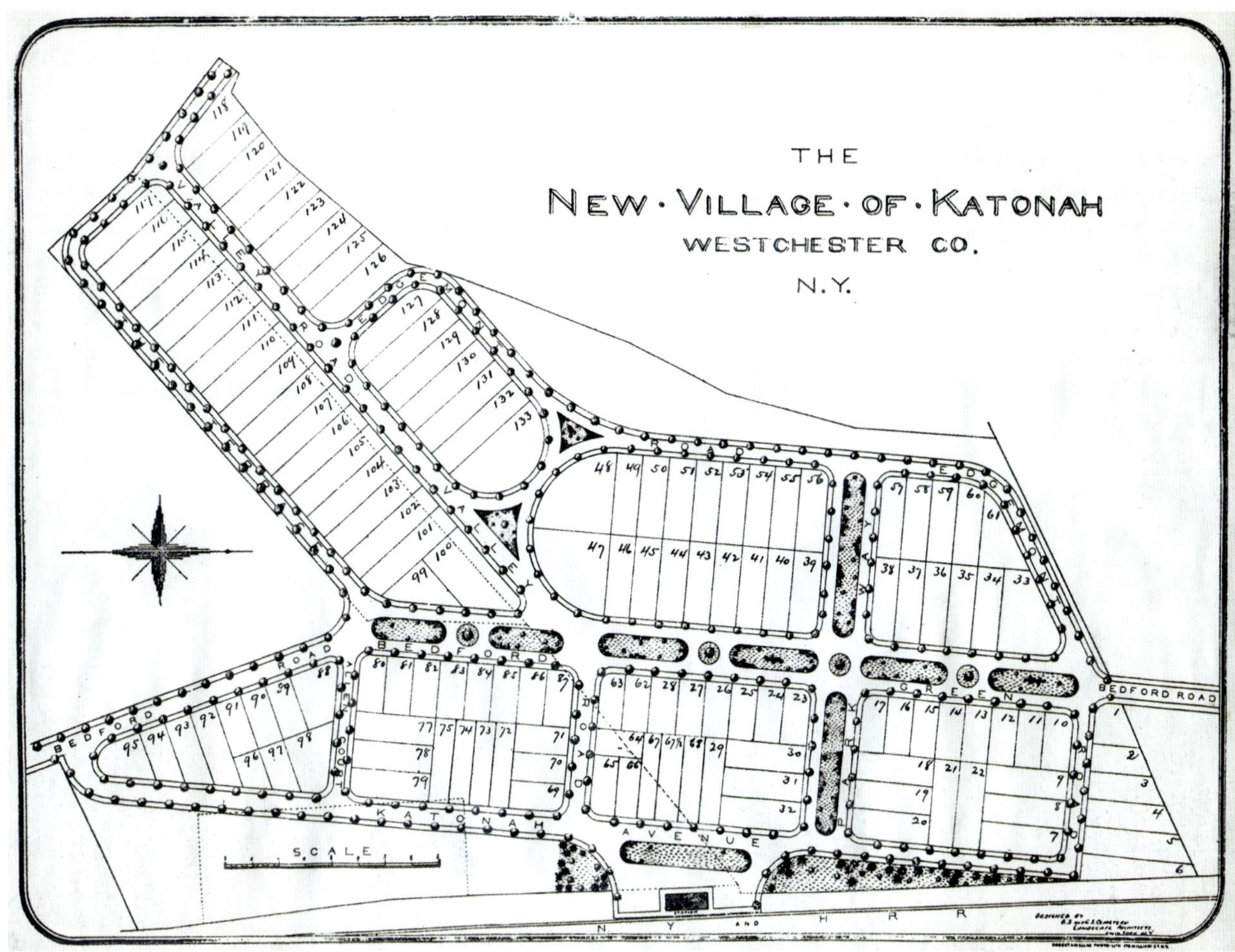

Figure 4.50: Katonah, New York. B.S. and G.S. Olmstead, 1897. Plan. Bedford Road is the boulevard in the center of the plan with a planted median. *Courtesy of Collection of the Bedford Historic Society*

The second main street was a block away, parallel to the first. Called Bedford Road, it was a broad residential street that also included two churches, the village library, and a schoolhouse. Most of the width of the street was filled with a large median in the center, planted with trees and grass (Figure 4.50).

A short but wide commercial street called The Parkway connected Katonah Avenue and Bedford Road.[26] Over the years, Bedford Road became a busy thoroughfare. In the 1980s, the Westchester DOT built a bypass from one of the busiest county roads near Katonah to Interstate 84 and the Saw Mill Parkway, making Bedford Road a quiet local street again. Why did we include Bedford Road? To underline the fact that before cars and traffic engineering took over the country, towns planned a variety of streets, including quiet places for strolling.

PROMENADE STREETS

Le Cours Mirabeau, Aix-en-Provence, France

Antoine Grumbach & Associates, 2002

Retrofit: Right-sizing the Roadway

Promenade Street

The Cours Mirabeau in Aix-en-Provence is a quintessential "great street."[27] There is no debating its attractions; it dates from an age when public works officials were heroes. This is what an eighteenth-century road-widening project looks like. By comparison, today's average traffic-capacity "improvement" looks to us like malpractice.

Two rows of mature plane trees line each side of the street (Figure 4.51). The resulting canopy is one of this French avenue's defining attributes. The first row of trees is eighteen feet from the building line. The second row is offset by an approximately thirty-foot walk. The buildings on each side are fifty to sixty feet tall.

In a city famous for its fountains, the Cours begins with a fountain at the place du General de Gaulle and extends for a quarter mile, before ending at a statue of King René. The Cours Mirabeau has long been known as a wide street with the majority of the street space dedicated to pedestrians, but, in fact, the division of the street into different parts has changed since its original construction. Before a rebuilding completed in 2002, significantly more space was dedicated to auto traffic. The redesign by Antoine Grumbach & Associates reassigned more than half the right-of-way to pedestrians (Figures 4.52 and 4.53). The retrofit eliminated two of the four travel lanes and extended the

sidewalks by thirteen feet on each side of the street.[28] The larger sidewalks contribute to the transformation of the boulevard from a primary transport route to a space dominated by pedestrians, cyclists, and market carts. Although still a direct route to the center of Aix, the pedestrian became the king of the new space. In addition to giving the pedestrian more room, the retrofit called for the installation of ramps around the fountains, with a slope of 5.6 percent.[29]

A natural traffic-calming device, the ramps and the narrowing of the travel lanes alter the character of the street, but not drastically. A lack of stoplights, traffic signs, and paint on the road further slow the traffic. The combination of these techniques makes the Cours Mirabeau a beautiful Complete Street, with a regional aesthetic all its own. The Cours provides shelter from the weather and is always full of people visiting local businesses and enjoying a stroll in the city.

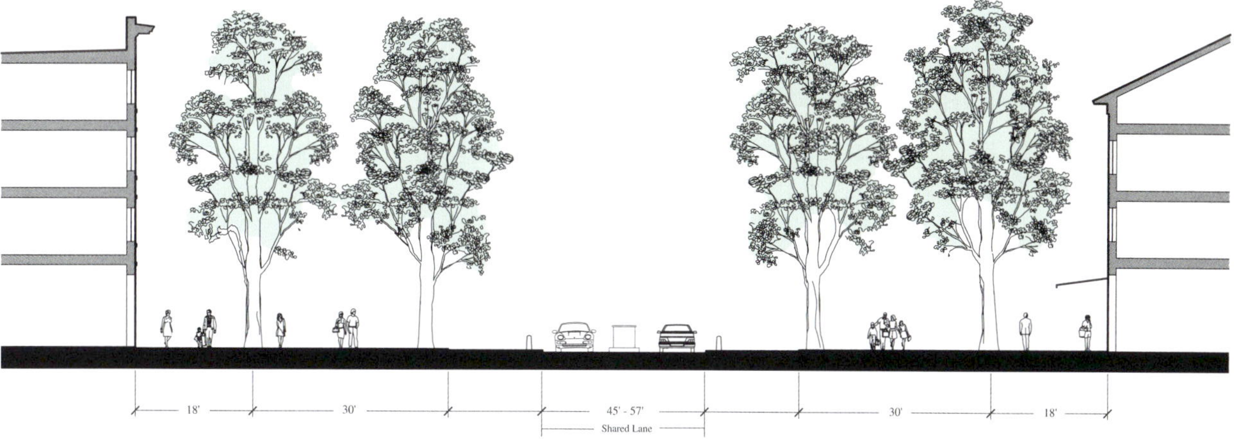

Figure 4.52: Cours Mirabeau, Aix-en-Provence, France. Antoine Grumbach & Associates, 2002. In its most recent makeover, the amount of space devoted to pedestrians was increased. *Courtesy of Andrew Georgiadis*

◄ **Figure 4.51:** Cours Mirabeau, Aix-en-Provence, France. Antoine Grumbach & Associates, 2002. Public works officials were once heroes, building places of such quality that they were postcard-worthy symbols of their towns. *Courtesy of Jason King*

▲ **Figure 4.53:** Cours Mirabeau, Aix-en-Provence, France. Antoine Grumbach & Associates, 2002. Section. Four rows of carefully spaced trees produce a high canopy over the space between the buildings. © *2013 Dover, Kohl & Partners*

Yorkville Promenade (Second Avenue), New York, New York

Massengale & Co LLC, Dover, Kohl & Partners, with H. Zeke Mermell, 2011

Retrofit: Reclaiming Street Space

Promenade Street

The following description is adapted from an entry in the competition *The Unfinished Grid: Design Speculations for Manhattan*, organized by the Architectural League of New York and the Museum of the City of New York.[30]

For so many reasons, we must reduce auto use in New York City. Studies completed for Mayor Bloomberg's administration showed that living on a high-traffic avenue in Manhattan is unhealthy, particularly for children. To add insult to injury, 80 percent of Manhattan households do not own cars, and only 20 percent of out-of-town commuters drive to work. The city's ugly, unhealthy avenues are more for the benefit of others than Manhattan's workers and residents.

Most Manhattanites live in small apartments and spend a great deal of time in public life. When the weather is nice, they consume expensive food and drink while sitting next to noisy, smelly streets made so that suburbanites can quickly and easily drive in and out of the city. The roads are one-way, the lanes are wide, and frequently there is no parking allowed on the avenues at rush hour, so that the speeding cars, trucks, and large express buses are sometimes just inches from their seats. The streets are suburban-style transportation corridors in the middle of America's most walkable, transit-rich city.

The New Yorkville Promenade

The Yorkville Promenade (Figures 4.54, 4.55, and 4.56) was our proposal for rebuilding Second Avenue after the completion of the new subway that was being built beneath it. Inspired by the famous Ramblas of Barcelona, the design gives the center of the wide avenue a new promenade for walking, biking, sitting, dining, and people-watching. Cafés and restaurants along Second Avenue would be licensed to have tables on the center island. Narrow traffic lanes and short-term parking lanes to each side let cars and deliveries come and go while eliminating speeding traffic from Second Avenue.

Construction would take advantage of the fact that work on the new subway line saw Second Avenue dug up

Figure 4.54: Yorkville Promenade, New York, New York. Massengale & Co LLC, Dover, Kohl & Partners, with H. Zeke Mermell, 2011. Aerial view looking south from 86th Street. Second Avenue today is a suburban-style arterial, encouraging suburbanites to drive into the city instead of taking the train, subway, or express bus. © 2011 Dover, Kohl & Partners / Massengale & Co LLC

between 96th Street and 63rd Street (where the Second Avenue subway will initially connect with the Q train, an existing line that runs from Queens to Coney Island, via Manhattan). The Promenade would be a special place that enlivens the Manhattan grid, like Broadway on the Upper West Side and Park Avenue on the Upper East Side, but with a vibrant street life unlike staid Park Avenue, where stores and trucks are prohibited.

West of Third Avenue on the Upper East Side, the introduction of Madison and Lexington avenues into the normal city grid produced shorter blocks that made the grid more interesting for pedestrians and thereby increased the value of the real estate (as discussed in East 70th Street, A Beautiful Block in New York, pages 18–23). Yorkville is east of Third Avenue. Its long blocks are less pedestrian-friendly, and that New York oxymoron—the elevated subway—blighted both Second and Third avenues, depressing real estate values and building quality for decades. But the last Upper East Side El was taken down in the 1950s, and more recently the area around Second Avenue has boomed in anticipation of the completion of the subway. The Yorkville Promenade would give it a linear neighborhood center unique in New York City, and the new subway line would also make it more accessible for tourists and other New Yorkers.

Update: In 2023, Billy Freeland made the construction of the Yorkville Promenade part of his campaign for Councilman from the Upper East Side. (Unfortunately, Freeland did not win.)[31]

Figure 4.55: Yorkville Promenade, New York, New York. Massengale & Co LLC, Dover, Kohl & Partners, with H. Zeke Mermell, 2011. Aerial view looking south from 86th Street. After reconstruction, Second Avenue could be a unique street in the New York grid, adding variety and claiming a local role. © 2011 Dover, Kohl & Partners / Massengale & Co LLC

Figure 4.56: Yorkville Promenade, New York, New York. Massengale & Co LLC, Dover, Kohl & Partners, with H. Zeke Mermell, 2011. Section. © 2011 Dover, Kohl & Partners / Massengale & Co LLC

WE SHAPE OUR INFRASTRUCTURE AND IT SHAPES US: WORKING WITH ENGINEERS

Elsewhere in the book, we make the point that engineers have been the *de facto* designers of the public realm since the Second World War. Anyone who takes part in the design of the public realm needs to know whom to talk to, because while it is common to call all engineers who work on road design and planning "traffic engineers," that is a mistake.

Civil Engineers have a Transportation Engineering branch consisting of three roles: Transportation Planner, Traffic Engineer, and Roadway Design Engineer.

The Transportation Planner takes responsibility for the big picture: forty-year plans for metropolitan planning organizations (MPOs), highway planning, rail planning, and the like. The Traffic Engineer is responsible for most things above the pavement, such as trees, signs, signals, and even striping (but not utilities). The Roadway Design Engineer designs everything from the pavement surface down: curbs, drainage, utilities, grades, slopes, etc.

Traffic Engineers and Roadway Design Engineers are Professional Engineers (PE). Transportation Planners can be either PEs or Certified Planners in the American Institute of Certified Planners (AICP). The American Association of State Highway and Transportation Officials (AASHTO) sets the standards, guidelines, and protocols for road design in the United States that are then adopted by local jurisdictions. The American Society of Civil Engineers (ASCE) represents PEs. The Institute of Transportation Engineers (ITE) began in 1930, sixteen years after AASHTO.

Over time, the ITE became more oriented towards cities and urban context than AASHTO. More recently, a coalition of transportation officials from fifteen of the largest cities in the United States founded the National Association of City Transportation Officials (NACTO) in 1996. As of August 2023, there are 26 Member Cities, 11 Transit Agency Members, and 52 Affiliate Members. Six of the cities are Canadian.[32]

The U.S. Department of Transportation (U.S. DOT) is a federal cabinet department in charge of transportation regulation. It includes the Federal Highway Administration (FHWA), the division of the U.S. DOT responsible for highways, and the Federal Transit Administration, which is in charge of transit systems.

In March 2006, the Congress for the New Urbanism and the ITE released the street design manual *Designing Walkable Urban Thoroughfares, A Context Sensitive Approach, An ITE Recommended Practice* to address "the challenges that New Urbanists face in creating streets that match the urban built environment." The 255-page manual gives engineers, planners, and designers guidance for interpreting AASHTO Green Book policy and "demonstrates for practitioners how CSS (Context-Sensitive Solutions) concepts and principles may be applied... in places where community objectives support walkable communities, compact development, mixed land uses and support for pedestrians and bicyclists." The manual, sponsored by the FHWA and the U.S. Environmental Protection Agency, can be downloaded at CNU.org or NACTO.org.[33]

Broad Street, Chattanooga, Tennessee

Dover, Kohl & Partners, 2023

Retrofit: Right-Sizing

Avenue

Symbolically, Broad Street is the most important street in Chattanooga. As it reaches the heart of the downtown, it has one of the widest cross-sections in the city; its long, straight vista towards the Tennessee River is semi-terminated by the silhouette of the Tennessee Aquarium. Broad Street occupies a natural position as a main street, situated in the valley between two ridges; on some old maps it is labeled "Valley Street." The scale and formality of the street scene attracted the city's first high-rise commercial buildings, including masterpieces by renowned local architect R. H. Hunt, such as the James Building (1907) and the Maclellan Building (1923). Hunt "designed every major public building in Chattanooga from 1893 to 1935" (Figure 4.57).[34]

Figure 4.57: Broad Street, Chattanooga, Tennessee, 1938. *Courtesy of Chattanooga Public Library*

Figure 4.58: Broad Street, Chattanooga, Tennessee. Aerial view looking north. © *2023 Dover, Kohl & Partners*

To adapt to changing times, the corridor has been the subject of several retrofits over the last century. As we write the second edition of *Street Design*, planning is underway for a new, more mature Broad Street.

During the postwar period, the City increased the amount of street space devoted to moving cars, and then increased it again, to three lanes of traffic in each direction. At the same time, they tore down buildings to create surface parking lots and erect parking garages.

Despite this, Broad Street traffic volumes never justified the excessive space allocated to automobiles. Unlike parallel Market Street, which carries traffic across the Tennessee River, Broad Street stops short of the river; fewer than 10,000 cars per day travel the northern portion of Broad today. Along the northernmost seven blocks, trees were replanted in 1995, and on some of the blocks those trees have grown a sizable canopy (Figure 4.59).

During the last two decades of the twentieth century, a renaissance unfolded in downtown Chattanooga. Significant reinvestment continued in the early 2000s. Landmark public projects in the Riverfront District, including the Tennessee Riverwalk, the Aquarium, Riverfront Parkway, restoration of the Tivoli Theatre, and the park at Ross's Landing anchored the tourism and hospitality industries. Private investment followed, and this part of town saw a stabilizing office market and a gradual rise in housing and new hotels. Together these spurred signature annual events that have become fixtures of the popular image of Chattanooga. In recent years, however, the arty Southside and Northshore neighborhoods have competed for attention with the Riverfront District, and it has become clear that Broad Street will need to offer new and better experiences to stay relevant and attract people.

In 2015, a city-led effort to improve Broad Street removed one travel lane in each direction, first to create additional on-street parking, and then to set up temporary, parking-separated bike lanes. The changes undoubtedly upgraded the street, but with those changes the street still has too much space to cars. The concrete curbs defining the temporary bike lanes are inadvertent tripping hazards, and preexisting bumpouts created an unintentional slalom course for bicyclists moving along the corridor (Figure 4.60).

River City Partners and the City of Chattanooga carried out a new planning process for the neighborhood, largely online, during the Covid-19 pandemic. Not surprisingly, citizens voiced desires for a more pedestrian-friendly Broad Street that could accommodate events more easily, with added space for sidewalk cafes, better trees, wayfinding, and lighting. Dover, Kohl and Partners was then hired in late 2022 to come up with new designs for Broad Street from Aquarium Way to Martin Luther King Jr., Boulevard. From a hands-on design session in January 2023, and community input received from an

Figure 4.59: Broad Street, Chattanooga, Tennessee. Aerial view looking south along Broad Street, showing mature Willow Oak trees.

earlier online survey, three common themes emerged to guide the designers in reimagining Broad Street:

- First, Broad Street should become safer and more welcoming for walking, biking, and gathering. New designs need to slow down traffic speeds and make the street more beautiful.

- Second, Broad Street should be greener. Citizens challenged the city to add shade trees to Broad Street less about cars and more about people. They asked that Broad become a "park with a street running through it."

- Third, Broad Street should be livelier. The goal is to attract more "bustle" along Broad Street, with more cafes, more housing, and more 24-hour life and activity.

Testing these ideas, sketching scenarios, and visualizing possibilities, the Dover Kohl team created a series of alternative street configurations.

Option A features a promenade street where much of the walking space is in the center of the street, down a shaded allée of trees, as seen in Las Ramblas in Barcelona (Figure 4.61).

Option B recaptures space for pedestrians by right-sizing the number of vehicle travel lanes and parking spaces. Broad(er) sidewalks flank both sides of Broad Street (Figure 4.62).

The centerpiece of Option C is a "park street" that moves all the cars to one side of the street, "super-sizes" the sidewalk on one side of the street, and preserves existing trees (Figure 4.63).

Figure 4.60: Broad Street, Chattanooga, Tennessee. The well-intentioned bike lanes on Broad Street are partly separated from traffic, but the path uncomfortably veers back and forth and the concrete separators have proven to be a tripping hazard for pedestrians. *Courtesy of Kenneth García*

Figure 4.61: Broad Street, Chattanooga, Tennessee. Dover, Kohl & Partners, 2023. Perspective Section. Proposed promenade street, option A. *© 2023 Dover, Kohl & Partners*

Figure 4.62: Broad Street, Chattanooga, Tennessee. Dover, Kohl & Partners, 2023. Perspective Section. Proposed broad(er) sidewalks, option B. *© 2023 Dover, Kohl & Partners*

Figure 4.63: Broad Street, Chattanooga, Tennessee. Dover, Kohl & Partners, 2023. Perspective Section. Proposed "park street," option C. © *2023 Dover, Kohl & Partners*

Figure 4.64: Broad Street, Chattanooga, Tennessee. Dover, Kohl & Partners, 2023. Plan. Existing conditions. © *2023 Dover, Kohl & Partners*

Figure 4.65: Broad Street, Chattanooga, Tennessee. Dover, Kohl & Partners, 2023. Plan. Proposed promenade street, option A.
© *2023 Dover, Kohl & Partners*

Figure 4.66: Broad Street, Chattanooga, Tennessee. Dover, Kohl & Partners, 2023. Plan. Proposed "park street," option C.
© *2023 Dover, Kohl & Partners. Axel and Margaret Ax:son Johnson Foundation for Public Benefit*

GROWING A CAR-OPTIONAL NEIGHBORHOOD: TEN INGREDIENTS

Figure 4.67: The Bend, Chattanooga, Tennessee. Dover, Kohl & Partners, 2019. Computer Model Aerial view of the Bend, Chattanooga's first twenty-first century, "car-optional" neighborhood. Urban Story Ventures plans new workplaces, residences, hotels, entertainment venues, parks, and squares to replace lost smokestack industries. The new street network will reopen this part of the city's core to its riverfront for the first time in over one hundred years. © 2019 Dover, Kohl & Partners

We urgently need to reduce our dependence on cars. That means making more walkable neighborhoods where daily life does not *require* a personal car for each adult, for every trip. Here's how. Note that seven or eight of the items on this Top Ten list depend on better street design and only one relies on evolving technology:

1. Bring things closer together: Achieve **livable density** and practical mixed-use, in town.

2. Design slow, safe, highly walkable, bikeable **streets**.

3. Lay out **small blocks** in an interconnected web pattern.

4. Require street-oriented, **street-shaping architecture,** and green, comfortable **public spaces**.

5. Plant **street trees**. Then plant more.

6. Connect to the surroundings via high-quality **bike infrastructure**: Incorporate trails, cycle tracks, and elegant protected bike lanes in a network.

7. Optimize for **new mobility**: Anticipate ride-hailing, bikeshare, scooters, car-share, electric vehicles, autonomous vehicles, delivery bots, ultra-compact vehicles, and whatever is next.

8. **Right-size parking**: Have just enough, not too much. Eliminate minimum parking requirements.

9. Incorporate showers and covered, secure bike parking at many **workplaces**.

10. Develop around a transit-connected **mobility hub,** linking region and neighborhood.

Figure 4.68: The Bend, Chattanooga, Tennessee. Dover, Kohl & Partners, 2023. Illustrative Plan. Showing the network of streets, squares, and parks in the "car-optional" neighborhood. © 2023 Dover, Kohl & Partners

Figure 4.69: The Bend, Chattanooga, Tennessee. Dover, Kohl & Partners, 2023. Computer Rendering. A Rowhouse Street in the "car-optional neighborhood." © 2019 Dover, Kohl & Partners

INCLUSIVE STREETS

Good streets should be a basic ingredient in all neighborhood design, not just a special feature for wealthy neighborhoods. The fundamentals of good streets can be achieved on a reasonable budget, and good streets inevitably help create or sustain a neighborhood's economic and cultural value.

What's more, 40 to 50 percent of adult Americans do not drive, because they are too old, too poor, disabled, or simply choose not to. As we have pointed out, we have a transportation system based on the use of private vehicles that excludes almost half the population.

The street designer therefore has roles to play in achieving social inclusiveness, with influence over factors that affect whether the goal of mixed-income, welcoming, diverse communities will be realized. These include:

- containing costs by avoiding the use of oversized streets that consume excessive amounts of land and material;
- making medium- and high-density housing desirable, thereby lowering the per-unit land cost;
- lowering household transportation costs by making walking, biking, and transit feasible; and
- taking into account all ages and abilities.

The physical design must aim for excellence and walkability that is shared by all, irrespective of the luxury or modesty of the surrounding real estate.

Figure 4.70: West Second Avenue, Conshohocken, Pennsylvania. The best historic streets maintain a welcoming, high-quality, inclusive public realm irrespective of the luxury or modesty of the surrounding real estate. © 2011 Sandy Sorlien

Supply and Demand

Judging by the market, people prefer to live on attractive and functional streets, and so prices are on the rise. America simply does not have enough good streets to meet the demand. In 2012, Brookings Institution researchers Christopher Leinberger and Mariela Alfonzo established a list of five criteria for measuring walkability, then classified sixty-six places in the Washington, DC, area according to the criteria. The researchers found an astonishing degree of correlation between walkability and economic performance. They concluded, "Considering the magnitude of influence that walkability has on economic performance, a one-level (or approximately 20-point) increase in walkability (out of a range of 94 points) translates into an $8.88 value premium in office rents, a $6.92 premium in retail rents, an 80 percent increase in retail sales, a $301.76 per square foot premium in residential rents, and an $81.54 per square foot premium in residential housing values" [in 2012 dollars].[35]

Some may get the impression that good streets are only for the rich. They are not. Certainly, the solution to creating mixed-income neighborhoods and ensuring social inclusiveness is not to design grim, repellent streets—Americans built too large an inventory of those in the twentieth century. We need to increase the supply of decent, economically produced streets, and to address the need for affordable housing with the array of financial and legal tools now available to government and developers.[36]

Mixed-Income Neighborhoods: HOPE VI

In the 1990s, the U.S. Department of Housing and Urban Development (HUD) set out to remake hundreds of miserable public-housing projects across the country with the Homeownership and Opportunity for People Everywhere VI initiative (HOPE VI). Most of the old projects used the tower-in-the-parking-lot model promoted by Le Corbusier. These had superblocks that required closing city streets, because, as we saw in Chapter One, Corbusier did not like the "old-fashioned" street. The "projects" were also for low-income tenants only.

HOPE VI replaced the public housing tracts with mixed-income, mixed-use neighborhoods. Leaders of the Congress for the New Urbanism collaborated with HUD to integrate principles of traditional neighborhood design into HOPE VI housing policy. At the core of the initiative was the notion that good street design is integral to thriving neighborhoods. Well-designed streets connect communities, link residents to public transport, and provide safe and healthy places for people to gather. They can also revitalize neighborhoods by introducing a mix of housing types and a convenient mix of land uses.

HOPE VI infill and retrofit projects like the Laurel Homes Revitalization in Cincinnati, Ohio, and College Park in Memphis, Tennessee, demonstrate the role that effective design can have in the transformation of depressed neighborhoods (Figures 4.71 and 4.72). Both projects aimed to reestablish a network of streets that accommodate walking, cycling, and transit, as well as driving.

At Laurel Homes (now known as City West), public policy and a master plan with affordable housing have begun to connect previously unconnected streets, making the neighborhood walkable, safe, and interesting. The retrofit focuses on the renovation of existing housing and infill development to further define the new street network while also introducing a mix of new housing types. A community center and live/work retail space on Linn Street integrate commercial uses into the neighborhood. This puts daily destinations like grocery stores within walking distance for residents, allowing them to become less dependent on cars.

The street designer has roles to play in achieving social inclusiveness.

Figure 4.71: Betton Street, City West, Cincinnati, Ohio. Torti Gallas & Partners, 1999. Looking west on Betton Street. HOPE VI buildings are designed to face the streets, unlike the campus-style housing projects that turned away from the streets. Photograph looking west. *Courtesy of Torti Gallas and Partners, Inc. / Steve Hall © Hedrich Blessing*

Figure 4.72: College Park Drive, LeMoyne Gardens, Memphis, Tennessee. Torti Gallas & Partners, 1994. Photograph looking north. Reimagining the streets was central to the vision of HOPE VI plans. *Courtesy of Torti Gallas and Partners, Inc. / Steve Hall © Hedrich Blessing*

MAIN STREETS

Kensington High Street, London, England

Royal Borough of Kensington and Chelsea, 2004

Retrofit: Restoring the High Street

Main Street

By the mid-1990s, many business owners and residents in the center of London came to the consensus that the capital's roads were congested, ugly, and inefficient for cars and pedestrians alike.[37] Authorities had turned important local streets like Kensington High Street into auto sewers for drivers going to and from the center of London. The "London Plan" identified the High Street as one of the thirty-five most important centers in the city, but heavy iron railings along the sides of the road prevented residents from crossing the street, except at widely dispersed crosswalks. The crosswalks had pedestrian landings in the middle of the road with staggered barriers that herded people crossing the road into what Londoners disparagingly call "pigpens." Traffic signs on the High Street were numerous and large, visually overwhelming both the road and the commercial signs. Most importantly, with almost no parked cars or errant parking to slow drivers down, traffic was fast, noisy, and polluting. (London had many diesel trucks, taxis, and private cars with few pollution controls, contributing to some of the dirtiest air in Europe.)

The railings along the sidewalks, the many signs, and the highway-scale striping and notices painted on the roadbed were all ugly and visually disruptive, weakening the pedestrian's experience of the space of the otherwise well-designed street, with some of the most beautiful buildings in London (Figure 4.74). Daniel Moylan, a councillor for the Royal Borough of Kensington and Chelsea, led the fight to redesign and rebuild Kensington High Street. Moylan observed that the street had been designed "with the principal purpose of making it difficult for road accident victims to bring successful litigation against the highway authorities."[38]

Applying the K.I.S.S. Rule (Keep It Simple, Stupid)

Moylan coordinated a new design effort that aimed to balance the needs of cars, pedestrians, and cyclists (Figure 4.75). The road had a high traffic count (2,100–2,300 vehicles per hour), and local merchants wanted to keep traffic flowing. At the same time, they wanted to make the street more attractive to the pedestrians coming from the affluent surrounding residential streets and the Underground stops on the street. Moylan said, "Some people find it difficult to get away from the idea that streetscape design is fundamentally an exercise in engineering that excludes enhancement of the visual experience."[39] Moylan, however, pragmatically pointed out that Kensington High Street was in competition with shopping malls. "Nobody designing an indoor shopping mall ever feels he has to justify making it attractive," Moylan said. "The idea is absurd. Yet, our critics talked as if paying serious attention to the attractiveness of the redesigned street was a form of guilty frivolity."

Figure 4.73: Kensington High Street, London, England. Before: Heavy iron railings lined the sidewalks and the High Street felt like an "auto sewer." *Property of the Royal Borough of Kensington and Chelsea*

Figure 4.74: Kensington High Street, London, England. Before: Excessive, auto-scaled striping, signs, and railings took the joy out of the street experience. *Property of the Royal Borough of Kensington and Chelsea*

Figure 4.75: Kensington High Street, London, England. Royal Borough of Kensington and Chelsea, 2004. Compare with Figures 4.73 and 4.74. Today, the railings are gone, the traffic markings and signals are minimal, and traffic is slower, making a more pleasant experience for people walking. Bicycle parking in the median has proved extraordinarily popular.

> Nobody designing an indoor shopping mall ever feels he has to justify making it attractive. The idea is absurd. Yet, our critics talked as if paying serious attention to the attractiveness of the redesigned street was a form of guilty frivolity.
>
> — Daniel Moylan, Councillor, Royal Borough of Kensington and Chelsea

On the issues of safety and liability, the team followed practices already proven to work in the Netherlands and went to great lengths to ensure that they did not compromise safety when departing from Britain's design norms. Following an evidence-based approach, they left "plenty of opportunities for pulling back if need be—it is, after all, easy to add guard railing later if absolutely necessary."[40] Since the work was completed, however, accident rates have significantly declined.

Before starting work, Moylan's team decided that they wanted "simplicity, quality, and elegance," and came up with a list of principles and goals that are worth repeating:

- No guardrails or bollards.

- The removal of all but the most necessary street clutter.

- Rectilinear footways, as far as possible.

- No pavement build-outs [bumpouts] other than to bring footways back to a true line.

- Improved north–south pedestrian crossings and, as far as possible, no staggered crossings.

- The central island should have no guard railing. (With no fences to stop them, pedestrians go straight across.)

- More pedestrian space, but useful pedestrian space, where the pedestrians wish to be and go.

- In practice, this means it should be on the south side, not on the north (where, on Kensington High Street, it would have been easier to accommodate).

- More bicycle parking (put in the center to avoid using pedestrian space).

- The detailed design of dropped curbs should be based on traditional vehicle crossovers rather than on contemporary models.

- Tactile blister paving should be replaced by a more attractive alternative. (In Kensington, stainless steel studs were used. See Figure 4.76.)

Figure 4.76: Kensington High Street, London, England. Royal Borough of Kensington and Chelsea, 2004. A limited palette of materials and colors gives the street simple elegance. Stainless steel studs replaced the usual yellow tactile blister paving at crosswalks.

- The number of surface materials should be reduced to a minimum. (On Kensington High Street, they used only two for the sidewalks: York stone and granite.)

The final effect is underwhelming—which is exactly the point. It is the buildings and the space between the buildings that draw your attention, rather than busy details that pull our focus away (Figures 4.77 and 4.78). Moylan, with no training in design or even planning, had the wisdom and the confidence to oversee the design of what at the time was the best new road in London. Most elected officials are more influenced by the professionals whispering in their ear about standards and lawsuits. "In Britain the flat-earthers deny evidence and cry that the great god traffic would 'grind to a halt' if streets were shared and traffic lights were abolished," wrote Sir Simon Jenkins FRSL, journalist and Chairman of the National Trust. "Yet as Galileo told the Inquisition, *Eppur si muove*, and yet it moves."[41]

> The final effect is underwhelming—which is exactly the point. It is the buildings and the space between the buildings that draw your attention, rather than busy details that pull our focus away from those.

Figure 4.77: Kensington High Street, London, England. Royal Borough of Kensington and Chelsea, 2004. The center median in the busiest part of the High Street has different uses in different places: service parking, bicycle parking, pedestrian islands, and occasional turn lanes. The traffic lanes still handle large volumes of cars and trucks every day, but now they go more slowly.

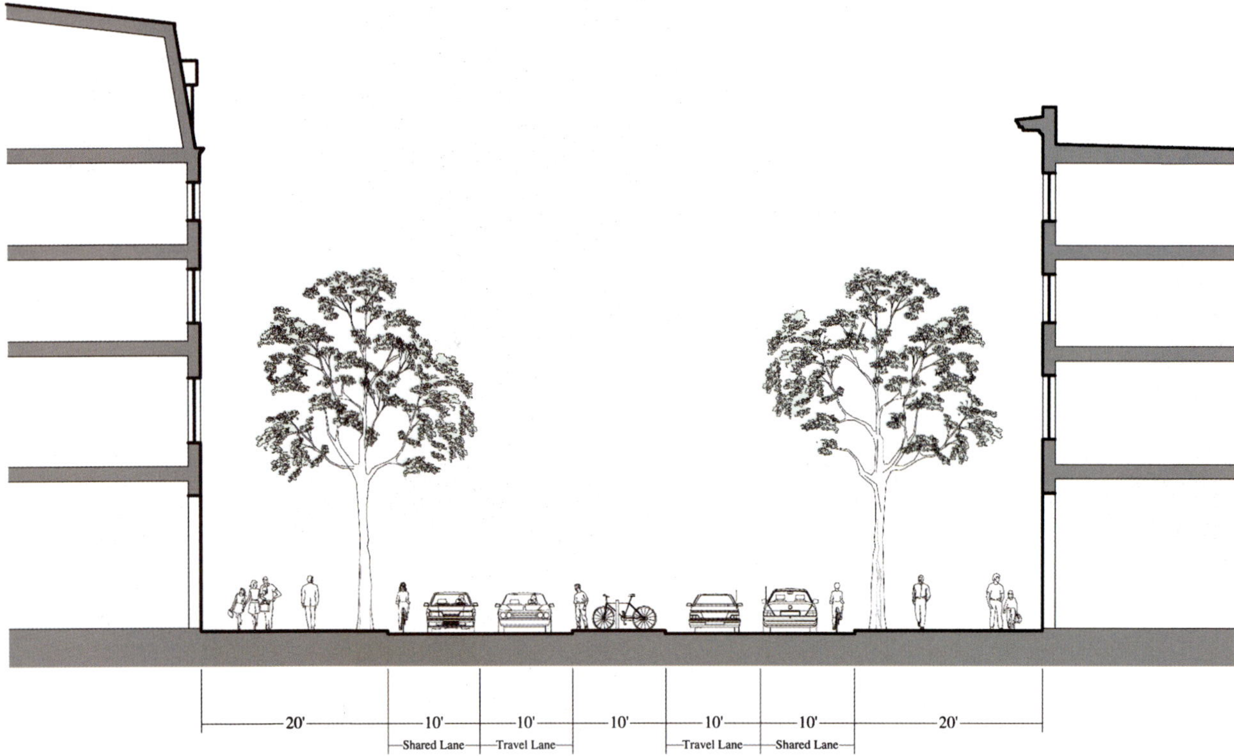

Figure 4.78: Kensington High Street, London, England. Royal Borough of Kensington and Chelsea, 2004. Section. © 2013 Dover, Kohl & Partners

Clematis Street, West Palm Beach, Florida

Dover, Kohl & Partners and Kimley-Horn, 2018–2020

Retrofit: Slow, Shaded, Flexible

Main Street

As cities mature, the demand mounts to evolve their most important streets. Clematis Street, the historical main street at the center of downtown West Palm Beach, has undergone several retrofits over the last century. Clematis Street connects east–west along the most crucial five blocks between the Lake Worth waterfront and the Rosemary Square neighborhood (Figure 4.79).

In the pioneer days that followed the City's founding in 1894, the wide street space made it easier to turn a horse-drawn carriage or a mule team. Upright, space-shaping buildings framed the crushed-shell surface of Clematis from the very beginning, lending urbanity even before the street was paved (Figure 4.80). At one point, the on-street parking on Clematis consisted largely of bicycle parking in the center of the street, where the bikes were eventually joined by early automobiles (Figure 4.81). By the time Clematis appeared in the controversial, largely

Figure 4.79: West Palm Beach, Florida. Dover, Kohl & Partners, 2019. Full site plan of Clematis Street. © 2018 Dover, Kohl & Partners

West Palm Beach, Fla.   Clematis Ave.

Figure 4.80: West Palm Beach, Florida. Photo looking down historic Clematis Street, West Palm Beach, Florida. *Courtesy of Florida Memory State Library and Archives of Florida*

Figure 4.81: Clematis Street, West Palm Beach, Florida. Historic photo. View of Clematis Street bicycle parking in the center of the street. *Courtesy of Florida Memory State Library and Archives of Florida*

Figure 4.82: Clematis Street, West Palm Beach, Florida. Historic photo. Model Ts parked diagonally on Clematis Street. *Courtesy of Florida Memory State Library and Archives of Florida*

unimplemented John Nolen plan for West Palm Beach in 1922–1924, the Model Ts were parked curbside, sometimes diagonally, sometimes backed-in (Figure 4.82).[42]

In Nolen's day, the planning profession was young, and lacked influence. (Trained as a landscape architect, Nolen was the first American to refer to himself as a professional city planner.)[43] Florida towns were growing quickly, and development regulations were almost non-existent. The results were often chaotic. Nolen's hotly debated, ultimately compromised plan for West Palm Beach dealt mostly with the extent, shape, zoning and—over Nolen's objections—racial segregation of the new growth beyond the edge of town. By then, however, Clematis Street was already an indelible, if small, feature at the center of his enlarged city map (Figure 4.83).

In the unpredictable, rapid-fire, *laissez-faire* boom of Florida's 1920s land rush, property owners could not count on government and rules to protect their interests. The mid-rise Comeau Building on Clematis (circa 1926, architects Harvey and Clarke) exemplifies how some of the best architects of the time dealt with the uncertainty. Upstairs offices in the Comeau Building offer spectacular views of the town, its waterfront, and distant Palm Beach; windows on all four sides also allowed for breezes in an era before air conditioning was commonplace. However, its owners knew their neighbors might someday build equally tall buildings on the immediately adjacent lots, and the Comeau's views, light, and ventilation would all be blocked on two sides. Their solution was to compose the building with a pedestal comprised of the bottom two floors, extending along the full width of the lot facing the main street, and then to top the pedestal with a slimmer tower of eight more stories, stepped back from the side property

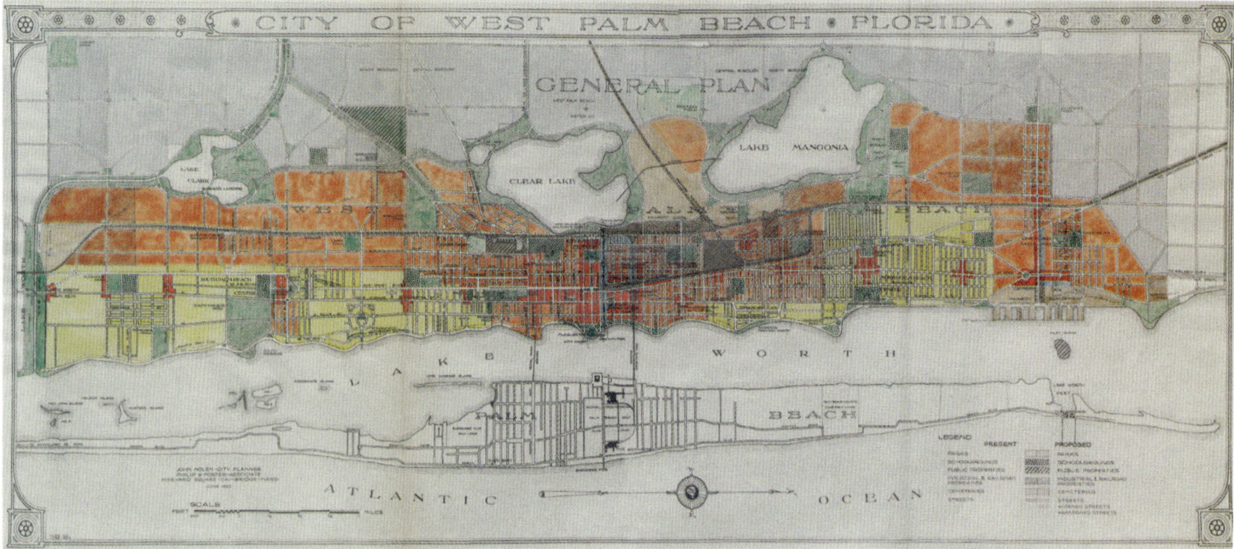

Figure 4.83: West Palm Beach, Florida. John Nolen, 1923. Master plan. West Palm Beach. *Courtesy of Florida Memory State Library and Archives of Florida*

lines. The gaps created by the step-backs lent assurance that their tenants would retain natural light and air from all four sides in the future.

The Florida real estate boom abruptly stalled after hurricanes in 1926 and 1927, and growth along Clematis slowed, though pedestrian life on the street continued to thrive. New structures from the Great Depression and postwar periods were smaller than the Comeau Building. But with maturity came larger department store buildings and five-and-dime stores, and Clematis was still the undisputed bright center of town.

All that changed after the postwar rush to the suburbs brought new competitive challenges to this main street, like so many others. Commerce shifted to suburban shopping centers and malls, and the public works priorities of the period were reflected in a gradual disfiguring of Clematis to make it a swifter thoroughfare, more efficiently emptying the downtown each evening at rush hour. First the city widened the car lanes and then made them one-way. The center of town declined, while whole neighborhoods nearby were razed for highway projects and painful urban renewal undertakings. Wide swaths of the city stood vacant when those projects halted or dragged on for decades.

By the early 1990s a significant makeover was clearly needed. The city restored the street to two lanes and two-way traffic, adding palm trees and decorative pavers. The reconstruction was carried out as an implementation step in a downtown revitalization plan and code created by

Duany Plater-Zyberk & Co (now DPZ CoDesign) and Treasure Coast Regional Planning Council, under the direction of Mayor Nancy Graham. This was a dramatic, effective step, and for its time it was a steep departure from the prevailing engineering ethos favoring speed and motoring capacity over all other considerations. A comeback ensued. Clematis was again the center of attention, as new developments filled in around it, and surrounding streets were upgraded. The street prospered, spawning a café scene and a full calendar of events. In 2014, the American Planning Association chose Clematis Street as one of "The Great Places in America."[44]

As the makeover approached twenty-five years old, however, even the 1990s redo proved to have been too cautious and car-oriented. Clematis still had narrow sidewalks, wide travel lanes that were fast, pavers that were difficult to keep clean, and more parking than needed. There were too few trees; sidewalks and storefronts along the southern side of the street were shaded by the buildings, but the north side of the street was bathed in glare. A Gehl Associates report in 2017 found three problems: too little space for pedestrians, too few places to sit, and too little shade.[45] In late 2017, a new retrofit was urged by the Community Redevelopment Agency (CRA) and the administration of Mayor Jeri Muoio.

Many main street retrofits are made to wake up a moribund downtown. The latest round of upgrades in West Palm Beach were a different case. To any observer who remembered the street in the 1980s, an economic

Figure 4.84: Clematis Street, West Palm Beach, Florida. Dover, Kohl & Partners, 2020. Computer Rendering. Concept illustration for the 500 block. © 2020 Dover, Kohl & Partners

revival was already vividly apparent; hundreds of new residential units and hotel rooms had been added to the neighborhood, and people had flocked to new businesses along its streets. The new retrofit was therefore to be carried out in the off-season (summertime, in this part of the subtropics) and in the heart of a thriving downtown without harming it. Whatever was to be constructed in each phase had to be started in late spring and completed by Halloween.

Besides schedule pressures, there were other questions to answer. Should the initial funding be concentrated in giving a single block segment of the corridor a full-blown upgrade, or spread thinly across all five blocks on dribbles of minor cosmetic improvements only? Was there support for reducing the amount of on-street parking, to free up space for more trees, walking space, dining, and seating?

There was little time before the 2018 summer construction season. The CRA and Dover-Kohl undertook an accelerated public process to answer those questions and make a plan. Going door-to-door for interviews with merchants and setting up a booth at the Saturday morning green market proved useful. During community meetings, keypad polling was used to probe public willingness to consider reducing the quantity of on-street parking; the polls revealed an appetite for that. Many participants reported that they had become accustomed to finding parking in nearby garages instead of trolling for a space along Clematis, and the new downtown residents were already walking there instead of driving. Several design options were considered at these gatherings, including an asymmetrical street configuration that was ultimately shelved in favor of a symmetrical one, in spite of the solar exposure. At these meetings, the public's preference was also confirmed for a curbless design, as originally suggested by Dover-Kohl designer Luiza Leite.

Buoyed by public enthusiasm, West Palm Beach chose to concentrate the funding and the 2018 summer of construction on wholly remaking the 300 block as a testing phase and proof of concept. The number of on-street parking spaces was cautiously reduced by about half. Once they saw the result, the initially nervous merchants endorsed a less cautious second phase; in 2019, the 200 and 100 blocks (three times the original work area) were reworked, and on-street parking was reduced even more. The city rebuilt the third phase on the 400 and 500 blocks during the 2020 Covid-19 lockdown, with

Figure 4.85: Clematis Street, West Palm Beach, Florida. Dover, Kohl & Partners, 2019. Photo looking west down 300 block. *Courtesy of South Florida Business Journal*

still less parking. Most of the spaces can be converted to outdoor dining areas.

In the new design, shade takes a high priority, with fifty-five new oak trees. Clematis is now slow and skinny. Trucks, buses, and the City's trolleys are able to pass one another, but not quickly. The effect on motorists was instantly noticeable; driving and parking maneuvers on Clematis require care and drivers have slowed down accordingly. Designer Kenneth García predicted this. He says, "The design of a street communicates to drivers how to navigate the space. A narrow street where you anticipate pedestrians lets you know that you are expected to drive slowly and cautiously."

By far, the most important change was getting rid of curbs. This added up to more than just reducing tripping hazards. The new Clematis feels like a festival street, and pedestrians clearly dominate. People on foot routinely cross midblock, whenever or wherever they feel safe.

Applying the K.I.S.S. Rule (see pages 388 and 513), the design team employed a neutral, gray-on-gray palette of lightly contrasting materials, reducing the sense of separation from one side to the other. There are two exceptions. The first is a high-contrast stripe of darker, textured material at the edge of the spaces used by automobiles, to aid the sight impaired and those with service dogs. This was especially important given the absence of curbs. The second exception was for high-visibility crosswalks with both darker and lighter pavers to draw the necessary attention.

The effect of the curbless design and the resulting flexibility is that Clematis functions at times like a main street, and at other times like a plaza. Raphael Clemente, executive director of the West Palm Beach Downtown Development Authority (DDA), significantly influenced the design and ran the public communications plan that supported businesses during the disruption of construction. He puts it this way: "The Clematis Street redesign has given Downtown West Palm Beach the best of both worlds, functioning equally well as a conventional street and as a pedestrian plaza. The ability to flex between the two is perfect for where our city is in its evolution as a place."[46]

In recent years, segments of Clematis Street have been closed to traffic from time to time, to host the community's weekly farmers market and various annual festivals. Retractable bollards along the edges of the crosswalks make closing the street easy and natural. Landscape architect Jonathan Haigh discovered during the early stages of construction that retractable bollards were only a few hundred dollars more expensive than fixed ones. The city then stipulated that in subsequent phases most bollards would be retractable, and sets of the bollards were installed on both sides of certain parking spaces. The retractable bollards allow the conversion of parking spaces into outdoor dining, cart vendor, or seating space—or back again—in a few moments (Figures 4.87 and 4.88).

The value of that flexibility came into full focus during the COVID pandemic and lockdown. Businesses worldwide, including the restaurants on Clematis Street, had to come up with ways to get running again, while respecting social distancing and avoiding confining customers indoors. West Palm Beach closed Clematis to car traffic, fully at times and partially at others, and allowed the businesses to expand into the street for open-air operations. To keep the scene organized, the West Palm Beach DDA rented tents and curtains and placed sticker dots on the street to show how far into the space restaurants could go; they branded the emergency program "Dining On the Spot," but it was about more than dining. The cycle gym moved its stationary exercise bikes outside to a former parking space, for instance.

The full street space was then completely open for walking, if only temporarily. The change reflected a global wave of similar moves, as cooped-up populations found themselves looking for safe ways to get out of their homes, stretch their legs, and be with other people. Main streets have become more important for strolling than at any time in a century. In response, the West Palm Beach DDA commissioned Dover, Kohl & Partners to create a booklet of instructions for "open streets"—that is, streets that are fully opened to pedestrians and closed to vehicles—for a range of conditions and streets throughout downtown.

Figure 4.86: Clematis Street, West Palm Beach, Florida. Dover, Kohl & Partners, 2019. Photo. Aerial view. © *2019 Dover, Kohl & Partners*

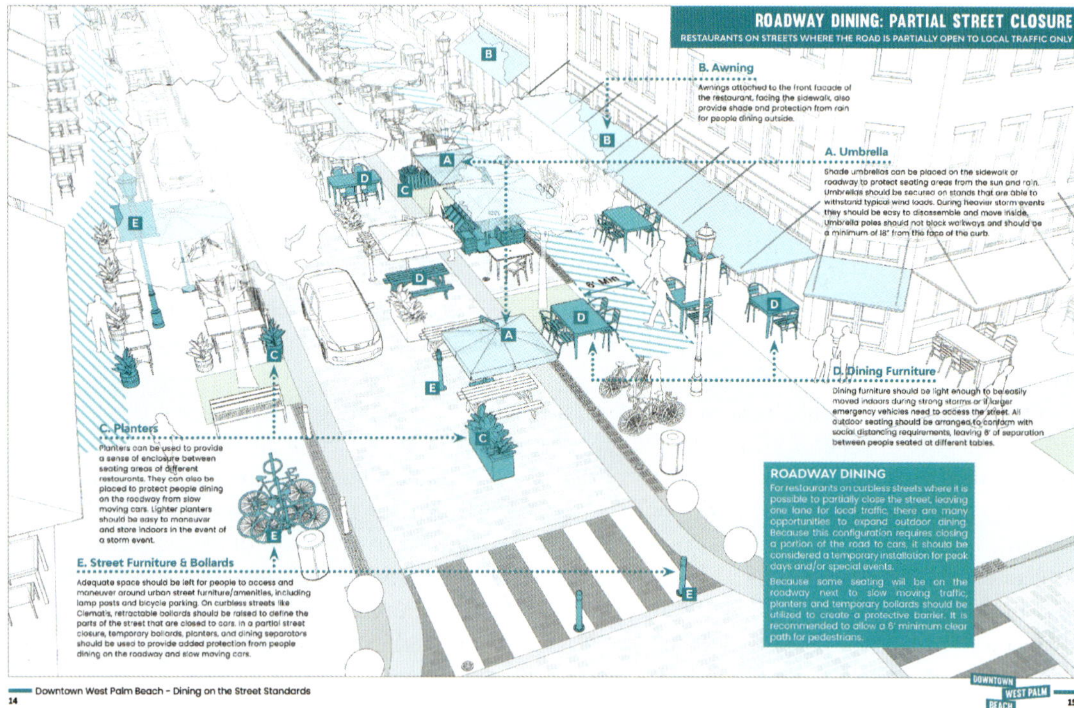

Figure 4.87: West Palm Beach, Florida. Dover, Kohl & Partners, 2020. "On-Street Dining Manual," created during the Covid-19 pandemic; page shows the layout for partial street closure, keeping one slow lane open for cars. © 2020 *Dover, Kohl & Partners*

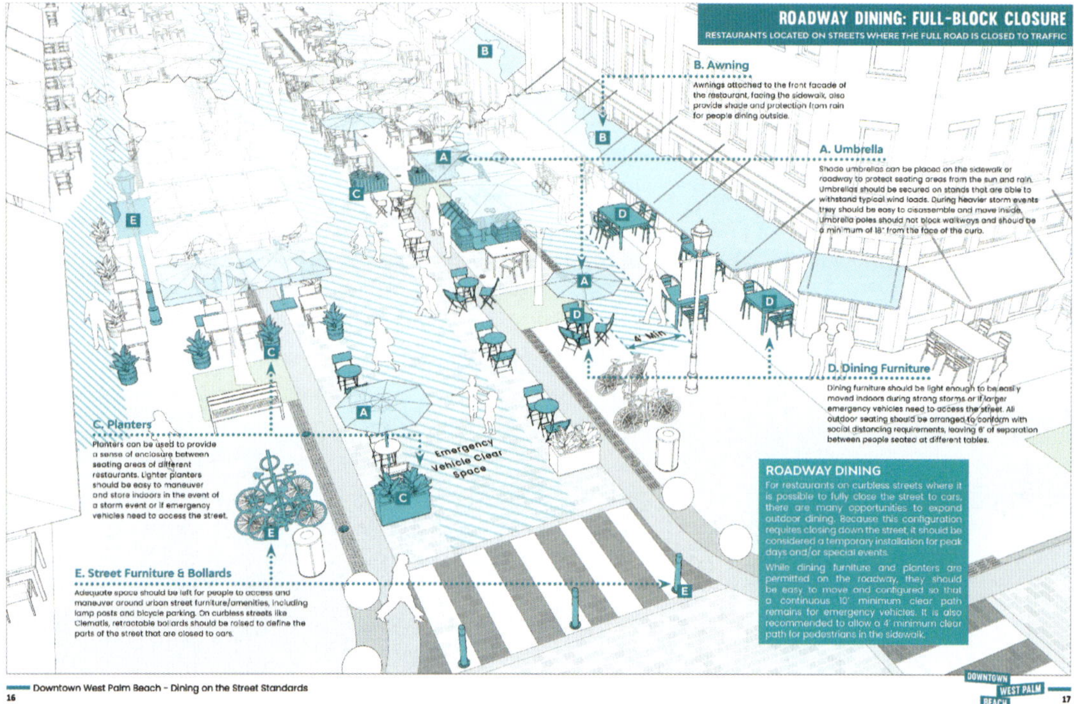

Figure 4.88: West Palm Beach, Florida. Dover, Kohl & Partners, 2020. "On-Street Dining Manual;" page shows the alternate layout for complete street closure, eliminating car traffic. Proponents call this approach "open streets" (rather than "closed") because the streets in their entirety are returned to being fully usable by people walking, after a century of being largely off-limits to pedestrians. © 2020 *Dover, Kohl & Partners*

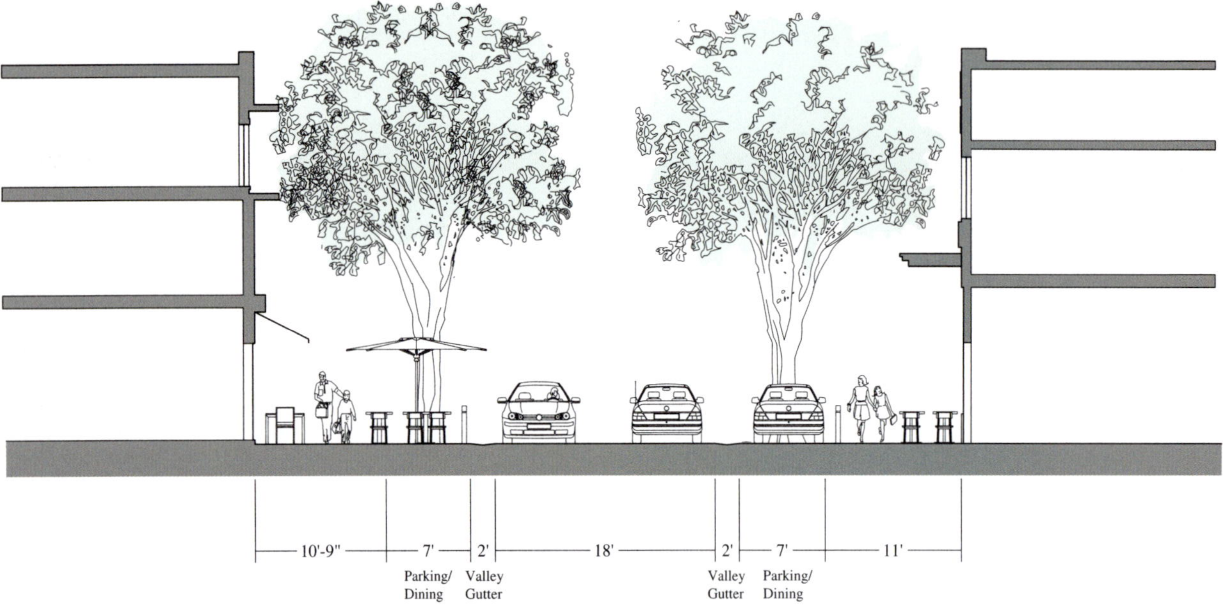

Figure 4.89: Clematis Street, West Palm Beach, Florida. Dover, Kohl & Partners, 2020. Section of Clematis Street. © 2019 Dover, *Kohl & Partners*

Worth Avenue and Hibiscus Place, Palm Beach, Florida

Sanchez & Maddux, Landscape Architects; Bridges, Marsh & Associates, Architects; 2010

Retrofit: Restoring Main Street

Main Street

Worth Avenue is a street for the superrich. We include it here because it is also one of the best American examples of a redesigned street that is as simple as Kensington High Street, without the signs, striping, warnings, multi-colors, multi-textures, and the like that dominate contemporary street design in the United States. Why should the rich get to have all the beauty? Of course, the rebuilding of the street was expensive, but that is not due to the lack of signs and striping.

Applying the Coco Chanel Rule to Main Street

The $15.8 million reconstruction (more than two-thirds of that for underground infrastructure) was spearheaded by the merchants on the street, who paid for much of the work with a bond (Figures 4.90 and 4.91). One of the driving forces behind the work was the Chanel boutique on the corner of Worth Avenue and Hibiscus Street, where the block in front of the store was rebuilt in a simple and elegant way as Hibiscus Place (Figures 4.92 and 4.93). Perhaps the designers were channeling Coco Chanel, the icon of early twentieth-century fashion and design, who famously said, "Simplicity is the keynote of all true elegance." She advised every woman before leaving the house to "look in the mirror and remove one accessory."

Worth Avenue is a shopping street, and so another of the basic ideas was to support the merchants. The noted New Urban shopping consultant Robert Gibbs said of

Figure 4.90: Worth Avenue, Palm Beach, Florida. Photo looking east on Worth Avenue before the renovation. Addison Mizner's apartment and studio were in the tall, arcaded building on the north side of the street. *Courtesy of Sanchez and Maddux, Inc.*

Figure 4.91: Worth Avenue, Palm Beach, Florida. Sanchez & Maddux, Landscape Architects; Bridges, Marsh & Associates, Architects, 2010. Photograph looking west on Worth Avenue, at Hibiscus Place.

Figure 4.92: Hibiscus Place, Palm Beach, Florida. Sanchez & Maddux, Landscape Architects; Bridges, Marsh & Associates, Architects, 2010. Photograph looking west on Hibiscus Place, at Worth Avenue.

Figure 4.93: Hibiscus Place, Palm Beach, Florida. Sanchez & Maddux, Landscape Architects; Bridges, Marsh & Associates, Architects, 2010. Photograph looking north on Hibiscus Place, from Worth Avenue.

the simple sidewalks on Worth Avenue, "Remember, the store owners want you to look at their windows—not the sidewalk or the streetscape" (see the "Secrets of Successful Retail" Chapter Two, pages 150–151). That is a lesson all designers who work on Main Streets should remember. The same principle applies to the street space in general. The experience of the public realm is what is important, not the streetscape or the new bumpout.

> The experience of the public realm is what is important, not the streetscape or the new bumpout.

In 1918, what is now Worth Avenue was a dirt road, with the private Everglades Club under construction at its western end. Standard Oil partner Henry Flagler had started to develop the island of Palm Beach approximately twenty-five years earlier when he extended his Florida railroad as far south as West Palm Beach on the Florida mainland. The Everglades Club was designed by the architect Addison Mizner, who had moved to Palm Beach and become the architect of choice for the rich building winter houses there. He later designed a building diagonally across Worth Avenue for his own office and studio, and behind it "Via Mizner" and "Via Parigi," small, picturesque shopping passages in the center of the large block between Worth Avenue and the next street over (Figure 4.94). The pedestrian passages expand the retail area of Worth Avenue and provide a convenient way to walk through the middle of the long block. Today, Via Mizner and Via Parigi (named after the Singer Sewing Machine heir and Florida developer Paris Singer) supply less expensive, smaller spaces than the buildings along Worth Avenue.

Mizner planted tall coconut palm trees on Worth Avenue. After these died in the lethal yellowing blight

Figure 4.94: Worth Avenue, Palm Beach, Florida. Sanchez & Maddux, Landscape Architects; Bridges, Marsh & Associates, Architects, 2010. Looking north, across Worth Avenue to Via Mizner.

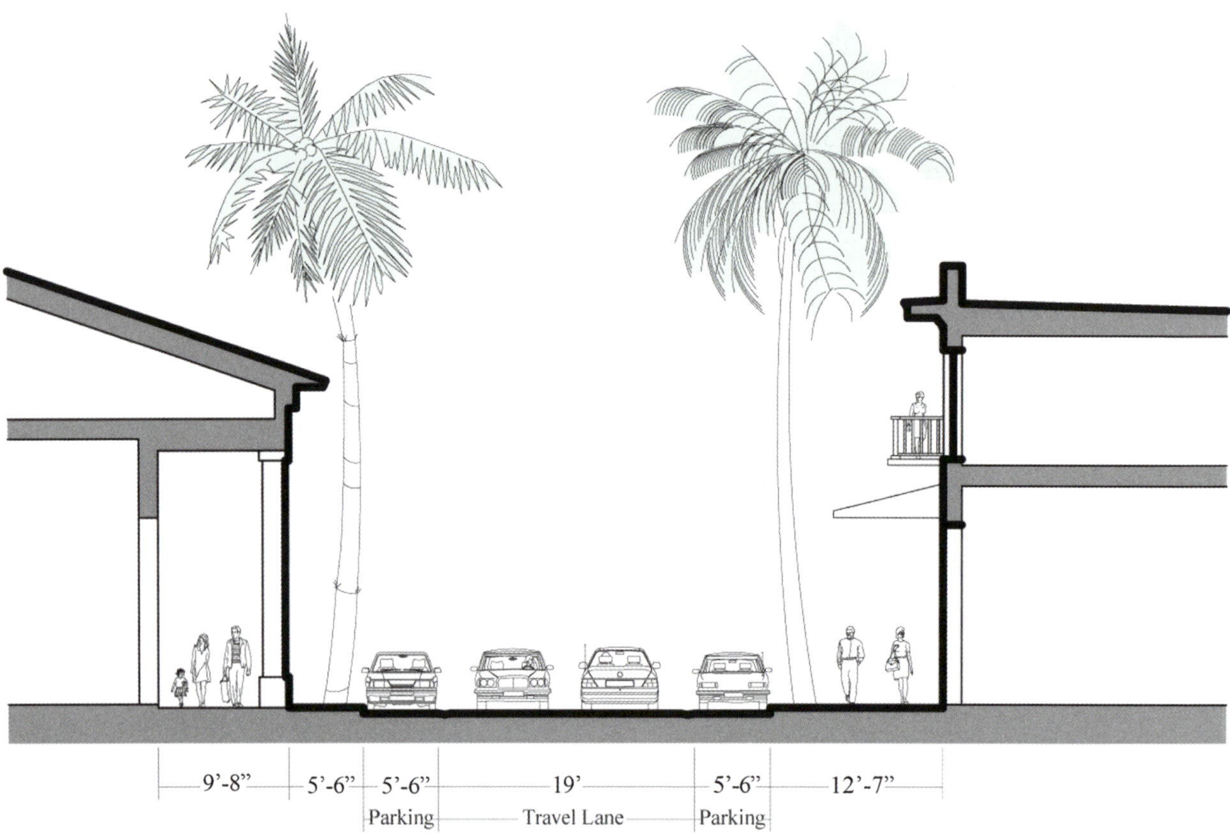

Figure 4.95: Worth Avenue, Palm Beach, Florida. Sanchez & Maddux, Landscape Architects; Bridges, Marsh & Associates, Architects, 2010. Section with data provided by the Town of Palm Beach. © *2012 Dover, Kohl & Partners*

of the 1970s, Palm Beach replaced them with shorter Christmas palms. Part of the recent renovation of Worth Avenue was to replace the Christmas palms with 32-to-40-foot-tall coconut palms. (This may sound strange to people from northern climes, but it is common practice in Florida and Southern California to plant fully grown palm trees that have been cultivated on tree farms.) At the same time, overhead power lines were buried and underground infrastructure was replaced.

The graceful trees are striking but restrained. Similarly, the tabby-shell concrete sidewalks may be more luxurious than plain concrete, but they are part of a muted palette that quietly works with other materials on the street, like the many stucco walls with coquina trim (the tabby shell also feels good under bare feet). Other visually restrained design moves were a drastic reduction in the number of signs and the quantity of striping, well-designed bumpouts, and the decision by

the clients to light the sidewalks during the evening with overflow light from the storefronts. At Hibiscus Place, parking lanes were removed. A well-proportioned island with a coquina fountain and coquina bollards was put in the center of the street to slow cars as they approach Worth Avenue. Because the cars were going slowly, a single, discreet directional sign could be used to guide traffic, without bold "Do Not Enter" signs on the inbound lane.

The slow speeds on Worth Avenue allowed other appealing choices that had fallen out of favor in downtown street design in America over the previous few decades. The sidewalk curbs are low, and the curbs around the grassy median are at the same level as the street and the grass as you approach from the north. At the other end of Hibiscus Place, a small planter extends into Worth Avenue at grade level. No curbs, bollards, signs, or striping prevent drivers from running over the planting

(Figures 4.92 and 4.93). And yet people rarely do. Even if they do, the plants are easy to replace.

Since there are no parked cars buffering people on the sidewalk from the traffic on Hibiscus Place, palm trees and coquina bollards prevent cars from veering onto the sidewalks. But the slow speeds of the cars, the lack of highway-scale graphics, and the placemaking touches in the design turn the roadbed into a space comfortably shared by walkers, cyclists, and motor vehicles.

For the most part, Worth Avenue follows Robert A.M. Stern's first rule of design: "Line things up." Since its blocks are long, however, there are midblock bumpouts to make people feel at ease crossing the street. One bumpout conveniently lines up with Via Mizner (Figure 4.94). All are much wider than the ubiquitous, formulaic bumpouts we have become accustomed to in new street construction. They are well proportioned and planted, so they are pleasing to look at, and there are benches where people can stop and rest without blocking the sidewalk.

Corner Conditions

One difference between new streets and historic streets is that we now want corner ramps for wheelchairs, disabled access, and even luggage with wheels. The formulaic, poorly designed, and narrow ramps built today can be both ugly and inconvenient. In their renovation of Worth Avenue, however, Sanchez & Maddux sloped the entire corner, a detail that works best with wide sidewalks. On the opposite side of Worth Avenue, where there is no cross street, they put an attractive bumpout, which visually breaks down the long block, reassuring the pedestrian that he or she has not been forgotten. Design solves problems. Formulaic design, like the placement of a poorly proportioned bumpout at every corner, commonly creates problems.

Figure 4.96: Worth Avenue, Palm Beach, Florida. Sanchez & Maddux, Landscape Architects; Bridges, Marsh & Associates, Architects, 2010. Looking south at the intersection with Golfview Avenue. Also see Figure 4.98.

BUMPOUTS

Our respective offices are in Manhattan and Coral Gables. All the sidewalks in the intersections on Broadway near the New York office have 7½-foot corner radii: that is, where the sidewalks come together at the corners they are rounded off with a quarter-circle with the 7½-foot radius used throughout most of Manhattan. In Coral Gables, the office is near the intersection of Sunset Drive and Red Road. In Bob Gibbs's terms (see "Secrets of Successful Retail," page 150), that is the local Main and Main corner. The corner radii there are approximately 25 feet. They are big so that large trucks can easily round the corner, even though the wide road gives lots of room for trucks to maneuver. The effect of the large radius is to make the already wide street even wider. Standing on the corner, the other side of the street seems far away, and you understand why street specialists working on bad suburban streets want to use bumpouts, also known as bulbouts, pork chops, and curb extensions (Figures 4.97 and 4.98). Unfortunately, this suburban solution has become both formulaic and ubiquitous in urban situations.

There is a place in good street design for bumpouts. For example, a bumpout might be appropriate when there is a midblock pedestrian crossing on a long block (as in Figure 4.94). Another type of problem is solved with the bumpout in Figure 2.129. The way they get pasted onto every wide intersection for "pedestrian improvements," however, is poor design. Most of the time, bumpouts are like lipstick on a pig—cosmetic efforts that do not address the real problem, which is that the street is too wide and dominated by cars and trucks (Figure 4.97). The balance on American streets between walking, driving, and other forms of transportation like bicycles and streetcars must be recalibrated.

A problem with bumpouts is that when they are pushed out into the street the traffic engineer wants to make the corner radius even larger, because a bumpout now intrudes into the area where the trucks turn. That is good for trucks and lazy drivers in any motor vehicle and bad for pedestrians. If the entire roadway is narrowed without bumpouts, the corner radius can be kept small, because the roadway is still available for turning.

In keeping with the K.I.S.S. Rule of Good Placemaking (Keep It Simple, Stupid), the public realm is usually most harmonious and comfortable when the curbs parallel the buildings (remember the Stern Rule,

page 405). Talented designers like Camillo Sitte and Léon Krier can play with this (Figure 4.176). Visually, ordinary bumpouts tend to take on a life of their own, disrupting the continuity and unity of the space. Throw in multi-color bus lanes, bike lanes, and inelegant crosswalk markings and the effect is magnified. Without exception, the beautiful streets of the world do not have this cacophony of elements.

In the redesign of Worth Avenue, one of America's most successful shopping streets (Figure 4.91), the designers took care to make the street visually harmonious. They removed most road signs, chose a limited color palette, carefully aligned the new palm trees, and used a light concrete rain gutter that visually lines up with the bumpouts—especially noticeable when the parking spaces are full, as they usually are. These bumpouts were also designed to be a visually pleasing width, unlike the formulaic, narrow bumpout we usually see (Figure 4.98).

No walkable city or town should need bumpouts "taming" the streets, because the streets should not need taming. Streets in cities and towns need narrow traffic lanes, slow speeds, drivers who are alert and afraid of hitting pedestrians—and whatever it takes to make streets where people want to get out of their cars and walk.

In the end, context has to be included in discussions about bumpouts. Worth Avenue was once a small-town Main Street, but it is now a luxury shopping street for aging, wealthy Boomers—and they might appreciate the comfort of a well-designed bumpout like the one in Palm Beach. Watch pedestrians in walkable places like Boston and New York, and you will see that they do not wait on the sidewalk, and that they certainly do not need bumpouts. Even though the New York City DOT turned many of New York's avenues into fast-moving suburban arterials in the 1950s and 1960s, New York pedestrians think the public realm belongs to them, and they wait out in their road for the light to change. Thankfully, the city is once again trying to give New Yorkers roads worthy of their ambitions. No walkable city or town should need bumpouts "taming" the streets, because the streets should not need taming. Streets in cities and towns need narrow traffic lanes,

Figure 4.97: El Camino Real, Santa Clara, California. The road is wide, the buildings alongside the road do not enclose the space, and the scale of the auto-centric road is boring for the pedestrian. In this situation, the bumpout may make it psychologically easier for the occasional pedestrian to cross the street. But why would the pedestrian even be on the street here? This is not a detail that entices people out of their cars, and it is not a detail that would improve a walkable street. *Courtesy of the City of Los Altos*

Figure 4.98: Worth Avenue, Palm Beach, Florida. Sanchez & Maddux, Landscape Architects; Bridges, Marsh & Associates, Architects, 2010. In context, this bumpout works well. It is for an upscale shopping street most people will drive to, it marks a place to cross in a long block, it functions well with the arcaded sidewalk, it breaks the monotony of a long block, and it works visually with the design of the rest of the avenue. In other words, it is an example of a good designer solving problems with good design.

slow speeds, drivers who are alert and afraid of hitting pedestrians—and whatever it takes to make streets where people want to get out of their cars and walk.

London's "cycling mayor" Will Norman said, with British understatement, "You won't see many bumpouts in London." But many street activists in New York and other cities are at an age where they have young children, and they don't want their kids standing behind cars—"people won't see them," "they'll run out and get killed"—so they listen to the traffic engineers who say therefore "daylighting" (removing parking, trash cans, trees, and other visual obstructions from a large area surrounding intersections) is a good thing. Leading British traffic engineer Phil Jones says there are UK studies that show daylighting is safe, and yet the same studies conclude that daylighting encourages drivers to go faster.

Park Avenue, Winter Park, Florida

Dover, Kohl & Partners, 1998

Retrofit: Restoring Main Street

Main Street

In the 1880s, two businessmen born in Massachusetts bought a large tract of land in Orange County, Florida, where they platted the town of Winter Park. They slowly developed the town, which they sold in 1885 to the newly formed Winter Park Company, which, in turn, successfully promoted the town as a winter escape for northerners. Park Avenue was the town's Main Street, which from the beginning was meant to be urban. Even before the roadway was paved, the earliest buildings had a traditional, street-oriented relationship to the public space of the avenue and its adjacent Central Park.

Over time, the street was widened to accommodate diagonal parking (later altered to parallel parking), and its distinctive curvature was relaxed and broadened, to speed up traffic. Unlike many other main streets, which became virtually deserted in the mid-twentieth century, Park Avenue continued to occupy a place of importance to the people of the town as the central scene of civic life. Periodically, however, Park Avenue was threatened and further encroached upon. Serious proposals were floated to pave Central Park for more parking; a bland suburban-style City Hall was built on a key focal site, and, in the early 1960s, the big Winter Park Mall opened just a few

blocks away. Some predicted the mall would kill Park Avenue, but things turned out differently. The avenue hummed along and grew more beloved, while the mall began to languish within a decade. Customers chose the authentic main street.

Eventually, though, leaders of prosperous Winter Park decided to improve the quality of Park Avenue—to bring it into the class of what Allan Jacobs called "great world streets." The timing for reconstruction was right, because the century-old lateral sewer pipes under the avenue had begun to fail, one at a time. The new design needed to preserve the classic main street without overdoing it, restore features that had been lost over the years, and improve the experience for pedestrians.

The City of Winter Park hired Dover, Kohl & Partners, whose design team began building consensus for a careful refurbishment of the avenue, starting in 1994 with the "Winter Park in Perspective" conference and a subsequent charrette. Dover-Kohl studied the existing conditions and decided to narrow the excessively wide travel lanes to accommodate widening the sidewalks and slowing down traffic. The team also judged the tree canopy along the avenue to be insufficient; the lack of shade made walking uncomfortable. Unsightly overhead wires ran along the avenue, and the poles for span-wire traffic signals in the sidewalks crowded pedestrians and seemed out of place (Figure 4.99). At the same time, it was clear that Winter Park wanted something better than another visually obnoxious, overcooked "streetscape"

Figure 4.99: Park Avenue, Winter Park, Florida. Before: Existing conditions, circa 1996.

Figure 4.100: Park Avenue, Winter Park, Florida. Dover, Kohl & Partners, 1998. After the retrofit, new rules for development were instituted, bringing the public effort to remake the street and private investment in new buildings into one, coordinated ensemble. A more abrupt midblock crank (center left) replaced a swooping curve. *Courtesy of Rick Hall, HPE*

Figure 4.101: Park Avenue, Winter Park, Florida. Dover, Kohl & Partners, 1998. Restored as a signature public space, Park Avenue is host to the Winter Park Arts Festival each spring.

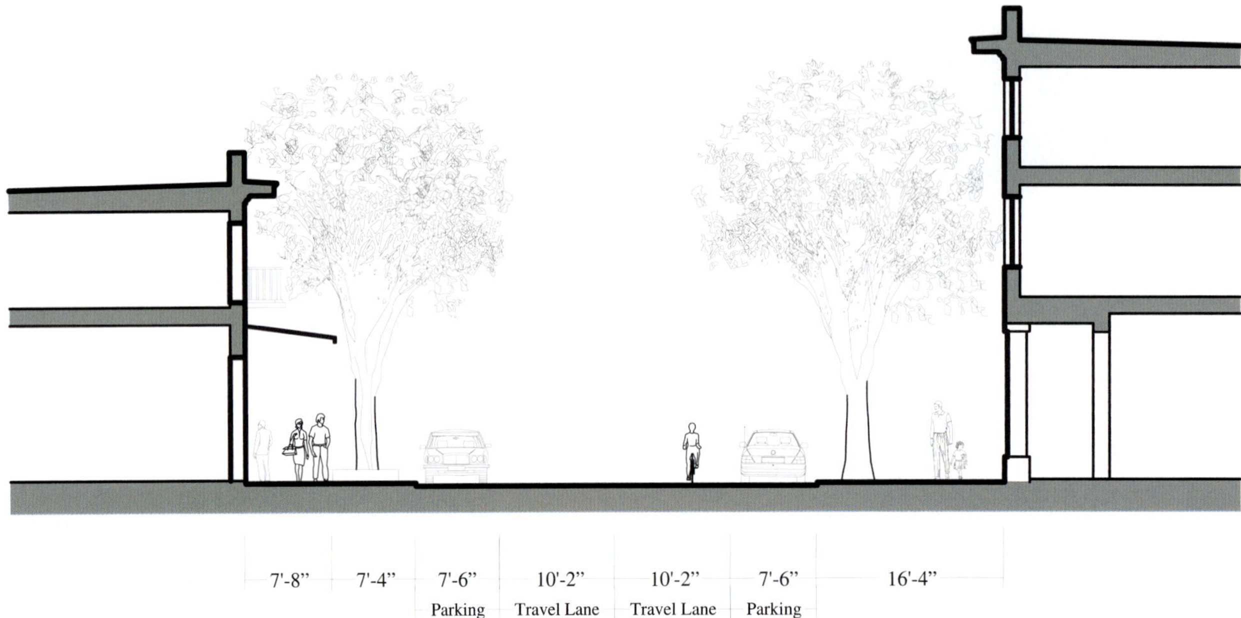

Figure 4.102: Park Avenue, Winter Park, Florida. Dover, Kohl & Partners, 1998. Section. © *1998 Dover, Kohl & Partners*

project with excessive amounts of the clichés Peter Katz calls the B's (benches, bollards, bricks, banners, and so forth).[47] The town's Community Development Director at that time, Don Martin, demanded instead a street design with "simple, understated elegance."

The reconstruction, largely completed by 1998, is considered a great success among local residents (Figure 4.100). First, the way Park Avenue changes in spatial character along its length, with a distinct beginning, middle, and end, was respected in the design (Figures 4.101 and 4.102). Each block was understood to be different, yet the corridor is still sufficiently united. The most challenging block was between Lyman Avenue and Comstock Avenue where, decades ago, engineers had reworked the street network to bend the alignment of Park Avenue into a high-speed curve. Here the geometry of the rebuilt street was significantly altered by changing that broad, swooping curve to a much more abrupt crank, tightening the scene visually, slowing the cars, expanding pedestrian space, and pulling new buildings into one's field of vision.

Sidewalks, intersections, and curb cuts were altered to meet ADA standards. The historic brick street surface, buried under generations of asphalt, was uncovered and restored, and new sidewalks poured with simple, easy-to-clean concrete. The old bricks in the roadway were combined with newer brick pavers in the crosswalks,

with a very slight difference in color and a muted pattern change to visually define the crosswalk in a subtle way (Figure 4.101). To address the perennial local problems with older brick streets, where the bricks shift and sag over time as their sand base is compacted or eroded, a more stable base of limerock, sand, and soil cement was made especially for Park Avenue.

Another common complaint about modern "streetscapes" is that the new trees do not grow to their proper height and full canopy; for Park Avenue, a customized tree-planting detail (copied from nearby Disney World, where the trees have thrived) has allowed the new oaks to grow to their full size. The overhead utilities on the avenue were buried. Mast-arm traffic signals, modeled on a century-old design rediscovered and revived by the landscape architect Jay Hood, replaced the span-wires. These employ a cable-stay to hold the arms steady instead of using fattened tubes, and the result is a visually welcome alternative to the standard mast arms used in most streetscape projects. Pedestrian signals were integrated at several intersections.

The City of Winter Park adopted new design rules to go with the refurbished street, and infill development following those rules has further sewn the segments of the street together. Galvanized by the success on Park Avenue, local leaders completed a grand plan for the long-term revival of Central Park under the direction of landscape

Figure 4.103: Rollins College entrance, Park Avenue, Winter Park, Florida. Dover, Kohl & Partners and Chael, Cooper & Associates, 1998–2002. *Courtesy of Chael, Cooper & Associates / Corey Weiner*

architect Forest Michael, and Rollins College installed a new monumental entrance and admissions hall to visually terminate one end of the avenue, both designed by architect Maricé Chael (Figure 4.103).

Main Street, Great Barrington, Massachusetts

massDOT, 2013

Retrofit: Rebuilding Main Street

Main Street

RETROFITTING A MAIN STREET THAT IS ALSO A STATE HIGHWAY: A CAUTIONARY TALE

Great Barrington, Massachusetts, was in transition when we wrote our Case Study for the first edition: massDOT had announced it would rebuild the town's Main Street and the Selectboard was leading public discussions. A postscript below looks at the results.

In May 2012, *Smithsonian* magazine named Great Barrington the best small town in America.[48] We like it too. One of the reasons we like it is because we enjoy spending time on the town's Main Street, a still-functioning town center and shopping street that is one of the best and most beautiful in New England (for historic photos of the street, see Figures 4.104 and 4.105). By the time this

book comes out, that may be less true, because if all goes as planned, the Massachusetts Department of Transportation (massDOT) will have cut down the Bradford Pear trees that line the street during a $5.5 million renovation (Figures 4.106 and 4.107).

Regardless of how that plays out (see the update on page 420), the reasons why Great Barrington's Main Street is threatened reflect the way most DOTs work, and how other professionals involved as DOT consultants frequently approach street design and tree planting. The proposed changes remind us that transportation and traffic engineers are now the default urban designers for a lot of America's public realm. Scenarios similar to Great Barrington's DOT disaster have gone on across America for decades.

Great Barrington's Main Street has all the elements necessary for making a street where people want to get out of their cars and walk. The central blocks are lined with three-story, mixed-use buildings—mainly loft-style brick buildings from the late nineteenth and early twentieth centuries characteristic of New England, with stores on the ground floor. Perpendicular to Main Street is another retail street, Railroad Street (see Figures 2.127 and 2.130). Parking is on-street and in lots behind Main Street.

> Great Barrington's Main Street has all the elements necessary for making a street where people want to get out of their cars and walk.

Figure 4.104: Main Street, Great Barrington, Massachusetts. Historic photo from the nineteenth century. The brick building on the southwest corner of Main Street and Railroad Street is 304 Main Street, the oldest "commercial block" on Great Barrington's Main Street. The building is also visible in Figure 2.127. In the one-hundred-plus years between the two photos, massDOT paved and widened the road. *Courtesy of James Mercer*

Figure 4.105: Main Street, Great Barrington, Massachusetts. Historic photo from the nineteenth century. Like Figure 4.104, a nineteenth-century view of Main Street, this one looking south. Planted along the still-dirt road, the trees grew well, forming a strikingly beautiful canopy. *Courtesy of the Special Collections Department / Iowa State University Library*

▶ **Figure 4.106:** Main Street, Great Barrington, Massachusetts. Photograph looking north on the east side of Main Street in the spring of 2012, from Bridge Street. By that time, massDOT had announced its plan to cut down the Bradford Pear trees, which are fuller when the green leaves are out. For the After view, see Figure 4.110.

Figure 4.107: Main Street, Great Barrington, Massachusetts. Photograph looking north from the corner of Main Street and Castle Street (opposite Bridge Street), in the spring of 2012. The road and the traffic lanes are wide, too wide visually unless there are mature trees. By that time, massDOT had announced its plans to cut down the Bradford Pear trees they had once planted.

Great Barrington is where the story of the *Alice's Restaurant* Thanksgiving "Massacree" began, and there is still a countercultural element in this part of the Berkshires, which has its own local currency, called BerkShares.[49] Locals and tourists alike frequent many of the stores and restaurants in the town, including a whole-wheat pizza place called Baba Louie's, a fancy cheesemonger with a slow-food café out back (Rubiner's and Rubi's, respectively), two excellent supermarkets (one a cooperative) that feature local produce, a great hardware store, and a regionally-important restored theater called the Mahaiwe. Matt Rubiner tells anyone who asks that one of the reasons he opened his shops in Great Barrington was the beauty of the downtown, which attracts customers the way it attracted him.

Main Street is also U.S. Route 7, which is why it comes under massDOT's jurisdiction. On its long passage from Long Island Sound in Connecticut to the Vermont–Canadian border, Route 7 has many different characters. In southern Connecticut it is a six-lane divided highway known locally as Super 7; in northern Connecticut, it is a wide two-lane road that winds along the scenic Housatonic River; and in many places along the route, it is the old Main Street in various cities and towns. In Great Barrington, Route 7 outside of the downtown is also an ugly modern arterial that attracts auto-centric shopping and

development of the kind that has harmed many downtowns by taking away stores and services.

MassDOT's first plan for Great Barrington followed its usual formula for downtown streets, as seen in a number of towns around Massachusetts. Fortunately, Great Barrington residents involved in local government committees eliminated a number of the formulaic elements in massDOT's design, like concrete sidewalks stamped with colored-brick patterns and ostentatious pedestrian crossings. Unfortunately, though, massDOT and its consultants have convinced the town that it will be best to cut down the Bradford Pear trees that were planted on their advice in the 1960s.

After all the problems of the demolition and messy reconstruction of Main Street are over—a process that will disrupt the businesses along the street and cost them sales—the result will be a place that in all meanings of the word will be less attractive: both locals and tourists will be less attracted to coming to Great Barrington and getting out of their cars. Keeping the pear trees and phasing in other trees over time would serve the town far better. If the town *is* going to take the drastic step of cutting down all the mature trees at one time, then it should demand more from massDOT in return. All over the country, progressive towns looking to the future and thinking about climate change are using the techniques of suburban retrofits to transform their arterials and transportation corridors into walkable places. Great Barrington could use some of those as well. We add a few ideas at the end of this essay.

Placemaking

In Chapter One, we wrote about techniques urban designers use to make streets like outdoor rooms, fashioning a public realm where people want to be. People enjoy being in these rooms when they are well-proportioned, comfortable spaces. If the space is wide and uncontained—as it is north and south of downtown on Route 7, where the road travels through sprawl and shopping centers—people are more comfortable in their cars than on foot.

MassDOT's plan for Great Barrington paid lip service to placemaking principles, and the state agency is more sensitive to the local importance of main streets than most state Departments of Transportation. Unlike the neighboring New York State DOT, massDOT has rarely gone through small towns simply removing all the trees along main streets and reducing sidewalks to uncomfortably narrow strips. But their first priority in the Great Barrington plan is still traffic flow and infrastructure engineering. Like most DOTs, massDOT has developed an impressive ability to procure funding, even in times of recession. Once it has those funds, one of its highest priorities becomes to spend the money on what it does best. If push comes to shove in public meetings, massDOT is ready to change the aesthetic parts of the design but strongly resists compromising on what is important to it, which is traffic controlling the roadbed and traffic flow.

A real Department of Transportation—as opposed to a Department of Traffic—might say the highest and best use of the funds in Great Barrington would be to rebuild the streetcar line that once ran along Route 7 from Canaan, Connecticut to Pittsfield, Massachusetts, or to restart train service to New York City, a possibility that private groups are discussing.[50] Instead, massDOT is saying that it is important to change the camber of the road from 1 percent to 2 percent, but urban designers or town planners with money to spend in Great Barrington would not begin by making a 1 percent change in the grade of the road, cutting down trees that are one of the town's greatest assets, or tearing up pleasant sidewalks, as massDOT plans to do in their Main Street

"reconstruction." They would recognize that the blocks between Great Barrington's Town Hall and the Post Office are one of the town's greatest strengths and that the money would be best spent to the north and south, where the town's fabric and walkability quickly fall apart. Great Barrington could use many more projects like the River Walk it built along the Housatonic River, where it flows downtown. It would not be hard to find better uses for this money.

If engineers were poets, they would never write, "I think that I shall never see/a poem lovely as a tree." They dislike trees near roads for a variety of reasons, beginning with their preference to remove any "hazardous object" from the side of the road. They realize there will be things next to the road in the center of towns, so they compromise by accepting a small tree, which will do less damage to a speeding car than a larger one. Overlooked is that pedestrians behind large trees will suffer fewer injuries, and might actually escape unharmed. Engineers also want to control variables; they do not like messy tree branches that can interfere with power lines. Last but not least, they use modern construction and compaction methods that damage tree roots.

The trees selected by massDOT's consultants can never grow to the height or the width required for a leafy canopy over the sidewalk or a "wall" of tall trees along the road (Figure 1.31). MassDOT uses consultants who think it is old-fashioned to use traditional street trees to shape a traditional American main street. The consultants use functional arguments about disease to support their preferences but frequently promote change for the sake of change.

They will tell you that the Bradford Pear trees on Main Street are brittle, which is true. They will not tell you that there is no such thing as a no-maintenance, mature tree. And they won't tell you that the trees they advise planting will never grow as large as the Bradford Pears or that a number of small-town main street studies show mature trees have a significant economic value for shopping streets. That is particularly true for Great Barrington because its Main Street is unusually wide, and the buildings that line it are not big enough to comfortably shape and visually contain the space between the buildings. The mature Bradford Pear trees are just large enough to do that, however, as well as jobs like providing shade in the summer.

In their work with massDOT, the town committees have thus far been unable to get a simple row of street trees of the type Great Barrington needs and has traditionally had. The town's Department of Public Works weighed in, calling for trees that would not block the town's cherry picker from getting to the second and third floors of the buildings along the street. This stipulation ignores the fact that the trees make a better life for the people on the second and third floors, just as the tree selection ignores the economic value of trees for the town's merchants and the fact that the small trees will block the view of the storefronts. Has there been a cherry-picker problem in the past, or is the DPW simply overthinking their part of the problem?

The DPW also complains about cracked sidewalks and places where the concrete slabs have been slightly raised by tree roots. Real bricks can alleviate that problem, as opposed to the brick-pattern stamped concrete panels that massDOT prefers. But traditionally, cracked sidewalks have not been a problem for New Englanders: in Nantucket, the town values what it has, so Nantucketers live with the uneven sidewalks instead of letting lawyers, engineers, and other specialists introduce monofunctional solutions. They balance function, construction, and beauty, and the result is one of the most beautiful and popular places in America (see the case study on pages 143–151).

A DOT LIMERICK

Figure 4.108: Pleasant Street, Nantucket, Massachusetts. A much-loved American Elm tree.

There once was a tree in Nantucket,
With none of its roots in a bucket.
"That can't be," said the State DOT,
But no car has ever yet struck it.

In Great Barrington, the specialists undervalue the trees' contribution to the beauty of Main Street and diminish it. They might tell you that beauty is in the eye of the beholder and that what's important is dotting all the i's and crossing all the t's in the limited criteria on their list of priorities. With the CNU ITE street design manual, *Designing Walkable Urban Thoroughfares: A Context Sensitive Approach,* this is less of a problem, because in the manual the ITE officially endorses context-sensitive design and calls for mature trees in the right downtown contexts. Great Barrington's Main Street discussion shows that these changes can take time to trickle down to the field, however.

The Future of Main Street

New Englanders used to be known as frugal people. "If it ain't broke, don't fix it," was a New England motto. When towns spent money on big changes, they demanded value. Since World War II, our road building and interstate highway program has been the most expensive public works project in the history of the world.* DOTs have become used to doing whatever they want with our roads, as well as controlling the funding to do so. They like plans along the lines of massDOT's $22 billion Big Dig in Boston, where the DOT tore down a functioning highway to improve traffic flow on the highway, at tremendous cost. That does not work well in the current state of the American economy, and it is not the best way to make our cities, towns, and neighborhoods more walkable.

Some say that all politics is local. All community design should be, too. When we look at the needs of Great Barrington, we can see ways to make the current plan better for the town, its citizens, and its merchants. The town's Main Street is a wide one that needs traditional street trees for its civic and economic well-being. Traditionally, new trees would have been phased in over time, because new trees grow slowly in New England. If the town decided to do that, the trees would have to be chosen so that they would make a harmonious streetscape with the Bradford Pear trees. Most of the trees in the massDOT plan have different shapes and sizes than the existing trees and will never form a canopy. They are chosen with the idea that they will be replaced again in ten or fifteen years.

If the town wants to go ahead with the DOT plan to dig up Main Street, then it should get a bigger bang for its buck. The money DOTs spend is our money, and we should put the needs of our communities ahead of a narrow focus on making spaces for cars. The time has come to build complete streets that put people, public transportation, bicycles, and placemaking back in the mix.

In Holland, the street might be made into a shared space, as welcoming and safe for pedestrians as cars. Massachusetts probably is not ready for that (even though Harvard and MIT students at the other end of the state are famous for plunging into the traffic on Massachusetts Avenue without looking). But there are other things that could be done, once one starts thinking about the space between the buildings as the public realm, rather than as a vehicular thoroughfare.

Great Barrington's Main Street is very wide. Narrowing the roadway could slow traffic (making the street safer), allow space for bike lanes, reduce maintenance costs, and even reduce construction costs. Narrowing the roadway means moving the curbs, and the DOT will probably argue that is too expensive. But if the reconstruction involves replacing the underground utilities that is not true (and that cost was never mentioned when the road was widened).[51]

An alternative to moving the curb is to dig a trench between the sidewalk and the road to create a place for growing healthy street trees. If the trees are planted eight feet out from the sidewalks on both sides of the street, bike lanes with pervious surfaces can be put between the road and the sidewalk, the drains can be left where they are, the road will be narrower and shaped by a beautiful allée. Since the road has two wide parking lanes and five wide traffic lanes, narrowing the road and decreasing vehicular space will improve it. The bike lanes will be protected from the traffic by parked cars and the trees, and even the water that goes into the drains and then the Housatonic River two blocks away will be cleaner. It's all good.

---

*Charles Marohn, the founder of Strong Towns (and the author of the excerpt in Chapter One from his book *Confessions of a Recovering Engineer*), calls the postwar financial system America used to fund roads a Ponzi scheme because local governments traded a near-term cash advantage for unfunded long-term financial obligations. See the Strong Town website for more information: https://www.strongtowns.org/contributors-journal/charles-marohn.

STREET TREES

The experience of revisiting great streets while writing this book confirmed many times over something we already knew: lining a street with parallel rows of classic street trees is one of the easiest and most productive steps in making a great street. Look at the old photos here of the Main Streets in Nantucket and Great Barrington, with their glorious canopies made by classic elms (Figure 2.73). Or look at the contemporary photos of Queens Road in Charlotte, North Carolina, which has a forest of mature oaks planted in seven rows (see the opening spread at the beginning of the book).

Here is a short description of how to make these beautiful allées today. First, select a tree species that will grow to a majestic height and form a canopy over the street. This generally means avoiding new hybrids. "On the whole, old species drawn from regional stock or old hybrids are better than the new hybrids for creating elegant canopies," landscape architect Douglas Duany says. "Many hybrids invented after 1950 right up to today develop problems over time, and—quite intentionally—few match the shapes of the old trees."[52] You can see these newer hybrids on roads all over America. Some grow tall but not wide. Others have dense balls of branches that never reach out across the road the way traditional street trees do. Some, loved by many current professionals, never grow tall *or* wide.

The new hybrids are sometimes recognizable by their names. This is not a hard and fast rule, but when a tree has a complicated and silly name like Espresso Kentucky Coffee Tree or Big Tooth Rocky Mountain Glow Maple (actual names chosen almost at random while looking at the list of trees at a tree farm), then it is a new hybrid that is likely to have an anti-traditional shape. Prewar hybrids have more straightforward names like Marshall's Seedless (a green ash that can be a good street tree).

We learn the characteristics of new hybrids from long observation in the field. The problems and weaknesses of the newer trees are therefore unknown until we have experience with them over time. That is why it is safer to choose more familiar hybrids—including ones like American sycamores, which are related to the London Plane trees we see in Great Britain and on the great boulevards of Paris and Barcelona. (The London Plane in turn is thought to be a sycamore hybrid that first appeared in Moorish Spain, where Middle Eastern sycamores and American sycamores were planted in close proximity. Another theory says they were developed in the royal gardens of England.) Time has taught us that among the trees that produce great canopies, the American sycamore and the London Plane are two of the heartiest (Figure 4.109).

Figure 4.109: Exmouth Market, London, England. London Plane trees in London. View looking southwest towards Farringdon Road. At the southwest end of Exmouth Market, what might have otherwise been an ordinary tangle of intersecting roads becomes a small shaded square, under the high canopy of trees that emerge directly from the pavement.

We learn the characteristics of new hybrids from long observation in the field. The problems and weaknesses of the newer trees are therefore unknown until we have experience with them over time.

Second, older street trees remain healthier and more resistant to disease if they are planted in groups. Therefore, they should be planted in ditches rather than boxes. In nature, most street trees are flood-plain trees that grow together in what are called "single stands." Flood-plain trees have the ability to survive periods when the oxygen supply to their roots is low. In the city, that helps them to resist the stress of compact urban soils. Equally important, in their single-stand natural habitat, the roots intermingle. That helps the trees communicate and protect one another against disease; a nutrient system produced by the trees feeds them when disease attacks. The nutrients are weakest directly under the tree, which means the planting boxes we typically use harm the trees in three ways: first, by restricting the spread of their roots and limiting the amount of nutrients and water the trees can get from the earth below them; and second, by preventing them from getting nutrients from other trees. Last but not least, any large tree needs to be able to spread its roots over an area larger than the tree box as it matures, or the tree is vulnerable to toppling over during storms with high winds.

Because it is best—aesthetically and ecologically—to think of trees in forests rather than as individual trees, street trees should be planted in ditches that approximate their condition in the woods. Dig and loosen or "decompact" a ditch the length of the block, place the trees in the ditch on a stable foundation, and backfill it with local dirt as quickly as possible. Water and gently tamp the soil but never compact it. There are various methods for lining the sidewalls of the ditch to prevent the roots growing horizontally near the surface. That prevents conflicts with utilities and roots raising the sidewalk, but may also limit the reach of roots looking for water. There are construction details for preventing compaction around the tree, and products such as Silva Cells have functional advantages over using tree boxes or the techniques described above. Silva Cells provide enough space for the roots and let the trees communicate while controlling where the roots go. Experience shows street trees planted in Silva Cells can be as healthy as trees with no restrictions.

A community that wants an allée may find that the professionals involved in the design oppose planting trees in a row or allowing street trees that make a canopy over the street, dismissing the idea as impractical and old-fashioned. Here, therefore, is a short summary of the benefits of a high deciduous canopy for an urban street.

On a wide road like Main Street in Great Barrington, Massachusetts, the trees are relied on to visually define a space that is too wide for the buildings to shape. The high canopy also solves solar problems: in the winter, bare branches reduce glare and sharp winter winds; and in the summer, the leaves shade the asphalt, which is best climatically and aesthetically. In addition, the high canopy shields the upper floors of buildings from viewers on the street and from traffic noise and pollution. Add to that the advice from retail experts like Robert J. Gibbs, who say to plant only trees that will rise above the storefronts, to avoid costing the merchants sales.

Unfortunately, the trees most engineers, arborists, and landscape architects propose cannot solve those problems, nor make the classic American tree-lined street we all picture in our mind's eye. It is important to understand that when an engineer or arborist proposes replacing a mature allée with a random mixture of popcorn trees and small ornamentals (Figure 1.31), they are proposing a radically different street. The justifications for such proposals vary, depending on the agendas of the professionals, but the unfortunate fact is that most of the professions involved in street design do not value the classic tree-lined street. If we agree with the urbanist's belief that these classic streets and their stunning canopies make great places, however, we can weigh the criteria differently and we may arrive at very different conclusions.

Traffic and roadway engineers traditionally have not liked large trees, because speeding cars are damaged when they hit them. (This is starting to change: the *CNU ITE Street Design Manual*, sponsored by the Federal Highway Administration and the Environmental Protection Agency, says that in urban contexts large trees are contextually appropriate. Drivers should not be speeding on urban streets, and trees simultaneously have a traffic-calming effect and protect pedestrians from errant cars.) Engineers also know that there is more public interest in saving mature trees than smaller ones and that it is easier to rebuild streets if the drivers of the heavy equipment used to do so do not have to be careful about killing the trees in the process. Thus, the engineers can kill two birds with one stone by replacing trees whenever they do major roadwork. For all these reasons, they often periodically replace street trees before they have matured.

Trained to control variables, engineers prefer to place trees in concrete planting boxes, so that they can contain the roots. This method has the added advantage, from the engineer's point of view, of limiting the root spread and thus the size of the tree: beyond a certain size, the tree starts to fail. As we have seen, this approach also limits the trees' resistance to disease. Last but not least, if the trees are in a box, the engineers can specify their preferred method of compacting soil when they work around the tree.

Arborists have different reasons for opposing traditional allées. They typically argue for replacement on the basis of disease and blight—and now climate change—as the primary reasons for putting in new tree species. Urbanists resist this way of thinking for several reasons. As already pointed out, the mixture of trees recommended by many arborists will never produce the great canopied street that makes the risk worth taking. Second is the fact that the risk is often small. Older Americans remember the terrible effects Dutch elm disease had on our towns and cities, because the elm was our greatest street tree. But while these blights can be terrible, they are rare.

In the long run, all trees age and have to be replaced eventually, and in that process, there is more than one way to introduce variety. The traditional method is to phase new trees in over time and, as discussed, to plant them in stands so that they can support each other. In the rare situation where disaster strikes, experienced gardeners will tell you that working with Mother Nature always involves dealing with problems and change. In the end, having a great stand of trees that has to be replaced is better than never having a great street.

Remember that the same professionals who say disease is a terrible problem may periodically recommend replacing all the trees for other reasons. Drastic tree-clearing is a philosophy that developed with Modernism. Modernist design frequently advocates the opposite of what was traditionally done and looks for a functional basis on which to build a new approach: as a profession, arborists tend to share that attitude. The cynical point out that arborists make money by replacing trees, but you do not have to be a cynic to acknowledge that while we have hundreds of years of knowledge about the old tree species, there is no significant body of knowledge about the new hybrids' problems over time. The dominant philosophy in contemporary landscape architecture is also a Modernist attitude that on principle opposes traditional design and placemaking. For decades, New Urbanists, who try to focus on what works rather than what is new or old, have been complaining that many landscape architects will not plant two trees in a row, because that is the traditional way of designing streets. Travel the world looking at our great tree-lined historic streets, and you will find that only a handful of them have a diversity of tree type, and that almost all have trees planted in parallel, symmetrical rows.

Some New Urbanists suggest introducing variety by using trees of similar size and shape on the same block, but that is not good for single-stand trees. Perhaps a greater number say it is better to introduce variety every other block or two rather than within each block, for both functional and aesthetic reasons. They value harmony and coherent unity and believe that a forced mixture of trees will never be as elegant as the great allées that were once the default standard for American street design.

Update: Great Barrington Redux

"The best time to plant a tree was twenty years ago. The second-best time is now."

— African proverb

Great Barrington's Main Street should be a place where people want to get out of their cars to shop, eat, and socialize—under a canopy of trees. But a year or two after the publication of the first edition of *Street Design*, massDOT and the town went ahead with what the publication *Better Cities and Towns* called "The massDOT Chainsaw Massacre."[53]

"The best time to plant a tree was twenty years ago," the Zambian-born economist Dambisa Moyo wrote in the book *Dead Aid* (apparently quoting an African proverb). "The second-best time is now."[54] In other words, it is not too late to fix the problems the recent rebuilding of Main Street brought to town. Great Barrington's Main Street has lost the curb appeal that helped make it *Smithsonian*

Figure 4.110: Main Street, Great Barrington, Massachusetts. mass-DOT, 2016. Photograph looking north on Main Street. The east side of Main Street after mass-DOT cut down the Bradford Pear trees and installed new sidewalks, lights, and street furniture. Compare this photo with the Before view in Figure 4.106.

magazine's best small town in America. "That's just aesthetics," some will say—including a few who contributed to the design decisions that make the new Main Street so ugly—but what real estate brokers and developers call "curb appeal" is not just aesthetics. It has economic and social value. Curb appeal describes good design, which affects how we feel about our cities and towns.

Studies by groups like the Yale School of Forestry and the National Association of Realtors show that majestic street canopies like the one Great Barrington used to have increase retail sales and real estate values. Surveys show people like street trees. The book *The Happy City* establishes that beautiful, mature trees increase our day-to-day happiness, and a growing body of research in cognitive science is beginning to record the data behind these effects.

After massDOT finished its reconstruction, the number of teenagers hanging out on Main Street visibly declined. Restaurant business went down too (several stopped serving lunch), and stores were less crowded. On the other hand, out-of-town-restaurants that people drove to had more customers. The town needs a post-occupancy survey, comparing both sales receipts and public attitudes before and after the rebuild.

For now, we can talk about some of the design details, which is what this book is about. We know from studies, surveys, and cognitive testing, that people like simple, harmonious, well-proportioned streets with good materials. The cheap concrete and concrete bricks, the oddly-shaped planters with attention-grabbing plantings, and the curbs that move up and down and in and out are poor urban design and weak placemaking. The highway-scaled lamps and oversized light poles are appropriate for exurban commercial auto sewers, but not for places where people walk next to them. And so on and so forth, right down to all the stripes and arrows in the roadway, bolder and brighter than the ones they replaced. The simpler and less expensive sidewalks in the town before the rebuilding were places where people were more inclined to walk.

Great Barrington's Main Street has lost the curb appeal that helped make it *Smithsonian* magazine's best small town in America.

The concrete in the new sidewalk was more pleasing to the senses. Then, the curbs jump in and out, the concrete bricks are frequently arranged in a way that draws attention to the fact that they do not align with neighboring shapes like the tree pits, the trees have a variety of shapes and sizes, the new pipe rail railings look cheap and are visually disruptive, and the new lights are large and more appropriate for a highway than a town center. It is the type of design that makes Fred Kent say, "Streetscape is a dirty word."[55] A less expensive sidewalk, with fewer materials and better-quality concrete, would make a better place for people.

The highway-scale pole in Figure 4.111 is typical of the lack of thought given to placemaking. The pole and the bolts holding it in place are inappropriate for a town center where people walk, as is the electrical box. Even if the plants grow in, the details will still be unattractive, just like the odd raised curbs. The galvanized metal is the wrong material for lights in a town center, and they look even stranger when painted "historic" poles are sometimes introduced. The yellow high-pressure sodium bulbs used are considered out of date for town centers as well. Many American municipalities retrofit their streets with LED lights, which use less energy, and can be adjusted for a more pleasing light.

National retailers know that replacing the wrong light in their stores with the right "warm" bulbs significantly increases sales per square foot, because people are more attracted to places with warm light. The same principle makes us more likely to stand in front of Baba Louie's Pizza at night to catch up with the neighbors.

In Figure 4.111, we can see the large corner radius between Elm Street and Main Street (meaning the curve that connects the curb on Elm Street to the curb on Main Street). That may have been one of the justifications for the raised planting bed, since large corner radii can interfere with pedestrian ramps. Large corner radii speed up and allow large trucks to turn the corner without going up on the curb. But high speeds are inappropriate here, and trucks do not need large corner radii when entering a wide street (Main Street) from a one-way street (Elm). The sidewalk detail is poor urban design because it takes a large piece of the sidewalk away from the pedestrian, making it narrow, and produces an awkward, ugly, and uncomfortable place.

Also odd is the fact that people on foot now need to push what is commonly called a "beg button" to legally cross the street. On a one-way, low-volume street like Elm Street, it is unusual to give priority to cars like that, particularly when the busier, wider Main Street has the opposite treatment, giving pedestrians the right to cross at all times.

The good news is that legally, Great Barrington owns Main Street. In most states, the state controls what happens to state highways as they go through towns, but as one can see from signs at each end of town, Great Barrington owns, and, in theory controls, the street between the bridge over the Housatonic River at the north end of town and the police station at the southern end.

We say "in theory" because massDOT controls the funding. But Massachusetts, almost uniquely among states, has a traffic code that differentiates between the design of town centers and state highways, and Great Barrington's Board of selectmen and its citizens have the legal right to make the street work for the town. If surveys and tax receipts show that the new street is making commerce and public life worse, logic says that the town should do what it can to make Main Street a place where people want to be.

Figure 4.111: Main Street, Great Barrington, Massachusetts. Photo. The galvanized metal pole and the electrical box are not appropriate for places where people are walking nearby.

Main st
Elm
ONE WAY

Figure 4.112: Stone Avenue, Greenville, South Carolina. Dover, Kohl & Partners, 2011. Watercolor rendering, aerial view. Before: Existing conditions in 2011. Decades of flight to the suburbs and road widening left the North End with substantial vacant property ready for infill and revitalization. © 2011 Dover, Kohl & Partners

Stone Avenue, Greenville, South Carolina

Dover, Kohl & Partners, 2011

Retrofit: From "Auto Sewer" to Neighborhood Main Street

Main Street

Just outside downtown Greenville, the city's North End streetcar suburb is a busy place again. A generation of new owners is restoring historic houses and starting up creative businesses.

The primary street in the North End, Stone Avenue, was historically a residential corridor, but it was widened in the late 1940s as a U.S. highway route. Over time, the avenue's auto-oriented design caused the corridor to decline; residential properties were retrofitted for office and low-level retail uses, and Stone Avenue became an auto sewer. In recent years, however, business owners and homeowners in the surrounding neighborhoods began incremental, grassroots efforts to reclaim the street as the centerpiece of the neighborhood.

The flaws in urban design that plague Stone Avenue are not unfamiliar. At a glance, it appears that it is unsafe to bike and walk along Stone Avenue and North Main Street, and it probably is. The street also has narrow sidewalks, overly wide traffic lanes, almost no appropriate street trees, a lack of building enclosure, a limited mix of land uses, and, most damaging of all, high-speed traffic. But a few key physical changes are bringing the street back to life (Figures 4.112–4.115).

The Stone Avenue Master Plan (2011) called for creating a mix of housing, retail, office space, civic and public open space, all located within a short walk of one another, as the first step to making this part of the city more livable. With City encouragement, private

Figure 4.113: Stone Avenue, Greenville, South Carolina. Dover, Kohl & Partners, 2011. Watercolor rendering, aerial view. After: The intersection of Main Street and Stone Avenue would reclaim its role as a physical and economic hub for the city, with new buildings positioned to frame the street spaces. As of 2024, a number of the buildings depicted in the rendering have been built (following the urbanism, but not the architecture). © 2011 Dover, Kohl & Partners

investors have since begun enthusiastically adding the missing elements on a number of North Main Street and Stone Avenue addresses. Several blocks within the plan's study area have been redeveloped, adding immensely to the real estate value (and tax base).

> The public-realm ideas have not been implemented, at least not yet. Instead, they were met by stubborn resistance and outright ridicule by engineers at the South Carolina Department of Transportation.

However, the plan also called for ensuring that an interconnected street system binds the addresses together, giving pedestrians easy access to their destinations. Narrowing the width of travel lanes to ten feet was a priority, to reduce traffic speeds and make the street easier to cross (Figure 4.116). Finally, the plan said the streets that connect these destinations must be designed for pedestrian use, with generous sidewalks, shade trees, and street-oriented buildings; traffic speeds must be lowered to make people feel safe using the street. Unlike the private components of the Stone Avenue plan, the public-realm ideas have not been implemented, at least not yet. Instead, they were met by stubborn resistance and outright ridicule by engineers at the South Carolina Department of Transportation. On the adjacent segment of Main Street, perfunctory bike lanes were installed, after great debate, but these are the curiously standard kind in which the doors of parked cars swing into the path of bikes (earning the derisive nickname "door lanes" among cyclists).

Figure 4.114: Stone Avenue, Greenville, South Carolina. Before: Existing conditions in 2011. Curb cuts for driveways are nearly continuous, and the scene is discouraging to pedestrians. © 2011 Dover, Kohl & Partners / UrbanAdvantage

Figure 4.115: Stone Avenue, Greenville, South Carolina. Dover, Kohl & Partners, 2011. Computer simulation. After: The plan calls for a step-by-step transformation of the street, converting to a narrower cross-section with on-street parking, street trees, and proper sidewalks faced by storefronts. The center turn lane is designed as a textured "safety strip," which allows the turns and keeps the street wide enough for emergency vehicles to navigate around traffic when necessary, but keeps the ordinary motoring speeds slow on all other occasions. © 2011 Dover, Kohl & Partners / UrbanAdvantage

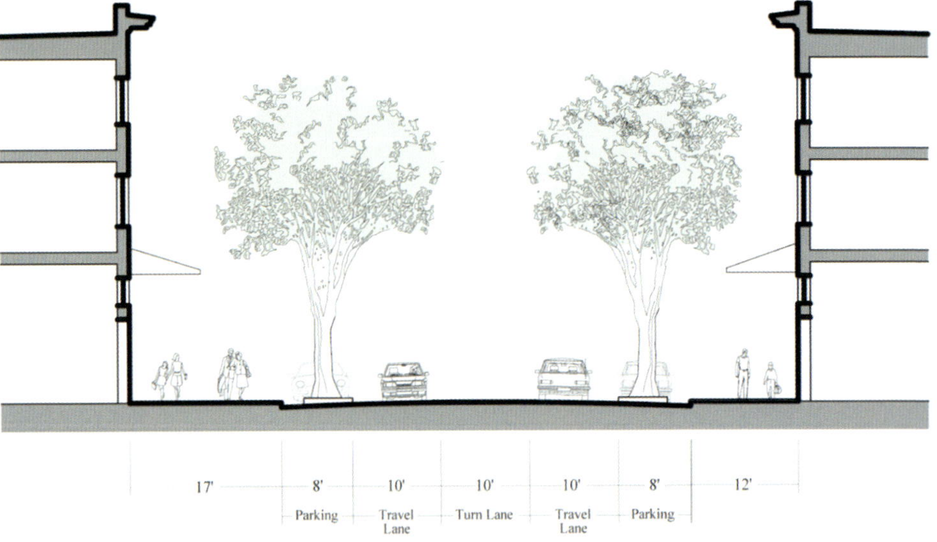

Figure 4.116: Stone Avenue, Greenville, South Carolina. Dover, Kohl & Partners, 2011 Section. © 2013 Dover, Kohl & Partners

ROUNDABOUTS, MONUMENTS, AND SLOW SPEEDS

The design of many of the streets in our most walkable cities, towns, and neighborhoods should begin and end with making places where people want to be. Ironically, many of the "Complete Streets" Americans build are incomplete when it comes to both placemaking and beauty. Too many designers of Complete Streets still give the car top priority, on a road that puts traffic flow first. This may seem fitting for suburban environments where most people drive and few people walk. But today we have suburban-style traffic-calming techniques that do not work for the creation or restoration of walkable places being used in towns, cities, and new traditional neighborhoods in the name of Complete Streets. As a nation, we are still learning how to make real complete streets that promote walkability as much as traffic flow. Figure 4.117 shows a roundabout in Okemos, Michigan, that Cleveland, Ohio's City Planning Commission uses to illustrate Traffic Calming in their Online Glossary. On the whole, the glossary is very good,[56] but there are a few reasons why the photo is not the best choice to illustrate traffic calming in the walkable parts of Cleveland (or Okemos, for that matter).

- There are many cars on the road today that can go through the roundabout at high speed: that is not traffic calming. The photo shows a design that puts the flow of cars above the comfort and well-being of the pedestrian. For a pedestrian crossing an urban street, the geometry of the corner radii should force the car to slow almost to a stop.

- What differentiates the design of a modern roundabout from old designs for traffic circles is that the "splitter island" at the entrance to the circle should narrow the lane and direct the driver to the right, slowing the car. "Slowing" is relative, however.

- The traffic lanes are wide: that is good for speeding cars, bad for pedestrians crossing the roads, and bad for making the street a space where pedestrians feel comfortable.

- The yellow striping is visually aggressive—more appropriate for highway traffic than speeds in neighborhoods where people walk and cycle.

- The large signs are designed to be read at fifty-five miles per hour, alerting the pedestrian that he or she is not in a pedestrian space.

- The sheer number of signs warning us of hazardous road conditions for cars also tells us we are in auto-dominated space.

- Things that people like, such as trees, have been removed.

- It is generally better for placemaking if the circle is not a perfect circle and has something more interesting to look at than dying grass.

- The curves on the outside of the traffic circle have no visual relationship to the circle at the center because the traffic engineer was thinking about moving cars rather than creating a place. One goal should be shaping space: the simplest way to do that with circles is to make concentric circles. But the shapes do not have to be circles. What's important is that the shapes define and make comfortable spaces for the pedestrian as well as the car.

As a photo, Figure 4.117 can make a good first impression. Everything is shiny and new, the design is orderly, and the gray of the asphalt goes well with the white concrete and the green grass—and, at this scale, with the white and yellow striping. But for the pedestrian standing on a splitter island, the auto-based geometry of the island subtly tells that person that he is standing in a place made for cars rather than humans. Add the bold striping and the bold signs, and the message is no longer subtle, even if we are so used to auto-dominated design that we do not consciously have that thought.

Traffic engineers control the design of our streets, always given great authority by states, counties, towns, and cities. But traffic engineers think about engineering rather than design. If they are going to retain this power, they need to understand placemaking and civic art, not just vehicular capacity.

Conventional traffic circles have stop signs or stoplights on entry roads perpendicular to the circle. Looking into the history of roundabouts, one discovers that the primary reason Departments of Transportation and Departments of Public Works reflexively put them in is because we have many bad drivers who roll

Figure 4.117: Okemos Roundabout, Okemos, Michigan. Aerial view. A shiny, new roundabout makes a strong impression—but it is not necessarily good for pedestrians, cyclists, or residential character. *Courtesy of the Ingham County Department of Transportation and Roads*

through stop signs and cause collisions. If a roundabout is substituted for a stop-sign system at a circle or a regular intersection, T-bone and rear-end crashes go from significant numbers to almost zero. Auto collisions become angled collisions, usually at low speed, that are less serious than the T-bone and rear-end crashes. In addition, throughput rises, and cars do not have to stop and start, which is good for increasing gas mileage.

Roundabouts, in other words, exist primarily for the benefit of cars and their drivers. Roundabouts can be uncomfortable for inexperienced or cautious cyclists, and pedestrians are moved away from the circle for crossings. This is because drivers approaching the circle and in the circle are usually looking to their left rather than in the direction of pedestrians crossing on their right. For the engineer, the priorities of pedestrians are secondary to the free flow of traffic.

Traffic Circles & Civic Art

There was a time before the era of Organized Motordom when traffic circles and roundabouts served everyone. Columbus Circle was built in Manhattan in 1905, when cars and pedestrians still shared the roads. The center of the circle was reserved for streetcars, their riders, and other pedestrians, while cars circled around the perimeter. The first modern roundabout in England had a center island that was

a refuge for pedestrians. Designed by the noted town planners Parker & Unwin and built in Letchworth Garden City in 1909, it was a response to increasing traffic. Today, as we have seen, pedestrians are typically banned from crossing the circle, while pedestrian crossings are pushed away from the circle.

The quiet, flat backstreets of Coral Gables, Florida, can be good places for walking. Around the time we wrote the first edition, however, the city-built roundabouts there. They were some of the best-looking roundabouts in the country—with low, planted circles in the center; few or no signs; and subtly-colored splitter islands—but they seem to simultaneously speed up the cars and make the roads slightly less pleasant for walking.

Many of these streets have no sidewalks because they have such low levels of traffic that they were shared spaces long before the Dutch named the concept.[57] With long allées of live oaks and banyan trees that form canopies over the street and keep the harsh sun at bay, the streets in the flat landscape have lovely vistas that highlight the slight topographical changes where they do exist. Even here, though, the scale is made for the car. New bumpouts interrupt the vistas and hide the streets on the other side, while the splitter islands and engineering geometry make the streets' intersections feel like a place where the pedestrian is unwelcome. The change can seem subtle, but one effect is that when a pedestrian at one of the roundabouts sees a car approaching, he or she knows that the driver probably will not see them until the car is exiting the roundabout. And these, the attractive roundabouts of Coral Gables, are among the simplest and most lowkey in America.

The old streets in Coral Gables show how differently designers approached these issues one hundred years ago. Most of the planned new town was built at the scale of the car, and some of the roads are very wide. But the traffic circle at De Soto Plaza, for example, has a large, stone fountain with a tall monument in the center (Figure 4.118). It appears to have been intended as a shared space when the city was designed in the 1920s, because the fountain is attractive in both meanings of the word, and it has a space around it where pedestrians may once have felt welcome to venture. Model Ts of the time topped out at about 20 horsepower. But our modern cars fly through the plaza, and it can be nearly impossible for pedestrians to reach the fountain. (Dover Kohl Principal Kenneth García illustrated a concept for modifying the De Soto Plaza circle to peel back the excessive pavement, reduce speeds, welcome back people, and improve safety, all while incorporating just enough of the lessons learned from the experiments with modern roundabouts in recent decades. See the before-and-after comparison in Figures 4.119 and 4.120.)

In the quieter, shared-space streets nearby, there are sometimes large grassy medians with no curbs that calm traffic the way roundabouts do but that also work as Civic Art, making beautiful places where people walking feel comfortable. They give variety and richness to the plan and varied experiences to the pedestrian. Occasionally there are trees in the middle of the road. Drivers manage to avoid hitting them or suing the city over them, despite the worst fears and warnings of traffic engineers of what will happen if you expect intelligent behavior from drivers.

We are *not* saying modern roundabouts should never be employed in street design or neighborhood design. When a roundabout is used for beauty, placemaking, and safety rather than just as a traffic-control instrument, it can be a welcome addition. But while modern roundabouts can be appropriate in some circumstances, they are often built where they should not be. After being banned for decades for the wrong reasons, roundabouts have recently become the embodiment of the old adage that if the only tool you have is a hammer, everything looks like a nail. A well-designed, modern roundabout with minimal striping and signs and a strong sense of Civic Art should be in every urban designer's toolkit, but it should be used sparingly.

Things in the Middle of the Road

In the nineteenth century, it was common to put monuments in the middle of roads, and many of them are still standing in the twenty-first century (Figures 2.80

Figure 4.118: De Soto Plaza Fountain, Coral Gables, Florida. Denman Fink, 1925. *Courtesy of Kenneth García*

Figure 4.119: De Soto Plaza Fountain, Coral Gables, Florida. Denman Fink, 1925. Before: Existing conditions. Even though there is a piece of civic art, the roundabout still looks and feels like it is meant to encourage large traffic speeds and discourage pedestrians.

Figure 4.120: De Soto Plaza Fountain, Coral Gables, Florida. Dover, Kohl & Partners, 2017. After: Computer simulation, showing proposed redesign. This design uses safe crossings and gives the center of the circle more space around it for an active plaza. A circle that has less asphalt and more people. © *2017 Dover, Kohl & Partners*

and 4.121). Lawyers and engineers require that we give them more protection today, in the form of fences, wide foundations, bold striping, and signs. But many of the perils of monuments go away if we just slow cars down in walkable areas—which has the extra benefit of letting the pedestrians live longer. Drivers going twenty miles per hour should not need striping, signs, splitter islands, or a fence to avoid hitting either objects or people. Their cone of vision widens as they slow down and they also have more time to react (Figure 1.55). Alternatively, finding out which drivers are unable to avoid hitting ten-ton pieces of granite will enable us to get them off the road.

Many of the perils of monuments in the road go away if we just slow cars down—which has the extra benefit of letting the pedestrians live longer.

In our current world of Happy Motoring, ubiquitous traffic engineers, and overzealous lawyers, that approach might seem dangerous. Like so many other factors in making better streets and places, it depends on slowing down the car. But places like Seven Dials show that it can be done (Figure 3.66).

Figure 4.121: Court Square Fountain, Montgomery, Alabama. Frederick MacMonnies, sculptor, 1885. When Hall Planning and Engineering and Dover, Kohl & Partners redesigned Court Square in 2007, it became an early example of a modern shared space in America. See the Dexter Avenue case study on page 345. *Courtesy of Peter Fouts, Fouts Commercial Photography*

Figure 4.122: Madison Square, New York, New York. Circa 1899. Photograph looking north from below Twenty-third Street. The Dewey Arch at the intersection of Broadway and Fifth Avenue. *Library of Congress / Detroit Publishing Company Photograph Collection*

Main Street, Rosemary Beach, Florida

Duany, Plater-Zyberk & Company, 1995

Main Street

Rosemary Beach is one of two resorts designed by Duany Plater-Zyberk & Company (DPZ) on the Florida Panhandle after the success of Seaside, where Robert Davis multiplied the value of his land more than a hundred times. Less than ten miles to the east, Rosemary Beach was developed by Leucadia National, a large holding and investment company. Their first plan for the site was more conventional than Seaside's, without a town center and with less emphasis on walkability. But after Leucadia learned that comparable houses in the area were selling for less than a third of the price of Seaside houses, the company hired Duany Plater-Zyberk & Company to develop a town plan for their site.

Selling exclusively to the rich was not Davis's intention when he started Seaside. Leucadia National, on the other hand, wanted to maximize the return on its investment, and market analysis convinced the company that building its development in the form of a town would give it the highest returns. The scarcity of placemaking in American development during the last few decades means that many of our best designed places—new as well as old—get a significant sales premium.

By the time Leucadia hired DPZ in 1995, many developers and house builders had copied what they thought was the Seaside style of architecture. Few of the new site plans, however, were designed for walkability and community. The developments repeated the conventional Florida model, with prominent garages on the street, direct access to the houses from the garages, and little or no emphasis on creating a comfortable public realm or making places to socialize. It was assumed that instead of walking downtown for dinner, people would get in their cars and drive to a restaurant.

The DPZ plan for Rosemary Beach made a downtown that straddled the main road through Rosemary Beach, with restaurants, stores, offices, and public meeting places. They designed the streets to encourage walking to the town center and to the beach. The first block of Main Street, which leads down to the Gulf from the main square, is shown in Figure 4.123. Both at the top of the block and the bottom of the block the land is flat, which visually accentuates the slope of the street in between. Where the slope of the street changes, the angle of the street and the individual buildings along it do so as well, drawing one's attention to their decreasing height as they go down the hill, and thus drawing attention to the fall of the hill itself. The deflected vista at the end of the street means that no one seeing the street for the first time knows if the slope continues beyond the guesthouse that terminates the view. In fact, there is a flat lawn there, which makes a comfortable place to gather by the Gulf.

Figure 4.123: Main Street, Rosemary Beach, Florida. Duany, Plater-Zyberk & Company, 1995. Photo looking south towards the Gulf of Mexico. The vertical facades of the two buildings that deflect the vista visually focus attention at the end of the vista.

NEW STREETS IN THE NEW TOWN OF CAYALÁ, GUATEMALA / MARÍA SÁNCHEZ

Léon Krier and Estudio Urbano, 2002–Present

Main Street

Cayalá is an extension of Guatemala City, built according to traditional urban and architectural principles. The town is on a plateau some four kilometers east (2½ miles) of the historic center of Guatemala City. Deep ravines surround Cayalá, dividing the site into four quarters: Paseo Cayalá, Nogales, Socorro Bajo, and Socorro Alto. The quarters are connected by a four-kilometer (2½-mile) main street that runs along a natural ridge. Altogether, the site comprises approximately 538 acres.

Planning began in 2002, when Estudio Urbano invited renowned master planner Léon Krier to join the effort. The team persuaded the landowner-client to embrace our vision for a vibrant, humanist new development. Master planning began with a charrette in 2003.

The goal was to create a humane, mixed-use, pedestrian-oriented urban environment that fosters a strong sense of community and pride of place in its residents. It is meant to be an open city for an open society, welcoming to everyone, regardless of income, social class, age, religion, or race. It has successfully generated local employment opportunities and a vibrant commercial and civic life.[58]

Cayalá represents a dramatic change in the architecture and urbanism of Guatemala built in the last decades, which is afflicted by sprawl, high-rise buildings loosely arranged in the landscape, lack of pedestrian streets and squares, and the growth of the isolated and self-referential single-use walled developments (Figure 4.124).

Figure 4.124: 10ma Avenida, Guatemala. Léon Krier and Estudio Urbano, 2002–Present. Photo looking north. The scene illustrates what the designers call "vernacular geometry," with the crank of the street alignment producing both deflected and terminated vistas. *Courtesy of Vicente Aguirre*

Quarters and Neighborhoods

Each quarter in Cayalá has a distinctive center, well-defined edges, and interesting views to the ravines around them (a design opportunity usually wasted in Guatemala City). The quarters come together around a central square, an east–west main street, and a north–south street (Figure 4.125).

The distance from the center of a quarter to its edge is approximately equal to ten minutes of walking, an efficient, traditional pattern that allows people to easily fulfill their daily needs. To lower the number of cars on all streets, we incorporated a system of out-of-sight underground parking, with escalators in the arcades or in elegant free-standing pavilions that adorn the street.

Each quarter has four neighborhoods, designed around central squares reserved for walking. From the center of the square to the edge of the neighborhood is a five-minute walk. We created an irregular pattern of small urban blocks, a building height limit of three or four stories, and a walkable network of streets and boulevards.

Figure 4.125: 10ma Avenida, Guatemala. Léon Krier and Estudio Urbano, 2002–Present. Aerial photo. 10 Avenida is the main street (*calle principal*) in the center of the photograph. *Courtesy of Google Earth*

We ensured pedestrian priority in Cayalá through the scale and layout of the streets.

On the main street, we introduced the concept of "shared space." Our shared space works without physical divisions, signs, or traffic lights. The speed of vehicles is controlled by the urban and civic character of its geometric configuration, the architecture, the landscaping, the street furniture, the pavement, the planting, and lighting. We carefully connected and integrated public buildings and spaces to promote social interaction and reinforce a sense of community. The intention was to create a beautiful, secure place for residents and visitors to enjoy for generations to come.

There are many ways to walk through the town, which has streets, avenues, boulevards, alleys, walkways, squares of various sizes, parks, and gardens. All have trees that cast pleasant shadows. To express the level of importance and hierarchy of the streets, we sized them differently and used a variety of materials.

Following our team leader Léon Krier, we used natural principles of design in creating the streets, because beautiful urban landscapes come from understanding and imitating nature. In nature, there is no such thing as a straight line and patterns have individual character, size, and proportions that while being similar and repetitive are never identical clones.

The streets are irregular, forming intimate and inviting urban spaces instead of corridor-type streets. Wide lanes make it easier for cars to speed, while irregular streets, narrower in some parts, naturally slow drivers down.

Instead of four-way "cruciform" intersections, we incorporated T-junctions wherever possible, to avoid congestion and collision points and ease circulation. Pedestrian streets of lesser width go through blocks and connect with each other, creating a coherent, virtually car-free network in each neighborhood.

Like the main street and the central square, the boulevards, streets, and squares of the residential neighborhoods are paved wall to wall with cobbles and flagstones to reinforce the primacy of the pedestrian over the automobile. Green spaces and parking options are shared among neighbors, and wooden elements such as doors, gates, and screens provide privacy and visual interest.

The most interesting and authentic public spaces come from variety in the geometry, dimensions, and functional use of lots, which in turn results in blocks that possess a great typological and architectural variety. This variety and the front alignment of the different buildings to the streets and squares create a pleasant continuity and coherence in the neighborhoods.

The use of local materials and craftsmen has been of great benefit, as it improves the economy through the opportunity provided to the labor force, strengthens the sense of identity and generates employment.

We employed vernacular rather than Euclidean geometry in designing the streets, but then used Euclidean order to generate building forms. In our experience, combining is the most interesting and efficient way to create enduring, attractive, and affordable places (Figure 4.126).

Figure 4.126: Las Ramblas, Cayalá, Guatemala. Léon Krier and Estudio Urbano, 2002–Present. Photo looking south. Vernacular geometry is used in designing the streets, but buildings and objects are generated using Euclidean figures and order. *Courtesy of Estudio Urbano*

A Place of Hope and Happiness

The traditional city promotes health and well-being by giving priority to the pedestrian and public life. It nourishes our mind and social nature with squares and streets that function as meeting places where neighbors can become friends and perhaps feel like part of an extended family. It feeds our soul with beauty and gives us a home we can be proud of (Figure 4.127).

The careful design of every neighborhood contributes to the overall character of Cayalá. Traditional town planning of the sort employed here encourages residents to engage in public life, and contrasts dramatically with the uncompromising uniformity and antisocial structuring of many contemporary residential developments. To many Guatemalans, the new town has become a place of hope and happiness. It is one of the most visited places in our country (Figure 4.128).

Figure 4.127: 10ma. Avenida, Guatemala. Léon Krier and Estudio Urbano, 2002–Present. Photo looking south in 2015. Citizens gather to make "carpets" of colored powders, sand, and flowers on the *calle principal* for a traditional Christian procession. *Courtesy of Estudio Urbano*

Figure 4.128: 10ma Avenida, Guatemala. Léon Krier and Estudio Urbano, 2002–Present. Photo looking north. Christmas season in Cayalá. The new town has become one of the most visited places in Guatemala. *Courtesy of Grupo Cayalá*

RULE OF FRONTS AND BACKS

Street-oriented architecture is crucial to livable, walkable cities, and need not be difficult to achieve.

Like humans, buildings have faces and backsides with different purposes. In the real world (no matter what creative protests architects may offer to the contrary), buildings almost always have a *front* and a *back*. The front—the public facade—naturally belongs to the public space.

The architectural grammar of this building-to-street relationship is both recognizable and flexible. The front should speak to the community and shape the public realm (Figure 4.129). For example, the front should have the primary entry door or doors, and openings that allow occupants to overlook the space and provide natural surveillance. In this way, the organization of the building is legible at a glance to passersby, and the individual building communicates with the larger community. The late Dan Camp called this "the presentation face" of the building.[59] In a Main Street building, the front may be expressed with storefronts, awnings, signs, perhaps galleries or arcades over the sidewalk, perhaps balconies extending from upper floors. In a residential building or a house, the front may be expressed with a porch, stoop, dooryard garden, verandah, or the like. In a civic building, a main portico could be positioned on the front to communicate the relationship of the building to the street. There are hundreds of ways to design buildings that let the front facade do its job of establishing the public space, so the creative architect is not unduly burdened or constrained. But without these features, a building that turns its face away from the public space sends a message of disrespect. At a minimum, placemaking suffers, and one might also end up with one of those buildings in front of which people stare in confusion or pace up and down, looking for the door.

The right place for certain service and messy utilitarian functions is the back of the building and its lot, in the most private outdoor spaces (Figure 4.130). Necessary but unsightly features belong away from public view, like garage doors, garbage cans, loading docks, drive-through lanes, parking lots, and parking structures. When at all possible, backflow preventers, air conditioning compressors, and the myriad boxes and meters needed by telecommunications companies and electric

and water utilities should also be positioned away from the important spaces of the public realm. This is easiest when the master plan provides alleys or rear service lanes. In some other cases, where blocks are sufficiently small and streets plentiful, certain streets can be identified as what Andrés Duany and Elizabeth Plater-Zyberk call the "A" streets and be kept free of the back-of-house functions that are relegated to the "B" streets or alleys. Either way, the thinking about the Rule of Fronts and Backs best begins when the streets, blocks, and lots are first laid out, long before the individual building architect appears on the scene.

Figure 4.129: Glenwood Park, Atlanta, Georgia. Dover, Kohl & Partners and Tunnell-Spangler-Walsh, 2001. Rowhouses designed by Historical Concepts in 2006 face the public space with their fronts. *Courtesy of James Dougherty*

Figure 4.130: Glenwood Park, Atlanta, Georgia. Dover, Kohl & Partners and Tunnell-Spangler-Walsh, 2001. The garage doors and midblock parking areas for the rowhouses designed by Historical Concepts are on the alley side. *Courtesy of James Dougherty*

MARKET STREET, HABERSHAM, BEAUFORT, SOUTH CAROLINA / THOMAS LOW

Duany Plater-Zyberk & Company, 1997

Main Street

Adapted from the *Light Imprint Handbook,* (New Urban Press, 2008)

With a sizable town center, the new town of Habersham will serve as an urban hub for surrounding villages. Market Street and Le Chene Circle are located in the heart of the town center and contain a post office, a fire station, seven restaurants, and a few dozen live–work units that provide living space above street-level commercial space (Figures 4.131 and 4.134). Together, Market Street and Le Chene Circle help to distinguish Habersham as a community that values sustainable development.

Initiated in 1997, the design and construction of Habersham was a case study for Duany Plater-Zyberk & Company's Light Imprint initiative. Light Imprint provides a framework for the design of sustainable neighborhoods based on New Urban planning principles. Some of the infrastructure is based on low-tech practices for providing good environmental design. By approaching each site as a unique entity, a Light Imprint system of stormwater management creates compact, walkable neighborhoods. Habersham has over thirteen thousand linear feet of marsh frontage, making the site especially sensitive to pollution from stormwater runoff and yet an ideal candidate for showcasing Light Imprint strategies.

The 283-acre site is crossed by a number of small creeks that drain into the Broad River marshes. Seventy-three acres of the site are preserved for parks, common areas, and natural drainage basins. Mature vegetation along the marsh edge creates a natural windbreak and an inviting habitat for wildlife. Extensive ecological analysis and tree surveys were conducted at the beginning of the design process; as a result, wetland preservation and marsh buffers are an important part of the master plan.

Additional Light Imprint techniques found on Market Street and Le Chene Circle include narrow streets with sidewalks on one side, pedestrian spaces constructed with pea gravel (Figure 4.135), and large medians planted with trees. Because these pervious surfaces and their arrangement help to slow down, percolate, and cleanse the runoff on rainy days, they assist with proper stormwater management and drainage patterns. Such initiatives are present throughout Habersham but are adjusted according to context—whether in the center of town, where development is most dense, or at the suburban edge, where development is less dense (Figures 4.136 and 4.137).

Figure 4.131: Market Street, Habersham, Beaufort, South Carolina. Duany Plater-Zyberk & Company, 1997. Looking north.

Figure 4.132: Market Street, Habersham, Beaufort, South Carolina. Duany Plater-Zyberk & Company, 1997. Photograph looking southwest. *Courtesy of Habersham Properties*

Figure 4.133: Market Street, Habersham, Beaufort, South Carolina. Duany Plater-Zyberk & Company, 1997. Photograph looking northeast. *Courtesy of Habersham Properties*

Figure 4.134: Le Chene Circle, Habersham, Beaufort, South Carolina. Duany Plater-Zyberk & Company, 1997. Photograph looking southwest.

Figure 4.135: Le Chene Circle in Habersham, Beaufort, South Carolina. Duany Plater-Zyberk & Company, 1997. Looking southwest, showing pervious pea gravel path. The "Light Imprint" approach at Habersham does not call attention to itself as stormwater infrastructure, because it looks and feels like traditional Lowcountry placemaking.

Figure 4.136: Harford Street, Habersham, Beaufort, South Carolina. Duany Plater-Zyberk & Company, 1997. Looking west, towards townhouses in the center of Habersham.

Figure 4.137: St. Phillips Boulevard, Habersham, Beaufort, South Carolina. Duany Plater-Zyberk & Company, 1997. The transect observed: As one moves away from the town center, the street network and the physical designs relax into less formal arrangements.

LIGHT IMPRINT URBANISM / THOMAS LOW

Green Infrastructure and Community Design

Habersham is the birthplace of what became known as the "light imprint" technique.[60] Low Impact Development (LID) and Light Imprint both offer systems for greener, more environmentally friendly stormwater infrastructure. However, when adopted through standard ordinances LID can promote sprawl, since its primary solutions push apart the impervious surfaces cities always have and wrap buildings with more (albeit smaller) ponds and trenches. This takes more land than conventional practice for the same amount of development and disrupts walkability. Light Imprint uses many of the tools found in LID, including a range of more pervious surfaces, but Light Imprint is a system of context-sensitive treatments. Treating each site as a unique entity, a Light Imprint system of stormwater management creates compact, walkable neighborhoods.

The Light Imprint philosophy of stormwater management avoids expensive drainage systems and excessive pavement to let the water filter directly into the soil. Light Imprint addresses these issues through a transect-based stormwater management system that integrates community design with tools found in LID. Light Imprint introduces a tool set for stormwater runoff that uses natural drainage, traditional engineering infrastructure, and infiltration practices. These tools are to be used collectively at the sector, neighborhood, and block scale. Light Imprint offers context-sensitive design solutions that work together on the community level. The design solutions highlight tools that are appropriate from the rural to the urban condition.

In the *Light Imprint Handbook*, a transect-based matrix organizes over sixty tools and resources in a simple, useful form.[61] Choosing the right Light Imprint tools to use in a given situation relies on using the matrix, based on variables. These variables include soil hydrology, slope condition, climate, initial costs, long-term maintenance factors, and Urban to Rural Transect zones. Once these variables have been analyzed, a customized palette of tools specific to the project's needs emerges.[62] For example, since Habersham is located near the Atlantic Ocean, heavy squalls can produce a large amount of rain there in a short time. The region is also prone to rainfall accumulations from tropical storms and hurricanes, making stormwater management a key consideration. Most of the street paving in Habersham is asphalt. It is a relatively cost-effective and readily available material. However, the impervious surfaces are minimized by keeping each street as narrow as is practical given its function.

Light Imprint: Materials and Configurations at Habersham

- Paving
  - Wood Planks
  - Crushed Stone/Shell
  - Asphalt
  - Concrete
  - Pea Gravel
- Channeling
  - Vegetative/Stone Swale
  - Slope Avenue
  - Shallow Channel Footpath
  - Concrete Pipe
  - Gutter
- Storage
  - Retention Basin with Sloping Bank
  - Retention Pond
  - Landscaped Tree Wells
  - Filtration
  - Wetland/Swamp
  - Filtration Ponds
  - Shallow Marsh
  - Surface Landscape
  - Natural Vegetation
  - Constructed Wetland
  - Green Finger

CORE STREETS

Madison Square, New York, New York

New York City Department of Transportation, 2008

Retrofit: Tactical Urbanism and Reclaiming Pedestrian Space

Intersection Redesign

Core Street

Commissioner Janette Sadik-Khan made New York City's Department of Transportation the most innovative and creative in America. With the blessing of her boss, Mayor Michael Bloomberg, Sadik-Khan was one of the first big-city DOT heads in the United States to take the virtually absolute power of the DOTs over our public realm and use that power to make things better for the pedestrians and cyclists rather than the cars.* That seems like common sense in Manhattan, where almost 80 percent of the residents do not own cars and only 20 percent of the workers commute by car, but until Sadik-Khan came along, the city's DOT was little different from other DOTs. Like practically everywhere else in the United States since World War II, New York City gave the car primacy over the streets of all five boroughs. In the 1950s and 1960s, the DOT narrowed sidewalks and sped up streets, making them one-way, with wider lanes and staggered traffic lights, creating suburban-style auto sewers in the middle of Manhattan (see Figure 5.58 on page 635). Sadik-Khan emphasized high-profile changes in the opposite direction—most notably by closing parts of wide New York streets to cars and setting up tables and chairs so that people could claim the former roadbeds as public space.

One night in 2008, work crews arrived at Madison Square at 11 pm and began putting out bollards and chairs to create new pedestrian plazas for Madison Square, where Fifth Avenue awkwardly meets the diagonal of Broadway (Figure 4.138). They painted the street brown where it was closed to cars, added large boulders and painted stripes to warn cars away, and by the time the morning traffic came, all the work was done. New Yorkers started using the spaces and chairs immediately. Even when the DOTs had what seemed like unlimited power to widen roads and remove pedestrians, the agencies rarely acted that quickly. But that raises important points as we move away from the auto-centric planning and street design of the last few decades. First, we should remember that all the DOTs that gravely say today, "We must be slow and cautious about change" acted quickly fifty years ago, when the national program was to widen roads. Second, America should take advantage of what has come to be known as Tactical Urbanism, which tries out and tests change with paint and other inexpensive materials.[64]

Most of the changes made by the New York City DOT during Sadik-Khan's tenure were popular—not that you would always know that from the press. "There are not only 8.4 million New Yorkers, but at times 8.4 million traffic engineers," Sadik-Khan once said. "And we're, you know, very opinionated."[65] Under Sadik-Khan, the DOT built 400 miles of bicycle lanes, including one in front of the Brooklyn building where Sadik-Khan's DOT predecessor lives with her husband, United States Senator Charles Schumer. The former commissioner was a leading figure in a campaign to remove the offending lanes, but the campaign ended unsuccessfully in court.[66] Meanwhile, the work completed in Madison Square raised property values in an already rising market.

The pedestrian crossing at the intersection of Broadway and Fifth Avenue on the north side of 23rd Street was the longest in the city. The length of two football fields, it was an ugly and hostile environment for anyone walking, with exhaust-stained, delaminating white and yellow plastic sticks marking traffic lanes, and highway-scale striping that assaulted the pedestrian's senses. The DOT's overnight transformation was an immediate and enormous improvement. The ugly plastic sticks are still

---

*Sam Schwartz, aka "Gridlock Sam," deserves an Honorable Mention. He was Traffic Commissioner from 1982 to 1986, but he had less power than Sadik-Khan. When the Department of Traffic became the Department of Transportation, Schwartz was second-in-command, with titles of First Deputy Commissioner and Chief Engineer from 1986 to 1990. Over the years, Schwartz tried to introduce congestion pricing and bike lanes to New York City, but was unfortunately unsuccessful.[63]

Figure 4.138: Madison Square, New York, New York. NYC DOT, 2008. Satellite view. Tactical urbanism at its best. *Courtesy of Google Earth*

there in a new arrangement, but they are not as dispiriting when you know they are temporary and in service of the public good. (Postscript: But how temporary are they? See page 446.)

Sadik-Khan instituted a program of pedestrian plazas.[67] Most began as similar tactical interventions. Temporary elements gave the neighborhood an idea of how the new plaza would feel. Inexpensive materials, such as brown epoxy paint with gravel that covers the closed portion of the street dominate. Boulders from public works excavations serve as traffic barriers and temporary seating. Lightweight folding chairs and tables like the ones used in Parisian parks, let individuals and groups arrange the furniture as they liked—in a circle, or perhaps two alone, in the sun. Ideally, the plaza quickly becomes a popular place to sit.

When Tactical Urbanism Becomes Permanent

Taking street space from the car and giving it back to the people was a radical step after sixty-plus years of Happy Motoring. Presenting the changes to the public as a temporary *fait accompli*—a done deal, but reversible—was a brilliant maneuver, because from the moment New Yorkers had a chance to use the spaces, it was clear they loved them. The inevitable criticism from some quarters was overwhelmed by public acclaim for the changes at Madison Square. Visiting the square today, we can see not only that the changes were a good idea, but also that—by correcting a few problems illuminated by the test run—the permanent solution can be even better.

In the first edition, we showed a sketch plan with ideas for improving the temporary plan. Ten years later, with the "temporary" plan still in place and new actions by the DOT and new ideas for the surrounding streets bubbling up from different sources, we present a new design (Figure 4.141). First, the DOT modified their original plan for Broadway at 24th Street (Figure 4.140) to make the "first shared street" in New York (more on this below).[68] Then, the group Transportation Alternatives received New York City's support for a "Broadway Linear Park" plan to remove traffic from the street and make it a place for walking and biking.[69] Our proposal responds to both ideas.

Only a traffic engineer could love the way Broadway currently ends at 25th Street. The next two blocks look like highway off-ramps were air-dropped into the Manhattan grid (Figures 4.138, 4.140, and 4.141). Attempting to turn the machine-space geometry of the first block into a shared space doesn't work. Compare what the DOT calls the first shared-space street in New York to what was probably the city's second example, described below in "A Chicane Is Not a Shared-Space Place." In the next block, 24th Street has another large curve, leaving a large no-man's land between the Broadway curve and the curve of 24th Street. The left-over space is painted brown, marking it as a place for people, but it is not a space where anyone wants to stop as they walk through. Surrounded by garish striping and bold warning signs for the traffic, it is a suburban-style place designed for cars, not people. Making a simple T-intersection and a stop sign or traffic light would make it feel like a normal New York street again.

Between 23rd and 24th streets, turning Broadway into a walking street would mean the pedestrian plaza could be attached to Madison Square (squaring the square), no longer separated from the park by cars, buses, and trucks. Below 23rd Street, the new plaza would fill the street. That is better than the current design, which diminishes the triangular "flatiron" shape made by Broadway cutting diagonally through the New York grid. The Flatiron Building is triangular precisely because it was designed (and named) for the intersection, and the test plaza visually removed some of its reason for being (Figure 4.139).

Figure 4.139: Madison Square, New York, New York. NYC DOT, 2008. Tactical urbanism at its best. Looking south on Broadway from the north side of 23rd Street. Overnight, boulders were placed in the roadbed to close large parts of the roadbed to cars, closed sections of the road were covered in sand-colored, textured paint, and lightweight, outdoor furniture was scattered around. New Yorkers responded immediately and enthusiastically.

Figure 4.140: Madison Square, New York, New York, NYC DOT, 2008. Looking south on Broadway from 23rd Street. A temporary solution. The wide traffic lane, the bold reflective signs, the high-speed striping, and the broad, fast curve—alien to the Manhattan grid—create an area where the pedestrian feels unwelcome.

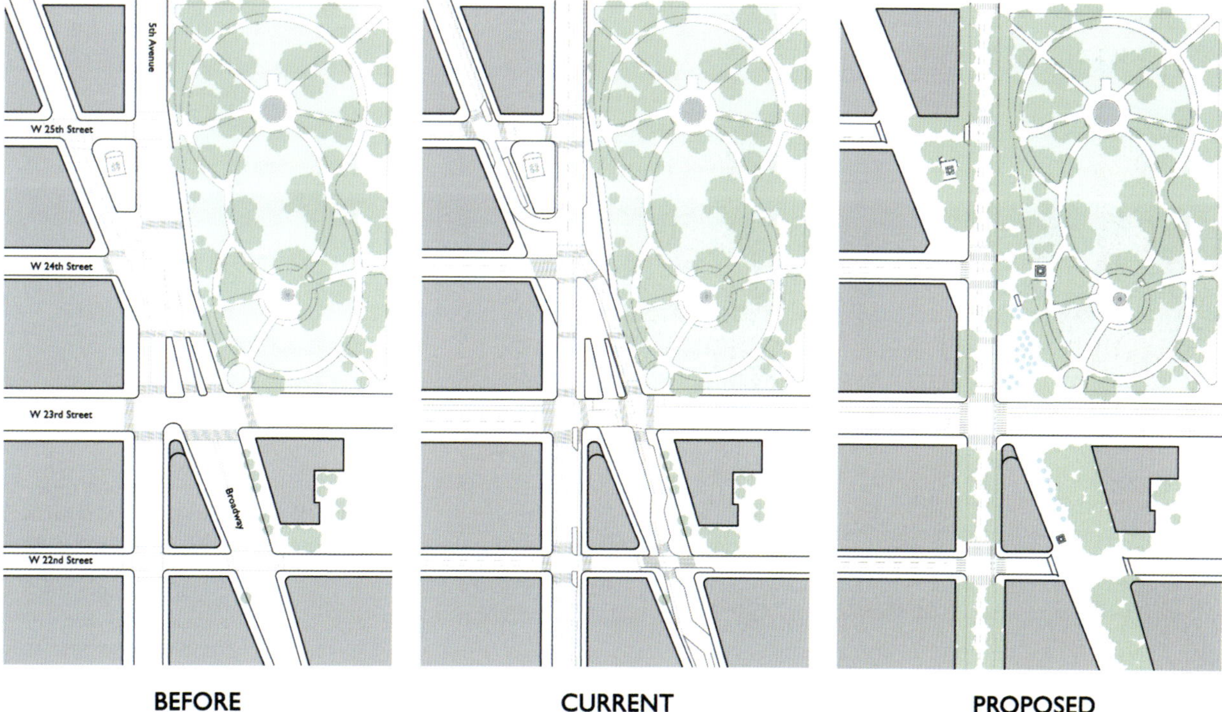

Figure 4.141: Madison Square, New York, New York. Plan view, Before, Current, and Proposed. © 2024 Dover, Kohl & Partners / Massengale & Co LLC

Figure 4.142: Madison Square, New York, New York. View looking south on Fifth Avenue and Broadway (on the left), circa 1910. In the center of the photo, between Fifth Avenue and Broadway, is the Flatiron Building, designed by Daniel Burnham and completed in 1902. For another historic photo of Madison Square, see Figure 4.122. *Library of Congress / Detroit Publishing Company Photograph Collection*

> Tactical Urbanism is praised for making changes at low cost. It gives people the chance to see what the changes will produce, tests the design, and allows improvements in the final product.

Weaving car lanes with lots of large warning signs are sub-urban, inappropriate for Manhattan. Designed by traffic engineers like a curved pipe for throughput, the lanes are separated from the grid that gives the city its character, and their highway-scale signs and stripes are machine scale, rather than human scale.

Until Commissioner Sadik-Khan started the DOT's bold land grab, all of Manhattan's existing "squares" were green parks, like Union Square and Madison Square. Her department's Tactical Urbanism showed that New Yorkers are also eager for paved squares like the ones we see in other parts of the world. Tactical Urbanism is praised for making changes at low cost. It gives people the chance to see what the changes will produce, tests the design, and allows improvements in the final product. The disappointing factor at Madison Square is that seventeen years later we still have "temporary" changes.

A CHICANE IS NOT A SHARED-SPACE PLACE

A space is only a place if people want to be there, we said at the opening of the book. After *Street Design* was published, the DOT added New York's "first shared-space street" to the short block by Madison Square shown in Figure 4.140. We can confirm that it is a space where drivers feel more comfortable than everyone else. That is because the design emphasizes traffic calming, not placemaking. Putting that another way, it is a product of traffic engineering, not a piece of urban design.

A later shared-space block built by the DOT just south of Union Square is similar. Figure 4.143 shows concrete tubs with shrubs in them filling some of the space. This disrupts the harmony of the space, and people like harmonious places. The street is no longer designed to make it comfortable for drivers to go as fast as possible, but as in most examples of traffic calming, the car comes first. Cutting the street up into pieces that have little or nothing to do with human perception or feeling disrupts the harmony.

Nothing between the sidewalks says, "This is a place for humans."

Like the shared street at Madison Square, the University Place "shared street" is only a block long, because one block is long enough to accomplish that. People walking or cycling, on the other hand, want a safe, comfortable network leading to destinations. That does not exist at Madison Square or Union Square. The NYC DOT website has block plans for each project.[70]

The rational mind sees a virtue in slowing traffic by disrupting the space, but when we experience a place, our reactions are fuller and more complex. In early Dutch efforts to reclaim the streets for people, the slowest streets frequently had similar traffic chicanes, with plantings that had no relationship to the overall space of the street. But the Dutch, who are decades ahead of us in these experiments, rarely use chicanes for shared-space city streets now. Recent conversations with some of the best street designers in the Netherlands confirm this.

Figure 4.143: University Place, New York, New York. NYC DOT, 2019. Photo of another NYC DOT shared-space block. Looking north towards Union Square. Drivers are more comfortable in this space than people walking or cycling. Nothing between the sidewalks says, "This is a place for people."

LAFAYETTE STREET AND FOURTH AVENUE AT ASTOR PLACE AND COOPER SQUARE, NEW YORK, NEW YORK / CLAIRE WEISZ

WXY architecture + urbanism and Piet Oudolf, 2016

Retrofit: Restoring the Balance

Core Street

> The reconstruction of Astor Place—and the reinstallation of the East Village's beloved Alamo—provides a terrific example of how well-designed public space can create a more unified, better functioning public sphere. Fluid, attractive and walkable spaces like Alamo Plaza are crucial as we work together to create a greener, healthier New York City.
>
> — Commissioner Mitchell Silver,
> New York City Parks

Every city has high-profile streets and parks. However, public spaces that are harder to define—but equally vital to contemporary American urban design—are "crossroads." Crossroads remain in the subconscious as attractors. Many public plazas today come from a typology dependent on the adaptive reuse of crossroads. These accidental public spaces result from the opportunities created by the history and traces of the street system. There are reasons why this unplanned condition became natural sites for new public spaces and plazas in North American cities. Understanding one instance does not necessarily lead to understanding the value of other plazas, but each tells a story about the potential of reimagining streets as we decrease our dependence on fossil fuels and move towards cities for walking, cycling, and new modes of delivery. The long history of change at Astor Place tells a story about the form of the city and its transformations in response to social use.

In Manhattan, spaces like Times Square, Herald Square, and Astor Place became exemplars of the American public realm. In the post-pandemic era, we must grapple with adaptation and help the public see spots of reclamation for pedestrian use as models for the entire city. Astor Place and Times Square were the first green-lit, permanent projects under former New York City Department of Transportation Commissioner Janette Sadik-Kahn's leadership. Although distinctly different in scale of impact, each project held problems and possibilities for adaptive reimagining as we make people rather than automobiles the design drivers of our urban space.

Figure 4.144: Astor Place, New York, New York. WXY architecture + urbanism and Piet Oudolf, 2016. Aerial photo of the rebuilt Astor Place. Looking north towards Fourth Avenue. The historic IRT Subway entrance is in the center of the new pedestrian plaza. *Courtesy of Claire Weisz*

Making crossroads into viable public spaces requires layers of public and disciplinary interactions and approvals. Subway lines, bus systems, buried utilities, underground vaults, and other existing infrastructure combine with the work of guerilla artists, transportation engineers, civil engineers, planners, landscape architects, and architects. An opportunity arose at Astor Place because the area was undergoing grade work for a third water tunnel, creating the potential to reconfigure the streets themselves after decades of advocacy and study by many groups. Since Astor Place reopened in 2016, it is possible to compare the area's character pre-transformation with the reconfiguration and then its success or failure with the change in commerce and patterns of use during the Covid-19 pandemic. The question of how "well-designed" streetscapes reflecting proportion, function, walkability, and the integration of natural systems hold up to the disruption of social systems and the adoption of more advanced technologies along with the widening wealth gap is a critical one for urban places.

"This space means a lot to me" DOT Commissioner Polly Trottenberg said as she opened the redesign of Astor Place to a handful of community members, transportation advocates, and city employees, asking them all to join her in spinning the iconic black cube sculpture Alamo by Bernard "Tony" Rosenthal. As a young college graduate in 1986, Trottenberg "once sold used books under the Astor Place Cube, back when you could still make money selling books." The end of that era, when used bookstores and publishers clustered in the surrounding blocks, is part of this area's history of public adaptation.

The Alamo was the first permanent contemporary outdoor sculpture installed in New York City, albeit on a traffic island. It became a symbol of Astor Place in the Seventies as an essential public space. As recounted by Trottenberg, the neighborhood served crowds of students from Cooper Union, NYU, and other schools making their way into the world. Nearby were the Public Theater, and The Carl Fisher Building.

Figure 4.145: Cooper Square, New York, New York. WXY architecture + urbanism and Piet Oudolf, 2016. Photograph looking south. Before the retrofit, this was a street. *Courtesy of Claire Weisz*

The Astor Place / Cooper Square Reconstruction was the bleeding edge of an urban design effort plugged into one of the most significant infrastructure projects that NYC has taken on—the Third Water Tunnel. As incongruous as it sounds, no direct funding stream for permanent public space was tied to rezoning or park funding. Policies had been against the design of a walkable city, despite advocacy from a small pedestrian unit in the DOT and many people on the community boards that straddle the area. But staff and agency leaders cooperated to cobble funding from other policy initiatives such as "amenity packages" and economic development funding.

Other factors arose that made it possible to reclaim these street geometries for public spaces. Cooper Union built its engineering building over the roadbed of Stuyvesant Street between Third and Fourth Avenue. The *Village Voice* sold its building to the Grace Church School, and New York University converted loft buildings into writing labs and classrooms. The city recognized there was a vital role for public space in the creation of green infrastructure, especially in response to deluge rain events. Still, at that time, the means of approving rain gardens and inlets kept the funding restricted to areas where the consent decree required stormwater retention. This meant that green infrastructure in Manhattan had to be argued for as a pilot program. The nature of the customized funding for Astor Place meant there were opportunities for a certain amount of design license, but this was restricted to using a "toolkit" being developed at the time by the Department of Transportation, with elements like benches, plantings, fencing, drinking fountains, paving, and lighting approved by both the Department of Parks and Recreation in anything designated a park. This once-in-a-lifetime infrastructure funding created the opportunity to solve two decades of advocacy for a public solution to new pedestrian space and logical crossings.

The design goal at Astor Place was to let users self-select their routes. The most significant change was at the southern end of the new pedestrian-only zone. The reconfigured curbs make a clear and safe way to walk to the Cube while adding routes to the historic Astor Place subway entrance. The stepped seating allows students to hang out in a place that felt like nature had reclaimed the historic roadbed for its porous and ecological benefits. It was a fantastic feeling and, from a mental health perspective, it provided a calming influence.

Moving to the historic Cooper Triangle Park, south of the Cooper Union's Foundation Building, with its new super-wide sidewalk and additional row of trees, the shocking difference was the opening of a former dead end and the creation of a park to walk through. This opening up of the space around the Cooper Union building also makes a much better setting for the new student building by Morphosis across the Bowery. There is plenty of room for future experiments and installations by Cooper Union, Parsons, or NYU students—or for one of the theater groups in the West Fourth Arts block to use the seating and accessible area for performances. The plan made the park more extensive and accessible. We widened the sidewalks and created a green infrastructure system between the sidewalks and the bus lines that buffered the buses from the square. The street is much narrower and one-way, going north. The whole extent of Cooper Square, street and plaza, supplies a more porous surface through trees and lushly planted beds by Piet Oudolf, master horticultural designer of the High Line and The Battery.

Complaints of conflicts before the new design between skateboarders, residents, students, and commuters in the area had created a false impression. But we designed new areas that were too bumpy for skateboarding and provided enough room to skateboard away from building entrances and pedestrian paths. We increased the number of places to sit in this historic and pivotal corner. From the "zipper bench," you can walk across to the piece of Cooper Square called Fourth Avenue, which now goes only north. We made an enormous plaza was at the north end of Astor Place—most famous for where the Mud Truck parked—where the Astor Place subway kiosk was formerly crammed between Lafayette Street and Fourth Avenue. Its most prominent feature is a 120-foot-long planter that operates as a large-scale piece of green infrastructure supporting trees and filtering stormwater. At both ends, there are low granite steps and, in the middle, the zipper bench. The giant granite steps widen at the entrance to the subway to make a place for waiting. This long-elongated pedestrian zone to Wanamaker Place, where Lafayette and Fourth Avenue meet, is in constant use.

▶ **Figure 4.146:** Astor Place and Cooper Triangle. WXY architecture + urbanism and Piet Oudolf, 2016. Plan of completed design. Astor Place is located to the north and Cooper Square to the south. Cooper Union's Foundation building sits between the two. *Drawing courtesy of Claire Weisz*

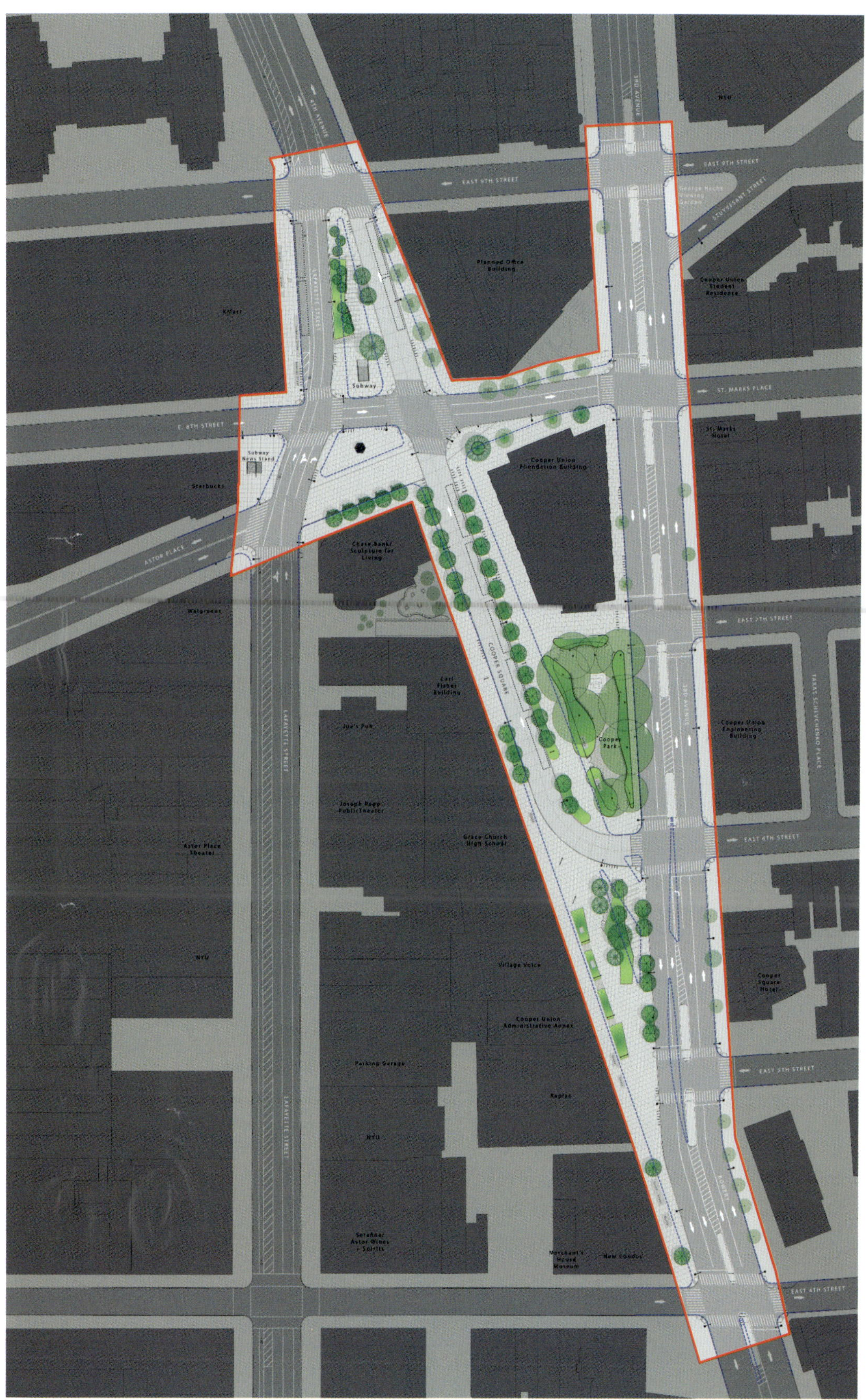
4TH AVENUE
THIRD AVENUE
NYU
EAST 9TH STREET
STUYVESANT STREET
George Hecht
Viewing
Garden
Planned Office
Building
Cooper Union
Student
Residence
KMart
Subway
ST. MARKS PLACE
St. Marks
Hotel
E. 8TH STREET
Subway
News Stand
Cooper Union
Foundation Building
Starbucks
Chase Bank/
Sculpture for
Living
ASTOR PLACE
EAST 7TH STREET
Walgreens
Cast
Fisher
Building
COOPER SQUARE
Cooper
Park
Cooper Union
Engineering
Building
3RD AVENUE
TARAS SCHEVENKO PLACE
Joe's Pub
LAFAYETTE STREET
Joseph Papp
Public Theater
Grace Church
High School
EAST 6TH STREET
Astor Place
Theater
NYU
Village Voice
Cooper
Square
Hotel
Cooper Union
Administrative Annex
Parking Garage
EAST 5TH STREET
Kaplan
NYU
BOWERY
Scrafmar
Astor Wines
+ Spirits
Merchant's
House
Museum
New Condos
EAST 4TH STREET

Figure 4.147: Cooper Square, New York, New York. WXY architecture + urbanism and Piet Oudolf, 2016. Aerial view looking north towards Lafayette Street from Third Avenue. *Courtesy of Claire Weisz*

Converting this web of streets to public plazas was a success based on metrics alone. The two-year construction project created two new pedestrian plazas and expanded and renovated two others, bringing 42,000 square feet of new pedestrian space to the neighborhood. The redesign incorporated an existing subway station and created a safer vehicular and pedestrian traffic configuration. It introduced larger sidewalks, 16,000 square feet of planting areas, 60 new trees, in-ground irrigation systems, 6,700 square feet of permeable pavement, 2,100 square feet of curbside rain gardens for improved drainage, and racks for more than 100 bikes. The Cube was restored at long last, and Jim Powers, an unofficial artist who decorated the bottoms of light poles with tile messages, had his work restored as well.

What held together this laundry list of improvements were ultimately the design decisions. By reimagining streets but not erasing history, we restored the balance of power to a point where places where people lived and worked were not designed for only one mode of transportation. Returning to Mitchell Silver's assessment, each of the contributing elements was there for its function and to act as a cohesive whole—an additional aspect of the design balancing this cohesiveness.

Today, this set of reshaped streets and plazas functions much better than the old wide traffic lanes and narrow medians. The subway below and buses crisscrossing above that used the medians as on-street storage complicated figuring out the right approach to reapportioning the street to the sidewalk, but it was not impossible. Many designed spaces depend on the quality of the details. In public space design, this starts with the ability of the surfaces to hold up over time, understanding planting diversity, considering landscape maintenance, and developing beautiful yet comfortable furniture systems that keep a crossroad's character intact. All urban streetscapes must respond to changes in use, social value, and climate conditions. Adapting important crossroads like Astor Place is the mandate of cities to go beyond pavement selection and understand the gift of space and the crossing of streets as a key tool for the regenerative potential of cities.

Figure 4.148: Astor Place, New York, New York. WXY architecture + urbanism and Piet Oudolf, 2016. One of the "zipper" benches referred to in the text. *Courtesy of Claire Weisz*

NINTH AVENUE, MEATPACKING DISTRICT, NEW YORK, NEW YORK

Adapted from Marvel and Ken Smith Workshop, 2019

Retrofit

Core Street

The Meatpacking District Business Improvement District (BID) is one of many gentrified New York neighborhoods that were once industrial or shipping centers close to the docks along the Hudson and East rivers.[71] Several of the areas kept their cobblestone streets long after the city paved over the cobblestones in residential neighborhoods and office districts. Today, BIDs and developers have made the streets and their surroundings chic. In addition to the Meatpacking District, think of SoHo (the area South of Houston Street that Robert Moses would have demolished if Jane Jacobs and others had not stopped him), TriBeCa (the Triangle Below Canal Street), and DUMBO (Down Under the Manhattan Bridge Overpass).[72]

Many who did not grow up in New York forget that the city was not always as expensive as it is now. As late as the 1990s, the East Village had many abandoned buildings and large numbers of squatters in those buildings.[73] At the same time, the Meatpacking District was home to wholesale butchers and meatpackers, prostitutes, and infamous S & M clubs. Those days are long gone: today the area is a walkable center of expensive, high-style boutiques and the usual galleries, hotels, bars, and cafes on beautiful stone streets.[74] As a group, BIDs have done a better job of making beautiful New York streets than the NYC DOT's. As we go to press, however, there are signs this may change. We say more about this is Chapter Five.

Figure 4.149: Ninth Avenue, New York, New York. Marvel and Ken Smith Workshop, 2019. Photo looking south towards West 14th Street. Hudson Street and Ninth Avenue funnel together here, making a wide street. The Meatpacking BID filled in half the block, leaving a narrow Ninth Avenue where Hudson Street used to be.

Figure 4.150: Ninth Avenue, New York, New York. Marvel and Ken Smith Workshop, 2019. Photo looking north from Little West 12th Street in the Meatpacking District. The historic paving stones and their muted color palette bring both visual richness and cohesive harmony. When most New York City streets were paved with cobblestones, it was common to provide smooth crosswalks. Today, smooth stones can also be used for bike lanes in the streets.

JANE JACOBS SQUARE, NEW YORK, NEW YORK

Bleecker Street between Christopher Street and West 10th Street

Massengale & Co LLC and Dover, Kohl & Partners, with H. Zeke Mermell, 2011

Retrofit: Shared Space

Core Street

Jane Jacobs Square was a proposal for a competition organized by the Institute for Urban Design called *By the City / For the City*.[75] Entrants could pick a site in New York City and suggest an improvement. We chose an interesting fissure in the Manhattan grid where Bleecker Street crosses Christopher Street and West 10th Street in Greenwich Village. Before the New York City Commissioners' Plan of 1811 platted a single, unifying grid over most of the island, the rural hamlet of Greenwich Village had a few small grids and individual roads laid out by various landowners as they thought best. Two of the grids come together in this block of Bleecker Street, making a comfortable triangular space that is unusual in Manhattan. Today, the space is dominated by the cars speeding down Bleecker Street. Jacobs Square reclaims it as a shared space for cars, bikes, and pedestrians.

The intersection of Bleecker and Christopher streets marks one of the original centers of the old Greenwich Village, where a public well once stood. It is still one of the centers of life in the Village, but the narrow sidewalks and ugly highway-scale graphics on the street claim the roadway for cars while telling people to stay over on the sidewalk where they belong (Figure 4.151). Anyone who walks through it can see its potential to be one of the great public places in New York—if we simply do away with its present 1960s-style, auto-dominated design.

The proposed redesign (Figure 4.152) uses principles of placemaking to create a square shared by walkers, cyclists, drivers, and diners:

- The sidewalk on the east side of the square is expanded to the west, forming a triangular "attached plaza" that covers the majority of the square.

- The roadway, now parallel to the buildings on the west side of the square, is reduced to a narrow traffic lane and a lane for parking and loading for the stores on the square.

- A speed table on the south side of West 10th Street raises the roadbed to the level of the sidewalks and the rest of the plaza.

- Trees along the sides of the road act like bollards, defining the boundaries of the street.

- A fountain on the north side of Christopher Street reduces ambient noise and recalls the eighteenth-century well once on the site.

- Bleecker Street no longer goes straight through: cars either stop and then turn right on Christopher Street or stop at Christopher before turning left. In the second scenario, Christopher Street either becomes a two-way street, or a one-way, eastbound street.

- A monument to the Commissioners' Plan of 1811 on the blank wall of the building on the southwest corner of Christopher and Bleecker streets visually terminates the view down Bleecker, while the fountain on axis with the southern leg of Bleecker points to the continuation of the street.

- Tables and chairs in the square, along with the new shade trees and programming for the public (restaurant and café tables, chess boards, and so on), encourage people to use the space.

Some of the tables and chairs would be like those used in Bryant Park and by the New York City DOT in outdoor spaces—lightweight furniture the public can easily rearrange. The plan would go a step further by embracing the storefronts lining the square to involve them in programming the space: two parts of the square would be reserved for a restaurant and a café, each with tables.

Cars passing by would be tamed by the speed table, the narrow roadway, the trees bordering the road, the cars and trucks entering and leaving from the parking spaces, and—most important—by the shared-space roadbed, now jointly used by cars, trucks, cyclists, and walkers.

New Yorkers are ready for spaces like the new Jacobs Square. They have small apartments and live much of their lives in public space and Third Places.[76] Their enthusiasm for the new squares like Astor Place and the Tactical Urbanism pop-up squares recently built around the city by the New York City DOT demonstrates their desire for more outdoor spaces that are not dominated by cars.

Jacobs herself lived nearby at 555 Hudson Street, and we like to think she would have embraced our idea. In *The Death and Life of Great American Cities,* she complained that long streets with unterminated vistas tire pedestrians and recommended improving the streets by adding trees, bridges over the street, or even buildings in the middle of streets.[77] Visit Jacobs Square and you will find exactly the type of tiring vista Jacobs disliked. Our design interrupts the unbroken vista, turning the square into a quiet oasis instead of an auto sewer.

Compare busy Bleecker Street with the quieter parallel street, West 4th. They run in opposite directions, making a "one-way couplet" beloved by transportation engineers. But West 4th (Figure 5.10) is less of a through street: while Bleecker runs south from the intersection of Hudson Street and Eighth Avenue to Sixth Avenue, where it turns east and heads towards NoHo and the Lower East Side, West 4th Street ends in a T-intersection at a small Village street, Jane Street. West 4th is narrower than Bleecker Street, and slightly less straight, with an almost continuous tree canopy over the street. There are no long, uninterrupted vistas on West 4th Street, but as mentioned, that is not true on Bleecker Street. This is particularly noticeable looking downtown from the north end of Jane Jacobs Square.

Bleecker Street is also a major thoroughfare for drivers going from the northwest corner of Greenwich Village to points south and east. Taxis and trucks speed on the wider, one-way street. An unprotected bike lane frequently feels unsafe. West 4th Street has no bike lane, but it feels like a shared-space street, where cyclists ride in the middle of the street (Figure 5.10). Customers waiting for tables at the sidewalk cafés along the street pick up on that and stand in the street while they wait. Drivers accordingly adjust their speed to accommodate the pedestrians and cyclists.

Filling Bleecker Street with a public space where it crosses Christopher Street is a logical next step in redressing the balance between pedestrians and cars in the neighborhood.

Two eighteenth-century grids come together at the intersection of Bleecker and Christopher streets, marks one of the old center of Greenwich Village, where a public well once stood.

Figure 4.151: Jane Jacobs Square, New York, New York. Massengale & Co LLC and Dover, Kohl & Partners, 2011. Watercolor rendering, aerial view. Before: Highway-scale striping and speeding cars currently dominate the potentially lovely space. © *2011 Dover, Kohl & Partners / Massengale & Co LLC*

▶ **Figure 4.152:** Jane Jacobs Square, New York, New York. Massengale & Co LLC and Dover-Kohl & Partners, 2011. Watercolor rendering. After: The aerial view shows one option. This version has the attached plaza on the east side of the square. © *2011 Dover, Kohl & Partners / Massengale & Co LLC*

Figure 4.153: Jane Jacobs Square, New York, New York. Massengale & Co LLC and Dover, Kohl & Partners, 2011. Before: Looking south on Bleecker Street from West 10th Street. The historic center of Greenwich Village. Also see Figure 5.11.

Figure 4.154: Jane Jacobs Square, New York, New York. Massengale & Co LLC and Dover, Kohl & Partners, 2011. Watercolor rendering. After: Looking south from West 10th Street at the new Jane Jacobs Square. Rendering by Gabriele Stroik. © 2018 Massengale & Co LLC,

Figure 4.155: Filosofgangen (Vester Voldgade), Copenhagen, Denmark. Looking north in 1880. On the left is Copenhagen's old "western rampart" (Vester Voldgade). *Copenhagen Museum / Wikimedia Commons / Public domain*

VESTER VOLDGADE, COPENHAGEN, DENMARK

COBE and GHB, 2012

Retrofit: Asymmetrical Promenade

Core Street

Vester Voldgade ("West Rampart Street") is interesting both for its current condition and its original form. The street is called "Voldgade" because like the original boulevards in Paris, it was built where an old city wall stood. As in Paris, allées were planted on the old ramparts, but in Copenhagen the result was different. Parisian boulevards became tree-lined, symmetrical streets (planted in patterns composed of squares—see page 99). In an old photograph of a section of Vester Voldgade then called "Filosofgangen" (Philosopher Path), we can see that there was a street next to the tree-covered ramparts, with all the trees on the rampart side of the road in what looks like a more naturalistic planting.[78]

Over time, the city removed the ramparts. By the late twentieth century, Vester Voldgade was a primary urban arterial delivering traffic to the center, a harsh place for pedestrians or cyclists. But a recent redesign of the street brought back the asymmetrical promenade, this time putting it on the eastern side of the street, where it can catch the afternoon sun.

The retrofitted Vester Voldgade has an interesting place in the current discussion about protected cycling lanes. For decades, Copenhagen has been a world leader in the creation of pedestrian streets and protected lanes, using a different model than the one followed in Amsterdam and the Netherlands. The Dutch idea of shared space, made famous by the great traffic engineer Hans Monderman and the widely visited streets of Amsterdam, puts all users on streets where cars go slowly enough to safely share the road. Copenhagen's model had more separation of use. Copenhagen made some streets car-free and remodeled others into transportation corridors that made space for people walking, cycling, and driving.[79]

The biggest and most famous of the pedestrian streets is the Strøget, similar to the German and Italian pedestrian zones started around the same time in the early 1960s.[80] Under the leadership of Alfred Wassard and Jan Gehl, Copenhagen moved forward with careful studies of public preferences and habits that led to more streets dedicated to walking. At the same time, the city built streets for what traffic planners call "multi-modal transportation." Typically, the traffic lanes on major streets were narrowed and the extra space between the buildings was used for wider sidewalks and broad bike lanes. The cycle lanes were a little higher than the roadway but a little lower than the sidewalk: the slight level changes, low curbs, and subtle paving patterns physically and visually separated the different areas. Frequently, the designs gave as much or more of the space to the use of pedestrians and cyclists as to motor vehicles.

The transportation corridors look utilitarian and not infrequently harsh. Unless the space between the buildings is very large, trees are left out in favor of more space for walking, cycling, and driving. Many of the corridors have little or no parking. In America, the default model for a protected cycling lane has bright red, blue, or green paint, next to a boldly-painted "door zone" separating the bike lane from parked cars (New York cycling enthusiasts love the sight of "fresh Kermit," meaning a newly-painted green lane). We used to call that a "Copenhagen Lane," but the actual lanes in Copenhagen are visually simpler and more efficient in their use of space. Only the slight difference in level between the street and the bike lane prevents a driver from going into the bike lane or even

Figure 4.156: Vester Voldgade, Copenhagen, Denmark. COBE and GHB, 2012. The new promenade on the east side of the street.

parking there. "We wouldn't do that in København," the Copenhagen City Architect said to one of the authors as they stood looking at a typical protected bike lane in New York City. She thought the lane was overdesigned and ugly, and the "mixing zone" at the intersection over-engineered and dangerous.

Vesterbrogade, a busy street near Vester Voldgade that accommodates cars, bicycles, and pedestrians is a typical example of the Copenhagen model (Figure 4.158). Vester Voldgade is a hybrid of that model, combining a high-style version of the Copenhagen Lane with the wide, pedestrian-friendly promenade. We include it because we think it is beautiful and because it illustrates the point that Copenhagen increasingly has a variety of cycle lanes in the city's large network of bicycle lanes and tracks. Some narrow streets in quieter areas are shared streets. Similarly, large transportation corridors in Amsterdam and other Dutch cities have separate, protected bike lanes, typically with less utilitarian design than seen at Vesterbrogade.

When Copenhagen and COBE Architects looked at redesigning Vester Voldgade, it was an important urban arterial, with a high level of traffic flow and many drivers impatient to get to their destination. On some blocks, the new design took away two of the three existing traffic lanes and, unusually, changed the street from two-way to one-way. That is because the traffic lane was narrowed, a

Copenhagen bike lane was added, and most significantly, the wide promenade was added on the sunny side of the street, creating an asymmetrical street of the type we promote in *Street Design* for some situations.

Compared to Vesterbrogade, Vester Voldgade is "designey," with special details that call attention to themselves. Examples include the unusually broad stone curbs and obviously varied stone paving. Compare Vester Voldgade to a street like Brabanste Turfmarkt in Delft, an asymmetrical street with more low-key details (Figure 5.50), and you can see how high-style Vester Voldgade is. Having visited many of the world's great streets over the years, we can say that all of them are simple: what is important is the unity of the space. Those are the streets that make the great outdoor rooms that support public life. Vester Voldgade has sleek, sharp modern details that contrast *slightly* with the surrounding buildings that shape the street, but only slightly. Walking on Vester Voldgade feels good. The design encourages people to sit down and hang out. On top of that, the new street is a greenway that connects to three new squares and that comes out at the square in front of City Hall, where it crosses the Strøget. We need more streets like Vester Voldgade, and more experiments in how to make the space between the buildings better for all of us, and not just the drivers.

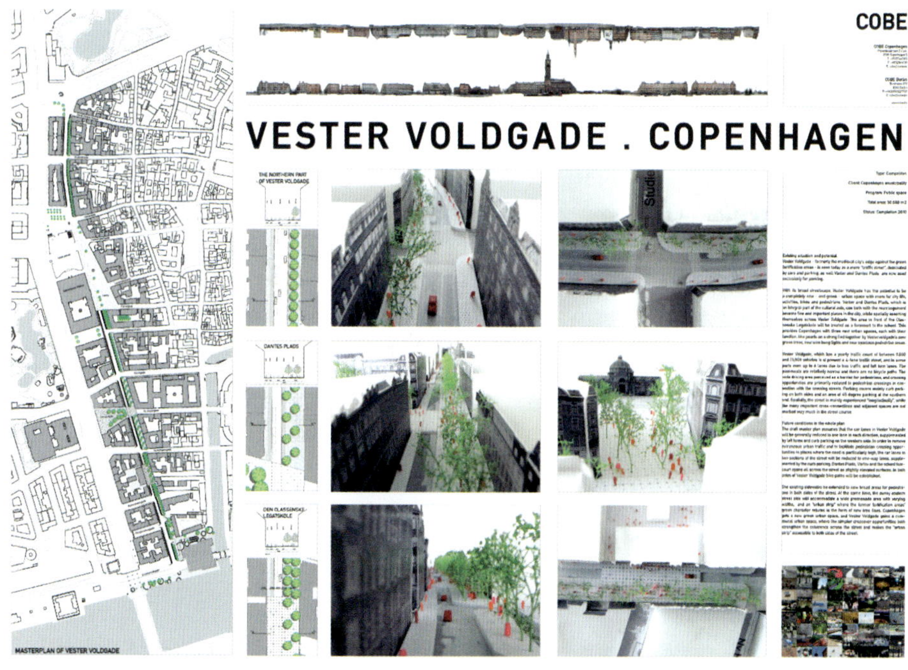

Figure 4.157: Vester Voldgade, Copenhagen, Denmark. COBE Architects, 2010. Competition entry. © *2010 COBE Architects*

Figure 4.158: Vester Voldgade, Copenhagen, Denmark. COBE and McKnight Hall, 2012. *© 2013 Michael MacKenzie / Wikimedia Commons / CC BY 2.0*

Figure 4.159: Vester Voldgade, Copenhagen, Denmark. COBE and McKnight Hall, 2013. *Courtesy of LYTT Architecture*

Figure 4.160: Vester Voldgade, Copenhagen, Denmark. COBE and McKnight Hall, 2013. *Courtesy of LYTT Architecture*

Figure 4.161: Vester Voldgade, Copenhagen, Denmark. COBE and McKnight Hall, 2013. Section. *© 2024 Dover, Kohl & Partners*

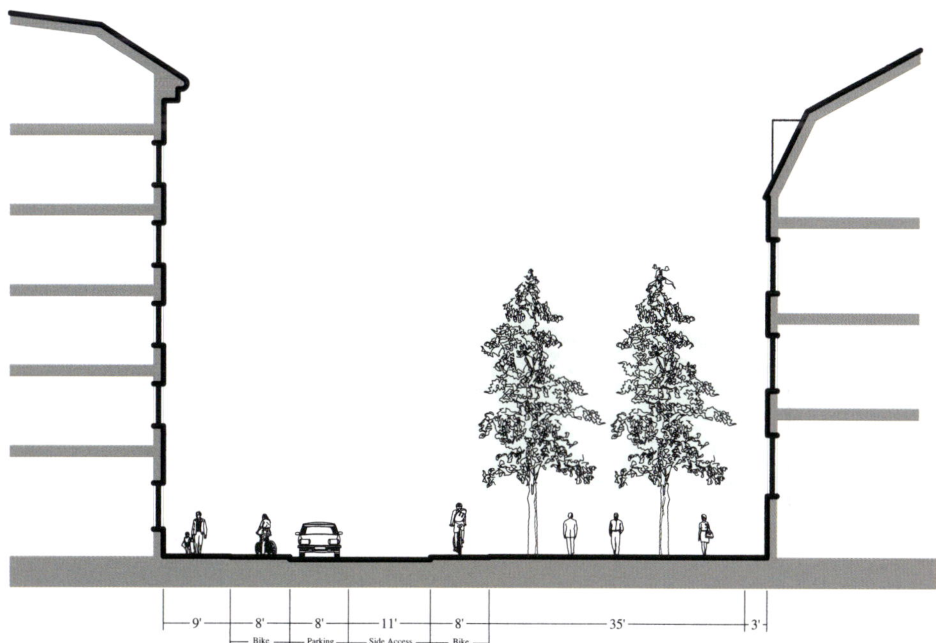

MARIAPLAATS, UTRECHT, THE NETHERLANDS

Retrofit: Restoring the Balance

Utrecht is a small, beautiful Dutch city with some of the best new street designs in the Netherlands. The city even removed a highway in the city that had replaced a canal and rebuilt the canal.[81] Alongside the historic districts, Utrecht has some starkly modern buildings, but the rebuilt pedestrian and shared streets typically follow a more traditional aesthetic.

Figure 4.162 might make you think something like, "Only old European cities can have streets that look like this, with so many layers of history." But the photo of Mariaplaats in the 1960s (Figure 4.164), when the Netherlands was putting cars first and replacing canals with highways, shows that this is not true. Vester Voldgade and Mariaplaats lead us to paraphrase Duke Ellington: There are two kinds of design—good design and the other kind.

Figure 4.162: Mariaplaats, Utrecht, the Netherlands. Photo of Mariaplaats in 2014. New paving and trees in a former parking lot. A street that looks like it has centuries of history behind it.

Figure 4.163: Mariaplaats, Utrecht, the Netherlands. Ronald Tamse and Werner de Feijter, 2015. Photo of Mariaplaats in 2019. The newest design for Mariaplaats removed the curbs, making the island less defined. Carts are rolled into the island for outdoor markets. *Courtesy of Ronald Tamse*

Figure 4.164: Mariaplaats, Utrecht, the Netherlands. Photo of Mariaplaats in the 1960s. Before the Netherlands became a walking and cycling paradise, its cities were choking in cars. *Het Utrechts Archief, Fotodienst GAU/Creative Commons Attribution-Share Alike 4.0 International*

FREDERIKSBERG ALLÉ, COPENHAGEN, DENMARK

Frederiksberg Allé is one of the main streets in a wealthy part of Copenhagen. The protected cycle lane divides the unity of the public space much less than the New York model engineered for the free flow of cars. Cyclists and pedestrians both appreciate the beautiful ride and the summer shade.

Figure 4.165: Frederiksberg Allé, Copenhagen, Denmark. Photo, looking east, across Askårdsvej. The "Copenhagen Lane" is not the only type of bicycle lane built in Copenhagen.

SWEDISH STREETS

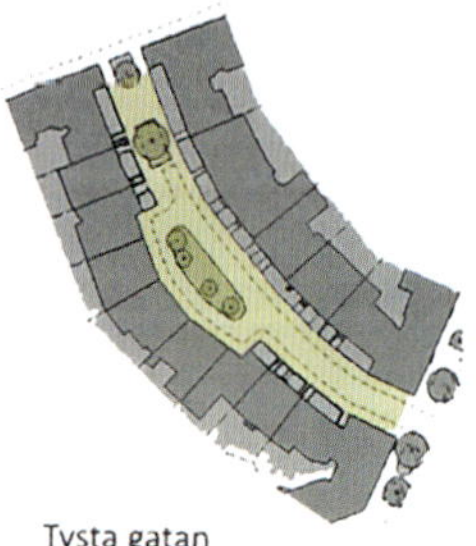

"Every larger street shall be planned in a form of its own. The shape of each street shall be a work of art and designed such that its match does not exist anywhere else. This can be achieved through the grouping of the buildings, their overall design, through gardens, the placing of sculptures, et cetera, and this is what the city planning artist can do for the street."

Tysta gatan

Bjurholmsplan

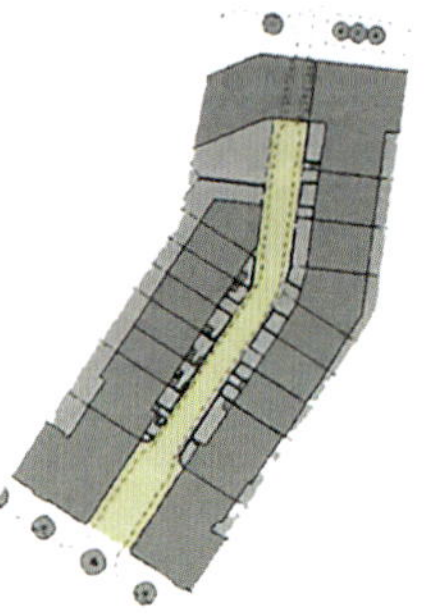

Danderydsgatan

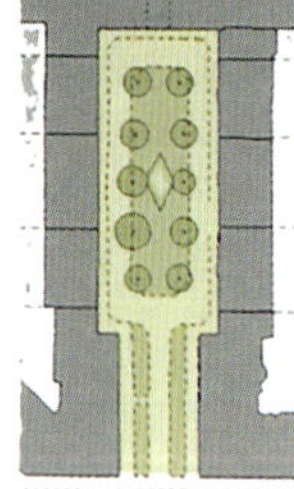

Breitenfeldsgatan

Tre Liljor

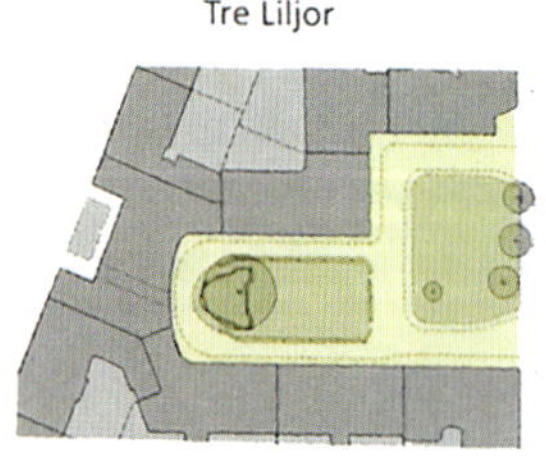

Figure 4.166: A plate from *Swedish Grace: The Forgotten Modern* (Axel & Margaret Ax:son, 2015), edited by Peter Elmlund and Johan Martelius.

CENTRAL BRANDEVOORT, THE NETHERLANDS / PAUL MURRAIN

KK Architects and Paul van Beek landschappen BNT, 1996–Present

Main Street & Neighborhood Streets

Brandevoort is a new town on the outskirts of Helmond in the Dutch province of Brabant. When completed, it will house 17,000 residents on a 365-hectare (900-acre) site. A project from the studio of Rob Krier and Christophe Kohl, Brandevoort is emblematic of their belief in acknowledging and celebrating the familiar, modifying where necessary, innovating in many ways, but always remaining firmly embedded in the traditions of place that have evolved over many centuries (Figure 4.167). The community of the town of Helmond specified its desire for this type of design.

Both Krier and Kohl studied the typologies of other local towns, and characteristics of Dutch facades were cataloged. Street, block, plot, and type are the tools of their trade.

The Urban and Architectural Codes: Working in Tandem

In so far as it is feasible today, Brandevoort manifests the New Urbanist dictum of "one code, many hands." So, in both product and process, the fine-grained adaptation of the building stock is a striking feature of its streets. There are never two adjacent buildings by the same architects, with exceptions only being made for important sites—for example, where Krier and Kohl required specific compositions for the termination of a vista. The other architects were at ease with this, aware that they were designing for a place where creativity was channeled towards a common language.

Figure 4.167: De Plaetse, Brandevoort, the Netherlands. KK Architects, 1996–Present. A view of a typical residential street in 2009. *Courtesy of Paul Murrain*

> In so far as it is feasible today, Brandevoort manifests the New Urbanist dictum of "one code, many hands."

It is the common language of the code that gives the "variety within a pattern" that typifies most places people find desirable. The street codes in Brandevoort are produced for different locations within the central area, including the primary streets that meet at its heart (and adapt an old country road), a secondary street system, and a specific type that runs around the perimeter, overlooking drainage canals and native vegetation.

Street Composition

After the elements of the code were established, the crucial stage of composing full street elevations by manipulating these elements in peer review was arguably the most unusual aspect of the project. Krier and Kohl referenced traditional Dutch townhouse facades for inspiration; the three-bay elevation is dominant on the three-story buildings in central Brandevoort that have dormers in the roof (Figures 4.168–4.171). Traditionally, two-story dwellings and narrower plots have two bays, but, again, symmetrically composed with immense variety. On occasion, the gable faces the street. Within this underlying pattern the possibilities for variety are many.

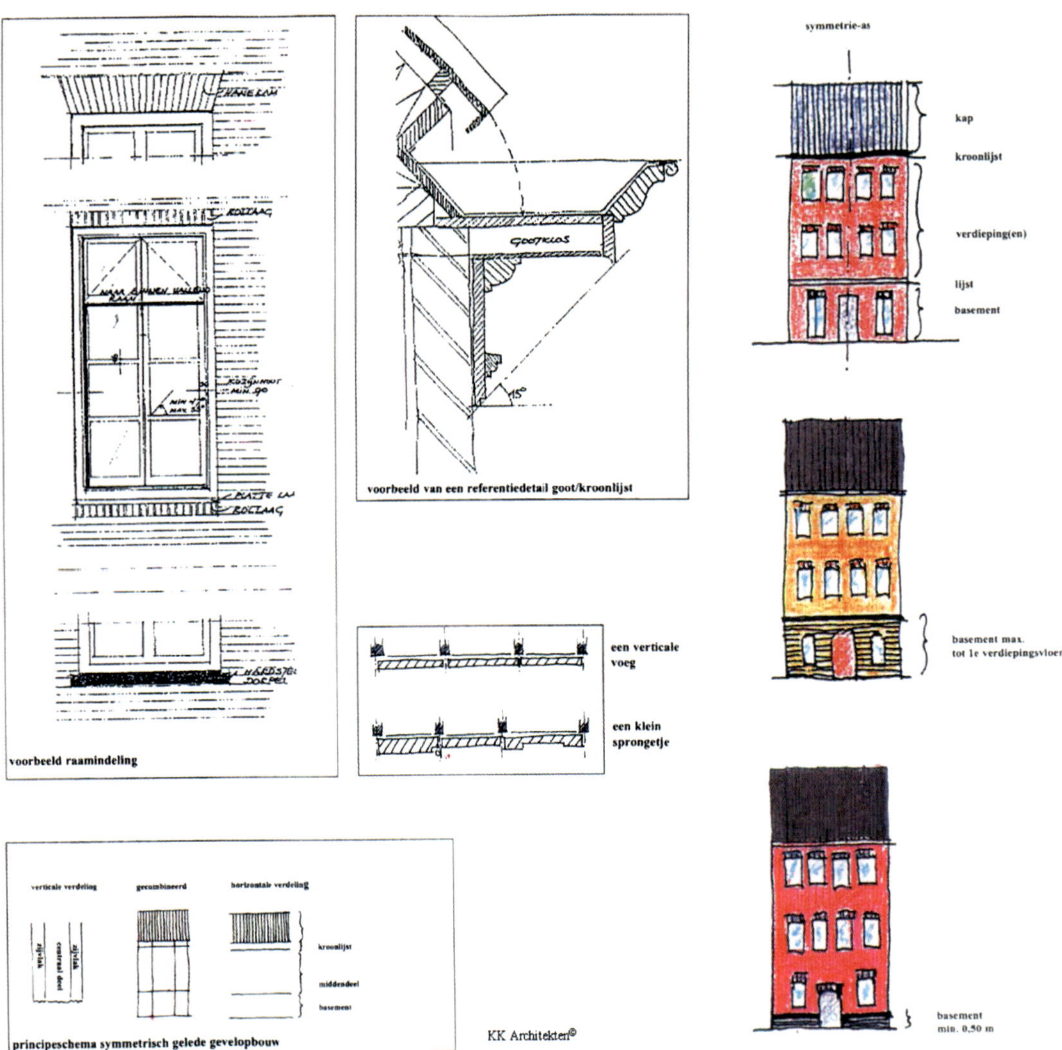

Figure 4.168: De Veste, Brandevoort, the Netherlands. KK Architects, 1997–Present. Urban codes for the rowhouse facades defined in the "image quality plan." *Archives KK Architects, Berlin*

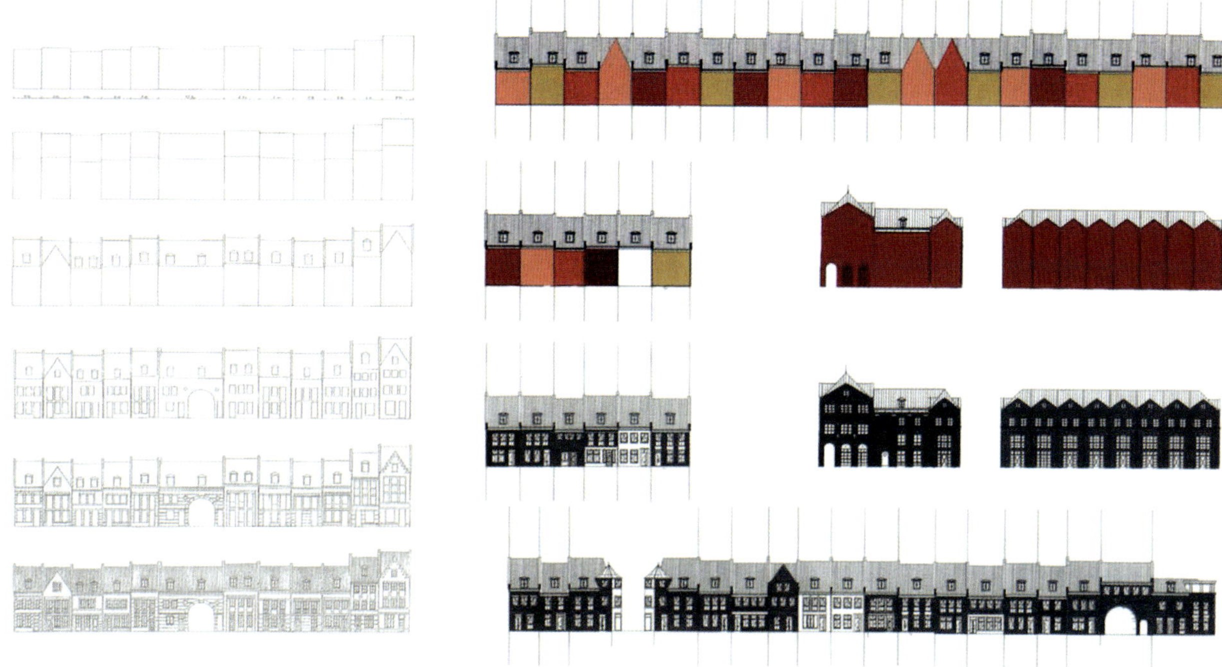

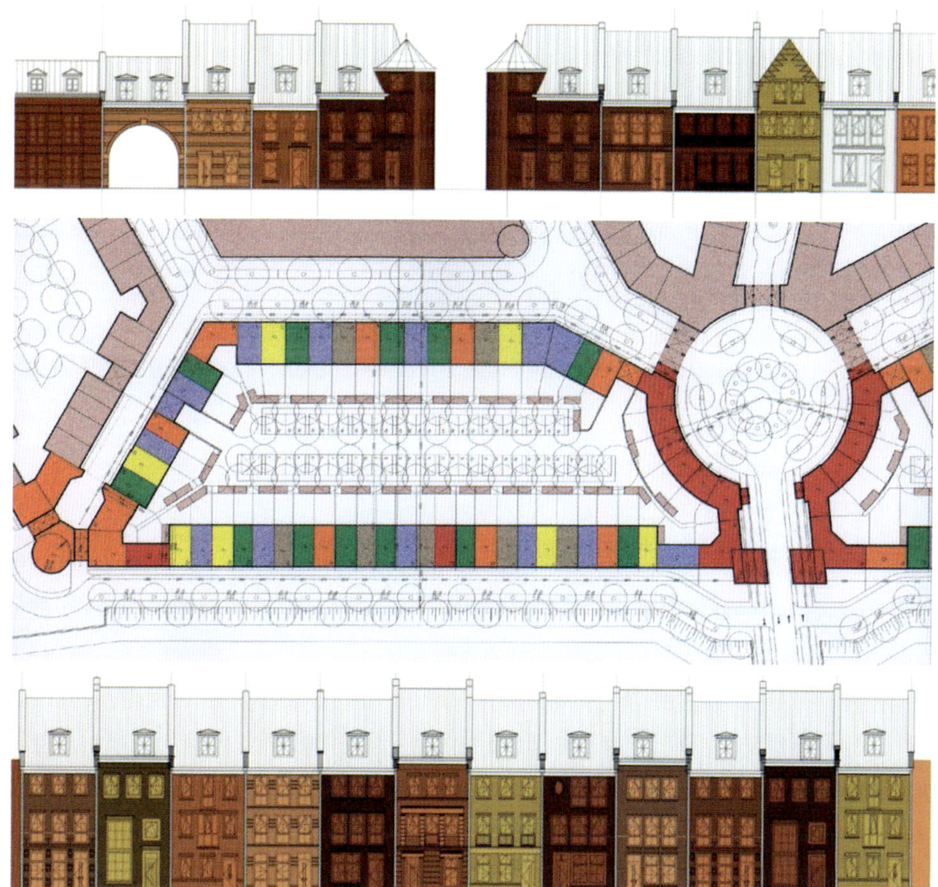

▲**Figure 4.169:** De Veste, Brandevoort, the Netherlands. KK Architects, 1997–Present. Sketch made during the facade design process for the groups of rowhouses. Within a subtle range, designers vary heights, widths of roofs, dormers, fenestration, materials, and ornament. *Archives KK Architects, Berlin*

Figure 4.170: De Veste, Brandevoort, the Netherlands. KK Architects, 1997–Present. Drawings made for the coordination of plan and elevation during the rowhouse design process. *Archives KK Architects, Berlin*

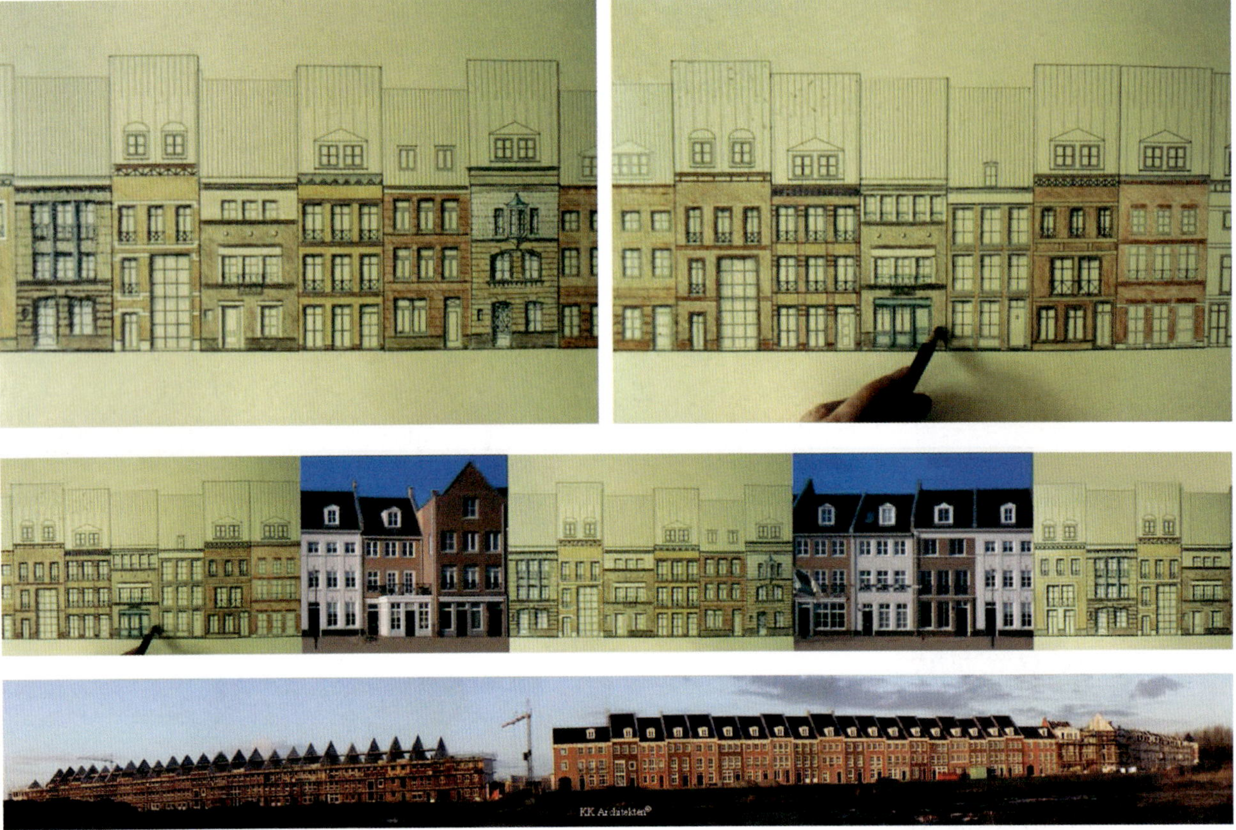

Figure 4.171: De Veste, Brandevoort, the Netherlands. KK Architects, 1997–Present. Study sketches. The underlying pattern permits immense variety despite the repetition of the type, using subtle manipulations of a few basic elements. *Archives KK Architects, Berlin*

Characteristically, the greatest freedom of expression is in the fenestration, which is again held together compositionally by the underlying patterns. The Dutch are at ease with displaying the interior of their dwellings to the street outside, even with the smallest unfenced privacy strip (Figure 4.172). Historically, the proportion of window to wall is significant, particularly on the first two floors of the three-story types. But this often involves the use of mullions and fine-grained glazing bars that operate as a filter when compared with sizable, single panes of glass.

The higher density streets of apartment buildings in the heart of central Brandevoort break these underlying patterns to a greater extent, being different and more recent building types (Figure 4.173). And yet even here it is evident that street composition played a significant role. Corners are afforded more freedom to allow building continuity at a variety of angles in the deformed grid, resulting in a number of prominent hipped roofs. (These, at times, can appear a little contrived.)

For the pedestrian, the streets of central Brandevoort serve up a superb urban experience because the horizontality of the street perspective is arrested by the powerful vertical emphasis expressed in the narrow plot subdivision. There are vertical proportions in facades and a subtle variety of reveals behind a relatively consistent frontage line. The ordered visual richness and overall permeability invite the pedestrian forward.

Figure 4.172: De Veste, Brandevoort, the Netherlands. KK Architects, 1997–Present. Completed facades. *Archives KK Architects, Berlin*

Figure 4.173: Biezenlann/Koolstraat, Brandevoort, the Netherlands. KK Architects, 1997–Present. A street with a mixture of apartments and rowhouses. *Courtesy of Paul Murrain*

LONGMOOR STREET, POUNDBURY, ENGLAND / LÉON KRIER

Léon Krier, 1998

Main Street & Neighborhood Streets

Longmoor is one of the six radial streets of Middle Farm Quarter fanning out from its center, Middlemarsh Square in Poundbury, in Dorset. It is 250 meters (820 feet) long, between 7 and 12 meters (23 to 39 feet) wide, and rises on a low gradient toward Burraton Square, the western gate to the Quarter (Figures 4.174 and 4.175). This first quarter of Poundbury is so named because of the old Duchy of Cornwall farm, which for two centuries dominated the rolling countryside with its majestic beech-tree crown. Half-moon shaped, the quarter nestles into a natural topographic bowl, the market square occupying the bottom. The Village Hall and the Market Tower[82] form the visual foci for the radiating streets. They also dominate the long prospect of the preexisting Cambridge Road, the suburban road linking Middle Farm eastwards to the historic center of Dorchester, seat of the regional government. Longmoor Street is, in fact, the continuation of the Cambridge Road axis. The Market Square has thus become an urban heart not just for Middle Farm but for the neighboring Cambridge Road suburb as well.

The West Dorset District Council and the residents of the suburb refused initially all vehicular or pedestrian connections with the new Middle Farm urban quarter. We designed and built connections, nevertheless, erecting bollards to checkmate the opposition. When, after years of refusal, Cambridge Road residents expressed the wish to be linked, the Middle Farm Residents Association refused the offer. The bollards still stand as monuments to the motorized-cul-de-sac-attitude. Indeed, the urban network of Poundbury is at various points absurdly blocked by bollards to this day, totem poles of an immoderate "Health and Safety" cult. In fact, we control speed ranges by design and geometry rather than by signage and obstacles. There has not been an accident in nearly twenty years.[83]

The street-section of Longmoor is of variable widths, and the facades are at times interrupted by garden walls, allowing large trees to spread their leafy crowns into an otherwise rather narrow public space. The setting back of some houses articulates at once the street and mews frontages, avoiding the excessive repetition of house or garage alignments. A subtle curving of the street axis allows for a dramatic sequential deployment of private and public buildings. The nonparallel runs of facades, curbs, and sidewalks caused the planning authorities the same anxieties as my refusal of cul-de-sacs. The sheer countless versions of the master plan of this phase resulted not from design indecision but from the necessity to overcome unrelenting bureaucratic chicanery, eager to enforce strict obedience to suburban design guidelines, while realizing a piece of true city. At some point, my line-drawing perspectives caused such hand-wringing despair among authorities and consultants that my clients had them redrawn by graphic artists from the development and real-estate milieu. The little horrors still adorn the office walls of the Duchy of Cornwall. Now only surviving in a poor 35-mm slide, Carl Laubin's beautiful original oil painting based on my sketch (Figures 4.176 and 4.177) was meticulously dissected by a palace-crisis-committee. It then issued a painstakingly detailed report instructing Carl on how to edulcorate his masterpiece or face its extinction.

The same committee replaced my open Market Hall with the Village Hall, whereas we had envisioned the latter on a different square on higher ground dominating Ashington Street, another major radial street overlooking the open countryside. The monumental gable and perron would have created another civic magnet for this quarter, marking a prominent symbolic articulation on the public park at the town edge.

The character of the street and two squares is decidedly urban, while both public and private buildings are held in a vernacular tone; more formal architecture and paving are reserved for the Civic Center of the New Town of Poundbury, Queen Mother Square. The architecture of the Tower and Village Hall and the Burraton House Employment Center, on Burraton Square, are without rhetoric or figural adornments, their monumentality being achieved by sheer size of the whole and detail, creating a strong contrast with the modest shopfronts, houses, and garden walls (Figure 4.178). The master plan premise of variously sized, shaped, and used lots has been generally followed on Longmoor; however, the ground floors of the entire street were

Figures 4.174 and 4.175: Longmoor Street, Poundbury, England. Léon Krier, 1989–1992. Sketch plans. Successive sketches trace the evolution of the Longmoor Street design. *Courtesy of Léon Krier*

originally to be occupied by shops. The West Dorset District Council opposes the idea of commercial high streets in Poundbury for fear of unwanted competition with the Historic Center of Dorchester. We are therefore only allowed to have continuous commercial frontage on the quarters' main squares (Figures 4.179 and 4.180).

Only natural building materials and lime-based render are used for building elements. Curbstones are made of local Purbeck stone and are laid in unequal lengths and widths. At the time our builders were able to buy local stone more cheaply than international products. Roads are paved in blacktop, with sidewalks, and mews paved in clear gravel. Market Square and Burraton Square are paved as "shared spaces." Given the fact that of the varied lot shapes, sizes, and uses, the vernacular urban geometry, and the volumetric compositions themselves ensure a rich formal articulation, I had desired a greater uniformity in building materials than what actually became policy. After nearly twenty years of aging, however, the effects of weathering and planting have attenuated the impression of an initially too-rich palette of color and materials. This is a quality we are able to achieve now at once, thanks to the lessons learned.

Figure 4.176: Longmoor Street, Poundbury, England. Léon Krier, 1992. Perspective. Study sketch looking east. The perspective drawing shows the more open market hall and tower originally envisioned for the focal point of the village. *Courtesy of Léon Krier*

Figure 4.177: Longmoor Street, Poundbury, England. Painted by Carl Laubin, as originally envisioned by Léon Krier, 1992. View looking east towards the open market hall proposed by Krier. *Courtesy of Léon Krier / Carl Laubin*

Figure 4.178: Longmoor Street, Poundbury, England. Léon Krier, 1989–1992. Perspective. Study sketch looking west. *Courtesy of Léon Krier*

Figures 4.179 and 4.180: Market Square, Poundbury, England. Sketches. Léon Krier, 1992. Studies showing the public and private buildings framing the square. *Courtesy of Léon Krier*

Figure 4.181: Longmoor Street, Poundbury, England. Léon Krier, 1992. Comparative study—Traditional. The architectural language matters; in this depiction, an immersive environment is created from traditional forms in harmony. *Courtesy of Léon Krier*

Figure 4.182: Longmoor Street, Poundbury, England. Léon Krier, 1996. Comparative study—Modernist. In this sketch, Krier illustrates the same spaces shaped into another kind of immersive environment—still consistent, but now one of dissonant, individualized forms. *Courtesy of Léon Krier*

Figure 4.183: Longmoor Street, Poundbury, England. Léon Krier, 1996. Comparative study—mixed styles. In this version, Krier combines the two into an absurd mixture; "anything goes" is not an adequate instruction for placemaking. *Courtesy of Léon Krier*

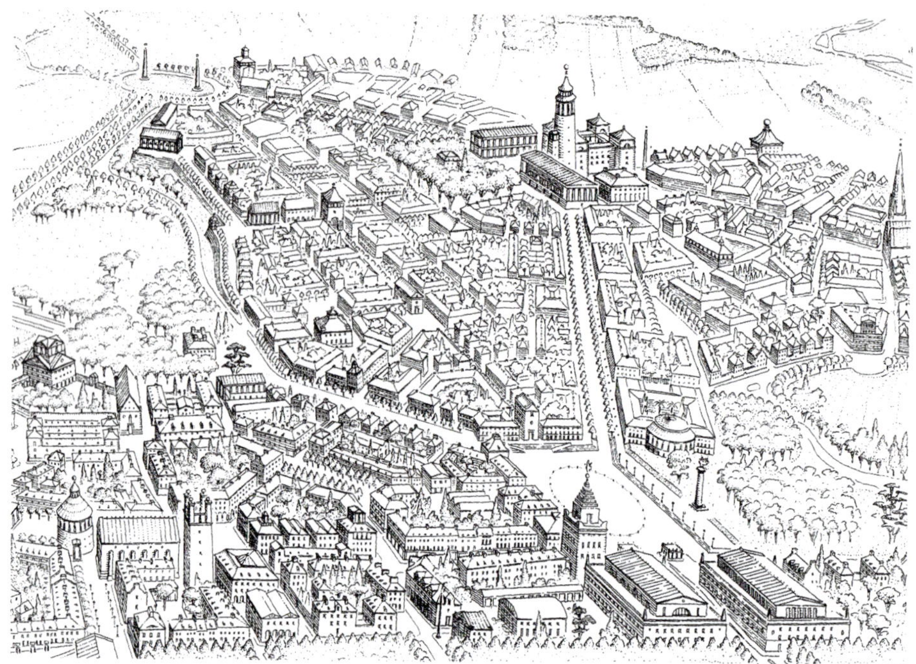

Figure 4.184: Poundbury, England. Léon Krier, 1989. Aerial perspective. When Krier was young, with little built work, he was mainly known through his drawings. Published in books, they spread new ideas about urban design. They are also sold in galleries, along with drawings by Michael Graves, Aldo Rossi, and others. *Courtesy of Léon Krier*

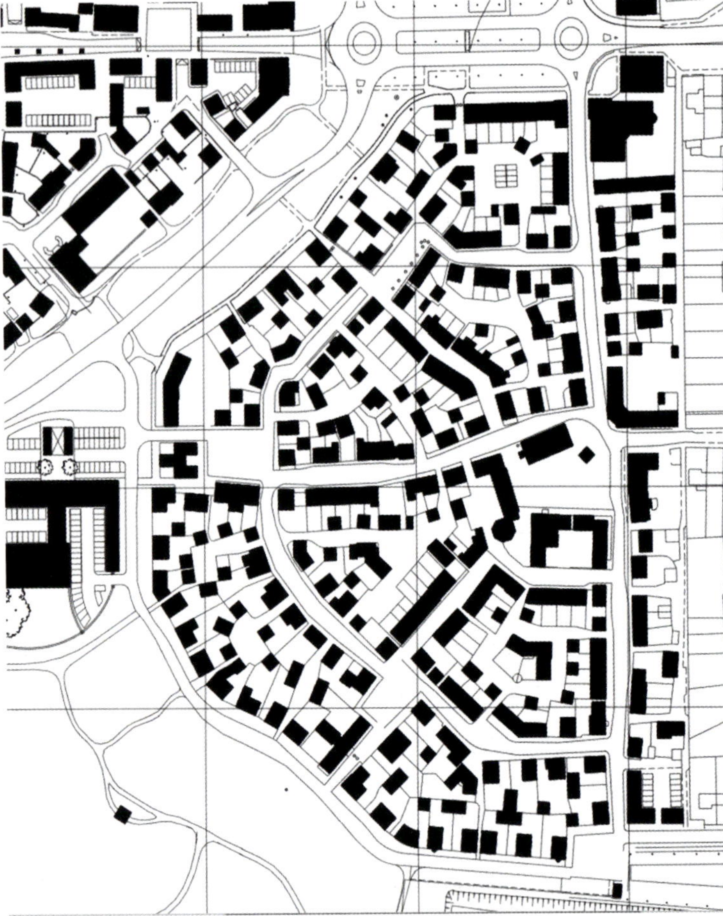

Figure 4.185: Middle Farm Quarter, Poundbury, England. Léon Krier, 1992. Plan. The Middle Farm Quarter was the first part of Poundbury to be built. *Courtesy of Léon Krier*

NANSLEDAN, NEWQUAY, CORNWALL, ENGLAND

ADAM Urbanism, Léon Krier, Colum Mulhern, 2013–Present

High Street

Newquay follows on the lessons and success of Poundbury. Like Poundbury, it is an urban extension developed and built by the Duchy of Cornwall in modern local vernacular. Construction began in 2013: 4,000 units will be built over fifty years on 540 acres. In 2024, the new Prince of Wales announced that he would increase the percentage of social housing at Nansledan from 30 percent of the project to 40 percent and add 24 houses for the homeless.[84]

Figure 4.186: High Street, Nansledan, Newquay, Cornwall, the United Kingdom. ADAM Urbanism, Léon Krier, Colum Mulhern, 2013—Present. Watercolor Rendering. The proposed High Street. *Courtesy of ADAM Architecture*

Bartram Street & Brasfield Square, Glenwood Park, Atlanta, Georgia.

Dover, Kohl & Partners and Tunnell-Spangler-Walsh & Associates, 2001

Rowhouse Street

Glenwood Park broke with the development industry's old habits and introduced the Atlanta region to new ways of doing things. Lots of things. It is a brownfield reclamation project, a traditional neighborhood, an infill development, a green-building trendsetter, a transit-ready development, a mixed-use/mixed-income/mixed-tenure neighborhood center, a small-is-beautiful business model—and an exemplar of walkable streets (Figures 4.187–4.189).

The twenty-eight-acre site, formerly home to a concrete plant, was organized into a neighborhood around an elliptical park. It was developed by Green Street Properties. The urban form of the neighborhood reflects the

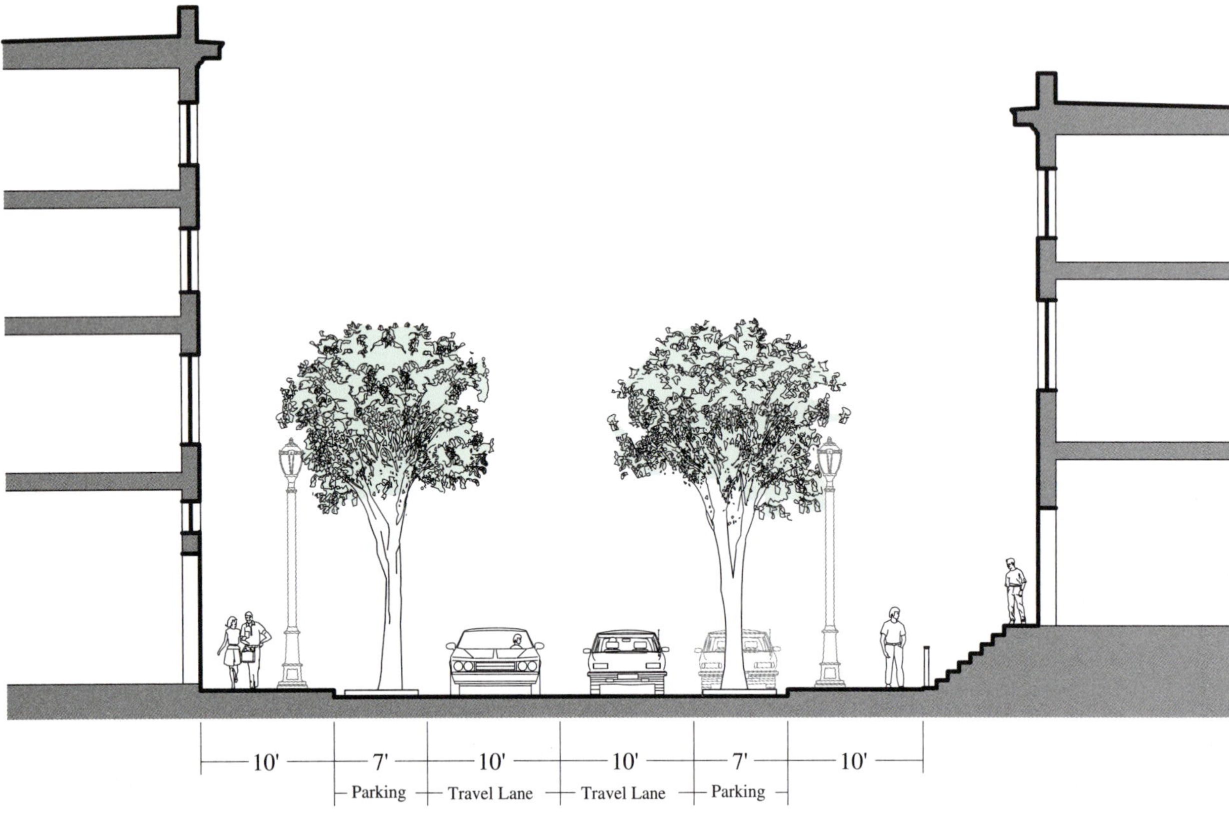

Figure 4.187: Bartram Street, Glenwood Park, Atlanta, Georgia. Dover, Kohl & Partners and Tunnell-Spangler-Walsh, 2001. Section. © *2013 Dover, Kohl & Partners*

Figure 4.188: Brasfield Square, Glenwood Park, Atlanta, Georgia. Dover, Kohl & Partners and Tunnell-Spangler-Walsh, 2001. Perspective. Rendering of the concept for the square. © *2001 Dover, Kohl & Partners*

Figure 4.189: Bartram Street, Glenwood Park, Atlanta, Georgia. Dover, Kohl & Partners and Tunnell-Spangler-Walsh, 2001. Aerial perspective. Detail of overhead view, showing the elliptical park at the end of Bartram Street. © 2001 Dover, Kohl & Partners

developers' goal to maximize the mixing of daily activities. Unlike every other local example of production building at the time, dwelling types, small businesses, and recreation were blended together in a fine-grained way.[85] The project emphasizes walkability, diversity, quality over quantity, and the importance of the public realm.

The developers of Glenwood Park set out to accommodate vehicular traffic without letting that control the design. It was, primarily, to be a place for people. This meant they had to wage a protracted and expensive technical battle with government officials over street details and dimensions. Eventually, the city permitted Green Street Properties to build narrower streets with smaller turning radii at the intersections. One result is that motorists drive slowly and carefully within the neighborhood—as they should. Another good outcome is that the City of Atlanta has steadily improved upon its standards and regulations, and it is gradually getting easier to implement traditional urbanism there.

Bartram Street serves as the main north–south axis of the community, connecting the elliptical park to Glenwood Avenue. The intersection where Bartram Street crosses Garrett Street—essentially Glenwood Park's main street—is really the psychological center of the neighborhood, and here Brasfield Square terminates Bartram Street in a formal, but small, civic space (Figure 4.190).

Brasfield Square itself is faced by mixed-use and live-work buildings on narrow lots. The high degree of spatial enclosure and the varied building facades make Brasfield Square the neighborhood's most memorable address. Note that here, too, is one part of Glenwood Park that demands a great deal of care and requires motorists to remain alert. Around the square the dimensions are tight and the turns are challenging (although few European motorists would be fazed by them). If you look closely enough at some of the corners, you will occasionally find black tire marks on the curbs. Those marks are worn proudly by Glenwood Park; they indicate that some speeder got a gentle reminder that this is a neighborhood where kids are playing and seniors are strolling, so they'd better slow down and drive through on the neighbors' terms.

North of the Square, Bartram Street necks down and is lined by rowhouses modeled after the predominant building type and architecture in Chicago's Lincoln Park neighborhood. Both sides of the street have parking, flanked by planting strips, sidewalks, stoops, and dooryards. Farther along, the view suddenly opens up to a roomy, slightly sunken ellipse, producing a contrast between the narrow space and the wide one that effectively dramatizes both. Connected yet varied spatial sequences like this create a genuine sense of the significance of the public realm in Glenwood Park (Figure 4.191).

Figure 4.190: Bartram Street interrupted by Brasfield Square, Atlanta, Georgia. Dover, Kohl & Partners and Tunnell-Spangler-Walsh, 2001. Photograph looking north on Bartram Street. *Courtesy of James Dougherty*

Figure 4.191: Brasfield Square, Atlanta, Georgia. Dover, Kohl & Partners and Tunnell-Spangler-Walsh, 2001. Photograph looking southeast. Upright, urbane live/work units face the square. *Courtesy of Justin Falango*

Figure 4.192: Bartram Street, Atlanta, Georgia. Dover, Kohl & Partners and Tunnell-Spangler-Walsh, 2001. Photograph looking north towards Faith Avenue.

GARDEN STREETS

Grace Park Way, Habersham, Beaufort, South Carolina

Duany, Plater-Zyberk & Company, 1997

American Garden Street

In the center of Habersham, lots are small and streets are edged with curbs. Away from the center, at the neighborhood edge, lots for the single-family houses are bigger and the streets gently become more suburban and rural near the marshfront. Grace Park Way (Figure 4.193) is in this looser section of the Traditional Neighborhood District's transect. It is notable that, even here, streets remain in a connected pattern, and a strong building-to-street relationship is legible, while lot sizes and the scale of houses vary.

Grace Park Way has a series of cottages on fourteen medium-sized building lots with shallow front yards. The street is narrow, without curbs, and shady. The pavement has an approximate width of eighteen feet, which allows two-way traffic to pass but discourages fast driving (Figure 4.194).

Even from the suburban areas within its range of Rural to Urban Transect zones, Habersham is walkable (Figures 4.195–4.197). A range of businesses and neighborhood institutions is within walking distance, so a car is not mandatory for every errand; Grace Park is less than half a mile away from the center of Habersham.

Figure 4.193: Grace Park Way, Habersham, Beaufort, South Carolina. Duany Plater-Zyberk & Company, 1997. Photograph looking northeast towards Mum Grace Park. The cottages and the curbless street are in the T3, or Suburban, transect zone.

Figure 4.194: Grace Park Way, Habersham, Beaufort, South Carolina. Duany Plater-Zyberk & Company, 1997. The developer chose an unusual number for the speed limit so that people would notice it and remember.

Figure 4.195: St. Phillips Boulevard, Habersham, Beaufort, South Carolina. Duany Plater-Zyberk & Company, 1997. Photograph looking north across St. Phillips Pond.

Figure 4.196: Mum Grace Park, Habersham, Beaufort, South Carolina. Duany Plater-Zyberk & Company, 1997. The lower density edge.

Figure 4.197: South Park Way, Habersham, Beaufort, South Carolina. Duany Plater-Zyberk & Company, 1997. At the neighborhood edge, the Light Imprint street is curbless and winding, saving mature trees.

Crystal Lake Drive, Hammond's Ferry, North Augusta, South Carolina

Dover, Kohl & Partners, 2005

American Garden Street

Hammond's Ferry is an infill development in the heart of downtown North Augusta, South Carolina. A new neighborhood on the Savannah River, the planned area includes two hundred acres that stretch one mile along the river, directly across from Augusta, Georgia.

Eventually, Hammond's Ferry will have shops, offices, and apartments in mixed-use buildings, and between 800 and 1,500 attached and freestanding houses. It was built by the North Augusta Riverfront Company, an affiliate of Leyland Alliance.

Crystal Lake Drive (Figures 4.198 and 4.199) is emblematic of the attitude the neighborhood founders adopted for Hammond's Ferry. It is a traditional neighborhood street, constructed around a sequence of pedestrian experiences and designed to show the houses well. During the planning process, Dover-Kohl visualized the

Figure 4.198: Crystal Lake Drive, Hammond's Ferry, North Augusta, South Carolina. Dover, Kohl & Partners, 2005. Section. © 2013 Dover, Kohl & Partners

Figure 4.199: Crystal Lake Drive, Hammond's Ferry, North Augusta, South Carolina. Dover, Kohl & Partners, 2005. Elevations along Crystal Lake Drive. © 2005 Dover, Kohl & Partners

experience of walking from Georgia Avenue, across the new trail that girds the historic downtown, through the neighborhood to the riverfront park, along the esplanade, and back again. Crystal Lake Drive literally follows that sequence as it makes the crucial connection between the older neighborhoods of downtown North Augusta and the new settlement on the riverfront. In the past, downtown North Augusta was cut off from its riverfront. The idea for Crystal Lake Drive was to reinforce the new, direct connection between the two in a memorable way.

For most of its length, Crystal Lake Drive is lined with a range of different types of housing—some houses are grand and large, others are small and modest, and there are rowhouses and apartment buildings as well (Figures 4.200 and 4.201). Small businesses have taken root within the growing neighborhood, and this becomes evident as the street intersects with Railroad Street. Subtle "cranks" or turns in the street tame traffic and create a village character by closing and opening the vista. For example, at one point, the street takes a substantial Z-turn and the space widens out into a small neighborhood

green. The green is unusually shaped, almost like two guitar picks placed back-to-back. This gives the houses unique addresses and views.

Much of Hammond's Ferry felt "grown in" early in the construction process (Figures 4.202–4.205). This is an example of what can be done with brand-new streets to establish both community character and a market position early on. Almost from the beginning, the sequence of experiences along the street showcased two key ideas about Hammond's Ferry. First was a pleasant new connectedness between the river and the rest of the town. Second is the neighborliness. Most homes have porches, stoops, balconies, and verandas within "conversational distance"[86] of the sidewalk. Everywhere, doors and windows face the public spaces, supplying "eyes on the street"[87] for a sense of safety and community.

The roadbed is generally twenty-four feet wide. When cars park in the informal, unmarked spaces on both sides, drivers going in opposite directions occasionally have to yield. The street widens into a slightly splayed form just before reaching the riverfront park and esplanade, opening sight lines to and from the waterfront.

Figure 4.200: Hammond's Ferry, North Augusta, South Carolina. Dover, Kohl & Partners, 2005. Aerial view of the present-day neighborhood. *Courtesy of North Augusta Riverfront*

Figure 4.201: Hammond's Ferry, North Augusta, South Carolina. Dover, Kohl & Partners, 2005. Rendering of the North Augusta Riverfront. In the past, downtown North Augusta was cut off from its riverfront; Hammond's Ferry unites the two. An aerial view shows the original vision for the riverfront square (since redesigned by others). © 2005 Dover, Kohl & Partners

Figure 4.202: Crystal Lake Drive, Hammond's Ferry, North Augusta, South Carolina. Dover, Kohl & Partners, 2005. Photograph looking southwest towards Savannah River. Front porches, not garage doors, face the street space. *Courtesy of Joseph Kohl*

Figure 4.203: Crystal Lake Drive, Hammond's Ferry, North Augusta, South Carolina. Dover, Kohl & Partners, 2005. Photograph from Railroad Avenue looking southwest. Mixing businesses, houses, and apartments feels natural at Hammond's Ferry, but it is rare today in newly developed neighborhoods. *Courtesy of Joseph Kohl*

Figure 4.204: Crystal Lake Drive, Hammond's Ferry, North Augusta, South Carolina. Dover, Kohl & Partners, 2005. Photograph looking north towards Railroad Avenue. Informal on-street parking means drivers going in opposite directions occasionally have to yield, keeping speeds down. *Courtesy of Joseph Kohl*

Figure 4.205: Crystal Lake Drive, Hammond's Ferry, North Augusta, South Carolina. Dover, Kohl & Partners, 2005. Photograph looking northeast. Where sizable houses face Crystal Lake Drive at Arrington Street, the other side of the same block has a grouping of modestly sized cottages. The street designs lend themselves to variety in households and building types. *Courtesy of Joseph Kohl*

Newpoint Road, Newpoint, Lady's Island, South Carolina

Vince Graham, Robert Turner, and Gerald Cowart, 1991

American Garden Street

Newpoint Road was the first street built in the break-through development of Newpoint—an early neo-traditional neighborhood across the Beaufort River from Beaufort's historic Old Point (Figure 4.206). Designed and largely developed before the term "New Urbanism" had been widely adopted, Newpoint Road has several impor-tant physical qualities that made it a model alternative to the typical suburban streets of the time. First, great care was taken to save the trees on the property; the street was threaded in between particularly impressive oaks. Second, the architecture along the streets in Newpoint was largely inspired by the Lowcountry coastal vernacular style seen, for example, on nearby Craven Street in Beau-fort. Together, the big trees and traditional architecture made Newpoint Road feel especially natural in its setting, to the point where it seemed almost instantly as if it had always been there. Locals started calling Newpoint Road the neighborhood's "historic district" about two years after the houses were built.

Third, although the house lots at Newpoint were not platted in an especially compact size because of local regulatory requirements at the time, some visual sleight of hand makes the scene feel neighborly and close; the lots are deep and narrow, with the narrow side along the street, and the houses are pushed forward on their lots and close to one another. In one stroke this design decision made the backyards more spacious and brought the houses within what developers Bob Turner and Vince Graham called "conversational distance from the sidewalk" (Figure 4.207).

Fourth, the calm visual impression and slow traffic speed on the street result from some simple, clever detail-ing. There is a moderate curb-to-curb width, an eight-foot planting strip, and a five-foot sidewalk. But the pavement surface is an old-school, exposed-aggregate asphalt that is slightly rough and matte-textured, unlike the shiny butter-smooth paving used in most residential subdivisions. This has the effect of discouraging speeders, yet it also makes Newpoint seem older and more natural.

Figure 4.206: Newpoint Road, Lady's Island, South Carolina. V. Graham, R. Turner, and G. Cowart, 1991. Looking west. The developers carefully built around the old trees, so Newpoint Road feels like an old street. In person, the new street looks like an upgrade compared to the scraggly, second-growth woods that covered much of the flat site.

Figure 4.207: Newpoint Road, Lady's Island, South Carolina. V. Graham, R. Turner, and G. Cowart, 1991. The porches on Newpoint Road are within "conversational distance" of the sidewalk.

Figure 4.208: Newpoint Road, Lady's Island, South Carolina. V. Graham, R. Turner, and G. Cowart, 1991. Looking east. Traffic calming as a by-product of placemaking: drivers have to slow down to go around the gazebo that interrupts the straight street and terminates the vista.

Last, since Newpoint Road is long and straight, the developers interrupted the straightaway with a small diversion around a gazebo (Figure 4.208). The gazebo itself is architecturally understated but highly visible, with its red roof and white columns silhouetted against the green backdrop. Urbanists familiar with the Lowcountry will immediately recognize the precedent at St. Philips Church in Charleston; the radically different scale and level of monumentality in these two examples show how some design ideas are adaptable, independent of scale and transect zone (Figures 2.149 and 2.150). Motorists see the gazebo from far down the street and slow down to go around it. When the developers were first requesting permits to build Newpoint, engineers and inspectors were not sure what to call this unconventional feature. Turner and Graham settled on "horizontal speed bump"; that sounded familiar enough, and they got their permits.

Newpoint became an influential prototype for a revamped vision of suburbia, one in which the character of the public spaces between buildings mattered more to home buyers than the quantity of space inside the house or the acreage of the lot.

West Drive in Warren Park, Rogers, Arkansas

Dover, Kohl & Partners, 2022

American Garden Street

Warren Park is an infill development planned as a pair of new traditional neighborhoods in the suburbs of Rogers, Arkansas. The two neighborhoods join along a slight ridge within an east–west easement that contains an overhead power line—the kind that cannot easily be buried or relocated. This easement aligns with an existing street (West Drive) to the east, currently running between sprawling neighboring subdivisions and a planned future link to the Razorback Greenway, the spine of the popular regional trail system.

The challenge in extending the street was threefold. First, Dover, Kohl & Partners needed to find a way to incorporate and work around the unsightly power line but not draw undue attention to it. Second, the street called for an upgraded level of accommodation for runners, walkers, and people on bikes making their way to the Razorback Greenway. Third, as the seam between two neighborhoods in Warren Park, the new extension of West Drive called for a unique treatment that would lend further prestige and distinctiveness to the addresses along it.

The first part of the solution was to jog and offset West Drive a half-block north powerline, putting it in a midblock alley behind houses instead of in front of them. The dense combination of houses and street trees provides multiple layers of screening from within the street or along the sidewalk. This simple and strategic adjustment of the block and street network focuses attention on the trees and front porches instead of the power line more than one hundred feet away.

The second part of the solution was to modify a classic American Garden Street, swapping out a row of on-street parking and a conventional sidewalk, replacing them with a wide multi-user trail instead, and then adding a third tree line so that trees frame both the trail and the carriageway, each running in its own allée.

Figure 4.209: Warren Park, Rogers, Arkansas. Dover, Kohl & Partners, 2022. Master plan. © *2022 Dover, Kohl & Partners*

Figure 4.210: West Drive, Warren Park, Rogers, Arkansas. Dover, Kohl & Partners, 2022. Perspective. Photorealistic rendering of West Drive. Example of American Garden Street, modified by adding a third tree line to frame the trail with its own allée. © *2022 Dover, Kohl & Partners*

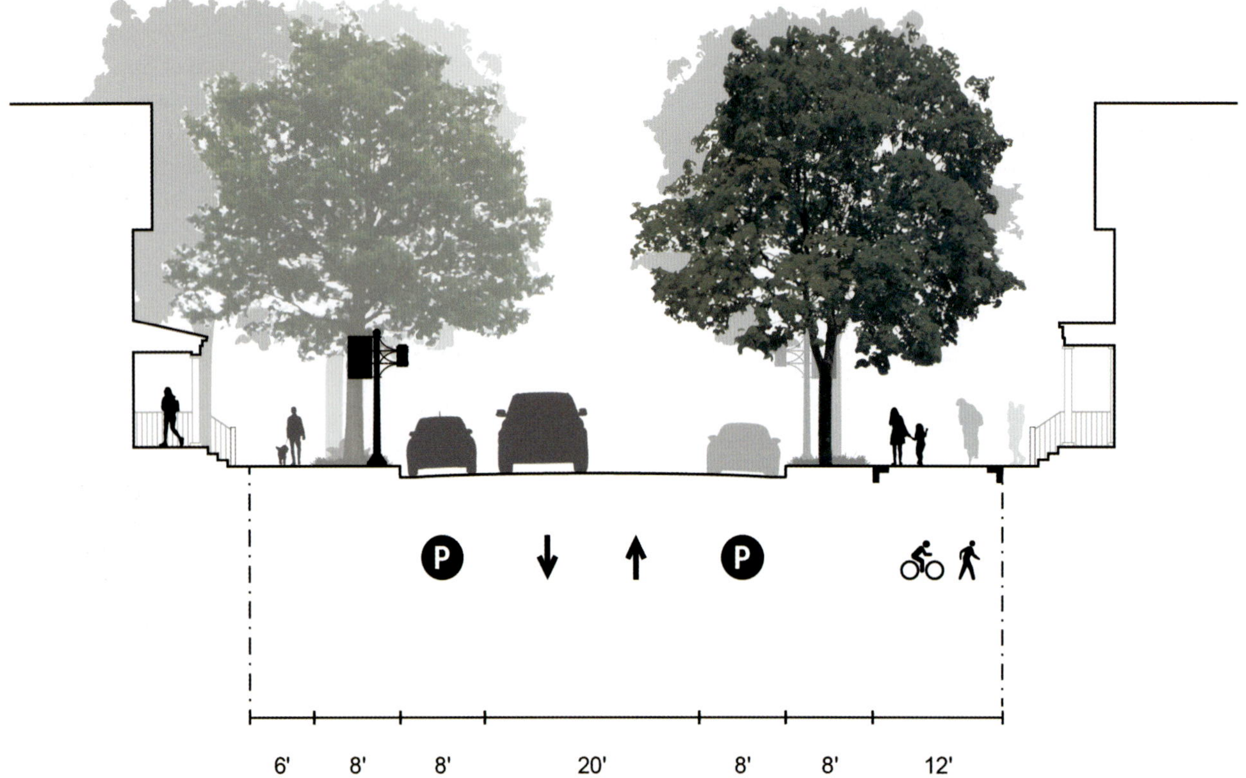

Figure 4.211: West Drive, Warren Park, Rogers, Arkansas. Dover, Kohl & Partners, 2022. Section. © *2022 Dover, Kohl & Partners*

GALT HOUSE DRIVE, THE NEW TOWN AT ST. CHARLES, MISSOURI / LAURA LYON

Duany Plater-Zyberk & Company, 2002

American Garden Street

The New Town at St. Charles is a New Urban development located far enough across the Missouri River from St. Louis that it is outside the metropolitan center. What was once an old sod farm began to come to life as a town in 2003. In 2013, New Town is ten years old, and it is also my home. I have been lucky enough to have arrived after plantings from the tree farm have grown in and are bountifully in tune with all four seasons. With undulating soils and the Big Muddy nearby, streets were drawn and plowed, cut and filled. Then, each street was framed so carefully, down to inches, to ensure a deep relationship to its users.

In New Town, the pattern of development gradually shifts from urban to rural. If you wander too far from the town's center, you will find agricultural crops growing within platted blocks, waiting for their future. But in the core, the inner sanctum, there is a framework of streets carefully and thoughtfully built out by the Town Architect and Town Founder. This net of roads, drains, and sidewalks—pervious and not—all works like a siphon, gathering people from their homes, cars, and parks, while moving water, wildlife, and seasonal color throughout the town.

I am fond of one street in particular: Galt House Drive. It is long in view but not in length (Figures 4.212–4.214). It reaches for all to be enveloped in its cross-section, but it is not wide in girth. It is stately and welcoming. It rises and falls from the tallest homes to the deepest fountains. In its proudest moments, it carries children to play, fishermen to the rocky piers, teens to sun on the banks of the canal, and mothers to the market, all while awaiting the community celebrations that each season brings.

The proud townhomes (recalling an eighteenth-century era of dignity) cast reflections in the grand fountains. The chapel peers from across the amphitheater, stealing the vista high and low, while the pedestrian is enveloped in a topologically whimsical cross-section. As the land reaches its highest point, the porches loom over, softened by cascades of ivies and blooms billowing out of garden walls and stairwells. The fountains bend to the windbreak of the view. In high rains, the waters lap at the walk. But this street also comes alive as the spine of the community, its social network. It is the prized place of honor for any parade-watcher. In winter, its bollards coated in ice and snow, this street waits quietly, majestically. Then its trees fill with blossoms and birds, and it awakens all the neighbors from their icy slumber and leads them back into spring. This is welcome. This is home.

Figure 4.212: Galt House Drive, New Town, St. Charles, Missouri. Duany Plater-Zyberk & Company, 2002. View towards the southeast, overlooking New Town Lake. *Courtesy of Lawrence Duffy*

Figure 4.213: Galt House Drive, New Town, St. Charles, Missouri. Duany Plater-Zyberk & Company, 2002. The streets in New Town are the backdrop for homegrown seasonal celebrations. *Courtesy of Lawrence Duffy*

Tupelo and West Grove Streets, Seaside, Florida

Duany, Plater-Zyberk & Company, 1981–Present

American Garden Streets

Urbanism takes time. Fifty years might pass before the last streets, parks, or buildings drawn in a master plan are completed. Or, during those years, there might be two new master plans. Most of us involved in urban design will be dead before any of our plans are fully built out. The Seaside plan was drawn in 1981. When *Progressive Architecture* visited in 1984, forty buildings had gone up, which is less than 10 percent of the final number.[88] Decades later (when we wrote this is 2012), houses were still rising. In 2024, the downtown continues to evolve and mature.

The *Progressive Architecture* critic Daralice Boles was impressed by the urbanism of Seaside when she wrote about it in 1985, although at that stage there was more *promise* of urbanism than the real thing. It is noticeable that not one of the photos in the seven-page article was a street view. In fact, although there were a few photos taken from Tupelo Street—the first street built—the roadway was carefully cropped out of every photo that included it.

One reason for the lack of street views was that Tupelo did not yet have any continuous streetwalls shaping it. The trees were too small, and the houses were so scattered that there was only one place with three houses in a row or two houses across the street from each other. Another view in the magazine shows Tupelo Street from behind, looking east over the scrub that covered the site. When that photo was taken, individual, freestanding houses rose above low scrub oaks that were barely higher

Figure 4.215: Tupelo Street, Seaside, Florida. Duany Plater-Zyberk & Company, 1981. Photograph looking south on Tupelo Street in 1983. Robert Davis carried out a number of experiments on Tupelo Street, the earliest street at Seaside. In the first year, seen here, there were palm trees and a clay roadbed. An unusually harsh winter killed many of the palm trees (which normally can grow on the "Redneck Riviera"), and the clay roadway was dusty once the traffic from buyers and potential buyers picked up. © 1983 Steven Brooke Studios

than the level of their raised floors, built on wooden pilings in the sand.

When the developer Robert Davis inherited the land, his grandfather had already clear-cut most of the site to build summer houses for the employees of the Pizitz Department Stores in Alabama. Many years later, the land had no plants or trees taller than three feet. The scrub was buffeted by salt air and winter winds off the Florida Gulf, so that in some years the oaks would grow an inch or two, but in other years they shrank.

Seaside's landscape code, one of the first *xeriscape* plans in the United States, emphasized the preservation

Figure 4.216: Tupelo Street, Seaside, Florida. Duany Plater-Zyberk & Company, 1981. Photograph looking north on Tupelo Street in 1984. In the second year, Robert Davis first planted tupelo trees and then sycamores as street trees but eventually settled on the same oaks that had originally covered the site. In 2014, the Seaside Community Development Corp. planted new palm trees downtown. © 1984 Steven Brooke Studios

Figure 4.217: Forest Street, Seaside, Florida. Duany Plater-Zyberk & Company, 1981. Looking east on Forest Street, towards Savannah Street, in 2011. Helped by the houses on the land breaking the wind and concentrating the rainwater, the Sand Live Oaks and shrub oaks grew rapidly. The oyster shell road surface seen in Figure 4.216 was replaced by bricks in 1985 because the oyster shells were too dusty in the summer when cars drove over them.

Figure 4.218: Tupelo Circle, Seaside, Florida. Duany Plater-Zyberk & Company, 1981. Looking east on West Grove Avenue in 2011. As the trees matured, they began to conceal some of the houses, reminiscent of tropical hammock street scenes in Coconut Grove, Miami, Florida.

of existing oaks and only allowed the planting of a few tree types that could withstand the climate without high maintenance. Eventually, the hundreds of houses and other buildings built for the resort buffered the wind and concentrated the rainwater on the site. Over time, the trees slowly grew, and the narrower east–west streets like Forest Street developed a natural tree canopy from the old scrub. Today, the streets can be quite lush, but they took many years to get to that stage.

In the two early pictures by the photographer Steven Brooke, one can see the original palm trees (in Figure 4.215) and the oyster shell roadbed (Figure 4.216). There are palm trees on the Florida Panhandle, but the first winter after the palm trees were planted at Seaside was a particularly hard one, and several of the trees died. Seaside replaced the palm trees in 1983 with sand live oaks, which today hide many of the houses on Tupelo and the surrounding streets (Figures 4.217 and 4.218). The effect is similar to the tropical hammock ecosystem that Davis knew from his former home in Coconut Grove, Florida, when he was a developer in Miami.

> Over time, the trees grew, so that narrower streets developed a natural tree canopy. Today, the streets can be quite lush, but they took many years to get to that stage.

Figure 4.219: Boathouse Close, I'On, Mount Pleasant, South Carolina. Duany Plater-Zyberk & Company and Dover, Kohl & Partners, 1995. Aerial view.

Boathouse Close, I'On, Mount Pleasant, South Carolina

Duany, Plater-Zyberk & Company and Dover, Kohl & Partners, 1995

Close

The notorious cul-de-sacs in suburbia aggressively disconnect and isolate people from their town. The bulb-shaped spaces at their ends fill no placemaking purpose but are simply vehicle turnaround areas. The overuse of the cul-de-sac pattern in America has given us too many dead-end streets and too few connecting routes to school, work, and stores.

On the other hand—although this is one of those medicines that should only be taken in small doses, once in a while—a dead-end street can be a useful feature in a plan. This is particularly true when site constraints make a fully connected street impractical. Traditional urbanism offers a range of solutions preferable to the suburban cul-de-sac, including the occasional close and the pedestrian court.

Closes have buildings arranged around a small square or green, with a lane or road fronting the houses. The central space, faced by the fronts of the buildings, is a "socially useful" public space (unlike the cul-de-sac's asphalt bulb).[89] The close itself is usually sized to accommodate the turning radii of fire trucks and moving vans, but the result does not look like automotive space.

The close can be a valuable addition to a neighborhood plan layout for several reasons. First, it offers a way to insert additional lots in what would otherwise be lost space in an irregular, oversized, or inefficient block, maximizing frontage and therefore lots; maneuvers of this sort that add interest and valuable curb appeal can be important in making the overall plan more financially feasible. Second, given its low traffic volume, the drive lane in the close can be designed to the lower-cost standards of driveways. Third, groupings of residences on a close lend themselves particularly well to varying lot sizes, unit sizes, and levels of luxury within the group.

Most of the streets in the new village of I'On are part of an interconnected web, without dead-end streets, but Boathouse Close is an exception to that rule (Figures 4.219–4.221). The close opens to Eastlake Road, on an otherwise unusually long block that gains a vista to the lake and some spatial relief from the close. At the opposite end, the close is anchored by a small but iconic civic building—the boathouse itself—and a view to the lake.

Figure 4.220: Boathouse Close, I'On, Mount Pleasant, South Carolina. Duany Plater-Zyberk & Company and Dover, Kohl & Partners, 1995. View looking west.

Figure 4.221: Boathouse Close, I'On, Mount Pleasant, South Carolina. Duany Plater-Zyberk & Company and Dover, Kohl & Partners, 1995. Photograph looking towards Eastlake Road. A neoclassical house (B. J. Barnes, 2001) on Eastlake Road shapes the space.

Three houses frame that view; each house has a front porch facing the Close. The result of this opening to the view is that there is a practical and psychological connection to the lake for residents who might live a block or several blocks away from the actual waterfront. The central space of the close is slightly splayed, eighteen feet wide at one end and thirty-four feet wide at the other. The tapering of the close as it moves away from the lake improves the view of the lake for the houses farthest from it, and the foreshortened perspective makes the space seem larger than it actually is. Facing the eastern end, a neoclassical house designed by B. J. Barnes on Eastlake Road formally concludes the space.

CHUCK MAROHN'S TAKE: THE DIFFERENCE BETWEEN A STREET AND A STROAD / CHUCK MAROHN

The highway system was a replacement for the railroad, and the railroad (as its name implies) is a road on rails—an efficient system for connecting one place to another. When highways were roads that connected two places, they also functioned well. The problem is that we have changed what a "road" is. We do not build roads or streets anymore. We have introduced elements of streets into roads, and vice versa. So our roads are no longer efficient connections, and our streets are financially unproductive. A street should be part of a network that allows you to get around in a place. Streets not only move cars; they also accommodate parking, walking, biking, people in wheelchairs, and people roller-skating. They serve as a framework for capturing taxable value from businesses, houses, office buildings, and hotels. But when we reconfigure our streets to have the characteristics of roads—as *stroads*—we are no longer able to capture the value or share the space. A modern stroad (Figure 4.222) is about the least safe traffic environment you could be in, too, with high-speed designs mashed up with turning traffic, stop-and-go traffic, sudden lane changes, and obnoxious signage. This ridiculously unsafe design is accepted as "normal" just because it was allowed to become ubiquitous.

STREETS, ROADS, STROADS, AND THOROUGHFARES / JOHN MASSENGALE

I grew up on, believe it or not, Goodwives River Road, an eighteenth-century street originally called Bell Avenue. When my Connecticut town became a railroad suburb of New York in the nineteenth century, the Town Selectmen decided that Goodwives River Road would sound like a more appealing place to live, so they changed the name. We like "roads" in New England, and there are many old roads we like living on.

England, too, has many roads that are favorite places. Some are beautiful country roads, lined with hedgerows and ancient trees. In the city, London has streets that were once country roads but are now good urban places, like Fulham Road, Brompton Road, and the King's Road—the hippest place in London during the Swinging Sixties. They sometimes have stretches with too much traffic, but less than 100 years ago they were shared-space streets. Chuck Marohn's point about the difference in character between streets, roads, and what he calls "stroads" is a valuable one, but it is a modern view based on the "thoroughfares" that DOTs and Functional Classification give us.

In other words, before Organized Motordom claimed control of our streets and redefined them to promote traffic flow, there was a large overlap between streets and roads.

In other words, before Organized Motordom claimed control of our streets and redefined them to promote traffic flow, there was a large overlap between streets and roads. Another London example is Shepherd's Bush, a quiet metropolitan suburb developed in the nineteenth century with block after block of terrace houses (rowhouses). The main streets are Uxbridge Road, Goldhawk Road, and Askew Road, built with small and mostly independent shops, pubs, and restaurants. Residential Street names end with "Garden," "Gardens," or "Road."

Terminology and Jargon

The dictionary tells us that the space between the curbs on a street is the road (engineers call it "the roadway"), while the space between the buildings, including the sidewalk and the road, is called the street. These precise definitions are not widely known, but it is not uncommon for someone in the city to say the sidewalk is "next to the road," or for someone in the country to think of walking out to the mailbox on the street. The ambiguous character of nature and public space in the suburbs further blurs the distinctions.

Many New Urban engineers and SmartCode writers do not agree with what I'm saying here. The prevailing school of thought says that if we want planners and engineers to respect us and our work, we have to adopt their precise language. Therefore, streets are only in the city, while roads are only in the country. Rather than saying all streets are roads, they say all streets and all roads are "thoroughfares," an archaic name that strikes most non-professionals as jargon. But as we move away from Functional Classification, the jargon can limit creativity. When we talk about "pedestrian mobility," we do not picture—or make—beautiful places.

While working on the second edition of *Street Design*, I asked a few friends who are accomplished writers and editors about planning jargon. When I introduced the phrase "the public realm," a *New Yorker* editor asked, "What's that?" He said the magazine would not allow it without an explanation. A friend with nine published books, four of them once on the *New York Times* bestseller list, had a one-word response to the word "pedestrianize": "Ugh." he wrote.

I like the use of the word "stroad" at the right time in the right context, but I also have some favorite "roads" around the world.

◄ **Figure 4.222:** Stroad, Anywhere, USA. A modern stroad makes pedestrians second-class citizens, in an unsafe environment. This ridiculous situation is accepted as "normal" because it was allowed to become ubiquitous. © *2012 Andy Boenau*

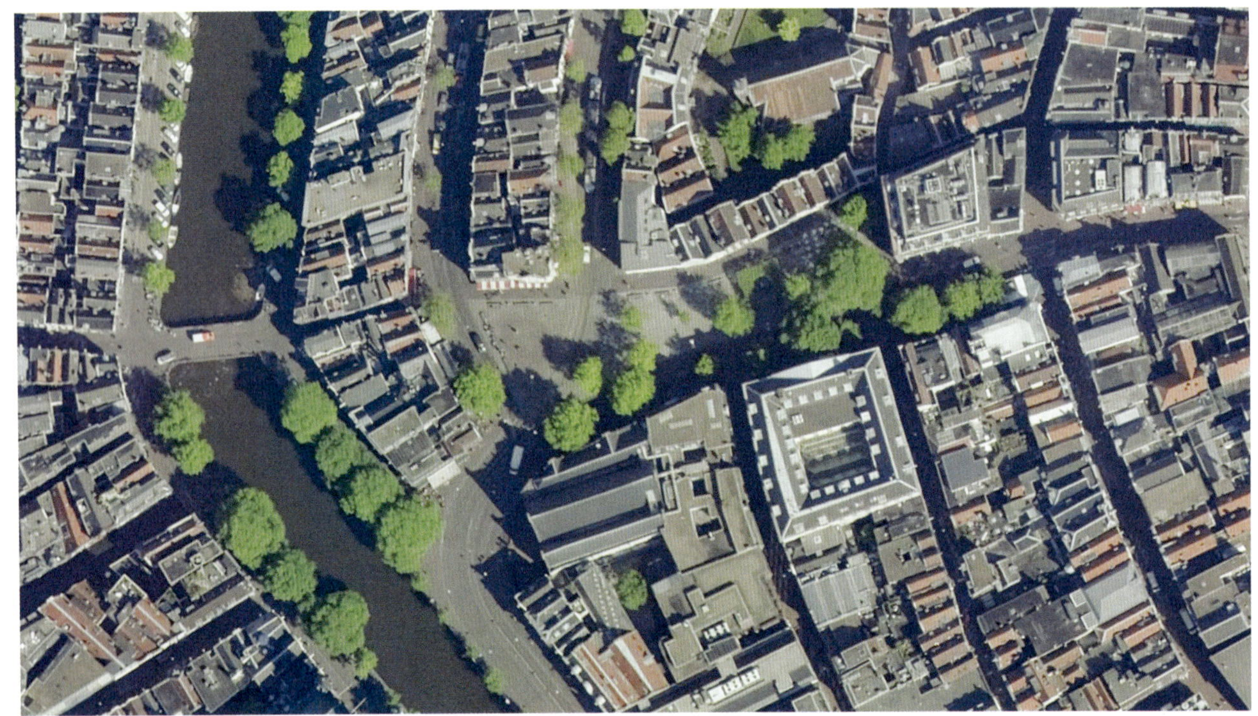

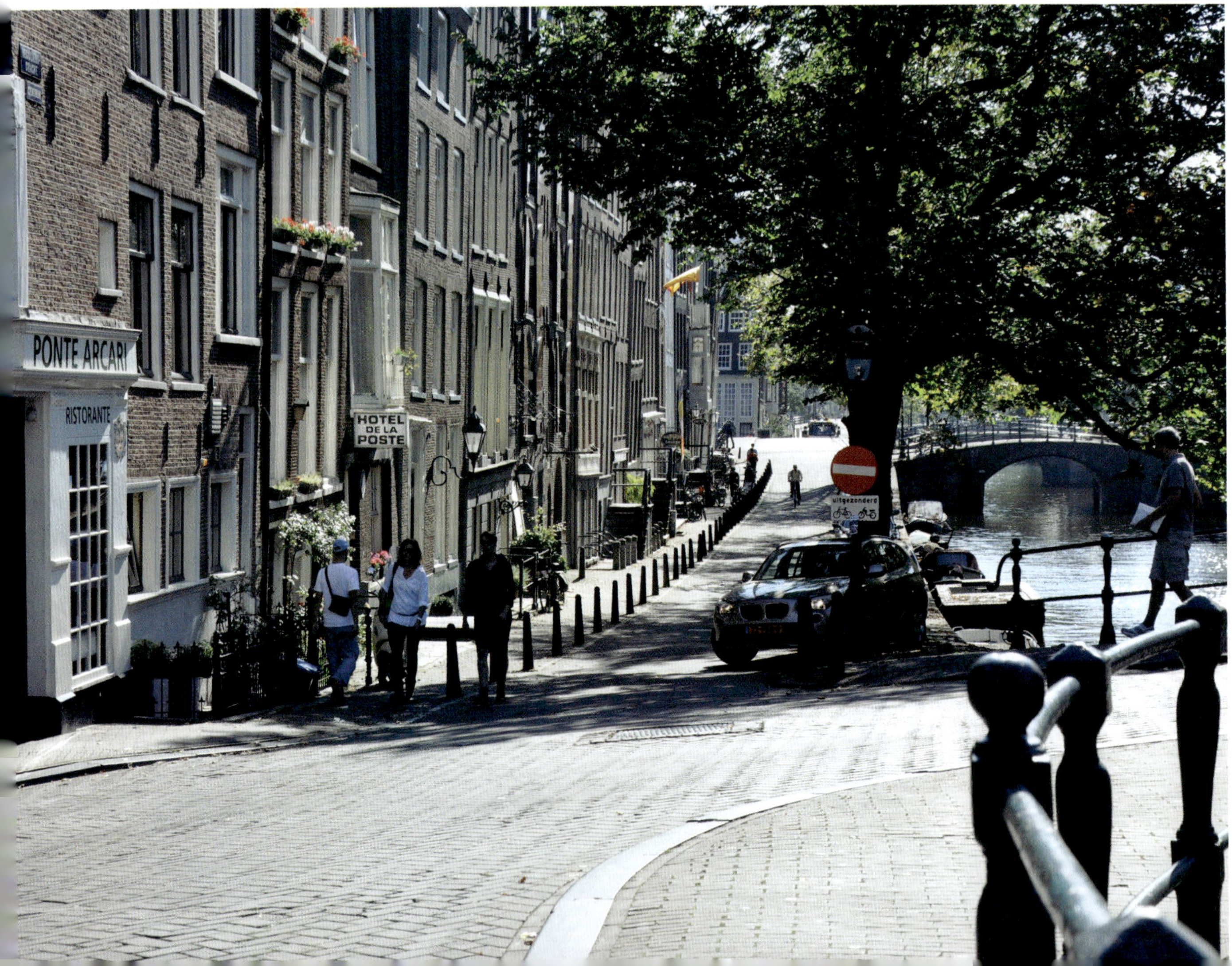
PONTE ARCARI
RISTORANTE
HOTEL
DE LA
POSTE
uitgezonderd

SLOW STREETS

The Spui, Amsterdam, the Netherlands
Shared Space

The corner in Amsterdam where Huidenstraat crosses the Singel canal on its way to the square called "the Spui" is an almost perfect urban place—which is one of the reasons why we put a photo of it on the cover of the first edition of this book (now Figure 5.1). Solid seventeenth- and eighteenth-century brick houses line both sides of the canal, most of them in a local Classical vernacular with simple trim and harmonious proportions that relate to the Classical music Amsterdammers love: 1 to 2, 2 to 3, 3 to 4, and so on. The houses are mainly red brick, but some have light-colored brick and some have dark brick. Most are three bays wide (three windows across), but there are also two-bay houses, and some are four or five bays wide. With the exception of the wonderful shop windows, all the windows are vertical and come from just a few "families" of windows. The window trim and the rest of the trim and moldings are usually, but not always, light stone or wood painted white. All in all, the houses produce a wonderful combination of the order and variety that creates good urbanism.

The similar but different houses line up along the street to make a solid street wall, punctuated by the rhythm of the vertical windows and the vertical houses with Dutch gables that make a varied skyline. The brick, stone, and painted wood are pleasing, and the street and sidewalks in front of the houses are a cheaper but also pleasant brick. Stone steps and low stone curbs in front of the houses separate the walks and the narrow street. Amsterdammers do not drive much, but they walk and ride bicycles a lot: the shared-space street is dominated by the bikes first and the pedestrians second. Cars and trucks come last although their drivers occasionally get impatient with the arrangement (Figure 5.52).

▶ **Figure 4.223:** The Spui, Amsterdam, the Netherlands. Satellite view. The canal on the left is the Singel, shown in Figure 4.226. *Courtesy of Google Earth / Aerodata International Surveys*

◀ **Figure 4.224:** Reguliersgracht, Amsterdam, the Netherlands. Photograph looking south on the Reguliersgracht from the Herengracht ("gracht" means "canal"). The roads alongside the canals rise up as they approach the bridges, creating a pleasant, man-made topography. *Photo by Andy Boenau / with permission*

The trees along the canal and the houses are a similar height, about 1½ to 2 times the width of the street, making a comfortable space ordered by the regular rhythm of the trees. On the other side of the trees from the houses is the canal, which is roughly twice as wide as the trees are tall. The proportions of the space are beautiful, and water in the middle of the city speaks to us in an almost mystical way (Figure 4.226). But that just begins to describe what makes the spaces so pleasant.

Amsterdam is flat, often built from fill, but the bridges and the streets approaching the bridges go up and down, making a pleasing, man-made topography (Figure 4.224). Almost all are arched bridges, highest at the center so that canal boats can pass underneath. The surface of the bridge typically slopes down at each end, and the streets coming off the bridges continue to slope down because most streets alongside the canals are low so that boats can easily dock there. The regular rise and fall create beautiful rhythms that delight the eye. The bend in the canal at the Huidenstraat adds another layer of delight.

After Huidenstraat crosses the Singel, it turns into a passage called Heisteeg, just wide enough for walkers and bicycles. Lined by human-scale buildings similar to the ones on the canal but with fewer stores, Heisteeg is approximately 75 feet long. Each side of the passage opens to well-designed, small-scale cafes and food shops before the passage ends at the northwest corner of the Spui, one of Amsterdam's most informal, most beautiful, and best-used squares (Figure 4.223).

Spuistraat comes into the square at that point. Lined with cafes, bars, and restaurants, it continues along the west end of the square, where there are outdoor tables that are heavily used day and night. The square has lightly trafficked roads along three sides and an important streetcar street (Nieuwezijds Voorburgwal) that cuts through near the western end. Most of the square is paved with stone blocks. The low-traffic streets along the edges have brick pavers, separated from most of the stone areas by low stone curbs. But the higher-traffic street with multiple streetcar lines narrows down to one traffic lane as it passes through the square, before changing from asphalt paving to the same stone paving as the no-traffic areas in the rest of the square.

"Spui" is the Dutch word for a particular type of sluice used in their canals. Where the square is now, there was once a body of water at the southern limits of the city. When the Singel canal was built as a moat around the city in the 1420s, the area was inside the city boundary, but

Figure 4.225: The Spui, Amsterdam, the Netherlands. Photograph looking west towards Nieuwezijds Voorburgwal. Cyclists ride along the south side of the square in the late afternoon.

the square was not filled until 1882. The Spui was renovated in 1996, and today it is both a destination and a place one happily passes through on the way to other destinations (Figure 4.225). Some of Amsterdam's primary shopping streets are just to the east of the square, so if you live on the Singel Canal or to the west, you might frequently go through the square. There are also important destinations to the north and south.

The Spui has the University of Amsterdam's historic library, and the square has traditionally been a center for bookstores. Every week, book and art markets set up there, and cafes, restaurants, and bars with outdoor tables are on every side of it.[90] On a workday, just a few minutes after five o'clock, every café and bar is full. Unless there is a hard rain or it is bitterly cold, people spill out into the Spui, sitting and standing, enjoying each other's company.

▶ **Figure 4.226:** Wijde Heisteeg, Amsterdam, the Netherlands. View looking south from a bridge over the Singel canal. In the 1960s, Amsterdammers decided not to tolerate the way the car was taking over the public realm of their beautiful city. Today, the vast majority of intersections there have no stop signs, no stoplights, and few traffic signs. The high quality of life in Amsterdam flows from the design of its streets, offering lessons for many other cities.

Dorn Avenue, South Miami, Florida

Dover, Kohl & Partners, 1993

Retrofit: Right-Sizing the Road

Slow Street

Dorn Avenue was one of the first streets in Florida to be *narrowed* in modern times; its retrofit was an early step in South Miami's citizen-driven Hometown Plan for its downtown. Unattractive and excessively wide (Figure 4.227), Dorn Avenue also had a prominent location, visible from South Miami's Metrorail platform and the town's two most important streets: Sunset Drive (its Main Street) and the South Dixie Highway (U.S. 1). Because it had the makings of a dramatic before-and-after sequence, Dorn Avenue was chosen in 1992 as a small "100 percent model" project to demonstrate what was possible on the neglected side streets of downtown.

Dorn Avenue is a key pedestrian link between the transit station and nearby multifamily housing, yet the street's narrow sidewalks, blank walls, and lack of shade made walking along it in 1992 unpleasant. South Miami prohibited outdoor dining, and high parking requirements and deep setback rules discouraged infill and redevelopment.

In a rebuilding of the street (Figure 4.228), Dover-Kohl narrowed the travel lane and replaced head-in spaces with parallel parking, which allowed significant widening of the sidewalks. The city installed pedestrian-scaled lighting and planted street trees in rows. The road surface and sidewalks were paved in bricks (which in hindsight have not weathered well and have been difficult to maintain, but which did achieve the effect of visually unifying the carriageway and the sidewalks). Six-inch-wide trench drains between the parking spaces and the travel lane were initially experimented with to minimize the need for both gutters and striping, minimizing the glare of concrete in the bright subtropical scene (Figure 4.229), but lackadaisical maintenance led to the trench drains filling with leaves; eventually, they were replaced with conventional, regrettable concrete valley gutters, but no one seems to notice the downgrade in elegance.

The local chamber of commerce, the downtown merchants association, property owners, and the City collaborated to press for the reconstruction of the street. To pay for the improvements, the City matched the privately raised funds with revenue from a local-option gas tax—meaning that motorists helped pay for improvements benefiting pedestrians and transit riders.

Figure 4.227: Dorn Avenue, South Miami, Florida. Photograph looking north. Before: Existing conditions circa 1992. It seemed less a street than a wide, barren alley, faced by blank walls. However, the scene is very visible from the Metrorail station platform (top center), so it was selected as the pilot street-transformation project.

Today, the street trees seem crucial to the scene, but they almost got scratched off the plans during construction. Once demolition began, city crews discovered an unexpected but crucial water pipe running under the sidewalk on one side, right where one of the new tree lines was to be aligned. There was no budget for relocating the line. City officials were prepared to eliminate the trees, but Dover-Kohl convinced them to realign this segment of the street instead, creating an asymmetrical cross-section so the trees could be planted after all (Figure 4.230).

The reconstruction of Dorn Avenue was big news. Although local authorities had plenty of experience with road-widening projects and shopping-center parking lots, this was the first time that *traditional* urban street dimensions and details had been used in greater Miami since the 1920s. In the learning process that followed, they transformed an anonymous and forgotten dead zone into the social center of the town. Eventually, South Miami reconstructed Sunset Drive. Following the Hometown Plan, the city narrowed it from five lanes to three, and they have remade several more downtown streets since. More upgrades are on the way; as we write this, Plusurbia Design is investigating alternatives for a further overhaul.

The physical changes prompted an overhaul of the land-development regulations. A new form-based code supplanted the conventional zoning, replacing setback rules with build-to lines, reducing parking requirements, and repealing the prohibition on outdoor dining.

Owners and merchants responded immediately, extending awnings, knocking windows into formerly blank walls, and eventually building new mixed-use buildings in the traditional main street format for the first time in seventy years.

The economic story was as profound as the bricks-and-mortar tale. In 1992, annual retail rents hovered around $6 to $7 per square foot, historic buildings were slated for demolition, and there were boarded-up storefronts around downtown. Ten years later, rents topped $60 per square foot, historic structures had been rehabbed, and cafés with white tablecloths had sprouted up on the sidewalks. Today, this block glows brightly on tax revenue analysis maps as the center of the City's most productive area in terms of property tax collected per acre.[91]

Decades later, the progress in South Miami continues, albeit with imperfections and occasional reversals. Spurring the rehabilitation of existing retail buildings proved easier than attracting the new housing and lodging needed in the downtown, at least at first; today, however, hundreds of new downtown residential units are finally under construction. The parking-policy reforms were unwound, reworded, reinstated, repealed, and reinstated again as the local political pendulum has swung from year to year. More than thirty years later, design work is underway on another remake of the street, with an even better pedestrian realm to coincide with the construction of a new residential tower along the street.

Figure 4.228: Dorn Avenue, South Miami, Florida. Dover, Kohl & Partners, 2014. Photograph looking south. After: Today the street is the social center of the town. Informal sidewalk cafés line the narrowed street. *Courtesy of Kenneth García*

Figure 4.229: Dorn Avenue, South Miami, Florida. Dover, Kohl & Partners, 2014. Photograph looking south. After: The least possible amount of space is devoted to circulating cars; speeds are slow. *Courtesy of Kenneth García*

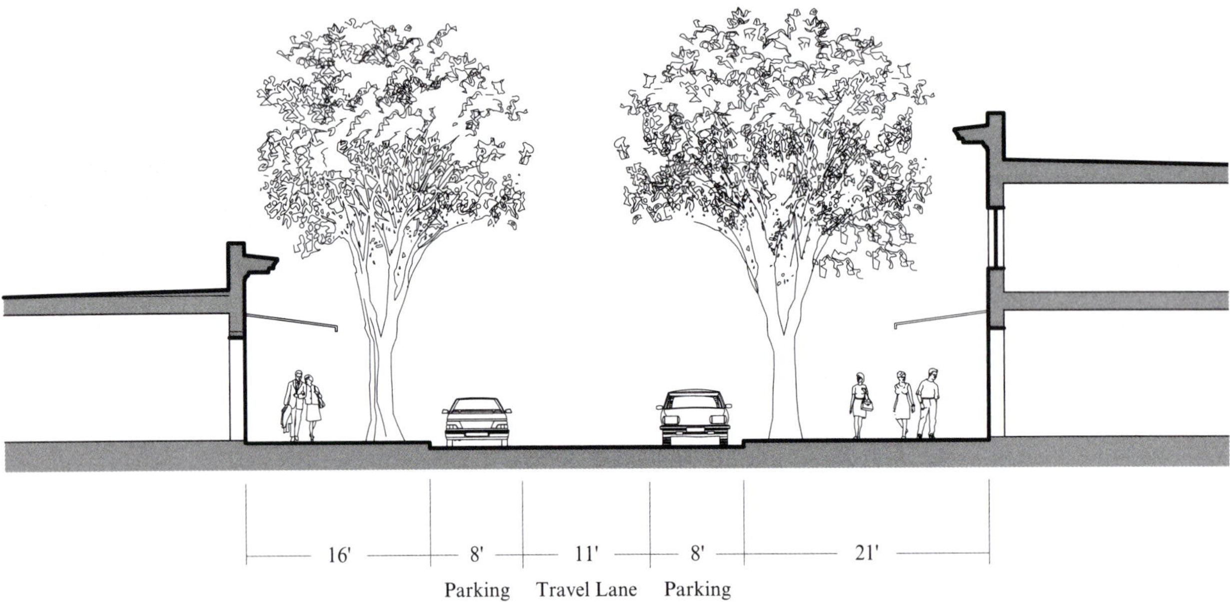

Figure 4.230: Dorn Avenue, South Miami, Florida. Dover, Kohl & Partners, 1993. Section showing the last-minute asymmetrical solution that allowed for street trees. © 2013 Dover, Kohl & Partners

Exhibition Road, London, England

Dixon Jones, 2012

Retrofit: Experimenting with Shared Space

Shared Space

Exhibition Road is another rebuilt street in the Royal Borough of Kensington and Chelsea (see the Kensington High Street, page 388). Lined with institutional buildings, Exhibition Road is a broad street, home to the Victoria and Albert Museum, the Science Museum, the Natural History Museum, the Royal Geographical Society, and Imperial College London. Like Kensington High Street, it had too much traffic, but the borough tried a different experiment on Exhibition Road than they had on the High Street, dividing most of the street down the middle. It gave one side exclusively to pedestrians and the other to both pedestrians and cars. The result is a mixed success (Figures 4.231 and 4.232).

Exhibition Road runs from the South Kensington Tube Station north to Hyde Park, where the 1851 Great Exhibition of the Works of Industry of All Nations was held in Crystal Palace. A lack of coherence on the street and in the surrounding area led to ambitious plans in the late twentieth century by Norman Foster, Daniel Libeskind, and others—which all foundered. In 2003, the architecture firm Dixon Jones won a competition organized by the borough to redesign Exhibition Road. They designed the shared space we see today.

The scheme begins at the intersection of Thurloe Street and Exhibition Road. The rebuilt road changes character as it goes north. First, it crosses two busy streets, Cromwell Road (the A4, one of the main routes in and out of London) and Thurloe Place (the A3218). In between these two streets is a short transitional block, where almost everyone is just passing through. North of Cromwell Road, the wide road runs straight as an arrow to Hyde Park. This long stretch is now one of the best-known shared spaces in England.

The innovations at the southern end of the street are more successful than the main stretch between the museums and other institutions. Thurloe Street and the adjacent plaza are small-scale, comfortable spaces that attracted people and new uses. Cafés with outdoor dining, a bookstore for the Victoria and Albert Museum, and other shops are always filled.

Cars can enter Exhibition Road here, but it feels more like a piazza than street. Low ventilating stacks for the Underground station and a tunnel leading to the station poke up in the middle of the space, helping the spatial definition. Built-in benches around the stacks are frequently occupied by those who do not want to have to spend money to spend time people-watching, and the few cars and vans that park in the square cluster around the stacks, giving some order to the parking. On the street, bold striping in light and dark grays sets the space off from the through traffic on Thurloe Place, at the top of the square.

HANS MONDERMAN, SHARED SPACE & THE MONDERMAN RULE

We quote Hans Monderman at the beginning of Chapter Five: "If you want vehicles to behave like they are in a village, build a village."[92] Monderman was a Dutch traffic engineer who saw that the signs, stripes, and signals his colleagues typically used to improve safety ironically served instead to lull motorists into an unsafe complacency. He pioneered modern concepts of shared space and reduced the detritus of modern traffic engineering.

Monderman was a conventional planner when he began his career, but as his experience grew, he became skeptical about the engineering "improvements" and came to believe there were many places in Dutch cities and towns where the best way to control traffic would be to remove curbs, signs, and markings, forcing the driver and the pedestrian to look each other in the eye when negotiating their way on the road

"The trouble with traffic engineers is that when there's a problem with a road, they always try to add something," Monderman says. "To my mind, it's much better to remove things." We call that "The Monderman Rule." Related design principles are the K.I.S.S. Rule (page 388), the Coco Chanel Rule (page 400), the "secret of successful retail' that says the "experience of the public realm is what's important" (page 150), and Mies van der Rohe's dictum that "Less is more."

> The trouble with traffic engineers is that when there's a problem with a road, they always try to add something. To my mind, it's much better to remove things.
>
> —Hans Monderman

Figure 4.231: Exhibition Road, London, England. Photograph looking south towards Cromwell Road. Before: Home to some of the most important cultural institutions in the city, Exhibition Road was dominated by the car. *Property of Royal Borough of Kensington and Chelsea*

Figure 4.232: Exhibition Road, London, England. Dixon Jones, 2012. Photograph looking north towards Cromwell Road. After: The super-graphic hyperstriping shouts, "Look at me." A row of Plane trees could divide the pedestrian space from the shared space, but the architects opted for thin metallic pylons instead. *Property of Royal Borough of Kensington and Chelsea*

The effect of these paving techniques on the long, wide piece of Exhibition Road to the north is less pleasing, because here the street feels vast and poorly shaped. The bold diagonal striping is visible over long stretches, and since there are no sidewalks, the supergraphic bumps into the institutions lining the road in an almost random and uncomfortable way. On the small piazza to the south, on the other hand, the paving pattern seems less repetitive and is more broken up by the ventilation shafts and the benches around them, the parked vehicles, and the number of people in the space. The outdoor tables on the southern block of Exhibition Road also cover the supergraphics at the edges, softening their effect.

A smaller pattern north of Cromwell Road would bring a more human scale to the vast space, and breaking it into smaller parts would help, too. A more traditional design would make borders along the edge and break the long space into smaller parts. Cars park in the center of the road, which has traffic on one side and pedestrians on the other. "Only the parked cars look comfortable," said Hank Dittmar, the former Chief Executive of the Prince's Foundation.[93]

An allée of London Plane trees would have been an effective way to shape and modulate the space, but Dixon Jones instead went for tall metal pylons down the center of the street. This out-of-scale and minimally detailed choice brings to mind two contrasting short passages into the Mount Street Gardens, a lovely midblock park a short distance away in Mayfair. One is a short block treated as a normal street, with building entrances, parked cars, and a Barclays Cycle rental station (now Santander Cycles, still popularly known as Boris Bikes). Even without trees, it is a beautiful small block. At the other end of the park is a short block of about the same size that terminates Carlos Place. Closed to traffic, it has an austere, minimalist design that is close to sterile. This block has a certain cerebral charm but little of the visual richness that makes the street at the other end of the gardens so charming. If its small, scrawny trees grow to a mature height, it may be a very good space, eventually. But the trees are in raised planters that may prevent them from growing, and the last time the authors visited the small space, like Exhibition Road, it suffered from too much fashionable ideology and too little placemaking.

A theme of this book concerns the perils of specialists making their specialties special. Engineers may overengineer, "pedestrian mobility" specialists draw special crosswalks that attract too much attention, and e-mobility

transportation planners design brightly colored cycle lanes that look like the most important thing on the street. Architects have given us many examples of overdesign, reinventing the street in a way that yells, "I'm special," even when it is austere and minimal.

By contrast, one of the greatest achievements of the other new road in the borough, Kensington High Street, is that borough Councillor Daniel Moylan made sure that everyone worked towards the common goal of a better Complete Street. Its clean, modern details are always understated, in the service of a simple street design, rather than for the purpose of making an architectural statement. Architectural expression can serve placemaking, but other values frequently speak more loudly.

Swift Street, Buena Vista, Colorado

Dover, Kohl & Partners and Peter Swift, PE, 2004

Village Street

"Swift Street" is an ironic name, because it is among the *slowest* new streets in the United States. It was named after transportation engineer Peter Swift, a tireless advocate of safer, saner streets where pedestrians are welcome.

Swift Street and the adjacent Swift Circle are part of the thirty-eight-acre South Main neighborhood in Buena Vista, Colorado (Figure 4.233). Dover-Kohl designed South Main for the developers Jed and Katie Selby as an extension of the small town's historic main street, which was lengthened to reach the new whitewater park, a regional recreation destination. The plan for the neighborhood reconciles Buena Vista's two existing street grids, joining them at a new square on the Arkansas River.

Swift is a narrow and intimate street. It was designed to contrast with the wider, more formal South Main Street, which frames long axial views of the surrounding peaks and gives the neighborhood with a legible orientation, rooting it in the town's larger context of mountain, valley, and river. Swift Street, on the other hand, is all about the short, close-up views across the skinny space and the experience of being in it. Swift collaborated on the design of all the thoroughfares in the South Main neighborhood and calls them his "proudest achievement," crediting the town staff and fire officials for their open-mindedness and "can-do attitude." The street was originally conceived as a *woonerf*, or pedestrian-dominated shared space,[94] but later plans were simplified to a more straightforward and

Figure 4.233: South Main, Buena Vista, Colorado. Dover, Kohl & Partners, 2004. Aerial rendering of the neighborhood. *© 2004 Dover, Kohl & Partners*

Figure 4.234: Swift Street, South Main, Buena Vista, Colorado. Dover, Kohl & Partners and Peter Swift, PE, 2004. Photograph looking northeast from South Main Street. The street is flanked by mixed-use buildings at S. Main Street; a small turret on the house where Swift Street bends in the distance is a focal point. *Courtesy of Justin Falango*

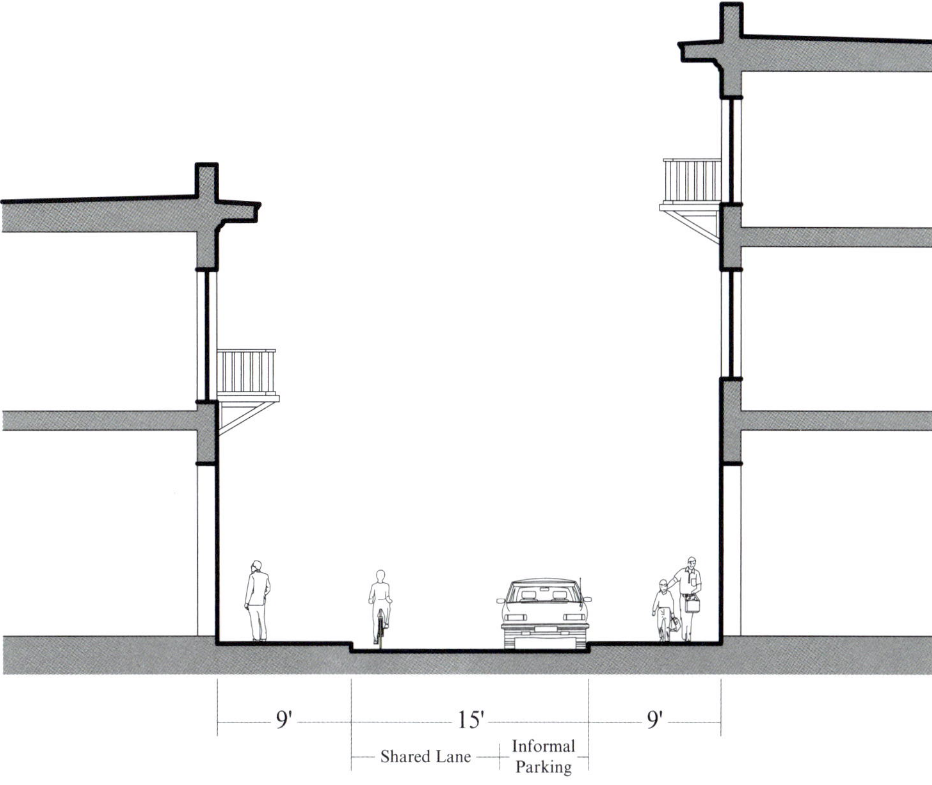

Figure 4.235: Swift Street, South Main, Buena Vista, Colorado. Dover, Kohl & Partners and Peter Swift, PE, 2004. Section. Street segment with parking. © *2013 Dover, Kohl & Partners*

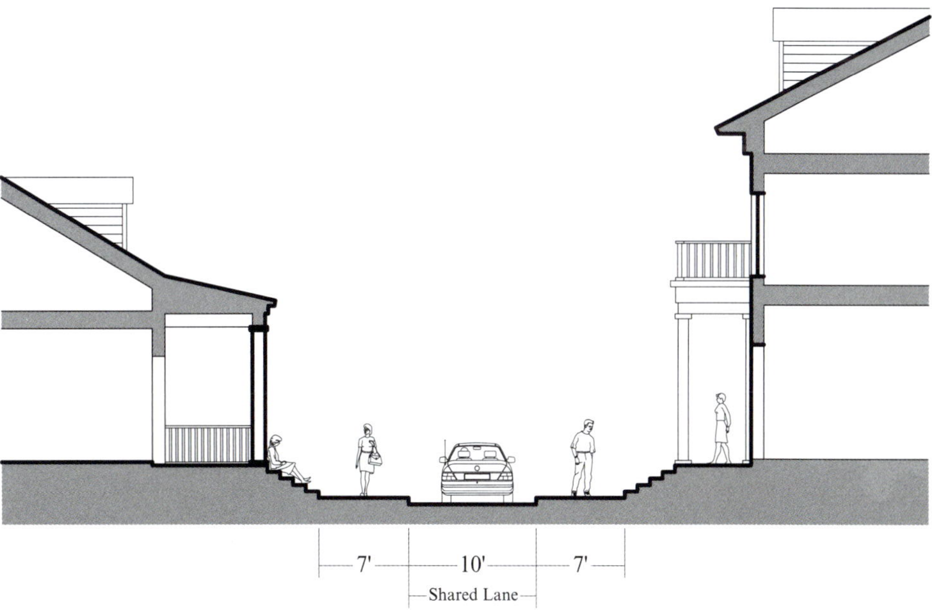

Figure 4.236: Swift Street, South Main, Buena Vista, Colorado. Dover, Kohl & Partners and Peter Swift, PE, 2004. Section. Street segment without parking. © *2013 Dover, Kohl & Partners*

Figure 4.237: Swift Street, South Main, Buena Vista, Colorado. Dover, Kohl & Partners and Peter Swift, PE, 2004. A form-based code controls both the streets and the buildings. *© 2004 South Main*

Figure 4.238: Swift Street, South Main, Buena Vista, Colorado. The roadbed is paved with local glacier stones, creating a connection to its rugged surroundings. *Courtesy of Kennley M Selby*

traditional street, albeit a narrow one. The result could be an example for many other American towns.

Swift Street intersects with South Main, creating a convenient location for a corner store, a local restaurant, and live/work units (Figure 4.234). To accommodate informal parking for the mix of uses emerging in the neighborhood, Swift Street briefly becomes wider and two-way where it crosses South Main Street (Figure 4.235). The street section then shifts, narrowing to a total curb-to-curb width of twelve feet to establish an ultra-slow-speed outdoor room that is quiet, safe, and easy to navigate (Figure 4.236). The roadbed is paved with hand-set local glacier stones, giving the street a visceral connection to its rugged surroundings (Figures 4.237 and 4.238). The choice was intended to keep the speed down—pedestrian advocate Dan Burden says seven miles per hour feels about right there—but the rough stone road surface also creates a kind of local distinctiveness to boot. Striping, signage, and other traffic-control devices are practically moot at these slow speeds and are kept to an absolute minimum. A narrow strip of flat, smooth concrete is incorporated in the wider section to keep the stones in place, with the added benefit of a few seconds of smooth riding for cyclists.

Sound environmental design was a high priority for the extension of downtown Buena Vista, where water is at a premium. In addition to implementing limited and water-efficient landscaping standards, the Selbys carefully considered other aspects of the local climate. The overall design corresponds with the building traditions of the region by responding and adapting to wind and weather patterns. The architecture maintains the sense of place with the Western mining-town vernacular but also shows adaptations of the traditional building types; the local architectural style is cross-pollinated with influences from that of other dry, high-altitude places. Materials such as adobe and stone provide insulation, high floor-to-ceiling heights promote air circulation, and shaded exterior spaces encourage outdoor activity.

The quality of the street space results from the combination of architectural design and details within the right-of-way. To get this ensemble right, a form-based code carefully controls both the street and the buildings (Figure 4.238). Every lot has build-to lines or zones that determine when stoops or porches are required. On the regulating plan, a "Privacy Side" symbol designates the side of the building where windows must be at least six feet above the average ground elevation, providing a measure of privacy for neighboring buildings. The primary

roof-ridge orientation for each lot is also predetermined in the code to optimize sun exposure and allow buildings to capitalize on the potential use of solar energy. These building regulations are accompanied by tight instructions from the urban designers on the configuration of the thoroughfares, via the Street Standards in the form-based code.

Twain Avenue, Davidson, North Carolina

Dover, Kohl & Partners, 1997

Yield Street

The St. Albans neighborhood[95] is an infill development near Davidson College, devised in response to the form-based Land Plan and code adopted by the town in the 1990s. The neighborhood has a large park, civic buildings, live/work units, large houses, cottages, and rowhouses.

Twain Avenue is a gently sloping street that connects a small neighborhood center to the formal Faulkner Square, another landmark in the new neighborhood (Figures 4.239–4.241). It is faced with rowhouses that terrace down the slope. Although the rowhouses may appear to resemble one another from the street—like facing like—they are actually different in plan, size, and cost. On one side of the street, the row of houses serves as a "liner" that screens the church parking lot, with semidetached garages on the alley. On the other side, the rowhouses are smaller park-under units that fit on shallower lots.

The street itself is as narrow as regulations would permit. It is a "yield street," meaning in this case there is two-way traffic but only one travel lane shared by motorists going in both directions. Drivers must sometimes pull aside to let oncoming cars pass, so they stay alert and go slowly. A travel lane and parallel parking on one side of the street have a combined width of twenty feet curb-to-curb, reflecting current fire regulations. Keeping the carriageway narrow saved enough room for street trees in a five-and-a-half-foot planting strip. Sidewalks are five feet wide, flanked by dooryards and stoops.

The neighborhood was created in an uneasy union of high-minded New Urban principles with the old habits of corporate builders accustomed to suburban subdivisions. This can be done successfully, but there are a handful of compromised details that would have turned out better had the builders been willing to follow tradition more closely. For example, the porches on Twain Avenue were oddly tucked within the building envelope behind

Figure 4.239: Twain Avenue, St. Albans, Davidson, North Carolina. Dover, Kohl & Partners, 1997. Photograph looking southwest towards Caldwell Lane. The street is framed by rowhouses that vary in size, cost, and floor plan. *Courtesy of Dover, Kohl & Partners / Douglas Boone*

Figure 4.240: Faulkner Square, St. Albans, Davidson, North Carolina. Dover, Kohl & Partners, 1997. Photograph looking southeast. The square is an elongated rectangle with narrow, tree-lined streets on each side. The elderly woman in this photo referred to the trees as her "lifeline," saying simply because they were planted, she gets her daily exercise and knows her neighbors.

Figure 4.241: St. Albans Lane, Davidson, North Carolina. Dover, Kohl & Partners, 1997. Photograph looking northeast towards St. Alban's Episcopal Church. The St. Alban's Parish steeple anchors two streets that intersect to make a small neighborhood center.

DORPSSTRAAT, ROTTERDAM, THE NETHERLANDS

BGSV and Rod'or, 2008

Bicycle Street (Fietsstraat)

Figure 4.242: Rotterdam, the Netherlands. BGSV and Rod'or, 2008. Photograph looking northeast. A Bicycle Street (the Dutch "fietsstraat") gives people on bikes priority over motorists, and the cyclists set the pace. The carriageway is divided into three narrow lanes instead of two: smooth travel lanes on either side and a middle lane that has a rough texture. Motorists have to drive over the rough texture to pass, forcing slow speeds. *Courtesy of BGSV bureau voor stedenbouw en landschap*

Figure 4.243: Rotterdam, the Netherlands. In the Dutch version, unsanctioned markings sometimes have variations on *"Auto te gast"* ("Your car is a guest"). © 2015 *Ben.manibog / Wikimedia Commons / CC BY 4.0*

HENNEPSTRAAT, UTRECHT, THE NETHERLANDS

Living Street (Woonerf)

Figure 4.244: Utrecht, the Netherlands. Photograph looking south towards Griftstraat. Living Streets (the Dutch "woonerfs" or "woonerven") have a maximum speed of four miles per hour. The speed limit applies to cyclists as well as cars, matching the average walking speed. With a narrow and usually curbless design, living streets become shared spaces where children can play and neighbors can interact safely. *Courtesy of Rebecca Albrecht*

the front facade, with flimsy aluminum railings, ultimately an unsatisfying experiment on the builders' part. One wishes the rowhouses had more traditional projecting stoops and porches like many in Savannah.

Another unfortunate detail is that the dooryards don't have the individual character they would display in Savannah, Alexandria, or Georgetown. There is one maintenance contractor, one plant palette, and a bland uniformity. This streetscape creates an exceedingly consistent plantings that gives off the faint scent of condo-style common management (Figure 4.239). Something about the lack of spatially defined side property lines betrays the dooryard space as condo-complex common area—which, of course, it is. The planting "strip," while green, subtly disengages the houses from the street. These criticisms aside, Twain Avenue and the other streets in the St. Albans neighborhood represent proof that the basic recipe of combining street-oriented buildings, street trees, and restraint in the amount of space given to the car can make streets that unify communities instead of isolating them.

PEDESTRIAN STREETS AND PASSAGES

Lincoln Road, Miami Beach, Florida

Morris Lapidus, 1960

Retrofit: Experimenting with Pedestrian Space

Pedestrian Street

In a very short time (urbanistically), the character and function of Lincoln Road profoundly changed. What was a dense grove of mangroves at the beginning of the century was cleared by the visionary developer Carl Fisher for the construction of Lincoln Road. In the 1920s and 1930s, the wide, auto-oriented street flourished (Figures 4.245 and 4.246).[96] Saks Fifth Avenue, Harry Winston Jewelers, and other tony retailers catered to well-heeled socialites; many of the Beach's VIPs were enthusiastic about motorcars and had ties to the rapidly growing transportation industries, so they were known at the time as "the Gasoline Society." The success of the shopping street continued until the middle of the century, when it experienced a rapid decline in popularity among tourists and locals. Much of the previously vibrant retail began to leave the road for other, more up-to-date locations, including suburban shopping malls.

In a radical move for the time, famous architect Morris Lapidus proposed to close Lincoln Road to car traffic, "pedestrianizing" the commercial street (Figures 4.247 and 4.248). Lapidus's design for an unusually wide pedestrian mall, slightly more than one hundred feet across, was completed in 1960.[97]

At first, eliminating cars did more harm than good; Lincoln Road went from struggling to moribund. The attempt to compete with suburban malls by being more like them backfired, and Lincoln Road went from bad to worse. By the mid-1980s, vacant storefronts were the norm, and every night Lincoln Road became dark and eerie. Gradually, artists took up low-rent space on the street, and a gallery scene emerged, catalyzed by the presence of the South Florida Arts Center. During this period, the rest of South Beach and the Art Deco District became a mecca for fashion photographers, and the area came back, one block at a time. Tourists soon returned.

Economically, Lincoln Road continued to struggle, but the street's edginess slowly attracted a following among the creative class. The City of Miami Beach toyed

Figure 4.245: Lincoln Road, Miami Beach, Florida. View looking east, circa 1936. The street was open to traffic and lined with tall palms. The asymmetrical Mediterranean Revival silhouette of the Van Dyke Building (1924) is visible in the background, at center right. *Courtesy of HistoryMiami, Archives and Research Center*

Figure 4.246: Lincoln Road, Miami Beach, Florida. View looking east, circa 1936. *Courtesy of HistoryMiami, Archives and Research Center*

Figure 4.247: Lincoln Road, Miami Beach, Florida. Morris Lapidus, 1960. Photograph looking west. After languishing for thirty years, Lincoln Road sprang back to life, with one of the highest concentrations of pedestrian activity to be seen anywhere in the American Sunbelt. *© 2013 Stephen A. Mouzon, Photographer*

Figure 4.248: Lincoln Road, Miami Beach, Florida. Morris Lapidus, 1960. Photograph looking east. The staid Gasoline Society finally gave way to hip café society; Lincoln Road was overhauled a second time by the City of Miami Beach, taking the kitschy "boomerang moderne" image up a further notch by restoring Lapidus' pavilions (center), adding even more glitzy fountains and pavements, and luring artists with subsidized rents. © 2013 Stephen A. Mouzon, Photographer

with the idea of reopening the street to traffic but backed off, opting instead to overhaul the finishes, fountains, and kitschy pavilion structures designed by Lapidus. After nearly forty years, when the surrounding neighborhoods were thoroughly revived, Lincoln Road finally took off. Instead of being a dim, forgotten nowhere, it is now a bright social center of the city. Crowds stroll and dine there every night, even in the off-season. The essential framework of the urban form (including building frontage lines and lot dimensions) has not changed much, but infill happened through the decades, reinforcing the street facade and the definition of the space.

The post-Covid, post-Amazon meltdown of bricks-and-mortar retail affected many streets. Anchor retailers like Books & Books (a fixture on Lincoln Road for thirty years) began closing during Covid. Lincoln Road remains popular with many tourists and diners, but anyone walking down the street cannot help but notice the number of empty storefronts. The City of Miami Beach intends to implement another round of improvements to the public realm in hopes of restoring the energy on Lincoln Road. In late 2024, the City enlisted Field Operations, the New York firm behind the High Line, to generate proposals for the latest makeover.

Stone Street, New York, New York

RBA Group and Signe Nielsen Landscape Architecture, 2000

Retrofit: Opening an Alley

Pedestrian Street

Stone Street is an old street in downtown Manhattan, near Wall Street. It has been called the oldest paved street in the city, although that seems to be an urban legend.[98] The original name was Brouwerstraat (Brewers Street), but it was renamed Stone Street in 1660, when it was paved with cobblestones. In the 1980s, the construction of a new headquarters building for Goldman, Sachs & Company split the street in two. A general level of unhappiness with that decision led to the present requirement that the closing of any street in downtown Manhattan must be approved by the city's Landmarks Preservation Commission.

In 1996, the Commission designated the derelict street part of a new Stone Street Historic District and worked with the Alliance for Downtown New York to transform it. "This street has such historic consequence," said Carl Weisbrod, president of the Alliance. "But it was really a back alley filled with graffiti, a garbage pit, used for low-level drug dealing."[99] The ensuing streetscape renovation, which was spearheaded by the Alliance, cost $1.8 million (Figures 4.249 and 4.250).

From building to building, Stone Street is approximately thirty feet wide. The restored buildings in the center of the block are fifty feet tall, with taller buildings at each end. The north end of the street is visually closed by the twenty-three-story Cotton Exchange at 3 Hanover Square (now apartments), designed by Donn Barber in 1923. Across William Street on the northeast corner of Stone Street is a tall 3½-story brownstone palazzo that was the original home of the Hanover Bank. The opposite corner of Stone Street has an interesting eleven-story private-banking building designed in a Renaissance Revival style by Francis H. Kimball in 1906, with a postmodern addition by Gino Valle; for many years, this was the home of Lehman Brothers. Near the southern end of Stone Street is a pair of Dutch Colonial facades designed in 1906 by the architect C.P.H. Gilbert.[100] Looming over the street is the ungainly behemoth designed by Skidmore, Owings & Merrill for Goldman Sachs. The RBA Group and Signe Nielsen Landscape Architecture designed the Stone Street renovation.

With the buildings on Stone Street taller than the street is wide, the space between them makes a comfortable outdoor room. The street and the sidewalks are stone, and the building facades are mainly stone and brick. The buildings have traditional vertical punched windows with deep shadows that emphasize the solidity of the walls and firmly contain the space. Most of the lots are narrow, so the horizontal repetition of the vertical building facades sets up a pleasing rhythm. There are a variety of building heights, and the styles are eclectic, but the double-hung windows used in most of the buildings other than the Goldman Sachs headquarters add some unity and harmony.

"They've put the stone back in Stone Street," the *New York Times* said in December 2000, referring to the cobblestones and bluestone sidewalks that replaced the cracked asphalt roadbed and concrete sidewalks.[101] Lampposts that look like gaslights were also installed. Before the renovation, the landlords of the buildings along this street—all of which also face neighboring streets—treated their Stone Street frontages as backs. Today, Stone Street is a successful "entertainment street."

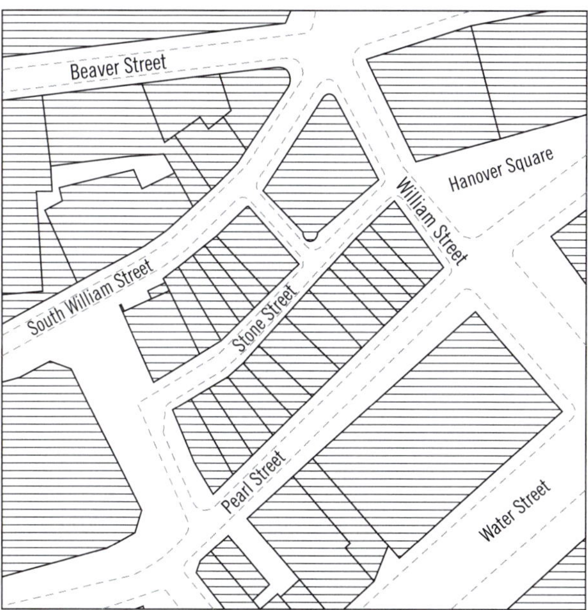

Figure 4.250: Stone Street, New York, New York. Figure-ground. © 2013 Dover, Kohl & Partners

◄ **Figure 4.249:** Stone Street, New York, New York. RBA Group and Signe Nielsen Landscape Architecture, 2000. Looking northeast on Stone Street. The former alley has become the main gathering spot for young Wall Street workers whenever the weather is warm enough after work.

Seventeen of the twenty-one restaurants on the block are owned by the Poulakakos family. Harry Poulakakos led a renovation featuring all-new storefronts along the alley. He had to argue with the New York City DOT, which expected the outdoor dining to fail and therefore wanted a street that could easily revert to traffic. But the Poulakakoses are kings of the street for as long as their restaurants succeed, and they were right that the street would become popular. Today, Stone Street is closed to traffic and filled with tables: it is one of the liveliest bar and restaurant scenes in lower Manhattan. On a nice afternoon or evening, the street overflows with people taking a break after work. Few of them probably realize that the historic-looking storefronts have been built since 1996.

Española Way, Miami Beach, Florida

Savino & Miller, 2002

Retrofit: From Vehicular Street to Shared Space to Pedestrian Passage

Also known as the Historic Spanish Village, Española Way is the only street on Miami Beach with a style of architecture and urban form that resembles what one might find in a Mediterranean city. Conceived as an artist colony, early on it took on notoriety for gambling and as a red-light district.

Today, Española Way is a prized address. It slowly evolved: most of the time it is a pedestrian-only environment, although the street can still be opened for deliveries and events on an occasional basis. The design of the outdoor space has seen multiple iterations, but the buildings have mostly been preserved.

In 2012 Española Way was selected as the Best Block in South Florida in a *Miami Herald* competition.

During the initial building boom on Miami Beach in the 1920s, the vehicular right-of-way on Española Way was nearly twice the size it had become by the early 2000s (Figure 4.251). Today, it is a pedestrian passage (Figures 4.252 and 4.253). The sidewalks have more than doubled in size, trees have been planted, and the historic facades are restored. The street is usually closed to car traffic via the use of movable planters and heavy bollards at its ends.

Like Aviles Street in St. Augustine, Florida, or Tweede Tuindwarsstraat in the Jordaan District of Amsterdam, Española Way is skinny, with a single central travel lane that measures 13 feet 4 inches (Figure 4.254). Large, 13-foot-wide sidewalks on both sides of the street are typically full of patrons dining, shopping, or walking to their apartments.

Figure 4.251: Española Way, Miami Beach, Florida. Before: Existing conditions in 1928. *Courtesy of HistoryMiami, Archives and Research Center*

Figure 4.252: Española Way, Miami Beach, Florida. Savino & Miller, 2002. Photograph looking west towards Pennsylvania Avenue. Wider sidewalks permitted trees and ornamental landscaping, and now the scene feels lush. *Courtesy of Jason King*

Figure 4.253: Española Way, Miami Beach, Florida. Savino & Miller, 2002. Photograph looking west towards Drexel Avenue. The widened sidewalks have been filled with tables from the restaurants, and most pedestrians walk in the center lane.

Figure 4.254: Española Way, Miami Beach, Florida. Savino & Miller, 2002. Section. © 2013 *Dover, Kohl & Partners*

PASEO PONTI, MIAMI DESIGN DISTRICT, MIAMI, FLORIDA

Duany, Plater-Zyberk & Company, 2011

Retrofit: Artful Shopping Destination

The Paseo Ponti is a multi-block, open-air pedestrian passage in the Miami Design District, described by the owner as "a creative neighborhood & shopping destination dedicated to innovative fashion, design, art, architecture & dining."[102] The developer, Craig Robins, bought many run-down buildings and empty lots in the area, eventually assembling 22.9 acres. He hired Elizabeth Plater-Zyberk at Duany, Plater-Zyberk & Company to make a form-based code for the streets and new buildings.

Duany Plater-Zyberk created the Paseo Ponti at the center of the district, linking the blocks. The Paseo runs north–south, across city streets, anchored by two civic plazas. The north–south orientation and shade trees protect it from the harsh Florida sun, making the project a walkable destination in a car-oriented city. At the intersections, raised paving at the level of the Paseo gives the pedestrians priority in the crossing. The design code emphasized simple modern buildings, giving the many architects involved room for individuality while keeping an urban harmony.

Figure 4.255: Paseo Ponti, Miami Design District, Miami, Florida. Duany Plater-Zyberk & Company, 2011. Photograph looking south. *Courtesy of Duany Plater-Zyberk & Company*

Figure 4.256: Paseo Ponti, Miami Design District, Miami, Florida. Duany Plater-Zyberk & Company, 2011. Photograph looking north. *Courtesy of Duany Plater-Zyberk & Company*

Figure 4.257: Paseo Ponti, Miami Design District, Miami, Florida. Duany Plater-Zyberk & Company, 2011. Photograph looking south towards NE 39th Street. *Courtesy of Duany Plater-Zyberk & Company*

Figure 4.258: The Alley, Montgomery, Alabama. 2WR/Holmes Wilkins Architects, 2008. *Courtesy of the City of Montgomery*

The Alley, Montgomery, Alabama

2WR/Holmes Wilkins Architects, 2008

Retrofit: Opening the Alley

Pedestrian Passage

The Alley was a service lane that opened onto Tallapoosa Street in the center of a block in downtown Montgomery. The sides of brick buildings line the alley—remnants from a time when even the most utilitarian structures might be built in a way that added texture and character to urban life. The city and local investors saw the opportunity to leverage that character and add a bright spot to downtown nightlife in Montgomery. The narrow passage—conveniently located between Biscuit Stadium and parking along Bibb Street—began as a place for a beer after the game, and a small portion of it was initially set aside for outdoor seating at the bar. Gradually, this forgotten space became the Alley (Figure 4.258), the liveliest part of the downtown pedestrian network. Initially shaped like an "L" and opening onto Tallapoosa and Commerce Street, the Alley now extends to connect Commerce and Coosa Streets with a "T" configuration.

According to Deputy Mayor Jeff Downes, the city and county invested $1.6 million in the common space at The Alley, while the private sector has invested a combined $25 million. This investment has paid off. Now the heart of downtown Montgomery's historic entertainment district, The Alley contains four restaurants, two bars, and an art gallery. Each opens to the alley space itself. The Alley Bar contains the "Back Alley" concert hall, a 5,000-square-foot party room and music venue.

An old, humble rooftop water tower was rescued and installed by the Tallapoosa Street entrance; the Alley's boosters then creatively lit the water tower and turned it into a unique combination of a signpost and a conversation piece, raising eyebrows among some preservationists and delighting others. Now the water tower is considered an iconic feature of the street and appears in most photographs of the Alley.

Today, The Alley is surrounded by three downtown hotels. The Entertainment Express rubber-tire trolley now circulates from the Old Cloverdale streetcar suburb and from Maxwell Air Force Base to the Downtown Entertainment District. The recent opening onto Coosa Street is very important because Coosa Street is an underused, two-sided architectural gem that is now attracting pedestrians for the first time in a long time—and a new restaurant is scheduled to open there as of this writing.

Winslow Homer Walk, New York, New York

Massengale & Co LLC and Dover, Kohl & Partners, with H. Zeke Mermell, 2011

Retrofit: Opening an Alley

Pedestrian Passage

The long blocks of midtown Manhattan are uncomfortable for pedestrians. To compensate, midtown has a tradition of passages like Shubert Alley that open the middle of the block, creating public spaces that create significant shortcuts for some trips. A newer tradition uses the bonus plazas given by New York City's 1961 zoning plan for pedestrian cut-throughs at buildings like the Equitable Building. New Yorkers familiar with the area know some midblock building lobbies go through the centers of the blocks.[103] For the *By the City / For the City* competition in 2011,[104] we looked at a midblock passage between 56th and 57th streets that was closed in the middle and poorly used by most passersby (Figure 4.261).

Directly opposite the Art Students League on West 57th Street, our proposed Winslow Homer Walk (Homer once taught at the school) brings the passage down to ground level and removes the high fence between the raised "Hooters Plaza" on 56th Street and the more formal raised plaza on 57th Street, which is connected to the building at 888 Seventh Avenue (Figure 4.262). Decades of studies have shown that raised plazas are less used than spaces at ground level, and the formal plaza at 888 is not welcoming.[105]

The existing plazas on 56th and 57th streets are both fifty feet wide, but there is a jog in the middle of the block where the fence currently separates the two spaces (Figure 4.263). In our design, we moved the diner in front of 888 on 57th Street around the corner so that it would face Homer Walk. The 56th Street side of the Walk is quieter, and we made that side wider, with more trees and more outdoor tables. A narrow two-story addition to 211 West 56th Street on the west side of the Walk gives a beautiful face to a boring utilitarian design from the 1960s and opens to the tables in the passage.

The long blocks of midtown Manhattan are uncomfortable for pedestrians. To compensate, midtown has a tradition of passages like Shubert Alley that open the middle of the block, creating public spaces that create significant shortcuts for some trips.

Figure 4.259: Parker Alley, Detroit, Michigan. View from pedestrian passage looking northwest. Echoing the laneways in Melbourne (see Figure 3.46) and The Alley in Montgomery, Parker Alley and adjacent Belt Alley were retrofitted in 2018 as part of the redevelopment of two blocks in downtown Detroit. Once the first passage was completed, the idea was extended to the next, and there are plans for more. *Courtesy of Mark Nickita, FAIA*

Figure 4.260: Parker Alley, Detroit, Michigan. View from pedestrian passage looking northwest. The passage is a public right-of-way adjacent to the Shinola Hotel. Narrow pedestrian passages are found in every climate and culture. *Courtesy of Mark Nickita, FAIA*

A new marble wall along the west side of the southern plaza, similar to the one on the other side of the plaza for the garage entrance for 888, makes a backdrop for a row of food carts adjacent to outdoor seating. New Yorkers love to take lunch breaks in quiet places like Homer Walk. The most-used of these, like Bryant Park or the paved plaza where Occupy Wall Street protesters encamped, have food vendors, free Wi-Fi, and benches and tables that can be moved by the users. In the end, the plan removed thirty-seven top-level parking spaces from the garages at 888 Seventh Avenue and 211 West 56th Street to bring the plaza down to ground level. At the same time, we added thousands of square feet of public and commercial space. That is the type of trade-off we

need more of as we move away from the auto-centric planning of the last fifty years, and less than a year after we drew our imaginary Winslow Homer Walk, the NYC DOT put up street signs, stoplights, and midblock crosswalks for something they call 6½ Avenue.[106] The mini-avenue marks a series of bonus plazas in the middle of the blocks between Sixth Avenue and Seventh Avenue, and between West 51st and West 57th streets. The next step would be to work with the owners of the lots and buildings 6½ Avenue passes through, because these "public plazas" can be quite barren. Some are next to popular restaurants, but the landlords do not let the restaurants put tables in the spaces.

In terms of density and transportation, New Yorkers live in the most walkable city in America, with carbon footprints similar to many Europeans. In terms of streets, squares, and places like Homer Walk and 6½ Avenue, however, New Yorkers need more places and more variety in the public realm, appealing places where people want to gather.

◄ **Figure 4.261:** Winslow Homer Walk, New York, New York. Before: Existing conditions, 2011. Rendering looking south from West 57th Street. The plaza on 57th Street is raised, which discourages use, and there is a wall between it and the outdoor space for Hooters on 56th Street. © 2011 Dover, Kohl & Partners / Massengale & Co LLC

◄ **Figure 4.262:** Winslow Homer Walk, New York, New York. Massengale & Co LLC and Dover, Kohl & Partners, with H. Zeke Mermell, 2011. After: Rendering shows the passage brought down to grade, opened to 56th Street, planted with trees, and lined with restaurants and food carts. © 2011 Dover, Kohl & Partners / Massengale & Co LLC

Figure 4.263: Winslow Homer Walk, New York, New York. Massengale & Co LLC and Dover, Kohl & Partners, with H. Zeke Mermell, 2011. Plan of proposed intervention. The diner that was moved to 57th Street when the office building was built is placed along Homer Walk. The office building gains a more important entrance on 57th Street but also has an entrance in the middle of Homer Walk. © 2011 Dover, Kohl & Partners / Massengale & Co LLC

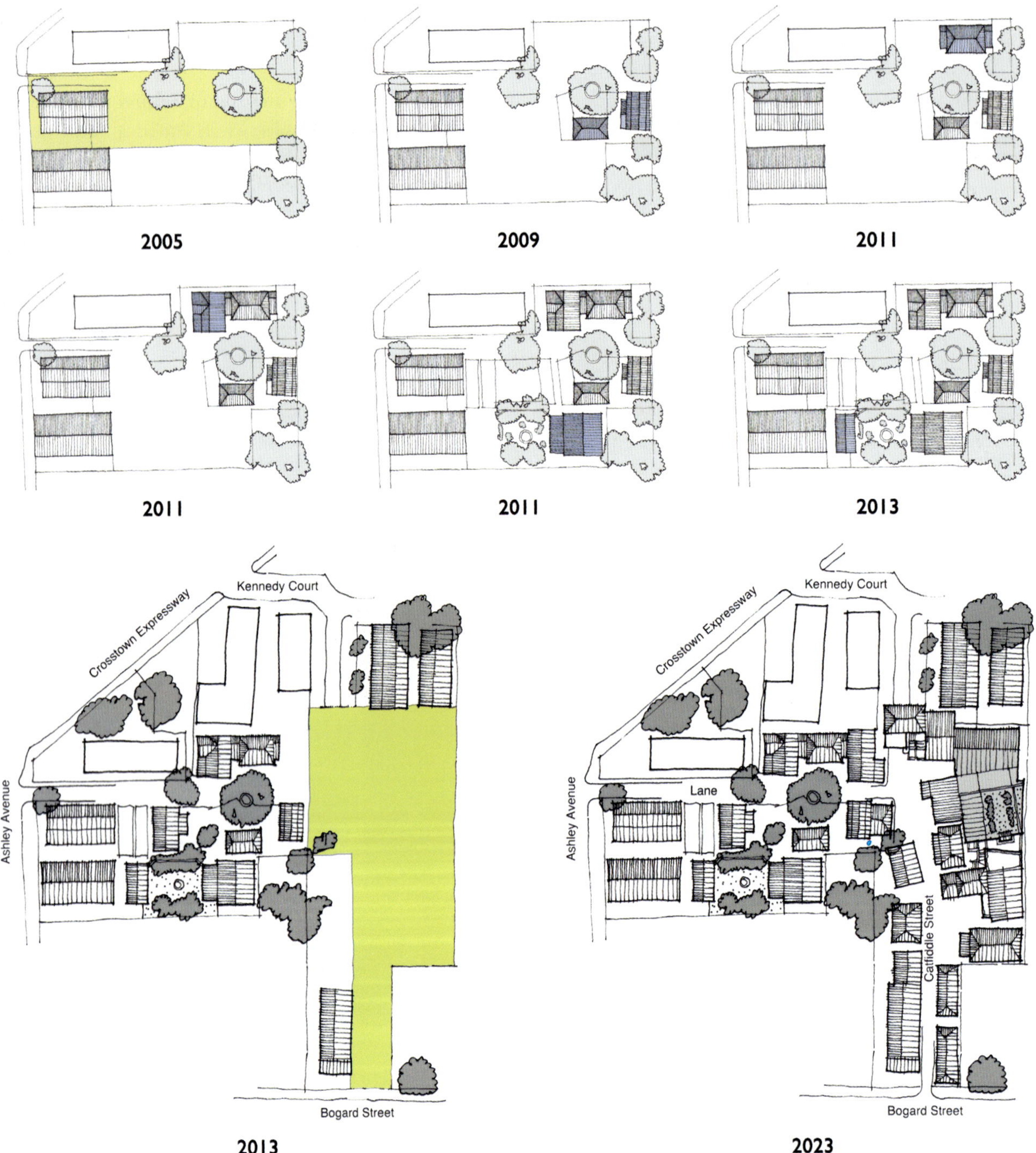

Figure 4.264: This series of plans shows a progression beginning with the first renovated house on Ashley Avenue (2005), the construction of two houses in the rear of the plot (2009), the expansion of the site on the north side and the construction of two more houses (2011). A semi-decrepit house was added on the south side (2011), a new cottage with a garden was built in the rear (2011), and a small outbuilding was constructed (2013) on the same lot. The project almost doubles in size with the purchase of the 0.4 acre lot on the east side (2013). The final plan shows Catfiddle Street complete (2023). *Courtesy of Witold Rybczynski*

CATFIDDLE, STREET OF DREAMS / WITOLD RYBCZYNSKI

Urban Ergonomics, 2016

Retrofit: Midblock Infill

To understand the Catfiddle Street project, it's necessary to go back to 2005, when Jerry Moran, a retired Air Force captain and Delta pilot, bought a property on Ashley Avenue in Cannonborough, a neighborhood north of Charleston's historic district. The hundred-year-old freedman's cottage needed work, but what attracted him was that the deep lot could accommodate three more houses. For the previous two decades, Jerry and his partners, George Holt a self-taught contractor and designer, and Cheryl Roberts who managed the business (she would tragically die the following year), had been restoring old properties in the city, and often in-filling new houses. That was the plan here.

Working with Andrew Gould, a recent Penn architecture graduate, George renovated the Ashley Avenue cottage and built two more houses, three and four stories tall with a footprint of less than three hundred square feet. Over the next eight years, Jerry enlarged the property, acquiring an adjoining lot and a partly derelict old house. George and Andrew built more new houses, including an Arts and Crafts cottage for Andrew's parents, and a yellow-ochre house with a stepped gable for Jerry's sister. Eventually there were eight houses, two old and six new, with space for two more. There was no grand plan; like Topsy the place just grow'd (Figure 4.264).

One of the new houses was for Reid Burgess, a bluegrass mandolin player from Brooklyn who was easing into a new career as a developer. He had always wanted a Palladian house, and with George and Andrew's help, he built a miniature villa based on the loggia of Palladio's Villa Saraceno. Later, he bought the derelict house from Jerry, and oversaw its extensive renovation. Reid, who had moved to Charleston and wanted to build a house for himself and his partner Sally, was on the lookout for a lot, not easy to find in a city that had become a hot real estate market. By chance, the empty lot immediately to the east of the Ashley Avenue compound was available. The half-million dollar price tag for less than half an acre was daunting, but Reid calculated that if he could subdivide the land into nine sellable lots—in addition to his own house—he could finance the purchase. Thus, the genesis of Catfiddle Street.[107]

George and Reid worked on the master plan. One of their models was the Arco degli Acetari in Rome, a medieval courtyard surrounded by picturesque houses

Figure 4.265: Ashley Avenue, Charleston, South Carolina. Looking east to Catfiddle Street. *Courtesy of Witold Rybczynski*

with exterior staircases and balconies, and walls painted in vivid tones of red, burnt orange, and ochre. George and Reid's plan included a new street that connected to the Ashley Avenue compound and to existing city streets. Not quite straight, the narrow lane, squeezed between tall houses, would appear almost accidental.

Reid had planned his and Sally's house around an atrium. Their home would be surrounded by its neighbors and invisible from the street, reached by climbing a narrow passage. Reid, who would sell lots to individual owners, wanted the other houses, which would be designed by different architects, to form a whole. But this would not be a themed development; although he and George would review the plans, there was no design code. If there was a precedent for Catfiddle Street, it was Clough Williams-Ellis' eclectic architectural fantasy of Portmeirion in Wales, which Reid and Sally had visited.

Reid and George selected the local architects of the nine houses: Jenny Bevan and Christoper Liberatos, Julie O'Connor, Kenny Craft, and the dean of Charleston classicists, Randolph Martz. George and Andrew designed one of the houses, and George and Reid were responsible for the final two. By then, more than a decade had passed since Reid had purchased the land. The built result puts me in mind of one of Christopher Alexander's lesser-known books, *A New Theory of Urban Design*. Alexander wrote that organic urban growth should have four qualities, it should be piecemeal, unpredictable, coherent, and full of feeling. Catfiddle Street, like the neighboring Ashley Avenue compound, exhibits all these conditions. But it also displays another essential quality: human agency. This involved not only design skills, which are very much in evidence, but also the series of contingent, quirky, and personal decisions that produced this little urban gem.

Figure 4.266: Catfiddle Street, Charleston, South Carolina, Urban Ergonomics, 2016. One of their models was the Arco degli Acetari in Rome, a medieval courtyard surrounded by picturesque houses with exterior staircases and balconies, and walls painted in vivid tones of red, burnt orange and ochre. Looking north to Jenny Bevan and Christopher Liberatos houses with Andrew Gould house on the left. *Courtesy of Witold Rybczynski*

Giralda Avenue, Coral Gables, Florida

Jaime Correa, Charles Bohl, and Jennifer Garcia, 2009

Cooper, Robertson & Partners, 2017

Retrofit: Pedestrian Street

For decades, popular restaurants lined Giralda Avenue, nicknamed "restaurant row." Coral Gables' subtropical climate invites diners to eat outside during most of the year if there is shade. Starting in the early twenty-first century, for one evening every month, the city closed an entire block to car traffic, and the restaurants extended their dining tables into the street space. Organizers called this event "Giralda Under the Stars."

For the rest of the month, the street was wide and barren (Figure 4.267). Analysis led to a simple conclusion: too much of the street space had been taken over by too many automobiles. Motor vehicles had sole possession of over 75 percent of the right-of-way, while pedestrians and customers for outdoor dining only had 25 percent of the space.

Urbanists Charles Bohl, Jaime Correa, and Jennifer Garcia proposed a retrofit to make the street more comfortable and better for outdoor dining. Their design (Figures 4.268–4.270) widened the sidewalks and added shade trees. To accommodate the expanding pedestrian realm, they proposed narrowing the central travel lanes and adjusting the parallel parking. Originally, the traffic lanes were more than twelve feet wide—too wide for an intimate neighborhood street.

The Garcia team proposed taking as much as 60 percent of the right-of-way for shoppers, diners, and strollers, and giving 40 percent to cars and trucks. Worth noting is that the number of travel lanes stayed the same, despite the dramatic increase in space for walking and dining.

An easy retrofit for any oversized street, narrowing travel lanes provides more space for people who are not in their cars. Having more people walking by

Figure 4.267: Giralda Avenue, Coral Gables, Florida. Before: Existing conditions, 2009. *Courtesy of Jennifer Garcia*

Figure 4.268: Giralda Avenue, Coral Gables, Florida. Jaime Correa, Charles Bohl & Jennifer Garcia, 2009. After Option 1: the street would have been narrowed, the street and sidewalk. *Courtesy of Jennifer García*

Figure 4.269: Giralda Avenue, Coral Gables, Florida. Jaime Correa, Charles Bohl & Jennifer Garcia, 2009. After: Option 2 in the original proposal interrupted the parking spaces with trees. *Courtesy of Jennifer García*

Figure 4.270: Giralda Street, Coral Gables, Florida. The "Giralda Under the Stars" monthly street closure, c. 2012. *Courtesy of Kenneth García*

Figure 4.271: Giralda Avenue, Coral Gables, Florida. Cooper, Robertson & Partners, 2017. The reconstruction made the street a popular pedestrian-only destination for art, people-watching, and dining. In this early photo, the walkway down the center of the street was still open for strolling. *Courtesy of Kenneth García*

restaurants and shops on the sidewalks is simply better for business.

The wildly successful "Giralda Under the Stars" experiments convinced Coral Gables and its Business Improvement District to go further. When the city contracted Cooper, Robertson & Partners to redesign Coral Gables' nearby main street, Miracle Mile, they added a retrofit of one block of Giralda Street. The final result on Giralda is more impressive than Miracle Mile, which was ostensibly meant to be the more important project; Giralda feels more like a place than Miracle Mile. The remade Giralda segment attracts almost as many pedestrians as Miracle Mile, which is four times as long. The Cooper-Robertson team redesigned the 600-foot-long segment of Giralda as a curbless, linear plaza, with stone pavers in a warm range of colors and, down the center, a narrow bronze trench drain. The stones are arranged in a pattern of colliding, concentric circles radiating from tall palms. Although the luxurious pattern is conspicuous enough to be immediately noticeable in the retrofit, it is not jarring or gaudy; it feels as though it might have been there all along.

Reportedly, the city's original idea was that the reconstructed "flexible" street would be open to car traffic during the day and then closed to cars on certain evenings and weekends. But construction took a long time, motorists found other routes, and as far as we know, the city has never reopened the rebuilt block to cars, and downtown Coral Gables is none the worse for it.

This block of Giralda Avenue—sometimes now referred as Giralda Plaza—has become popular beyond all expectations. According to the city, more than two million pedestrians visited this short block of Giralda in 2021.[108] The combination of food, drink, public art, and a walkable street has proven attractive to patrons and innumerable Instagram users.

Recently, the lucrative business opportunities for Giralda addresses have led to a modern-day case study on the "tragedy of the commons."[109] Local restaurants, eager to squeeze in more and more tables, successfully petitioned the city to allow them to absorb and block the center walkway. Narrow walking paths still exist on either side up against the storefronts, crisscrossed by staff waiting on tables, but the street no longer feels inviting in the same way—especially to anyone in a wheelchair. Planner/engineer Jessica Keller, a former city official who helped usher in the changes according to the Cooper-Robertson plan, says "Walkability has been significantly reduced, and the shades [the merchants have been allowed to add] are now walling off the space. It feels chaotic," adding, "I watched a family with a normal-sized stroller get stuck trying to go past… They turned around."[110]

Universal accessibility advocate Steve Wright commented, "Designed as a paradise for [both] walkers and wheelchair users, [it is] now pinched down to something totally off limits to people who use assistive mobility devices. Warned Gables officials about this for years. Dozens of emails and posts 100% ignored."[111]

Figure 4.272: Giralda Avenue, Coral Gables, Florida. Cooper, Robertson & Partners, 2017. Photo looking southeast after reconstruction. The stone paving pattern, with a series of intersecting concentric circles, lends a level of classic luxury in keeping with Coral Gables' rising affluence. On the other hand, the blinding, eyeball-height "Q-tip" light fixtures are unfortunate, but the intense glare from these fixtures is partly screened by café table umbrellas. *Courtesy of Kenneth García*

Figure 4.273: Giralda Avenue, Coral Gables, Florida. Cooper, Robertson & Partners, 2017. Photo looking east in 2024. The central walkway and the elegant paving pattern have been subsumed by more and more dining tables, a modern-day tragedy of the commons. *Courtesy of Jessica Keller*

Figure 4.274: Hobbs-Hanley Block, Baldwin Park, Orlando, Florida. Skidmore, Owings & Merrill, 1988. Photograph looking west. What if we replace the asphalt with trees, shrubs, and grass?

HOBBS-HANLEY BLOCK

Baldwin Park, Orlando, Florida

Skidmore, Owings & Merrill, 1988

Green Street

Suppose we build a classic American Garden Street—the staple but elegant kind that might have been named Elm Street or Maple Street in a streetcar suburb from 1900—but swap out the cars and asphalt for a shared lawn and shrubs? This is a *green street*. Baldwin Park, in Orlando, Florida, has such a street. Between Hobbs Alley and Hanley Alley, an unnamed one-block-long green street connects Trapp Lane to the lakeside drive called Fox Street.[112]

Except for the missing on-street parking and car lanes, the symmetrical configuration is typical of many American garden streets: the houses sit forward on narrow lots, with front porches within conversation distance of the sidewalks, shaded by regular rows of street trees. The street is open to pedestrians but not cars. Two alleys provide vehicles with access to the rear side of the lots.

The Hobbs-Hanley houses have raised first floors that give the interiors privacy and give anyone on the porch a sense of command over the street. The alley is higher than the green street, matching the interior floor levels, allowing for wheelchair access and making it easy to bring in groceries from garages and driveways.

The pattern gives a gentle increase in density. Ten houses occupy the land area (and share the lake view and generous shared garden) where only four or five might have been. The houses are close together and have only small private gardens and a minimum of private outdoor open space, but the street scene is lush and green.

Figure 4.275: Green Street. The basic components are: **1** Houses on Narrow Lots; **2** Elevated Finished Floors; **3** Front Porches; **4** Sidewalks; **5** Street Trees; **6** Linear Garden replacing what would typically be parking and driving lanes. © 2024 Dover, Kohl & Partners

PEDESTRIAN COURT AND ALLEY STUB

Like a close, a pedestrian court is a street type that can reconcile challenging geometry with good placemaking. Urban designers use pedestrian courts to add variety and value to a neighborhood by extending the public realm, lot frontages, and street connectivity. The court is a short, semi-public space for pedestrians only, in which the primary access to the front of the buildings is from walkways in the courtyard rather than from streets. It is a variation on the green street type.

Houses face onto the court, which is paired with a parallel alley stub that gives vehicle access and keeps garage doors and service areas away from the court. This allows adherence to the Rule of Fronts and Backs (see "The Rule of Fronts and Backs" in this chapter) and promotes "like facing like" (discussed in Chapter One),[113] even where block size or topography conspire against it. Used judiciously, the pedestrian court can connect sidewalks and bike paths where, for one reason or another, a full-blown street cannot go through.

The conceptual model for the Juniper Point neighborhood in Flagstaff, Arizona, abuts a "wash," a steep, wooded valley that was preserved as a natural park (Figures 4.276 and 4.277). There are spectacular long views across the wash, but under a conventional layout, only the houses on the outer layer would enjoy the view. The other households would have no psychological connection to the open space, and front-loaded lots would have to be wider and fewer in number. On the other hand, with a combination of alley stub and pedestrian court kept open at the end closest to the wash, the whole neighborhood shares the view. Driveways can be eliminated or consolidated, and higher density is feasible without negative consequences. The sidewalks of the street network within the neighborhood seamlessly and spatially connect to the network of trails in the park.

Variations on the pedestrian court street type lend themselves to concepts like cohousing[114] and pocket neighborhoods (Figure 4.278).[115]

Figure 4.276: Pedestrian Court, Juniper Point, Flagstaff, Arizona. Dover, Kohl & Partners, 2006. Computer model. The scenic view can be shared with the whole neighborhood, and the street and park are spatially united. *© 2024 Dover, Kohl & Partners*

Figure 4.277: Juniper Point, Flagstaff, Arizona. Computer model. Conventional Layout. This form increases impervious surfaces, produces less real-estate value, and hides the view so that it can be enjoyed by only a few households. *© 2024 Dover, Kohl & Partners*

Figure 4.278: New Oaks Pocket Neighborhood, Lake Wales, Florida. Maricé Chael and Victor Dover, 2021. Computer model. Variations on the Green Street and Pedestrian Court types can be used to organize cohousing, cottage courts, and pocket neighborhoods. Note the requirement for two of the buildings to have two "front" façades, so that both the interior courtyard garden and the adjacent street are faced by worthy frontages. © 2021 Dover, Kohl & Partners

SEA GARDEN WALK, ALYS BEACH, FLORIDA / MARIEANNE KHOURY-VOGT AND ERIK VOGT

Duany Plater-Zyberk & Company, 2003

Pedestrian Passage

Sea Garden Walk is a pedestrian street that integrates architecture and landscape into a harmonious whole and delineates the transition from an urban to a rural environment along its passage.

It is one of several north–south passages that run through the community and connect it to the beach at the south and a woodland preserve along the northern edge.

The design was overseen by the Alys Beach Town Architect, following a strict urban and architectural code provided by Duany Plater-Zyberk & Company. The primary building type is the attached courtyard house.

Sea Garden Walk encompassed two related goals. First, to extend into town the relationships of solid and void, of public and private space, formed by the residential courtyards and their *zaguans* (the traditional entries into the courtyards found in Spanish colonial towns). Second, to establish the primacy of a harmonious streetscape over individual architectural expression. To that end, the house designs contributed by over a dozen architects were assembled into larger elevations for the street, which were then edited and synthesized to form a coherent whole. The Town Architects then looked at the street as a public room harmoniously furnished with paving, landscaping, lighting, and other amenities. Street trees mark entryway axes, urns balance adjoining facades, and benches or fountains animate places of gathering.

A transect embedded in the master plan set a more urban character in the southern half of the development. Going from the south to the north, the character gradually changes to a more rural and naturalistic character. Lot sizes, building disposition, and house types all

follow suit, reinforced by all elements of the streetscape, designed or selected to fit appropriately within its transect zone. Grosvenor Atterbury and the Olmsted Brothers followed a similar design process at Forest Hills Gardens, although they made the designs themselves, rather than making a code for other designers to follow. See "The Transect Observed: Forest Hills Gardens" in Chapter Three, page 298.

At the southern end Sea Garden Walk (Figure 4.279), the street is a continuously paved surface, house-wall-to-house-wall, punctuated by elements of an urban landscape—formal tree groupings, overscaled planted urns, climbing vines on wall and roof trellises, and contained planters built into the house facades. As one moves north, the paving gradually narrows down to a broad walk bordered by landscaped swales containing more informal groupings of native plantings that weave into the existing scrub.

Finally, the northernmost end of the walk opens up to a man-made lake, designed as a picturesque landscape. Native oak, pine, and cypress trees blend into the backdrop of the woodland preserve. The walk divides into a narrow, paved path that circumscribes the lake edge and joins up with a boardwalk that snakes through the preserve.

Here, the architecture becomes episodic, subordinate to the predominant landscape. A pedestrian bridge and fountain work together to mark the southern edge of the lake and connect with an east–west path through the town (Figure 4.280). A terrace in the form of an exedra juts into the lake, its built-in wall benches shaded by a willow tree. The vehicular bridge spanning the lake is marked by a tower, its form inspired by the simple gabled houses of the town. From here, a prospect can be taken southward of the path, walk, and street that lead to the sea.

Figure 4.280: Sea Garden Walk, Alys Beach, Florida. Duany Plater-Zyberk & Company, 2003. At its northern end, the Walk opens up into less tightly wound space. (Construction photo.) *Courtesy of Kurt Lischka*

◄**Figure 4.279:** Sea Garden Walk, Alys Beach, Florida. Duany Plater-Zyberk & Company, 2003. At its southern end, the Walk has continuous paving in a tighter, more urban space. *Courtesy of Kurt Lischka*

QUICK SAFETY FIXES FOR A MORE WALKABLE CITY / JEFF SPECK

In my book *Walkable City*, I discuss about a dozen ways in which urban streets can be made safer quickly and cheaply, often just using a new topcoat and a few cans of paint. With this excerpt, I hope to convince city leaders to make the three quickest and cheapest fixes at the first opportunity. They are the restoration of one-way systems back to two-way traffic, the replacement of traffic signals with all-way stop signs, and the removal of centerlines. The paragraphs that follow are slightly modified from both *Walkable City* (2012) and its 2022 update.[116]

The One-Way Epidemic

In 1918, a flu pandemic killed more than seventy-five million people worldwide. A century later came Covid. Almost exactly in between these disasters, the United States was hit by another epidemic that, while less harmful to humans, laid waste to city after city from coast to coast. I am talking, of course, about the wholesale replacement in downtowns of two-way traffic with one-way traffic, a plight that few American cities escaped. Its impacts were profound and haunt us to this day.

The logic was simple enough: to stay competitive in the face of suburban out-migration, cities needed to retool themselves around the goal of moving suburbanites in and out of the downtown quickly. One part of this effort—the obvious part—involved building elevated interstates, with the near-suicidal outcomes that have been well documented. The other part, less discussed, involved the remaking of downtown street networks around free-flowing systems of one-way pairs. By replacing two-ways with one-ways, cities were able to introduce synchronized signals and eliminate the slowdowns caused by left turns across traffic.

Like the interstates, these retrofitted streets were effective at speeding commuters, enough so that there was no longer any reason to live downtown. They also turned what had once been a great urban asset—the public realm—into little more than a collection of surface freeways. Thoroughfares that once held cars, pedestrians, businesses, and street trees became toxic to all but the former. Freed of other uses, they effectively turned into automotive sewers.[117]

We are all aware that one-way multilane streets contribute to anti-pedestrian driving. Add to that the elimination of all friction from cars headed in the opposite direction, and the sheer momentum represented by two to four columns of unopposed traffic, and you can see why these streets quickly depopulated. It is difficult to name a midsized or larger American city that was not impacted by this technique, whether it takes the form of a largely one-way network—St. Louis, San Diego—or just a single one-way pair—Alexandria, Virginia, or Cornelius, Oregon. Indeed, driving west from Portland to the Oregon coast, I witnessed how a single DOT had managed to send a good number of a state's Main Streets onto life support with this one trick.

One-ways wreck downtown retail districts for other reasons beyond noxious driving, principally because they distribute vitality unevenly, and often in unexpected ways. They have been known to kill stores consigned to the morning path to work, since people do most of their shopping on the evening path home.[118] They also create a situation in which half the stores on cross streets lose their retail visibility, being located over the shoulders of passing drivers. They intimidate out-of-towners, who are afraid of becoming lost, and they frustrate locals, who are annoyed by all the circular motions and additional traffic lights they must pass through to reach their destinations.

Indeed, these circular motions call deeply into question the presumed greater efficiency of one-way systems. Sure, they move vehicles faster, but does that greater speed make up for the additional distances that motorists have to travel... especially lost motorists? While there are plenty of studies documenting the congestion-busting efficacy of one-ways, I have yet to see one that factors in the marginal congestion caused by circling.

> The logic was simple enough: to stay competitive in the face of suburban out-migration, cities needed to retool themselves around the goal of moving suburbanites in and out of the downtown quickly.

I was reminded of this fact on my first visit to Lowell, Massachusetts, on the day I was hired to work on their downtown. After twenty minutes spent lost—despite Google Maps—I finally had to call the Deputy City Manager, who talked me in for a landing. For a city planner with what I thought was a well-calibrated internal compass, this was a profound embarrassment. Later, as I got to know the city, I began to feel a bit better. The superimposition of a one-way network on Lowell's preindustrial cranked grid, interrupted by canals and rivers, had created one of the most discombobulating street networks in America. In my eventual report, I took great pleasure in documenting how the drive from the Memorial Auditorium to its designated parking lot, a mere 200 yards away, required a looping 5-turn odyssey of more than a mile.

At this point, some astute readers will be asking about Portland: it has a one-way grid, and it is doing great. What gives? Portland adds a major caveat to this discussion: if the grids are simple and the blocks small, corresponding to a dense network of relatively tiny streets, then one-way systems can function quite well—picture most of the residential cross streets in Manhattan. But Portland has a number of one-ways that are simply too big to invite walking, and so does Seattle, another small-block gem. When the streets get more than two lanes wide, it takes some pretty tall buildings to make them feel comfortable, buildings that most American cities do not have.

Take Savannah. In 1969, a one-way system was applied to many of the north–south streets in Oglethorpe's delicate grid (see page 269). Most still remain, and create perhaps the only significant impediment to pleasurably strolling this otherwise eminently walkable city. Recognizing this problem, the city government commissioned the architect Christian Sottile to study what happened to just one thoroughfare, East Broad Street, when it became a speedway. He dug into the tax rolls and counted the number of active (tax-paying) addresses located along the street in 1968 and then a few years later. He learned that, due to the conversion, the street lost almost two-thirds of these addresses.[119]

Happily, there is a flip side to that story. Worried about speeding as it built a new elementary school, the city returned East Broad Street to two-way. In short order, the number of active addresses shot up by 50 percent.[120]

Savannah's experience is not alone. Based on a few well-publicized successes, dozens of American cities are beginning to revert their one-way systems back to two-way traffic. These include Oklahoma City, Miami, Dallas, Minneapolis, Charleston, Berkeley[121]... and, soon, Lowell. Perhaps the best documented of recent reversions was accomplished in Vancouver, Washington. As told by Alan Ehrenhalt in *Governing* magazine, Vancouver had "spent millions of dollars trying to revitalize its downtown," but these investments "did nothing for Main Street itself. Through most of this decade the street remained as dreary as ever."[122] He continues:

> Then, a year ago, the city council tried a new strategy. Rather than wait for the $14 million more in state and federal money it was planning to spend on projects on and around Main Street, it opted for something much simpler. It painted yellow lines in the middle of the road, took down some signs and put up others, and installed some new traffic lights. In other words, it took a one-way street and opened it up to two-way traffic. The merchants on Main Street had high hopes for this change. But none of them were prepared for what actually happened following the changeover on November 16, 2008. In the midst of a severe recession, Main Street in Vancouver seemed to come back to life almost overnight.[123]

The success has continued, and business owners remain ecstatic. Twice as many cars drive past their businesses each day, and the once-feared traffic congestion has never occurred. Now, the head of Vancouver's Downtown Association, Rebecca Ocken, has some planning advice for other cities: "One-way streets should not be allowed in prime downtown retail areas. We've proven that."[124]

For small and midsized cities like Vancouver (population 162,000), she is almost certainly right. In larger cities, it depends. I, for one, am not about to revert Manhattan's Columbus and Amsterdam Avenues back to two-way traffic, but it is fair to say that New York would be an even more walkable place if that change were made. Bottom line: if your downtown lacks vitality and has one-ways, it is probably time for a change.

MONON BOULEVARD, CARMEL, INDIANA

Speck & Associates, Gehl Partners, RAI Landscape, 2012

Jeff Speck says Lancaster Boulevard (page 346) was an inspiration for Monon Boulevard, a short street that connects at each end to the existing Monon Greenway, a walking and cycling trail in a former rail right-of-way. The tracks were in the middle of the block in an industrial area. Speck created the 140-foot-wide boulevard, making it the centerpiece of the redevelopment of the industrial area. The slow roads on each side of the boulevard are eighteen feet wide, including parking.[125]

Figure 4.281: Monon Boulevard, Carmel, Indiana. Speck & Associates, Gehl Partners, RAI Landscape, 2012. An aerial view of Monon Boulevard. *Courtesy of Brandon Lust*

Oversignaled

Briefly discussed in the first edition of *Walkable City* is the safety improvement that all-way stop signs offer over traffic signals. Not mentioned was the data: when Philadelphia removed the signals from 472 intersections and replaced them with all-way stop signs, severe crashes dropped by 62.5 percent. In most streets, a city's investment in traffic signals—at six figures apiece—is an investment not in safety, but in high-speed driving, often from red light to red light.

And there is the rub. Because while green lights allow drivers to shoot through intersections at dangerous speeds, they are usually part of a system that actually delays trips across town. The high-speed driving is easily outweighed by time wasted idling at reds. As

Charles Marohn notes, "A city without traffic signals is a city where most of us arrive at our destinations sooner and all of us travel safer."[126]

That fact takes a while to sink in. Were traffic signals not invented to make urban driving safer and more efficient? That may have been the intention, but at a certain point, we have to acknowledge that they do the opposite.

I had a funny experience in my walkability study for downtown Albuquerque, when I recommended the removal of seventeen signals. The city took me up on nine. But people complained—change is always hard—so the city put three back. Then people complained even louder, because they observed the signals made traffic worse. The headline on the TV news put it this way: "City reinstalls downtown stop signs to reduce traffic pileup, speeding."[127]

All-way stops can be a bit confusing to the inexperienced, requiring a level of neighborly cooperation that we are spared by signals. For many Americans, the experience of taking their turn at a four-way stop may be the only moment during their day when they are asked to interact with, and therefore build trust with, their fellow community members. That might be a good thing.

The Streets We Need

Of course, replacing signals with stop signs is just a way station on the path to the Safety Apotheosis also discussed in Step 5: naked streets. Even as almost no progress has been made stateside in the last decade, the naked streets revolution continues apace in Europe. In the Netherlands, all-way stops are now considered bad form; they are ripping them out and replacing them with… nothing. Drivers have grown accustomed to approaching every intersection as a negotiation, and it has become apparent that signs just get in the way. Streets are properly understood as social spaces in which the most vulnerable participants dominate, and drivers proceed with care.

Sounds like a nightmare for drivers, right? Tell that to Waze: their analysis of the experiences of 65 million monthly users in 235 countries identified the Netherlands as "the most satisfying place in the world to drive."[128]

One aspect of naked streets has accidentally crept into American road design, however, and we should not neglect to use it: centerline removal. If you look around your town, you will find that many streets do not have yellow lines down the center; they are common only on the bigger ones, where they ostensibly make driving safer by keeping everyone in their lane. Except they do not. Risk homeostasis alert! Centerlines give drivers confidence, enough confidence to drive 7 mph faster, according to one British study. For a pedestrian, that 7 miles per hour roughly doubles the risk of death.[129] In my work, I remove centerlines wherever I can.

Considering the three street network modifications discussed above, it is fair to say that centerline and signal removal are no-brainers, except for any legal barriers that might exist. Removing signals often requires a "warrants study"; by all means, commission that if you must. Removing centerlines on busy streets in the United States can run afoul of the notorious MUTCD (the Manual on Uniform Traffic Control Devices); some federal funding can be tied to following this manual during repaving. One engineer friend has suggested an intriguing workaround: stripe the centerline by all means, but use latex house paint.

The largest item, two-way restoration, can get quite expensive if it requires the placement of new signal heads. But, done properly, it should not. As intersections of multilane one-ways become intersections of two-lane two-ways, it becomes possible to remove signals entirely in favor of all-way stops. This magical synergy can allow two-way restorations to pay for themselves, as signals removed from newly two-way streets can be deployed elsewhere in the city for the mere cost of relocation.

As an urban planner, one is wary of having a "rubber stamp" that one takes as a universal fix. Yet the above recommendations should likely be deployed wherever one-ways, signals, and centerlines proliferate—in other words, just about everywhere. Why? Because they reverse another rubber stamp, one that made our cities universally more dangerous and less walkable over the past half century. So, stamp away!

Note: Our colleague Jeff Speck is right. Europe is doing a much better job of making streets for people than we are here in the United States. For one thing, they don't put traffic engineers in charge, as we do. We will look at this again in Chapter Five. But let's not forget all the good streets new and old in America, like our last example in Chapter Four, the Rue du Grand Fromage, in the Cotton District in Starkville Mississippi.

Rue du Grand Fromage, The Cotton District, Starkville, Mississippi

Dan Camp, 2007

Village Street

Few who visited Dan Camp's Cotton District development in Starkville, Mississippi, will forget either the District or its flamboyant creator. Camp was a tall, funny, good ol' boy from Mississippi straight out of central casting.[130] His unique development the Cotton District, as Camp liked to point out, was "New Urbanism before New Urbanism."[131]

Camp was teaching mechanical drafting and other subjects in the Industrial Education Department at Mississippi State University when he began buying inexpensive land and run-down shacks near a former cotton mill in Starkville. He had the idea that he could build low-cost student apartments on the land, situated between the moribund town center and the university campus.

Figure 4.282: Dixie Court, The Cotton District, Starkville, Mississippi. Dan Camp, 2011. Student apartments *Courtesy of The Cotton District*

To further save money, he made the apartments small and developed a crew of workers, whom he taught to make bricks and fabricate windows and moldings for less than it cost to buy them. Space-saving techniques included traditional porches that doubled as exterior circulation and design touches like galley kitchens that also functioned as the hallway between the living room and the bedrooms.

Most of all, Camp built luxurious-looking traditional Southern buildings, combined with extravagant touches inspired by trips to France and Italy. Then he managed his development with his inimitable style, befriending generations of Mississippi State students. He became the largest landlord in town after the university and held the office of Mayor for several years.

He began by building a street of tiny rowhouses in the middle of a large empty block that was his least valuable land. That put him on the map, so to speak. La Rue du Grand Fromage, which includes stores, a restaurant where Camp liked to eat, and more student apartments, was one of his later streets. We suspect he named it after himself, the Big Cheese.[132]

Figure 4.283: Rue du Grand Fromage, The Cotton District, Starkville, Mississippi. Dan Camp, 2007. Photograph looking north from University Drive. A modern shared-space street with student apartments, restaurants, and coffee bars.

NOTES

1. As we write the second edition in the new post-Covid world, it seems like there's at least one new video of the rue de Rivoli on Twitter every week, as well as many photos. Streetfilms made a good video in September 2023 called "Paris vs NYC: What It's Like to Bike"—see https://www.youtube.com/watch?v=jCFKCfdXHJA&list=PLRW0byauSy-l9PTb_0Y6GxHN4pmKAkQVS. Streetsblog NYC wrote about some of the Paris videos: "STREETFILMS: *Regardez Ces Pistes Cyclables Impressionnantes Parisiennes!*" Streetsblog NYC, July 19, 2022, https://nyc.streetsblog.org/2022/07/19/streetfilms-regardez-ces-impressionnants-pistes-cyclables-parisiennes, and "Streetfilms: Paris Kicks New York's Ass as a Biking Capital," Streetsblog NYC, September 25, 2022, https://nyc.streetsblog.org/2022/09/25/sunday-streetfilms-paris-kicks-new-yorks-ass-as-a-biking-capital.

2. Giulia Carbonaro, "Americans Being Priced Out of Their Love Affair with Cars," *Newsweek*, February 7, 2024, https://www.newsweek.com/2024/02/16/americans-being-priced-out-their-love-affair-cars-1866374.html.

3. Robert D. Putnam, *Bowling Alone: The Collapse and Revival of American Community*, revised and updated (Simon & Schuster, 2020).

4. Culdesac Tempe is the first project built by the Culdesac development company, which calls it "the first car-free neighborhood built from scratch in the US." The company's website is https://culdesac.com/. Conor Dougherty wrote about Culdesac Tempe in *The New York Times*: "The Capital of Sprawl Gets a Radically Car-Free Neighborhood," October 31, 2020, https://www.nytimes.com/2020/10/31/business/culdesac-tempe-phoenix-sprawl.html. The first paragraph says, "Phoenix, that featureless and ever-spreading tundra of concrete, has been called 'the world's least sustainable city.' It has been characterized as a 'sprawling, suburbanite wasteland' and 'a monument to man's arrogance.' *The Onion* has darkly predicted that by 2050, 'most of Earth's landmass' will be swallowed by the encroaching Phoenix exurbs. The Walk Score index ranks the place as the second-worst big city in America for pedestrians, and traversing it has been described as 'a slog through a desert, plus the occasional McDonald's.'"

5. Bernard Rudofsky, *Streets for People: A Primer for Americans* (Doubleday, 1969).

6. William H. Whyte, *The Social Life of Small Urban Spaces* (The Conservation Foundation, 1980). A later edition can be purchased from the Project for Public Spaces at https://www.pps.org/product/the-social-life-of-small-urban-spaces. At the same time that he published the book, Whyte released a film called "The Social Life of Small Urban Spaces" that was a compilation of the film studies he made of public spaces in Manhattan. This can be seen online at https://archive.org/details/social-life-of-small-urban-spaces. Whyte's full name, William Hollingsworth Whyte, was the source of his nickname, "Holly."

7. William H. Whyte, *City: Rediscovering the Center* (Doubleday, 1988).

8. Jane Jacobs, *The Death and Life of Great American Cities* (Random House, 1961): 183.

9. Werner Hegemann and Elbert Peets, *The American Vitruvius: An Architects' Handbook of Civic Art* (The Architectural Book Publishing Company, 1922). In the 1980s, a mail-order bookstore that specialized in remaindered books fortuitously sold copies of *Civic Art* for $1 each.

10. Maurice Culot, ed., *Rational Architecture Rationelle: The Reconstruction of the European City* (AAM Edition, 1981).

11. Robert Davis, in conversation with John Massengale, October 2012.

12. Kurt Andersen, "Design: Best of the Decade," *TIME*, January 1, 1990. Andersen also wrote two other *TIME* articles about Seaside: "Design: Building a Down Home Utopia," *TIME*, April 18, 2005, https://content.time.com/time/subscriber/article/0,33009,1050544,00.html and "Oldfangled New Towns," *TIME*, May 20, 1991, https://content.time.com/time/subscriber/article/0,33009,972988,00.html.

13. Kaid Benfield's essay originally appeared on his NRDC blog. See "How Retrofitting a California Suburb for Walkability Is Spurring Economic Development," Smart Cities Dive, https://www.smartcitiesdive.com/ex/sustainablecitiescollective/how-retrofitting-california-suburb-walkability-spurring-economic-development/108556/.

14. See Transportation Alternatives, "Streets for New York's Next Generation," *Reclaim*, vol. 20(3), Fall 2014: 12-15, https://issuu.com/transalt/docs/reclaim-fall-2014. Also see Stephen Miller, "Envisioning a Safer Queens Boulevard Where People Want to Walk," Streetsblog NYC, November 25, 2014, https://nyc.streetsblog.org/2014/11/25/a-safer-queens-boulevard-must-be-a-place-where-people-want-to-walk; John Massengale, "Two Birds with One Design: Affordable Housing & The Boulevard of Death," There Are Two Kinds of Architecture, November 19, 2014, https://blog.massengale.com/2014/11/19/queensboulevard/; John Massengale, "Affordable Housing & The Boulevard of Death Followup," There Are Two Kinds of Architecture, December 2, 2014, https://blog.massengale.com/2014/12/02/qbfollowup/; and John Massengale, "QB Redux: I think that I shall never see…," There Are Two Kinds of Architecture, August 7, 2015, https://blog.massengale.com/2015/08/07/qbredux/. Also see https://street.design/slowny.

15. See Bill Parry, "DOT Begins Phase 2 of the Queens Blvd. Project in Elmhurst," AMNY QNS, August 4, 2016, https://qns.com/2016/08/dot-begins-phase-2-of-the-queens-blvd-project-in-elmhurst/. Also see the many stories in Streetsblog NYC, including images of safety improvements: https://nyc.streetsblog.org/search?s=Queens+Boulevard.

16. See page 610 in Chapter Five.

17. Bill Parry, "Transportation Alternatives Celebrates Completion of Queens Boulevard Redesign Project," AMNY QNS, November 16, 2021, https://qns.com/2021/11/transportation-alternatives-celebrates-completion-of-queens-boulevard-redesign-project/.

18. See HR&A, "Columbia Pike Commercial Market Study", October 2019, https://arlingtonva.s3.amazonaws.com/wp-content/uploads/sites/31/2019/12/Columbia-Pike-Commercial-Market-Study_Final-revised-dec-2019.pdf.

19. *Ibid.*

20. James Jarvis, "County Leaders Says Columbia Pike's Future Looks 'Bright' as Construction Nears Completion," ARLnow, February 9, 2024, https://www.arlnow.com/2024/02/09/county-leaders-says-columbia-pikes-future-looks-bright-as-construction-nears-completion/.

21. *Plan El Paso* is available online at https://www.elpasotexas.gov/planning-and-inspections/planning-division/planning/.

22. See the Case Study on Gran Via de les Corts Catalanes in Barcelona in Chapter Two, page 109.

23. See "A Village on the Move," Katonah Village Improvement Society, https://www.katonahvis.org/katonah-village-history/a-village-on-the-move/. Also see Robert Khederian, "How Katonah Was Moved for New York City's Water System," Streetsblog NYC, August 3, 2017, https://archive.curbed.com/2017/8/3/16089078/katonah-croton-nyc-water-history.

24. B. S. Olmstead (1803–1897) was a florist and landscape architect. See his obituary: "Benjamin S. Olmstead," *The Daily Standard Union*, Brooklyn, New York, May 4, 1897, page 3. George S. Olmstead was his son. See "Osborn Memorial Home Haven for Women in Declining Years," *The Daily News*, Tarrytown, New York, August 28, 1934, page 9. Both articles can be found at FultonHistory.com. An 1870s Census entry at Ancestry.com says the Olmsteads lived in Port Chester, New York, in the Town of Rye, West Chester County, New York (now Westchester).

The Osborn Memorial Home is the only other work we found attributed to the Olmsteads. Known today as The Osborn, it was designed by the distinguished architect (and father of etiquette author Emily Post) Bruce Price. The Osborn website has a short history of the large Georgian-style complex at https://www.theosborn.org/about.

Families in early America sometimes spelled their shared names differently, and Frederick Law Olmsted and B.S. Olmstead were both Connecticut natives and landscape architects, but there is no known connection between them.

25. MTA Press Release, "MTA to Purchase Grand Central Terminal, Harlem Line and Hudson Line for $35 Million," November 13, 2018, https://apps.cio.ny.gov/apps/mediaContact/public/view.cfm?parm=92FAF7B8-02CA-E021-4E8F8DF271CEA0AD_74898BB8-5056-9D2A-10F7020A8F081395.

26. Khederian, *op. cit.* Fifty-five of the households in old Katonah elected to move their houses to the new village. A contractor jacked up the houses and moved them on a set of temporary wooden tracks greased with soap. The move went up and down hills and across rivers and streams, but some residents stayed in their houses during the move.

27. Allan B. Jacobs, *Great Streets* (MIT Press, 1995).

28. See official correspondence of the Association pour la Protection des Demeures Anciennes et Paysages Aixois (APDAPA), La "requalification" du Cours Mirabeau; Projet municipal et Contre-project, http://apdapa.free.fr/pages/cours/cours_projet_fr.htm and https://apdapa-free-fr.translate.goog/pages/cours/cours_article01_defigurer.htm?_x_tr_sch=http&_x_tr_sl=fr&_x_tr_tl=en&_x_tr_hl=en.

29. See the works of architect and urban designer Antoine Grumbach at https://jeannebucherjaeger.com/artist/grumbach-antoine/.

30. The Architectural League of New York, "The Unfinished Grid, Design Speculations for Manhattan," https://archleague.org/publications/the-unfinished-grid/.

31. Julianne Cuba, "Upper East Side Candidate Billy Freeland One-Ups Competitor By Calling for Replacing FDR Drive," Streetsblog NYC, March 30, 2021, https://nyc.streetsblog.org/2021/03/30/upper-east-side-candidate-billy-freeland-one-ups-competitor-with-transportation-plan-calls-for-replacing-fdr-drive.

32. See the list of NACTO Member Cities at https://nacto.org/member-cities.

33. See the downloadable PDF at https://nacto.org/wp-content/uploads/designing_walkable_urban_thoroughfares.pdf.

34. "Reuben Harrison Hunt," Preserve Chattanooga, https://www.preservechattanooga.com/hunt.

35. Christopher Leinberger and Mariela Alfonzo, "Walk This Way: The Economic Promise of Walkable Places in Metropolitan Washington, DC," Walkable Urbanism, https://www.brookings.edu/articles/walk-this-waythe-economic-promise-of-walkable-places-in-metropolitan-washington-d-c/.

36. Emily Badger, "Why We Pay More for Walkable Neighborhoods," Bloomberg, May 28, 2012, https://www.bloomberg.com/news/articles/2012-05-28/why-we-pay-more-for-walkable-neighborhoods. Also see: Peter Katz, "Hope VI and the Inner City," Better! Cities and Towns; Melinda J. Milligan, Kevin Fox Gotham, and James R. Elliott, "HOPE VI, New Urbanism, and the Utility of Frames: A Reply to Melendez and Coats," https://journals.sagepub.com/doi/pdf/10.1111/j.1535-6841.2004.00095.x?casa_token=jR3xQjOIiPUAAAAA:zMJC9fXnRD7oleihujPWdBGFBM_6L6m8b3BXkyA5EqjhILU1p8hb73XyFc5Gxwa10LH3Upqk5p4; and Henry G. Cisneros and Lora Engdahl, eds., *From Despair to Hope: Hope VI and the New Promise of Public Housing in America's Cities* (Brookings Institution, 2009).

37. Daniel Moylan, "Committed to the Cause," *Green Places* (June 2004): 32–34.

38. *Ibid.*, 32.

39. *Ibid.*, 33.

40. *Ibid.*, 34.

41. Simon Jenkins, "Rip Out the Traffic Lights and Railings," *The Guardian*, February 29, 2008, https://www.theguardian.com/commentisfree/2008/feb/29/guardiancolumnists. City of Cambridge, Massachusetts, Department of Public Works, "Five Year Sidewalk and Street Reconstruction Plan," https://www.cambridgema.gov/theworks/ourservices/~/media/53AEF489A1144E4CB31B9324F48DBF91.ashx.

42. R. Bruce Stephenson, *John Nolen: Landscape Architect and City Planner,* (University of Massachusetts Press in association with Library of American Landscape History, 2015).

43. *Ibid.*

44. See https://www.planning.org/greatplaces/streets/2014/clematisstreet.htm.

45. See Gehl, "West Palm Beach Public Realm Action Plan", https://issuu.com/westpalmdda/docs/2017_west_palm_beach_public_realm_a.

46. Quoted from an email from Raphael Clemente on February 24, 2023.

47. See also Peter Katz, *The New Urbanism: Towards an Architecture of Community* (McGraw-Hill, 1994).

48. Susan Spano and Aviva Shen, "The 20 Best Small Towns in America," *Smithsonian*, April 30, 2012, https://www.smithsonianmag.com/travel/the-20-best-small-towns-in-america-of-2012-66120384/.

49. BerkShares were created by the Schumacher Center for New Economics, founded in Great Barrington in 1980 as the E.F. Schumacher Society. Schumacher was the author of *Small Is Beautiful: A Study of Economics As If People Mattered* and an advocate for regional, sustainable, and socially just economies. The Center owns Schumacher's personal library. For BerkShares, see "BerkShares, Our Currency for the Berkshire Region" at https://berkshares.org/ and "BerkShares: an alternative vision for thriving regional economies - Schumacher Center for a New Economics," https://centerforneweconomics.org/publications/berkshares-an-alternative-vision-for-thriving-regional-economies/. Beautifully designed, BerkShares are the only currency on Earth with an illustration of a skunk. For the Schumacher Center, see their website at https://centerforneweconomics.org/ and "Schumacher Center for a New Economics," Wikipedia, https://en.wikipedia.org/wiki/Schumacher_Center_for_a_New_Economics.

50. See the website for The Train Campaign, https://barringtoninstitute.org/traincampaign/. The Train Campaign was created by Karen Christensen. For more information on her many activities, including a podcast with John Massengale, see https://karenchristensen.org and https://karenchristensen.substack.com/p/the-secret-to-great-cities-and-towns: "The Secret to Great Cities and Towns, and How Traffic Engineers Got It Wrong," August 11, 2025.

51. See "Reconstructing the Street? Move the Curb!" in the Strong Towns blog, https://www.strongtowns.org/journal/2023/9/11/reconstructing-the-street-move-the-curb.

52. In conversation with John Massengale.

53. The title referred to the "Alice's Restaurant Thanksgiving Massacree" in Stockbridge, Massachusetts (near Great Barrington), as well as to the *Texas Chainsaw Massacre* movie. See "Alice's Restaurant Massacree" on YouTube, https://www.youtube.com/watch?v=WaKIX6oaSLs. The local *Berkshire Record* published several articles on our Main Street essay, including: John Massengale, "A Compromised Main Street," *Berkshire Record*, (June 29–July 5, 2012): A5, Part 2 (July 6–July 12, 2012): A4. Julie Ruth, "Main Street Renovation Examined in New Book," *Berkshire Record*, B1; and John Massengale, "Occupy Main Street," *Berkshire Record*, (September 9–September 15, 2016): A5, https://blog.massengale.com/2016/09/10/occupy-main-record/. "A Compromised Main Street" was also published online by Streetsblog: John Massengale, "MassDOT Mistake: How Not to Rebuild Main Street," Streetsblog USA, July 10, 2012, https://usa.streetsblog.org/2012/07/10/massdot-mistake-how-not-to-rebuild-main-street; and "Street Design in the *Berkshire Record*," Streetsblog USA, February 20, 2014, https://blog.massengale.com/2014/02/20/street-design-in-the-berkshire-record/.

54. Dambisa Moyo, *Dead Aid: Why Aid Is Not Working and How There Is a Better Way for Africa* (Farrar, Straus and Giroux, 2009). Also see a discussion of the origins of this proverb on Reddit: https://www.proverbshub.com/african-topic/planning/.

55. In conversation with John Massengale. Fred Kent founded the Project for Public Spaces, and later the Social Life Project. For the Project for Public Spaces, see https://www.pps.org/. For the Social Life Project and Kent, see https://www.sociallifeproject.org/author/fredkent/.

56. City Planning Commission, City of Cleveland, "Glossary of Terms," https://planning.clevelandohio.gov/cwp/glossary/glossary.php.

57. Hans Monderman (1945–2008) was a Dutch traffic engineer who pioneered modern concepts of shared space and reduced traffic-control markings after seeing that excessive signage and specialized space intended to improve safety ironically lulled motorists into unsafe complacency. See "Hans Monderman, Shared Space & the Monderman Rule" in this chapter.

Figure 4.284: Margaret E. French, "Letter to the Editor: Pedestrians' Woes Not Yet Ended," *Oakland Tribune*, September 3, 1928. Margaret French's commentary on the problems between cars and people on foot, written almost a century ago, is as true today as the day it was written. In fewer than six hundred words, Margaret French explains some of the problems traffic engineers caused when imposed on our streets a system that puts drivers first. *Courtesy of Peter Norton*

Oakland Tribune   MONDAY EVENING, SEPTEMBER 3, 1928

PEDESTRIANS' WOES NOT YET ENDED.

To the Editor of The TRIBUNE:

May not a "walker" have a little space in your paper to point out certain aspects of the "Stop-and-Go" system of regulating street traffic, which has now been in operation long enough to make possible fair criticism of its success?

Pedestrians, who formerly might dart in and out at their will—or rather their peril—are now controlled in their own interest and a few seconds of comparative safety provided for them. But the "transverse" vehicular stream is more dangerous than ever, owing to haste and a sense of their own right on the part of drivers, and the "foot passenger" caught in it may hope for very little consideration either on the street or in court. He may, therefore, reasonably expect maximum protection during the short duration of his own right to cross the street. He is, however, almost forced to feel in this matter that "Blessed are they who expect nothing." His right of way is subject to two forms of invasion under the mechanical signal system, both sufficiently menacing to deserve attention.

The first is the right hand turn within the central traffic district. Here the left-hand turn, which cuts a long swath in automobile traffic, has been eliminated; but the right hand turn remains and is iniquitous, no matter on which signal it is made, cutting across the pedestrian's path during the few moments when he might expect to be free from danger. A pedestrian leaving the curb on the proper signal is often so delayed by one vehicle after another cutting around the corner in front of him that he cannot possibly make his goal on time and is caught in highly dangerous alternative traffic. The city which has wrested from him the right to look out for his own safety (and wisely in the main) thus deceives him and leaves him in a veritable trap. This is particularly true where right-hand turns are permitted from, as well as into, the transverse traffic—or against a stop signal, since the pedestrian is thereby laid open to invasion both at the start and at the finish of his crossing. Exclusion of the right-hand turn against a stop signal is provided for by the uniform traffice ordinance within the central district. But all right-hand turns ought to be eliminated on the most congested corners.

A second form of invasion of the walker's very limited right of way at congested corners is from vehicles crossing from a street entering the main street "on the bias," or so as to form a "gore." As this traffic usually unites with the vehicular traffic of the main street, it crosses pedestrian traffic only. Often it is very heavy. On certain corners I have seen correctly moving pedestrians so invaded by a mass of automobiles and street cars launching against them as to be afforded no practical right of way at all. Were these to cross vehiclular traffic a third stop-and-go signal would have to be provided them, and pedestrians, who are the most common victims of accidents, ought to be similarly protected. Traffic from bias streets ought to be held back until the signal operates to hold back the pedestrians whom it menaces.

Both of these forms of invasion of the pedestrian's right of way are extremely dangerous and ought to be abolished on certain corners, at least, of the central traffic district.

With thanks for this hearing, I am,

MARGARET E. FRENCH.

58. An article about Ciudad Cayalá published in *The New York Times* in 2024 had a headline that captured a lot of attention: "New Utopian Enclave? Or a Testament to Inequality?" (Simon Romero and Jody Garcia, January 18, 2024, https://www.nytimes.com/2024/01/18/world/americas/cayala-guatemala.html). Within a week, a software engineer named Zach Caseres, who runs a website called Startup Cities, wrote a harsh critique of the *Times* article on his Substack site: "*The New York Times* is Wrong About Guatemala's New Town," January 24, 2024, https://www.startupcities.com/p/the-new-york-times-is-wrong-about.

59. Dan Camp is the founder, designer, and builder of the Cotton District infill development in Starkville, Mississippi, and a former mayor of Starkville. See https://www.cottondistrictms.com/. He died on October 25, 2020. *The New York Times* published a nice tribute to him and his work: Neil Genzlinger, "Those We've Lost: Dan Camp, Who Created a Mississippi Jewel, Dies at 79," *The New York Times*, November 9, 2020, https://www.nytimes.com/2020/11/09/obituaries/dan-camp-dead-coronavirus.html.

60. *Light Imprint* was named one of the "25 Great Ideas of the New Urbanism" by the CNU. See Robert Steuteville, October 31, 2017, https://www.cnu.org/publicsquare/2017/10/31/25-great-ideas-new-urbanism.

61. The *Handbook* is available in bookstores or by contacting Tom Low at info@civicbydesign.com.

62. This section on *Light Imprint* is taken in part from the *Light Imprint Handbook* (New Urban Press, 2008) by Thomas Low. For a further description of the Urban to Rural Transect, see "Context Matters" in Chapter One.

63. See Aaron M. Renn, "When New York City Tried to Ban Cars—The Extraordinary Story of 'Gridlock Sam,'" *The Guardian*, June 1, 2016, https://www.theguardian.com/cities/2016/jun/01/new-york-city-ban-cars-gridlock-sam-schwartz-manhattan. Also see Samuel I. Schwartz, William Rosen, *Street Smart: The Rise of Cities and the Fall of Cars* (Public Affairs, 2015) and Nicole Gelinas, *Movement: New York's Long War to Take Back Its Streets from the Car* (Empire State Editions, 2024): *passim* but especially 378–379.

64. For Tactical Urbanism, see Mike Lydon and Anthony Garcia, *Tactical Urbanism: Short-term Action for Long-term Change* (Island Press, 2015). Tactical Urbanism is related to what the Project for Public Spaces calls "The Lighter, Quicker, Cheaper Transformation of Public Spaces": https://www.pps.org/article/lighter-quicker-cheaper. Using tactical urbanism techniques, one can quickly test public reaction to new ideas and use that as the basis for more permanent designs. Bright paint and bold designs announce that the street is changing, but those designs don't always age well. Calling attention to the change rather than the place, they can disrupt the harmony of the space. The paints and plastic delineators can look grimy after a short time. And in the case of Madison Square, they are still in place seventeen years later.

65. Frank Bruni, "Bicycle Visionary," *New York Times*, September 10, 2011, https://www.nytimes.com/2011/09/11/opinion/sunday/bruni-janette-sadik-khan-bicycle-visionary.html. Also see Gelinas, *Movement, op. cit.*, and Sadik-Khan's own book, Janette Sadik-Khan and Seth Solomonow, *Streetfight: Handbook for an Urban Revolution* (Viking, 2016).

66. Ben Fried summarizes the long, sordid history in "Good Riddance to the Prospect Park West Bike Lane Lawsuit," Streetsblog NYC, September 22, 2016, https://nyc.streetsblog.org/2016/09/22/good-riddance-to-the-prospect-park-west-bike-lane-lawsuit.

67. See "Pedestrian Plazas," New York City DOT, https://www.nyc.gov/html/dot/html/pedestrians/nyc-plaza-program.shtml. For the work at Madison Square, see Sadik-Khan, *Streetfight, op. cit.*, 123–125.

68. Described in "NYC DOT Announces 'Shared Street' Completed in Manhattan's Flatiron District," NYC DOT Press Release, https://www.nyc.gov/html/dot/html/pr2017/pr17-059.shtml.

69. The Broadway Linear Park has a website: https://www.broadwaylinearpark.nyc/. Also see 2017 Pedestrian Intercept Survey https://sites.google.com/view/broadwaylinearpark/community-support?authuser=0.

70. For the shared-space street at Madison Square, see "NYC DOT Announces 'Shared Street' Completed in Manhattan's Flatiron District," NYC DOT Press Release, August 9, 2017, https://www.nyc.gov/html/dot/html/pr2017/pr17-059.shtml, and "Flatiron Shared Street," March 27, 2017, https://www.nyc.gov/html/dot/downloads/pdf/flatiron-shared-street-mar2017.pdf. For Union Square, see "University Place Shared Street," https://www.nyc.gov/html/dot/downloads/pdf/univ-place-shared-st-may2019.pdf. John Massengale critiqued the DOT's design in a blog post "A Chicane Is Not A Shared Space Place," There Are Two Kinds of Architecture, October 22, 2019, https://blog.massengale.com/2019/10/22/chicane/."

71. The BID's website is https://meatpacking-district.com/.

72. Most people think it stands for Down Under the Manhattan and Brooklyn… something. DUMBO does have a BID, but a developer and arts organizations bear more responsibility for the creation of what is now an Historic District. See https://dumbo.nyc/ and https://en.wikipedia.org/wiki/Dumbo,_Brooklyn for more information.

73. The East Village name was introduced by real estate agents in the 1960s, after hippies and artists followed the migration of beatniks from Greenwich Village (now the West Village) to Lower East Side. When Latins moved to the Lower East Side in large numbers, it became known as "Loisaida," a phonetic spelling of Spanish pronunciation of the name.

74. Leading to plenty of jokes about businesses prostituting themselves. Philip Johnson famously quipped that architecture is the world's second oldest profession. The new National Historic District is described in Wikipedia: https://en.wikipedia.org/wiki/Meatpacking_District,_Manhattan.

75. The results of the competition were published in Anne Guiney and Brendan Crain, eds. *By the City / For the City: An Atlas of Possibility for the Future of New York* (Multi-Story Books, 2011). Jane Jacobs Square is on page 457.

76. Ray Oldenburg, *Celebrating the Third Place: Inspiring Stories about the "Great Good Places" at the Heart of Our Communities* (Marlowe & Company, 2000): *passim.*

77. See "Walls Without Windows" in Chapter Five regarding feng shui and energy.

78. The Danish Architecture Center is appropriately located on one of the Copenhagen's most high-design streets. See "Vester Voldgade, Urban Spaces," Danish Architecture Center, https://dac.dk/en/knowledgebase/architecture/vester-voldgade-2/. Also see "Vester Voldgade," Wikipedia, https://en.wikipedia.org/wiki/Vester_Voldgade.

79. See John Massengale, "There Are Better Ways to Get Around Town," *New York Times*, May 15, 2018, https://www.nytimes.com/2018/05/15/opinion/there-are-better-ways-to-get-around-town.html.

80. See Gianni Longo and Roberto Brambilla, *For Pedestrians Only: Planning, Design and Management of Traffic-Free Zones* (Whitney Library of Design, 1977).

81. "It is being viewed as the correction of a historic mistake. More than 40 years after parts of the canal that encircled Utrecht's old town were concreted over to accommodate a 12-lane motorway, the Dutch city is celebrating the restoration of its 900-year-old moat," Daniel Boffey wrote in "Utrecht restores historic canal made into motorway in 1970s," *The Guardian*, September 14, 2020, https://www.theguardian.com/world/2020/sep/14/utrecht-restores-historic-canal-made-into-motorway-in-1970s.

82. The Market Tower is both the only unbuilt structure from Poundbury's first phase, as well as the only building designed by Léon Krier for the first phase.

83. For "Shared space was the nub of what Prince Charles had in mind in 1987 when he founded the experimental village of Poundbury on land that he owns in the English countryside," according to Hank Dittmar, the former Chair of the CNU and the former CEO of the Prince's Foundation for the Built Environment. See his essay on his favorite space in London, Seven Dials: Chapter Three, page 307. Architecturally, the village is often panned as a nostalgic exercise in faux-bucolic Englishness. But in prioritizing people over cars, says Dittmar, the winding streets and discreet signs used in Poundbury make it a model for high-density urban design. The bigger challenge, he says, is 'retrofitting places that were built before the automobile. The old idea for traffic was to separate pedestrians and motor vehicles, but what it has devolved to is guardrails that fence people in.'" Quoted by Michael Brunton, "Signal Failure," *TIME*, January 30, 2008, https://time.com/archive/6683576/signal-failure/.

In the early part of this century, Dittmar was in the thick of the battle between Prince Charles and the British architectural establishment. The fight began in 1984 when Charles gave what was expected to be a polite royal speech at a 150th Anniversary dinner for the Royal Institute of British Architects. Instead, he called a proposed extension of the National Gallery in London "a monstrous carbuncle on the face of an old friend." See Ian Mansfield, "30th anniversary of the 'monstrous carbuncle' speech," ianVisits, May 17, 2014, https://www.ianvisits.co.uk/articles/30th-anniversary-of-the-monstrous-carbuncle-speech-11696/, and Mira Bar-Hillel and Will Hurst, "Prince Charles: When I said monstrous carbuncle...," *The Standard*, April 12, 2012, https://www.standard.co.uk/hp/front/prince-charles-when-i-said-monstrous-carbuncle-6774545.html.

Sir Richard Rogers, whose wife ran a restaurant popular among politicians, became the leader of the opposition after deciding that Charles had cost him more than one commission. Dittmar told the authors that it was noticeable that if Charles criticized Rogers one day, the next day might find a critical article about Camilla Parker Bowles on the front page of one of Rupert Murdoch's British papers. The most famous of these became known as "Camillagate." One of Murdoch's editors obtained a recording of one of Prince Charles's calls and printed intimate comments he made to Parker Bowles.

From the earliest days of Poundbury, the architectural establishment was critical of the traditional architecture and urban design. Today, however, Britons hold the development in high esteem. See, for example, a 2016 article by *The Guardian*'s architecture critic Oliver Wainwright: "A royal revolution: is Prince Charles's model village having the last laugh?," *The Guardian*, October 27, 2016, https://www.theguardian.com/artanddesign/2016/oct/27/poundbury-prince-charles-village-dorset-disneyland-growing-community, and Michael Mehaffy, "How Charles was right (it's probably not what you think)," CNU Public Square, October 31, 2022, https://www.cnu.org/publicsquare/2022/10/31/how-charles-was-right-its-probably-not-what-you-think.

The former Prince's Foundation for the Built Environment is now the King's Foundation: https://kings-foundation.org/. Poundbury and newer developments by Charles like Nansledan, page 481, have become part of the Labour government's housing discussion. See Sean Coughlan, "King makes rare joint visit with PM and Rayner," BBC, February 10, 2025, https://www.bbc.com/news/articles/cew5rylqj2ko. In 2025, Nansledan received the ICAA's Gindroz Award for Excellence in Affordable Housing: see https://www.classicist.org/articles/the-icaa-announces-nansledan-as-the-2025-recipient-of-the-gindroz-award-for-excellence-in-affordable-housing/.

84. Brodie Owen, "Nansledan: Prince William to Build Homes for Homeless on Duchy Land," BBC News-South West, February 18, 2024, https://www.bbc.com/news/uk-england-cornwall-68330944. Also see note 83, about a joint visit to Nansledan by King Charles, UK Prime Minister Keir Starmer, and UK Deputy Prime Minister and Housing Secretary Angela Rayner.

85. Neighborhood founder Charles Brewer used a fictional town from an animated children's cartoon—Busytown from *The Busy World of Richard Scarry*—to explain his ideal development program: a bustling place where neighbors from all walks of life are in frequent contact and come to know one another.

86. We first heard the term "conversational distance" used by Vince Graham and Bob Turner, the developers of Newpoint (*q.v.*).

87. Jane Jacobs, *The Death and Life of Great American Cities* (Random House, 1961): 269.

88. Daralice D. Boles, "Robert Davis: Small Town Entrepreneur," *Progressive Architecture* 66 (July 1985): 111–118.

89. Duany Plater-Zyberk & Company, *The Lexicon of the New Urbanism*, 2002, https://www.dpz.com/wp-content/uploads/2017/06/Lexicon-2014.pdf.

90. A bar called Hoppe on the west side of the square, at the intersection of Spuistraat and the tiny lane leading to the Huidenstraat, is reportedly where Freddy Heineken got his start. A simple and unpretentious Amsterdam "brown café," it was called by *Newsweek* one of the ten best bars in the world.

91. Tax Valuation Maps by Prof. Philip Stoddard, Matthew Toro, Felipe Delgado and Florida International University GIS Center, c. 2014.

92. Fred Kent frequently quotes Monderman on this topic: the exact language can vary, but the concept is always the same. For example, see "Can a Street Be a Place?," Reports from the Editors, Planners Web, https://plannersweb.com/2010/05/can-a-street-be-a-place/. Also see Michael Brunton, "Signal Failure," *TIME*, January 30, 2008, https://time.com/archive/6683576/signal-failure/. "'My theory,' said Monderman, at a Congress for the New Urbanism (CNU) summit in London 'was if you want people to behave in a village, maybe you have to make it feel like a village.' Monderman's flowerpots reduced average traffic speed by 10%; using shared space cut it in half."

93. In conversation with John Massengale, December 2012.

94. A *woonerf* is "a street, also known as a home zone, shared zone, or living street, where pedestrians have priority over vehicles and the posted speed limit is no greater than 10 miles per hour. Physical elements within the roadway, such as shared surfaces, plantings, street furniture, parking, and play areas, slow traffic and invite pedestrians to use the entire right-of-way." The definition is from the USGBC's LEED-ND rating system: https://www.usgbc.org/credits/SSpc14.

95. It was originally called simply "the new neighborhood in Old Davidson," breaking with the convention among developers that each new subdivision is given its own annoying brand name, e.g., "Arcadia Ridge at Hunter's Glen" or "Fawn Creek at Farmers Run." (A favorite parody: Jerry's parents on *Seinfeld* live in "Del Boca Vista Phase III.") Eventually, neighborhood founder Doug Boone agreed to make a new sanctuary for St. Albans Episcopal Church, the landmark centerpiece of what has since become known informally as the St. Albans neighborhood.

96. See also Allan T. Shulman, ed. 53, *Miami Modern Metropolis: Paradise and Paradox in Midcentury Architecture and Planning* (Balcony Press, 2009): 233.

97. *Ibid.*, 238–239.

98. Albert Ullman, *A Landmark History of New York* (D. Appleton & Co., 1901): 19. Also Wikipedia article "Stone Street, (Manhattan)", https://en.wikipedia.org/wiki/Stone_Street_(Manhattan).

99. David W. Dunlap, "Turning an Alley into a Jewel," *New York Times*, December 6, 2000, https://www.nytimes.com/2000/12/06/nyregion/commercial-real-estate-turning-an-alley-into-a-jewel.html.

100. Christopher Gray, "A Nod to New Amsterdam," *New York Times*, December 27, 2012, https://www.nytimes.com/2012/12/30/realestate/streetscapes-on-south-william-street-a-nod-to-new-amsterdam.html.

101. Dunlap, *op. cit.*

102. Also see "Design District." DPZ CoDesign, https://www.dpz.com/projects/design-district/.

103. Jane Jacobs, *The Death and Life of Great American Cities* (Random House, 1961): 178: "Condition 2: Most blocks must be short; that is streets and opportunities to turn corners must be frequent."

104. Guiney and Crain, *op. cit.*, 347.

105. For example, see *The Social Life of Small Urban Spaces*, *op. cit.*

106. Matt Chaban, "Meet Me on 6½th Avenue: DOT Planning Public Promenade Through Middle of Midtown Towers," *New York Observer*, March 26, 2012, https://observer.com/2012/03/meet-me-on-6%C2%BDth-avenue-dot-planning-public-promenade-through-middle-of-midtown-towers/.

107. Catfiddle Street and its makers are described in Witold Rybczynski's book *Charleston Fancy: Little Houses & Big Dreams in the Holy City* (Yale, 2019). Also see, "Enduring Places: Little Houses and Big Dreams in the Holy City with Witold Rybczynski," Institute of Classical Architecture & Art, April 2, 2024, https://www.classicist.org/articles/enduring-places-little-houses-and-big-dreams-in-the-holy-city-witold-rybczynski/

108. In his 2022 State of the City address, Mayor Vince Lago reported that, despite the pandemic, retail space in downtown Coral Gables is 96% occupied and office space is 87% leased: "Mayor Lago Delivers State of the City Address," City of Coral Gables, September 22, 2022, https://www.coralgables.com/news/mayor-lago-delivers-state-city-address.

109. "The tragedy of the commons refers to a situation in which individuals with access to a public resource (also called a common) act in their own interest and, in

doing so, ultimately deplete the resource. This economic theory was first conceptualized in 1833 by British writer William Forster Lloyd. In 1968, the term "tragedy of the commons" was used for the first time by Garret Hardin in *Science* magazine." See https://online.hbs. edu/blog/post/tragedy-of-the-commons-impact-on-sustainability-issues#:~:text=This%20economic%20 theory%20was%20first,Garret%20Hardin%20in%20 Science%20Magazine.

110. Keller (@jkellerfsu) Tweeted "I stopped by #GiraldaPlaza in #CoralGables for the 1st time in months. Disappointing to see some of the changes. #Walkability has been significantly reduced by extending dining tables into the center walkway and these shades are now allowed walling off the space. Feels chaotic," X/Twitter, January 20, 2024, https://twitter.com/jkellerfsu/status/1748800983373 652186?s=19.

111. Wright reposted Keller's Tweet and commented, "Designed as a paradise for walkers and wheelchair users... now pinched down to something totally off limits to people who use assistive mobility devices. Warned Gables officials about this for years. Dozens of emails and posts 100% ignored," @stevewright64, X, January 21, 2024, https://twitter.com/stevewright64/status/1749058852803 260535?s=19.

112. Because the green street extends the view from the houses towards the lake across Fox Street, neighbors have taken to calling it "Fox Mews," an unofficial and groan-inducing bit of rhyming. But on city maps, the street has no name; houses were given addresses on either one of the alleys. For a time, a landmark pin appeared on Google Maps pointing to its "Shady Park Bench," but that label is gone, perhaps because we visited with cameras once too often.

113. Pedestrian courts and variations on the idea are employed to good effect in New Urban neighborhoods including The Waters, Montgomery, Alabama, and Rosemary Beach, Florida. The benefit of a public pedestrian space facing an important view is also vividly illustrated at Newpoint Green, Beaufort, South Carolina (*q.v.*). Older developments like Bronxville, New York and Forest Hills Gardens, Queens, New York that inspired New Urban projects have excellent examples of pedestrian courts. For Bronxville, see Chapter Three, page 303; Eloise L. Morgan, editor, *Building a Suburban Village, Bronxville, New York, 1898–1998* (Bronxville Centennial Celebration, 1998); Robert A.M Stern with John Massengale, *The Anglo-American Suburb* (Architectural Design Profile, 1981): 31; and Robert A.M. Stern, David Fishman, and Jacob Tilove, *Paradise Planned: The Garden Suburb and the Modern City* (Monacelli Press, 2013): 310–315. For Forest Hills Gardens see Sage Foundation Homes Company, *Forest Hills Gardens* (The Company, 1913); Susan L. Klaus, *A Modern Arcadia: Frederick Law Olmsted Jr. and the Plan for Forest Hills Gardens* (University

of Massachusetts, 2004); Stern and Massengale, *The Anglo-American Suburb*, *op. cit.*, 33; Robert A.M. Stern, Gregory Gilmartin and John Massengale, *New York 1900, Metropolitan Architecture and Urbanism 1890–1915* (Rizzoli, 1983): 427–429; and Stern *et al*, *Paradise Planned*, *op. cit.*, 133–135 and 140–144. For additional references, see the index and bibliography in *Paradise Planned*. Also see Paul Goldberger, "Design Notebook; An Honorable U.S. Tradition of Suburban Planning," *The New York Times*, November 12, 1981, https://www. nytimes.com/1981/11/12/garden/design-notebook-an-honorable-us-tradition-of-suburban-planning.html.

114. Kathryn M. McCamant, Charles Durrett, and Ellen Hertzman, *Cohousing: A Contemporary Approach to Housing Ourselves* (Ten Speed Press, 1988).

115. Ross Chapin, *Pocket Neighborhoods, Creating Small-Scale Communities in a Large Scale World* (Taunton Press, 2011).

116. Jeff Speck, *Walkable City: How Downtown Can Save America, One Step at a Time* (North Point Press, 2012); Jeff Speck, *Walkable City: How Downtown Can Save America, One Step at a Time*, 10th Anniversary Edition (MCD Picador, 2022); and Jeff Speck, *Walkable City Rules: 101 Steps for Making Better Places* (Island Press, 2018).

117. Andres Duany, Elizabeth Plater-Zyberk, and Jeff Speck, *Suburban Nation: The Rise of Sprawl and the Decline of the American Dream* (North Point Press, 2001): 64.

118. Just such a numbskull move devastated Calle Ocho, the main drag of Miami's Little Havana, in the 1970s. *Ibid.*, 161n.

119. Christian Sottile, "One-Way Streets: Urban Impact Analysis," commissioned by the City of Savannah (unpublished).

120. *Ibid.*

121. Melanie Eversley, "Many Cities Changing One-Way Streets Back," *USA Today*, December 20, 2006. The article is no longer online, but it is archived at https://forums. anandtech.com/threads/cities-changing-one-way-streets-back.1984432/.

122. Alan Ehrenhalt, "The Return of the Two-Way Street," *Governing*, March 16, 2010, https://www.governing.com/ archive/the-return-of-the.html.

123. *Ibid.*

124. *Ibid.*

125. "Monon Boulevard Corridor," Congress for the New Urbanism, https://www.cnu.org/what-we-do/build-great-places/monon-boulevard-corridor.

126. Charles Marohn, *Confessions of a Recovering Engineer* (John Wiley & Sons, Inc., 2021): 103.

127. "City reinstalls downtown stop signs to reduce traffic pileup, speeding," KQRE 13 News, https://www.youtube. com/watch?v=v3MLZkZdjjk.

128. John Chew, "This Is the Best Country to Drive in," *Fortune*, September 30, 2015, https://fortune.com/2015/09/30/best-country-drive-waze/.

129. Speck, *Walkable City Rules:* 168–169.

130. *The New York Times* ran his obituary in the print edition on November 18, 2020, Section B, Page 10. For the online version, see Neil Genzlinger, "Dan Camp, Who Created a Mississippi Jewel, Dies at 79," *New York Times*, September 11, 2020, https://www. nytimes.com/2020/11/09/obituaries/dan-camp-dead-coronavirus.html.

131. In 2013, Mississippi Public Broadcasting ran a short program on Camp and the Cotton District: "The Cotton District, Mississippi Roads," MPB, https://www.youtube.com/watch?v=wRK3lyk1r70.

132. Photos of the Rue du Grand Fromage are on the Cotton District website at https://www.cottondistrictms.com/rue-du-grande-fromage/.

Figure 4.285: Pacific Street, Brooklyn, New York. Dover, Kohl & Partners, Massengale & Co. LLC, 2021. View of the new continuous sidewalk looking north from Fourth Avenue. A continuous sidewalk crossing pacific street creates a slow, shared space block. Designed for BrooklynSpeaks Community Engagement around Atlantic Yards / Pacific Park.

Figure 4.286: Paseo Ponti, Miami Design District, Miami, Florida. Duany Plater-Zyberk & Company, 2011. Watercolor, pen, and ink. Bird's-eye study of the pedestrian passage on page 530. Rendering by Eusebio Azcué. *Courtesy of Duany Plater-Zyberk & Company*

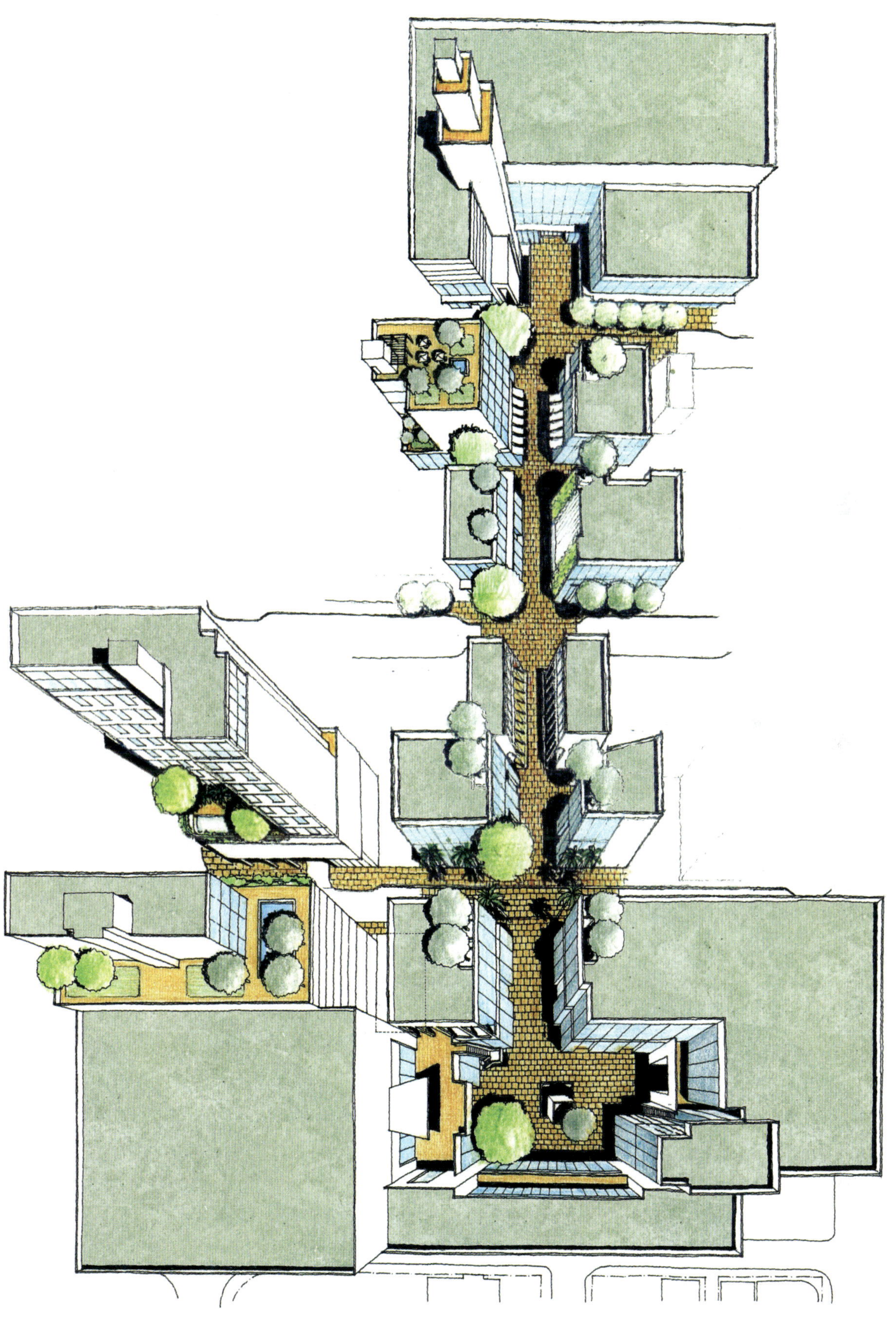

CHAPTER FIVE

WHOSE STREETS?

> The question of what kind of city we want cannot be divorced from the question of what kind of people we want to be, what kinds of social relations we seek, what relations to nature we cherish, what style of life we desire, or what aesthetic values we hold.
>
> — David Harvey, *Rebel Cities*

> Every family should have the right to spend their money, after tax, as they wish, and not as the government dictates. Let us extend choice, extend the will to choose and the chance to choose.
>
> — Margaret Thatcher, *Let Our Children Grow Tall*

IF WE BUILD STREETS FOR PEOPLE, we get a city for people. If we build streets for cars, we get a city for cars (paraphrasing Fred Kent, who was paraphrasing Hans Monderman). We have all grown up in the age of the automobile, so we are all used to the idea that cars should occupy most of the space on the street in our cities and towns. Since the first edition of *Street Design* came out, however, the world has come to see more clearly than ever many of the problems that come with putting driving first. We briefly summarize the most obvious ones on the next page in "Some Problems with Cars and Driving." But this book is about street design and public life, so we focus mainly on that. Even if science solves all climate change and pollution problems tomorrow, we—John and Victor—will still want to change the way American cities and towns use their streets. We believe in cities for people, with fewer cars and more public life.

Diminishing the public realm damages the common good. Traffic degrades street life and makes walking less safe. Nevertheless, here in America, cities and towns give most of their public space to the movement of machines rather than public life. We isolate ourselves in cars and live in suburban and exurban patterns that separate us from our neighbors. By every measure, Americans are more lonely and less connected to their fellow citizens than they used to be.[1] Our cars and how we use them are not the only causes of loneliness, but they contribute to it. Driverless vehicles have the potential to turn our cars into mobile cocoons that will only make the problem worse.

◄ **Figure 5.1:** Singelgracht, Amsterdam, The Netherlands. Looking south towards Wijde Heisteeg. A shared-space street on the west side of the Singel Canal.

Almost every street in America that is a good place for people was built before motor vehicles existed. But we wrote *Street Design* ten years ago because, in both Europe and America, attitudes about the proper design of streets were changing. Now, the Covid pandemic has accelerated change. Covid both pulled us apart and brought us together. All over the world, driving dramatically decreased after the lockdown started, practically overnight. Before long, air quality in cities and urban areas dramatically improved, providing some easy and easily visible answers to questions about driving and pollution that previously could only be studied through scientific modeling.[2] Most offices and stores closed, along with restaurants, bars, and cafés.

People replaced cars in the streets, sometimes filling them. Black Lives Matter groups marched on city streets across the country, growing into the largest protests in American history.[3] Riots on the street in Portland, Oregon, went on for months.[4] On a happier note (for some), when news organizations projected Joe Biden as the winner of the Presidential election on November 7, 2020, hundreds of thousands of people in cities across the country poured out onto the streets in spontaneous celebrations that were the biggest public demonstrations since VE Day in World War II. Donald Trump supporters later had their own marches and protests in the streets.[5]

In many cities, cyclists joined marchers in nightly demonstrations, riding along in advance and rear guards. On some nights in New York City, more than a thousand demonstrators on bikes turned out for rides from Brooklyn to Manhattan (and vice versa). "Jingling bike bells, the riders chanted, 'Whose streets? Our streets' and 'Say his name: George Floyd,'" the *New York Times* reported in July 2020. "Many had hand-drawn signs—including 'We Demand Change' and 'Defund the Police'—taped to their backpacks, side racks and handlebars. Others wore T-shirts that read 'Black Lives Matter.'"[6]

Early in the pandemic, politicians, urbanists, and activists began quoting Winston Churchill: "Never let a crisis go to waste."[7] Soon, they announced plans to change their city streets and transportation or accelerate

SOME PROBLEMS WITH CARS AND DRIVING

First, the American way of driving has put more carbon in the atmosphere than any other single source. Unless scientists invent some way of removing those gases, that carbon is permanently with us. We cannot continue adding greenhouse gases as we have if we want future generations to have good lives. Those future generations include many people alive now.

Second, putting cars first is killing people: in traffic collisions, because of pollution, and through a lifestyle that leads to obesity, heart disease, diabetes, and other illnesses. Despite the many advances of modern medicine, poor health is lowering American life expectancy for the first time in our history, while crashes kill over 40,000 Americans every year and wound and maim hundreds of thousands more. Around the world, 1.3 million people die in traffic every year, with another 20–50 million injured.[8]

With more than 8 billion people in the world and almost 300 million cars in America alone, depletion of the materials needed to make cars and delivery vehicles will only get worse. The earth has finite materials, so the production of automobiles, the maintenance of roads and highways for automobiles and trucks, and the manufacture of batteries will inevitably become more expensive, dependent on dwindling resources. (And do not forget the pollution that comes from vehicle manufacturing, road building, and the particulate matter from tires and brake pads on all vehicles, including electric ones.)

We wrote in Chapter Two that since Robert Moses built the Henry Hudson Parkway in 1934, the number of personal and commercial motor vehicles in the United States has increased by more than 11,000 percent (with 21,544 cars and 3,717 commercial vehicles then). The results are in the air (permanently, as we said).[9] The world needs city streets that encourage us to get out of our cars and enjoy the place where we are.

Figure 5.2: 16th Street NW, Washington, DC. In the first few months of the pandemic, large public marches included Black Lives Matter demonstrations and street slogans. © 2020 Planet Labs / Wikimedia Commons / CC BY 2.0

Figure 5.3: Wyckoff Street, Brooklyn, New York. Early in the pandemic. Black Lives Matter demonstrators on bicycles ride by a "streatery." When Covid removed most cars from our streets, we changed the ways we used our streets. *Courtesy of Amir Hamja*

Figure 5.4: Railroad Street, Great Barrington, Massachusetts. Looking east on July 4, 2020. Outdoor dining on an Open Street in the first year of the pandemic. For more on Railroad Street, see pages 179–182.

efforts already underway to reduce traffic and increase public life. Cities from New York to San Francisco created a variation on what is known as an "Open Street." Open Streets were not new (they began in Bogotá in 1974 and arrived in the United States in New York City in 1999),[10] but until the lockdown, they were not widely recognized outside of cities and towns that had large numbers of bicycle advocates. Before Covid, Open Streets in the United States were city streets that, on special occasions, were closed to cars and open to everyone else (especially bicycles.) The best Open Streets created during the pandemic were part of larger bicycle networks connecting different parts of the city.[11] At the same time, restaurants and cafes convinced cities and towns to let them open outdoor dining spaces. People were desperate for a taste of public life, businesses needed to make money, and "streateries" were born.

In the weeks before the pandemic, Paris Mayor Anne Hidalgo unveiled the 15-minute city—*la ville du quart d'heure*—in her reelection campaign. In retrospect, that was perfect timing for the lockdown that was coming (see the *Street Design* Foreword by Carlos Moreno, page xii, and a short discussion of the 15-minute city in this chapter on page 636).[12] But her argument was for places where residents and visitors could drive less. "There are more and more of us fighting for a different vision of the world," Hidalgo wrote in late 2019, "a world that takes care of our most precious resources: the air we breathe, the water we drink and the places we share."[13] Marco Granelli, a Deputy Mayor of Milan, said early in the pandemic: "We worked for years to reduce car use. If everybody drives a car, there is no space for people, there is no space to move, there is no space for commercial activities outside the shops."[14]

These circumstances and possibilities were not lost on Americans after the lockdown started. In New York, Senator Charles "Chuck" Schumer said, "My goal, my dream is to be like Amsterdam. I visited Amsterdam; bikes are a way of life there. And it is a beautiful way to live. New York has to aspire to that." Then he announced he had secured "an effing lot of money" to make safer streets in New York City.[15] We will see in this chapter that New York City is definitely not Amsterdam (but see the mention of Nieuw Amsterdam on page 612).

Before Covid, many European cities and towns were better than any American city in making streets that, at the very least, gave people walking and cycling as much right to the street as motor vehicle drivers. During the pandemic, Europe surged ahead while America debated whether or not streets were made for cars. As a result, it seems to us that Covid was less damaging to European social life than in America. In retrospect, we can see that the pandemic had a strong negative effect on social customs and public life in the United States. Many Americans seem to have forgotten how to act in public (that includes obeying traffic laws), and we are now perhaps lonelier than ever.[16]

MY MANHATTAN / JOHN MASSENGALE

Many things changed for me on July 17, 2011, when my beautiful BMW 540iT with self-leveling sport suspension self-immolated. Ever since, I've been carless in Manhattan (thank you, *Grosse Berthe*[17]). I'm happy to live in New York City, where I can get up in the morning and go about my day without getting in a car. Many car owners do not realize how liberating that can be.

I was born in New York, but I grew up in the suburbs, where I was a car lover from a young age.[18] I still love cars, but even if the world did not understand the direct connection between driving and climate change, many would agree that sprawl, Functional Classification, and traffic engineers have taken most of the fun out of driving. In any case, now I live on the Upper West Side of New York, where more than three-quarters of my neighbors do not own cars.[19] We have access to the best public transportation in North America. With trains, subways, ferries, buses, and bike share, I can quickly get anywhere in the city and the region without owning a car.[20] My neighborhood is one of the two or three densest in the

Western world, with 110,000 people per square mile (42,500/km²). That is twice the density of the Left Bank of Paris, the birthplace of the concept of the 15-minute city. I can easily walk, bike, bus, or take the subway to almost any store, service, or entertainment I can imagine. It is a wonderful way to live. Visiting friends assure me it is not normal in most of the United States.

Those are a few facts about the Upper West Side and New York City. More to the point, every time I go out the front door of my apartment building, I walk down endlessly entertaining, frequently beautiful streets that make me feel good. I can walk up the promenade on Riverside Drive to Columbia University (where I have a free student ID because I live in one of the Manhattan zip codes that qualifies for that[21]) and be struck every time by the beauty of the walk and the daily change in the trees, flowers, and bushes. I can get on my bike and coast downhill to the Hudson River and the East Coast Greenway (see page 232). With public transportation, I can quickly and easily be in Brooklyn or Greenwich Village, where the streets are so different that the experience is almost like taking a vacation. The city is full of interesting stores and great restaurants and coffee places that are not national chains. I know where to stand on the subway platform so that when I arrive at the other end, I will be where I want to be to easily access the exit I want. I know which blocks have buildings by Carrère & Hastings and McKim, Mead & White, and I know where the ugly blocks are and how to avoid them. I know where Babe Ruth lived, and I know that half an hour after the Yankee games end, I will hear the fans coming out of the subway. Everywhere I go, I rub shoulders with New Yorkers, and I love that. There are 8.5 million stories in New York, but I run into friends and acquaintances on the street. I love the energy on the street, and I love New Yorkers. I love New York.

Most New Yorkers make a trade-off: we live in small, expensive apartments in exchange for the rewards of life in the city. That is true for the rich, the poor, and people in between. In recent years, the extreme and growing income inequality of the twenty-first century has distorted and damaged that equation, but this is a book about street design, so I am only going to say here that the city I was born in, the city I moved back to, and the city I live in are not only for the rich. My neighborhood is under attack by forces that have been in play since Mayor Michael Bloomberg (the richest man in New York) said the city should be a luxury brand you have to pay for. Statistically, the income per capita

THE IMPORTANCE OF SHORT BLOCKS / JOHN MASSENGALE

Working on *Street Design* made me realize that the five apartments where I've lived in Manhattan were all in neighborhoods with short blocks. The offices I've rented have also been in neighborhoods with short blocks. I didn't always realize that at the time, but in retrospect, the blocks influenced why I chose those apartments and offices. Their neighborhoods are more lively, with more interesting stores and life on the street. Let's look at that.

"Short blocks are good" is not a new idea, of course. Jane Jacobs had a chapter in *The Death and Life of Great American Cities* called "The Need for Small Blocks." Since she mainly wrote about Greenwich Village, I was interested to see that she discussed the long blocks I regularly pass through when I walk to Central Park. She even diagrammed them in the book.[22]

For those who don't know Manhattan well, I have made a note summarizing the relevant blocks where I lived on the Upper East Side and the Upper West Side.[23]

Three of my five apartments were on the East Side between Lexington Avenue and Central Park, like the block of East 70th Street discussed in Chapter One. The short blocks on the Upper East Side resulted from adding Lexington Avenue and Madison Avenue to the original streets in the Commissioners' Plan of 1811.[24] Where I live now, facing Riverside Park, is less than 400 feet from West End Avenue (the first avenue to the east of my apartment). The next two blocks are each around 350 feet long. But beyond that, Robert Moses created superblocks with suburban buildings on the other side of Amsterdam, blocking 98th and 99th streets. That is why I take "the scenic route" farther south when I walk to Central Park. The old buildings there make more pleasing streets, but the long blocks cause problems too.

In *Death and Life*, Jacobs writes,

> Consider, for instance, the situation of a man living on a long street block, such as West Eighty-eighth Street in Manhattan, between

on the Upper West Side is so high that one has to call the neighborhood rich, but the neighborhood is more diverse than most American neighborhoods. Within a few blocks of my apartment is a 22.5-acre tower-in-park apartment complex built by New York City Housing Authority (NYCHA). One of the avenues near the housing project has a five-block, quarter-mile stretch with old, relatively inexpensive apartment buildings and Latin-owned stores that cater to working-class Latin population there. Other NYCHA housing projects lie between my apartment and Central Park, where there are many small, inexpensive, rent-regulated apartments in converted rowhouses and walk-up tenements. Those are from a time, not so long ago, when New York was a poorer place. Scattered on Broadway and side streets near me are small buildings bought by the city in the 1980s and 1990s and owned now by agencies that provide homeless shelters and specialized housing with social services. Due to the peculiarities of the New York housing market, about 10 percent of the units in my cooperative apartment building (another New York peculiarity) are

inexpensive rent-regulated apartments with tenants who moved into the building many decades ago.

Covid and the lockdown quickly changed New York's streets. There have always been times when traffic was low in New York (late at night, on holidays and Sundays, or after a snowfall, to name a few), but all traffic on Manhattan's streets decreased dramatically virtually overnight in March 2020, making the streets consistently empty. Most of the 3.58 million people[25] who commuted into the city stopped doing so, and some young people went home to live with their parents. Hundreds of thousands who owned second homes moved into them for the duration, making only periodic, short trips back. Hundreds of thousands more bought or rented houses outside New York. Life on the street was profoundly different.

Many in New York started doing what I frequently do: walking in the middle of the street, as people do in European historic centers or on streets in Third World countries. That changed their bodily experience of New York streets. There was time to slow down, look around from a new perspective, admire the buildings,

Central Park West and Columbus Avenue. He goes westward along his 800-foot block to reach the stores on Columbus Avenue or take the bus, and he goes eastward to reach the park, take the subway or another bus. *He may very well never enter the adjacent blocks on Eighty-seventh Street and Eighty-ninth Street for years* [emphasis added].

This brings grave trouble. We have already seen that isolated, discrete street neighborhoods are apt to be helpless socially. This man would have every justification for disbelieving that Eighty-seventh and Eighty-ninth streets or their people have anything to do with him. To believe it, he has to go beyond the ordinary evidence of his everyday life.[26]

The long blocks limit the choice of routes, while short blocks make many routes possible. That might seem obvious, but perhaps less obvious until you experience it, is that a choice of routes means that people walking do not have to spend much time waiting at red lights.

Longtime city dwellers know that when they walk in neighborhoods with short blocks and many routes, it is frequently possible to go either left or right at an intersection, and therefore rarely having to wait for a light to change. Contrast that with the experience of drivers in the city or the suburbs, where everyone spends long periods staring at red lights.

The experience of moving my office a few times taught me another lesson. I remember the first time I walked from a new office to the nearest supermarket. The most obvious route included an ugly, wide, noisy street. The distance was about a third of a mile, but the walk seemed long. Because the blocks are short, however, over the next few weeks I found more scenic routes. Taking different routes, I developed a three-dimensional mental map of the neighborhood. I soon came to see that the supermarket wasn't far away at all and thought of a trip there as fast and easy. I will come back to this phenomenon of how we come to understand new neighborhoods through their streets in the section about Slow New York, later in the chapter.

and appreciate the space between the buildings. We feel the street more immediately from the middle of an empty street than when we are on a sidewalk with cars buffering us from cars driving by. When the Congress for New Urbanism began in 1993, best practice said that cars parked on the street were good because they protect the pedestrian. However, they also visually and physically cut the space between the buildings—the public realm—into three pieces and give most of the public space to motor vehicles. That changes our relationship with public space and how we feel in it. We will come back to this later in the chapter when we talk about the experience of place, ideas like embodied cognition, and what Tony Hiss calls "simultaneous perception" (page 609).

Before the pandemic, that zone in the middle of the street was a dangerous place for people walking.

Figure 5.5: Riverside Drive, New York, New York. Frederick Law Olmsted, 1873; Calvert Vaux, 1881; Clarke & Rapuano, 1934. Photo looking north from 115th Street. Empty streets during Covid. Also see the Riverside Drive case study in Chapter Two (page 222).

WALLS WITHOUT WINDOWS

Most of the time, urbanists discourage "blank walls." But as with everything in design, that depends on the context and the circumstances. Both the northern wall of the main gallery at the Frick Collection and the south wall of the grounds at St. Patrick's Old Cathedral, for example, have no windows and few openings. They are very different, but they both shape places that feel good. Carrère & Hastings' sophisticated design delights the senses, using principles of Classical composition, Classical ornament, and good materials and proportions. The cemetery wall has Classical brownstone quoins at the corners, but its charm comes from the uneven settling of the wall, the patina on the bricks, and the texture and color of the individual bricks. They are two of the most beautiful window-free walls in New York.

Figure 5.6: Frick Collection, East 71st Street, New York, New York. Carrère & Hastings, 1914. Looking west on 71st Street towards Fifth Avenue and Central Park along the north face of the museum's main gallery. The gallery has no windows, but walking next to the long wall feels good because of the Classical composition and details, the materials, and the proportions.

Figure 5.7: Cemetery Wall, Basilica of St. Patrick's Old Cathedral, New York, New York. Joseph-François Mangin, 1815. Looking west on Prince Street from Mott Street.

It is a place for fast-moving, noisy machines that can easily kill us. "If you want to understand what bad feng shui is," Douglas Duany once said to me, "Stand on a New York City street corner after the light changes." That led to the following lines in *Salon* after *Street Design* first came out:

> John Massengale and I are standing in the middle of 1st Avenue at East 4th Street, in New York's East Village, and he does not like the feng shui. He points to the thick, white lines in the roadway, directing drivers toward a left turn. 'Automobile-scale striping,' he notes. 'It's telling you: 'This is not a place for you.''
>
> No doubt about that. Nevertheless, here we pedestrians are, waiting for a torrent of cars to be released from a red light at Houston Street, four blocks south. 'Something is telling your limbic system you're not supposed to be standing out here,' Massengale continues, as the lights change on 1st Street … 2nd Street … 3rd Street…
>
> A taxi's horn blares and we scamper to the sidewalk.

The middle of the street was dangerous for walking, by design. The pandemic gave us a glimpse of better streets for people, with quiet streets, Open Streets, and streateries. A dramatic expansion of city-sanctioned Open Streets and streateries early in the pandemic opened many New Yorkers' eyes to how much better city streets could be for public life if cars did not always come first. "How New Yorkers Want to Change the Streetscape for Good," the *New York Times* wrote in December 2020. "The pandemic led to a mass experiment in closing streets to cars. New Yorkers embraced the change and want permanent spaces for playing, dining and performing," said the subhead.[27]

> Of all the ways the pandemic reshaped New York City's streetscape, the most profound example

might have been found on Vanderbilt Avenue as it cut through brownstone Brooklyn. On weekends jazz bands played on the corners. Friends reunited on the median. Children zigged and zagged on their bikes as diners sat at bistro tables atop asphalt. The faint sound of cars could be heard in the distance. Just as the early days of the coronavirus forced New Yorkers inside, it eventually pushed them outdoors—for fresh air, for exercise, for eating, for relief—in what became an organic takeover and reimagining of the city's streets across its five boroughs. City officials handed over 83 miles of roadway to cyclists, runners and walkers, allowed nearly 11,000 restaurants to stretch onto sidewalks and streets and let retailers expand their storefronts beyond their front doors. People reclaimed the pavement and are, by and large, unwilling to give it back.[28]

That all sounds good, but looking back three years later, we can see that New York wasted the proverbial crisis. For reasons we do not need to discuss at length, the Mayor went back on many of his campaign promises for better streets for walking and cycling.[29] In 2024, many of the Open Streets are closed, and even the most successful, like Vanderbilt Avenue, may not stay open. Worthy of more discussion is how the NYC DOT responded. The DOT Commissioner—a former City Councilman long interested in bringing better streets with fewer cars to New York—was muzzled by the Mayor. At the same time, the five thousand career engineers in the department continued to act like the traffic engineers they are. The NYC DOT is neither all good nor all bad, but their heart is in making traffic flow, and they regularly illustrate the Monderman Rule (page 513) by adding unnecessary things.* The design moves they brought to the Open Streets were over-engineered to the extent that they were bad placemaking. In a nutshell, the engineers at the DOT are not urban designers or placemakers, but they are in control.

---

*Note 92 in Chapter Four repeated an interesting fact Monderman reported at a CNU conference in London. In his early days as a traffic engineer, Monderman experimented with putting large flowerpots in the street to slow drivers down (similar to the way New York's DOT uses flowerpots to channel traffic away from pedestrians). The flowerpots reduced traffic speed by 10 percent. But when Monderman removed the flowerpots and turned the roads into shared space streets, the average traffic speed was cut in half.

Figure 5.8: "21st Century Street," New York, New York. Computer simulation. Steven Nutter, 2011. Looking east on Fourth Avenue in Brooklyn. A winning entry for a competition in New York City to redesign Fourth Avenue in Brooklyn, imagining it as the 21st Century Street. Instead of making a comfortable place where Brooklynites might want to stroll or shop, the design focused on the roadbed, which it divided into pieces. The bicycle specialist who designed the entry made a special, visually dramatic bike lane that would have cut the space in two and forced cyclists to weave dangerously across an anti-urban turn lane. This emphasis on the intersection where the vehicles come together ignored the design of the public realm between the intersections. In all of this, it was representative of the work of specialists in the twenty-first century. *Courtesy of Steven Nutter*

Figure 5.9: Madison Avenue, New York, New York. Looking south from East 55th Street. America's most expensive shopping street, in the heart of America's most walkable city, is a bus and auto sewer. Pedestrians on the sidewalk are close to large, noisy, smelly express buses barreling along. The traffic signs and the bold graphics are highway-scale, telling the pedestrian, "This is machine space. This is not for you. Danger!"

Figure 5.10: West 4th Street at West 11th Street, New York, New York. Looking east on 11th Street. Many of the streets in Greenwich Village came before the Commissioners' Plan of 1811 for the Manhattan grid, with the result that in this part of the city, 4th Street crosses 11th Street (instead of being parallel and seven blocks south). With light traffic, narrow streets, and no bike lanes, West 4th Street frequently functions as a natural shared-space street.

Once upon a time, as discussed in Chapter Four (page 444), the NYC DOT was probably the most innovative DOT in the United States, far ahead of most residents of New York in imagining better scenarios. That was after Janette Sadik-Kahn became the City's DOT Commissioner in 2007. But since then, other DOTs have passed New York's, while New Yorkers came to believe that the city's streets could be even better. They saw the new plazas and CitiBike stations, and they liked what they saw. They read about climate change and wanted to change the city. Advocacy groups like Transportation Alternatives and Open Plans, the publisher of Streetsblog, and campaigns like The War On Cars (created by the founding editor of Streetsblog and others) effectively publicized new ideas.[30] Since then, New York has hired good Deputy Commissioners and Commissioners like Michael Replogle and Polly Trottenberg (the U.S. DOT Deputy Secretary under Biden), and there are passionate young activists in the DOT who want more bike lanes, pedestrian plazas, Open Streets, and shared-space streets—but the five-thousand engineers kept the department moving straight ahead, on course, like the proverbial battleship. Importantly, our mayors too frequently support maintaining the status quo on city streets, quietly abandoning policies they pledged to support before they took office.

The world has changed. Change accelerated everywhere during the pandemic, and streets in cities and towns all over Europe are dramatically better for walking and cycling than they used to be. But the NYC DOT mainly moves sideways. There is very little it does today that it did not do almost twenty years ago. The good news as we go to press is that the new Mayor Zohran Kwame Mamdani and his DOT Commissioner Mike Flynn (an excellent choice, see page 40) are saying all the right things.

A new program of "daylighting" intersections has a timetable that, as announced, will take 470 years to complete (although it is already behind schedule). Compare that to the past: "Manhattan alone widened 453 roadbeds in the years 1953–1958, and its borough president announced this was only a start," Jane Jacobs wrote in *The Death and Life of Great American Cities*.[31] "In the 1950s, nothing was too good for our gold-plated transportation system," Andrés Duany says. Now the DOTs cry poverty when changes are proposed. If you are wondering why I do not say more about congestion pricing here, it is because Mayor Michael Bloomberg started working on it in 2002 and first tried to implement it in 2007.[32] When Norman Mailer ran for Mayor in 1969, he proposed free transit and free bicycles for all while banning private cars in Manhattan.[33]

Figure 5.11: Hoyt Street, Brooklyn, New York. Hoyt Street between Atlantic Avenue and Pacific Street. New York's four-track, local-express subway system makes long trips easy. Door to door, without a car, I can go from my office in a beautiful seventeen-story, spec-office building near City Hall designed by Cass Gilbert to the restaurant shown here in under twenty minutes. On a good day, I can travel the six miles from home to my office in twenty-five minutes. The subway is cheap, I can read on it, and when I go to Greenwich Village (Figure 5.10) or Hoyt Street, it is like going on vacation.

Streets for People?

Whose streets? DOTs around the country design streets and highways that induce vehicular demand—and the sad fact is that New York City's DOT has too, even though, in terms of population density, transportation, and Walk Score, New York City is unique in America.[34] But Organized Motordom fought for control of America's streets a hundred years ago, and it won. Today, Motordom has a powerful system enforced by federal, state, and local DOTs: they essentially own our streets, with large budgets, traffic manuals, and legal precedents on their side. If cities, towns, and states do not do as the regulations and manuals say, lawyers working for the cities, towns, and states will testify in lawsuits that they did not follow "accepted practice." At least in prevailing theory, that means they lose if someone sues them. And no municipalities want to turn away the dollars from the State and Federal DOTs.

The lawyers and engineers will argue that the DOTs' accepted and best practices are about ensuring public safety. We know that is not true. If it were, speed limits across the country would be lower and road designs would make crashes rarer. Zero traffic deaths are possible, but best practice for Vision Zero means cars will have to go slowly, and DOTs do not want that. When they say "safer," they mean safer to go faster, in the context of more than 40,000 annual American deaths. Even in New York City, "pedestrian mobility" and cycling safety frequently come second to the priorities of out-of-town drivers.

Mayor Bill de Blasio committed New York to a goal of zero traffic deaths on New York City's streets around the same time my wife and I moved to Riverside Drive from the Upper East Side of Manhattan. The first New Yorker run over and killed after Hizzoner's announcement died a few blocks from our apartment, hit at the busy intersection of Broadway and West 96th Street,[35] a deadly crossing near on-off ramps for the Henry Hudson Highway.[36] An article in *New York* by the magazine's architecture critic, Justin Davidson, chronicled traffic deaths he is personally familiar with there, in what he called the "Triangle of Death." Most of what Davidson wrote about has happened since Mayor de Blasio backed off implementing Vision Zero (meaning, don't slow the cars down *too* much).[37]*

---

*The original goal was to have zero traffic deaths in New York City by 2024. When traffic deaths increased the first few years, the city quietly pushed the ten-year deadline back, and then later abandoned the idea of a deadline. In 2021, the DOT stopped reporting relevant annual statistics. A story on NPR in acknowledged the slow pace in America, "Vision Zero marks a milestone, but the goal of ending traffic deaths is still far off."

LEXINGTON AVENUE: BEFORE AND AFTER, ONE HUNDRED YEARS LATER

Figure 5.12: Lexington Avenue, New York, New York. Before: The northwest corner of Lexington Avenue and East 89th Street in 1913. Photographed during the construction of the East Side IRT subway line, which explains why the sidewalk has wooden planks rather than concrete. *Courtesy of the New York Transit Museum*

Figure 5.13: Lexington Avenue, New York, New York. After: The northwest corner of Lexington Avenue and East 89th Street in 2013. Lexington Avenue was widened twice, most drastically in 1960, when the DOT converted the road to one-way traffic and widened the roadway to speed up traffic. A comparison of Figures 5.12 and 5.13 reveals that the sidewalk was narrowed so much that the houses along Lexington lost both their stoops and the window wells for the basements. Underneath the avenue is the busiest subway line in America, and one block away are three commuter train lines that run north from Grand Central Terminal to Connecticut, Westchester County, and the Hudson Valley. The attitude behind the road widening is why America burns twice as much gas per person as the residents of Middle Eastern oil-producing countries, and why we have such ugly streets. (This pair of photos was inspired by a similar pairing in Streetsblog NYC by Aaron Donovan.)

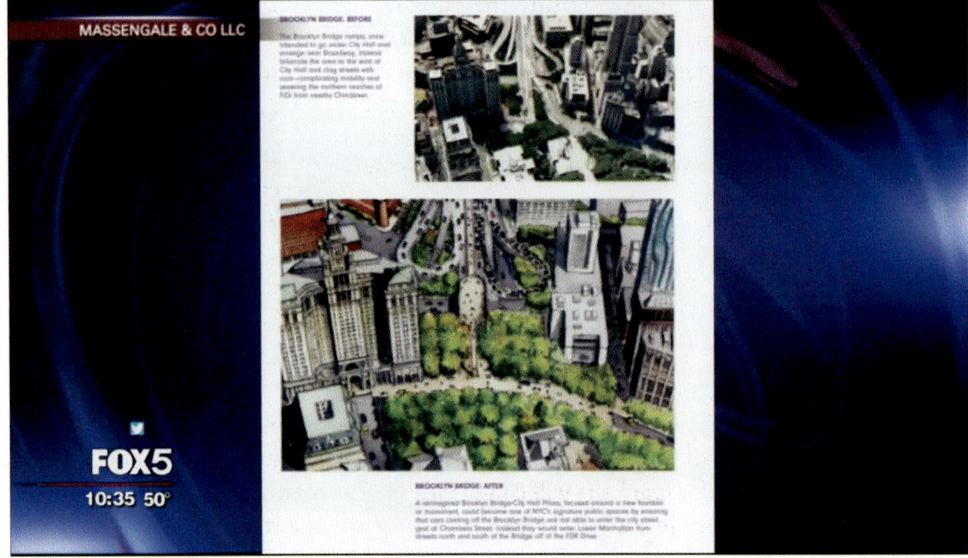

Figure 5.14: Car-Free Brooklyn Bridge Before & After on FOX 5 New York. Television screenshot taken April 21, 2019. The images are part of the Make Way for Lower Manhattan plan. See "Slow New York" on page 610 and Figure 5.28.

"The Department of Transportation boasts of having 'driven'—its word—'traffic deaths to historic lows,'" Davidson writes, "which is true if by 'history' you mean since last year."[38] That is because the year before the article was the worst year for pedestrian deaths since New York City committed to Vision Zero. In the previous nine years, motor vehicles killed an average of 125 pedestrians on the streets of New York, or one every three days. Altogether, 255 people died in New York traffic in 2022. Since the city officially committed to Vision Zero, the annual total has never gone below 200. That is because deaths are built into the system as practiced by 99.99 percent of American Departments of Transportation, including New York's. Nationwide, the profession accepts more than 40,000 deaths every year, despite all the elaborate and expensive safety devices in our cars. The deaths are seen as part of the cost of keeping traffic flowing at an acceptable speed. Putting it that way can make it sound shocking, but it is true (see note 143).

> Deaths are built into the system. More than 40,000 Americans die in crashes every year.

I understand that many Americans have no alternative to driving every day, and that many like driving. But that is not true for me and my neighbors. The streets we live on would be safer if our city followed more closely the practices of cities like Helsinki, Oslo, and even nearby Jersey City and Hoboken (I can see the last two from my apartment). Both New Jersey cities reduced pedestrian and cycling deaths to zero, but New York continues to prioritize moving cars into, out of, and through the city

(another form of induced demand).[39] There are a surprising number of New York streets where traffic makes walking both unpleasant and unsafe. There are not many New York City streets that are as good for biking and walking as thousands—or tens of thousands—of streets in Europe.

Only the political powers that be prevent large parts of New York from being a Vision Zero pedestrian paradise. These are: the New York State Legislature (which gained control over various aspects of New York City when the city almost went bankrupt in 1975);[40] the State and City DOTs; and to varying degrees, the mayors and the City Council.[41] In an article in Common Edge in 2023 I asked if New York's dangerous, unnecessarily polluted streets should be illegal. "I know that sounds ridiculous to many," I wrote, "but New York City policies and laws promise us safer streets, while drivers kill more people, and what doesn't kill us makes us unhealthy," I said, "with asthma, diabetes, heart disease, and maybe missing limbs."[42]

Earlier in the year, New York State Attorney General Letitia James looked at the problem and suggested that legal action might be reasonable. James wrote to a disgruntled developer who built a truck depot in Harlem that the facility may be a "public nuisance." According to state law, she wrote, that means it "interfere(s)" with or "cause(s) damage" to a "public place or danger(s) or injure(s) the property, health, safety, or comfort of a considerable number of persons."[43]

Explaining why she was considering banning the depot, James declared, "Trucks are associated with increased traffic delays, noise, and vibrations, and can endanger the safety of pedestrians and cyclists. Trucks also generate local air pollution."[44] Using the same criteria, many streets in New York City are public places that meet the public nuisance definition. Many city streets in Europe were once as dangerous and polluted as New York's. Why can't American cities make the same fixes as European cities? "We cannot solve our problems with the same thinking we used when we created them," Albert Einstein famously said.[45] Organized Motordom and traffic engineering caused the imbalance in our streets today. Why should we expect the profession to solve the problem? I closed the opinion piece with some news:

> Mayor Eric Adams has created a new position, a director of the public realm, Ya-Ting Liu, who will report directly to two deputy mayors. Functioning as a "traffic czar," she will coordinate city agencies that work in the public realm, including the departments of transportation, parks, climate, buildings, and environmental protection.
>
> I wish her all the luck in the world and hope her power quickly expands. Traffic engineers do not lead the design teams that make the beautiful, safe, low-traffic streets for people we see in Europe. Those streets are

21st-century paradigms for moving away from the car-dominated models of the 20th century still plaguing us.[46]

New York City should do whatever it takes to change our streets, how we make them, and how we use them. The city has been a model for the nation and can be again. No place in America is more ready for safe, slow, and walkable streets.

MY MIAMI / VICTOR DOVER

Miami-Dade County is not usually thought of as a place for walking or biking. In pop culture, greater Miami has been perpetually portrayed, glamorized, and engineered as a motorized region for drivers in a hurry. But there is more to the story.

When Alfred Hitchcock showed Ingrid Bergman's character living in Miami in the 1946 movie *Notorious*, she was behind the wheel of a Cadillac convertible— where else?—in her pivotal scene early in the film. She was speeding, and more than a little drunk, with the bayfront skyline behind her.[47] Miami, at least in the movies, is for motoring around in the sunshine with the top down.

Figure 5.15: Ocean Drive, Miami Beach, Florida. Looking north. New bike lanes separated by "armadillo" fixtures now parallel the sidewalk. The city has opened and closed the street to car traffic in various configurations over the last several years.

In the 1980s, the television pop-cop drama *Miami Vice* used the transitions between its gritty and pastel scenes to show the detectives roaring around the city in confiscated Ferraris, having a blast in high-fashion sunglasses.[48] More recent reality-TV shows reinforced that image, making it seem like everyone arrives in limousines to go club-hopping. Miami, at least on TV, is a city for driving fast and flashy.

In the real world of Miami today, however, everyone in the street moves in slow motion. There are some of us who drive relatively little, some who drive all the time, and a few who hardly drive at all, getting around via walking, biking, and transit. But none of us can quickly zip around, and none of us have as much fun as the glamorous movie and TV characters. We are more likely to be stuck in road rage amid epic traffic congestion, or walking too far in the sun to catch a bus, or waiting too long for a train, or enduring too much stress when we are on our own two feet or two wheels. Street design errors are to blame for much of the predicament.

It is not all bad. Some of our historic neighborhoods and retrofitted places are highly walkable and bikeable, and they are beloved, thanks to their combinations of street design, architecture, and street network. These point the way to better streets in Miami. Several of the exceptional streets are found in this book, including Española Way in Miami Beach, Giralda Street in Coral Gables, Paseo Ponti in the Design District, and Dorn Avenue in South Miami. Whole neighborhoods have also found ways to defy the stereotype, too; in the North Beach neighborhood of Miami Beach, for instance, 26 percent of households manage without a car.[49]

Miami has long had a potent design culture. In recent history, given the sprawl, it logically became a birthplace of the New Urbanism movement. Our spirited local conversation about city design has been continually encouraged by the University of Miami School of Architecture under the leadership of Elizabeth Plater-Zyberk, who was the Dean from 1995 to 2013. (She now runs the UM Master of Urban Design Program.) She and her partner Andrés Duany, two of the six founders of the Congress for the New Urbanism, opened Duany, Plater-Zyberk & Company in Coconut Grove in 1980, after leaving Arquitectonica. ("DPZ" is now officially "DPZ CoDesign.") UM and DPZ made Miami a place that influenced the national conversation, attracting students and practitioners from around the world. Many New Urbanists call the region our home; we find ourselves flying all over the country and the Caribbean to help other people figure out how to grow and fix their towns. But we also have a lot to fix *here*.

For one thing, walking or riding a bike in South Florida remains unnecessarily perilous. Our metro area ranked as America's fourteenth worst in the latest *Dangerous by Design* report, with 954 pedestrian deaths in the area during the period of 2016–2020.[50] As bad as that sounds, it represents an *improvement* in the rankings. We were in the top five for a long while.

Four of Florida's other metropolitan areas have worse numbers, meanwhile, maintaining the Sunshine State's reputation for killing pedestrians. In the state-by-state contest, Florida held the top position for pedestrian deaths relative to population for decades, but finally slipped to #2 for the first time—behind New Mexico—in the 2022 *Dangerous by Design* report.[51] As in other parts of the Sunbelt, many of the deaths in both Florida and New Mexico are blamed on signalized crosswalks being set too far apart on supersized, high-speed roads. Some of us are out there walking and biking, just the same.

I live just under two miles from work and usually use my old rattling bike or my newer e-bike to go to the office or to the University of Miami to teach my class. Experimenting over the years, I have realized that my commutes by bike take about *half* as much time as driving and parking my car. It feels good to be part of the solution rather than more of the problem, too. Some of my friends are known to ride around town wearing bright green t-shirts with "One Less Car" or "We Are Traffic" printed on the back.

My own experience suggests if we could just get street design right, walking and biking could be big here. The local saying goes "We get to live where other people go on vacation." We have flat terrain. It gets hot here in our subtropical paradise, but we have good biking weather seven months a year, and *phenomenal* biking weather for five. It does rain here, too, but most storms are of the twenty-minute variety, and then the sunshine comes right back. (Those are somewhat better weather conditions than Davis, California, the city that currently holds the U.S. title for the most bike commuters,[52] and much better weather than cycling capitals like Portland or Amsterdam.) Plus, although street trees are rare here, where we *do* have street trees, the routes stay green and shady year round. I am lucky because most of my ride from South Miami to Coral Gables is along

Figure 5.16: The executive parking lot at Dover, Kohl & Partners, 2015.

well-connected streets with a shady canopy, keeping the route direct and the heat in check. Most of my daily ride is tranquil. A short section of my route is stressful, when I am forced to mix it up with impatient motorists for a minute or two. But overall, my bike trips to and from work are relaxing and fun. How many American workers can say that their commute home after work is one of the best parts of their day?

> How many American workers can say that their commute home after work is one of the best parts of their day?

Commutes are not the only trips we could improve with better street designs in Miami, of course. At least 40 percent of all trips in the United States are under two miles in distance.[53] In flat South Florida, those trips should be easier to shift to walking or biking than they are; old engineering habits are the main obstacle. Our *transpocracy*—the interlocking set of agencies and authorities in control of transportation—is mainly set up to deliver an illusion of car convenience at the regional scale for car-based lives. That transpocracy is not focused on the safety and comfort of pedestrians or people on bikes, nor on promoting the vitality of neighborhoods, although it should be.

The Manhole Lane

My daily ride includes a segment on long, straight Red Road (SW 57th Avenue), a corridor we complained about in the first edition of *Street Design*. The Florida

Department of Transportation (FDOT) rebuilt 3.6 miles of Red Road in 2014. That was an opportunity to heal a historic two-lane straightaway adjoining the cities of Coral Gables and South Miami and the University of Miami campus. After some handwringing, FDOT reduced the lane widths and painted bike lanes on the shoulder on both sides, allowing one to ride right next to traffic (if one dares). That is better than nothing, but there was room for so much more. Early and often in the design process, the FDOT team heard from citizens about the need for protected bike lanes, slower design speeds, and street trees. They were shown how everything needed could fit within the right-of-way.[54]

The technical brush-off was swift. Project engineers blocked new shade trees from appearing on the plans to any meaningful degree. Astonishingly, the bike lanes abruptly vanish a block short of connecting to the Underline (now the centerpiece of the regional trail system), two blocks short of downtown South Miami's main street, and three blocks from the Metrorail station. Subtle details suggest that the street was designed by someone who does not regularly ride a bike; locals nicknamed our Red Road bike lane "the manhole lane," for instance. Most people on bikes along the corridor can still be seen riding on the sidewalk along that stretch instead of rumbling over the manholes in the bike lane next to loud SUVs.

Big Plans

When we first published *Street Design* in 2014, it looked as though things were just about to get better here. Greater Miami briefly topped national lists for having one of the "Fastest Growing Percentages of Bike Commuters." (Read that headline carefully: the *fastest growing* percentage is a great statistic. If you start with a sufficiently small number of people overall, it does not take that many new folks to grow the percentage at a rapid rate.) According to the American Communities Survey, the hardy group was indeed growing rapidly between 2001 and 2011,[55] but after that it appears to have leveled off. Looking at all trips, however, and not just commutes, both walking and cycling are still increasing in Miami-Dade, with the biggest increases occurring on weekends.[56] Mechanical counters showed the average number of pedestrians observed in fifty-five locations rising 216 percent between 2007 and 2018. The number of cyclists at those locations rose 207 percent. Some streets have gotten better at attracting people than others; in a subgroup of twelve of the locations where counters were installed, the pedestrian activity rose 482 percent.

Working with the City of Miami, Mike Lydon, Tony Garcia, and their firm Street Plans Collaborative had created an ambitious *Bike Master Plan* in 2009.[57] Following the plan, the city gradually painted several miles of both conventional on-street and protected bike lanes, mostly in the city's core. The cities of Coral Gables and Miami Beach adopted similarly ambitious bike/pedestrian plans in the years that followed. Those inspired still more plans for suburbs like South Miami and Pinecrest; while their visions were more modest, the plans contributed to the overall sense that progress was in the making.

As with many cultural awakenings, young people have been pointing the way. Celebrities have added new glamor to the two-hundred-year-old technology of bikes, too. In 2013, tennis star Serena Williams opted to pedal to her tournament in Miami instead of driving, and images of her pedaling under palm trees were flashed all around the world. I can't help thinking it was just faster than sitting in stalled traffic outside the stadium.

Critical Mass rides—the giant, unofficial, ostensibly leaderless rides that tie up streets all over the globe on the last Friday evening of each month—regularly draw hundreds of participants in Miami, most of them twenty-somethings. Critical Mass in the United States began in San Francisco and spread to New York and other cities as bona fide protests to draw attention to the imbalance of power in their streets. In those cities, I have seen Critical Mass riders hollering, shaking their fists, and going out of their way to anger blocked motorists and get noticed. Not here in the subtropics. Miami's tradition is more laid back, happy, and friendly. Riders wave, smile, sing, and yell "Thank you!" to the stalled drivers, and the cities provide police escorts.

In 2014, word got out that LeBron James and Dwyane Wade of the Miami Heat were joining in, and Critical Mass swelled to an estimated four thousand cyclists.[58] The star athletes became well known for riding to the arena for practice, too. King James also appeared at the front of a throng of cyclists on Miami streets in a popular Nike commercial.

Reform?

Under our metropolitan charter, Miami-Dade County bureaucrats, not our various incorporated municipalities, are ultimately in charge of everything related to local street design, traffic circulation, signage, markings, and more. The county has earned a tough reputation for saying no to anything similar to the better examples in this book. (One colleague quipped, "They'll approve traffic calming as long as no actual traffic is actually calmed.") Ever the optimist, I once suggested that starting engineering projects by assuming slower design speeds would save lives and create possibilities for correcting oversized lanes, planting street trees closer to the curb, and so on. The chief engineer answered immediately: "Over my dead body." He went on to explain that he simply would not survive having his bosses get angry calls from "the motoring public" if anything ever suggested the county might be intentionally slowing them down. I think about that story each time I hear advocates complain about the common use of the term "accident" among engineers and reporters. As they point out, we should be using "crash" or "collision;" it is hard to call them *accidents* when you knew perfectly well they were going to happen, and you *engineered* it to be likely.

Around 2014, however, I was asked by local leaders how to grow transit ridership in our region. I answered, *Put the person in charge of growing transit ridership in charge of street design.* My idea was that welcoming, tree-lined routes walkers and cyclists can use to get to the transit station or bus stop are not luxury frills. They are mission-critical equipment. They are just as important as the escalators in the train station, if not more important. If we want people to use the train or the bus, their last-mile walk to and from the transit stop has to be a good experience. A transit leader, I thought, will understand this and change the old street design habits.

BICYCLE LANES & COMPLETER STREETS

In *Copenhagenize: The Definitive Guide to Global Bicycle Urbanism*, Mikael Colville-Andersen identifies four main types of bicycle lanes. "One of these four basic fits every street in the Danish Kingdom," he says, "and, indeed, every street in every city in the world."[59] Traffic volume and speed determine which of the four should be used. The speeds used in Copenhagenize are a little higher than we use in the United States. Perhaps Europeans drive less aggressively than we do?

1. Cycle Tracks: "When the speed limit hits 50 or 60 km/h (approximately 35 mph)" Colville-Andersen writes, it is time for the cycle track, which is defined by physical separation from motor vehicle traffic. At that speed, the cycle track must be one-way, for safety at intersections. In Denmark, a small curb is typically used for the separation (see discussion below). But when the speed limit for cars hits 70 km/h (approximately 45 mph) or higher, "there is only one goal: getting cyclists as far away as possible from the motorized traffic." When cycle tracks are used in off-street contexts with no contact with cars, a two-way design is "suitable."

2. Bike Lanes: "When the speed limit hits 40 km/h (25 mph) or the volume of cars rises, painting the lane is an acceptable solution," Colville-Andersen says. Painted lanes are not as safe as Cycle Tracks, or the American protected lane (discussed below). For safety at intersections, bike lanes must be one-way.

3. Shared Lanes: Including Shared Space and Sharrows. Traffic must be light and traveling at less than 40 km/h, which is 25 mph. In England and America we expect this to be 20 mph.

4. Bicycle Streets: Optimized for bicycle traffic. "Cars are allowed, but have to follow the tempo of the cyclists."

The designs for all of these are evolving around the world, with variations from country to country.

Well-designed cycle tracks in a network of well-designed streets are the safest choice for streets with high volumes of motor vehicle traffic. The tracks come in many forms, different in urban design, street design, and safety. As we noted in Chapter Four (page 462), street designers in Denmark don't like the way the New York City Department of Transportation and now the National Association of City Transportation Officials have adapted what the NYC DOT first called "Copenhagen Lanes" (seen on page 635). Many Danish cycle tracks are on streets with large numbers of motor vehicles, bicycles, and pedestrians.

In the Netherlands, protected cycle tracks on normal city streets typically have more physical separation from the street than Danish or American examples. As noted, however, when traffic reaches 40 miles per hour or above, Danish practice also moves the cyclists farther away from the street than in the Copenhagen Lane, as best practice always should. The protected bike lane in New York City we illustrate on page 635 is typical of the new cycle tracks on many American "Complete Streets." You can see these in greater detail in NACTO's design manuals, the *Urban Bikeway Design Guide* and the *Urban Street Design Guide*.[60]

Jon Orcutt, the spokesman for Bike New York, helped create the modern American standard when he was a Deputy Commissioner of the New York City DOT. He says, "Parking-protected bike lanes are a good interim step for places that may not have political will to do away with parking, but there are very real issues…. Cities looking to implement the lanes should be very aware of that."[61]

The alternatives Victor wrote about in *My Miami* (page 583) are better models. In the New York solution seen in Figure 5.58, parked cars separate the cycle track from moving cars. That protects the cyclists, but the cars and the paint visually and functionally cut the street into pieces, fracturing the public realm in the space between the buildings. The over-engineered details combine with the under-designed placement of the cars in the road to make streets that are frequently ugly and unpleasant for pedestrians. Even though we get used to the designs, we still feel the degradation of the public realm. We need Completer Streets.

Those protected bike lanes are common on one-way Complete Streets. They have "mixing zones" that have safety problems when turning cars and the cyclists going straight ahead cross and come into conflict. In the first iteration of the design, particularly on one-way streets, left-turn lanes and left-turn lights made the mixing zones dangerous. Parked cars obstructed the drivers' view of the cyclists and drivers moving across the bicycle lane with a green turn signal frequently drove quickly. Collisions and deaths resulted.

The left-turn lanes in cities like New York, which previously had few turn lanes, also led to the death of pedestrians in the crosswalks on the side streets, even when the pedestrians had the right of way. Current best practice in traffic engineering forces drivers into slow, sharp turns. This is safer, but it is still a result of prioritizing traffic flow over bicycle and pedestrian safety. Note that bicycle tracks that put the cyclists behind parked cars also tell the drivers that the bicycles will stay away from "their part" of the road (making speeding feel more comfortable). In Denmark, where streets with cycle tracks tend to be two-way, cycle tracks once ended before the intersection, creating a zone where bikes mixed with cars turning right. Collisions and deaths resulted. The solution was to make cars come to a complete stop before crossing the cycle track.[62]

The cycle tracks on city streets in Denmark typically have curb separations: the cycle track is 2 to 2½ inches (5 to 7 centimeters) above the street level, and 2 to 2½ inches below the sidewalk. Most European cities and countries today put the cycle track on the street with barriers separating them from the street. These can be low but wide strips of stone or concrete, or taller barriers such as planters. Rubber curbing and "armadillos" are also common. Armadillos are small, rounded separators that should be bolted to the ground.

In the past, European cities have frequently put bicycle lanes on the sidewalk at the same level as the section reserved for people walking, without barriers between the two parts. This only works on wide sidewalks where there are few cyclists. With the rise of cycling and e-mobility around the world, pedestrians in many cities are increasingly experiencing sidewalk conflicts with cyclists and riders on motorized scooters and skateboards.

We like to think that the anti-urban, semi-protected American bike lanes we wrote about in the first edition of *Street Design* are fading into the past, but one can still see them all over America. We hasten to add that we don't think the new American standard for 2025 should be the permanent solution. First, it has safety

issues, discussed below. And it is over-engineered, usually built with lots of garish paint, cheap white-plastic sticks that look worse and worse over time, and the occasional Jersey Barrier, sometimes with reflective orange paint that also ages badly.

Sharrow markings and unprotected cycle lanes in the roadway are a compromise that cities and towns should only use on slow streets with little traffic. For decades, DOTs have experimented with sharrows because they are cheap and easy. But while simply stamping sharrow markings alone on a street might seem like a good first step, paint is not a barrier. Painting a picture on the pavement of a child on a bike does not change a dangerous environment to one safe enough for children to ride there.[63]

Peter Flax, in an article bluntly titled "Why Sharrows are Bullshit," said cities use sharrow markings when they "care enough about safety to do the very least." Citing two engineering reports, Flax said "sharrow markings are worse [for safety] than doing nothing."[64]

A better choice is to make shared-space Slow Streets with little traffic. Note that many American drivers are not ready for this treatment. For situations like that, raised crossings and "continuous sidewalks" like the one in Figure 4.285 alert the drivers that they are entering Slow Streets, unlike the streets they are used to.

Continuous sidewalks are best when designed for people walking rather than driving. By that, we mean that the fewer visual cues there are for the drivers—such as a change in materials and bright, multi-color paints—the better. Those cues signal to the pedestrian and the driver that the pedestrian is entering a machine space. A simple continuous sidewalk spanning the street, on the other hand, tells the driver he is leaving machine space.

Finally, cities that want increased levels of cycling in the future need to think about Bicycle Streets. Dutch cities, of course, have done that. We show one of their *Fietsstraaten* on page 521 and we include a Bike Street as one of the Essential Street Types.

FLORIDA'S STATEWIDE REFORM EFFORT / VICTOR DOVER

The true creator is necessity, who is the mother of our invention.

> — Benjamin Jowett's 1894 translation of Plato's *Republic*

We used to joke that Florida had two types of locations outlined in its Department of Transportation design manuals: "Rural," by which they really meant suburban, and "Urbanizing," by which they *also* meant suburban. And on Florida's roads, suburban design has magnified the hazards.

In *Dangerous by Design* and other recurring reports, Florida is always listed as a leading producer of fatalities in traffic. For decades, state transportation officials simply doubled down on the wider, faster roads approach, in a futile chase of congestion relief that never came, as if inducing even more driving might somehow reduce the body count. But in 2014, there was a sea change in state policy. Florida Secretary of Transportation Ananth Prasad declared a Complete Streets Policy, sending a memo to the entire agency announcing that this policy "will be integrated into the Department's internal manuals, guidelines and related documents governing the planning, design, construction, and operation of transportation facilities."

Demand for reform had been building for some time. Way back in 1980, the FDOT added planner Dan Burden to its staff roster; inventing his own title, Burden became America's first State Bicycle and Pedestrian Coordinator.

In 1984, urged on by Burden, Governor Bob Graham signed a state law that today we recognize as one of the first complete streets policies in the nation, requiring state engineers at least *try* to incorporate infrastructure for walking and biking in each new road project. In those days, engineers frequently found ways to wriggle out of meaningful departures from business as usual; when they did build them, the earliest bike lanes on state roads were perfunctory at best, of little use to cyclists.[65] But thanks to Burden, the conversation had begun.

In addition to its bookshelf of fifty-odd manuals controlling design on FDOT's own state facilities until recently, the agency promulgates a rulebook for *local* governments to follow on their streets, confusingly titled *The Florida Green Book*. The book was ridiculed

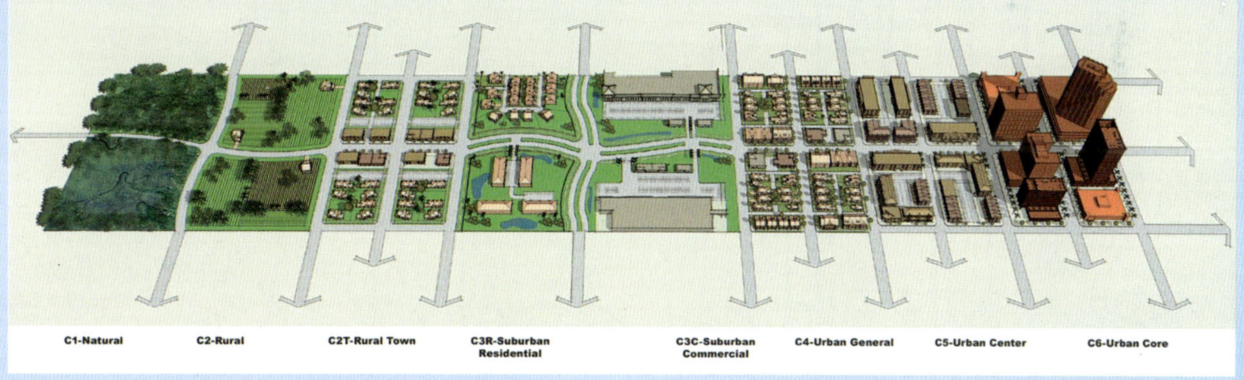

Figure 5.17: DeWayne Carver *et al.*, Florida Department of Transportation (FDOT) Central Office, *The Florida Design Manual (FDM)*, Tallahassee, 2017. Graphic depicting the eight Context Classifications now used by FDOT. *Courtesy of DeWayne Carver / Florida Department of Transportation / Public Domain*

by barnstorming New Urbanists in the early days of the movement because nothing permitted in it resembled the streets of the walkable, traditional neighborhoods so many people desire and need. In 2011, a revision committee including engineers Billy Hattaway and Rick Hall overhauled *The Florida Green Book*. They were finally allowed, after considerable negotiation, to add a new chapter for Traditional Neighborhood Developments, or TNDs. Legend has it that one serious member of their committee, from the engineering old school, held up the discussion for a while; he adamantly argued hard that TNDs, if allowed at all, should require big signs at their boundaries, warning of the dangers in the walkable streets that lie within. More of a cultural shift was needed.

At first, Secretary Prasad simply issued a two-paragraph memo that said, "the Department will routinely plan, design, construct, reconstruct, and operate a context-sensitive system of Complete Streets" and "The Department specifically recognizes Complete Streets are context-sensitive and require transportation system design that considers local land development patterns and built form."[66] This was helpful, but left district-level managers with little technical material to go on; the old official state manuals seemed to prohibit the very designs the Secretary was ordering them to produce.

But Secretary Prasad had hired Billy Hattaway to lead both FDOT District One and the statewide effort to change the system. This was Hattaway's second tour of duty at FDOT; he had previously served as State Roadway Design Engineer and left for the private sector after getting frustrated at the slow pace of reform at the agency. This time, Hattaway and Prasad had a plan. Hattaway toured the state to train staff at each of FDOT's seven district headquarters, installing a Bike/Ped Coordinator in each one. Then, in 2015, he had the Department partner with Smart Growth America and the National Complete Streets Coalition on an overhaul of the manuals themselves. This effort was dubbed "M2D2" (officially, the Multimodal Design and Delivery program); it was based upon a similar attempt within Michigan DOT the year before.

The M2D2 "task force" reviewed more than 50 sacred scrolls of transportation planning and engineering,

eventually identifying about a dozen manuals that were central to the roadway engineers' bad habits, including the notorious *Plans Preparation Manual*. (The task force was almost completely populated by DOT staffers, but Rick Hall, Kim Delaney and I were allowed to sit in; we were nicknamed "the civilians.")

By 2018, most of these archaic books had been replaced by the new *Florida Design Manual* (FDM) plus a couple of rewritten companion documents (including, notably, the *Lane Repurposing Policy*). Several updates have followed.

The breakthroughs are many, but the biggest one is this: Traffic engineering on state thoroughfares is no longer one-size-fits-all. Now, context matters.

FDOT's set of eight T-zones closely resemble the six T-zones in the Urban-to-Rural Transect popularized by Andrés Duany and the New Urbanists (see Chapter One, page 48). FDOT's Context Classification System assigns a "draft" context classification to each segment of each state thoroughfare; now every significant road project triggers a check of a long list of features to confirm whether the context classification is correct or needs upgrading. At least in the more urban context zones, the *FDM* gives engineers clear permission to say yes to components and dimensions they previously refused to consider. For the prior half-century, state officials just said no to slower design or target speeds, 10-foot travel lanes, shade trees close to the lanes, and the like—blaming their manual. Now they can say yes, and sometimes even will, if nudged.

The *Florida Design Manual* gives engineers clear permission to say yes to components and dimensions they previously refused to consider.

Rob Steuteville wrote, "If context-based street design works in the most automobile-dominated state, it can make a difference anywhere." DeWayne Carver, who shepherded the reforms from within FDOT's central office in Tallahassee, said, "This idea was not that hard for traffic engineers to get their heads around. I was honestly surprised at how quickly they adapted."[67]

For his leadership, Hattaway was named Public Official of the Year by *Governing* magazine.

WHAT IS THE MUTCD?

Many of the essays and sections in this chapter are about influences on street design and the design of public space, particularly from the fields of architecture and urban design. As we have said before, however, traffic engineers and transportation planners dominate the design of our streets. A century of work by Organized Motordom gives DOTs and DPWs legal control over streets. They also control the funds for building and maintaining most streets, and the cost of that construction and repair is a large part of many state, county, and municipal budgets.

> The *Manual on Uniform Traffic Control Devices for Streets and Highways* (MUTCD) could be called *Making Ugly Things Compulsory Details in Cities and Towns.*

It extends the power of the profession beyond basic engineering decisions like the details and dimensions of intersections by setting rules that determine how traffic signs, markings, and signals planned for sub-urban highways and auto sewers should be installed and maintained on neighborhood and city streets.[68] Managed by the Federal Highway Administration (FHWA), the MUTCD is frequently updated to keep up with changes in transportation policy, safety technology, and new research. The book now has more than 1,100 pages of over-engineered and frequently anti-urban objects that unfortunately litter the streets of America.

Street lights and traffic lights are a prime example. In images like Figures 2.32 to 2.34, we see city-scaled lampposts on New York's Park Avenue that fit the urban context and complement the surrounding buildings. In the most recent of the photos, Figure 2.33, galvanized metal posts have replaced the old painted and tradi-tionally styled iron posts. But the new posts are simple and still human-scaled. What one cannot see in the photo are that the avenue also has much taller galva-nized posts with long mast arms from which modern luminaires hang, indistinguishable from the street lights on suburban traffic corridors.[69]

History of Revisions

After various experiments around the country in the 1910s and 1920s, the forerunner of the American Association of State Highway Transportation Officials and the National Conference on Street and Highway Safety jointly published the first national edition of the MUTCD in 1935. The federal government took

Figure 5.18: Northeast 15th Street, Miami, Florida. Photograph showing the recently updated MUTCD as garishly applied in the field: Paint, paint, and more paint, highway-style. Even markings meant to increase the visibility of the crossings for people walking and biking inadvertently reinforce the feeling that these streets are first and foremost hostile, automotive environments. *Courtesy of Miami-Dade Department of Transportation and Public Works social media feeds*

over in 1966.[70] Counting updates between eleven official editions, the MUTCD has been revised at least 35 times. The latest version took full effect upon the federal government's publication of its final rule in December 2023.

Advocates consider it a significant improvement over preceding editions, though skepticism remains. Under a provision of the Bipartisan Infrastructure Law (BIL) passed in 2021, the MUTCD will be updated every four years.[71]

Much of the MUTCD is written as guidance, but engineers and lawyers at the local level tend to treat it as gospel. They will tell you that you do not *have* to follow the MUTCD "*unless* you want funding from the federal government or not to get sued." However, the reality is more nuanced.[72]

First, application of the MUTCD varies by jurisdiction. Federal law requires states to either adopt the federal government's version of the MUTCD (as Florida and seventeen other states have done), *or* to create their own "substantially conforming" version (following the examples of Texas, California, and eight other states), *or* to adopt the federal version with a supplement. New York, twenty-one other states, Puerto Rico, and the District of Columbia took that approach.[73]

Second, while some features are mandatory—stop signs shall be red, for example—much of MUTCD is indeed guidance. A major shift starting with the 2000 edition "organized content into *Standard*, *Guidance*, *Option*, and *Support* provisions, with those headings clearly labeled in every instance."[74]

Third, the authors of the MUTCD, at least on paper, did not intend to promote "one size fits all." They say up front that designers should always exercise judgment, site-specific in each case:

> The decision to use a particular device at a particular location should be made on the basis of either an engineering study or the application of engineering judgment. Thus, while this Manual provides Standards, Guidance, and Options for design and applications of traffic control devices, this Manual should not be considered a substitute for engineering judgment. Engineering judgment should

be exercised in the selection and application of traffic control devices, as well as in the location and design of roads and streets that the devices complement (MUTCD Section 1A.09).

Lastly, the MUTCD has an extensive and paperwork-intensive process for officially sanctioned experimentation. New techniques that prove useful, once tested, can be incorporated as accepted practices in a future version.

Note, however, the infrastructure law does not exactly let agencies *ignore* the MUTCD. Localities may use different rules for marking streets, but only if these rules are approved by the FHWA and follow all other important laws. Departments that use different design rules must be prepared to justify their choices when setting up signs and signals on roads. FHWA refers to the MUTCD not just as a useful publication, but as "the law."[75]

Complaints abound about the MUTCD and its application, even with recent refinements. For example, with traffic signals, which are notoriously overused, confusion remains; just because one is technically *allowed* to use something does not mean one is *required* to use it, but that is exactly what happens with signals. The MUTCD itself states, "The satisfaction of a traffic signal warrant or warrants shall not in itself require the installation of a traffic control signal."

Another example is the widely loathed "85th percentile rule." This is the practice, dating from the 1960s, of setting the speed limits based on the pace most drivers naturally choose when traffic is free-flowing—even if that is dangerously fast in areas we should be making safer for pedestrians.

The federal government says it was never intended as a "rule," and added new language to the MUTCD vaguely describing how various other factors might be considered in setting speeds, but critics argue the discredited 85th percentile method does not belong in a modern document at all.

Streetsblog editor Kea Wilson writes, "Some advocates say that the Administration is naive to believe that engineers will consider factors other than 85th percentile speeds—especially without clear guidelines telling them exactly how to set safer limits instead. And that is in part because, however much the FHWA says that the Manual of Uniform Traffic Control Devices" is not a road design manual… that carries the force of law, overworked engineers fearful of lawsuits certainly treat it that way, even when the text gives them nominal permission to consider options besides the status quo."[76]

Other countries have documents similar to the MUTCD. But engineers around the world hesitate to use the flexibility the rulebooks offer. In the United Kingdom, the *Traffic Signs Regulations and General Directions 2016* (TSRGD) document carries the force of law, but some of the less-is-more solutions it permits are rarely taken up by street designers. Engineer Phil Jones reports, "A few authorities experimented with crossings that just used the stripes, without the zig-zag markings on the approaches or the yellow flashing globes ('Belisha Beacons'). They work very well but our Department for Transport really don't like them and few places have used this approach, sadly."[77]

To my surprise, Miami-Dade *did* reorganize. The mayor at the time rearranged the county staff, merging the old Miami-Dade Transit Agency and the Department of Public Works into one new entity, the Department of Transportation and Public Works (DTPW), with one boss. Confidence rose.

But gradually it became clear that they had merged the agencies but kept the silos. A decade later, the mid-level traffic engineers still prioritize flow over safety. Level of Service analysis, though nationally discredited, still rules. Engineering decision-makers deep in the Traffic Operations silo (177 employees of the 4,122 in the department as a whole, in 2023) still operate in a car-first way; staffers in the larger department still refer to the group as "Highway," as in "I heard over in Highway they have a project to widen that road." It remains rare to get permission for slower design speeds, lane repurposing, road diets, protected bike lanes, street trees, or anything else.

The Bike/Transit Combination

Halfway through the 2010s, cyclists were finally allowed, officially, to bring their bikes aboard the Metrorail without first getting special authorization. A quarter-century earlier, when Miami-Dade's Metrorail service initially opened, the procedure was bizarre. One was expected to leave the bike at home, take the train downtown, fill out duplicate forms at county offices, and commute back home, then keep a card or a carbon copy of the paperwork on their person on any future date when they wanted to roll their bike onto a train. Once registered, they would be expected to enter only the last door of the last car of any given train.

When Metrorail's Orange Line extended to connect with Miami International Airport in 2012, though, transit officials realized space would be needed for luggage. They removed a couple of benches from each train car. Local cycling advocates like Eric Tullberg seized the moment. Tullberg successfully argued that the luggage space could double as bike storage space, and even managed to get stickers with bike symbols affixed to the wider turnstiles at stations, right next to the wheelchair symbol. Since then, we have been able to enter any door on any car or train. Folks with bikes are not exactly welcome yet, but at least we're not prohibited (or left guessing where on the station platform that last door might open, then scurrying toward it).

Government agencies are not known for creative self-promotion of their services. A group of us realized that

Figure 5.19: Posters for the "WHEELS FLORIDA" event, November 2015. Jennifer Garcia, Herb Hiller, and Victor Dover, 2015. *Courtesy of Dover, Kohl & Partners*

the new extension of Metrorail to the airport had barely been advertised. The upgrade meant more than most people had noticed; to facilitate the branching of the system so half the trains could run to the airport, Miami-Dade Transit had divided the rolling stock, shortened the trainsets, and scheduled the trains to move through many stations twice as frequently. Notoriously long waits at the station platforms vanished, overnight. It was finally becoming practical to combine bike rides with transit trips; I began referring to my own bike as my "horizontal elevator," moving me to and from the transit stations. Bikes and transit and trails combine powerfully, each extending the range of the other. But no transportation agency was actively promoting that combo in our region.

At age 86, Herb Hiller, the longtime cycling advocate credited with essentially inventing bike tourism in Florida, came up with the idea of a week of special events in November 2015 to shine a spotlight on the bikes/trails/transit combination. We called it WHEELS. There were bike-in volunteer activities, rides to and from Metrorail to show how easy it is to access the transit stations by bike, fun runs on new trails, a history lecture, a symposium with national speakers, contests, and tactical urbanism installations, video productions, and more. DTPW, Tri-Rail, FDOT, and a score of nonprofit organizations, foundations, and companies collaborated to stage the WHEELS events.

Driving everywhere for everything is not our only history, and WHEELS offered a chance to remind people of this. The origins of the League of American Bicyclists can be traced to Miami's Coconut Grove neighborhood, for example. Kirk Munroe, who moved to the Grove to write books for young readers, cofounded the group and held membership card number one in what was then called the League of American Wheelmen. The League has been unchallenged as the leading organization in the Good Roads and cycling movements for over one hundred forty years, and one of Miami's own was its first "Commander." Munroe organized The Wheelmen's Winter Rendezvous in Miami-Dade County each year, hosting well-to-do bike enthusiasts from up north for long, bone-shaking rides over gravel dikes on hard rubber tires. WHEELS attendees got to learn about Munroe and ride together on one of his routes.

Most importantly, WHEELS brought thousands of people into contact with everyday cycling and into dialogue with each other about biking and walking in Miami-Dade. During that time you could not help but feel change was coming. Sometimes ideas take a while to make it all the way down our long peninsula and sink in, though.

Making Progress?

Miami-Dade was the last large metro area in America to introduce genuinely protected bike lanes. We still have a total of only 4.8 miles of them![78] Compare that with greater Minneapolis, with a similar population but at least four times more protected bike lanes in the core city alone.[79]

In our suburbs, the rare streets that do have separated bike lanes are still designed like 100 percent automotive environments with a bit of bike stuff added on at the edges. Supportive elected officials waged a difficult battle to upgrade a 0.79-mile stretch of SW 211th Street with protected bike lanes, for instance, and in 2020 the county finally built it. But all five adjacent car lanes remain wide—noticeably wider than even the car lanes on Florida's Turnpike—and fast. As for the curb-separated bike lanes, they are interrupted, they swerve, and they are left unprotected for so many storm-drain inlets, curb cuts, driveways, and bus stops, the protection amounts to surprisingly little. The slim medians that separate the bikes from the cars "protect" less than 27 percent of the length of the bike lanes. People on foot on the sidewalks and people on bikes in those lanes won't find shade, either, since the only trees are skinny palms and most of those are in the medians out in the center of the corridor. Despite all the political effort, SW 211th Street remains an unnerving place meant for cars (see Figure 5.20).

Just a few little improvements can make a meaningful difference, however. My morning route now includes a short segment of bike lanes with slightly better protection in South Miami, on SW 64th Street. As soon as it opened, I revised my regular route to take advantage of it, and the sense of increased safety and reduced stress was immediately noticeable.

Gables Greenways and the Bikelash

For its part, in 2017 Coral Gables started gearing up to implement the *Bike and Pedestrian Master Plan* the City Commission had adopted in 2014, promising

Figure 5.20: Protected bike lane on SW 211th Street, Cutler Bay, Miami-Dade County, Florida. Photograph looking west, showing how the corridor remains focused on fast driving, not on cycling or walking. The slim medians separating the bike lanes are interrupted abruptly and often; less than 27 percent of the length of the route is actually "protected."

Figure 5.21: Salzedo Street, Coral Gables, Florida. City of Coral Gables, Bike Walk Coral Gables, and Dover, Kohl & Partners, 2017. Computer simulation. The "Gables Greenways" initiative envisioned ways to modify oversized streets, adding street trees and protected bikeways.

Figure 5.22: University Drive, Coral Gables, Florida. City of Coral Gables, Bike Walk Coral Gables, and Dover, Kohl & Partners, 2017. Photograph looking west. The "Gables Greenways" demonstration project included temporary protected bike lanes on University Drive and Salzedo Street. They worked well but were quickly dismantled. *Courtesy of Kenneth García*

twenty-seven miles of new bike lanes, sidewalks, and crosswalks.[80] They followed up with a thicker and more comprehensive *Multimodal Transportation Plan.*

The "Gables Greenways" effort offered a preview of how good—and how bad—it could get. The program included a demonstration project with protected bike lanes on University Drive and Salzedo Street.[81] The separations between the temporary bike lanes and car traffic were accomplished with upmarket planters and cascading flowers, in keeping with Coral Gables' City Beautiful reputation, and it was all quite elegant compared to most other Tactical Urbanism installations. Traffic-wise and safety-wise, the setup worked well and looked good, too.

But before long, Coral Gables ran into a buzzsaw of internet-stoked, anti-bike protest. The first round of "bikelash"[82] befell the Gables Greenways bike lanes. One business owner along Salzedo Street, accustomed to lightly-used on-street parking on both sides of the two streets adjoining his corner lot, claimed that removing a handful of spaces would ruin his business. At a public hearing on the matter, he was joined by a chorus of red-faced speakers who voiced variations on "I don't bike, everybody knows it's too hot here, so why should anybody else be biking?" and "It scares me to see bikes when I'm driving," and essentially "My [speedy] drive is more important than someone else's safe bike ride." The former mayor asked the city engineers if they could come up with a type of bike infrastructure that is "only there" for recreational rides at off-peak times and on weekends, but otherwise disappears to make more room for rush hour driving and parking. After that, the City administration backed off completely on the Gables Greenways effort.[83]

A second round of bikelash came in 2020 when Gables public works officials tried to implement the

North Alhambra Circle bike lanes described in their *Bike and Pedestrian Master Plan*. Although the plan had been adopted following months of public outreach, committee work, and hearings, and the construction would have been paid for by a $597,000 state grant, the residents along North Alhambra Circle organized loud opposition. Once again, the complaints largely amounted to "I don't bike, why should you?" and homeowners claimed they had never heard about the plan.[84]

Opponents of the North Alhambra Circle bike lanes and similar projects decided to make it personal and targeted ferocious attacks on the deputy public works director assigned to the projects. They did not just make her feel unwelcome; this engineer was pushed out for doing the very job she was hired to do. Her boss, the city manager and another professional engineer, commented at a meeting of the City Commission, "She is a bike person, and that is very big for her, and I understand that. But this is *a city that does not do bikes*. If you like bikes that much maybe you might go to Seattle. But this is not the city."[85] Coral Gables then shelved the North Alhambra bike lanes.

Meanwhile, in the neighboring City of Miami, city commissioners voted to reinstate downtown parking minimums in 2022, just as hundreds of cities around the country were doing the opposite. At the meeting, Commissioner Manolo Reyes made a comment similar to that of the Coral Gables city manager. His reasoning for voting to repeal parking reforms, he said, was that Miami is "not a pedestrian and bicycle city."[86]

Jessica Keller, the planner and engineer recruited to implement the Gables plans, reflects on the episodes this way: "You cannot overestimate the importance of public engagement. Doing it properly costs money. You cannot have too much public engagement. Coral Gables found out too late just how important it is."[87]

Put Safety Up for a Vote?

One way to tamp down the clamor for better streets is to pile procedural obstacles so high few advocates can hurdle them. Despite the setback on North Alhambra, resident Roberta Neway and her neighbors a mile away on South Alhambra Circle continued pressing for the planned bike lanes on their section of the street. The City rushed a new procedure into place: On any given street, before traffic calming and bike infrastructure can be installed,

now there must be a vote of all property owners along the street, with a 51 percent majority voting in favor. (The original proposal was for a nearly impossible two-thirds majority, but cooler heads prevailed.) Neway was undeterred. She went door to door for weeks, patiently informing as many neighbors as she could. When the official vote was tallied in October 2022, 62 percent of the owners on South Alhambra who bothered to vote had voted in favor.[88] Now it remains to be seen if the transpocracy will find some new excuse for not rebuilding the street. At the end of 2025, nothing has been built.

There are several fundamental problems with the vote-by-the-homeowners approach. First, it assumes that the only citizens who need or care about the streets are the people living right on them. That is not the case. Our streets are public rights-of-way for a reason. Riding or walking down South Alhambra, there are citizens whose trips originate outside the neighborhood and end on the other side of it at schools, workplaces, and other destinations. Some would benefit from the bike lanes.[89]

Second, such a vote implies the only people qualified to decide are the property owners. What about renters, underpaid workers tending the homeowners' gardens, workers changing the beds in the neighborhood hotel, disabled persons using mobility devices, or students? All of them might welcome an inexpensive alternative to driving.

Third, and most importantly, it sets a bad precedent to insist that safety should be put up for a vote among the neighbors. Engineer and writer Andy Boenau puts it this way, "Voting is a double-edged sword. Americans have been trained to associate the words voting and democracy as inherently righteous in any context. It is like the old joke about a Sunday School teacher asking a question (any question), and a little kid answers, 'Jesus.'" Town planners have been led to believe 'more voting' and 'more democracy' are always the correct answers to any question about street design. But for one hundred years, the majority has raced to vote against safe streets."

Boenau continues, "I have no expertise in rocket science. People like me should never be deciding how NASA performs quality control on satellites. Similarly, most people have no expertise in driver behavior, multimodal accessibility, or civil engineering. Their observations should be treated like they are users of a system, not authorities on design. The average person is more concerned about personal convenience than public safety. Your life and my life should never be up for discussion, let alone a vote."[90]

Ocean Drive: On Again, Off Again, On Again

South Beach residents are justifiably proud of having the most urbane, walkable area in the region. But Ocean Drive, the most famous street in South Beach, also symbolizes the cycle of innovation, backlash, retreat, experimentation, and compromise.

When I first got to know Ocean Drive (Figure 5.15) in 1986, the street was wide, drab, beige, and dowdy. In the early phase of the neighborhood comeback, the city adjusted the lane widths to make room for more sidewalk space. It worked. Soon after that, the Drive became shiny, hip, and pleasant, a favorite spot for fashion photographers.[91] The retrofit accelerated the revitalization of the Drive and of the Art Deco District as a whole. Crowds returned to South Beach, so gradually pressure mounted for another makeover on Ocean Drive.

Recently, advocates began asking the city to do more to reduce the dominance of cars on Ocean Drive—and defenders of the status quo responded to each proposal with a counteroffensive. The City of Miami Beach converted the street into a one-way operation and installed separated bike lanes on the Drive. Then they removed them. Then they reinstalled them, in a new version. For a short while during the Covid crisis, the street was only open to pedestrians and people on bikes. Then it was reopened to traffic. Now it is opened, and closed, and opened again intermittently. With each change, one faction howls and the other cheers. The experiments continue.

"Quick-Build"

It is all taking too long. Like other cities with energetic leaders in the Tactical Urbanism movement, Miami-Dade initiated a "Quick-build" program, to implement accelerated pilot projects and demonstrations in sync with a growing urban population expecting better bicycle, pedestrian, and transit infrastructure. There was considerable fanfare behind the announcement in 2018. One of the program advisers, Tony Garcia of Street Plans Collaborative, is a founder of the global Tactical Urbanism movement (and literally wrote the book on *Tactical Urbanism*).[92]

The program started with a goal of constructing two to four projects within a year. A contest was held to crowdsource a set of projects. The response was overwhelming; over five hundred project ideas were submitted from around Miami-Dade County. Given the response, the project team expanded the number of projects they hoped to complete to eighteen. Projects ranged from protected bike lanes and dedicated bus lanes to simple crosswalks. The team designed and vetted all eighteen projects with the community within six months of the initial selection.

Miami-Dade County could not follow through. Unlike peer cities vividly presented in Garcia's book, the County had no internal process to accelerate the permit process. When the county engineering staff finally reviewed projects, they lacked basic knowledge about active transportation best practices and were ill-equipped to design streets for people walking, biking, and taking transit. The Quick-build projects did not end up being quick, and hardly anything has been built. The few projects that *were* built were done without any formal county approval. The program faded quietly from there. Asked to describe the impact of the program, Garcia was frustrated but hopeful. "While the program was short-lived, it taught us that there is incredible demand for safer streets in Miami-Dade. The successful projects had a strong group of engaged residents who advocated for their implementation. As our walkable urbanism continues to develop, that constituency will continue to grow. I think it is only a matter of time before the county has to change. The question is, how long will that take?"[93]

Wins Among the Losses

All these stories paint a clear picture: Greater Miami's street designs have not kept pace with designs from our peer cities. Education explains the gap. On the other hand, there have been *some* wins. The pedestrianization of Giralda Avenue is an astounding success (Figure 4.273). Lincoln Road (Figure 4.248) finally "worked." In one of the earliest road diets, Sunset Drive in downtown South Miami was narrowed from five lanes to three lanes, thanks to the courage of insistent elected officials. (That was pushed through despite the county public works director's initial reaction, "I hate to *lose pavement* on a street like Sunset Drive.")

Figure 5.23: Ludlam Trail, Miami, Florida. Dover, Kohl & Partners, and AECOM, 2020. Computer simulation. Thanks to the success of the Underline, other ambitious trail projects have gone from being far-off visions to being funded for construction. The six-mile Ludlam Trail crosses several street corridors at grade, so level crossings were devised, moving the local conversation forward on street design.

The Underline is another win. The term "trail" does not do the Underline justice. It is a ten-mile linear park, a showcase for native landscape plantings, a pair of paths for walking and biking, and, according to the organizers, "a living art destination." Conceived by master fundraiser and promotional genius Meg Daly, the Underline has been fashioned from the neglected leftover space beneath the elevated Metrorail tracks, along the former maintenance path. It stretches from the Miami River downtown south through the Brickell and Coconut Grove neighborhoods, and will soon reach Coral Gables, South Miami, and Downtown Kendall. One of the interesting, underreported aspects of the Underline story is that it required rethinking street designs and crosswalks along busy corridors like SW First Avenue and U.S. Highway One.

The Underline, Ludlam Trail, and Miami River Greenway will eventually link up to form what architect Maricé Chael named The Miami Loop, a twenty-five mile system of high-quality urban trails.[94] The Ludlam Trail will cross city streets at grade in eleven places, so just like with the Underline, the planning forced a rethinking of

street and crosswalk designs. In a region obsessed with driving, the outpouring of interest in the Ludlam Trail is encouraging. The County bought the land in the former rail corridor, and construction is funded. More than four hundred people joined in one Zoom meeting in late 2020, as planning continued during the Covid lockdown—another win.

Arriving at a consensus on connecting *to* the trail with sidewalks and protected bike lanes has proven more elusive, however. A fierce not-in-my-*front*-yard back-lash stopped South Miami from extending the SW 64th Street bike lanes to connect to the Ludlam Trail, at least for now; that would have linked the University of Miami campus to the trail and to the western half of the Miami Loop. Advocates have not given up.[95]

The Commodore Trail may be the next battleground. Unlike the Underline and the Ludlam Trail, which generally run perpendicular to vehicular rights-of-way, the Commodore Trail runs along and within the right-of-way. For more than sixty years, the trail has appeared on maps and plans, but there is still not much built infrastructure for more than three miles of its five mile length.

Figure 5.24: Miami, Florida. The Commodore Trail, along South Miami Avenue. Dover, Kohl & Partners, 2023. Before: Photograph of existing conditions at the historic Vizcaya estate, looking west within the Farm Village.

Figure 5.25: Miami, Florida. The Commodore Trail, along South Miami Avenue. City of Miami, Dover, Kohl & Partners and Chen Moore Associates, 2023. After: Computer simulation of a proposed redesign as a raised plaza at Vizcaya, looking west across the corridor. *Courtesy of Dover, Kohl & Partners*

The shady route connects from Coral Gables through Coconut Grove to the Rickenbacker Causeway—at least on maps. Along some stretches, the path is very evident as a multi-user trail, but in most spots there is just a narrow sidewalk, or a dusty shoulder, or nothing at all. Nevertheless, this is still the most direct route for runners and cyclists. That means it is also the most popular course for people training for the Miami Marathon and triathlons, despite the lack of accommodation. Each month more than 30,000 cyclists pass through the area where the Commodore Trail connects (on paper) to the Rickenbacker Causeway. The nonprofit Friends of the

Commodore Trail built a coalition advocating for the trail's full implementation, twisted the arms of the County and City of Miami leaders to run an outreach program that has drawn robust, enthusiastic participation, and our firm has prepared a detailed master plan. The next step is to see if the County will, finally, prioritize safety over flow. Street design will be the key to all of it.

Choices: Car-Dependent, Car-Lite, and Car-Free?

We are not a monolithic population. Even in auto-oriented Florida, some households are reducing car dependence. Some are breaking completely free.

"I arrived in Miami expecting to buy a car within a few months," Mark R. Brown wrote in *Human Speed*, after seven years of working in Baltimore and Dallas. "But that didn't happen. I got so used to life without car payments, gas payments, and parking fees that I kept putting a car purchase off. Work was accessible by bike, and all of my errands and social engagements were accessible either by Miami's Metrorail or a short car-share ride." He adds, "Cycling around Miami Beach is a joy…. The scenery. The ocean. The palm trees. The unique architecture and eclectic neighborhoods. I've traveled all over the country and no place rewards getting out of a vehicle like Miami."[96]

My colleagues Kenneth and Jennifer Garcia settled here after marrying and set a powerful example. As soon as they could, they sold their car. Kenneth rides his bike to our office, occasionally switching to transit or ride-share on the days when biking won't work well; Jennifer bikes or walks to her office in Coral Gables. They rent a car from time to time to visit relatives upstate. With the money they saved each month from car payments, car insurance, gasoline, parking, and maintenance, they invested instead in residences in one of the nicest, most walkable, mixed-use neighborhoods in Coral Gables—one for them, and another to rent out at a handsome premium.

Many couples the same age took the opposite path, with the opposite math: The high costs of two cars leave those couples with so much less income available for housing, they feel forced to drive far from work to find a home they can afford, and while they are spending a lot of their life driving, Kenneth and Jennifer are walking to Giralda Street to eat at an outdoor café, walking their toddler to swim lessons, and collecting rent.

STREET DESIGN AND THE POWER OF POLITICAL WILL

But if you're looking for Mr. Riley's most distinguishing mark, you can find it in the title of one of his speeches—"The Mayor As Urban Designer." He believes "the lasting mark of a civilization is the city." Americans may have left cities by the millions, he notes, "but we need our cities more than ever."

—*Baltimore Sun*, January 3, 1994

Mayor Joseph P. Riley's work in Charleston exemplifies the principles of traditional Western urbanism and the politics of the Progressive Era (see page 14 in Chapter One). Riley holds that elected officials are the chief urban designers of any city, yet those principles and their obligations also extend to citizens. The former Chair of the architecture school at the University of Notre Dame, Carroll William Westfall, writes about people's responsibility in the traditional city to work within "a civil order" and build cities that express the aspirations of their community.[97] More simply, Westfall writes,

> The city is the most important thing people build. We strive to build the good city to enable us to live the civil life that assists all of us to enjoy our natural rights and pursue our happiness, to use the words of 1776.
>
> At the center of the good city is the commons where people come to know the character of others. This knowledge requires more than the casual, occasional, and superficial encounters that suburbs, and now increasingly cities, offer. A commons can have any number of forms. The most familiar now is a street where people intermingle framed by a variety of residential, commercial, and institutional buildings at the center of a neighborhood or town and where diverse individuals feel at home. Traditional architecture and urbanism have always provided the guidance necessary to build a commons.

After the events of the last seventy-five years that will sound overly idealistic to many, but before the Modernism of the second half of the twentieth century,

Figure 5.26: Columbus Boulevard, Coral Gables, Florida. The most enjoyable streets in the subtropics are shaded by the tree canopies intersecting overhead. *Courtesy of Kenneth Garcia*

those were common beliefs—and that is when we built the cities, towns, and neighborhoods we love. In the nineteenth century, the French National Assembly stated that every citizen of Paris had the right to "justice and beauty."[98] In the twenty-first century, facing large crises, people seem to be going in two directions: more individualistic and concerned about personal freedom, and more attracted once again to experience and the common good. The latter group frequently emphasizes the importance of both happiness and beautiful experiences. In *The Architecture of Happiness*, Alain de Botton explains why he thinks beauty and our sense of the good life are intertwined.[99]

As we saw with the case study on Clematis Street in West Palm Beach, improving street designs requires political will and leadership. The details of projects like Clematis can easily take years or even decades to achieve the necessary consensus among bureaucrats, technicians, property owners, and citizens at large. But West Palm Beach Mayor Jeri Muoio gave clear instructions that the best solutions must be found within ninety days, so the street reconstruction could begin before available funding and yet another construction season slipped away. All parties responded, right on time (see page 392).

The tenure of Philip Stoddard, PhD., mayor of South Miami from 2010 to 2020, reflected what one can accomplish with a mix of political will, doing one's homework, and persistence. One of his wins was one of the first residential solar panel mandates in the country. The SW 64th Street protected bike lanes (see page 594) were built on his watch. Stoddard issued clarion calls about the need for adaptation to sea level rise that resonated across the country; he was named one of "America's greenest mayors"[100] and singled out for recognition by the editors of *Politico*.[101]

He got a lot done, but in his earliest years in office, Dr. Stoddard was stymied on some of his favorite transportation-related initiatives, including fixing parking policy. He encountered the skeptical pushback new ideas tend to attract. Stoddard began to realize this had been a pattern with his predecessors, too. For example, Dover, Kohl & Partners suggested eliminating minimum parking requirements in downtown South Miami as far back as 1992, but the proposal had repeatedly stalled.

Stoddard is a research biologist, a healthy skeptic himself, and an academic; he loves data and the scientific method, plus he has the stomach for reading long, technical reports (especially if they challenge what he previously thought to be true). So he borrowed, and almost inhaled, a copy of the late UCLA professor Donald Shoup's two-inch tome *The High Cost of Free Parking*. Armed with new facts, Stoddard's administration eventually got the parking reforms done, but it took until 2015, twenty-three years after they had first been brought up.

Mayor Stoddard took it upon himself to study theories of change after that, rereading Machiavelli and a stack of business books and research into institutional cultures. Change happens, he learned, in a specific sequence. What Stoddard realized can be hard for designers to remember: "You have to sell the problem first, and *then* the solution," Stoddard says—just as Shoup had.[102] Change also requires perseverance: "They tell you no six times before they say yes when you ask the seventh time."

Street Design in a Divided World

Every change to a street is a political act, and at this moment, we the People are frequently divided. After Houston's newly elected mayor reportedly ordered a halt to "all projects with roadway diet (i.e., narrow lane to 10′), lane reduction, and on-street bike lanes" in 2024, *Governing* magazine's headline read "Street Design Is Increasingly Politicized":

> The shift in Houston's transportation focus
> comes as street redesigns are increasingly
> politicized around the country. They are
> becoming a more frequent site of conflict
> between state and local governments. Republican
> leadership at the state level in Texas has taken
> a particularly strong stance against reducing
> vehicular lanes. In one instance, the Texas
> Department of Transportation stepped in to
> block a long-planned—and voter-approved—
> road diet in San Antonio several years ago.
>
> The state GOP's official platform says:
> "We oppose anti-car measures that punish
> those who choose to travel alone in their own
> personal vehicle, and oppose any measure to
> impose 'road diet' mandates designed to shrink
> auto capacity and/or intentionally clog vehicle
> lanes to force deference to pedestrian, bike and
> mass transit options (whose users do not pay
> gas tax)."
>
> A new transportation law in Florida also
> increases public meeting requirements for
> projects that repurpose automobile lanes, as
> part of an effort to "prevent localities from
> agenda-motivated lane reductions to force
> people out of their cars," according to Florida
> GOP Gov. Ron DeSantis.

ART, ARCHITECTURE, URBANISM, AND THE PUBLIC REALM

In the second half of the twentieth century, Americans radically changed how they looked at and used the public realm. The United States went from a place with communities built around public space connected by mass transportation to a nation where tens of millions of Americans moved from walkable cities and towns to single-family houses isolated in auto-dependent sprawl. They happily moved, but that is not the point here: for better or for worse, the day-to-day lives of many Americans dramatically changed. Their houses gained large, "feature" laden interiors and private backyards, but their connection to public life weakened.[103]

At the same time, the attitudes of many artists and architects towards art in the public realm changed. Consider the important modern sculptor Richard Serra, who lived from 1938 to 2024. Roberta Smith's obituary for Serra in the *New York Times*, combined with Michael Kimmelman's memorial remembrance in the same paper, give a glimpse into how much attitudes changed. Smith, the paper's co-chief art critic, called Serra one of the era's greatest sculptors. "Brilliant, uncompromising and endlessly argumentative," she wrote, he often seemed ready for a fight.[104] "Ready to fight" was how he frequently approached his work in the public realm. Kimmelman, the *Times* architecture critic and former chief art critic, quoted Serra: "I think if work is asked to be accommodating, to be subservient, to be useful to, to be required to, to be subordinated to, then the artist is in trouble."[105]

An obvious illustration of that egocentric philosophy is one of Serra's most famous works, the Tilted Arc sculpture in front of the Jacob K. Javits Federal Building in Lower Manhattan. Workers in the building, including the Chief Judge, thought the sculpture was both ugly and intimidating. Not only did the 120-foot-long, 12-foot-high, 15-ton unfinished plate of rusting Cor-Ten steel obstruct their way in and out of the building every day; many feared it would fall on them. Writing about Jackson Pollock, Serra and Tilted Arc, Kimmelman said,

> All these decades later, a wide swath of the
> public today continues to be baffled and
> occasionally galled by Pollock, just as it didn't
> get Serra for years. "Tilted Arc," the giant steel
> sculpture by Serra, was still a fresh wound. . .

Public officials had removed it from a plaza outside the courthouses in Lower Manhattan in 1989. Fellow artists objected to the removal, but office workers who ate their lunches in the plaza implored City Hall… Serra still wore his fury like a badge of honor.[106]

The contemporary artist and architect James Wines called the sculpture "Plop Art" and "the turd in the plaza" Many others at the time said the public had the right to remove the sculpture.[107] I don't think it is the function of art to be pleasing," Serra said, "Art is not democratic. *It is not for the people* [emphasis added]."[108]

Wines founded an art and architecture practice in SoHo called "Sculpture In The Environment," later known as "SITE." His comment reflected the New York art world of the 1970s, as well as the rise of Postmodernism, which was more contextual and more involved with public taste than the earlier art community Serra lived in. Serra was fervently anti-traditional; Wines liked to play with traditional taste.

Before the Modernism Serra grew up in, painters and sculptors were less "challenging." Some were notoriously egotistical or unconventional, but when they made public art, they worked to create beautiful works and beautiful places. When that changed in the twentieth century, the public frequently pushed back, with the help of public intellectuals like Holly Whyte and Jane Jacobs. New Urbanists (and this book) emphasize that any building on a public street has a public role to play. Unlike a work of art in a museum, gallery, or private collection, the buildings that shape streets and the skyline are part of the public realm and public life.

"As a public object a building carries with it what any person or act carries, which is an obligation to serve a public interest and contribute to the common good or, at a minimum, not to disrupt them," Carroll William Westfall says. "This obligation extends to whoever hires the architect, whether a private person, a corporation, or a public body. What builders and architects do is 'touched by a public interest,' as common law puts it, no less than the errand run by a young person, the work of captains of industry and finance, and the activities of public officials…."[109]

Many architects today continue to disagree, however. After a brief period of teaching Postmodern and traditional urbanism, many of the most prestigious architecture schools in the United States returned to embracing egocentric art school attitudes, as seen in Serra's work. An interview conducted in 2011 by *New York* magazine's architecture critic Justin Davidson with five architects affiliated with Columbia University's Graduate School of Architecture, Planning and Preservation (GSAPP) illustrates this:

Bernard Tschumi (former Dean of the GSAPP): "Every one of those buildings [we are discussing] is a bad "citizen"—in a good way. Before 2000, everything was about being contextual, and buildings were supposed to be good citizens. And when somebody from out of town asked me what new architecture to see, I had a hard time giving them an answer. Now I can tell them about all these exciting new buildings that break the pattern and don't play the typical New York game of the podium with the tower on top. So suddenly we have buildings that no developer in their right mind would build—but they did."

Justin Davidson: "One example, Bernard, might be your Blue condo, a glass tower in varied shades of, yes, blue that looms over the brick tenements of the Lower East Side. Does violating New York traditions make a building less New Yorky?"

Tschumi: "No. New York can take it."

Justin Davidson: "Has any building gone beyond what New York can tolerate?"

Winka Dubbeldam (former GSAPP faculty, principal at Archi-Tectonics): "I wish!"

Robert A.M. Stern (former GSAPP faculty, then Dean of Yale): "Well, the buildings that entertain Bernard's friends, who jet in from wherever, don't really make any contribution except as big art objects. The city can take them, but what are they telling us? They don't offer any new insights about how people live, or about the relationship to the street or to the sky. Just a new curtain wall, and a strange one at that. To be a good citizen is to work with the city and not against it."

Gregg Pasquarelli (GSAPP faculty, partner at SHoP Architects): "I disagree. Like other kinds of art, great buildings contradict everything else. They make us think. They start conversations, so people talk about what it means to fit in, what it means to have courage. It's okay for some buildings not to work."

Tschumi: "Maybe that's what a city is: confrontation and complication. In New York, the name of the game is to have one's own envelope. When you arrive from the airport and you look at the skyline, you see this incredible variety—a symphony of envelopes."[110]

Few New Yorkers would endorse the idea that what they need is more confrontation and complication. Living in New York can feel like drinking coffee with an extra shot of espresso all day. Stress levels can be high, and when you go out on the street, sirens are blaring and drivers are loudly honking at pedestrians. Going into Central Park for relief, you find yourself in the shadow of Pasquarelli's 1,428-foot-tall tower on Billionaires' Row.[111] Pasquarelli also designed the tallest building in Brooklyn, a supertall residential tower commonly known as "the Batman Building" and frequently referred to as "evil." "But perhaps the most popular comparison is to the Tower of Sauron in the evil land of Mordor from 'The Lord of the Rings,'" a story in the Real Estate section of the *New York Post* said.[112]

The Batman Building is a tower people love to hate, and in 2024, the developer defaulted on the mortgage.[113] But many in and around the world of New York architecture promote architecture and urbanism like the Batman Building. A prominent architecture critic in New York wrote on Bloomberg's City Lab, "The Batman building, the Tower of Sauron—the nicknames write themselves. And why not? Better this than another squared-off tower that simply fiddles with the ratio of white solid to blue glass. Maybe I should hate it for its bigness, its blackness, its thrust—but I don't."[114] And the former Dean of the Cooper Union School of Architecture says the design is "impeccable," calling the way it is "redefining" the Brooklyn skyline "unprecedented."[115]

"Unprecedented" is high praise in the world of high-fashion architecture. Professors of architecture at Cooper Union and Columbia University are not interested in aspiring to the traditional "good city" described above by Bill Westfall. They want to be artists, doing unprecedented work. The result was described by Rem Koolhaas when he wrote about his own firm's work: "The work we do is no longer mutually reinforcing, but I would say that any accumulation is counterproductive, to the point that each new addition reduces the sum's value."[116]

THE EXPERIENCE OF PLACE

In the first chapter, we wrote that everyone who walks into the Piazza San Marco in Venice instinctively knows it is beautiful. That is central to one of our main themes. *Street Design: The Secret to Great Cities and Towns* looks at design theory and includes histories of cities and streets, but a primary concern of the book is understanding how we humans experience place, so that we can translate understanding of that experience into making and preserving places where people want to be.

We are all affected by our surroundings, regardless of how aware we are of the experience. The documentary movie *Motherload*—an excellent, wide-ranging film about cycling, cultural change, and climate change—shows the joy people can experience when they are out walking or exercising in nature:

> When we listen to our bodies, it's clear, we're hardwired to move. And to connect, not only with nature, but with each other. These instincts have served us and the planet well, for millennia. Let's hold on tight to them. Let's show the world, and our kids, what it means to be fully, truly, human.

The film's description of listening to our bodies accompanies a scene that takes place on a beautiful hilltop in Marin County, California. That is obviously a different experience than walking in a city or town. In those situations, we watch other people. We walk by interesting buildings and pass through invigorating spaces. Stores catch our attention. We enjoy the street and public life. But a scene in *Motherload* effectively shows how different that experience is when we are sitting in a car.*

Street Whisperers

Good urban designers are street whisperers, with an awareness of how a place makes them feel and a talent for

---

*The scene in *Motherload* effectively demonstrates how cars cut us off from our surroundings. When the scene opens, we see a therapist walking along the street, directly addressing the camera and the audience. While he's talking, he climbs into a car and closes the door, leaving the audience on the street. He continues talking, but we can no longer hear him. When the camera moves inside the car, he discusses how much sensory input from the street he has lost. Cars are private spaces that insulate us from the world. It is not surprising that Road Rage against others can be one of the results. Considering only the view *from* the car, we see that cars are bad for touring the city. The roof hides the sky and the tops of buildings, and the car doors and the hood block the view too. And the faster we go, the less we see (page 53). Car rides are most enjoyable in wide-open landscapes, where we see everything at a distance, including the sky.

translating that understanding into design. Their understanding may be more intuitive than conscious because the design process frequently works that way. Design connects unconscious creativity with problem-solving. Some of the factors that need to be incorporated in the design may be mundane, but good design can bring solutions that are simultaneously pragmatic and transcendent. On the other hand, theory, ideology, the intellect, the ego, and even fear can all get in the way of good design.

Creating good places often requires that designers reject and overcome the philosophy and principles they were taught in school, whether that school was an architecture program at a university, an urban design program, an engineering school, or a business school. Conversely, some of the qualities of street design we've looked at here might sound like intellectual concepts, but they are not. For example, why should we believe that streets proportioned according to the rules of harmonic proportion feel the best? The answer has multiple parts: because they do, because people skilled in placemaking and urban design have understood this for thousands of years, and because new scientific studies are beginning to reinforce what centuries of wisdom and human experience have already told us.

When trying to put this into words, metaphors and similes are popular. "Music is liquid architecture, and architecture is frozen music," Johann Wolfgang von Goethe famously said in the eighteenth century, writing that "the influence that flows upon us from architecture is like that from music."[117] In the twenty-first century, former Beatle Paul McCartney says music is "a bunch of frequencies" that "affect us:"[118]

> Even when they're organized into a musical composition, there's no reason really why they should affect us. But they do. And you can hear sometimes a piece of Classical music, there's no lyrics, so there's no reason for it to touch your heartstrings, but it does. It's just like, oh my god, I'm going to cry. How? I don't know. It's magical.[119]

"It's magical," or as Goethe said more formally, "Beauty is a manifestation of secret natural laws, which otherwise would have been hidden from us forever."[120] That sounds abstract and perhaps unconvincing to modern ears, but Aristotle, Vitruvius, Leonardo da Vinci, Leone Battista Alberti, William Shakespeare, Edmund Burke, John Keats, Ralph Waldo Emerson, Walt Whitman, William Morris, Augustus Saint-Gaudens, Charles McKim, Louis Sullivan, Frank Lloyd Wright, Christopher Alexander, and many other artists, architects, writers, philosophers, and poets all said the same thing in different ways. "Beauty is the promise of happiness," Stendhal wrote (and Edmund Burke may have too).[121]

> In the twenty-first century, many architects and urbanists want to reclaim beauty and public space. Reflecting that, they talk about places and buildings people love. To that, we add the idea of making things *with* love.

In the visual arts, this phenomenon was frequently understood as bringing a visible manifestation of a divine or unseen supernatural world into the material world. John Muir,* who emphasized the role our senses play in perceiving the natural world, also kept journals recording "the sublimity of Nature" in "an aesthetic and spiritual notebook."[122] Today, the young are more likely to talk about "vibes" and "being in the moment." Some say "everything is energy," which is not far removed from what some quantum physicists are saying about energy and reality.[123] "The pursuit of truth and beauty is a sphere of activity in which we are permitted to remain children all our lives," Albert Einstein said.[124]

Consider the experience of Jill Bolte Taylor, the Harvard brain scientist whose stroke became the basis for the first "viral" TED talk and a book called *Stroke of Insight* that was on the *New York Times* bestseller list for sixty-three weeks.[125] "Our right human hemisphere is all about this present moment," Taylor says in her TED Talk:

---

*Muir, of course, was an environmentalist, co-founder of the Sierra Club, and "Father of the National Parks."

◄ **Figure 5.27:** Broadway, New York, New York. Looking south towards the Financial District in the early spring. The broad sidewalk benefits from lean-in trees in City Hall Park bring constant change to Broadway. Below the park, the relatively narrow street lined by the tall buildings of the Financial District is known as the Canyon of Heroes. On special occasions, Presidents, astronauts, championship sports teams, and others honored by the public ride in open vehicles from Battery Park to the steps of City Hall while the city gives them a tickertape parade.

It's all about 'right here, right now.' Our right hemisphere, it thinks in pictures and it learns kinesthetically through the movement of our bodies. Information, in the form of energy, streams in simultaneously through all of our sensory systems and then it explodes into this enormous collage of what this present moment looks like, what this present moment smells like and tastes like, what it feels like and what it sounds like. I am an energy being connected to the energy all around me through the consciousness of my right hemisphere. We are energy beings connected to one another through the consciousness of our right hemispheres as one human family. And right here, right now, we are brothers and sisters on this planet, here to make the world a better place. And in this moment, we are perfect, we are whole, and we are beautiful.[126]

Taylor experienced all that after losing the connection to the left side of her brain, which is linear and logical. Her brain forced her to be in the present, so she stopped worrying about the past and the future. "In the moment," her surroundings flooded her senses with information about the world around her.[127] That sounds New Agey, but remember that neuroscience is moving in a similar direction. We know now that it is not just our brain and our eyes that give us information about the world around us. It's our bodies, our gut, and our hearts too. And of course, we have said that throughout history. "What does your gut say?" "Trust your heart." "That resonated with me."

Neuroscientists like Sergei Gepshtein, the founding director of the Collaboratory for Adaptive Sensory Technologies at the Salk Institute, partially explain the experience of place by talking about "embodied cognition."[128] Scientists now broadly accept embodied cognition as a concept, but the science still needs testing and verification. "We have more than five senses, but fewer than one hundred," Gepshtein says,[129] and precise measurement of how our senses react to specific stimuli is still in the early stages.

Not only is it difficult to isolate the many different factors in a real-world test, but it is also difficult to find funding for the complicated, time-consuming tests. Neuroscience is not high on the priority lists of Federal funding sources or Big Pharma, and there are questions of what criteria to use in the tests. Gepshtein is a Board member of the Academy of Neuroscience for Architecture (ANFA), the leading research body in the field.[130] The architect and author Mark Hewitt is on ANFA's Board of Advisors and the author of *Draw in Order to See*, one of the best books on the relationship between embodied cognition and architectural design.[131] Gepshtein, Hewitt, and John Massengale were the organizers of the small conference in Sweden "Neuroscience & Measuring the Experience of Place," where they proposed testing qualities drawn from traditional architecture and urbanism and from Modernist design.[132] Attempts to find adequate funding were unsuccessful, but it is worth mentioning some of the most promising research.

One approach is to connect individuals to mobile EEG equipment that measures electrical activity in different parts of the brain.[133] That can be combined with electrodes monitoring skin conductance, which rises when the wearer is aroused. Another approach is to film the subject, focusing on body language and facial expression, updating Holly Whyte's Street Life Project, published in the landmark book *The Social Life of Small Urban Spaces*. Artificial Intelligence can tirelessly watch many hours of the films from public spaces, and AI is learning to analyze body language and human expressions.[134]

Ann Sussman's and Justin Hollander's work with eye-tracking shows that when there is a lack of richness, or a lack of order, the eye does not know where to look and the viewer loses interest.[135] That is not surprising: in the words of the architect, educator, and author Robert A.M. Stern, "When the eye is bored, the mind is bored." Sussman and Hollander also provide some evidence for the idea that traditional architectural design is frequently anthropomorphic.[136]

In Chapter One and the case studies in subsequent chapters, we discussed qualities and characteristics that make streets good places for walking and congregating. Neuroscience is beginning to corroborate conventional wisdom about the elements that influence people's experience, but there is still more conjecture and theory than reproducible science. From the ratio of the width of the street compared to the height of the buildings enclosing the space, to the Principle of Sublimation, to our reaction to materials and colors, we are surrounded by influences that affect us without our conscious attention. Many we integrate into our experience almost by osmosis, certainly without consciously thinking about them most of the time. We can categorize these from personal observation and from the recorded observations of designers

MATERIALS MATTER

Materials matter: many Modernists who dislike traditional design and details still like traditional materials like cobblestone paving and iron fences (see Figures 4.148 and 4.150). Chain link fences and galvanized metal railings are less pleasing to the senses than iron railings. In Christopher Alexander's terms, they have less "life."[137] Wood never really dies, and wooden clapboards have more life than aluminum or vinyl siding.* A wooden floor with a natural finish has more life than a wood floor sealed with polyurethane (polyurethane is plastic).[138] Plaster is more pleasing than sheetrock: horsehair plaster and wood paneling are one reason why genuine Colonial houses feel better than New Colonial houses in subdivisions. Good Modernist architects know that dead materials like sheetrock can be made more "lively" by angling and tilting walls. Pay attention to these things when you see them: they might make you happy.

---

*Wood also gets a pleasing patina as it ages. Aluminum and vinyl siding become grungy (they also hide any water damage and rot that might be behind the siding, but that is a different story.) Luxury car makers know how to make plastic that is pleasing to the senses, but building materials manufacturers have not learned this lesson.

from Classical times until now, in books such as *The Ten Books on Architecture* by Vitruvius and Frank Ching's *Architecture: Form, Space, and Order*.[139]

Lacking scientific proof that a traditional design principle like the Rule of Threes is "hard-wired" in humans,[140] or why the fractal designs generated by Classical architecture (the most generative architectural language we have) are attuned to human nature, we look to descriptions that ring true, like Tony Hiss's explanation of "simultaneous perception" in *The Experience of Place*.[141] Today, many then look inside themselves, like Jill Bolte Taylor.

Beauty and Love

In the second half of the twentieth century, it was common for designers to proclaim that beauty is in the eye of the beholder. The saying dates from the late nineteenth century, but it was adopted by Modernist artists and architects who wanted to break with traditional conventions and establish new, more personal standards for art and architecture. Dissatisfaction and discomfort became virtues, replacing harmony and beauty (pages 603 to 605). Master chefs talk about love being an essential ingredient in cooking, and the best New Urban places are communities where developers and designers combined the creation of places they would like to live in themselves with the making of places others will love too. The first New Urbanists were students in architecture school who said to their teachers, "Look, this ordinary city street is better than the superblocks you're showing us." The first New Urban development was Seaside, consciously created by Daryl and Robert Davis for their children, grandchildren, and generations of families they had not met yet.

Seaside, Vince Graham's I'On, and South Main in Buena Vista, Colorado (built by Jed Selby, Kennley Selby, and Katie Urban) are the opposite of the sprawling subdivisions in movies like *E.T. the Extra-Terrestrial* and *Poltergeist*. Production builders built those non-places, subdivisions that look like no love went into their design and construction of the "spreadsheet" architecture and planning—which is usually correct. Design fees are minimal, construction costs are kept low, and the owners of the large production building companies rarely live in their own projects, usually preferring architect-designed houses in expensive suburbs.[142]

Seaside, I'On, and South Main are places built with love that are now loved. "In order to be authentically sustainable, buildings and places have to be cared for and loved over generations," Chris Alexander wrote, "Beautiful buildings and places are more likely to be loved, and they become more beautiful, and loved, through the attention given to them over time. Beauty is therefore, not a luxury, or an option, it includes and transcends technological innovation, and is a necessary requirement for a truly sustainable culture."[143]

This takes time. Rome wasn't built in a day. Paraphrasing G.K. Chesterton, we don't love Rome because Rome is beautiful and great. Rome is beautiful and great because Rome was loved.[144]

SLOW NEW YORK / JOHN MASSENGALE

To some degree, this book began one quiet Sunday morning when I was walking downtown in Manhattan. I came upon a street the New York City Department of Transportation had "improved" with paint, boulders, and plastic sticks, and I intuitively thought something along the lines of, "This is ugly. New York deserves better than this." Calling my friend and colleague Victor Dover, whose firm Dover, Kohl & Partners was a New Urban planning firm involved in making better streets, I told Victor what I was seeing and texted him some photos. At the annual CNU congress a month or two later, we decided to write a book together.

In a bigger context, I called Victor because we had both been working on making better streets for decades.

We knew each other through the CNU, founded in 1993 by architects, developers, traffic engineers, and urban activists who came together with a desire to make better cities, towns, and neighborhoods. As we say in the first line of this book, the design of cities, towns, and neighborhoods begins with the design of streets, and that was how the design process always began on charrettes. Making better places frequently meant rejecting what planning norms, our education, and our professions told us was best practice, and New Urbanists collectively shared the lessons they learned individually. Victor and I met at the I'On charrette conducted by Dover, Kohl & Partners and Duany, Plater-Zyberk & Company for the developer Vince Graham (page 502). This book is a direct result of what we, John and Victor, learned over the years on charrettes and at CNU congresses from our New Urban colleagues, including the traffic engineers like Walter Kulash and Rick Hall and the developers like

Figure 5.28: Car-Free Brooklyn Bridge, New York, New York. Massengale & Co LLC, 2019. Watercolor. After: Aerial view from the Manhattan side of the bridge, which lands by City Hall and the New York Municipal Building. See "Financial District Slow Zone" on page 612. Showing Massengale & Co's proposal for a car-free Brooklyn Bridge and a new City Hall Plaza. Earlier versions of the plan can be seen in the *Make Way for Lower Manhattan* booklet (which can be downloaded from a link in note 154). © *2019 Massengale & Co LLC, rendering by Gabriele Stroik Johnson*

Graham and Robert Davis (pages 338 and 498). *Street Design* contains countless shared lessons from hundreds of colleagues.[145]

While writing *Street Design*, Victor, Dover-Kohl's Director of Design James Dougherty, and I drew up some quick entries for a competition called "By the City/For the City: Making a Better New York," organized by the Institute for Urban Design.[146] We published them in the first edition of *Street Design*, and I later developed one of the ideas, Jane Jacobs Square, a little further.[147]

That was before New York City adopted Vision Zero policies in early 2014 (in the same month *Street Design* came out).[148] Since that time, New York has adopted over twenty-five Slow Zones, with Slow Streets, and instituted a citywide, twenty-five-mile-per-hour speed limit.[149] However, in the United States, "twenty-five-miles-per-hour" means, "You will never get ticketed if you don't

Rome wasn't built in a day. Paraphrasing G.K. Chesterton, we don't love Rome because Rome is beautiful and great. Rome is beautiful and great because Rome was loved.

go faster than thirty-five miles per hour," and a lot of the old guard at the NYC DOT don't like the new limits.[150] And it is not hard to avoid getting a ticket in New York City.[151]

The Slow Zones do not make drivers go slowly. Not surprisingly, the retrofitted streets in the Slow Zones could be used to illustrate the Monderman Rule: "The trouble with traffic engineers is that when there's a problem with a road, they always try to add something" (page 513). When creating a Slow Zone, the NYC DOT adds paint, boulders, plastic sticks, warning signs, speed bumps, bumpouts, highway-scale graphics, and the like to *tell* drivers to slow down, rather than making streets

Figure 5.29: City Hall Park, New York, New York. Before: An aerial view of the current conditions of the Brooklyn Bridge landing at City Hall Park in Manhattan: the Before view for Figure 5.28. City Hall is at the center of the photo. The New York Municipal Building is the tower across Centre Street from City Hall and the Tweed Courthouse on the left. City Hall Park. Robert Moses built the ramps to and from Brooklyn Bridge at the top of the photo. *Courtesy of Google Earth*

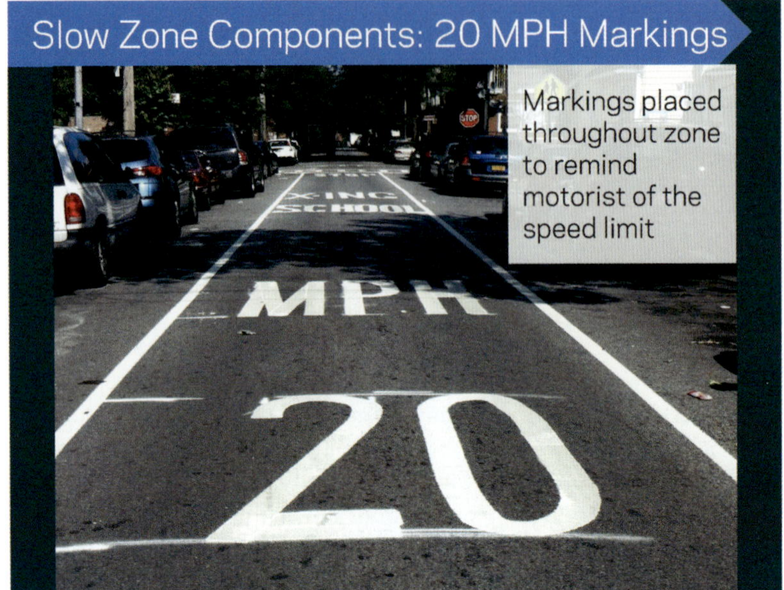

Figure 5.30: Slow Street Example, New York, New York. NYC DOT, 2014. Poster child illustration of the Monderman Rule (page 513). The graphic is large enough to be read by a driver going 75 mph. The reductions do not reduce deaths and injuries to zero, as promised by the city's Vision Zero pledge. Nor do they make streets more pleasant for walking. *Courtesy of NYC DOT, image from public records, U.S. Department of Transportation*

where people will drive slowly. The roadbed and traffic lanes are as wide as ever, now with big, bold, painted graphics (Figure 5.30). Streets like those will never lead to zero traffic deaths, and from the point of view of the pedestrian over on the sidewalk, little has changed, except that the machine space in the middle of the road is often uglier than before. New York City streets used to have less paint and fewer signs than they have now. Standard sub-urban elements like turn lanes didn't arrive until the twenty-first century.

As a New Yorker and an urbanist, I think the residents of New York deserve more. I don't want to sound like a broken record, but New York is the American city with the fewest cars per person, the highest Walk Score, and the best public transportation. The city has a large number of groups and organizations advocating for alternatives to private cars and has more bicycle riders and bike lanes than any other American city.

New good streets movements like "Move NY" and "Charleston Moves" (for Charleston, South Carolina) are catchy titles, but people like Vince Graham and I think

> Streets like those will never lead to zero traffic deaths, and from the point of view of the pedestrian over on the sidewalk, little has changed, except that the machine space in the middle of the road is often uglier than before.

we need to shift the emphasis to new approaches like "Slow New York" and "Still Charleston." The Move NY plan would reduce traffic, and Charleston Moves is a great organization that works for better streets, reduced traffic, and transportation alternatives to cars.[152] We just want to shift the message and the mindset around city streets from Go to Slow.

Graham lives without a car in downtown Charleston. He likes "Still Charleston" because it can mean both, "Be still Charleston," and "We're still Charleston—so why do we have so many cars?" I like "Slow New York" because the name says we should slow New York traffic to make a great Slow City, with Slow Streets. Of course, it also references La Cittaslow Internazionale, the Italian movement for "bringing balance to cities and improving lifestyles."[153]

Financial District Slow Zone

My biggest Slow New York project began when the DOT Commissioner suggested to the Financial District Neighborhood Association (FDNA) Board that I might be a good team member for their project Make Way for Lower Manhattan (luckily for me).[154] FDNA was working with Kate Ascher's Cities team at Buro Happold. The Financial District is New York's oldest and most historic place. Laid out by the Dutch in the seventeenth century, Nieuw Amsterdam

MENTALLY MAPPING THE STREETS / JOHN MASSENGALE

When I moved my office to Fulton Street, an old street in the northern end of the Financial District, I began regularly walking streets I had only visited a few times in the decades I lived in and around New York City. Slowly but surely, I gained a new understanding of how it all came together as a place. The design of the cities begins with the design of streets—understanding a city requires knowing how it's streets connect.

The office was at the corner of Fulton and Nassau streets, two of the oldest streets in the city. They cross at one of the highest points in Lower Manhattan.[155] From there, Nassau Street goes downhill to the south to the intersection of Wall Street and Broad Street. South of Wall Street are man-made "canyons," narrow, seventeenth-century streets now filled with towering skyscrapers from the American Renaissance. The canyons focus our eyes on "the city at eye level": all the old buildings have rich, human-scaled first stories.[156] The wider streets, too, have beautiful bases, like the long curving wall of the Standard Oil Building along Broadway as it opens to the colonial-period Bowling Green. The rhythm of the tall arched windows on the *piano nobile* engages us as we walk along the unfolding wall. The fifteen-story podium shapes the street, while the Mausoleum of Helicarnassus at the top of the tower pierces the skyline. We simultaneously have our feet on the ground and our minds in the sky.

From the canyons, we glimpse extraordinary structures high above. McKim, Mead & White's New York Municipal building has a fantasy based on the Monument of Lysicrates (Figures 1.17 and 5.44). Like the Mausoleum of Helicarnassus on top of the Standard Oil Building (Figure 5.31), they capture our attention and our imagination. The Financial District is not a place to go if the sight of capitalism and its monuments upset you, but it is home to great buildings and streets. When you know it well enough to picture it as a three-dimensional whole, you understand what a spectacular place it is, from the street to the sky.

For more than a hundred years, it was the entrance to America that greeted the tens of millions who arrived in the great harbor by ship (as seen in Colin Campbell Cooper, Figure 5.34). For those who built America's new financial center, it was an announcement that the United States would no longer be the provinces, but the rival of Europe. Complemented by the Statue of Liberty, next to Ellis Island, they announced, "Welcome to the New World."

Figure 5.31: Standard Oil Building, 26 Broadway, New York, New York. Carrère & Hastings, 1928. The *New York Times* wrote that 26 Broadway was "to oil what 1600 Pennsylvania Avenue is to politics."[157] A late addition to downtown. The Mausoleum of Helicarnassus on the top of the tower was an important addition to the skyline. *Courtesy of New York Public Library*

"New York Harbour is loveliest at night perhaps," Rupert Brooke wrote,[158]

> On the Staten Island ferry boat you slip out from the darkness right under the immense sky-scrapers. As they recede they form into a mass together, heaping up one behind another, fire-lined and majestic, sentinel over the black, gold-streaked waters. Their cliff-like boldness is the greater, because to either side sweep in the East River and the Hudson River, leaving this piled promontory between. To the right hangs the great stretch of the Brooklyn Suspension Bridge, its slight curve very purely outlined with light; over it luminous trams, like shuttles of fire, are thrown across and across, continually weaving the stuff of human existence. From further off all these lights dwindle to a radiant semicircle that gazes out over the expanse with a quiet, mysterious expectancy.

Walking around the "canyons" of Lower Manhattan, we look up at a terrace and see ourselves there. We remember trips on the ferry and imagine the whole.

Figure 5.32: Nassau Street Shared Space, New York, New York. Massengale & Co, 2017. Watercolor. Looking north towards the New York Municipal Building and the new civic space at the entrance to the Brooklyn Bridge. © 2019 Massengale & Co LLC, rendering by Gabriele Stroik Johnson

has old, European-style streets (so when some New Yorkers complain about bicycles and say in response to the proposed plan for the district "We're not Amsterdam"… they are forgetting their history).

The district was America's first national capital, presided over by George Washington. In the nineteenth and twentieth centuries, it was the financial capital, with the world's most famous skyline. When world travel was by boat, that skyline was important; tens of millions of future Americans first set foot on American soil after sailing into the great harbor and seeing the famous sight. The skyline view from the harbor supplied what Henry James called in *The American Scene* (1907) "the happily-excited and amused view of the great face of New York," adding, "Standing up to the view, from the water, like extravagant pins in a cushion already overplanted," [the buildings] seem "stuck in as in the dark, anywhere and anyhow."

In the twenty-first century, many investment houses, banks, and corporations moved to midtown Manhattan. Developers converted old office buildings to apartments, and FiDi (a new name, replacing "Wall Street") gained 75,000 residents. At the same time, the old World Trade Center site and the South Street Seaport (on opposite sides of FiDi) became large tourist attractions. In the

words of the *Make Way for Lower Manhattan* report, "A 21st century vision for Lower Manhattan would change the way that cars and people interact on city streets, making the lives of pedestrians safer and more pleasant in a "slow-street" district where the needs of residents, businesses, and tourists can all be met."[159]

Shortly after I joined the FDNA team, the local residents made some decisions about the streets. In a nutshell, they wanted,

- Slow, shared-space streets
- Curb space given to garbage collection, delivery services, and parklets
- A reduced number of on-street parking spaces, with no on-street parking over 30 minutes

Looking at the big picture, it was clear that the shared-space streets would work best between Broadway and Water Street, and south of Chambers Street and City Hall Park all the way to the Battery (see Figures 5.14 and 5.28). Many of the blocks already feel like shared space. A few of the streets, like the long block of Broad Street below Wall Street and a few blocks of Wall Street, are closed to traffic. Quite a few of the streets south of

Fulton Street are narrow and crowded with double-parked cars and delivery vehicles. Many blocks have scaffolding covering the building and obstructing the sidewalks, and the influx of new residents has caused garbage-collection problems: commercial buildings and residential buildings don't share garbage collection in New York City, with the result that the narrow sidewalks can be full of large piles of garbage bags.[160] That leaves little room for all the office workers, tourists, and residents—some pushing baby carriages and shopping carts—so many people walk in the street. As a result, there are only a handful of streets in the area where drivers go as fast as twenty miles per hour.

To acquaint myself with the streets, I walked every block with a map and colored pens in hand, marking places that were comfortable for walking, and places that were uncomfortable (like the block that had a parking garage entrance open to the street and seventy-five feet wide). Many sections of the blocks were neutral, and got no marks. The final map suggested a network of streets best for walking, as well a variety of blocks that might be best for things like service streets for the volume of trucks a dense neighborhood like FiDi needs.

Walking the blocks, I noticed a problem: the community wanted one big walkable neighborhood, but City Hall Park and City Hall (and the multitude of subway stops there) are cut off from FiDi by the cars constantly going to and from the Brooklyn Bridge (Figure 5.29). At my office on the opposite corner of City Hall Park (The Broadway Chambers at Broadway and Chambers Street, the first New York building designed by Cass Gilbert), we listen all day to sirens from emergency vehicles fourteen floors below that cannot get through the congestion caused by impatient drivers from Brooklyn and Long Island (congestion pricing has slightly reduced that problem—but compromises in the plan mean there are still too many cars around City Hall). I can see the bridge from my desk, and I realized one day that the on-ramps and off-ramps Robert Moses built in the 1950s could be used to keep all the traffic away from City Hall Park (see Figure 5.28). And that led to more ideas. The street between City Hall and McKim, Mead & White's beautiful New York Municipal Building could become a civic space extending City Hall Park.

Underneath the Municipal Building are four subway lines that come up in a wonderful vaulted space made with Guastavino tiles.[161] Going anywhere to the west or south now requires crossing the no-man's-land created by the traffic coming off the bridge fewer than forty feet away. But if that traffic goes away, there is an immediate connection to Nassau Street, one of the oldest streets in New York City. Nassau Street goes south to the intersection of Broad Street and Wall Street, coming out at Federal Hall. It could make the northern half of a tourist trail to the old city green and the harbor, perhaps called the Nassau Trail or the Knickerbocker Trail (Figure 5.32). The Lenape tribe was in Lower Manhattan before the Dutch, but their trail was Broadway, on the other side of City Hall Park. Their name for what the Dutch named *Brede weg* (Broad way) was *Wickquasgeck*. (Manhattan was originally *Manahatta*.)

Figure 5.33: Broad Street, New York, New York. Lithograph printed in color. View looking north in 1797. George Washington's Inauguration as the first President of the United States was on the balcony of the Federal Hall on Wall Street. Today, the space in front of Federal Hall is one of the great public spaces in America. See Figure 1.1, which shows the new Federal Hall at the top of Broad Street circa 1905. *Courtesy of New York Public Library*

Today, the space in front of Federal Hall is one of the great public spaces in America. See Figure 1.1.

The FDNA plan was popular, gaining support from the local Councilwoman, the Community Board (several FDNA Board members were or are on the Community Board), local and citywide media, local residents at visioning meetings, and Mayor Bill de Blasio's Sustainability Office, which gave the DOT $500,000 for a study building on the FDNA plan. It was the latest in five or six downtown plans calling for pedestrianization, going back to Mayor John V. Lindsay's administration in the 1960s (that one was supported by David Rockefeller, the unofficial mayor of downtown at Chase Bank, and his brother Nelson, the actual governor). In 2008, the Bloomberg administration released a multi-year study called "A Street Management Framework for Lower Manhattan: The Downtown of the 21st Century."[162] Anyone reading the various plans has to be struck by how similar they are. They all propose fewer cars, and they all envision similar pedestrian networks downtown, beginning with the north–south axis of Nassau and Broad streets. None of them have happened (although Nassau Street was closed to traffic for months). Political machinations behind the scenes killed the Make Way plan: at one point the DOT contracted with an engineering firm for another Nassau Street study, but most of the $500,000 is still on the books at the agency. Even Covid was not enough to kickstart any action.[163]

I love New York. I cannot imagine a better American city for a Slow Zone with slow, shared-space streets, nor a better place to try that in New York than in the Financial District (New Amsterdam). The residents want it, their political leaders support it, their Community Board supports it, the U.S. Senate Majority Leader says he wants New York to be more like Amsterdam, and that he will get "an effing amount of money" for it, and yet here we are, stuck with the status quo. Congestion pricing will be a step forward, but it took us seventeen years to get it. You can read more about Make Way for Lower Manhattan at fidinewyork.org.[164]

Car-Free Brooklyn Bridge & the Brooklyn-Queens Expressway

In the 1950s, Robert Moses demolished buildings, blocks, and streets at both ends of the Brooklyn Bridge to make way for ramps to and from the bridge. On the Manhattan side, he created the ramps seen in Figure 5.37. At the opposite end, he cleared land in one of the oldest parts of Brooklyn to connect the Brooklyn Bridge and the nearby Manhattan Bridge to his new Brooklyn-Queens Expressway (also known as the Brooklyn-Queens Expressway or BQE, and now part of Interstate 278). Riding a bike one day in what feels like a Demilitarized Zone between the two bridges, I was struck by three things (and almost by a car).

First, the ride was dangerous—I am lucky I *wasn't* struck by a car. The bike lane weaves from one side of the road to the other, sometimes going in the middle of the street between impatient drivers rushing between

Figure 5.34: Colin Campbell Cooper, "Hudson River Waterfront," circa 1917, New-York Historical Society. Oil on canvas. For more than a century, millions of Americans, immigrants, and tourists arrived in New York City by ship, greeted by one of the most distinctive and beautiful skylines in the world. When Cooper painted this view from the harbor, the Woolworth Building (792 feet) and the Singer Tower (612 feet) were the two tallest buildings in the world. The economic boom of the Roaring Twenties saw the addition of several more iconic towers to the Lower Manhattan skyline. *Courtesy of New York Historical Society*

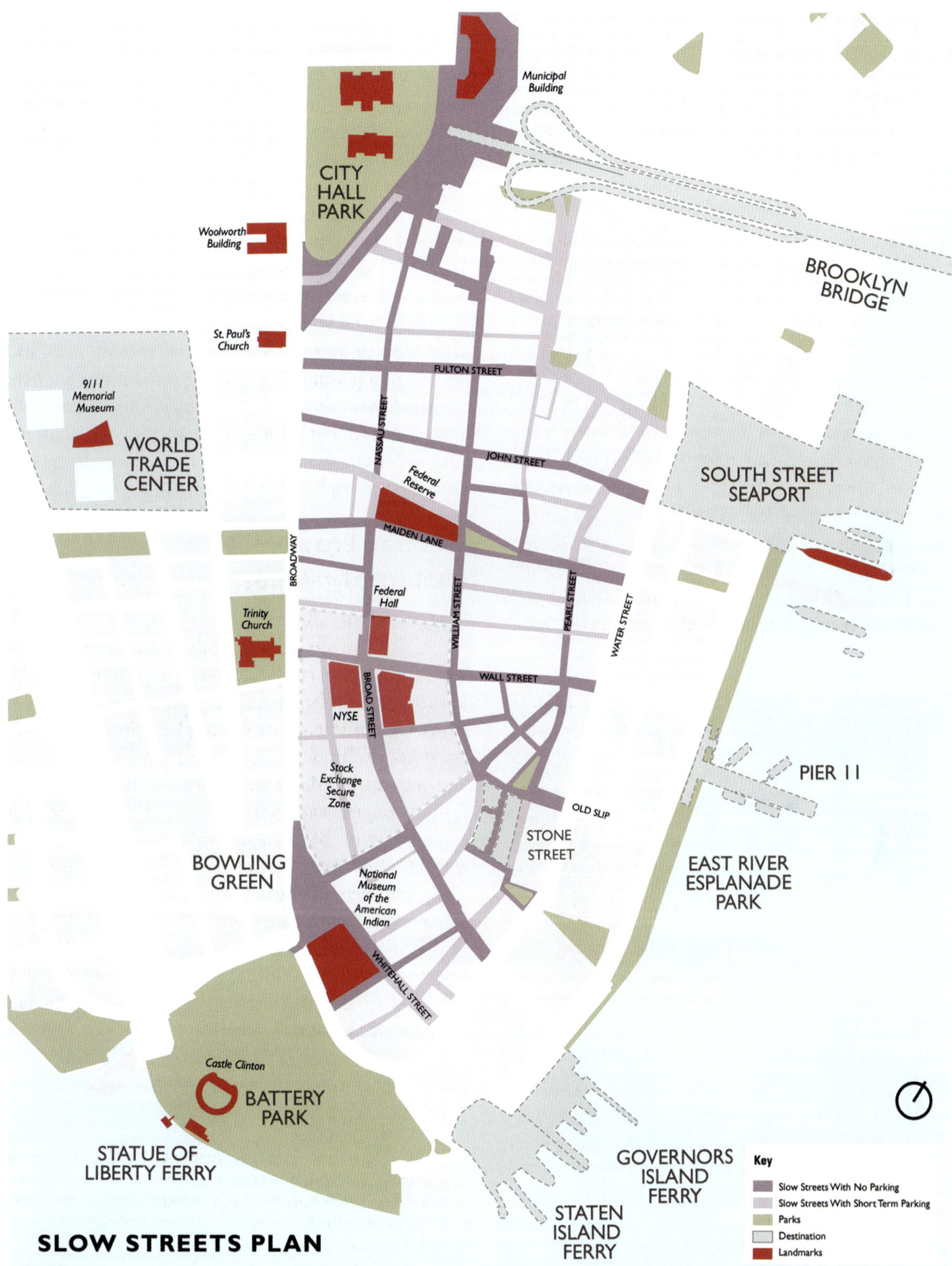

Figure 5.35: Slow Streets Plan, Lower Manhattan, New York, FDNA, 2019. Make Way for Lower Manhattan Plan Diagram. The Slow Streets networks. All the streets between Broadway and Water Street, from Chambers Street south to Battery Park, would be shared-space streets. *Courtesy of Buro Happold*

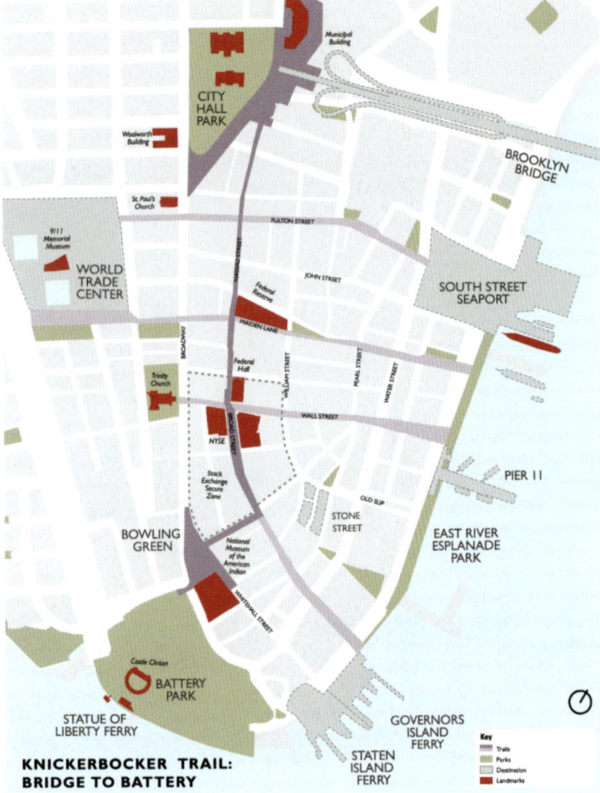

Figure 5.36: Tourist Trail Plan, Lower Manhattan, New York, FDNA, 2019. Make Way for Lower Manhattan Plan Diagram. Nassau Street and Broad Street become the north–south backbone of the plan, while three east–west routes cross the district. All the streets between Broadway and Water Street, from Chambers Street south to Battery Park, would be shared-space streets. *Courtesy of Buro Happold*

the bridges and the highway. Second, the ramps and the elevated highway create what Jane Jacobs described as a "border vacuum"—a dead space in the urban fabric created by large buildings, large infrastructure, or large empty spaces. On a trip to New York, John Norquist[165] pointed out how much parts of the elevated BQE look like the old Berlin Wall.[166] Three, this all takes place in what could be one of the prime locations in Brooklyn, situated between Brooklyn Heights, Brooklyn's civic center and parks, and the expensive loft district known as Down Under the Manhattan Bridge Overpass (DUMBO).

At about the same time, New York was talking about what to do with a portion of the BQE the city owns that was (and still is) falling down.[167] I thought there were many advantages to tearing it down: doing that would cost billions of dollars less than repairing the highway; the

fix would be permanent; New York would have fewer cars and trucks and less pollution; the blight would be gone, the city would suddenly own acres of land in a prime location convenient to mass transit that could be used for affordable housing; the streets would be safer and more pleasant for the residents; *and we could make Brooklyn Bridge car-free!*

There are links in the notes and at Slow New York (slownewyork.city) to articles and more images, so I will only say a few words here.[168] As I write this, New York City has had congestion pricing for several months, but after compromises in the execution of the plan, the Brooklyn Bridge is still like an open hydrant that pumps cars into Manhattan day and night. If we are serious about changing that, there are better ways to spend billions of dollars than pushing cars through Manhattan, Brooklyn, and Queens.* (Even if we aren't serious about changing that, there are still better ways to spend billions of dollars.) The benefit of swapping the highway for prime housing sites is obvious. And the end result would be one of the greatest walks in North America, going from Brooklyn's Borough Hall to New York's City Hall and Municipal Building, across one of the most iconic structures in New York. New Yorkers would use it to commute on foot or by bicycle. For tourists, it would be more popular than the Highline. Starting the transformation would be simple: the NYC DOT could put jersey barriers out overnight and have a car-free bridge the next day.

Slow Zones and the Slow City

New Yorkers are waiting for fewer cars in their neighborhoods and more places like a car-free Brooklyn Bridge. If the FDNA plan for a large Slow Zone with Slow Streets had gone ahead, it would be enormously popular. Other parts of the city would quickly call for their own Slow Zones and slow neighborhoods—finally. The Bloomberg administration and Commissioner Janette Sadik-Kahn started a revolution in 2007, but the evolution of the revolution has been disappointingly slow—especially when we look at what European cities have done.

____

*The BQE/I-278 creates an "asthma belt" in Brooklyn and Queens. However, I-278 goes through four of the five boroughs, coming from New Jersey through Staten Island and continuing on into the Bronx. It also induces traffic in Midtown and Lower Manhattan. "If you build it, they will come."

Figure 5.37: Brooklyn Bridge Landing, Brooklyn, New York. Before: Aerial photo of the auto sewer at the Brooklyn end of the Brooklyn Bridge. The Brooklyn-Queens Expressway (BQE) comes out from underneath the Brooklyn Heights Promenade (top left), connecting to ramps for the Brooklyn Bridge and the Manhattan Bridge (just visible in the lower right). In the 1950s, Robert Moses demolished several blocks of buildings in historic Brooklyn to make room for the ramps. The new buildings around the bridge in the photo are on sites Moses cleared, although there are many more outside the photo.

Figure 5.38: Brooklyn Bridge Landing, Brooklyn, New York. After: Watercolor sketch suggesting how the city-owned land under the BQE might be re-used for parkland, housing, and civic buildings after closing the Brooklyn Bridge to traffic and removing the BQE. © 2020 Massengale & Co LLC, rendering by J.J. Zanetta

Figure 5.39: Car-Free Brooklyn Bridge, Brooklyn, New York. Photorealistic rendering of a car-free Brooklyn Bridge. Looking west on the southern roadway, redesigned for walkers and runners. The northern roadway would be for bicycles and other self-powered wheeled vehicles. © 2020 Massengale & Co LLC, rendering by Zeke Mermell

London, England introduced their Congestion Charge Zone (CCZ) in 2003. I happened to be there at the time.[169] Apart from the CCZ, London's streets were a mess, with lots of old-style auto sewers. Experiments like the Kensington High Street retrofit (page 388) and Walthamstow's "Mini-Holland" hadn't happened yet. But in 2024, London has many Low Traffic Neighborhoods (LTNs), and every one of them is more effective at slowing traffic than any New York Slow Zone.[170] In 2024, one-hundred-and-twenty-five miles of the streets controlled by Transport for London (TfL) had twenty-mile-per-hour limits, and London boroughs had adopted twenty-mile-per-hour limits on almost all their streets. TfL data shows that twenty-mile-per-hour limits inside London's congestion zone led to a 25 percent reduction in deaths and serious injuries (and London's statistics were already much better than New York's).[171]

London's example suggests that New York can think bigger than one neighborhood. A group called Streetopia Upper West Side has been working towards this in upper Manhattan, and Greenwich Village is a natural Slow Zone in lower Manhattan. Why not remake the entire island of Manhattan, after extending congestion pricing all the way north? We have discussed the Commissioners'

Plan of 1811 for Manhattan a few times in *Street Design* (pages 9 and 25). The simplicity of the grid could make New York one of the easier places to begin moving away from streets that all put cars first to a better balance of streets for traffic and people. Add to that all the information and data about cars and transportation in Manhattan that we have mentioned before, including the low rate of car ownership and use, the transportation system, the city's walkability, and the growing use of bicycles.

To rebalance how New York uses its streets, the city could begin by focusing on the numbered north–south avenues (all one hundred feet across or wider) and the regular grid of one-hundred-foot cross streets. Then it could redesign the avenues where moving cars is a priority and change the priority on other avenues.* All avenues where cars are prioritized should be two-way, with narrower lanes and wider sidewalks. That would make the avenues safer and slower than today.

---

*Most New York avenues are boulevards, with "endless sections" leading to unterminated vistas (as discussed in Chapter One). But because New York calls them "avenues," we do too.

Figure 5.40: Brooklyn Bridge Pier Elevation with New Pedestrian Deck, from the Car-Free Brooklyn Bridge Proposal. Old elevation with addition by Massengale & Co LLC, 2020. The deck can be seen in Figure 5.39.

Until the 1950s and 1960s, almost all avenues in New York were two-way, including the narrower avenues like Lexington Avenue and Manhattan Avenue. Some are still two-way here and there (as are the wide cross streets). When the city converted avenues to one-way streets, New York began on the outside of the island and worked its way to the center. The last two avenues made one-way were Fifth and Sixth avenue. Similarly, in the future, the avenues where cars have the highest priority should be

on the east and west sides of the island. An ideal plan would coordinate that with turning the Henry Hudson Parkway and the East River Drive into beautiful riverside boulevards.

The avenues should have three functional types: one would still be devoted to moving cars around the city, but these avenues would become two-way. Like the wide cross streets, they could have a design speed of 25 miles per hour. European examples show that slow, two-way streets are the safest transportation corridors. Exercising next to noisy, polluting cars and trucks is unhealthy and unpleasant. The second type of avenue should be two-way and car-free, redesigned for bicycles and buses—with wide sidewalks on each side for walking. Finally, there should be "shared-space" avenues and promenades for pedestrians and slow cyclists. Particular blocks, such as along Fifth Avenue to the east of Rockefeller Center, might be only for pedestrian. When the avenues are no longer auto sewers, the traffic engineer's endless section can be discarded. People walking like variety, as we know. For them, changing the street every 600–800 feet or so (a furlong) works well (see page 159).

Many of the wide cross streets like 23rd Street and 96th Street could remain two-way streets where fewer cars go more slowly but still have priority. Some of the wide cross streets should have higher priority for buses and cyclists in protected lanes.[172] Over time, custom designs would evolve. There is little traffic on West 106th Street, because it ends at Central Park on the east and only goes as far as Riverside's small local road on the west—granite steps lead down to the main boulevard (Figure 2.193). One possible design for West 106th Street could be along the lines of Vester Voldgade (Figure 4.160) in other words, dividing the street in half and removing cars from the southern portion. The vast majority of Manhattan's cross streets are sixty feet wide rather than one hundred feet across. Most could become shared-space Slow Streets, where cars move at a speed safe for people on foot or on bikes, which means less than 20 miles per hour. The design techniques that support shared-space places also support good urban design and placemaking, making them streets where people want to get out of their cars and stroll. Some could become School Streets, closed to traffic.

New York has been a leader in street design before and should lead now.

Figure 5.41: Broadway, New York, New York. NYC DOT, 2023. Looking south towards West 26th Street and Madison Square. Another illustration of the Monderman Rule: an over-designed, over-engineered, bicycle transportation corridor and outdoor dining area that will extend from Times Square to Madison Square and then perhaps south to Union Square. With five colors of paint, Jersey barriers, MUTCD-compliant orange traffic barrels, and random bumpouts that serve little purpose, the street is ugly Tactical Urbanism. But remember that some of the "temporary" tactical installations in New York have lasted almost two decades. *Courtesy of NYC DOT, image from public records*

Before we had heard of Covid, City Council Speaker Corey Johnson and other New York politicians were talking about "breaking the car culture" in the city.[173] To do that, we need to stop treating every avenue like a suburban transportation corridor.* Instead, the city continues to tinker with designs that seemed a decade ago. We are moving too slowly, while our cars go too fast.

---

*"C'mon in," our streets say to suburban drivers. "Our streets are just like yours." New York City has adopted congestion pricing and officially pledged to lower traffic and pollution from cars, but old habits die hard.

IF YOU KNOW IT IS USEFUL, AND FEEL IT IS BEAUTIFUL, REPEAT IT

Adolf Loos, the famous Modern Viennese architect, is best known for his essay "Ornament and Crime."[174] "The evolution of culture is synonymous with the removal of ornament from utilitarian objects," he wrote in the article, where he said it is "the artist's task to find a new formal language for new materials. Everything else is imitation."[175] But when an architectural colleague criticized Loos's design for the three-legged Thebes stool shown in Figure 5.43 as too traditional, Loos reportedly said, "If you know it is useful, and feel it is beautiful, repeat it" (Figure 5.43).[176] We agree. Innovation is good. So is learning from history. That applies to every level of design, including street design and urban design.

One of the main messages in *Street Design* is that humans know how to make great cities and towns: we have been doing that for thousands of years. Cities for people start with streets for people. What is important is that they are places that support city life, not how original

Figure 5.42: Passeig des Born, Palma de Mallorca, Spain. Looking south from the Plaça dei Rei Joan Carles I. A beautiful place for people: the NYC DOT design seen in Figure 5.41 could be similar. The beauty of the street is not in the buildings; they are good, but so are the buildings along Broadway. What makes the street beautiful is the proportion of the center to the side, the plane trees (cousins of our London Planes and Sycamores), the stone paving, the sculpture, and the regular simplicity of the whole. God is in the details in the hand-made tree wells and the sphinxes standing guard. *Courtesy of Andrés Duany*

or unprecedented they are. Many design schools today teach that creativity begins with the designer's personal vision or artistic process, but that process has more failures than successes when it comes to creating public space. Contrast those failures with tens of thousands of streets around the world like East 70th Street in New York City (page 18) and Railroad Street in Great Barrington, Massachusetts (page 179) that show how easy it can be to make a good place.

What is important is that they are places that support city life.

That does not mean we should simply copy what has been done before, although that can work. History is full of examples of Classical buildings that have been literally reproduced. Figure 5.44 shows the Choragic Monument of Lysicrates, which was measured and drawn in the eighteenth century by the English architects John Stuart and Nicholas Revett and then published in their book *The Antiquities of Athens* (1762). Reproductions of the monument stand in parks around the world.[177] Designers

like the nineteenth-century American architect William Strickland kept the form recognizable but transformed the function and the construction. Strickland famously based the cupola on his Greek-Revival-style Merchant Exchange Building in Philadelphia on the monument, helping to make Philadelphia "truly the Athens of America," a local newspaper reported.[178] The original monument, however, did not sit on the roof of another building, with windows to let light down into the interior of a trading room.

American architects trained at the École des Beaux-Arts used the design to crown City Beautiful skyscrapers. McKim, Mead & White's bold version of the Choragic

Figure 5.43: Left: Three-legged Thebes Stool, Adolf Loos, circa 1903. Right: Three-legged Thebes Stool, Liberty of London, circa 1890. Three-legged and four-legged stools discovered in early nineteenth-century archaeology expeditions sponsored by the British Museum inspired copies and variations for over 100 years. The original furniture came from Egypt's Eighteenth Dynasty (1550–1300 BC). This Liberty stool was one of three designs closely based on the ancient stools found by the museum that were patented by the company in 1884 (see note 176). Loos made his versions almost twenty years later. When an architectural colleague criticized his design as too traditional, Loos reportedly said, "If you know it is useful, and feel it is beautiful, repeat it."

Monument of Lysicrates atop the New York Municipal Building (Figure 1.17) makes the temple part of a stepped composition scaled for the city's skyline. In other words, some architects literally copied the design, while other architects used it for inspiration, creating new interpretations.

Some designers thought they were copying the monument, but when we compare their work to the original today, we know that one is from ancient Greece and the other from Victorian England (for example). In his book *The Impecunious House Restorer: Personal Vision & Historic Accuracy*, the American furniture scholar John T. Kirk uses photographs to document the use and furnishing of "period" rooms in house museums in different decades, illustrating how much our interpretation of historical accuracy changes over time.[179] Once again, we can usually guess when the rooms were made. In some cases, the curators worked from contemporaneous paintings of the rooms when they were new, but the look of the room still changes every decade or two.[180]

Munich has wonderful examples of designs copied from other places. One hundred miles north of the Alps, Munich is the capital of Bavaria. On the other side of the Alps is sunnier, warmer Italy, a place many Bavarians are drawn to. When the Kings of Bavaria expanded Munich in the nineteenth century, they created two grand boulevards that shifted the center of the city to Max-Joseph-Platz, in front of their palace (the Residenz). Then they placed architectural souvenirs from Florence on the boulevards in the new heart of the city. The longer boulevard is the Ludwigstrasse. Before it arrives at Max-Joseph-Platz, Ludwigstrasse splits in two, going around a monument that visually terminates the axis. Called the *Feldherrnhalle* (Field Marshalls' Hall), it looks like the Florentine Loggia dei Lanzi. Ask most *Münchners* or tourists about the loggia and they will tell you it is an exact copy of the Florentine original (it isn't), complete with Bavarian lions that look like the Medici lions in Florence.

To the east of the Feldherrnhalle is the Residenzstrasse, which continues to Max-Joseph-Platz. The oldest part of the royal palace is the narrow street, with a long façade painted to look like an Italian palazzo. A little before the square, the light, painted facade transforms into a heavy, rusticated stone palazzo. That is the first indication that in the square sits a large version of Pitti Palace. Again, many Münchners believe that it is an exact copy of a

Florentine landmark. On the opposite side of the square, facing the palace, is a copy of the *Ospedale degli Innocenti* (the Foundling Hospital, one of the most famous buildings in Florence, designed by Fillipo Brunelleschi). Most architecture professors today (and dating back to before World War II) would call these three buildings "pastiche," which is one of the harshest, most critical words in their vocabulary. But the architects and rulers who built them would not agree that pastiche is bad.

In music, a *pastiche* or *pasticcio* is a work with multiple composers. Wolfgang Amadeus Mozart and Georg Friedrich Handel are among those who composed *pasticcii*. Leo von Klenze was the Royal Architect, considered one of the best German architects of the nineteenth century. The manner in which his pasticcio Palazzo Pitti design joins the painted palazzo next to it, and the way that his rooms inside the Königsbau relate to the German-style rooms in the old palace rather than the Italian-style facade, were precisely what he and his client wanted. Most importantly, they wanted the residents of Munich to enjoy the feeling of a palpable connection to Italy in their city (Figure 5.46).

America has had similar aspirations, sometimes for philosophical reasons. Thomas Jefferson* modeled the Virginia State Capitol on the ancient Roman Maison Carrée in Nîmes, France, although only the portico is like the original (the Roman temple did not have exterior walls or double-hung windows). Being a believer in the ideals of the Enlightenment, Jefferson wanted America to have an architectural connection to the Classical world, to reinforce the democratic ideals of education, rationality, and civic responsibility that he saw in Classical architecture. When he designed the Lawn at the University of Virginia—frequently voted the most beautiful place in America—he used the ten pavilions as "textbooks" to teach the students Classical design. They were the built equivalent of the plates in the architectural treatises Jefferson owned.[181] For each pavilion he drew one page showing the pavilion facades and floor plans. On the reverse side (Figure 5.47), he made notes on dimensions, details, and materials.

---

*When President John F. Kennedy held a dinner at the White House to honor forty-nine Nobel Prize winners, he famously said, "I think this is the most extraordinary collection of talent, of human knowledge, that has ever been gathered together at the White House, with the possible exception of when Thomas Jefferson dined alone."[182]

In the twenty-first century, we no longer have to follow the twentieth-century Modernist dogma that said the rejection of history and the architectonic expression of technology are the only acceptable ways to build. Modernism produced many great buildings, but it produced many more mediocre buildings, and worse: in the words of Andrés Duany, "in any other business the win-loss ratio would be unacceptable." The combination of Modern planning and transportation also had a low rate of success in creating good places, giving us sprawl, strip shopping centers, shopping malls, suburban office parks, vast single-family subdivisions, auto sewers, and what Jane Jacobs called "urban removal." At the very least, we can learn from the examples of Jefferson, Charles McKim, and Stanford White, and traditional urbanism. Most of the cities, towns, neighborhoods, blocks, and streets we love came from those traditions.

Good streets—the spaces between the buildings—require good architecture. They are outdoor rooms in which the walls and perhaps the "ceiling" (the sky and the skyline) are usually more important than the "floor." Good streets for people can be simple, like East 70th Street (page 18), and they can be more complex, like the Boulevard de Rochechouart (134–136 pages). In the last one hundred years, we have turned city streets into harsh, anti-urban auto sewers, with endless boring sections, and wide dangerous roadways. We can reclaim the streets with urban design, with the principles and the concrete examples of good streets, including the streets in this book. East 70th Street shows the benefits of short blocks. The richness of Rochechouart comes from the rich combination of its parts, which change as we walk along it. The Corso Cavour (page 159) is another street made comfortable and interesting by the changing experience.

Maximilianstrasse and Ludwigstrasse in Munich have copies of buildings we know from other places: they give us happy memories that add to the pleasure of being on the street. The Feldherrnhalle is what designers call "multivalent," meaning that the viewer can experience it in multiple ways. The monument decisively terminates the vista, with a structure that is pleasing to look at as an object at the same time that it reminds us of another place and the associations of that place. Unlike the architects of the buildings that shape the streets and their experience, street designers rarely make literal copies of other streets in the space between the buildings (although traffic engineers usually do).

An American promenade like Commonwealth Avenue (137–138 pages) might have elements and principles derived from Spanish ramblas, but the details are different, and so the experience is different.

Design Solves Problems

When engineers say "the Devil is in the details" (page 40), they mean it can be hard to make the formulas laid out in traffic manuals fit the existing conditions in cities and towns. The difference between "the Devil is in the details" and "God in the details" is design. Design can follow principles, but good design is flexible, adaptable, and accommodating. That is a crucial difference between traffic engineering and street design. Building code requirements are frequently challenging for architects, as traffic manual standards are for street designers. "If technical issues prevent you from making something beautiful," the architect Michael Graves used to say, "you shouldn't be a designer."

New Urbanism succeeded in the beginning by countering the status quo in planning and development with beautiful drawings showing places better than people imagined. The status quo for new development was sprawl, ugly buildings, and auto sewers. People expected new development to be bad, in other words, because when they looked around them, that was usually what they saw. When local residents were instead shown designs for walkable, mixed-use places they liked, they frequently supported those on land that otherwise would have ended up as single-family sprawl subdivisions. The perspective drawings visualizing something better were a crucial part of the process.

How were the drawings for those places created? In the same way that most good design is done. The process is not linear. It involves different parts of the brain, embodied memories, and even the use of the body to create visions for a particular place. The body is involved in the drawings because they are hand-drawn. In the book *Draw In Order To See*, the architect Mark Hewitt explains that through a combination of experience, repetition, and talent, designers develop the ability to pull together information, form-making, spatial imagination, and memories of drawings they have made and places they have visited to create a design.[183] During the design process, thoughts loop back and forth between multiple parts of the brain.

At one point, the thoughts might go from the left brain, where logic and language lie, to the more emotional and creative right brain. Memories of physical places and objects drawn stored in the body contribute: a door in that corner will feel like *this*; window details are like *that*. The process is influenced and guided by thoughts more than controlled. This contributes to the common idea that designs and compositions are received gifts. It is a process that works best with relaxation, when brain waves are in Alpha.

Figure 5.44: Choragic Monument of Lysicrates, Epimenidou 3, Athens, Greece. 334 BC. The English architects James Stuart and Nicholas Revett measured and recorded ancient Greek monuments before publishing engravings of the structures in *The Antiquities of Athens* (1762). The Monument of Lysicrates is the oldest recorded example of the Corinthian order. Thirty-three feet tall, the monument was reproduced many times, both as a freestanding reproduction and in freely interpreted versions on the tops of buildings. The design of the Soldiers' & Sailors' Memorial seen in Figure 2.192 was based on the Choragic Monument of Lysicrates and is approximately one hundred feet tall.

Figure 5.45: The San Remo, 145 & 146 Central Park West, New York, New York. Emery Roth, 1930. View from the southwest looking towards Central Park after a snowstorm. Crowned by variations on the Choragic Monument of Lysicrates, the twin towers are modeled after the Giralda Tower in Seville, Spain.[184] © 2012 Jun / Wikimedia Commons / CC BY SA 4.0

Figure 5.46: Königsbau, Max-Joseph-Platz, Munich, Germany. Leo von Klenze, 1835. A souvenir of Italy in the Bavarian capital, it is one of three copies of famous Florentine buildings built by the Bavarian kings in the heart of the city within eyesight of the others. Most residents of Munich believe it is an exact copy of the Pitti Palace in Florence, even though each floor is significantly different from the original. The interiors of the Königsbau are similar to earlier rooms in the palace, with little direct influence from Italy. © 2014 Wikiolo / Wikimedia Commons / CC BY SA 4.0

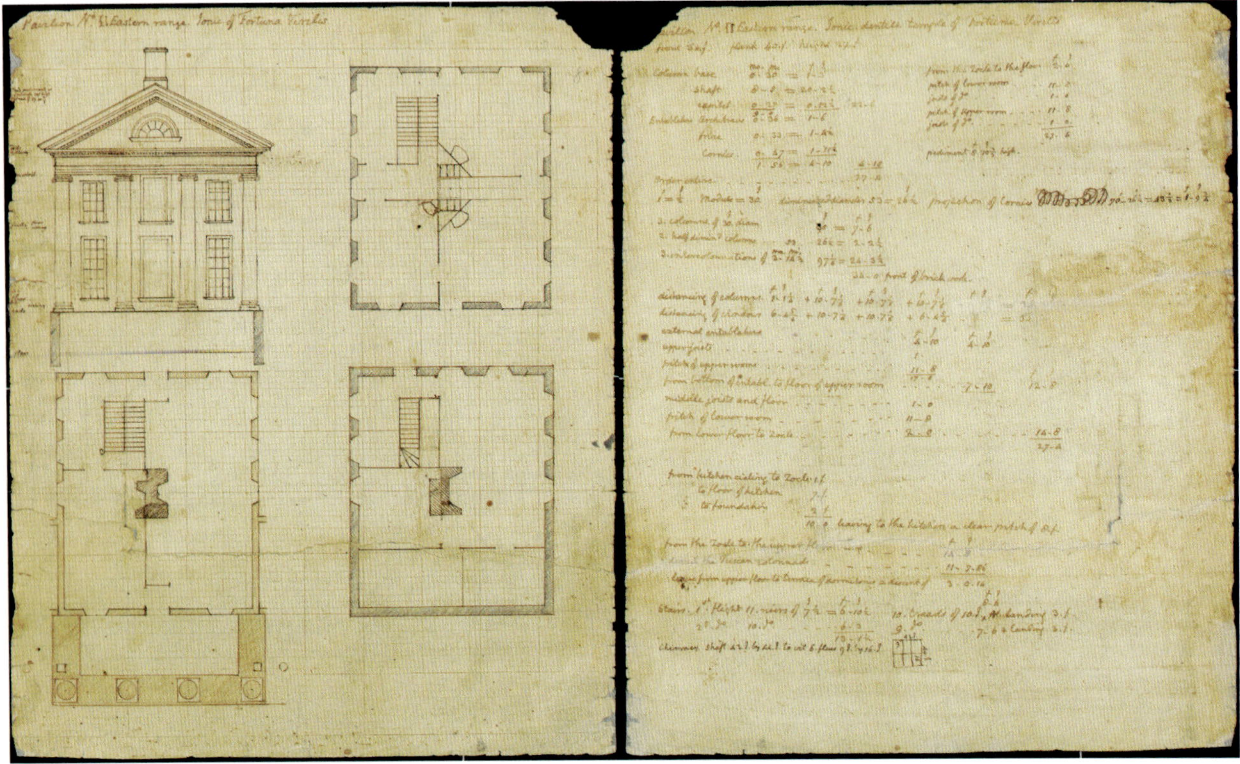

Figure 5.47: Pavilion II, The Lawn, University of Virginia, Charlottesville, Virginia. Thomas Jefferson, June 1819. Sketch design by Jefferson for the pavilion's facade and floor plans, ink on paper. Jefferson designed the ten pavilions to teach students the use of the Classical Orders. His description at the top of the drawing says this illustrates the Ionic Order, as used at the Temple of Fortuna Virilis in Rome. Jefferson never got closer to Italy than Nîmes, in Southern France, which he visited to see the Maison Carrée: "Here I am, Madam, gazing whole hours at the Maison Carrée, like a lover at his mistress."[185] He knew the Temple of Fortuna Virilis from Classical treatises like Palladio's *Four Books of Architecture. Courtesy of University of Virginia Library Special Collections*

We have written about designs that are literal copies. Recreating a New York City block is not hard in a greenfield development, for instance. If you can create good traditional brownstones on each side of the street and then line up similar curbs, reasonable sidewalks, good street trees, and well-designed stoops, you will have a good block. But we also talked about how infrequently we find exact copies in urban design. Many designers want to put their own taste into the design, and many put their taste into the design without realizing it. Commenting on design in the early twentieth century, architects said "No one can design like Stanford White [or Louis Sullivan or Frank Lloyd Wright] because no one can draw like him." You must have the equivalent of his hand and eye to be able to draw it, and you have to understand what White drew in order to recreate it.

A good designer can draft a form-based code that will enable many to build a good New York-style block on a flat site from scratch, following the formula in a manual. But if the project is a retrofit or sits on complex topography, design skills and instincts may be required, because the street is too wide or too narrow. What if the street curves, or the local zoning prohibits characteristic elements? That is when design is needed to solve problems. Here a good designer will do a better job than ninety-nine out of one hundred engineers robotically following formulas.

To those who want to be street designers, we believe that if you study and understand the streets in this book, you will be on your way to designing good streets. If you visit the streets and let your body feel the space, you will be one step closer to your goal (so to speak). Then you can copy them, modify them, use them as jumping-off points, or invent something new. What's important is to design streets where people want to be and life will flourish.

ASYMMETRICAL STREETS

Most of the streets we show in *Street Design* are symmetrical: the left side of the street is the same as the right side, with the same uses on each side ("like faces like") and the same dimensions. The left and right curbs are equidistant from the centerline of the street, and the buildings are too. There are exceptions, of course, and designing an asymmetrical street can be part of the fun of urban design.

The section through Riverside Drive constantly changes (*Street Design* never advocates "the endless section"), but with Riverside Park to the west of the Drive and buildings to the east (sometimes behind the islands), Riverside Drive is never symmetrical (see Figure 2.193). Léon Krier played with asymmetry in his plan for Longmoor Street (page 474), and Douglas Duany describes some of the asymmetry in Orvieto's main streets (pages 159–165).

We have complained about engineers' designs oblivious to the asymmetry introduced that subtly, or sometimes not so subtly, erodes the harmony of the public realm. Examples include the visual imbalance created by protected bike lanes that push parked cars, heavy infrastructure, and bold striping out towards the middle of the street (see "Bicycle Lanes & Completer Streets" on page 587) and the NYC DOT's plans for making long stretches of Broadway between Times Square and Union Square car free. People have been talking about making that narrow section of Broadway car free for many years, and we support that—but not in the way the DOT proposes, with an ugly bikeway weaving down one side of the street (Figure 5.41).[186]

Good design solves problems. Bad design can create problems. In Chapter Four, we wrote about complications that came from the asymmetrical redesign of Exhibition Road in London (Figure 4.232). The experience of walking along Exhibition Road is inferior to the experiences we find on the beautiful asymmetrical streets illustrated here.

Figure 5.48: Pont Street, London, England. View looking west towards St. Columba's Church. One block has an asymmetrical cross-section, adding the double row of trees on the south side to form Cadogan Square.

Figure 5.49: Rue Saint-Sulpice, Paris, France. Looking east on the rue Saint-Sulpice between rue Bonaparte and rue des Canettes. The broad, shady sidewalk is as wide as the street in front of it. *Courtesy of Michael Ronkin, Designing Streets for People*

Figure 5.50: Brabantse Turfmarkt, Delft, the Netherlands. Looking south on Brabantse Turfmarkt from Burgwal. Traffic on the two-way street is so light that it is easy to imagine the street divided down the middle like Exhibition Road (Figure 4.232).

Figure 5.51: Mount Vernon Street, Boston, Massachusetts. Looking east on Mount Vernon Street, near the top of Beacon Hill. Setting the houses back on one side gives them more sun in the cold New England winter.

A STREET IS A TERRIBLE THING TO WASTE

When we wrote the first edition of this book, we were struck by two things. First, changes—both good and bad—were happening on American streets. Second, several European cities were decades ahead of any American city or state in terms of making streets for anything other than driving: public life, walkability, and cycling. They were safer too, with far fewer traffic deaths.[187] We decided to ignore the bias among American urban designers and planners against using European streets as models.

At that time, Americans were less concerned with climate change than they are now. But the national discussion about sprawl, loneliness, health problems, traffic jams, and the economic cost of sprawl grew every year. A majority of American residents said they wanted to live in more walkable cities, towns, and neighborhoods.[188] European streets had many lessons, so we got a travel grant from the Richard H. Driehaus Charitable Trust[189] and went to Europe together to visit streets. We visited streets we knew and streets new to one or the other or both. Slowly the book came into shape, with a combination of lessons from Europe and America. We did not write about any streets we had not personally seen, although some of our guest authors wrote about streets that neither of the two of us had visited. That is why most of the streets in the book are in the United States or Europe.

The most famous exemplars of change in 2011 and 2012 were Amsterdam and Copenhagen. Both were cycling centers and centers of street reforms in their countries. The history of the reforms went back to the late 1960s and early 1970s, but the two countries developed different approaches. In cities and town centers, the Dutch favored shared-space streets, while the Danes had transportation corridors that gave as much priority to pedestrians and cyclists as motor vehicles (pages 461–462).[190] Frequently forgotten today is that at the same time, Italian cities began emphasizing *Centri Storici* (historic centers) with fewer cars allowed, shared-space streets, and pedestrian malls. Transportation histories ignore that story, because the changes were made in the name of historic preservation rather than transportation. Dutch traffic engineer Hans Monderman wrote about and publicized the advantages of shared-space streets. Italian authorities did not.

Figure 5.52: Antwerp, Belgium. Photographed from Grote Markt looking northwest towards Antwerp City Hall. *Courtesy of Peter Katz*

Figure 5.53: Via della Dogana Vecchia, Rome, Italy. Looking north from Via degli Staderari. A shared-space street in the Roman *Centro Storico*.

Our trips were eye-opening and inspirational. We saw the urban paradise described in Chapter Two (page 96) and streets in Paris and London so clogged with traffic that we avoided walking on them when we could (page xv). Today, the situation is very different. London and Paris are two of the world leaders in reducing traffic and supporting alternatives, including walking and cycling.

In 2003, London was the second major city in the world to introduce congestion pricing. London had heavy-duty auto sewers and flyover highways in some parts of the city but slowly changed its car and road culture. Kensington High Street (page 388) and Exhibition Road (page 513) are high-profile examples of changes in London streets.

In 2016, London Mayor Sadiq Khan offered an economist named Will Norman—who was the Director of Global Partnerships at Nike—the job of Cycling Commissioner.[191] Norman said he would only take the job if he could be the Cycling *and Walking* Commissioner. Khan agreed, and Khan and Norman soon announced that together they would make London the world's most walkable city.[192]

Norman works for the Greater London Authority, which is to say "London," not to be confused with the City of London. The City is the oldest part of London and one of the thirty-three boroughs that make up Greater London.[193] Khan and Norman link walking, cycling, street design, congestion pricing, and public transport together across the city to move forward with fewer cars on the road and cleaner air.[194] After experimenting with Ultra Low Emissions Zones, in 2023 Khan made all of London an Ultra Low Emission Zone (ULEZ).[195]

Greater London controls 360 miles of streets in the city, but the other 9,500 miles belong to the boroughs.[196] London, therefore, made a citywide cycling network on its own streets. But Norman's group and TfL also worked with the boroughs to create a "Cycleways" network that combines London's old Cycles Superhighway network on the main roads with the Quietways network that was in the boroughs. Despite the bureaucratic problems involved, London's Cycleways system is a more complete and better-designed urban cycling network than any in America. "Urban" is a key word, since America only has six cities larger than five million people, but London's network is also better than those in San Francisco, Portland, Minneapolis, Austin, Washington, and Boston.[197]

Khan's Transport Plan for London promotes cooperation with boroughs to create Low Traffic Neighbourhoods. "LTNs help to make streets around London easier to walk and cycle on by stopping cars, vans, and other vehicles from using quiet roads as shortcuts."[198] The first three opened in 2014, when the Mayor of London at that time, Boris Johnson, funded a £100 million grant for three boroughs to create "mini-Hollands" that included slower streets and "Dutch-style cycling infrastructure."[199] After Covid arrived in London, the city expanded the budget for LTNs to create more space for walking and cycling safely during the pandemic. Today, there are nine LTNs, with more in the works.

In central London, neighborhoods like Mayfair and Marylebone were built by aristocratic landowners whose estates still own them in the English system of leasehold. The estates are involved in the maintenance and design of the sidewalks and streets. Over the last ten years, they have tended to make the streets and intersections visually simpler. Many T-intersections in Mayfair have no traffic lights of any kind, no stop signs or yield signs, no painted crosswalks, and no supergraphics on the pavement telling the driver to yield or turn. A single line across half the intersection (if the street is two-way), tells the driver that vehicles on the cross street have the right-of-way. These streets have lower crash rates and traffic deaths than formulaic Slow Zone streets in America (also compare the

Figure 5.54: Orford Road, Walthamstow, London, England. A street in one of London's "Mini-Hollands." Looking west at the Walthamstow Village Square. Streets in London's Low Traffic Neighbourhoods (LTN) reduce traffic passing through the neighborhood and create comfortable places for walking and cycling. *Courtesy of Transport for London*

Figure 5.55: Orford Road, Walthamstow, London, England. Between the Village Square and East Street, looking west. The narrow street is no longer overwhelmed by traffic and parked cars. *Courtesy of Transport for London*

feeling to the view of the historic center of Greenwich Village in Figure 5.10). With stone sidewalks, they are also very comfortable to walk along. Removing all the MUTCD detritus makes the public space between the buildings more harmonious and welcoming. Seeing such a simple intersection, where the curbs line up with the buildings and the trees line up with the curbs, makes one realize how visually disruptive a formulaic bumpout can

be. "You won't see many bumpouts in London," Will Norman says, "We don't like them."[200]

More and more streets in London, in all the boroughs, look like that. We—John and Victor—are happy about that. When it comes to street design, we endorse Frederick Law Olmsted's principle of subordination, Mies van der Rohe's "Less Is More," the KISS principle, the Coco Chanel Rule, and the Monderman Rule (which

Figure 5.56: Motcomb Street, London, England. Corner Condition: Most of the great streets of the world have an understated, muted palette of colors, materials, and textures, rather than a riot of bright stripes and loud traffic signs.

says that the problem with traffic engineers is that they do the opposite of the KISS principle). To all of these, we add one more, from the Anglo-German author of *Small Is Beautiful*: "Any intelligent fool can make things bigger, more complex, and more violent. It takes a touch of genius, and a lot of courage, to move in the opposite direction." (Figure 5.58)[201]

Although there are plenty of old-school traffic engineers in England, the profession has produced many good designers, including Phil Jones and David Milner. Milner worked on projects like Poundbury and Nansledan for Prince Charles and the Prince's Foundation.[202] Jones founded Phil Jones Associates, now PJA, a large influential practice. He was part of the team that produced the UK *Manual for Streets* for the Department for Transport (DfT), "a comprehensive guide to the design of urban, residential and lightly trafficked streets" that is much different, and better, than a US traffic manual.[203] It endorses the need for placemaking and concepts like innovation. In Section 2.6, under "Risk and liability," it says,

2.6.3 In fact, imaginative and context-specific design that does not rely on conventional standards can achieve high levels of safety. The design of Poundbury in Dorset, for example, did not comply fully with standards

Figure 5.57: Intersection of Green Street and North Audley Street, London, England. Looking east on Green Street. For Americans, what is striking about this photo of a busy "t-junction" in Mayfair is what we don't see: traffic lights, stop signs, yield signs, reflective plastic sticks, red paint, green paint, supergraphics painted on the street, bumpouts in the street… or any of the engineering detritus American DOTs tell us we must have if we want safe streets. And these streets are far safer than American city streets. Compare to Figure 4.153, which shows the historic heart of Greenwich Village, in America's most walkable city. *Courtesy of Google Earth*

INCOMPLETE URBAN STREET

Figure 5.58: Eighth Avenue, New York, New York. NYC DOT, 2009. Looking north towards the intersection with 18th Street. A few good photos can be worth a thousand words: compare this image with Figure 1.58.

When the NYC DOT began rolling out this Complete Street design on Manhattan avenues in 2007, in the context of what other American DOTs were building, the design seemed progressive and forward-looking. In 2025, knowing how European cities like London and Paris have recently transformed their avenues and boulevards, Manhattan avenues look like suburban-style, one-way arterials in the middle of America's densest urbanism.

This is still the DOT's only design. The same street is one of several options in the National Association of City Transportation Officials (NACTO) *Urban Street Design Guide*. It is the most likely choice for the old-school traffic engineers who run the nation's DOTs when designing a one-way city street.

Jeff Speck argues on page 552 that one of the easiest ways to make streets better is to make them two-way. Virtually all Manhattan streets were two-way until New York rebuilt them in the 1950s and 1960s. Even so, before twenty years ago, turn lanes were rare in Manhattan, because they prioritize traffic over walking and cause pedestrian death.

The majority of public life in Manhattan takes place on the wide north–south avenues, but most of the public space there is given over to machines and their "throughput." The over-designed, over-engineered streets are ugly. The decaying plastic sticks, the bold paint (now yellow and white), and the parked cars and pedestrian island sitting asymmetrically in the street contribute to the ugliness. They cut the street into pieces, claiming most of the street for automobiles and trucks, and kick the pedestrian to the side of the road.

We know that ugly streets discourage walking and that people avoid them. Beautiful streets encourage lingering. Lingering, we encounter our neighbors. Many of the historic and new streets illustrated in this book and its Catalog of Ten Essential Street Types show better models for large streets that better balance the needs of all people on the street, not just the drivers. They support city life as well as urban transportation.

and guidance then extant, yet it has few reported accidents. This issue was explored in some detail in the publication *Highway Risk and Liability Claims*.[204]

The accompanying photograph in the *Manual for Streets* shows a blind corner designed to slow people.[205] Poundbury has no stop signs and was built, of course, by the current King of England. It is very different to walk around Poundbury than a Toll Brothers subdivision next to a regional mall (see Figures 4.176 and 4.177). But it has been at least as profitable as the cookie-cutter Toll Brothers developments.

Jones was also instrumental in an important change to the National Highway Code. Following his lead, the UK DfT rewrote the law to establish a "Hierarchy of Users" at intersections, with a "Hierarchy of Responsibility": *"...those in charge of vehicles that can cause the greatest harm in the event of a collision bear the greatest responsibility to take care and reduce the danger they pose to others."*[206] In the city, suburbs, and country, intersections are where drivers, passengers, cyclists, and pedestrians are at greatest risk. In the UK now, if a driver turns into a side street and hits a cyclist, the driver is at fault. If a cyclist hits and injures a pedestrian, the cyclist is at fault. The United States should have a law like this.

PARIS AND THE 15-MINUTE CITY

Amsterdam, Copenhagen, London, and Paris have all followed different paths to greater walkability and mobility. Paris is now the home of the 15-minute city. The model set out in Carlos Moreno's book of the same name calls for a city in which everyone can meet daily needs within 15 minutes of their home by walking or biking, and most services with 15 minutes on public transport services.[207] Moreno presents this as a critique of Modern planning, which separated sleeping, eating, working, and playing. A word cloud in *The 15-minute City* confirms that many of his influences are planners and urban designers like Jan Gehl, Peter Calthorpe, and Andrés Duany.[208]

The 15-minute city is different than the New Urban five-minute neighborhood or quarter, however. The latter is created by drawing a quarter-mile radius around a neighborhood center, while the 15-minute city centers on each person's home ("chez moi"). In a fine-grained city like Paris, each person's 15-minute city is different. Paris is a

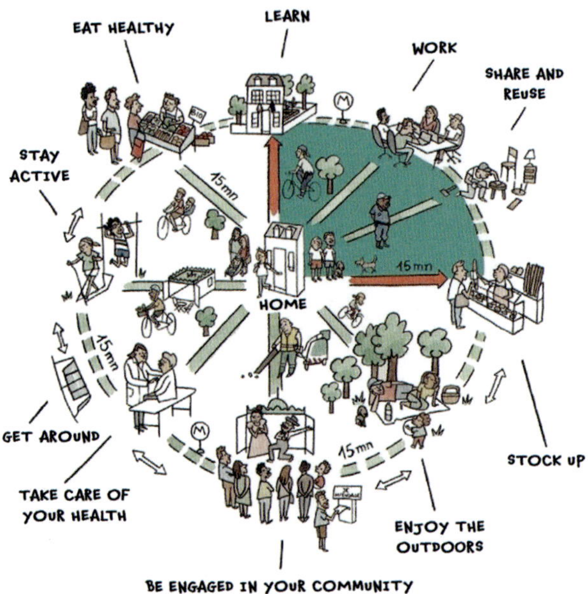

Figure 5.59: 15-Minute Paris. A diagram of the concept from Carlos Moreno. Unlike the New Urban Five-Minute PedShed, which is drawn with a quarter-mile radius around a neighborhood, the 15-minute city centers on each individual and household. In the 15-minute city, everyone can reach essential services within 15 minutes of their home by walking, biking, or public transit. *Courtesy of Micaël/Micaeldessin.com*

city divided into 20 *arrondissements*, each with a town hall and other civil services. Public schools are located every 400 meters (around 435 yards), and small, locally-owned shops are numerous. There are so many local centers that the fabric of centers can sometimes be continuous, part of an interconnected, complex organism.

When we (John and Victor) visited Paris while writing the first edition of *Street Design*, the 15-minute city was 10 years away in the future. The changes we saw in Parisian streets and plazas were mainly physical. Plazas that had been ugly parking lots were beautifully restored with no surface parking but large garages underneath. Car navigators led drivers to available parking spaces, probably inducing demand. "We have a parking space for you," the system said. "We will guide you straight to it."

Figure 5.60: Rue Charles Baudelaire, Paris, France. A recent photograph of the School Street (bottom), compared to Google Street view from earlier in the decade. *Courtesy of Google Earth / Richard Bono / Melissa and Chris Bruntlett*

Figure 5.61: Rue de Rivoli, Paris, France. View looking northeast, October 2022. People on bikes now dominate the once traffic-choked corridor. *Courtesy of Margaret Flippen*

Traffic clogged many parts of the city. Some famous boulevards were so crowded that we avoided them. Paris was beautiful, but heavy traffic made it hard to see that. Today, traffic is greatly reduced. There is less parking, many former traffic lanes are now bicycle lanes (Figure 5.61), some streets are closed to traffic, the highway along the Seine is gone, and large vehicles cost more to park than smaller ones.

Some friends and colleagues tell us Paris suddenly has the best bicycle lanes in the world. Others say this is a transitional period, with poor designs that should be replaced by something better. The first group may be colleagues who favor utility over beauty. On Twitter and Instagram, the hashtag *#saccageParis* ("raging Paris," but it also means "sacking Paris") is popular, a rallying point for complaints about changes to the Paris streetscape (such as tearing out historic elements like benches installed by Baron Haussmann), new glass towers, and rude cyclists on the streets.[209] "Politicians want to make Paris a cycling city, but no one is following any rules," a longtime resident of the Marais told the *New York Times*. "It's becoming risky just to cross the street!"[210]

Many cities have experienced this in the post-pandemic era. Drivers run red lights, electric bike riders speed on sidewalks and park paths. Some "e-mobility" users seem oblivious to how much they can scare pedestrians. After a desocializing isolation period, some percentage of the world has forgotten how to act in public. Since the authors last visited Paris before the pandemic, we can't make first-hand observations, but we agree with the philosophy of the 15-minute city. Until Organized Motordom changed the world, all cities were 15-minute cities. Paris, a great city, is well suited to this restoration of a more balanced public realm.

The sponsor of the 15-minute city, Mayor Anne Hidalgo, promoted the concept when she began running for a second term as Mayor in late 2019. When Covid hit Paris in March, 2020, Hidalgo and Moreno picked up the pace. Taking advantage of the number and proximity of public schools in Paris, they closed streets on blocks with schools, creating safe, public spaces where people could gather. Before long, the city rebuilt the school streets, conveniently located for most of the population.[211]

The new school streets did many good things. They created safe spaces in front of schools, where both children and parents could socialize. Soon after the lockdown started, the school streets gave tens of thousands of people new public spaces. Children saw a different vision of what a city can be. "The middle of the street used to be the most dangerous spot. Now the middle of the street is the most pleasant spot, where people want to be," a Dutch resident of Paris said to Streetfilms. "With shade, with greenery, with life."[212]

At the same time, Paris emphasized the rapid transformation of car lanes into bicycle lanes.[213] If you were on #BicycleTwitter, you saw hundreds of photographs and videos of the rue de Rivoli showing cars confined to one lane, with most of the street given to bicycles moving in two directions, as seen in Figure 5.61.[214]

Early in the pandemic, architects and urban designers sitting at their computers started turning out elaborate designs for transforming Paris streets. There were long, straight blocks with new, multicolor winding paths. You saw computer simulations of streets with so many trees that they looked like the Fontainebleau Forest, a UNESCO biosphere reserve. The usual designs by specialists to make their specialty special, in other words. Thankfully, the city has ignored those and reverted to the Parisian tradition of simple streets, simply made.

To understand typical Parisian intersections before Covid, use the "See more dates" option in Google Streetview to look at *ronds-points* (traffic circles) in the sixteenth *arrondissement*. They were frequently like the new intersections in London (Figure 5.57), with no traffic lights, traffic signs, or crosswalks and very little paint on the street. What paint there is tends to be faded and half the size our MUTCD calls for. Once again, remember that Parisian streets have far fewer crashes and fatalities than American streets. More paint, we say again, makes the street feel like a place for machines, and makes the view from the sidewalk ugly.

Pedestrians avoid ugly streets when they can. We know that. Parisians take pride in what they are sure is the most beautiful city in the world (which is why some changes have produced the #saccageParis revolt). As we pointed out in Chapter Three (page 277), Paris has the most stringent and comprehensive form-based building codes in the world. The buildings that shape Parisian streets have uniform heights, uniform setbacks (usually none), and repetitive façades. Most Parisian buildings are built from the same local stone, and the spaces between the buildings are very simple. But no one returns from a trip to Paris saying, "Oh, it was so boring."

▶**Figure 5.62:** Main Street, Staunton, Virginia. Looking west at sunset. *Courtesy of Staunton CBC*

Figure 5.63: Commercial Street, Provincetown, Massachusetts. Looking northeast. Locals and visitors long ago turned the holiday town's main street into a de facto shared space. © *mararie* / *Wikimedia Commons* / *CC BY 2.0*

SAFER, BETTER STREETS FOR BETTER CITIES AND TOWNS

We are not suggesting that all streets should be like the streets of Paris. The street-design revolution that spread across Europe over the last ten years provides countless examples of interesting, innovative streets that balance the needs and desires of people walking, cycling, and driving. Small Spanish cities have outstanding new streets.[215] Switzerland and Austria have developed stunning *Begegnungszonen* (meeting zones). These include shared-space streets that might change to pedestrian-only spaces for a few blocks before changing back again. Suddenly, Munich is a new German cycling capital, with many bike tours available for visitors.

America has great streets too. We love them, but in the last ten years, they have had less love lavished on them than their European counterparts. American streets are controlled by construction-oriented Departments of Public Works and the engineering-dominated Departments of Transportation, which are primarily Departments of Traffic—their passion is still moving cars around. In the 1950s and 1960s, the DOTs worked miracles and got things done. But seventy-five years later, we know all the problems that came with Organized Motordom.

Figure 5.64: Open Street, the Bronx, New York. People dancing in the streets. "What is the city but the people?" *Courtesy of Brittainy Newman / The New York Times*

Tens of millions of Americans want walkable places where they can live, work, and play.[216] Many of them live in neighborhoods and towns that were built before the car: there are hundreds of these places, and they are low-hanging fruit ready for improvement. These are places with good Walk Scores, like downtown Provincetown, Massachusetts (Figure 5.63, Walk Score 96), historic Santa Fe, New Mexico (page 204, with Walk Scores ranging 77 to 85 in different neighborhoods around the downtown), and the Germantown neighborhood in Columbus, Ohio, where the Walk Scores vary from 88 to 94.[217] With a little tender loving care, the Walk Scores in all of them could be 97, like South Beach in Florida, or 98, like the Upper West Side of Manhattan.

We know how to make them more walkable—and more livable—again. Begin with streets that are great public spaces where people want to get out of their cars and walk, and then give them places to walk to. Make beauty and safety more important than traffic flow. Build new neighborhoods like Culdesac in Phoenix, Arizona and The Bend in Chattanooga, Tennessee, where people don't have to climb into their cars every morning.[218] Work with the fact that until a hundred years ago, our cities and towns were places where people lived, worked, and flourished without cars. Enable Americans to make 21st-century, 15-minute places for all who want them. For "What is the city but the people?"—William Shakespeare, *Coriolanus*, Act 3, Scene 1.

NOTES

1. The increase in driving time does not allow time for more human interactions. Even though time alone for Americans has predated Covid, the pandemic has increased the amount of time we spend alone. See Note 16.

2. Air quality improved across the world: see Jessie Yeung, "Pandemic lockdowns improved air quality in 84% of countries worldwide, report finds," CNN, March 16, 2021, https://www.cnn.com/2021/03/16/health/world-air-quality-report-intl-hnk-scn/index.html. Also see "Air quality improvements from COVID lockdowns confirmed," *United Nations News*, https://news.un.org/en/story/2021/09/1099092 and "Impact of COVID-19 Pandemic on Air Quality: A Systematic Review," National Institute of Health, National Library of Medicine, https://www.ncbi.nlm.nih.gov/pmc/articles/PMC8871899/. "Not Back to Normal," Breaking Ground, July 22, 2020, https://breakingground.us/not-back-to-normal/.

3. Maneesh Arora, "How the coronavirus pandemic helped the Floyd protests become the biggest in U.S. history," *The Washington Post*, August 5, 2020, https://www.washingtonpost.com/politics/2020/08/05/how-coronavirus-pandemic-helped-floyd-protests-become-biggest-us-history/.

4. Sergio Olmos, Rick Rojas, and Mike Baker, "From Antifa to Mothers in Helmets, Diverse Elements Fuel Portland Protests," *New York Times*, July 19, 2020, https://www.nytimes.com/2020/07/19/us/portland-protests.html.

5. Brandy Zadrozny and Ben Collins "As vaccine mandates spread, protests follow—some spurred by nurses," NBC News, August 11, 2021, https://www.nbcnews.com/tech/social-media/vaccine-mandates-spread-protests-follow-spurred-nurses-rcna1654. Also see Berkeley Lovelace Jr., "Trump campaign rallies led to more than 30,000 coronavirus cases, Stanford researchers say," CNBC, October 31, 2020, https://www.cnbc.com/2020/10/31/coronavirus-trump-campaign-rallies-led-to-30000-cases-stanford-researchers-say.html.

6. Troy Closson and Sean Piccoli, "Thousands Join N.Y.C. Bike Protests: 'It's Like Riding in the Cavalry'," *New York Times*, July 2, 2020, https://www.nytimes.com/2020/07/02/nyregion/Floyd-bike-protests-new-york.html.

7. There is no evidence that Churchill ever said that, but the quote is commonly attributed to him. Never let a good quote and attribution go to waste. Also see, Susannah Black, "Not Back to Normal, An interview with John Massengale on urbanism after the pandemic," Breaking Ground, July 22, 2020, https://breakingground.us/not-back-to-normal/

8. A podcast from Streetsblog suggests the number is much higher: "One in 32 people around the world die from car crashes, car-related air pollution, and car-related lead exposure every year. But even the astonishing number doesn't tell the whole story." The podcast is referenced at "The Brake Podcast: How Many People Does Car Culture Kill, Exactly?," Streetsblog USA, March 12, 2023, https://usa.streetsblog.org/2024/03/12/how-many-people-does-car-culture-kill-exactly. Veronica O. Davis, *Inclusive Transportation* (Island Press, 2023). In a chapter called "Should There Be a War on Cars?" another Island Press book says we shouldn't put cars first: "Prioritizing cars creates traffic congestion… [because cars] are not efficient ways to move people…. The more we invest in non-car modes, such as public transportation, bicycles, and pedestrian infrastructure, the more we can bring healing and restoration to communities." See Veronica O. Davis, *Inclusive Transportation: A Manifesto for Repairing Divided Communities* (Island Press, 2023). "The War on Cars" is also the title of a popular podcast from Aaron Naparstek, Doug Gordon, and Sarah Goodyear found at https://thewaroncars.org/.

9. See "If We Stopped Emitting Greenhouse Gases Right Now, Would We Stop Climate Change?" The Conversation, July 4, 2017, https://theconversation.com/if-we-stopped-emitting-greenhouse-gases-right-now-would-we-stop-climate-change-78882.

10. See the origin of Open Streets: https://www.openplans.org/our-history; "Livable City: The Beginning of Open Streets: Bogotá, Colombia Changes the Game," Livable City Non-profit, March 13, 2019, https://www.livablecity.org/the-beginning-of-open-streets-ciclovia-changes-the-game/.

11. San Francisco, Oakland, and Portland all had networks. San Francisco has the best big-city DOT in America, the San Francisco Municipal Transit Authority (SFMTA), run by Jeffrey Tumlin. Tumlin was previously Interim Director of the then-new Oakland Department of Transportation and before that was the director of strategy at Nelson\Nygaard Consulting Associates, "a transportation planning and engineering firm that focuses on sustainable mobility." See "Jeffrey Tumlin, SFMTA Director of Transportation," https://www.sfmta.com/people/jeffrey-tumlin.

12. Three introductions to the 15-minute City: Laurie Winkless, "The '15-Minute City': What They Are And How To Build Them," *Forbes*, September 30, 2022, https://www.forbes.com/sites/lauriewinkless/2022/09/30/the-15-minute-city-what-they-are-and-how-to-build-them/?sh=2fc8d8376bdd; Elizabeth Plater-Zyberk, "The 5-minute neighborhood, 15-minute city, and 20-minute suburb," CNU Public Square, January 8, 2024, https://www.cnu.org/publicsquare/2024/01/08/5-minute-neighborhood-15-minute-city-and-20-minute-suburb; and Paul Tullis, "Has Paris Become the Healthiest City in the World?," *Town & Country*, April 27, 2022, https://www.townandcountrymag.com/leisure/travel-guide/a39715752/paris-bicycle-health-news/.

13. Hidalgo wrote that in September 2019 and introduced the 15-minute city in the weeks before the French Presidential election on March 15, 2020. "Anne Hidalgo: How Paris Is Becoming a Green City," *Time*, September 12, 2019, https://time.com/5669067/paris-green-city/.

14. Laura Laker, "Milan announces ambitious scheme to reduce car use after lockdown," *The Guardian*, April 21, 2020, https://www.theguardian.com/world/2020/

apr/21/milan-seeks-to-prevent-post-crisis-return-of-traffic-pollution.

15. Dave Colon, "Schumer: NYS Will Get 'An Effing Lot of Money' for Street Safety and Cycling," Streetsblog NYC, November 24, 2021, https://nyc.streetsblog .org/2021/11/24/schumer-new-york-state-will-get-an-effing-lot-of-money-for-street-safety-and-cycling.

16. D. Larivière-Bastien *et al.*, "Children's perspectives on friendships and socialization during the COVID-19 pandemic: A qualitative approach," National Library of Medicine, https://www.ncbi.nlm.nih.gov/pmc/articles/PMC9111596/. Many adults still suffer too. Covid highlighted the problems of loneliness when isolation is extreme. Also see Kenan Malik, "We think loneliness is in our heads, but its source lies in the ruin of civil society," *The Guardian* March 24, 2024, https://www.theguardian.com/global/2024/mar/24/we-think-loneliness-is-in-our-heads-but-its-source-lies-in-the-ruin-of-civil-society. Judy Kugel, "An epidemic of Americans alone in a crowd," *Boston Globe*, August 22, 2024, https://www.bostonglobe.com/2023/08/22/opinion/american-elderly-loneliness-epidemic/. Adam Piore and Jason Laughlin, "Four years of COVID: What it taught us about loneliness," *Boston Globe*, March 9, 2024, https://www.bostonglobe.com/2024/03/09/metro/covid-pandemic-anniversary-massachusetts/. Public life and public discourse have coarsened, and road rage and illegal driving seem higher than ever in many parts of America. NYPD has worked to diminish the raised number of fraudulent plates this time, see Gersh Kuntzman, "Wednesday's Headlines: Fake Plates Crackdown Edition," Streetsblog NYC, March 13, 2024, https://nyc.streetsblog.org/2024/03/13/wednesdays-headlines-fake-plates-crackdown-edition.

17. A personal note that gives some sense of how much I love cars—my 5-series station wagon was easily the biggest car I ever owned, so I named her after a very large German siege howitzer from World War I: see "Big Bertha (howitzer)," Wikipedia, https://en.wikipedia.org/wiki/Big_Bertha_(howitzer).

18. From the age of 11, I subscribed to *Road & Track* and *Car and Driver* and managed to influence my parents to buy cars like a Triumph TR4 sports car in British Racing Green and a Rover 2000 TC, advertised in America as "the car for your teenage son to have an accident in." The car came with twin carbs, magnesium wheels, and a five-speed manual transmission with a stubby shifter. Driving was fun then, when there were few traffic jams and not much sprawl. Connecticut was still a state of walkable towns and cities connected by railroads. I had a bicycle, and my parents didn't expect to drive me everywhere. Once I got my driver's license, the roads outside the town centers were quiet, and relaxing to drive on. Gas cost 17¢ a gallon, and no one I knew had heard of climate change. Now our cars are where we develop road rage, flipping off the guy who had the audacity to slip in ahead of us. Cars are our five-thousand-pound, gas-guzzling cocoons, insulating us from contact with strangers and neighbors alike. One way to make driving fun is to make us less dependent on driving on arterials and crowded highways.

19. The Upper West Side of Manhattan has .26 cars per household: see the Hunter College Urban Planning Infographic Report at https://www.hunterurban.org/wp-content/uploads/2024/06/Car-Light-NYC-Infographics-May-2024.pdf. Also see the Nurture Nature Foundation Balanced Transportation Analyzer at https://nurturenature.org/pages/balanced-transportation-analyzer and the Center for Neighborhood Technology's Housing and Transportation Affordability Index at https://htaindex.cnt.org.

20. Some say cars represent freedom, a concept we understand. But consider this list of government requirements for owning and operating a motor vehicle: The driver must be over 16; the driver must take a driving test and a written test to obtain a license; the driver must regularly pay for and renew the license; the driver must obey speed limits; the driver must be insured; the driver may not drink and drive; the driver may not be high while driving; the driver may not text or talk on a hand-held phone while driving; the driver may not park in no-parking zones; the driver may not run red lights; etc. The car must be insured; the car must display license and registration; the car must be regularly inspected and maintained; the car must pass annual emissions testing (in most states); the car is subject to annual fees; the car must be a model approved for use in the United States (includes expensive crash testing); the car must have expensive emissions controls and air bags; the emissions controls, safety equipment, and crash testing add thousands of dollars to the price of the car; car occupants must wear seat belts; and tires and brakes must be in good condition.

21. That gives me access to a beautiful urban campus with a large, open-stack library designed by James Gamble Rogers that, during term-time, is open 24/7. Columbia has the best architecture library in America, designed by McKim, Mead & White, and there are many places on campus to work on public computers with fast internet. Most of my *Street Design* revisions were written in Milstein Library at Barnard College (a women's college that is one of four undergraduate colleges at Columbia). I even get to audit one class per semester—all at a great university I can walk to, and all for free.

22. Jacobs, *op. cit.* Chapter Nine, "The Need for Small Blocks" starts on page 178 in the 1961 paperback. The diagrams and the discussion about the Upper West Side blocks are on pages 179–184.

23. Madison is between Park and Fifth Avenues. Lexington Avenue is between Third Avenue and Park Avenue, originally called Fourth Avenue. The avenues on the Upper West Side were renamed early in the development of the area (page 223). Eighth Avenue is Central Park West, Ninth Avenue equals Columbus Avenue, Tenth is Amsterdam, Eleventh is West End Avenue, and Riverside Drive, which swings to the west and the east, is between West End Avenue and where Twelfth Avenue would be if it existed (page 223). My other apartment was in Tribeca, not far from my current office.

24. The blocks of the Upper East Side are discussed on page 25.

25. Dave Colon, "Report: 85,000 More Damn Cars Will Enter Manhattan's Core Daily in 2023," Streetsblog NYC, October 18, 2021, https://nyc.streetsblog.org/2021/10/18/report-85000-more-damn-cars-will-enter-manhattans-core-daily-in-2023.

26. Jacobs, *op. cit.*

27. Matthew Haag, "How New Yorkers Want to Change the Streetscape for Good," *New York Times*, December 18, 2020, https://www.nytimes.com/interactive/2020/12/17/nyregion/nyc-open-streets.html. The Prospect Heights Neighborhood Development Corporation runs two of the best and most popular Covid-era Open Streets, Vanderbilt Avenue and Underhill Avenue, and maintains an excellent online list of articles about Open Streets founded since the Covid lockdown began: see "In the News," Prospect Heights Places, https://prospectheightsplaces.com/inthenews/.

28. *Ibid.*

29. Dave Colon, "In 2023, Mayor Adams Basically Erased the 'Streets Master Plan'," Streetsblog NYC, January 2, 2024, https://nyc.streetsblog.org/2024/01/02/year-in-review-in-2023-nycs-ambitious-streets-master-plan-was-just-pretty-paper-and-maps. In 2025, aides and associates to Adams were arrested for taking bribes, including bribes to stop the construction of a bike lane: see Kevin Duggan, "'Classic Bribery': How a Powerful Brooklyn Family Crashed and Burned Over a Simple Bike Lane'," Streetsblog NYC, August 22, 2025, https://nyc.streetsblog.org/2025/08/22/road-diet-fat-cash-how-mcguinness-blvd-blew-up-brooklyn-bike-lane-foes, and Kevin Duggan, "Just Absurd': Adams Calls Unsafe McGuinness Blvd. Compromise a 'Win' Despite Lewis-Martin Bribery Indictment," Streetsblog NYC, August 22, 2025, https://nyc.streetsblog.org/2025/08/22/just-absurd-adams-calls-unsafe-mcguinness-blvd-compromise-a-win-despite-lewis-martin-bribery-indictment.

30. See more information on these advocacy groups at https://transalt.org/ourstory, https://nyc.streetsblog.org/about, https://www.openplans.org/our-history, and https://thewaroncars.org/about/.

31. *Death and Life, op. cit.*, 364.

32. As we go to press, New York City has congestion pricing, and it is a success. But Governor Kathy Hochul weakened it, first by delaying it, and then by lowering the top charge by 40 percent. To track the history and the current status of congestion tracking in New York City, check these three links: https://www.nytimes.com/search?query=congestion+pricing, https://gothamist.com/search?q=congestion+pricing, and https://nyc.streetsblog.org/search?s=congestion+pricing. Also see "Financial District Slow Zone" and "Slow Zones and the Slow City" in this chapter.

33. Mailer wanted to build a monorail around the perimeter of Manhattan that would connect parking lots where private cars could be left. His running mate was Jimmy Breslin—they were undoubtedly the first All Pulitzer Mayoral team in New York (or anywhere). Breslin was a hard-drinking journalist who liked to walk from bar to bar, so he supported Mailer's idea. See Steve Dunleavy, "Banning Cars In Manhattan Not A Bad Idea," *New York Post*, December 15, 1999, https://nypost.com/1999/12/15/banning-cars-in-manhattan-not-a-bad-idea/. Breslin described their campaign in a cover story for *New York Magazine* entitled, "Mailer-Breslin Seriously?" May 5, 1969, https://nymag.com/docs/07/11/mailerbreslin.pdf.

34. Walk Score says that San Francisco is the most walkable city in America, but their rating does not include transportation options, where New York ranks number one. Almost every other rating available online puts New York first. For example, see Alicia Underlee Nelson, "The Most Walkable Cities: 10 U.S. Metros Where You Don't Need a Car," Rent.com Blog, September 18, 2023, https://www.rent.com/blog/most-walkable-cities-in-usa/. Walk Score explains their rating system here: https://www.walkscore.com/how-it-works/. Many U.S. cities and towns have neighborhoods with high Walk Scores. My Upper West Side neighborhood has a Walk Score of 98, which Walk Score says is only "the 23rd most walkable neighborhood in New York"; https://www.walkscore.com/. "Walk Score," Wikipedia, https://en.wikipedia.org/wiki/Walk_Score.

35. John Massengale, "Walkability: A Street is a Terrible Thing to Waste," *New York Conference of Mayors Municipal Bulletin*, Summer 2014, https://blog.massengale.com/wp-content/uploads/2014/10/MassengaleNYCOM.pdf.

36. Statistics from the DOT show that the streets around city highway exits, bridges, and tunnels where drivers enter Manhattan are particularly dangerous—as in my neighborhood. New York neighborhoods next to highways also have the worst air quality. But when we get an opportunity to fix that, we spend billions of dollars to keep people driving in and out.

37. Joel Rose, "Vision Zero marks a milestone, but the goal of ending traffic deaths is still far off," NPR, January 12, 2024, retrieved on January 13, 2024, from the WYSO public radio website: https://www.wyso.org/2024-01-12/vision-zero-marks-a-milestone-but-the-goal-of-ending-traffic-deaths-is-still-far-off. For recent stories on Vision Zero in New York City, see "NYC's Vision Zero Nears Decade Mark Without Evident Effects," National Criminal Justice Association, April 28, 2023, https://www.ncja.org/crimeandjusticenews/nyc-s-vision-zero-nears-decade-mark-without-success; "Child & Pedestrian Fatalities Rose in 2024 While 132 Vehicles Received 100 Safety Camera Tickets with Two Exceeding 500: New Data from Transportation Alternatives & Families for Safe Streets," Transportation Alternatives, January 27, 2025, https://transalt.org/press-releases/child-amp-pedestrian-fatalities-rose-in-2024-while-132-vehicles-received-100-safety-camera-tickets-with-two-exceeding-500-new-data-from-transportation-alternatives-amp-families-for-safe-street; and "Lessons from Vision Zero in New York

City," Transportation Alternatives, February 12, 2024, https://projects.transalt.org/lessons-from-vision-zero-new-york-city. Also see page 611 ("…the old guard at the NYC DOT don't like the new limits") and note 150.

38. Justin Davidson, "The Upper West Side's Zone of Pedestrian Death," Curbed, January 24, 2023, https://www.curbed.com/2023/01/upper-west-side-zone-pedestrian-death.html.

39. If you build it, they will come. "What Is Induced Demand?," Planetizen, https://www.planetizen.com/definition/induced-demand.

40. Jeff Nussbaum, "The Night New York Saved Itself from Bankruptcy," *The New Yorker*, October 16, 2015, https://www.newyorker.com/news/news-desk/the-night-new-york-saved-itself-from-bankruptcy. The situation led to one of the most infamous headlines in New York's history: "Ford to City: Drop Dead." The large tabloid headline was published on October 29, 1975: https://www.nydailynews.com/2015/10/29/ford-to-city-drop-dead-in-1975/.

41. In recent years, the City Council has pushed the mayors to do more. David Meyer, "WHAT NEXT?: Four Things Mayor Adams Must Do to Save Vision Zero," Streetsblog NYC, February 15, 2024, https://nyc.streetsblog.org/2024/02/15/what-next-four-things-mayor-adams-must-do-to-save-vision-zero.

42. John Massengale, "Should New York City Streets be illegal?" Common Edge, March 7, 2023, https://commonedge.org/should-new-york-city-streets-be-illegal/. Also published in Streetsblog and City and State.

43. Quotes from Julianna Cuba, "'Park Your Fleet': Harlem Residents Decry Newly Opened Truck Depot on Site of Proposed Housing," Streetsblog NYC, January 19, 2023, https://nyc.streetsblog.org/2023/01/19/park-your-fleet-harlem-residents-decry-newly-opened-truck-depot-on-site-of-proposed-housing.

44. *Ibid.*

45. Or at least that quote is famously attributed to Albert Einstein. The closest verifiable quote comes from a *New York Times* story about a telegram Einstein sent to President Franklin Delano Roosevelt in 1939 "'to let the people know a new type of thinking is essential' in the atomic age 'if mankind is to survive and move toward higher levels.'": see "Atomic Education Urged By Einstein," *New York Times*, May 25, 1946, https://timesmachine.nytimes.com/timesmachine/1946/05/25/100998236.html?pageNumber=11.

46. Massengale, "Should New York City Streets Be Illegal?" *op. cit.*

47. Or at least a rear-projection of it. "The cast never left Los Angeles." Martin Purvis, "*Notorious*—Alfred Hitchcock (1946)," The Film Sufi, February 19, 2014, http://www.filmsufi.com/2014/02/notorious-alfred-hitchcock-1946.html.

48. Andres Viglucci, "The Vice Effect: 30 years after the show that changed Miami," *Miami Herald*, September 29, 2014, https://www.miamiherald.com/article2266518.

html. "The Beach backgrounds in those early *Vice* episodes were desolate for a reason, [Michael] Kinerk recalled, and not because streets were closed for filming by the city film office. There was no city film office. There simply weren't many people on the street." …[Kinerk said] "They were filming all over Miami Beach," he said. "They could film in the middle of the street. There was literally nobody there. There were no cars parked in the street."

49. Goodkin Research, for City of Miami Beach, *Plan NoBe: North Beach Master Plan*, adopted October 19, 2016, https://www.doverkohl.com/plan-nobe.

50. Ebony Venson, Abigail Grimminger, and Stephen Kenny, *Dangerous By Design*, Smart Growth America, July 2022, https://smartgrowthamerica.org/dangerous-by-design-2022/. The complete PDF can be downloaded from https://smartgrowthamerica.org/wp-content/uploads/2022/07/Dangerous-By-Design-2022-v3.pdf.

51. *Ibid.*, 31, The report puts it this way: "This year, the rankings for the deadliest states for pedestrians changed slightly. Previous #1 Florida—where it should be noted that overall deaths still increased significantly in 2020—was surpassed by the increase in New Mexico, which is now the most dangerous state for pedestrians. No state that improved their position in this top 20 list achieved that feat because they reduced their fatality rate. All 20 have grown more deadly with a higher fatality rate compared to their average rate for 2011–2015."

52. "List of U.S. cities with most bicycle commuters," Wikipedia, https://en.wikipedia.org/wiki/List_of_U.S._cities_with_most_bicycle_commuters.

53. Of all trips 2 miles or less, 2 percent are by bike. 26 percent are on foot. *National Household Travel Survey*, U.S. Department of Transportation, Federal Highway Administration, January 8, 2010, https://bikeleague.org/sites/default/files/2009_NHTS_Short_Trips_Analysis.pdf.

54. I know, I was there—Victor.

55. "The ACS asks only about commuting. It does not tell us about bicycling for non-work purposes." See "2010 Bike Commuting Data Released," The League of American Bicyclists, September 24, 2011, https://bikeleague.org/2010-bike-commuting-data-released/.

56. The Corradino Group, "Miami-Dade 2018 Bicycle & Pedestrian Data Collection," Miami-Dade Transportation Planning Organization, 2018, pages 25 and 31, https://miamidadetpo.org/library/studies/miami-dade-bicycle-and-pedestrian-data-collection-2018-06.pdf.

57. From the plan: "A review of the City of Miami's existing network conditions revealed that the corridors, most of which are County or State owned, are designed primarily for auto mobility. The major corridors within the city allow for a high volume of swift moving traffic which results in the isolation rather than connection of the City's neighborhoods. Existing motor vehicle speeds do not provide for a safe environment for bicyclists along these important thoroughfares. Furthermore, the existing conditions

research revealed the lack of bicycle facilities, parking, and the unbalanced geographical distribution of what has been implemented." The Street Plans Collaborative, *City of Miami Bicycle Master Plan* (City of Miami, 2009), https://www.miami.gov/files/sharedassets/public/v/1/transportation/2009-city-of-miami-bicycle-master-plan-1.pdf.

58. "LeBron and Wade Participated in Miami's Critical Mass Bike Event," WPLG 10, February 28, 2014, https://www.youtube.com/watch?v=VaxwOTW57M4.

59. Colville-Andersen is the founder of the Copenhagenize office, https://www.copenhagenize.eu/, and the author of *Copenhagenize: The Definitive Guide to Global Bicycle Urbanism* (Island Press, 2018).

60. *Urban Bikeway Design Guide*, Third Edition, National Association of City Transportation Officials (Island Press, 2025) and *Urban Street Design Guide*, National Association of City Transportation Officials (Island Press, 2013) and https://nacto.org/publication/urban-bikeway-design-guide/.

61. See Aaron Short, "A Bill of Goods: Why Are We Putting Up With Parking-Protected Bike Lanes?" Streetsblog USA, December 3, 2019, https://usa.streetsblog.org/2019/12/03/a-bill-of-goods-why-are-we-putting-up-with-parking-protected-bike-lanes.

62. See Colville-Andersen, *op. cit.*, passim. Also see Short, *op. cit.* For more on one-way Danish cycle tracks, see Troels Andersen, "One-way cycle tracks," Cycling Embassy of Denmark, https://cyclingsolutions.info/one-way-cycle-tracks/. In the First Edition of *Street Design*, we had photographs of different types of bike lanes. See page 375. On page 374 we quoted Enrique Peñalosa, the former mayor of Bogotá, Colombia: "A true bicycle network is one that can be used by a child."

63. Dave Snyder of the People for Bikes nonprofit organization was an early proponent. In 2023 Snyder wrote "We Were Wrong About Sharrows," saying they "accomplish something pernicious which I did not anticipate. They allow officials to take credit for doing something for bicycle safety without impacting car traffic, even though that something is next to nothing. It's just pretending, and it's worse than being honest about priorities. It's insulting to the public to encourage bicycling by painting bike symbols on the street but doing nothing to actually make riding a bike any safer." See Dave Snyder, "We Were Wrong About Sharrows," PeopleForBikes, January 19, 2023, https://www.peopleforbikes.org/news/we-were-wrong-about-sharrows.

64. Flax is the former editor-in-chief of *Bicycling* magazine. He also wrote, "To the politicians and engineers stuck in the middle, sharrows seem like a devilishly perfect compromise — a way to placate the pro-car populists while still being able to claim you did something. . . . There's only one problem: Sharrows are make-believe safety infrastructure." See Peter Flax, "Op-Ed: Why Sharrows Are Bullshit," Streetsblog USA, November 17, 2021, https://usa.streetsblog.org/2021/11/17/op-ed-why-sharrows-are-bullshit.

65. DeWayne Carver, who today is FDOT's Criteria Publications Manager, once pointed out that the skinny, unprotected bike lanes right next to high-speed traffic were more like a stealth road widening, nicknaming them the "cell-phone recovery zone," meaning drivers had extra asphalt for regaining control when they drifted out of their lane while in the middle of composing a text message or social media post.

66. See Florida Department of Transportation "Complete Streets" Policy: https://fdotwww.blob.core.windows.net/sitefinity/docs/default-source/roadway/completestreets/000-625-017-a.pdf?sfvrsn=5f76a980_2.

67. Robert Steuteville, "Florida's Success with Context-Based Street Classification," CNU Public Square, July 17, 2023, https://www.cnu.org/publicsquare/2023/07/17/florida%E2%80%99s-success-context-based-street-classification.

68. *Manual on Uniform Traffic Control Devices for Streets and Highways* 11th Edition, United States Department of Transportation Federal Highway Administration, December 19, 2023, https://mutcd.fhwa.dot.gov/pdfs/11th_Edition/mutcd11thedition.pdf. Title 23 of the Code of Federal Regulations, Part 655.603 states that the MUTCD is the national standard for all traffic control devices installed on any street, highway, or bicycle trail open to public travel: See https://mutcd.fhwa.dot.gov/knowledge/faqs/faq_general.htm#genq2. Page 1, Section 1A.01 to 1A.02 in the Manual states, "Infrastructure elements that restrict the road user's travel paths or vehicle speeds, such as islands, curbs, speed humps, and other raised roadway surfaces, are not traffic control devices. Transverse or longitudinal rumble strips are also not traffic control devices. Operational devices associated with the application of traffic control strategies such as fencing, roadway lighting, barriers, and attenuators are shown in this Manual for context, but their design, application, and usage are not specified since they are not traffic control devices."

69. The article "Street Light" in Wikipedia, https://en.wikipedia.org/wiki/Street_light, has illustrations of streets lights new and old, as well as a long list of citations to other sources.

In 1906, a scant generation after electrified street lighting appeared, the much-admired lampposts on the Passeig de Gràcia in Barcelona (page 647) were introduced by Pere Falqués, the municipal architect. Today, they are a famous signature of the street. "While almost every other detail of the boulevard has changed since then, the beloved lampposts and benches remain," Trià s wrote to the authors. "Barcelona's fixtures were commissioned designs, created by professional artists and architects, not by engineers or lighting experts. They were designed to be permanent parts of the civic world." The job ahead for advocates of placemaking and good streets is to achieve a level of artistry in the designs that will engender loyalty and love among the local residents.

Figure 5.65: Streetlight on the Passeig de Gràcia, Barcelona, Spain. Pere Falqués, 1906. Ramon Triàs points to street lights and lampposts in his native Barcelona as an exemplar of using design to create objects that go beyond the utilitarian expression of technology and function (see note 69 in this chapter). A natural role for civic art is to absorb various elements of the design into context-specific placemaking. The technology's new components should evolve and be integrated into the rest of the street design, so that they add to, rather than subtract from, community character—perhaps even to the point where they become part of the city's fabric and spirit.
Courtesy of Paolo Rosa

70. "The Highway Safety Act of 1966 granted authority to the Secretary of the newly formed U.S. Department of Transportation to establish the national standards for traffic control devices. This meant that compliance with the MUTCD would become mandatory throughout the United States. This action recognized that providing clear, consistent, uniform messaging to road users on a national level is an effective, key component in fostering a safe travel environment. The MUTCD was codified in law for the express purpose of promoting the "safe and efficient utilization" of the nation's system of roads—for all its users." "The Evolution of MUTCD," *Manual on Uniform Control Devices*, FHWA, https://mutcd.fhwa.dot.gov/kno-history.htm.

71. The Bipartisan Infrastructure Law (passed into law November 2021), also known as the Infrastructure Investment and Jobs Act (IIJA), established a rhythm for more frequent MUTCD revisions. As of March 2024, the FHWA website lists 47 known errors in the December 2023 version of the MUTCD. "Manual on Uniform Traffic Control Devices for Streets and Highways," *Manual on Uniform Control Devices*, FHWA, https://mutcd.fhwa.dot.gov/.

72. Fear of being sued and losing funding are the prime enforcement muscles behind the MUTCD. "There is no 'MUTCD Police Department' that is responsible for identifying violations of the MUTCD." Gene Hawkins, PhD, PE and Kelly Laustsen, PE, Kittelson & Associates, "Legal

Aspects of the MUTCD," MUTCD Resources for Practitioners & Attorneys, 2024, https://mutcd.kittelson.com/legal-aspects-of-the-mutcd/.

73. Practitioners should become familiar with any unique mandates and options that apply where they are working. "MUTCDs & Traffic Control Devices Information by State," *Manual on Uniform Control*, FHWA, https://mutcd.fhwa.dot.gov/resources/state_info/index.htm.

74. Practitioners be advised: Learn the difference between the terms *shall, should,* and *may.* A common complaint is that engineers interpret all three as mandatory absolutes, but "should" and "may" mean there are options that can be explored. "Evolution of the MUTCD," *op. cit.*

75. "Frequently Asked Questions - General Questions on the MUTCD," *Manual on Uniform Control*, FHWA, https://mutcd.fhwa.dot.gov/knowledge/faqs/faq_general.htm#genq2.

76. Kea Wilson, "Feds, Advocates Talk About What's In The New MUTCD (And What Isn't)!," Streetsblog USA, December 19, 2023, https://usa.streetsblog.org/2023/12/19/feds-advocates-talk-about-whats-in-the-new-mutcd-and-what-isnt.

77. Email from Phil Jones to Victor Dover, March 24, 2024.

78. Email from Miami-Dade Transportation Planning Organization, March 11, 2024.

79. City of Minneapolis, "Getting Around," https://www.minneapolismn.gov/getting-around/.

80. See The Street Plans Collaborative, City of Coral Gables Bicycle Master Plan, https://street-plans.com/city-of-coral-gables-bicycle-master-plan-coral-gables-fl/.

81. "Gables Greenways," Bike Walk Coral Gables, August 25, 2017, https://bikewalkcoralgables.org/updates/2017/7/23/gables-greenways.

82. The term "bikelash" may have originated with Gersh Kuntzman when he was editing *The Brooklyn Paper* in 2009, Aaron Naparstek wrote us in March 2024. The term made the cover of *New York* magazine in early 2011 and was used by Naparstek in a presentation to the Harvard Kennedy School of Government in October of that same year. See Matthew Shaer, "Bikelash: How a Little Squabble Over Bike Lanes Became the Newest Urban Culture War," *New York,* March 28, 2011.

83. Opposition to walking can get as fierce and indignant as outrage over biking. In 2019, the same former mayor flimsily invoked historic preservation as his reason to oppose adding more sidewalks to Coral Gables. "We are a mature city," he said, "we've been here for ninety-some years, we are set, our streets are set, our homes are historic. . . I'm not going to be part of efforts to change the nature of our city. I'm just not going to be part of it." Video recording of City Commission meeting, November 15, 2019, www.coralgables.granicus.com.

84. Aaron Leibowitz, "There's a New Bike Lane Proposal in Coral Gables. Residents Are Furious Again," *The Miami Herald,* January 4, 2020, https://www.miamiherald.com/news/local/community/miami-dade/coral-gables/article238904138.html#storylink=cpy. From the article: "In the city planning industry, the backlash to bike lanes is known as "bikelash," [public works director] Santamaria said—though he perhaps underplayed the unique zeal it takes on in his community."

85. Video recording of Coral Gables City Commission meeting, March 10, 2022, www.coralgables.granicus.com.

86. James Brasuell, "Miami City Commissioners Vote to Reinstate Downtown Parking Minimums," Planetizen, March 31, 2022, https://www.planetizen.com/news/2022/03/116687-miami-city-commissioners-vote-reinstate-downtown-parking-minimums.

87. Interview with Jessica Keller, PE, January 12, 2024.

88. Based on public records requested by Jessica Keller, PE, and interview with Robert Neway, March 16, 2024.

89. Some might even work at Dover, Kohl & Partners.

90. Interview with Andy Boenau, PE, March 16, 2024.

91. Thanks in part to *Miami Vice* crews who repainted and relit old buildings as they went from location to location; and thanks in part to the successful efforts of Barbara Capitman and the Miami Design Preservation League, who promoted the Art Deco District; and thanks in part to the pioneering real estate investments of Tony Goldman, who re-established South Beach as hot property.

92. Mike Lydon and Tony Garcia, *Tactical Urbanism: Short-Term Action for Long-Term Change* (Island Press, 2015), http://tacticalurbanismguide.com/guides/tactical-urbanism-the-book/.

93. Interview with Tony Garcia, March 13, 2024. A revamped version of a quick-building program was launched by Miami-Dade County in 2023 in preparation for the NACTO conference scheduled in Miami for May 2024; we may yet see progress.

94. Mari Chael, "Prioritize cycling and connectivity with the Miami Loop," *The New Tropic,* May 4, 2015.

95. In the moment, the elected officials chose pacifying the adjacent homeowners over safety. Finding himself on the losing side of a City Commission vote regarding the bike lanes, former mayor Philip Stoddard chastised the other commissioners who shut the project down: "You will have blood on your hands." Victor Dover watched the meeting.

96. Mark R. Brown, *Human Speed: A City Planner's Evolution from a Daily Driver to a Multimodal Traveler* (Car Free America, 2023), https://carfreeamerica.net/human-speed/.

97. For more on Westfall's discussion of the Good Life in the Good City, see Samir Younés and Carroll William Westfall, *Architectural Type and Character: A Practical Guide to a History of Architecture* (Routledge, 2022): 28–29, and Carroll William Westfall, "The Architect as Citizen," *Traditional Building,* https://www.traditionalbuilding.com/opinions/architect-citizen. "The architect as citizen seeks beauty as the counterpart to justice" is one idea in the essay.

98. See Anthony Sutcliffe's Paris, *An Architectural History* (Yale University Press, 1993): 99. The National Assembly was debating "Manhattanization."

99. Alain de Botton, *The Architecture of Happiness* (Pantheon, 2006); John Massengale, "Building for Beauty," *The Wall Street Journal*, November 19, 2006, https://blog .massengale.com/2006/11/19/wsj-building-beauty/.

100. Gillian Neimark, "How A Florida Wildlife Biologist Became One of The Greenest Mayors In America," *Good*, February 14, 2018, https://www.good.is/articles/phillip-stoddard-florida-mayor.

101. Stoddard was #47 on the "Politico 50" list of "thinkers, doers and visionaries transforming American politics." *Politico*, September 13, 2016, https://www.politico .com/magazine/politico50/2016/philip-stoddard-harold-wanless/.

102. Donald Shoup, *The High Cost of Free Parking* (American Planning Association, 2005). There is also an Updated Edition, published by Routledge in 2011.

103. A best-selling book about this is *Suburban Nation: The Rise of Sprawl and the Decline of the American Dream* (North Point Press, 2000) by Andrés Duany, Elizabeth Plater-Zyberk, Jeff Speck.

104. Roberta Smith, "Richard Serra, Who Recast Sculpture on a Massive Scale, Dies at 85," *New York Times*, March 26, 2024, https://www.nytimes.com/2024/03/26/arts/richard-serra-dead.html.

105. Michael Kimmelman, "For Richard Serra, Art Was Not Something. It Was Everything," *New York Times,* March, 27, 2024, https://www.nytimes.com/2024/03/27/arts/ design/richard-serra-death-appraisal.html.

106. *Ibid.*

107. The Wikipedia article on Tilted Arc has a number of quotes pro and con: "Tilted Arc," Wikipedia, https://en.wikipedia .org/wiki/Tilted_Arc.

108. Sean O'Hagan, "Man of steel: an interview with Richard Serra," *The Guardian*, October 4, 2008, https://www .theguardian.com/artanddesign/2008/oct/05/serra.art.

109. Westfall, "The Architect as Citizen," *op. cit.*

110. Justin Davidson, "The Greatest Building: If I Had To Pick One Tower, It Wouldn't Be the Empire State Building," *New York Magazine,* January 11, 2011, https://nymag .com/news/features/greatest-new-york/70475/. All the architects in the article except Robert A.M. Stern were on the architecture faculty at Columbia University's Graduate School of Architecture, Planning and Preservation, which has an outsized influence on architecture, planning, and urban design in New York City (Stern, rumored to be the youngest tenured faculty member in the history of Columbia, left Columbia in 1998 to become Dean of the Yale School of Architecture).

In 2009, members of the Columbia architecture faculty invited Stern and Elizabeth Plater-Zyberk (then the Dean of the University of Miami's School of Architecture) to a small, invitation-only meeting at the school's Avery Hall to discuss the future of architecture after the financial crisis of 2008. At one point in the meeting, faculty star Liz Diller said to Stern and Plater-Zyberk, "You have both done a good job of dealing with public taste, but we don't want to deal with public taste. We want to be artists" [like Richard Serra].

At that time, Amanda Burden was in her first term as Mayor Michael Bloomberg's New York City Planning Commissioner, just beginning to promote "dynamic, world-class" Modernism that paired luxury developers with Starchitects to interest the global 1% in buying real estate assets in New York City in places like Billionaires' Row. Burden was instrumental in getting developer Bruce Ratner to hire Frank Gehry to design Atlantic Yards, and in enabling the Supertall towers designed by Starchitects on Billionaires' Row and at Hudson Yards. Burden and the Bloomberg administration gave the Highline invaluable support, and it became "an unaffordable housing generator," surrounded by many new towers designed by Starchitects. (See John Massengale, "Big Real Estate's Continuing Stranglehold Over New York City," Common Edge, February 14, 2022, https://commonedge.org/big-real-estates-continuing-stranglehold-over-new-york-city/ and John Massengale, "My Meals With Bob: My Mini Memoir in Common Edge," There Are Two Types of Architecture, April 12, 2022, https://blog.massengale .com/2022/04/12/ramsa/.)

As New York boomed in the years following the financial crisis, Diller's firm benefited from the pairing of luxury development and self-proclaimed avant-garde architects. Her firm grew rapidly and received several large commissions, including the remake of the Highline and the design of The Shed, a theater in Hudson Yards at the end of the Highline. Not everyone shared the love: see Ben Davis, "The Shed Is a Shiny Billionaire's Box of Dreams. Can It Be an Important Cultural Center Too?," artnet, April 4, 2019, https://news.artnet.com/ art-world-archives/the-shed-review-1508725, and Tiana Reid, "The Shed Sucks: A Dispatch From New York's Latest Cultural Megaspace," *The Nation*, May 23, 2019, https://news.artnet.com/art-world-archives/ the-shed-review-1508725: "The Hudson Yards cultural center proves—yet again—that art is inseparable from commerce."

Many quotes in the story sound like something Howard Roark might have said in *The Fountainhead* (Ayn Rand, 1943). The Wikipedia article on *The Fountainhead* cites architects who have called the novel an inspiration for their work: "Architect Fred Stitt, founder of the San Francisco Institute of Architecture, dedicated a book to his "first architectural mentor, Howard Roark." According to architectural photographer Julius Shulman, Rand's work "brought architecture into the public's focus for the first time." He said *The Fountainhead* was not only influential among twentieth-century architects, but moreover "was one, first, front and center in the life of every architect who was a modern architect." Some of New York's developers have been known for their egos. During his 2016 presidential campaign, real estate developer Donald

Trump praised the novel, saying he identified with Roark." *The Fountainhead*, Wikipedia, https://en.wikipedia.org/wiki/The_Fountainhead. Trump's daughter Ivanka wrote in her autobiography, "*Walking down Fifth Avenue, I look up and wonder: What piece of the skyline will be mine?*" See Ivanka Trump, *The Trump Card: Playing to Win in Work and Life* (Simon & Schuster, 2009): 99. Also see John Massengale, "Who's Afraid of Virginia Woolf," There Are Two Types of Architecture, March 18, 2004, https://blog.massengale.com/2004/03/18/virginia-woolf/.

111. Justin Davidson, "The Supertalls Have Walled In Central Park Even from deep inside the park, the supertalls are impossible to not see," Curbed, April 3, 2024, https://www.curbed.com/article/supertalls-scale-towers-central-park-height-view.html.

112. Emily Lefroy, "New Brooklyn Tower divides NYC with its 'evil' 'Sauron' vibes," *New York Post*, May 10, 2023, https://nypost.com/2023/05/10/brooklyn-tower-divides-nyc-with-its-evil-sauron-vibes/.

113. Hannah Frishberg, "Developer of Brooklyn's tallest skyscraper defaults on $240M loan—93-story building faces foreclosure," *New York Post*, April 2, 2024, https://nypost.com/2024/04/02/real-estate/developer-of-brooklyns-tallest-skyscraper-defaults-on-240m-loan/. Also see Mary K. Jacobs, "Here's what it's like living in an empty, luxury Brooklyn high-rise," *New York Post*, March 12, 2025, https://nypost.com/2025/03/12/real-estate/what-its-like-living-in-an-empty-luxury-brooklyn-high-rise/.

114. Alexandra Lange, "Downtown Brooklyn Gets the Gotham City Treatment," *Bloomberg City Lab*, September 25, 2023, https://www.bloomberg.com/news/features/2023-09-25/brooklyn-tower-and-100-flatbush-lead-brooklyn-s-art-deco-revival.

115. Lefroy, *op. cit.*

116. In 2025, a planning student went to Columbia's GSAPP Career Office and asked for help getting a summer job with a New Urban office in New York City. "We can't help you," the career advisor said, "We don't like New Urbanism." Told to the authors, by the student. For Koolhaas, see G. La Giorgia, "Market v. meaning," Architecture Week, September 5, 2009, http://www.architectureweek.com/2007/0905/design_3-1.html/.

117. It is commonly accepted that Goethe said "Music is liquid architecture, and architecture is frozen music," but the closest written documentation today comes from the book *Goethe's Conversations with Eckermann*: "I've found a page among my papers, where I describe architecture as frozen music," Goethe says. "And there's something in that, you know; the state of mind produced by architecture is similar to the effect of music." See Johann Peter Eckermann, *Gespräche mit Goethe in den letzten Jahren seines Lebens*, https://www.projekt-gutenberg.org/eckerman/gesprche/titlepage.html. Also see "Goethe on architecture and music: source," Reddit, https://www.reddit.com/r/askphilosophy/comments/13j7n2h/goethe_on_architecture_and_music_source /.; Daniel Walden, "Frozen Music: The Synthesis of Music and Mechanical

Theory in Vitruvius' *De Architectura*," *Sunoikisis Talk Final Draft*, April 27, 2012, https://research-bulletin.chs.harvard.edu/wp-content/uploads/2012/04/SURS.Walden.paper_1.pdf; and Feby Susan Philip, "Architecture is Frozen Music," *Elemental*, April 29, 2021, https://www.elemental-architects.com/post/architecture-is-frozen-music.

118. "Paul McCartney on the Magic of Music," *Beatlemania Anonymous*, YouTube, https://www.youtube.com/watch?v=vAwtkzSo_0I. Yo-Yo Ma told Stephen Colbert that "music is energy" on The Late Show with Stephen Colbert: "Like Humans, Nature Is Unbelievably Creative and Utterly Destructive – Yo-Yo Ma," https://www.youtube.com/watch?v=6gvLTBhtf4M.

119. *Ibid.*

120. See Homer, "Goethe: 'Beauty is a manifestation of secret natural laws, which otherwise would have been hidden from us forever'," The Socratic Method, January 21, https://www.socratic-method.com/quote-meanings-and-interpretations/johann-wolfgang-von-goethe-beauty-is-a-manifestation-of-secret-natural-laws-which-otherwise-would-have-been-hidden-from-us-forever-2.

121. Alain de Botton translated and quoted Stendhal in *The Architecture of Happiness* (Pantheon Books, 2006): 78. He actually said, "La beauté n'est que la promesse du bonheur." See Stendhal, *De L'Amour* (Garnier Frères, 1826): Chapter XVII, https://www.gutenberg.org/cache/epub/60882/pg60882-images.html. See also John Massengale, "Building for Beauty," *The Wall Street Journal*, November 19, 2006, https://blog.massengale.com/2006/11/19/wsj-building-beauty/. "Beauty is the promise of happiness" is frequently attributed to Edmund Burke, although there is no evidence today that he said that verbatim. See, however, Burke, *A Philosophical Enquiry into the Origin of Our Ideas of the Sublime and Beautiful* (Thomas McLean, 1823): *passim*. For an online version, see https://archive.org/details/philosophicalinq00burk/mode/2up.

In his Nobel Prize acceptance speech, Aleksandr Solzhenitsyn discussed an "enigmatic remark" by Dostoevsky that "beauty will save the world." An English translation of his speech can be found at https://www.nobelprize.org/prizes/literature/1970/solzhenitsyn/lecture/.

122. Denis C. Williams, *God's Wilds: John Muir's Vision of Nature* (Texas A&M University Press, 2002). Muir famously said, "When we try to pick out anything by itself, we find it hitched to everything else in the universe." See *My First Summer in the Sierra* (Houghton Mifflin, 1911):110. In *The Most Astounding Fact*, a video by Max Schlickenmeyer, 2012, https://www.youtube.com/watch?v=9D05ej8u-gU, Neil deGrasse Tyson said, "We are all connected; to each other, biologically, to the Earth, chemically; to the rest of the universe, atomically."

123. See "What Is Quantum Physics?" California Institute of Technology, https://scienceexchange.caltech.edu/topics/quantum-science-explained/quantum-physics.

124. Albert Einstein is quoted at: https://www.goodreads.com/quotes/63087-the-pursuit-of-truth-and-beauty-is-a-sphere-of.

125. From Taylor's website, "In 2008, Dr. Jill gave the first TED talk that ever went viral on the Internet, which now has over 27.5 million views. Also in 2008, Dr. Jill was chosen as one of *Time* magazine's '100 Most Influential People in the World' and was the premiere guest on Oprah Winfrey's 'Soul Series' webcast." https://www.drjilltaylor.com/.

126. Jill Bolte Taylor, "My Stroke of Insight," TED Talk, February 2008, https://www.ted.com/talks/jill_bolte_taylor_my_stroke_of_insight.

127. John Massengale talked about this in the On Cities Podcast: see https://blog.massengale.com/2023/08/29/me-on-the-on-cities-podcast/. For the On Cities site, see https://www.voiceamerica.com/show/4119/on-cities.

128. "Sergei Gepshtein," Salk Institute, https://www.salk.edu/scientist/sergei-gepshtein/.

129. Quoted by Mark Hewitt, in conversation. A non-profit group called the Heartmath Institute promotes "heart-brain coherence," using music. See their website https://www.heartmath.org/.

130. "Sergei Gepshtein," Academy of Neuroscience for Architecture, https://anfarch.org/users/sergei-gepshtein.

131. Mark Hewitt, *Draw in Order to See: A Cognitive History of Architectural Design* (Oro Editions, 2020). Michael Crosbie's review of the book is a good short introduction to it: Michael J. Crosbie, "Reaffirming the Essential Role of Drawing in Design" Common Edge, January 6, 2021, https://commonedge.org/reaffirming-the-essential-role-of-drawing-in-design/.

132. John Massengale, "Neuroscience & Measuring the Experience of Place," Place Science, October 27, 2017, https://www.place-science.com/engelsberg/. John Massengale and Mark Hewitt will contribute an article called "Measuring the Experience of Place in City Streets" to Derek Clements-Croome, Ali GhaffarianHoseini, and Charles Walker, editors, From *Neuro-architecture to Neuro-cities: Call for Book Chapters* (Springer Nature, 2026).

133. Some recommendations from William Browning, author with Catherine Ryan and Joseph Clancy of *14 Patterns of Biophilic Design* (Terrapin Bright Green, 2014, revised 2024), https://www.terrapinbrightgreen.com/reports/14-patterns/: Jenny Roe and Layla McCay, *Restorative Cities: Urban Design for Mental Health and Wellbeing* (Bloomsbury Visual Arts, 2021). Roe did early work with mobile EEG looking at differing outcomes from a walk through three environments (park, suburban and urban); Julio Cesar Bermudez, author of *Transcending Architecture: Contemporary Views on Sacred Space* (Catholic University of America Press 2015) uses advanced mobile EEG to study how architecture influences the mind and spirit, with differing responses to sacred and secular spaces. The website Spiritual Understanding & Architecture, Empirical Inquiries has experiments and references by Bermudez and others: https://www.building-spiritual-understanding.net/; another author at the Spiritual Understanding website, Zakaria Djebbara, was the editor of *Affordances in Everyday Life, A Multidisciplinary Collection of Essays* (Springer, 2022) also uses advanced mobile EEG to demonstrate subconscious responses to indoor spaces. For more on Browning, see "Bill Browning: The Buildings People Love Have a Strong Biophilic Component," Venetian Letter, January 23, 2025, https://venetianletter.com/2025/01/23/bill-browning-the-buildings-people-love-have-a-strong-biophilic-component/.

134. William H. Whyte, *The Social Life of Small Urban Spaces* (The Conservation Foundation, 1980). Working with the New York City Planning Commission in the late 1960s, Whyte used still cameras, movie cameras, and notebooks to record public life in an objective and measurable way.

135. Sussman and Hollander, *Cognitive Architecture, Designing for How We Respond to the Built Environment* (Routledge, 2021). Also see a 2023 publication of eye-tracking research in the *Journal of Applied Sciences*: Hernan J. Rosas, Ann Sussman, Abigail C. Sekely, and Alexandros A. Lavdas, "Using Eye Tracking to Reveal Responses to the Built Environment and Its Constituents," November 6, 2023, https://www.mdpi.com/2076-3417/13/21/12071.

136. *Ibid.* Also see Colin Ellard, *Places of the Heart: The Psychogeography of Everyday Life* (Bellevue Literary Press, 2015), Sarah William Goldhagen, *Welcome to Your World: How the Built Environment Shapes Our Lives* (Harper-Collins Publishers, 2017), Sarah Robinson, *Architecture is a Verb* (Routledge, 2021), and Davide Ruzzon, *Tuning Architecture with Humans: Neuroscience Applied to Architectural Design* (Mimesis International, 2022).

137. Christopher Alexander, *The Nature of Order* (Center for Environmental Structure, 2002). Four volumes. Alexander writes. "I believe we are on the threshold of a new era when the proper understanding of the deep questions of space, as they are embodied in architecture will play a revolutionary role in the way we see the world and will do for the world view of the 21st and 22nd centuries, what physics did for the 19th and 20th."

138. When painted with a high-quality paint a "cementitious" material called Hardie Board can be a pleasing, lower-cost choice than wood.

139. A partial list includes: Vitruvius Pollio, *The Ten Books on Architecture* (Harvard University Press, London: Oxford University Press, 1914); Andrea Palladio, *Four Books of Architecture* (MIT Press, 1997); Camillo Sitte, *The Birth of Modern City Planning* (Rizzoli, 1986); Werner Hegemann and Elbert Peets, *The American Vitruvius, An Architect's Handbook of Civic Art* (Princeton Architectural Press, 1988); John Belcher, *Essentials in Architecture* (Princeton Architectural Press, 1988). More recent books include Christopher Alexander, *A Pattern Language: Towns, Buildings, Construction* (Oxford University Press, 1977); Steven Semes, *The Architecture of the Classical Interior* (W. W. Norton & Company, 2004); Stein Eiler Rasmussen, *Experiencing Architecture* (MIT Press, 1964);

Frank Ching, *Architecture: Form, Space, and Order* (John Wiley & Sons, 1979). Create Street's *Of Streets and Squares, Which places do people want to be in and why?* (Cadogan, 2019) from the English group Create Streets: it surveys and summarizes research and architectural theory from many sources, all included in a useful bibliography. Download available at: https://issuu.com/cadoganlondon/docs/of_streets_and_squares_26_march_wit?e=32457850/68741701.

140. Semes, *op. cit.* The Rule of Three says when things are grouped in threes, they are more appealing to the eye, appearing more natural. There is also the Rule of Thirds, which divides surfaces into thirds. Photographers frequently use the Rule of Thirds, which some call a simplification of the Golden Section.

141. See in particular the first chapter: Tony Hiss, *The Experience of Place* (Knopf, 1990). A more recent book is Steve Bass's *Beauty Memory Unity: A Theory of Proportion in Architecture* (Lindisfarne Books, 2019). Bass has related videos online: Steve Bass, "Beauty Memory Unity: A Theory of Proportion in Architecture & Design," YouTube, https://www.youtube.com/watch?v=wtbKBXFDdPs, and Keith Critchlow, "The Art of the Ever-True," YouTube, https://www.youtube.com/watch?v=6V1qwLOUKrI.

142. Yes, that is a generalization. But talk to architects and planners who work in that field and they will usually confirm the generalization.

143. Alexander, *The Nature of Order, ibid.*

144. G.K. Chesterton, *Orthodoxy* (The Bodley Head, 1908); 121. Digital copy at https://www.gutenberg.org/ebooks/130.

145. Notably, the Saturday night entertainment at the first Congress was two hours of listening to traffic engineers. All later congresses had parties on Saturday night, but the CNU added elected officials, retail consultants, investment experts, and the like over the years. The focus of the congresses was sharing knowledge about how to design, get approvals for, build, and sell better places.

146. See Brendan Crain, "By the City/For the City: Making a Better New York," *Next City*, April 13, 2011, https://nextcity.org/urbanist-news/by-the-city-for-the-city-making-a-better-new-york. The exhibit catalog was published 2011.

147. "Slow New York," Street Design, https://street.design/slowny.

148. New York City has information about Vision Zero at https://www.nyc.gov/content/visionzero/pages/#.

149. The NYC DOT maintains a list of Slow Zones here: https://www.nyc.gov/html/dot/html/motorist/slowzones-list.shtml. Also see the city's *Street Design Manual*: https://www.nycstreetdesign.info/. Regarding the 25 mph speed limit, a story at cbsews.com says, "The new default speed limit applies to all streets where no other limit is posted. Highways like the FDR, West Side Highway and Riverside Drive will still have higher limits and school zones will have lower limits. The 25 mph limit also does not apply to the following streets and parkways: Webster Avenue and Mosholu Parkway in the Bronx; Hylan Boulevard and Richmond Terrace on Staten Island; Fort Hamilton Parkway and Ocean Avenue in Brooklyn; and Utopia Parkway and Bell and Springfield boulevards in Queens." See "NYC's New 25 MPH Speed Limit Goes Into Effect On Most City Streets," CBS News, November 7, 2014, https://www.cbsnews.com/newyork/news/nycs-new-25-mile-per-hour-speed-limit-takes-effect/. Note: In 2024, the speed limit on Riverside Drive everywhere but on the viaduct over the 125th Street valley is 25 mph.

150. Stephen Miller, "Queens Blvd Gets 'Slow Zone' Label, But Speed Limit Remains the Same," Streetsblog NYC, May 2, 2014, https://nyc.streetsblog.org/2014/05/02/queens-blvd-gets-slow-zone-label-but-speed-limit-remains-the-same. A New York City Vision Zero story from John Massengale: I once attended a meeting between the New York City DOT Commissioner, the Deputy Borough President of Queens, the Executive Director of Transportation Alternatives, several of the top engineers at the DOT, and a few additional people from Transportation Alternatives and the Deputy Borough President's office. The Commissioner and Vision Zero were both new to New York. The Commissioner gave a short presentation on Vision Zero and Queens Boulevard, "the Boulevard of Death" (see page 353). After explaining that 60 percent of the traffic deaths in New York City happen on 10 percent of the streets, she talked about Vision Zero and the Mayor's pledge to reduce traffic deaths in New York City to zero by 2024 (10 years later). That made the redesign of Queens Boulevard a high priority New York City DOT, because it was one of the most deadly streets in New York. The Commisioner then turned to the engineer in charge of the traffic lights in the city and said, "So Joe, what do you think we should do?"

"I don't think we should do anything," he said, "because traffic is flowing well."

Then he explained that the lights were set so that cars going 45 miles per hour would not have to stop very often. If the DOT reset the target speed at 25 or 35 miles per hour, the engineer said, rear-end collisions would increase.

Three takeaways from this: First, and I have seen this in other interactions between the engineers and the Commissioners, there is a belief among DOT staff that Commissioners come and go, and that the engineers need to maintain department and professional standards. Second, as we said (page 583), don't expect the profession that caused the problem to solve the problem. Third, in a semi-public meeting with important elected officials present, the engineer was willing to ignore what the Commissioner had explained was an important goal for the new Mayor and his Commissioner: dramatically slowing traffic to eliminate traffic fatalities.

151. "Reckless Driving Casualties Rising as NYPD Enforcement Lags," Streetsblog NYC, July 14, 2009, https://old.nyc.streetsblog.org/2009/07/14/ta-report-reckless-driving-casualties-rising-as-nypd-enforcement-lags/. Also see Eve Kessler, "New App Helps Reckless Drivers Thumb Their Noses at City's Speed Cameras," Streetsblog NYC,

May 13, 2022, https://nyc.streetsblog.org/2022/05/13/new-app-helps-reckless-drivers-thumb-their-noses-at-citys-speed-cameras.

152. "Executive Summary," Move NY, https://iheartmoveny.org/executive-summary and "About Us," Charleston Moves, https://charlestonmoves.org/about-us/. Also Molly McArdle, "This Transportation Engineer Won't Give Up on Moving New York City," Next City, May 2, 2016, https://nextcity.org/features/move-ny-plan-transportation-plan-gridlock-sam-schwartz.

153. La Cittaslow website: https://www.cittaslow.org/content/association. Cittaslow developed from the Slow Food organization The Slow Food website: https://www.slowfood.com/our-history/. Slow Zone projects are already shown in Chapter Four: Yorkville Promenade, Winslow Homer Walk, Queens Boulevard.

154. FDNA and Buro Happold created a book that was printed in small numbers for distribution in the neighborhood. A PDF of the presentation can be downloaded at https://docs.wixstatic.com/ugd/9e36dc_5d4721ac7f9f4c3da0622f9302feebb5.pdf: *Make Way for Lower Manhattan* (Financial District Neighborhood Assocation, 2019). One of the FDNA Board members wrote an article that documented the history and many of the goals: Catherine McVay Hughes, "A Greener FiDi—Make Way for Lower Manhattan: Shared Streets Project," *The Sallan Foundation*, July 2, 2019. The FDNA website is at https://www.fidinewyork.org/.

155. "Mental Mapping," There Are Two Types of Architecture, November 11, 2023, https://blog.massengale.com/2023/11/11/mapping

156. Hans Karssenberg, Jeroen Laven, Meredith Glaser, and Mattijs van 't Hoff, Editors, *The City at Eye Level* (Eburon Academic Publishers, 2016): *passim*. Download the ebook at https://issuu.com/stipoteam/docs/ebook_the.city.at.eye.level_english#google_vignette.

157. David W. Dunlap, "Commercial Property: Unusual Spaces; An Oak Board Room, Anyone?," *New York Times*, December 22, 1991, https://www.nytimes.com/1991/12/22/realestate/commercial-property-unusual-spaces-an-oak-board-room-anyone.html.

158. Rupert Brooke, "Arrivals," *Letters from America*, Preface by Henry James (Charles Scribner & Sons, 1916): https://www.gutenberg.org/files/6445/6445-h/6445-h.htm.

159. *Make Way, op. cit.*: 28.

160. New York City has something called Local Law 11. See Penelope Green, "Our Lives, Under Construction Three hundred miles of protection envelops New York City. What can we make of it?," *New York Times*, January 2, 2020, https://www.nytimes.com/2020/01/02/style/scaffolding-new-york-city.html. FDNA found that both the scaffolding and the garbage on the sidewalks are big concerns for FiDi residents. The FDNA website has a website about this: https://vimeo.com/322157109.

161. The space should be on this list of places vaulted with Guastavino tiles: Logan Ward, "8 Majestic Guastavino Tile Vaults from Around the Country," *Preservation*, Summer 2020, https://savingplaces.org/stories/7-majestic-guastavino-tile-vaults-from-around-the-country. Perhaps the close proximity to the traffic coming off the bridge is why it is not.

162. "A Street Management Framework for Lower Manhattan: The Downtown of the 21st Century," 2008, https://www.sallan.org/pdf-docs/Snapshot-ss_trb_09_lmsm_paper.pdf.

163. Dave Colon, "FiDi Shared Streets Advocates Press DOT to Show 'Urgency' on Neighborhood Makeover," Streetsblog NYC, July 1, 2022, https://nyc.streetsblog.org/2022/07/01/fidi-shared-streets-advocates-press-dot-to-show-urgency-on-neighborhood-makeover and Carl Glassman, "Putting Pedestrians First in FiDi? City's Search for Answers to Start Anew," *Tribeca Trib*, August 12, 2022, https://www.tribecatrib.com/content/putting-pedestrians-first-fidi-citys-search-answers-start-anew.

164. *Make Way for Lower Manhattan, op. cit.* Also see Vincent Barone, "Imagine Manhattan's Financial District embracing a 'slow streets' approach," *amNY*, March 12, 2019, https://www.amny.com/transit/financial-district-slow-streets-1.28429784; Sydney Pereira, "FiDi Group Wants To Open Lower Manhattan Streets To Pedestrians," *Patch Media*, March 19, 2019, https://patch.com/new-york/downtown-nyc/fidi-group-wants-open-lower-manhattan-streets-pedestrians; Mark Alan Hewitt, "Slow streets proposed for New York's Financial District," *The Architect's Newspaper*, April 8, 2019, https://www.archpaper.com/2019/04/slow-streets-financial-district/; and Dave Colon, "FiDi Shared Streets Advocates Press DOT to Show 'Urgency' on Neighborhood Makeover," Streetsblog USA, July 19, 2022, https://nyc.streetsblog.org/2022/07/01/fidi-shared-streets-advocates-press-dot-to-show-urgency-on-neighborhood-makeover.

165. John Norquist, "Tear It Down!" *Democratic Leadership Council's Blueprint Magazine*, September 1, 2000, http://www.preservenet.com/freeways/FreewaysTear.html.

166. Considering that greenhouse gases from motor vehicles are America's biggest contribution to global warming, it is surprising that the big environmental groups have not been more active on this issue. In April 2024, the Brooklyn-Queens Expressway Environmental Justice Coalition finally took a strong stand. Bill McKibben has been one of the leaders of climate change since 1989. Find more in the biography on his website: https://www.billmckibben.com/bio.html. Also see Mary Frost, "Feds approve $5.6M grant to reconnect neighborhoods split by BQE," *Brooklyn Daily Eagle*, March 12, 2024, https://brooklyneagle.com/articles/2024/03/12/feds-approve-5-6m-grant-to-reconnect-neighborhoods-split-by-bqe/. For a photograph of Mayor Norquist next to the BQE, see "What's the Difference Between BQE and the Berlin Wall?" There are Two Types of Architecture, February 10, 2026, https://placemakers.com/codes-study/.

167. John Massengale, "Traffic flow and status quo: There are better ways to spend $10 billion than rebuilding the BQE," *Daily News*, January 15, 2023, https://www.nydailynews.com/2023/01/15/traffic-flow-and-status-quo-there-are-better-ways-to-spend-10-billion-than-rebuilding-the-bqe/; John Massengale, "Buck the BQE: The city should not replace a key stretch of the cantilevered Brooklyn highway," *Daily News*, November 25, 2019, https://www.nydailynews.com/2019/11/25/buck-the-bqe-the-city-should-not-replace-a-key-stretch-of-the-cantilevered-brooklyn-highway/; Ariama Long, "The whole story behind the BQE reimagining," *Amsterdam News*, August 26, 2021, https://amsterdamnews.com/news/2021/08/26/whole-story-behind-bqe-reimagining/.

168. Streetsblog NYC wrote an article called "TO END THE MESS: Make the Brooklyn Bridge Car-Free," Streetsblog NYC, https://nyc.streetsblog.org/2020/07/09/to-end-the-mess-make-the-brooklyn-bridge-car-free. The drawings can be seen in the PDF by Massengale & Co LLC called "Change the Car Culture – Car Free Brooklyn Bridge": https://urbanist.massengale.com/BrooklynBridgeChange.pdf. For more articles, see Slow New York, https://slownewyork.city or https://photos.massengale.com/slow-ny/.

169. I was staying in a hotel on the street called High Holborn, which continues Oxford Street to the east. The Central Line, one of the original lines in the Underground system, is a busy line that runs underneath Oxford Street and High Holborn and extends beyond the city limits to the west and the east (one of only two lines that does that). But the Central Line was closed for service at that time, and traffic on High Holborn was slow. It was faster to walk to Oxford Street than to take a bus. The day the CCZ started, however, traffic flowed like water and buses zipped along. I have been a fan of congestion zones ever since.

170. Peter Walker, "'Mini-Holland' schemes have proved their worth in outer London boroughs," *The Guardian*, June 26, 2018, https://www.theguardian.com/environment/bike-blog/2018/jun/26/mini-holland-schemes-have-proved-their-worth-in-outer-london-boroughs.

171. Noah Vickers, "London: Crashes drop by 25% after 20mph limit introduced." *BBC*, February 14, 2023, https://www.bbc.com/news/uk-england-london-64637389.

172. Fourteenth Street in Lower Manhattan is already a Bus Priority street, with limited private automobile traffic.

173. "Corey Johnson for Mayor: Will He 'Break' or Merely 'Bend' the Car Culture?" Streetsblog NYC, October 14, 2019, https://nyc.streetsblog.org/2019/10/14/corey-johnson-for-mayor-will-he-break-or-merely-bend-the-car-culture; Danny Pearlstein, "Opinion: Escaping Our Car Culture by Bus," Streetsblog NYC, November 22, 2023, https://nyc.streetsblog.org/2023/11/22/opinion-escaping-our-car-culture-by-bus.

174. First published in French as "Ornement et Crime," *Les cahiers d'aujourd'hui N° 5*, June 1913, https://www.edition-originale.com/fr/litterature/editions-originales/collectif-les-cahiers-daujourdhui-n-5-1913-68409. The "Theban Stools" were based on the design of historical Egyptian stools well known at the time. "'Dear friends, I want to let you in on a secret: there is no modern furniture!' is the provocative beginning to a 1924 essay by Loos entitled 'The abolition of furniture'." See Christopher Menz, "'There is no modern furniture!': Adolf Loos and the Viennese apartment of Jakob and Melanie Langer," *Art Journal*, January 5, 2013, https://www.ngv.vic.gov.au/essay/there-is-no-modern-furniture-adolf-loos-and-the-viennese-apartment-of-jakob-and-melanie-langer/.

175. Jeanne Willette, "Adolf Loos (1870-1933), Part One," arthistoryunstuffed, January 11, 2019, https://arthistoryunstuffed.com/adolf-loos-1870-1933-part-one/.

176. The statement by Adolf Loos is reported by Stefanos Polyzoides, Dean of the School of Architecture at the University of Notre Dame. Relevant architectural trivia: The School's building, designed by John Simpson and opened in 2018, has a replica of the Choragic Monument of Lysicrates in the form of a lantern crowning a tower. The Notre Dame school administers the Richard H. Driehaus Prize that annually honors major contributors in the field of contemporary traditional and classical architecture. Winners of the prize receive a bronze model of the Choragic Monument of Lysicrates. In the School's Hall of Casts, where students learn by sketching plaster components of the Classical orders, an eight-foot miniature of the monument is prominently displayed for study.
Regarding the Theban stools, the British Museum dig discovered them in the ancient Egyptian city of Thebes, five hundred miles south of Cairo. Thebes is now part of the city of Luxor: the Luxor Temple and other famous ruins are there. For more information on the stools and the nineteenth century's fascination with Egyptian stools see Donato Esposito, "From Ancient Egypt to Victorian London: the impact of ancient Egyptian furniture on British art and design 1850–1900," *The Journal of the Decorative Art Society*, 27 (2003), 81–93. To see one of the three-legged stools owned by the museum, go to https://www.britishmuseum.org/collection/object/Y_EA2481. For a four-legged stool, see "Thebes Stool for Liberty Antiques," Michael Pashby Antiques, https://www.michaelpashbyantiques.com/shop/seating/thebes-stool-for-liberty-co/.

177. Wikipedia has a list of examples. See "Choragic Monument of Lysicrates," Wikipedia, https://en.wikipedia.org/wiki/Choragic_Monument_of_Lysicrates.

178. "Philadelphia Merchant's Exchange," *ushistory.org*, https://www.ushistory.org/tour/merchants-exchange.htm.

179. John T. Kirk, *The Impecunious House Restorer: Personal Vision & Historic Accuracy* (Knopf, 1984): *passim*.

180. Owen Barfield, an Inkling and anthroposophist, was an important influence on fellow Inklings C.S. Lewis and J.R.R. Tolkien. His primary focus was on the evolution of consciousness, which for purposes of discussion he divided into three historical stages. The first was most widespread at the time of the ancient world. He called it "original perception" and said ancient peoples were

more fully a part of the natural world (much as American Indian culture describes). Studying the literature of the time, he believed that ancient Greek had no word for blue because Greeks of the time did not see blue. Homer's "wine dark sea," he says, is an example. More recent linguistic studies agree with Barfield's judgment. Mark Hewitt discusses some of the recent studies in his lecture "The Body Breathes Emotion: how Greek architecture moves us."

181. See Calder Loth, "Specimens for the Architectural Lectures: Jefferson's Orders at the University of Virginia," *ICAA*, April 27, 2021, https://www.classicist.org/articles/specimens-for-the-architectural-lectures-jeffersons-classical-orders-at-the-university-of-virginia-with-calder-loth/. The discussion of Pavilion II begins at 11 minutes and 57 seconds. Also see "The Lawn, Academical Village," *Albert and Shirley Small Special Collections Library*, University of Virginia Library, https://small.internal.lib.virginia.edu/collections/featured/the-cabell-family-papers-2/cabells-at-uva/cabells-and-founding/the-lawn/.

182. Other guests at the dinner included James Baldwin, Van Wyck Brooks, Pearl Buck, Ralph Bunche, John Dos Passos, Robert Frost, Robert F. Kennedy, Arthur M. Schlesinger, Lionel Trilling, and Diana Trilling. "Remarks at a Dinner Honoring Nobel Prize Winners of the Western Hemisphere (161)," *Public Papers of the Presidents: John F. Kennedy, 1962*, April 29, 1962, https://www.jfklibrary.org/asset-viewer/archives/jfkwhp-1962-04-29-b#?image_identifier=JFKWHP-AR7188-A.

183. This section owes much to Hewitt's book, *op. cit.*, and discussions with him. Any errors are the fault of the authors.

184. The tower at McKim, Mead & White's Madison Square Garden was also modeled after the Giralda Tower. At the San Remo, the round temples conceal the wooden water towers that were required for residential buildings in New York. See Susan Xu, "10 Fun Facts About The San Remo Luxury Apartment Building In NYC," Untapped New York, June 20, 2017, https://untappedcities.com/2017/06/30/10-fun-facts-about-the-san-remo-luxury-apartment-building-in-nyc/.

185. Richard Armstrong, "Thomas Jefferson and the Maison Carrée," *The Engines of Our Ingenuity*, Cullen College of Engineering, University of Houston, https://engines.egr.uh.edu/episode/2977.

186. For example, see Ethan Kent, "Streets as Places to Come Together: The Next Evolution for the Transportation Revolution," Social Life Project, May 11, 2022, https://www.sociallifeproject.org/streets_as_places_transportation_revolution/; Jeff Speck, "The Great Green Way," *New York Daily News*, January 10, 2019, https://www.nydailynews.com/2013/04/14/the-great-green-way-4/; and Transportation Alternatives' call for "Broadway Linear Park," https://www.broadwaylinearpark.nyc/home. The NYC DOT has closed the street to cars for Open Streets several times: https://flatironnomad.nyc/event/car-free-earth-day-2025/.

187. See Sri Taylor, "US Auto Crash Death Rate Highest in Study of Rich Countries," *Insurance Journal*, July 6, 2022, https://www.insurancejournal.com/news/national/2022/07/06/674788.htm. "US death rates from motor vehicle crashes rates are the highest among 29 upper-income countries, according to a government study that adds to the country's poor public-health record among developed nations.

Annual US fatalities from car, truck, and motorcycle collisions were 11.1 per 100,000 people in 2019, researchers from the US Centers for Disease Control and Prevention said, 2.3 times higher than the average in wealthy countries. While 21 other countries saw a decrease in crash-related mortality from 2015 to 2019, the pace of US deaths remained the same."

188. National Association of Realtors and American Strategies, *Community Preference Survey: Americans Prefer to Live in Mixed-Use, Walkable Communities*, October 2013, https://cdn.nar.realtor//sites/default/files/reports/2013/2013-community-preference-analysis-slides.pdf?_gl=1*to8pld*_gcl_au*NDY3NTk5MTE0LjE3MTU4OTAyODc.

189. The Driehaus Foundation was founded by Richard Driehaus. See "The measure of one's personal holdings is of less importance than the impact of our collective aspirations made real," for an article about Driehaus and his life at https://driehausfoundation.org/about.

190. Andy Boenau, "Opinion: How American vs. Danish Pedestrian Laws Stack Up," Streetsblog USA, August 17, 2023, https://usa.streetsblog.org/2023/08/17/opinion-how-american-vs-danish-pedestrian-laws-stack-up.

Also see John Massengale, "There Are Better Ways to Get Around Town," *New York Times*, May 15, 2018, https://www.nytimes.com/2018/05/15/opinion/there-are-better-ways-to-get-around-town.html

191. "Dr. Will Norman, Walking and Cycling Commissioner," Mayor of London, https://www.london.gov.uk/who-we-are/what-mayor-does/mayor-and-his-team/dr-will-norman.

192. "London set to become the world's most walkable city," *Mayor of London Press Release*, Mayor of London, July 19, 2018, https://www.london.gov.uk/press-releases/mayoral/mayor-launches-londons-first-ever-walking-plan.

193. Also known as "the Square Mile," the City is London's financial center. In typical British fashion, it is larger than a square mile.

194. "New Data Reveals Mayor's Policies Have Improved Air Quality in London Significantly Since 2016," Greater London Authority, April 24, 2023, https://www.london.gov.uk/new-report-reveals-dramatic-improvements-londons-air-quality-2016.

195. "Ultra Low Emission Zone Expands London-Wide in a Landmark Moment," Mayor of London, August 29, 2023, https://www.london.gov.uk/Ultra%20Low%20Emission%20Zone%20expands%20London-wide%20in%20a%20landmark%20moment%20for%20the%20capital.

196. "Red Routes," Transport for London, https://tfl.gov.uk/modes/driving/red-routes?intcmp=2187.

197. Jaime Dunaway-Seale, "The Most Bike-Friendly Cities in the U.S. (2023 Data)," Clever Real Estate, July 17, 2023, https://listwithclever.com/research/most-bike-friendly-cities-us/#ranking-50.

198. "Low Traffic Neighbourhoods," Transport for London, December 15, 2020, https://madeby.tfl.gov.uk/2020/12/15/low-traffic-neighbourhoods/.

199. "Transforming Cycling in Outer Boroughs: The Mini-Hollands Programme," Greater London Authority, March 3, 2021, https://www.london.gov.uk/programmes-strategies/transport/cycling-and-walking/transforming-cycling-outer-boroughs-mini-hollands-programme.

200. Zoom call with the authors, March 24, 2024.

201. E. F. Schumacher, *Small Is Beautiful: A Study of Economics As If People Mattered* (Blond & Briggs, 1973).

202. The Prince's Foundation was run for many years by Friends of Street Design Hank Dittmar, Ben Bolgar, Richard John, and Paul Murrain. It is now the King's Foundation. Scroll down on the following webpage to see the planning projects: https://www.kings-foundation.org/practice.

203. For the UK Department for Transport Highway Code, see https://www.gov.uk/guidance/the-highway-code. For a downloadable PDF of the UK *Manual for Streets* (Thomas Telford Publishing, 2014), https://assets.publishing.service.gov.uk/government/uploads/system/uploads/attachment_data/file/341513/pdfmanforstreets.pdf.

204. *Manual for Streets, op. cit.*, https://assets.publishing.service.gov.uk/government/uploads/system/uploads/attachment_data/file/341513/pdfmanforstreets.pdf.

205. *Ibid.*, Figure 2.5: "Typical road and street types in the Place and Movement hierarchy."

206. "Phil Jones Instrumental in Change to the Highway Code," Phil Jones Associates, August 2, 2021, https://pja.co.uk/2021/08/02/phil-jones-instrumental-in-change-to-the-highway-code/.

207. Carlos Moreno, *The 15-Minute City: A Solution to Saving Our Time and Our Planet* (Wiley, 2024): *passim.*

208. *Ibid.*, 74.

209. "Le Mouvement #SaccageParis," Saccage Paris, https://saccage-paris.com.

210. Liz Alderman, "As Bikers Throng the Streets, 'It's Like Paris Is in Anarchy,'" Paris Dispatch, *New York Times*, October 2, 2021, https://www.nytimes.com/2021/10/02/world/europe/paris-bicycles-france.html. For more positive coverage, see Streetsblog NYC's ongoing coverage at https://nyc.streetsblog.org/search?s=paris and videos at https://www.youtube.com/@StreetfilmsCommunity/search?query=paris.

211. Streetfilms video: "Paris Streets Prioritize Pedestrians, Kids, and Micromobility," Streetfilms, March 5, 2025, https://www.openplans.org/blog/paris-streetfilms.

212. *Ibid.*, at 1:04 minutes.

213. Alderman, *op. cit.*

214. Kea Wilson, "Cycling Through COVID-19: Paris," Streetsblog USA, October 13, 2022, https://usa.streetsblog.org/2022/10/13/cycling-through-covid-19-paris-france.

215. "Parisians have voted in favour of pedestrianising 500 more streets in the French capital, bolstering City Hall's ongoing campaign to reduce car usage and enhance air quality. A referendum held on Sunday saw nearly 66 per cent of voters approve the measure to create more car-free zones … . This latest vote marks the third such referendum in recent years, following a 2023 decision to ban e-scooters and a 2024 move to significantly increase parking fees for large SUVs." Benoit Van Overstraeten, "Parisians vote to ban cars from 500 more streets," *The Independent*, https://www.independent.co.uk/news/world/europe/paris-car-ban-pedestrian-streets-parking-b2720978.html.

216. Conor Dougherty, "The Capital of Sprawl Gets a Radically Car-Free Neighborhood," *New York Times,* October 14, 2021, https://www.nytimes.com/2020/10/31/business/culdesac-tempe-phoenix-sprawl.html /.

217. Two articles about Charlotte, North Carolina are interesting: Patrick Sisson, "Can Car-Free Living Succeed in Cities Built Around the Automobile?," Bloomberg, March 14, 2023, https://www.bloomberg.com/news/articles/2023-03-14/can-car-free-living-succeed-in-cities-built-around-the-automobile and Mathew Wald, "A Southern Success Story for Public Transportation," *New York Times*, April 5, 2010, https://archive.nytimes.com/www.nytimes.com/gwire/2010/04/05/05greenwire-a-southern-success-story-for-public-transporta-52742.html. Charlotte has old, walkable neighborhoods, but the city spent a lot of money making car-first streets and patterns of development. Nevertheless, the Queen City has four residential neighborhoods with Walk Scores that range from 78, in Dilworth, to 89, in the Fourth Ward. "Can't we have a neighborhood with a Walk Score over 95?" a resident asked

"In a first-of-its-kind national survey, 18 percent of US adults express interest in 'car-free' living, and an additional 40 percent are open to the idea. That is in addition to 10 percent of US households that currently live without a car." See Robert Steuteville, "Nearly one in five is interested in car-free living," CNU Public Square, January 23, 2026, https://www.cnu.org/publicsquare/2026/01/23/nearly-one-five-interested-car-free-living.

218. For Culdesac Tempe, see Chapter Four, note 4. For The Bend, see pages 383–384. Also see "Choices: Car-Dependent, Car-Lite, and Car-Free?" page 601.

Figure 5.66: Concept of a Market Street. © 2022 *Dover, Kohl & Partners*

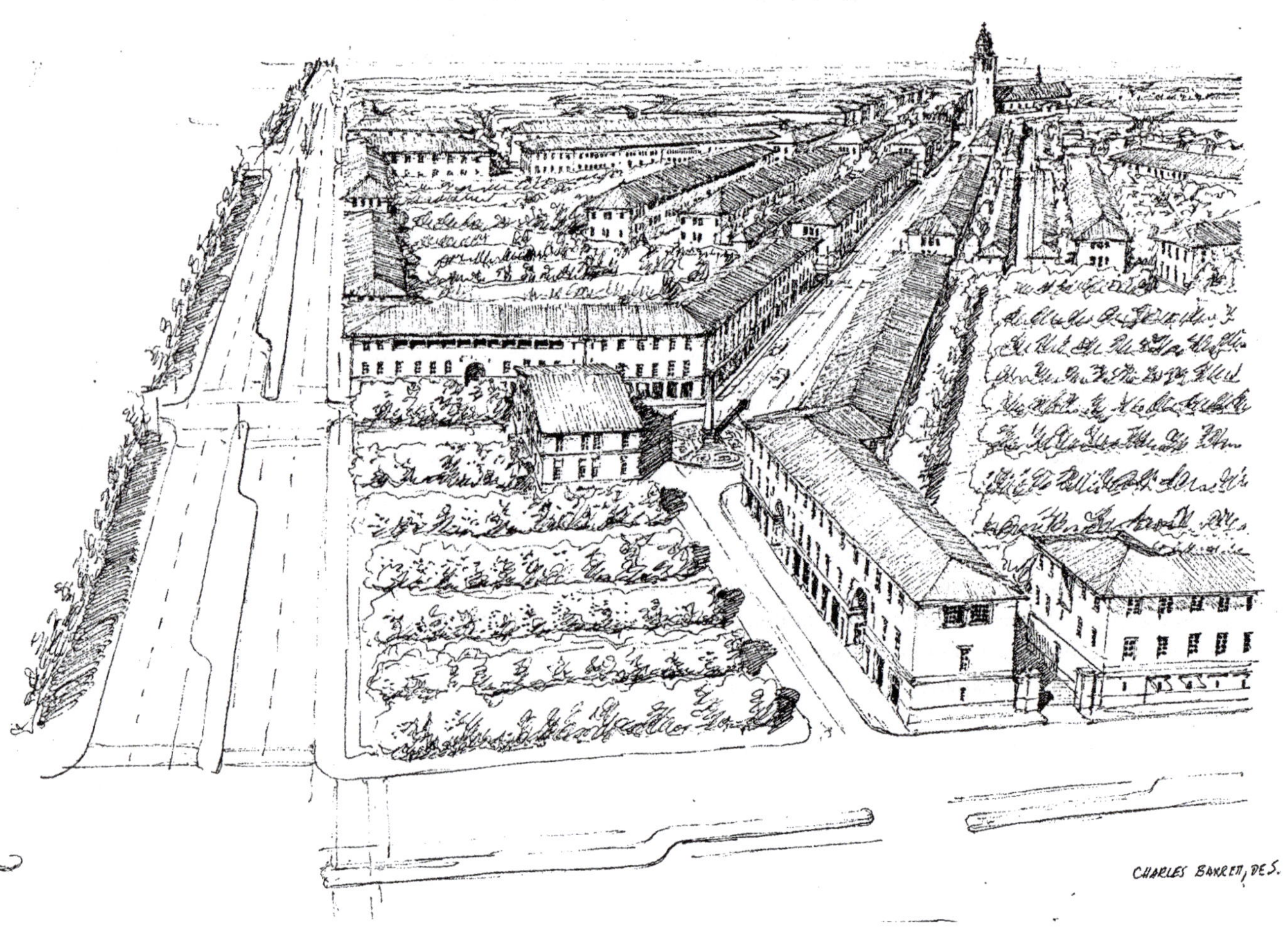

Figure: Avalon Park, Orlando, Florida. DPZ CoDesign, 1997. Aerial study. Pen and ink rendering by Charles Barrett. *Courtesy of DPZ CoDesign*

AFTERWORD

Elizabeth Plater-Zyberk

PRESUMABLY THE READER is wondering what yet another voice could add to the enlightenment experienced in the preceding pages. The following thoughts may serve only as a reminder of the marvel that is in your hands, and how useful its knowledge can be.

Street Design is an edifying compendium and the subtitle is apt: *The Secret to Great Cities and Towns*. Indeed, there is not one secret but many. You have been initiated into a society of aficionados that understands these secrets, especially the importance of spatial enclosure and its dimensions, the harmony possible in combining natural and built components (trees, paving, buildings), and the importance of beauty, a concept often absent in the built environment today.

Less of a secret now for you, dear reader, is the book's central premise—that streets are the chief component of placemaking in cities and towns, and of community identity and appeal. To imagine streets as "communicators" at a time when human encounters challenge us, reaffirms the possibility of shared experience and goals. Where else but in the public realm of streets and sidewalks can we find the unifying solidarity of the comings and goings of daily life—the opportunities to greet, give way in courtesy, and exchange in commerce—for people of all ages and occupations?

More ground than the book's title suggests has been covered here, in glimpses providing a taste of additional pleasures to be derived from deeper dives elsewhere— the history of cities, the engineering of infrastructure, or a deeper understanding of a city about to be visited— Philadelphia, Charleston, London, Barcelona, among those here included.

What all these examples share—no matter where they are in the world, or when they were constructed, or which principle they are illustrating—is that they were all intentional designs by humans who understood the interaction of two-dimensional and three-dimensional perception. These are not the result of happenstance, as is sometimes suggested. Across the globe, across time, and across the transect, from narrow urban passages to scenic rural roads, repetitive patterns suggest universal principles of design. Dimensions, orientation, controlled views, materials hard and soft, accommodation of various types of mobility, often repeat. But their specificity to context removes them from the domain of formula, to engender character and a sense of place—a specificity that will forever challenge AI or whatever rote intention seeks easy replication.

Dover and Massengale have come to know and love these places and admire the people who made them. They have focused on what these places can teach us about universally admired excellence—from centuries ago to contemporary endeavors battling against modern quantitative conventions. The selection of photographs, elucidating diagrams and tables, bespoke drawings for scale-comparisons, and especially the enticing color illustrations (all thankfully at a larger scale than the first edition), makes this effectively a handbook for designers, community leaders, civic activists, teachers, and students of any age.

The first edition of *Street Design* landed 10 years ago. The contemporary projects gathered there reflected the history of the first decade and a half of the New Urbanism. The writers of its essays are among the leading thinkers and practitioners of the movement. In the ensuing decade much has transpired. Some of the projects have indeed been realized—many with great success owing in large part to encouragement by publication in the first edition. These have added to the list of glorious victories against mindless convention. There is much to cheer about and to deploy in continuing battles.

Seeing photographs of the recent victories (Lancaster) or of the historic examples that have been added to this edition (Riverside Drive), one can't help but think of many more instances that could benefit from a better use of public transportation funding. Which might bring the reader to ask, how do we modify the nearly century-old conventions of roadway design and building? Nothing less than culture change is needed! Certainly a tall order, but not impossible, as exemplified in Florida, now allowing (although not yet requiring) context-sensitive street design—a significant victory over regulations that have long prevented correspondence of engineering and urban design.

At this writing, we are at a special moment, as federal funding for infrastructure re-orients to support long-term resilience. The legislation recognizes walkable communities as one of the leading effective responses for emissions reduction. While decarbonizing the economy puts a large focus on changes in technology, it is the behavioral response to be induced by an amenable public realm that encourages walking, transit use, and compact living, that is the guiding principle of this book.

Beyond the urgency of the moment, education of the next generation also must be addressed. This book should be required reading for students of urban design, architecture, and landscape architecture. And an understanding of its content should be part of the licensing requirements for those professionals as well as civil, traffic, and transportation engineers.

Street Design is dense with useful material—so much in fact that it may not facilitate memory at the time of need. So here is a final reminder of the riches to which at a given moment you may wish to return in the preceding pages: the taxonomy of classifications (including for reference, the merely functional); the role of urban design and its components; the aggregation of streets as networks; the lucid diagrams about connectivity; the walkability index; bicycle encouragement; retail secrets; the role of street trees and what they need to thrive; circles and roundabouts; good and better; among many others. *Street Design* has everything you need to know about embellishing human experience in a shared public realm!

INDEX